Soil Management:
Building a Stable Base for Agriculture

Soil Management:
Building a Stable Base for Agriculture

Jerry L. Hatfield and Thomas J. Sauer, Editors

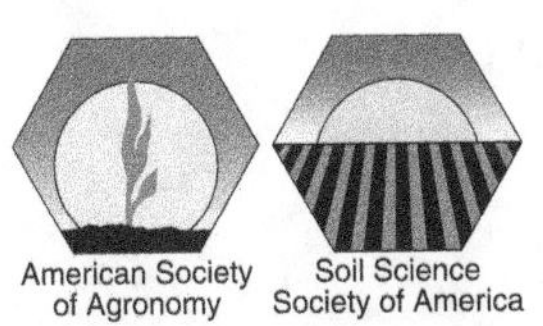

American Society of Agronomy
Soil Science Society of America
5585 Guilford Road, Madison, WI 53711, USA

ISBN: 978-0-89118-853-7
Library of Congress Control Number: 2011905781

Cover photo courtesy of John A. Kelley, USDA-NRCS,
http://SoilScience.info

Cover design: Patricia Scullion

Contents

Challenges facing society over the next several decades include food security, loss of soil and water resources and associated ecosystem services, energy security, and climate. Soil is at the root of these issues. Embracing science-based soil management practices that directly respond to these challenges is both crucial and essential for the long-term sustainability of soil resources. That is the essence of this book. The chapters found herein are relevant, current, and written to respond to these global issues. The strength of this book is its ambitious goal of taking the best science and presenting it without oversimplification, but achieving an accessibility that will be appreciated by readers from related disciplines or those who are not Ph.D.s. It is not a textbook, but it is a knowledge-packed book for learning that is readable for a vast audience.

Because the Soil Science Society of America and the American Society of Agronomy support the management of soil as fundamental to life, this book speaks to a priority message of our sciences.

Charles W. Rice, Soil Science Society of America President , 2011
Newell R. Kitchen, American Society of Agronomy President, 2011

Management of the soil encompasses a wide range of practices with the express purpose of improving the capability of the soil to perform the various functions. Improvement in soil management practices will result in increased soil functionality and will be realized only if these practices are based on scientific principles. In assembling this book, we had two primary goals. The first goal was to gather the information about the emerging challenges in soil management from a number of different perspectives. The second goal was to develop this book to help a wide range of readers understand the need and opportunities in improving soil management. As editors and active researchers in the area of soil management, we are aware of the need to enhance our soil and protect our environment for future generations.

We feel that we must continue to enhance our soils and prevent further degradation of the soil resource. If we consider the need to increase the world's food supply over the next 40 yr to meet the demands of the rapidly expanding population, we will have to develop soil management practices capable of increasing and sustaining the production necessary to meet these demands. A focus on the efficiency of the current production systems will be necessary to maximize what we produce from a given amount of water, nutrients, carbon, and light. Our personal concern is that we have not focused enough on how to improve our soils and soil management practices. The demands to produce food, feed, fuel, fiber from our soil resources demands soil with the capability of supplying water, nutrients, oxygen, and microbial diversity.

We express our sincere thanks to all of the chapter authors and their unselfish efforts to share their knowledge by preparing these chapters as a summary of the current state of knowledge. Without their dedication to advancing scientific knowledge, it would be impossible to develop this collection of materials covering the diverse set of topics within this book. We also thank the ASA and SSSA headquarters staff for their help and assistance in preparing this final product and the Societies for supporting and encouraging us to undertake this task. Finally, we thank you as the reader for taking the time to increase your knowledge of the challenges in soil management.

Jerry L. Hatfield
Thomas J. Sauer

Andrews, A.A.   USDA-NRCS, ENTSC, 200 E. Northwood, Ste. 410, Greensboro, NC 27401

Archer, D.W.   USDA-ARS, 1701 10th Ave. SW, P.O. Box 459, Mandan, ND 58554

Brouder, S.   Agronomy Dep., Purdue Univ., 915 W. State St., West Lafayette, IN 47907-2054 (sbrouder@purdue.edu)

Calderón, F.J.   USDA-ARS, Central Great Plains Research Station, 40335 County Rd. GG, Akron, CO 80720-0400 (francisco.calderon@ars.usda.gov)

Clapperton, M.J.   Earthspirit Consulting, 8337 Lamar Trail, Florence, MT 59833 (earthspiritconsulting@gmail.com)

Gish, T.J.   USDA-ARS Hydrology and Remote Sensing Lab., Beltsville, MD 20705 (timothy.gish@ars.usda.gov)

Grant, C.A.   Agriculture and Agri-Food Canada, Brandon Research Ctr., Brandon, Manitoba, Canada

Hatfield, J.L.   USDA-ARS National Lab. for Agriculture and the Environment, 2110 Univ. Blvd., Ames, IA 50011 (jerry.hatfield@ars.usda.gov)

Hao, X.   Agriculture & Agri-Food Canada, 5403 1st Ave. South, Lethbridge, Alberta, Canada T1J 4B (xiying.hao@agr.gc.ca)

Helmers, M.J.   Iowa State Univ., Dep. Agricultural and Biosystems Engineering, Ames, IA 50011

Hernandez-Ramirez, G. The New Zealand Inst. for Plant and Food Research Limited, Soil Water and Environment Group, Plant and Food , Research Lincoln Gerald St., Lincoln, 7608, New Zealand (Guillermo.Hernandez@plantandfood.co.nz).

Hillel, D.   29 Yefe-Nof St., Zichron Yaacov, Israel 30900 (d_hillel@netvision.net.il)

Johnson, J.M.F.   USDA-ARS, 803 Iowa Ave., Morris, MN 56267 (Jane.johnson@ars.usda.gov)

Kang, S.   Oak Ridge National Lab., P.O. Box 2008, MS 6407, Oak Ridge, TN 37831 (kangs@ornl.gov)

Karlen, D.L.   USDA-ARS, National Laboratory for Agriculture and the Environment, 2110 Univ. Blvd., Ames, IA 50011 (Doug.Karlen@ars.usda.gov)

Kaspar, T.C.   USDA-ARS, National Laboratory for Agriculture and the Environment, 2110 Univ. Blvd., Ames, IA 50011 (Tom.Kaspar@ars.usda.gov)

Kelley, J.   4018 Starboard Court, Raleigh, NC 27613 (jkelley13@nc.rr.com)

Kirkham, M.B.   Dep. of Agronomy, 2004 Throckmorton Plant Sciences Center, Kansas State Univ., Manhattan, KS 66506-5501 (mbk@ksu.edu)

Kladivko, E.J.   Agronomy Dep., Purdue Univ., West Lafayette, IN 47907 (kladivko@purdue.edu)

Kover, J.L.   USDA-ARS, National Laboratory for Agriculture and the Environment, 2110 Univ. Blvd., Ames, IA 50011-3120 (John.Kovar@ars.usda.gov)

Kustas, W.   USDA-ARS Hydrology and Remote Sensing Laboratory, 10300 Baltimore Ave. Bldg. 007, Room 104, BARC-West, Beltsville, MD 20705

Kyei-Baffour, N.   Dep. of Agric. Eng., Kwame Nkrumah Univ. of Science and Technol., Kumasi, Ghana (nkyeibaffour@hotmail.com)

Larney, F.J.   Agriculture & Agri-Food Canada, 5403 1st Ave. South, Lethbridge, Alberta, Canada T1J 4B1 (francis.larney@agr.gc.ca)

Levy, G.J.   Institute of Soil, Water and Environmental Sciences, Agricultural Research Organization, The Volcani Center, Bet Dagan, Israel (vwguy@volcani.agri.gov.il).

McKee, L.G.   USDA-ARS Hydrology and Remote Sensing Laboratory, 10300 Baltimore Ave. Bldg. 007, Room 104, BARC-West, Beltsville, MD 20705

Menalled, F.D.   Dep. of Land Resources and Environmental Sciences, Montana State Univ., Bozeman, MT 59717 (menalled@montana.edu)

Mullen, R.W.   130 Williams, OARDC-Wooster, The Ohio State Univ., Wooster, OH 44691 (Mullen.91@osu.edu)

Nielsen, D.C.   USDA-ARS, Central Great Plains Research Station, 40335 County Road GG, Akron, CO 80720-0400 (david.nielsen@ars.usda.gov)

Ofori, E.   Dep. of Agric. Eng., Kwame Nkrumah Univ. of Science and Technol., Kumasi, Ghana (oforiegh@yahoo.com)

Paulitz, T.C.   USDA-ARS, Root Disease and Biological Control Research Unit, Pullman, WA 99164-6430 (paulitz@wsu.edu)

Prueger, J.H.   USDA-ARS National Laboratory for Agriculture and the Environment, 2110 Univ. Blvd., Ames, IA 50011 (john.prueger@ars.usda.gov)

Reicosky, D.C.   USDA-ARS, North Central Soil Conservation Research Lab., 803 Iowa Ave., Morris, MN 56267 (don.reicosky@ars.usda.gov)

Rożej, A.   Lublin Univ. of Technology, Dep. of Land Surface Protection Engineering, Nadbystrzycka 40B, 20-618 Lublin, Poland (a.rozej@wis.pol.lublin.pl)

Russ, A.   USDA-ARS Hydrology and Remote Sensing Laboratory, 10300 Baltimore Ave. Bldg. 007, Room 104, BARC-West, Beltsville, MD 20705

Ryan, M.R.   Dep. of Crop and Soil Sciences, The Pennsylvania State Univ., Univ. Park, PA 16802 (mrr203@psu.edu)

Sands, G.R.   Univ. of Minnesota, Dep. Bioproducts and Biosystems Engineering, St. Paul, MN 55108

Sauer, T.J.   USDA-ARS National Laboratory for Agriculture and the Environment, 2110 Univ. Blvd., Ames, IA 50011 (tom.sauer@ars.usda.gov)

Singer, J.W.   USDA-ARS, National Laboratory for Agriculture and the Environment, 2110 Univ. Boulevard, Ames, IA 50011 (jeremy.singer@ars.usda.gov)

Smith, R.G.   Dep. of Natural Resources and the Environment, Univ. of New Hampshire, Durham, NH 03824 (richard.smith@unh.edu)

Stępniewska, Z.   John Paul II Catholic Univ. of Lublin, Kraśnicka 102, 20-551, Lublin, Poland (stepz@kul.pl)

Stępniewski, W.   Lublin Univ. of Technology, Dep. of Land Surface Protection Engineering, Nadbystrzycka 40B, 20-618 Lublin, Poland (w.stepniewski@pollub.pl)

Stigter, C.J.   Agromet Vision, Bondowoso, Indonesia and Bruchem, the Netherlands (cjstigter@usa.net)

Strock, J.S.   Univ. of Minnesota, Dep. Soil, Water, and Climate and Southwest Research and Outreach Center, Lamberton, MN 56152 (jstrock@umn.edu)

Topp, E.   Agriculture & Agri-Food Canada, 1391 Sandford St., London, Ontario, Canada N5V 4T3 (ed.topp@agr.gc.ca)

Van Pelt, R.S.   USDA-ARS, Wind Erosion and Water Conservation Research Unit, 302 West I-20, Big Spring, TX 79720

Wienhold, B.J.   USDA-ARS, 305 Entomology Hall, Univ. of Nebraska, East Campus, Lincoln, NE 68583 (Brian.Wienhold@ars.usda.gov)

Walker, S.   Dep. of Soil, Crop and Climate Sciences, Univ. of the Free State, Bloemfontein, South Africa (walkers@ufs.ac.ca)

Weyers, S.L.   USDA-ARS, 803 Iowa Ave., Morris, MN 56267

Wilhelm, W.W.   deceased, formerly USDA-ARS, University of Nebraska, Lincoln

Williams, J.   Dep. of Agribusiness, Plant and Animal Sciences, 525 S Center St., Rexburg, ID 83460 Brigham Young Univ.–Idaho, Rexburg, ID (williamsj@byui.edu)

Zobeck, T.M.   USDA-ARS, Wind Erosion and Water Conservation Research Unit, 3810 4th St., Lubbock, TX 79415 (zobeck@ars.usda.gov)

Section 1
Framing the Soil Resource Problem

An Overview of Soil and Water Management: The Challenge of Enhancing Productivity and Sustainability

Daniel Hillel

During the greater part of its existence on earth, the human species subsisted as roving bands or clans engaged in hunting, gathering, and scavenging. From its origin in Africa, the species known as *Homo sapiens* gradually spread to all other habitable continents, adapting to the various ecological conditions and to the vicissitudes of changing climates. The momentous transition from the traditional nomadic mode of life to sedentary farming first took place in the Fertile Crescent, part of the region known today as the Near East, during the so-called Neolithic Age, which began some ten millennia ago.

Subsequently, agriculture in various forms has developed and spread throughout most of the habitable continents. The attainment of relative food security by means of farming and grazing (along with sanitation and medical advances) has enabled the gradual growth of population, from perhaps a few million in Neolithic times to well over six billion today. Current and foreseeable trends are projected to increase the globe's population to some nine or even ten billion by the middle of this century. So far, food production has managed to keep pace, by and large, with population growth (although not in all regions—notably not in parts of Sub-Saharan Africa).

However, the increase of food production has been achieved at a great ecological cost: the progressive encroachment on extensive land and water resources and the eradication of their native biota. Agriculture has indeed appropriated a goodly fraction of the terrestrial domain's net primary productivity, leaving less and less of it to sustain natural ecosystems. Among the most egregious processes of degradation resulting from the spread of agriculture are land denudation and soil erosion, nutrient depletion, salination, and waterlogging, as well as diversion and pollution of surface and subsurface water resources. All these processes amount to an unsustainable exploitation of the Earth's natural resources and its biotic communities. Now looms the prospect of anthropogenic climate change, which threatens to further disrupt established patterns of food production.

The future of agriculture hangs in the balance. What can be done to rectify the destructive practices of the past and present and to ensure the sustainability of agriculture in the face of rising demands, a changing climate, and the growing scarcity and cost of conventional energy sources?

The first requirement is to intensify production on lands that can be improved and managed efficiently, while relieving the pressure on vulnerable or fragile lands that are not, or cannot be, so managed. As currently practiced, much of the world's agriculture is unsustainable, as well as exceedingly inefficient in terms of its energy balance, conservation of resources, and

D. Hillel, 29 Yefe-Nof St., Zichron Yaacov, Israel 30900 (d_hillel@netvision.net.il).

doi:10.2136/2011.soilmanagement.c1

productivity. Hence there is great latitude for improving both the economic efficiency and ecological sustainability of production and resource utilization.

Modes of Cultivation

Agricultural soil management generally consists of a series of practices that include cultivation, planting, fertilization, pest control, irrigation, drainage, and erosion control. The more efficiently these practices are performed and optimized, the more productive and sustainable can agriculture become.

Cultivation or *tillage* is defined as the mechanical manipulation of the soil to improve conditions affecting the various stages of crop production. Three principal aims are generally attributed to tillage: control of weeds, incorporation of organic matter into the soil, and preparation of a favorable seedbed to facilitate germination and seedling establishment. An additional rationale is sometimes claimed for tillage, namely the conservation of soil moisture by enhancing infiltration and inhibiting evaporation. Not all of these traditional claims can be substantiated in many cases.

A distinction must be made between primary tillage and secondary tillage. *Primary tillage* is typically performed by means of moldboard or disk plows, which slice and pulverize a layer of topsoil along parallel furrows and invert it so as to cover the surface residues of plant matter remaining from preceding seasons. An alternative mode of primary tillage is the use of subsoilers and chisels, which break and loosen the topsoil without inverting it. All such methods of primary tillage are generally designed to penetrate to a depth of at least 20 cm, and sometimes to a depth as great as 50 cm.

Secondary tillage is often performed subsequent to primary tillage to repeatedly loosen the topsoil (generally to a depth of no more than 20 cm) and thereby cut the roots of weeds that grow between the rows during the crop growing season. In other cases, secondary "light" tillage is performed in lieu of primary tillage in soils that are naturally loose and require no deeper tillage at all. The implements used for secondary tillage are disk harrows, spike harrows, sweeps, rotary hoes, and various other tools that work the soil to shallow depth and also help to disrupt crusts where they occur. Often, however, such implements, while effective in the short run, ultimately contribute to the degradation of soil structure by excessively pulverizing and grinding down the soil's natural aggregates.

In recent decades, the advent of chemical herbicides has reduced the importance of tillage as the primary methods for the eradication of weeds, although the high cost of such chemical treatments and their residual environmental effects limit their application, especially in developing countries. At the same time, the formerly prevalent practice of inverting the topsoil to bury manures and plant residues has become a less important function of tillage in modern field management. Plant residues can, and in many cases should, be left over the surface as a *stubble mulch* to protect against evaporation and erosion.

An essential task of agriculture is soil structure management because it affects water infiltration and runoff, wind erosion and evaporation, gas exchange and soil aeration, as well as the planting and germination of crops. However, tillage practices suitable in one location may become harmful in another. Arid-zone soils with low organic matter contents and unstable aggregates are particularly vulnerable to compaction, crusting, and erosion by wind (during dry spells) and by water (during rainstorms). The precise effects of various modes of tillage must be defined for each set of conditions to ensure the sustainability of soil management.

Tillage operations are especially consumptive of energy. The amount of earth-work involved in repeatedly loosening, inverting, and then recompacting the topsoil is indeed very considerable. In a typical field of 1 ha, the topsoil to a depth of only 30 cm weighs no less than 4000 tons. In an extensive farm of 1000 ha, the mass of soil thus manipulated in each cycle of operation may exceed 4 million tons. The consumption of energy, as well as the wear and tear of tractors and implements, increases steeply as the depth of tillage increases. With the rising cost of fuel, the costs of tillage also increase progressively. Much damage is done to soil structure by the repeated passage over the soil of heavy tractors and other machinery, and such damage, which induces erosion while inhibiting infiltration, aeration, germination, and root development, is difficult to rectify.

Current trends in soil management are aimed at minimizing tillage operations and overland traffic, both to reduce costs and to avoid soil compaction, while tailoring each operation to its specific zone and objective. This approach, in principle, underlies the methods variously termed *minimum tillage*, *precision tillage*, and even *zero tillage* or *no-till*. The tendency is to adopt a system of "precision farming," which consists of a balanced combination of practices designed to optimize nutrient supply, tillage, water use, and pest control. Instead of treating a large block of land uniformly, the effort is to define the soil's inherent heterogeneity so as to adapt all operations to the space-variable and time-variable requirements for applying pesticides, fertilizers, seeds, and water.

The entire set of practices thus described replaces the old tradition of "clean cultivation" of the entire top layer of the soil, which in the past entailed burning or plowing-in the stubble of previous crops and disrupting the natural structure of the soil, thus making it more vulnerable to erosion. Instead, special equipment is used that is designed to sow seeds into narrow slits while retaining the residues on the surface between the rows. The organic matter remains at the surface, called *mulch*, helping to conserve moisture and protect the soil against erosion. The problematic aspect of zero tillage is that it relies on the use of herbicides instead of mechanical cultivation to eradicate weeds that might otherwise compete with crop plants for moisture, nutrients, space, and sunlight.

The Practice of Irrigation

Irrigation is the supply of water to agricultural crops so as to permit farming in arid regions and to offset periodic droughts in semiarid regions. Wherever rain-fed farming is a high-risk enterprise due to erratic rainfall, irrigation can help to ensure stable and reliable production.

Irrigation has long played a key role in feeding populations and is expected to play a still greater role in the future. Although irrigated land amounts to only some 17% of the world's cropland, it contributes about 40% of the value of agricultural production. That vital contribution is most important in arid regions, where the natural supply of water by rainfall is least even as the demand for water, imposed by the bright sun and dry wind, is greatest.

Ideally, irrigation not only raises the yield of specific crops but also prolongs the effective crop-growing period in areas with extended dry seasons, thus permitting multiple cropping each year where only a single crop could be grown otherwise. With the security provided by irrigation, additional inputs needed to intensify production (e.g., pesticides, fertilizers, improved varieties, physiological treatments, environmental controls, and soil amendments) can become economically feasible. Irrigation reduces the risk of such expensive inputs being wasted by crop failure due to lack of water.

The practice of irrigation consists of applying water to the part of the soil that serves as the root zone, for the immediate and subsequent use of the crop. However, the initiation and continuation of irrigation in a given area tends to induce a series of processes that can profoundly affect both the on-site and the related off-site environments. Over time, some of its potentially self-destructive effects may make the very practice of irrigation unsustainable.

Off-Site Aspects of Irrigation

The first requirement of an irrigation project is a dependable supply of fresh water and the means to deliver it to the site to be irrigated. The second requirement is the availability of suitable land and soil in which to grow the crops under irrigation. The third requirement is an outlet for the safe disposal of wastewater from the irrigated land. Every irrigation project therefore consists of withdrawing water from some source (e.g., a lake, an aquifer, or a river) and diverting it to the site of irrigation. In the process, inevitably, that water is denied to some other site that had been its natural recipient. In the typical case of a river valley project, changes take place both in the upstream and in the downstream sections of the riverine domain.

River flow is generally time-variable. More often than not, the season of peak irrigation demand coincides with the period of low river flow. Hence, an irrigation project typically requires the construction of engineering structures—dams and canals—designed to regulate the flow so as to ensure adequate storage and supply throughout the growing season.

Dam construction is a problem in itself. Appropriate locations for dams are relatively rare. The ideal topographic, geologic, and climatic conditions are seldom found in the proximity of target irrigation projects, and less-than-ideal conditions may make the construction and maintenance of dams economically prohibitive and environmentally unsustainable. Unfavorable topography may require very massive construction and may result in the submergence of very large areas, with consequent damage to natural ecosystems, displacement of long-established human populations and infrastructure, and loss of historical cultural heritage and scenic sites. Some areas are inherently vulnerable to potential natural disasters such as earthquakes. Porous or fractured substrata may cause great losses of water by practically uncontrollable seepage, and a dry climate may impose additional losses of water by evaporation. A case in point is the Aswan Dam and its associated lake in Egypt, located in the midst of the Earth's driest desert, where evaporative losses of water may be in the range of 12 to 16 billion cubic meters per year—some 20% of the inflow.

Additional losses of water by evaporation and uncontrolled seepage occur in the system of conveyance from a dam to the fields. Generally, this conveyance takes place via canals and ditches, which, in the interest of minimizing costs, are often dug into the ground and left unlined (i.e., not underlain with concrete or some other impervious and scour-resistant material). Even where closed conduits (i.e., metallic, ceramic, or concrete pipes) are installed, in time they tend to develop leaks and to incur considerable losses of water and hydraulic pressure.

Water storage behind dams is also subject to silting, which, gradually over a period of some decades, reduces a dam's capacity and may eventually clog up its storage basin entirely. Silt accumulation in reservoirs and canals is especially rapid in regions where the upper watersheds of the river have been denuded of their natural vegetative cover by overgrazing and further destabilized by excessive cultivation, thus subjecting the catchments to accelerated soil erosion.

Thus far, we have listed processes that may take place upstream of an irrigated area. In addition, irrigation also entails a series of processes that occur downstream. The first of these is the diminution of the river flow resulting from the extraction of the water used for irrigation. Consequently, all the associated riparian ecosystems along the riverbanks, floodplain, and estuary are deprived of vital water supplies and thereby impoverished. Natural wetlands that were originally biologically diverse and highly productive may be subject to periodic or permanent desiccation, and even to the eradication of vital species or biotic habitats, especially spawning fisheries. Where a river discharges naturally into a freshwater lake, the deprivation of the lake's inflow resulting from the river's diversion for irrigation may cause the lake to shrink markedly, and in places drastically. The shrinkage of the Aral Sea in central Asia is a prime example of the sort of environmental disaster that can be caused by large-scale irrigation development.

These environmentally damaging processes are often exacerbated by the downstream disposal of irrigation-generated wastes. Drainage from irrigated lands may be laden with salts, as well as with residues of the fertilizers and pesticides that are often applied in excess to the irrigated crops. If the drainage from the bottom of the root zone percolates toward an aquifer, it may gradually contaminate its groundwater and make the aquifer unusable for humans. In extreme cases it may even become unsuitable for irrigation. Nitrates, as well as chlorides, may accumulate in groundwater underlying irrigated lands, to the extent of posing a health hazard to communities relying on wells. Other agents contained in the drainage from irrigated agriculture, as well as from households and industries, are various toxic and carcinogenic elements and compounds from pesticides.

Where the drainage from irrigated lands is channeled through ditches or pipes and discharged into the river downstream, it may pollute the water there and make it unusable for people as well as harmful to natural fauna and flora. The alternative to discharging the drainage into the river is to convey it to the sea, to lagoons or wetlands (whose biota, however, may also be vulnerable to the pollutants), to environmentally isolated evaporation basins in the desert, or to very deep aquifers where the pollutants are diluted. All of these alternatives may be quite expensive to carry out, and perhaps unsustainable in the long run. The Kesterson Reservoir in the Central Valley of California is an example where the discharge of drainage from irrigated areas into wetlands has resulted in damage to wildlife because of the accumulation of waterborne toxic elements.

On-Site Aspects of Irrigation: The Hazard of Soil Degradation

The twin processes of degradation that typically affect irrigated lands in river valleys and low-lying coastal plains, specifically those in arid regions, are waterlogging and salination. Waterlogging results from the tendency of irrigators to apply a volume of water to the soil in excess of the amount of soil water taken up by the crop. In part, this is a matter of necessity, to prevent the root zone from accumulating salts.

Irrigation water is never entirely pure; hence, the application of irrigation necessarily adds water-borne salts to the soil. Moreover, many arid-zone soils and subsoils contain natural reserves of salts, which are also mobilized by irrigation. Since crop roots typically exclude most salts, the salts left in the root zone tend to accumulate to the detriment of the crop, unless leached from the soil and driven downward by drainage.

Maintaining an optimal balance of water and salts in the root zone is a delicate dynamic process. Water flowing downward below the root zone eventually reaches the water table and augments the groundwater saturating the substrata. If the water table remains deep and the groundwater beneath it has its natural outflow, the balance of water and salts in the root zone can remain favorable to crop growth. If, however, the water table is shallow and the rate of natural groundwater drainage is slow, the addition of irrigation water from above will cause the water table to rise. Sooner or later (perhaps within a few decades) the water table approaches the soil surface. The root zone then becomes saturated with water and deprived of oxygen. Since most crop plants require oxygen in the soil for their roots to respire, the saturation of the root zone itself restricts crop growth and, in the case of sensitive plants, may cause total crop failure.

The deleterious effect on soil productivity resulting from waterlogging is further exacerbated by soil salination, which typically occurs in arid regions wherever the groundwater is brackish and the water table rises toward the surface. Instead of percolating downward and leaching away the salts, the groundwater seeps upward to the soil surface in response to evaporation-induced suction gradients. As the groundwater evaporates, it leaves salts behind. So much salt can accumulate at and near the surface as to render the soil sterile to most crops. In this manner, well-intentioned irrigation projects can, quite inadvertently, cause the degradation of originally productive soils.

A further scourge of irrigation is the phenomenon of soil sodicity (also called alkalinity), a condition caused by the specific effect of sodium ions adsorbed onto the electrostatically charged clay particles. Such particles normally attract a swarm of cations in the aqueous phase (the ambient solution) that surrounds the particles. These cations are exchangeable, in the sense that they can be replaced by other cations whenever the composition of the ambient solution changes.

Divalent cations such as calcium ions tend to compress the ionic swarm surrounding the particles, thus allowing the particles to approach one another sufficiently to clump together and to form flocs (a process termed *flocculation*, which contributes to the formation of a desirable soil structure). In contrast, monovalent cations such as sodium tend to diffuse farther away from the particle surfaces and thus to thicken the hydration envelope surrounding each particle. This, in turn, causes swelling and dispersion of the soil flocs (a process termed *deflocculation*, which in effect thickens the hydration envelopes surrounding the clay particle and causes them to draw apart from one another). The latter process destroys soil aggregates and restricts the soil's permeability to water and air. When wet, a sodic soil becomes a slick and sticky mud; when dry, however, it hardens to form a tough crust. This condition reduces the entry of water and air into the soil and also forms a barrier to the emergence of germinating seedlings and to the penetration of their roots.

Sustainability of Irrigation

The sustainability of irrigation is never to be taken for granted. Waterlogging and salination of soils, along with other degradation processes, not only have caused the collapse of irrigation-based societies in the past, but are indeed threatening the viability of irrigation at present. The problem is global in scope. Decimation of natural ecosystems, deterioration of soil productivity, depletion and pollution of water resources, and conflicts among sectors and states over dwindling supplies and rising demands have become prevalent and pressing problems closely linked with the practice of irrigation in various regions.

Irrigated agriculture can be sustained in the long term only if and where certain stringent requirements are met. The requirements are effective prevention of upstream, on-site, and downstream damage to the environment.

Although there are cases where the degradation is already severe and remediation may seem impractical or prohibitive, in most cases irrigation projects can and should be made sustainable, provided the correct diagnosis and actions are undertaken.

To be and remain viable, irrigated agriculture must strive for a balance between the immediate need to maximize production and the ultimate need to ensure continued productivity in the future. It must also strive to achieve a harmonious interaction with the external environment, which includes both natural ecosystems and other human enterprises. More specifically, irrigation projects should ensure that water supplies of adequate quality are and will continue to be available, the salt balance and hence the productivity of the land can be maintained, the drainage effluent can be disposed of safely, public health can be safeguarded, and the economic returns can justify the costs of initial investment and subsequent continuing upkeep.

The key to ensuring the sustainability of irrigation is the timely installation and continuous operation of a drainage system to prevent waterlogging and to dispose safely of excess salts. All too often drainage creates off-site problems beyond the on-site costs of installation and maintenance, since the discharge of briny effluent (e.g., into a stream or a lake) can degrade the quality of water along its downstream course. Where access to the open sea is feasible, solving the problem is likely to be easier than in closed basins or in areas far from the sea. In those cases, the disposal terminus eventually becomes unfit for human use, as well as for wildlife, hence the importance of reducing the volume and salinity of effluents by such means as improving the efficiency of water use, a task that in itself can bring economic and environmental rewards. Modern irrigation technology offers the opportunity to conserve water through reduced transport and application losses, coupled with increased yields per unit volume of water.

Modern Irrigation

In recent decades, revolutionary developments have taken place in the science and art of irrigation. A more comprehensive understanding has evolved regarding the soil–crop–water regime as affected by climatic, physiologic, and soil edaphic factors. These conceptual developments have led to technical innovations in water control that have made possible the maintenance of near-optimal conditions of soil moisture, aeration, and nutrients throughout the growing season.

Foremost among the innovations are techniques of high-frequency, low-volume, partial-area applications of water and of fertilizers directly into the rooting zone at rates calibrated to satisfy crop needs. Properly applied, these irrigation methods can raise yields while minimizing waste (by runoff, evaporation, and excessive seepage), reducing drainage requirements, and promoting the integration of irrigation with essential concurrent operations such as fertilization and pest control. The use of brackish water has become more feasible, as has the use of sandy, stony, and steep lands previously considered unirrigable. Additional potential benefits include increased crop diversification and cropping intensity (i.e., the number of crops that can be grown in succession each year).

The traditional method of irrigation consisted of flooding the land to some depth with a large volume of water so as to saturate the soil completely, then waiting some days or weeks until the moisture stored in the soil was nearly depleted before flooding the land once again. In this low-frequency, high-volume, total-area pattern of irrigation, the typical cycle consisted of periods of excessive moisture alternating with periods of insufficiency. Optimal conditions occurred only briefly in transition from one extreme condition to the other.

In contrast, the newer methods of irrigation are designed to apply a small, measured volume of water at frequent intervals precisely to where the roots are concentrated. The aim is to reduce fluctuations in the moisture content of the root zone by maintaining optimal conditions continuously, without subjecting the crop either to oxygen stress from excess moisture or to water stress from lack of moisture. Moreover, applying the water to the base of the plants (at the soil surface or just below it, along the crop rows) has the effect of keeping much of the soil surface between the rows in a dry state, thus helping not only to reduce evaporation but also to suppress the proliferation of weeds.

Since the high-frequency irrigation systems can be adjusted to supply water at very nearly the exact rate required by the crop, the irrigator no longer needs to depend on the soil's ability to store water during long intervals between irrigations. Hence the water-storage properties of the soil, once considered essential, are no longer decisive in determining whether a soil is suitable for irrigation. New lands, traditionally believed to be unsuitable for irrigation, can now be brought into production. Examples are coarse sands and gravels, where moisture storage capacity is very low and where the conveyance and spreading of water by surface flooding would cause too much seepage.

Of particular interest is the method of *drip irrigation* (also called *trickle irrigation*) and its many variants, such as *microsprayer irrigation*. The idea of applying the water slowly, literally one drop at a time, at a rate that is absorbed by the root zone continuously, is not entirely new. However, what has made it practical was the development of low-cost weathering-resistant plastic tubing and variously designed emitter fittings. System assemblies are now available that are capable of maintaining sufficient pressure in thin lateral tubes to ensure uniform discharge of water throughout the field, with a minimum of clogging. A variant of the system is the subsurface placement of perforated or porous tubes that can ooze water continuously with practically no loss due to evaporation.

The application systems described have been supplemented by ancillary equipment such as filters, timing or metering valves (enabling the irrigator to predetermine the quantity, duration, and frequency of irrigation), and even equipment to inject fertilizers and pesticides into the water supply. Mechanical clogging by suspended particles in the water supply can be prevented by proper filtration, whereas chemical clogging by precipitating salts and biological clogging by algae can be reduced by slight acidification and algicidal treatment of the water. Numerous trials in varied locations have resulted in increased yields of both orchard and field crops (perennial as well as annual), particularly in adverse conditions of soil, water, and climate. Drip irrigation is also applied widely for greenhouses and gardens and lends itself readily to labor-saving automation.

A variant of high-frequency irrigation applicable to large-scale mechanized farms is the so-called *center-pivot irrigation system*. It consists of a central vertical pipe that delivers water to a horizontal rotating tube fitted with sprinklers or drippers. The horizontal tube circles continuously, and can thus irrigate a large area automatically. The speed of rotation can be adjusted so that every spot of land (and every plant) receives an increment of irrigation at high frequency (e.g., several times daily) throughout the growing season. The water-application rate can be adjusted according to the variable (weather-determined) evaporative demand and to the stage of crop growth. The circular pattern of center-pivot irrigation systems can be observed from the air by airline passengers flying over large sections of the U.S. Great Plains.

Drainage Requirements

The term *soil drainage* refers in a general sense to the outflow of water from soil. More specifically, it can serve to describe the artificial removal of excess water or the set of management practices designed to prevent the occurrence of excess water. The removal of free water wherever it tends to accumulate over the soil surface, achieved by appropriately shaping the land surface, is termed *surface drainage*. The removal of excess water from within the soil, generally by lowering the water-table or by preventing its rise, is termed *groundwater drainage*, and it is an integral aspect of sustainable soil management.

Groundwater drainage is generally achieved by means of ditches, pipes, or "mole channels," into which groundwater flows as a result of the hydraulic pressure gradients existing in the soil. The drains themselves are made to direct the excess water, by gravity or by pumping, to an outlet, which may be a stream, a lake, an evaporation pond, or the sea. In some places, drainage water may be recycled, or reused, for agricultural, industrial, or ecological purposes.

Water-Use Efficiency

Any concept of efficiency is a measure of the output obtainable from a given input. Irrigation efficiency, or water-use efficiency, can be defined in different ways, however, depending on the nature of the inputs and outputs considered. For example, one can define as an economic criterion of efficiency the financial returns in relation to the money invested in the installation and operation of a water-supply system. The difficulty is that costs and prices fluctuate from year to year and vary widely from place to place, so they are not universally comparable. Another criterion for the relative merit of an irrigation system is an agronomic one, namely the added yield resulting from irrigation per unit of land area, or per unit volume of water applied.

A widely used expression of efficiency is the *crop water-use efficiency*, which is defined as the amount of vegetative dry matter produced per unit volume of water taken up by the crop from the soil. Because most of the water taken by plants in the field is transpired, while generally only a small amount is retained, the plant water-use efficiency is in effect the reciprocal of what had long been known as the *transpiration ratio*, defined as the mass of water transpired per unit mass of the dry matter produced by the plants.

What irrigation engineers term *irrigation efficiency* is defined as the net amount of water added to the root zone divided by the amount of water taken from some source. As such, this criterion of efficiency can be applied to large regional projects, to individual farms, or to specific fields. In each case, the difference between the amount withdrawn from the source and the net amount of water added to the root zone represents the loss incurred in conveyance and distribution.

The *agronomic efficiency of water use* (WUE) can be defined as follows:

$$WUE = P/W \tag{1}$$

where P is crop production, either in terms of total dry matter or the marketable product, and W is the volume of water applied.

Since only a fraction of the applied water is actually absorbed and used by the crop, we need to consider the various components of the denominator W as follows:

$$W = R + D + E_d + E_s + T_w + T_c \tag{2}$$

Here R is the volume of water lost by runoff from the field, D is the volume drained out of the root zone by deep percolation, E_d is the volume lost by evaporation during delivery and application of water to the field, E_s is the volume evaporated from the soil, T_w is the volume transpired by weeds, and T_c is the volume transpired by the crop. All these volumes pertain to the same unit area and time period.

Water-use efficiency can be maximized by increasing crop yield (P) and by reducing the amount of water use (W). Crop yields can be increased by using high-potential varieties that are well adapted to the local soil and climate and by optimizing agronomic practices. The latter includes proper timing and performance of planting and harvesting, tillage, fertilization, and pest control. Of the items included in the term W, the one that generally should not be reduced is the transpiration by the crop. In the open field, little can be done to limit transpiration without curtailing growth and yield. This is because the unrestricted absorption of carbon dioxide, required to maximize photosynthesis, depends on the stomates remaining open, a condition that generally allows unrestricted transpiration as well. (Differences do exist, however, between C3 and C4 plants with respect to their physiological responses to water stress.)

Plant diseases and pests, as well as competing weeds and deficient fertility of the soil, may depress yields without a proportionate decrease in water use. All management practices can thus influence water-use efficiency, and none can be considered in isolation from the others. Given the expected increase in the world's population and the necessity to alleviate human deprivation while maintaining biodiversity and environmental quality, and given the limited availability of suitable land and water resources, there is an imperative need to improve the efficiency and ensure the sustainability of soil and water management in all regions of agricultural production.

Contemporary Issues

Several problems beset the present and threaten the future of agriculture. The first of these is the progressive expropriation and degradation of land areas and their ecosystems by the voracious demands of expanding populations, most especially in fragile tropical regions. The alternative to this threat is the intensification of production via sustainable resource management of favorable lands, coupled with the release of degradable lands and restoration of their naturally functioning ecosystems. The necessary intensification of land use must be based on the protection of soil and water resources against abuse and degradation by means of effective methods of conservation and quality enhancement.

Important advances have been made in recent years in the development of efficient soil management, including conservation tillage (minimum tillage, or even zero-tillage), precision application of nutrients to prevent contamination and eutrophication of the environment, maintenance of protective mulch coverage over the land surface to minimize evaporation and erosion by wind and rain, and the practice of agroforestry to restore the biotic community and its ecosystem services.

Dramatic advances have been made in the practice of irrigation, including the high-frequency, low-volume application of water precisely calibrated to answer the time-variable and space-variable needs of growing crops while minimizing the waste of water by evaporation, runoff, and excessive seepage (as well as the restriction of root-zone aeration, leaching of nutrients, and the scourge of salination) which have beset irrigated agriculture throughout history.

Advances have also been made in the control of the aerial environment of crops, by means of plastic covers to control and optimize light intensity and composition, temperature, atmospheric humidity and composition (e.g., CO_2 concentration), as well as to control fungal and insect infestation. The objective of all such environmental measures, coupled with genetic improvement of the yield potential of crops, must be to greatly increase yields per unit of land area and of resource investment (including energy), while ensuring the integrity and sustainability of the larger environment.

2

Challenging Balance between Productivity and Environmental Quality: Tillage Impacts

D.C. Reicosky, T.J. Sauer, and J.L. Hatfield

The increasing pressure to provide food security, enhance environmental quality, and address societal problems creates challenges for agriculture and requires we consider how to change our current systems to become more sustainable. Wiebe (2003) stated, "Not only is the contemporary food system inherently unsustainable, increasingly it is damaging the environment." There have been adverse effects in all parts of the world on soils, water, and biodiversity. Poor management of our agricultural systems has contributed to human-induced climate change, and, in turn, human-induced climate change threatens agricultural productivity. In many developed countries, access to quality food is taken for granted, and farmers and farm workers are poorly rewarded for acting as stewards of the Earth's land area used for agricultural production. There is little emphasis on the conservation ethic. More troubling, the environmental degradation caused by intensive agriculture will likely worsen as the global population grows to eight or ten billion in the next three decades.

Modern agriculture is no longer approached as a single issue and is a business that includes far more than just production of food. We have to learn how to pay farmers to not only produce food, animal feed, fiber, and biofuel, but to value the environmental services they impact during crop production. We must consider the environmental issues of biodiversity and water, the economic issues of marketing and trade, and the social concerns of gender and culture. All of this must be done in an economically, environmentally, and socially sustainable manner. We need a fundamental reevaluation of agricultural knowledge, science, and technology transfer to achieve a sustainable global food system. Challenges include giving farmers better access to knowledge, technology, and credit and bringing the necessary information and infrastructure to rural areas.

Much of the environmental damage from agriculture is directly related to the intensive tillage methods used in present-day agricultural production systems. The environmental damage takes a number of forms: erosion and salinization of soils, deforestation as more land is brought into cultivation, fertilizer runoff that ultimately creates enormous "dead zones" around the mouths of many rivers, loss of biodiversity, fresh water scarcity, and agrochemical pollution of water and soil (Lal et al., 2007).

D.C. Reicosky, USDA-ARS, North Central Soil Conservation Research Lab, 803 Iowa Ave., Morris, MN 56267 (don.reicosky@ars.usda.gov); T.J. Sauer and J.L. Hatfield, USDA-ARS National Laboratory for Agriculture and the Environment, 2110 University Boulevard, Ames, IA 50011. The USDA is an equal opportunity provider and employer.

doi:10.2136/2011soilmanagement.c2

In short, we in developed countries create unprecedented abundance while ignoring the long-term consequences of our actions. This is reminiscent of previous agricultural societies—the Greeks, Babylonians, and Romans—who all destroyed soil and habitat in their efforts to feed growing urban populations and collapsed as a result (Lowdermilk, 1953; Gebregziabher et al., 2006; Lal et al., 2007). Historical lessons from impacts of intensive agriculture and plow tillage provide modern civilization learning opportunities that can lead to a brighter future. Hopefully, we can quickly learn and act to avoid repeating history.

Objective

Agriculture uses natural resources—sun, soil, water, and air—to produce food. We in developed countries still have a poor understanding of the properties and conditions of these critical resources. Intensive agriculture can lead to deterioration in the quality of these natural resources and their ability to support food production for an expanding population. The consequences of intensive agriculture on environmental quality require reevaluation and an enhanced educational effort to bring together a better understanding of the unintended human impact on our food production systems (Tinker, 1998; Wiebe, 2003). The contemporary buzzword "sustainable development" has become overworked and needs to be related to productivity and environmental quality (Du Pisani, 2006). The

objective of this chapter is to identify key issues that need to be addressed to maintain agricultural production for the expanding population with minimal impact on environmental quality by addressing agricultural, social, economic, and political issues as they interact to provide a quality of life acceptable to all. Emphasis will be placed on the importance of soil management, as illustrated in Fig. 2|1, particularly the unintended consequences of tillage, as the controlling factor in maintaining this delicate balance between agricultural productivity and environmental quality (Lal, 1998; Grandy et al., 2006; Robertson and Swinton, 2005; Sanchez et al., 2004; Sandor and Eash, 1995; Troeh et al., 1999).

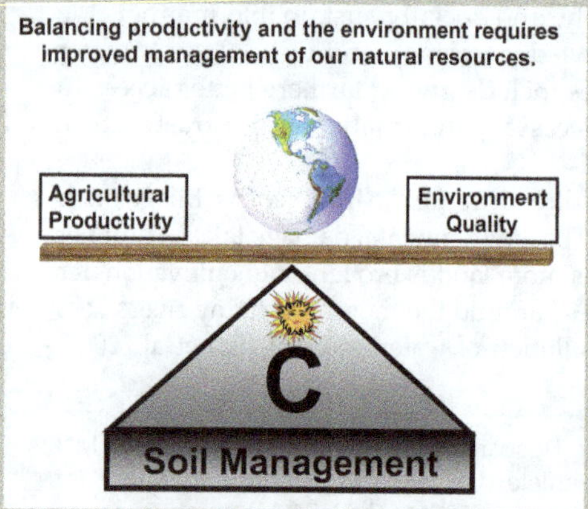

Global Population Growth and Land Available for Food Production

One way to increase food production is to increase the amount of land devoted to agricultural production. Agriculture has a finite amount of land suitable, with today's technology, to support the ever-increasing global population and the accompanying required increase in the capacity of food production systems. About 71% of the Earth's surface area is covered by oceans, and 29% (148.9 million km^2) is land area (Weast, 1981). Ice-free lands include desert and wastelands, forests and savannas, pasture and rangeland, swamps, bogs and lakes, and urban areas (FAOSTAT, 2009; Weast, 1968, 1981). As of 2006, about 37% of Earth's land area was used for agriculture. About one-third of this area, or 11% of Earth's total land (~163 million ha), is used for arable crops. The balance, roughly 26% of Earth's land area, is pastureland, which includes cultivated or wild forage crops for animals and open land used for grazing (FAOSTAT, 2009).

Population growth is expected to increase, and the world population is projected to reach ten billion by 2050, which decreases the per capita arable land. The estimated global population in 2008 was 6.707 billion people and is presently expanding at the rate of ~1.2% per year (U.S. Census Bureau, 2007). These numbers indicate that annual individual food and living requirement must be grown on

Fig. 2|1. Our world rests on balancing agricultural productivity and environmental quality through improved soil management and carbon cycling.

~0.21 ha of arable land. This is the same area as a square lawn area 46 m on each side. If we include the area of permanent crops and permanent pastures as an area to produce food and living space, then each individual has ~0.74 ha. These estimates are similar to those of Ramankutty et al. (2002), who stated, "the cropland base diminished from ~0.75 ha person^{-1} in 1900 to ~0.35 ha person^{-1} in 1990. This loss of croplands was not globally uniform, and more than half the world's population, living in developing nations, lost nearly two-thirds of their per capita cropland base." More intensive agricultural production will have to meet the increasing food demands for this increasing population, especially because of an increasing demand for land area to be used for biofuels. Thus, the long-term soil degradation as a result of intensive plow tillage must be addressed, and heeding its implications for long-term food security must significantly influence our agricultural management practices to protect soil quality for our expanding population. Overall, the global food production system is increasingly vulnerable to regional disruptions because of our increasing reliance on expensive technological options related to fossil fuels to increase agricultural production.

The FAO (FAOSTAT, 2009) categories of agricultural land (total of 4,967,579,500 ha) may not be precise quantitatively because they do not reflect land that has gone out of production or other land that may be potentially used for future production. The potential for increased production on good-quality cropland is considerably more limited; however, there does appear to be some potential to bring new land under cultivation, even though it may be a relatively small amount. Bringing new land under production must be balanced by losses of land area used for production. As the population grows and urban areas expand, some farmland is paved over. The impact of urban sprawl is obvious in many developed countries, especially in the United States. On a worldwide basis this sprawl takes up only 3% of the land area, but it will eventually have an effect on global agriculture (Heimlich and Anderson, 2001; Gardner, 1996; Buringh, 1989).

The global climate change attributed to agriculture may result in some redistribution of agricultural production more favorable to the crop plants. The primary concern of many experts is that land currently in production, may eventually be degraded by soil loss or contamination to such a degree that can be no longer used to grow crops (Larson et al., 1983; Lal, 1989, 1998; Pimentel et al., 1995; Lal et al., 1998). Gardner (1996) stated, "While the world's major grain producers have over expanded into marginal land in recent years, damaging large areas of land in the process, they now are pulling back the land that can be cultivated with a resulting loss in grain production." However, these definitions and estimated areas indicate the relatively small land area per capita available for food production to cope with the expanding global population.

From the global soil map and soil climate GIS information, Beinroth et al. (1994) stated that "about 29.45% of the Earth's ice-free land surface is too dry for sustainable human habitation). Advanced water and energy supply and irrigation techniques have enabled some use of these arid lands. About (15.46%) of the land is the cold tundra zone, which are not easily amenable to normal agriculture." Eswaran et al. (1997) noted that "there are other constraints, which prevent the use of soils for agriculture. Saline and alkaline soils, for example, occupy 2.4% of the land surface, and soil acidity affects 14.1% of the total land. There are sloping lands, sandy soils, soils with low water and/or nutrient-holding capacity, soils with high organic matter (peats), etc. Some limitations are considered permanent or cannot be corrected by low to medium-level inputs."

Smil (1987) stated that "a country with less than 0.07 ha of arable land per person cannot feed its population in the absence of very intensive agriculture. This is a population supporting capacity of about 14 persons per ha of land, which itself is an unrealistically high number. The fossil energy inputs required to produce sufficient food at this level would be excessively high for any meaningful output. With countries that have per capita land area numerically close to 0.07 ha; this does not imply that their agriculture production is designed to support 14 or more persons. The concern is the 0.07 ha estimate is now widely used by United Nations organizations as an optimistic reference for evaluating agricultural land's carrying capacity."

Effect of Climate Change on Agriculture

The IPCC report (IPCC, 2007) stated, "Global environmental change concerns us all." Scientists have assembled the evidence for climate change and emphasized its anthropogenic causes. In a remarkably short time, scientists have concluded that warming of the climate system is "unequivocal" and that there is a "very high confidence that the globally averaged net effect of human negativity since 1750 has been one of warming." The much greater problem for farmers is destabilization of weather patterns. We face not just a warmer climate, but climate chaos: droughts, floods, and stronger storms in general (hurricanes, cyclones, tornadoes, hail storms)—in short, unpredictable weather of all kinds. Farmers depend on relatively consistent seasonal patterns of rain and sun, cold and heat; a climate shift can spell the end of farmers' ability to grow a crop in a given region, and a single storm can destroy an entire year's production. Eswaran et al. (1997) concluded, "Environmental degradation, water shortages, salinization, soil erosion, pests, disease, and desertification all pose serious threats to our food supply, and are made worse by climate change. But many of the conventional ways used to overcome these environmental problems further increase the consumption of finite oil and gas reserves and environmental degradation as we attempt to maintain agricultural productivity for the expanding population." Science may lay out the possibilities but cannot affect the solutions that require social and economic interaction and policy implementation.

Global warming can affect crop yields in a variety of ways, both positively and negatively (Rosenzweig and Hillel, 1995). The increase in atmospheric carbon dioxide concentration increases the efficiency of photosynthesis and thereby enhances plant growth and increases water-use efficiency. The higher temperatures associated with global warming will increase the length of the growing season, and thus, agricultural production may become feasible in areas presently too cold. The increased temperatures may cause plants to mature faster, but any yield loss may be offset using double cropping techniques. Another concern is that pests and diseases often thrive in higher temperature conditions and may contribute to lower yields. Using models, Rosenzweig et al. (1993) estimated that if temperatures increased by 2°C, then maize and rice yields will increase about 8%. One major negative is the frequency of extreme meteorological events such as unseasonable frosts, hurricanes, tornadoes, heavy rainstorms, and droughts, which can all disrupt crop production and cause lower yields.

Another concern is that global warming may cause agricultural production to shift from one geographical area to another. There may be production time lost as we adapt to this shift in climate change, and another secure food source may be required during this transition stage. There are concerns that problems in establishing the institutional and infrastructure needed to move the food from the production area to the consuming areas can be considerable (Rosenzweig and Hillel, 1995). Rosenzweig and Hillel (1995) noted, "While the overall, global impact of climate change on agriculture production may be small, regional food deficits may increase, due to problems of distributing and marketing food to specific regions and groups of people."

A recent comprehensive report covering this subject is one of a series of products from the U.S. Climate Change Science Program (CCSP, 2008), with special emphasis on agriculture covered by Hatfield et al. (2008). This report builds on extensive scientific literature and a series of recent assessments of the historical and potential impacts of climate change and climate variability on managed and unmanaged ecosystems and their constituent biota and processes. The general time horizon for this report is from the recent past through the period 2030–2050, although longer-term results out to 2100 are also considered. Assessment of the effects of climate change on U.S. agriculture, land resources, water resources, and biodiversity are provided. These insights are mainly focusing on effects of climate on cropping systems, pasture and grazing lands, and animal management. The report discusses the nation's ability to identify, observe, and monitor the stresses that influence agriculture, land resources, water resources, and biodiversity, and evaluates the relative importance of these stresses and how they are likely to change in the future.

Briefly, the CCSP (2008) report summarizes the effects of climate change on U.S. land and water resources, agriculture, and biodiversity and provides the following conclusions:

1. Climate changes—temperature increases, increasing CO_2 levels, and altered patterns of precipitation—are already affecting U.S. water resources, agriculture, land resources, and biodiversity.

2. Climate change will continue to have significant effects on these resources over the next few decades and beyond.

3. Many other stresses and disturbances are also affecting these resources in the form of multiple environmental drivers (e.g., land use change, nitrogen (N) cycle changes, point and nonpoint-source pollution, wildfires, invasive species) that are also changing.

4. Climate change impacts on ecosystems will affect the services that ecosystems provide, such as cleaning water and removing carbon (C) from the atmosphere, but we do not yet possess sufficient understanding to project the timing, magnitude, and consequences of many of these effects.

5. Existing monitoring systems, while useful for many purposes, are not optimized for detecting the impacts of climate change on ecosystems and natural resources.

Even under the most optimistic CO_2 emission scenarios, important changes in sea level, regional and subregional temperatures, and precipitation patterns will have profound effects on both agricultural production and environmental quality (Smith et al., 2008). They state:

Management of water resources will become more challenging. Increased incidence of disturbances such as forest fires, insect outbreaks, severe storms, and drought will command public attention and place increasing demands on management resources. Ecosystems are likely to be pushed increasingly into alternate states with the possible breakdown of traditional species relationships, such as pollinator/plant and predator/prey interactions, adding additional stresses and potential for system failures. Some agricultural and forest systems may experience near-term productivity

increases, but over the long term, many such systems are likely to experience overall decreases in productivity that could result in economic losses, diminished ecosystem services, and the need for new, and in many cases significant, changes to management regimes.

The CCSP report addressed the broad agricultural problems through consideration of cropping systems, pasture and grazing lands, and animal management (Hatfield et al., 2008). They state:

"The many U.S. crops and livestock varieties are grown in diverse climates, regions, and soils. No matter the region, however, weather and climate factors such as temperature, precipitation, CO_2 concentrations, and water availability directly impact the health and well-being of plants, pasture, rangeland, and livestock which all translate to productivity. For any agricultural commodity, variation in yield between years is related to growing-season weather; weather also influences insects, disease, and weeds, which in turn affect agricultural production."

Hatfield et al. (2008) stated:

The primary conclusions for the agricultural sector were:

- With increased CO_2 and temperature, the life cycle of grain and oilseed crops will likely progress more rapidly. But, as temperature rises, these crops will increasingly begin to experience failure, especially if climate variability increases and precipitation lessens or becomes more variable.

- The marketable yield of many horticultural crops—e.g., tomatoes, onions, fruits—is very likely to be more sensitive to climate change than grain and oilseed crops.

- Climate change is likely to lead to a northern migration of weeds. Many weeds respond more positively to increasing CO_2 than most cash crops, particularly C3 "invasive" weeds. Recent research also suggests that glyphosate, the most widely used herbicide in the United States, loses its efficacy on weeds grown at the increased CO_2 levels likely in the coming decades.

- Disease pressure on crops and domestic animals will likely increase with earlier springs and warmer winters, which will allow proliferation and higher survival rates of pathogens and parasites. Regional variation in warming and changes in rainfall will also affect spatial and temporal distribution of disease.

The importance of water in agricultural production cannot be overstated. While a very critical issue in balancing agricultural productivity and environmental quality, climate change impacts on water deficits are a critically important aspect of agricultural production (Postel, 1989). Agriculture's demand and use of water continues to expand to meet the food needs of our expanding population. On land, the amount and frequency of rainfall determine the success of crops, as well as the survival of natural and agricultural ecosystems. Precipitation varies by both season and geographic area, with large impacts on both agricultural productivity and environmental quality. As one result, highly specialized ecosystems have developed, from deserts to rain forests. In the event of global warming, regional rainfall patterns may shift. Similarly, the removal of forest cover for agricultural production may alter rainfall distribution because of reduced evaporation of water from plants. Changes in patterns of precipitation could have dramatic effects, positive or negative, on all life on Earth.

Irrigation has underpinned the advancement of human societies for several thousand years by enabling farmers to apply water when and where needed, turning many of the Earth's sunniest, warmest, and most fertile lands into important crop-production regions (Postel, 1989). Today, farming accounts for some 70% of global water use. In recent years, however, forces have begun to slow irrigation's expansion and to raise questions about agriculture's heavy claim on the world's rivers, streams, and underground aquifers. As air temperatures rise and rainfall patterns shift with the onset of climate change, water supplies will increase in some areas and decrease in others. Whatever the outcome for particular regions, adjusting the global irrigation base to the changes in water availability will be costly. Moving rapidly to greater efficiency and equity in the use of water is the surest way to avert shortages and lessen irrigation's ecological toll. Stepped-up research into breeding strains of crops that are more salt tolerant and drought resistant would help prepare for a likely future of increased shortages and hotter, drier climates in some important food growing regions. The delicate balance between agricultural productivity and environmental quality rests on balancing the water budgets of many countries as much as slowing the rapidly growing populations (Postel, 1989).

Environmental Issues: Resource Management versus Resource Conservation

Soil Quality and Soil Degradation

Soil quality is the fundamental foundation of environmental quality. Soil quality is largely governed by soil organic matter (SOM) content, which is dynamic and responds effectively to changes in soil management, primarily tillage and C input (Lal, 1987, 2003). Maintaining soil quality through carbon management can reduce problems of land degradation, decreasing soil fertility, and rapidly declining production levels that occur in large parts of the world needing the basic principles of good farming practice (Lal, 1989, 1995, 2003).

Soil quality is a way of expressing the potential productivity of the soil based on qualitative attributes. While there are many attributes of soil quality, the one that contributes the most to "good properties" of the soil is the carbon content. While a typical soil carbon content can have a range of up to 5% in mineral soils, the small amount of carbon is very critical for optimizing the soil physical, chemical, and biological properties. The overlapping interaction in the role of soil carbon and impacting soil properties is illustrated in Fig. 2|2. The critical role of soil carbon in maintaining these properties is also linked to agricultural global warming issues that may have contributed to the increase in atmospheric CO_2 (Lal, 1995, 1999a, 2003, 2004).

There is a global and urgent need for restoring the quality of degraded soils.

The urgency is highlighted by large areas of land degraded to some extent (Lal, 1987, 1989, 1998; Oldeman et al., 1991), and the need is underscored by the world's finite soil resources and high population pressure. The per capita arable land area is rapidly shrinking, especially in densely populated China and India, where arable land resources are limited and the soil degradation risks are alarmingly high. Critical strategies for soil restoration must be identified on the basis of the concepts of soil resilience. Soil resilience refers to the ability of soil to resist or recover from an anthropogenic or natural disturbance (Greenland and Szabolc, 1994; Lal, 1997). Soils differ in resilience depending on several inherent characteristics. Highly resilient soils have high buffering capacities and high rates of recovery or restoration. Fragile soils, in contrast, are unstable, cannot recover to the initial state, and may have lost some of their initial characteristics (Greenland and Szabolc, 1994). Soil resilience is closely linked to soil quality through soil carbon, because resilient soils have high soil quality and vice versa. Therefore, soil resilience can be assessed from the rate of change of soil quality with time (Lal, 1994).

Knowledge of soil resilience is essential to adopting appropriate restorative measures. The choice of restoration techniques depends on numerous factors, including the on-site and off-site impacts, socioeconomic issues of the farm household, and institutional support and policy issues. However, biophysical processes affecting soil quality decline are extremely important in the choice of technical measures. Equally or more important are the magnitudes of the past erosion and the current rate of soil erosion (Lal, 1987, 2003). The soil erosion must be controlled while alleviating the soil quality-related constraints caused by the past soil erosion. Because of the complexity of the problem and numerous interacting factors involved, it is appropriate to adopt an agroecosystem approach to soil quality restoration (Lal, 1989).

Contributing to a better understanding of tillage impacts on the environment, Warkentin (2008) defined soil tilth as the "condition" of a soil in relation to the habitat provided for seed germination and plant growth. Soil structure of surface horizons was perceived for many years as tilth of the seedbed and plowing or tillage to achieve it.

The early social and technical history of tilth was largely the history of how to plow and how to plow well (Lal et al., 2007). With the advent of soil science laboratories, the dominant concern soon became measurement of static properties expected to be related to tilth and stability of structure: aggregate-size distribution, soil bulk density to calculate porosity, grain-size distribution, shape and size of aggregates, and stability of aggregates (Mikha and Rice, 2004; Olchin et al., 2008; Shepherd et al., 2001; Blanco-Canqui and Lal, 2004; Warkentin, 2008). The importance of organic matter and clay content in soil structure was generally recognized; however, all these soil properties were related to changes in soil caused by cultivation, and cultivation impacts on tilth and crop yield (Tisdall and Oades, 1982). Tisdall (1996) stated, "the concept of hierarchical arrangement of different aggregate sizes, and the bonds responsible for stability, drew attention to different void sizes and to the soil functions of each. The unique role of soil in ecosystems led to considering soil structure as defining the habitat for soil biological, physical, and chemical functions such as decomposition, water routing, etc. Structure is now a concept centered on voids, and the term soil architecture became more appropriate; the spaces and surfaces of the spaces were more important than the solids of walls and roof." Today's soil structure research is again productive in concepts and applications and encourages less intensive tillage (Jastrow et al., 1996; Six et al., 1998, 2002; Warkentin, 2008).

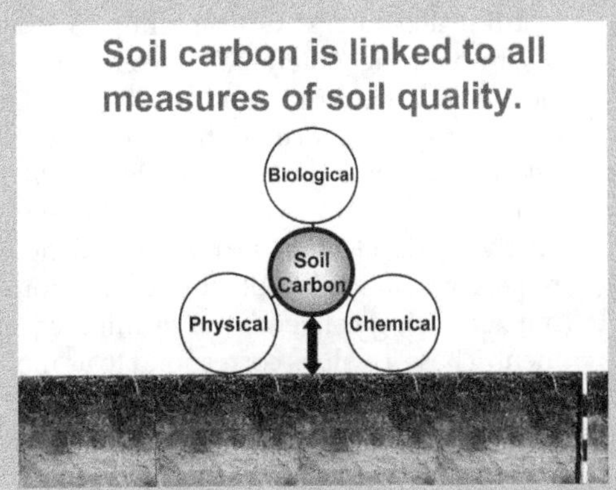

Fig. 2|2. The interdependence of soil physical, chemical, and biological properties on soil carbon.

Soil Erosion

One of the most degrading forces acting on the soil is erosion. Soil erosion is caused by intensive agricultural production. Soil that is loosened by tillage is more easily transported by wind or water, increasing the rate of erosion. In annual production systems, the soil surface is left bare for months at a time, leaving a soil susceptible to erosion (Lal, 1987, 1995, 1998, 1999b, 2003).

Lal (1998) stated

Principal processes that lead to decline in crop yield due to erosion are: (1) reduction in effective rooting depth, (2) loss of plant nutrients and soil organic carbon (SOC), (3) loss of plant available water and available water-holding capacity (AWC), (4) loss of land area, and (5) damage to seedlings. There are also numerous indirect effects of accelerated soil erosion that are primarily due to the loss of resources during the season, for example, loss of fertilizer and agrichemicals, delayed sowing or re-sowing, supplemental irrigation required due to the loss of water runoff, additional machinery cost, etc. Other soil degradation processes, including compaction, acidification, toxic contamination, and salinization largely relate to specific regions in some countries.

On a global scale, the on-site effects of soil erosion are undermining the productivity of about 33% of the world's cropland (Pimentel, 1993; Pimentel et al., 1987, 1995; Brown and Young, 1990).

In the 1950s, when the Soil Conservation Service (now known as the Natural Resources Conservation Service) began defining "tolerable" rates of soil erosion from agricultural land, hardly any data on rates of soil formation were available. The agency thus determined the so-called soil loss tolerance values, or "T values," on the basis of what farmers could do to reduce erosion without "undue economic impact" using conventional farming equipment. These T values correspond to as much as 25.4 mm of erosion in 25 yr. But, recent research has shown the soil erosion rate to be far faster than the rate at which soil rebuilds (Montgomery, 2007; Heimsath et al., 1999). What was once "tolerable" is now "intolerable" based on what we have learned about soil formation.

Soil erosion is a major global issue because of its adverse economic and environmental impacts (Lal, 1987, 1998, 2003; Pimentel et al., 1987, 1995). Lal (1998) stated:

Economic impacts on productivity may be due to direct effects on crops/plants on-site and off-site, and environmental consequences are primarily off-site due either to pollution of natural waters or adverse effects on air quality due to dust and emissions of greenhouse gases. Off-site economic effects of erosion are related to the damage to civil structures, sedimentation of water ways and reservoirs, and additional costs involved in water treatment. On-site effects of erosion on agronomic productivity are assessed with a wide range of methods, which can be broadly grouped into three categories: agronomic/soil quality evaluation, economic assessment, and knowledge surveys.

Agronomic methods involve greenhouse and field experiments to assess erosion-induced changes in soil quality in relation to productivity. There is also a need to assess on-site impact of erosion in relation to soil loss tolerance, soil life, soil resilience or ease of restoration, and soil management options for sustainable use of soil and water resources (Lal, 1998). Restoration of degraded soils is a high global priority (Oldeman et al., 1991). If about 1.5×10^9 ha of soils in the world prone to erosion can be effectively managed to control soil erosion, it would improve air and water quality, sequester C in the soil at the rate of about 1.5 Pg yr^{-1}, and increase food production (Lal, 1995, 1998). Erosion-caused losses of food production are most severe in developing countries and elsewhere in the tropics.

Soil erosion is a complex process that depends on soil properties, ground slope, vegetation, and rainfall amount and intensity. Changes in historical land use, associated with intensive tillage, are widely recognized as accelerating soil erosion, and it has been long recognized that erosion in excess of soil generation would eventually result in decreased agricultural potential (Lal, 1998; Lal et al., 2007; Pimentel et al., 1987). Montgomery (2007) concluded that "erosion rates from conventionally plowed agricultural fields averaged 1–2 orders of magnitude greater than rates of soil production, erosion under native vegetation, and long-term geological erosion." He concluded that "hill slope soil production and erosion evolve to balance geologic and climate forcing, whereas conventional plow-based agriculture increases erosion rates sufficiently to prove unsustainable." Montgomery (2007) concluded further that "no-till agriculture produces erosion rates much closer to soil production rates and therefore could provide a foundation for sustainable agriculture." No-till, zero till, and direct seeding involves leaving crop residue on the soil surface instead of plowing it under. The seeds are inserted directly into the soil by a specialized drill. The layer of organic matter left on the soil surface acts as a surface mulch to promote infiltration and reduce runoff and erosion. In the 1970s, few farmers in the United States used no-till techniques, but by 2000, 16% of the cultivated plants in the United States used no-till methods (Derpsch, 2001). Although no-till practices have been increasingly adopted in North and South America, approximately 5% of the global cropland is managed by using no-till (Lal et al., 1998). Only a fraction of U.S. no-till cropping systems are permanent no-till. The imbalance between agricultural soil loss and formation under both native vegetation and geologic time scale is that, given enough time, continued soil loss will become a critical problem for global agricultural production under conventional upland farming practices. With little new land that could be brought under sustained cultivation (FAO-STAT, 2009; Brown, 1994; Brown and Young, 1990; Chen, 1990; Fischer and Heilig, 1998; Gardner, 1996; Greenland et al., 1998; Young, 1999) and the projected increase in the global population (Ehrlich et al., 1993), the issue of long-term conventional (high-intensity tillage) agricultural sustainability will become an increasingly pressing issue.

Soil translocation from tillage operations has been identified as a source of soil erosion, which at specific landscape positions can be greater than the soil loss tolerance levels (Lindstrom et al., 1992; Govers et al., 1994; Lobb et al., 1995, 2007; Poesen et al., 1997). Soil translocation or tillage erosion is the net movement of soil downslope in response to the action of mechanical implements. The soil is not directly lost from the fields by tillage translocation or tillage erosion; rather, it is moved away from the convex slopes and deposited in concave slope positions. The loosened soil is redistributed as a result of the tillage tool and gravitational forces interacting. Lindstrom et al. (1992) showed that soil movement on a convex slope in southwestern Minnesota could result in a soil loss of approximately 30 t ha^{-1} yr^{-1} from annual moldboard plowing. Lobb et al. (1995) estimated soil loss in southwestern Ontario from a shoulder position to be 54 t ha^{-1} yr^{-1} from a tillage sequence of moldboard plowing, tandem disk harrow, and a C-tine cultivator. In this case, tillage erosion, as estimated through resident cesium-137, accounted for at least 70% of the total soil loss. Tillage speed increases the rate of tillage erosion nonlinearly (Lobb et al., 1995; Lindstrom et al., 2000). Schumacher et al. (1999) concluded that tillage erosion resulted in more soil loss in the shoulder position, while soil loss from water erosion occurred primarily in the mid to lower backslope position. The decline in soil productivity was greater when both processes were combined compared to either process acting alone. The net effect of soil translation from the combined effects of both tillage and water erosion was increased spatial variability of crop yields, which led to a decline in overall soil productivity (Schumacher et al., 1999).

Agriculture's Impact on Greenhouse Gases

Agricultural production also interacts with the environment on a global scale, and especially on global climate change. It is clear that agricultural production creates greenhouse gases (Reicosky et al., 2000) and releases CO_2 during and after tillage (Reicosky and Lindstrom, 1993, 1995; Ellert and Janzen, 1999). If agriculture production grows to keep pace with the food demand, it will likely add further to the greenhouse gas problem. There is a relatively high degree of consensus among scientists that the average temperature of the Earth is increasing. There is some disagreement about the extent to which global warming has already occurred and about the extent to which human activities are responsible for the increase in greenhouse gases that cause global warming. Agriculture is a relatively small player contributing to greenhouse gases (Lal, 1999a). The three main greenhouse gases from agriculture that may cause global warming are CO_2, methane (CH_4), and nitrous oxide (N_2O) (Reicosky et al., 2000). The increased mechanization in intensive agriculture in addition to the petrochemicals used raises concern that energy use in agriculture will become an environmental problem. Chen (1990) reported that agricultural production accounts for only 3.5% of commercial energy use in developed countries and 4.5% in developing countries.

Agriculture affects the condition of the environment in many ways, including impacts on global climate change through the production of greenhouse gases (Robertson et al., 2000). In 2004, the USEPA estimated that agriculture contributed approximately 7% of the U.S. greenhouse gas emissions (in carbon equivalents, or CE), primarily as CH_4 and N_2O. While agriculture represents a small but relevant source of greenhouse gas emissions, it has the potential, with new practices, to also act as a sink, tying up, or sequestering, CO_2 from the atmosphere in the form of soil C (Lal, 1999a). Lal (1999a) stated, "Estimates of the potential for agricultural conservation practices to enhance soil C storage range from 154–368 million metric tons (MMTCE), which compare to the 345 MMTCE of reduction proposed for the U.S. under the Kyoto Protocol." Thus, agricultural production systems can be manipulated for the dual benefits of reducing greenhouse gas emissions and enhancing C sequestration, which contributes to increased productivity and environmental quality.

Tillage versus No-Tillage?

Tillage and Its Role in Crop Production

Tillage has many roles in crop production (Cornish and Pratley, 1987; Titi, 2003; Reicosky and Allmaras, 2003). The concept of tillage systems combines various aspects of tilling, planting, managing residue, and applying pesticides and fertilizers. Because of the number and diversity of components in tillage systems, it is difficult to give any one system a meaningful name or very precise definition (Reeder and Westermann, 2007). Systems can be identified according to their ultimate objective, whether it is conventional or conservation tillage, and sometimes they are described by the primary tillage implement used (e.g., whether it's a moldboard plow or a chisel plow). The name problem often is compounded because the definitions differ between

geographic regions. Different names may be used to identify a similar tillage system in different parts of the country. Listing all the operations in the system results in the most accurate description as described in Reicosky and Allmaras (2003).

Tillage can have a major impact on soil functions and soil quality (Karlen et al.,1994a, 1994b; Reicosky, 1997). Warkentin (2001) discussed how alteration of soils by tillage changes the sustainability of soil functions. Soil tillage presents an enigma in thinking about soil sustainability in ecosystems. There is a several-thousand-year history of gradually increasing the disturbance of ever-greater volumes of soil (Lowdermilk, 1953; Gebregziabher et al., 2006; Lal et al., 2007) that gave the perception these tillage changes increased crop production. Specific soil uses, specific soils, characteristics of the site, and whether short or long-term changes are being evaluated determine the effects of tillage in agroecosystems. The largest effects of tillage are increased recycling rates and long-term decreases in porosity and diversity of habitat. The only benefits of tillage in annual crop production appear to be a temporary improvement of water and oxygen conditions in a seedbed and the destruction of competing plant species (Warkentin, 2001). Environmental concerns such as water quality, diversity of habitat, storage of carbon and nitrogen, and water partitioning are all negatively affected by tillage.

The most widely recognized function of tillage is seedbed preparation. Seed placement requires some form of tillage; even in no-till crop production systems, some soil must be disturbed to place the seed. Tillage has been used for thousands of years to release nutrients from the soil through accelerated mineralization of organic matter and to incorporate nutrients found in manures and crop residues. More recently, equipment has been developed to inject manures, commercial fertilizers, and other amendments into the soil. Tillage, in the form of cultivation, has been used extensively in the past to control weeds and insects. The recent development and use of commercial pesticides has greatly reduced the use of tillage. The widespread adoption of no-till/zero till systems in North America has been largely attributed to the availability of cost-effective glyphosate for weed control. With increasing interest in limiting or eliminating pesticide use in crop production, a return to more intensive tillage systems can be expected if soil erosion concerns are addressed. Tillage is also used to manage soil moisture (e.g., hilling row crops) and soil structure (breaking up soil crusts and alleviating soil compaction). Possibly the best example of a farm management practice which is not thought of as tillage but results in significant soil disturbance is the harvesting of root crops such as potatoes and sugar beets.

More farmers have been adopting no-till farming to capture efficiencies in crop production, saving money, time, and energy; to stop the loss of valuable topsoil by erosion; and to curb the runoff of sediment, fertilizers, and pesticides into rivers, lakes, and eventually oceans. Despite the benefits of no-till, adoption worldwide remains low at less than 7% (Derpsch, 2001). A balanced evaluation of the sustainability of no-till agricultural systems describing advantages and disadvantages is warranted and must address challenges, including different equipment, pest management strategies, crop rotation, and fertility management, all which contribute to a steep learning curve for farmer adoption. The future of no-till farming will address global climate change, population growth, hunger and food nutrition, energy conservation and biofuels, environmental degradation, endangered species, pesticide use, genetically modified organisms and crop diversification (Phillips and Young, 1973). Huggins and Reganold (2008) stated, "No-till is not a cure-all; rather, no-till is a component of a larger vision of sustainable agriculture that is continually evolving and where diversity of farming methods from no-till to organic is healthy." Future no-till farming will need to employ more diverse pest management strategies, including biological, physical, and chemical measures to lessen the threat of pesticide resistance. Greater diversity of economically viable crops would also advance no-till farming and its adoption. There is a need to move away from intensive annual tillage, primarily monoculture farming, such as current wheat- and corn-based production, toward integration of perennial crop production practices, and precision technologies into no-till and conservation tillage systems.

In discussing various soil management practices that impact soil erosion, it is essential to understand the difference between conventional agriculture (conventional tillage) and conservation agriculture (conservation tillage with minimum soil disturbance). There are regional differences in the meaning of "conventional" that need to be clarified when discussing site-specific characteristics. Typically, conventional tillage over the last 30 yr has consisted of moldboard plow, disk harrow, and field cultivator before planting. Reicosky and Allmaras (2003) described the difference between various tillage management systems presently used in North America. In no-till or zero till the soil is left undisturbed from harvest to planting except for nutrient and/or seed injection. Planting or drilling is accomplished in a narrow seedbed or slot created by coulter, row cleaners, disk openers, in-row chisels, or roto-tillers. Weed control is accomplished primarily with herbicides. Cultivation may be used for emergency weed control.

While this definition of no-till/zero till/direct seeding may not be globally accepted, it serves to illustrate the conservation aspects of no-till come from minimum soil disturbance, continuous crop residue cover, and the use of diverse rotations and cover crops to protect the soil surface from erosion forces. The effect of no-till in reducing soil erosion rates that approximate the long-term rates of soil production suggests a more sustainable system (Montgomery, 2007).

No-till has the potential to deliver benefits that are increasingly desirable in a world facing population growth, environmental degradation, rising energy costs, and climate change, among other daunting challenges. The terminology being developed for such systems is *Conservation Agriculture* (CA) (FAO, 2009). Conservation agriculture implies conformity with all three principles supporting CA defined by FAO (2009) as: " 1) minimum soil tillage disturbance, 2) diverse crop rotations and/or cover crops, and 3) continuous plant residue cover." Others are promoting the integration of crop and livestock production and controlled traffic into the vision of CA. The foundation underlying the three main principles is how they interact with and contribute to soil carbon, the primary determinant of soil quality. Conservation agriculture includes concepts of no-till, zero till, and direct seeding as the ultimate form of CA. But no-till is not a cure-all; rather it is part of a larger system, requiring higher-level management, evolving vision of sustainable agriculture, in which a diversity of farming methods is considered healthy. Ultimately, all farmers should integrate all aspects of conservation agriculture with no-till on their farms for sustainability.

Tillage and the Environment

The primary environmental benefits of all soil management practices are, first and foremost, improvements in soil and water quality. Of all farm management practices, tillage may have the greatest impact on the environment. Reicosky (2008) stated:

> Tillage, by affecting crop production, affects the environment: crop productivity affects the production and consumption of CO_2, the production of biomass above and below ground, the uptake of soil water and its transpiration, and the efficiency of cropping inputs such as fertilizer and pesticides. In addition to its effects on crop production, tillage also affects a variety of soil biophysical properties and processes that impact the environment.

Tillage releases CO_2 and mixes soil and crop residue to allow rapid decomposition of SOM (Reicosky and Lindstrom, 1993, 1995; Reicosky, 1997, 2002; Ellert and Janzen, 1999). In this way, tillage is a "double negative," rapidly releasing carbon from the soil and contributing to the increase in the atmospheric CO_2 and an enhanced greenhouse gases effect. Tillage under windy conditions loses soil carbon faster than under low wind speeds (Reicosky et al., 2008). Reicosky et al. (2008) stated that tillage "affects wind, water and tillage erosion, leaching and runoff, greenhouse gas emissions, pesticide sorption and degradation, as well as other biophysical processes. Tillage intensity, by affecting the amount of crop residue on the soil surface and how that residue is distributed on and anchored to the soil, and by affecting the size of soil aggregates and their stability, has a large impact of wind and water erosion." Tillage, through the action of soil disturbance and the downward force of gravity, causes the slow progressive downslope movement of soil, i.e., tillage erosion (Lobb et al., 1995). Soil erosion results in the redistribution of soil within fields and losses from fields. Typically, in cultivated topographically complex landscapes soil loss is most severe on hilltops.

Huggins and Reganold (2008) concluded that "no till has the potential to deliver a host of benefits that are increasingly desirable in the world facing population growth, environmental degradation, rising energy costs and climate change." But they also stated that "no-till is not a cure-all; because such a thing does not exist in agriculture." Huggins and Reganold (2008) considered no-till as part of a larger component of sustainable agriculture in which the different farming methods from no-till to organic and combinations are considered healthy. All farmers should integrate all aspects of conservation agriculture, and no-till if feasible, on their farms for environmental protection (Reicosky and Saxton, 2007a,b; Reicosky, 2008).

Intensive tillage influences SOC dynamics and storage. Studies have shown that the adoption of no-till leads to an accumulation of SOC at or near the soil surface (0–10 cm; see West and Post, 2002; Deen and Kataki, 2003). There have been several meta-analyses and scientific literature reviews on no-till vs. conventional tillage on SOC in world soils (e.g., Six et al., 2002, 2004; West and Post, 2002; Alvarez, 2005), in which various forms of conventional tillage were considered (conventional tillage may involve noninversion primary tillage such as disk plowing or using a heavy cultivator, such as practiced in the Canadian prairies).

Furthermore, reducing tillage increases soil carbon sequestration compared with conventional moldboard plowing (Deen and Kataki, 2003). One of agriculture's main greenhouse gas mitigation strategies is soil carbon sequestration (Lal et al., 1998; Izaurralde et al., 2001) wherein crops remove CO_2 from the atmosphere during photosynthesis, and nonharvested residues and roots are converted to soil organic matter, which is 58% carbon. About one-half of the overall potential for U.S. croplands to sequester soil carbon comes from conservation tillage, including no-till (Dick and Durkalski, 1997; Lal et al., 1998).

Soil carbon sequestration has benefits beyond removal of CO_2 from the atmosphere. No-till cropping reduces fossil fuel use, reduces soil erosion, and enhances soil fertility and water-holding capacity. Beneficial effects of conservation tillage on SOC content may be short-lived if the soil is plowed even after a long time under conservation tillage (Pierce et al., 1994; Gilley and Doran, 1997; Stockfisch et al., 1999; Garcia et al., 2007; Quincke et al., 2007). Stockfisch et al. (1999) concluded that "organic matter stratification and accumulation as result of long-term minimum tillage were completely lost by a single inversion tillage in the course of a relatively mild winter." Several experiments in North America have shown more SOC content in soils of conservation tillage compared to plow till seed beds (Doran, 1980, 1987; Doran et al., 1987; Rasmussen and Rohde, 1988; Tracy et al., 1990; Havlin et al., 1990; Kern and Johnson, 1993; Lafond et al., 1994; Reicosky et al., 1995; Reicosky, 2001a,b). Liebig et al. (2005) reported "continuous cropping and no-tillage resulted in carbon accumulation of 0.27 Mg C/ha/yr, a value specific to the lower rainfall area of the U.S. Northern Plains."

Soil plays a key role in maintaining a balanced ecosystem and in producing quality agricultural products (Organization for Economic Co-operation and Development, 2001). There can be a significant time delay between recognizing soil degradation and developing conservation strategies to maintain soil health and crop productivity. The intensity of rainfall, degree of protective crop cover, slope, and soil type are the controlling factors of water erosion. The process of wind erosion is also controlled by climate (soil moisture conditions), crop cover, and soil type and involves detaching and transporting soil particles (mainly silt and fine sand) over varying distances. Loss of topsoil by any type of erosion also contributes to the loss of nutrients. Soil tillage practices can also contribute to erosion by moving soil on hilly landscapes, removing soil from the hilltop to the bottom (Lobb et al., 1995, 2007).

Without permanent no-till, many of the agronomic and environmental benefits are not realized (Grandy et al., 2006). Years of soil regeneration can be lost to a single inversion tillage event (Stockfisch et al., 1999). Social and agronomic challenges as well as political challenges continue to limit both the extent and rate of no-till development. These challenges should become an agricultural research priority that emphasizes integrated systems approach and long-term dynamics. Interdisciplinary solutions must address these complex problems so that no-till systems can be more extensively adopted and permanently maintained, thereby yielding

their full agronomic, economic, social, and environmental potential. Although some carbon is sequestered (Smith et al., 2001), accelerated water erosion is responsible for net emission of about 1 Gt C yr^{-1} (Lal, 2003). Leaving crop residues after grain harvest increases the carbon content of soil and controls erosion, but the benefits are lost if the biomass is plowed under because microorganisms quickly degrade residue C to CO_2 (Reicosky et al., 1995). Essential nutrients as part of SOC disappear with its depletion. Thus, farmers require more fertilizer, irrigation, and pesticides to preserve yield while water quality can deteriorate when less SOC is available for natural filtering.

No-till agriculture reduces the loss of the SOC pool (Dalal et al., 1995; Sa et al., 2001; West and Post, 2002), with surface residues conserving soil water and inhibiting weeds. Soil C enhancement improves agronomic productivity (Bauer and Black, 1994) and resource-use efficiency of impoverished soils. The beneficial effects of enhanced SOC cannot be fully replaced by increased levels of fertilizer, especially in soils of the tropics (Kanchikerimath and Singh, 2001). No-till, in combination with mulching and crop rotation to enhance the SOC pool (Angers et al., 1995; Jenkinson, 1991; Smith and Powlson, 2000; Drinkwater et al., 1998; Uhlen and Tveitnes,1995), is also a viable strategy for sustainable management of soils of the tropics.

Tillage causes the release of the labile fraction soil organic matter from within the aggregates; plus the incorporation of the aboveground biomass leads to increased SOC availability for decomposition (Shepherd et al., 2001). As result, microbial activity increases, leading to accelerated CO_2 emissions (Reicosky and Lindstrom, 1993; Ellert and Janzen, 1999) and N mineralization rates. Although some of the nitrate produced from tillage is taken up by the plants, the release of this N is often poorly synchronized with plant nitrogen needs, which usually do not peak for eight or more weeks after tillage, making the mineralized N highly susceptible to loss via leaching and denitrification. Associated with these changes in soil aggregation and organic matter availability are increases in soil temperature and oxygen concentration that further stimulate microbial decomposition. This process can be reversed by eliminating tillage, but the recovery of the aggregates and aggregate-associated carbon pools takes several years longer than the destruction that occurs with a single tillage event (Pierce et al., 1994; Stockfisch et al., 1999; Garcia et al., 2007; Grandy et al., 2006; Quincke et al., 2007). Clearly, tillage has an immediate and striking effect on soil fauna and biological processes (Calderón et al., 2001; Jackson et al., 2003), and even intermittent tillage of no-till systems may undermine efforts to restore soil physical and biological processes and sequestered carbon in the systems (Six et al., 2004).

Several years are often required before soil aggregate stability increases in no-till systems. As a result, compaction may limit growth and denitrification rates may be high due to anaerobic sites (Ismail et al., 1994; Six et al., 2004). Grandy et al. (2006) found substantial changes in soil aggregation within 31 d following tillage, suggesting that tillage per se, rather than the indirect effects of bare soils or plant community changes, destroyed soil aggregates. A handful of studies in different geographical regions demonstrated changes in the distribution of organic matter between soil size fractions and depths following tillage of long-term no-till soils (Tiessen and Stewart, 1983; Pierce et al., 1994; VandenBygaart and Kay, 2004). Soil aggregation changes rarely have been studied following this conversion from no-till. Soil aggregation may be the single best indicator of the agronomic and environmental effects of tillage because it influences soil structure, soil permeability, and water-holding capacity, as well as soil organic matter turnover in nutrient cycling (Jastrow et al., 1996; Grandy et al., 2002; Shaver et al., 2003).

Historical data have demonstrated the degradation of accelerated soil erosion in agricultural societies to the extent that episodes of severe soil erosion were associated with the rise and decline of civilization in the Middle East (Lowdermilk, 1953; Gebregziabher et al., 2006; Lal et al., 2007). Montgomery (2007) analyzed historical data and concluded that the erosion rates from conventionally plowed agricultural fields averaged one to two orders of magnitude greater than the rates of soil formation or long-term geological erosion. Losing soil faster than it can be generated is not a sustainable system and will eventually result in decreased agricultural production. Soil erosion rates in conventionally plowed fields can erode through a typical hillslope soil profile over time scales that are comparable to the longevity of some major civilizations. Montgomery (2007) stated that "no till agriculture produces erosion rates much closer to soil production rates and therefore could provide a foundation for sustainable agriculture."

In the past several decades, scientists have determined that measuring the soil concentrations of certain isotopes that form at a known rate permits direct quantification of soil production/formation rates. Applying this technique to soils in temperate regions Heimsath et al. (1999) found soil production rates ranging from 0.02997 to 0.08001 mm (0.00118–0.00315 inch) per year. As such, it takes 300 to 850 yr to form 25.4 mm of soil in these places. Montgomery's (2007) global compilation of data revealed an average rate of 0.01701 to 0.03607 mm (0.00067–0.00142 inch) per year—equivalent to 700 to 1500 yr to form 25.4 mm of soil. With natural soil production rates of centuries to millennia per 25.4 mm and soil erosion rates of millimeters per century under plow-based agriculture, it would take just several hundred to a couple of thousand years to plow through the soil in these regions. This simple estimate predicts remarkably well the life span of major agricultural civilizations around the world. With the exception of the fertile river valleys along which agriculture began, civilizations generally lasted 800 to 2000 yr, and geo-archaeological studies have now shown a connection between soil erosion and the decline of many ancient cultures (Lowdermilk, 1953; Gebregziabher et al., 2006; Lal et al., 2007).

No-till offers some economic advantages to farmers. The number of passes over a field needed to establish and harvest a crop with no-till typically decreases, requiring 50 to 80% less fuel and 30 to 50% less labor than tillage-based agriculture, significantly lowering production costs per acre (Phillips and Phillips, 1984). Although specialized no-till seeding equipment can be expensive, running and maintaining intensive tillage equipment is no longer necessary, lowering the total capital and operating costs of machinery required by up to 50% (Phillips and Young, 1973; Phillips and Phillips, 1984). With these savings in time and money, farmers can be more competitive at smaller scales, or they can expand and farm more land, sometimes doubling farm size using the same equipment and labor. Many farmers appreciate that the time they once devoted to rather mundane tillage tasks they can instead spend on more challenging aspects of farming, family life, or recreation, thereby enhancing their quality of life.

Future no-till agriculture will need to employ more diverse pest and weed management strategies, including biological, physical, and chemical measures to lessen the threat of pesticide resistance. Crop rotation is already helping no-till's war on pests and weeds by helping to break up the weed, pest, and disease cycles when one species is grown continuously (Calegari et al., 2008). The capacity to grow a diverse selection of economically viable crops would advance no-till farming and make it more appealing to farmers. Experts continue to debate the merits of growing fuel on farmland, but if society decides to expand biofuel crops, we will need to consider using no-till with diverse crop rotation to produce them sustainably (Moebius-Clune et al., 2008). Development of alternative crops for bioenergy production on marginal lands, including perennials such as switchgrass, could complement and promote no-till farming, as would perennial grain food crops currently under development.

Soil Carbon, Nitrogen, and Nutrient Cycling

The importance of the soil carbon cycle is often overlooked in traditional agricultural studies because the primary focus is on the crop yield, which is not subject to known carbon limitations, and on those nutrients such as nitrogen that do limit productivity. The decomposition portion of the carbon cycle governs many agronomic processes that occur below ground and manifest themselves above ground. Microorganisms control the decomposition of C, and their activity regulates nutrient cycling in soils. Even though the consequences of their activities can be quite obvious, the presence of most microorganisms is usually taken for granted in cropping systems. In studies of ecosystems, microorganisms are generally considered not as organisms, in an autoecological sense, but as disembodied rates and pools of nutrients. A better mechanistic understanding of many ecosystem processes could be obtained from knowledge of microbial species composition, physiology, and metabolism; of the factors controlling microbial populations and activities; and of the spatial and temporal distribution of microorganisms.

Soil humus comprises a large and stable pool of soil organic matter (SOM); hence a better understanding of the fate of C in soil humic fractions can provide valuable information for the development of alternative tillage practices that may lead to long-term soil C sequestration. Murage and Voroney (2008) reported tillage effects on the dynamics of native C (C3-C) and corn-derived C (C4-C) in fulvic acid (FA), humic acid (HA), and humin fractions:

> Humic substances were extracted from soils cropped with corn for 11 yr and managed under either 55 yr of conventional tillage (CT) or no-tillage (NT). No-tillage resulted in higher proportions of C4-C in the upper 5 cm and generally lower C4-C proportions below 5 cm than CT. Up to 31, 27, and 34% of C4-C were assimilated into FA, HA and humin fractions, respectively, indicating that even the humin fraction, often described as passive, old, or resistant, acted as a sink for recently added C, and that it is heterogeneous with some young components. Recovery of large proportions of C3-C in the humic fractions demonstrated their importance in the long-term stabilization of SOM. Within each sampling depth, there were no unique differences in the distribution of C3-C among the three humic fractions, suggesting similar turnover of C3-C in all the fractions.

These results show the subtle impact of intensive tillage on new and old carbon cycling in our production systems that awaits further research (Angers et al., 1997; Huggins et al., 2007).

While the adoption of no-till can lead to the accumulation of SOC in the surface soil layers, a number of recent studies have shown that this effect is sometimes partly or completely offset by greater SOC content near the bottom of the plow layer under full-inversion tillage (FIT) (Six et al., 2002; West and Post, 2002; Baker et al., 2007; Blanco-Canqui and Lal, 2008; Christopher et al., 2009). Angers and Eriksen-Hamel, (2008) reviewed the literature in which SOC profiles had been measured under paired no-till and FIT situations. Full-inversion tillage is inferred as moldboard plow as primary tillage likely followed with some form of secondary tillage (Reicosky and Allmaras, 2003). Angers and Eriksen-Hamel (2008) found profiles of SOC had to be measured to at least 30 cm, and in most studies, SOC content was significantly greater under no-till than full-inversion tillage in the surface soil layers. At the 21- to 25-cm soil depth, which corresponds to the mean plowing depth for the data set (23 cm), the average SOC content was significantly greater under FIT than no-till. The relative accumulation of SOC at depth under FIT could not be related to soil or climatic variables. Furthermore, the organic matter accumulating at depth under FIT was present in relatively stable form, but this hypothesis and the mechanisms involved require further investigation. Significant differences in SOC stocks between FIT and no-till situations occur at the soil surface but also at depth, which further highlights the importance of taking into account the whole soil profile when comparing soil C stocks (Baker et al., 2007; Blanco-Canqui and Lal, 2008; Christopher et al., 2009). The accumulation of SOC at depth in FIT situations is occurring at the bottom of the plow layer, but to some extent also below it. The greater SOC content at depth under FIT did not completely offset the gain under no-till in the surface horizon, with the net result that the average SOC stocks were greater under no-till than under FIT. The extent, mechanisms, and factors controlling SOC stabilization at depth require further investigations for all types of tillage implements, especially inversion tillage.

Sanchez et al. (2004) investigated the impact of cropping system management on C and N pools, crop yield, and N leaching in a long-term agronomic experiment in southwest Michigan. Four management types ranging from conventional to transitional organic were applied to two crop sequences with and without legume cover crops. Using compost as a fertility source and reducing the use of herbicides and other chemicals resulted in long-term changes in soil organic matter pools. Mineralizable N varied within the rotation, tending to increase after soybean and decrease after corn production in all systems. Corn yield was closely associated with 70-d N mineralization potential, being greatest for first-year corn with cover crop and least for continuous corn without cover crop under all management types. Fall nitrate level and nitrate leaching were higher for commercially fertilized corn than for any other crop or for compost-amended corn. This unique long-term agriculture experiment shows how a production system integrating reduced chemical inputs and a well-designed crop rotation can produce higher yield and lower leaching than a comparable conventional system. Organic C and N storage increased up to 43 and 33% in the integrated compost and transitional organic systems, which decreased the need for additional fertilizers and should tend to improve soil structure and physical condition. Legume cover crops were particularly important within wheat stubble and resulted in a 13% increase in first-year corn yield.

The results of Sanchez et al. (2004) are general and can be applied to corn-based agroecosystems anywhere. In general, agroecosystems that make better use of short- and long-term C and N pools will tend to be more productive and environmentally sustainable than systems that rely on heavy applications of chemical fertilizers and herbicides. Applying a properly structured diverse crop rotation to soils under limited tillage, utilizing legume cover crops where appropriate, taking an integrated pest management approach to weed management, and supplementing fertility with animal waste products all tend to increase labile soil organic C and N. Sequestering C in soils has a significant impact on the global C cycle and enhances the mineralizable forms of C and N, resulting in greater soil N supplying and recycling capacity. This may be useful in all production systems but is essential where synthetic fertilizers are not the primary N source (Russell et al., 2006; Khan et al., 2007). Widespread adoption of the strategies suggested in this study have the potential to improve soil and water quality without affecting yield and are likely to contribute to a cleaner environment on a global scale.

Biomass and Bioenergy Concerns

Biofuels have huge potential for renewable energy development. With the global population growing, the demand for food and energy is intensifying. Despite our best efforts, intensive agricultural practices are still compromising the natural resource base that we rely on for food production (Larson, 1979). Increased land area required for producing biomass for energy use may result in less land available for food production and, hence, less food production, possibly endangering global food security (Giampietro et al., 1997). Using a modeling analysis, Wolf et al. (2003) showed that when a high input system of agriculture is applied, 55% of the present global agricultural land area is needed for food production to the year 2050. The remaining 45% can be used for other purposes, such as biomass production. On the other hand, if a low input system is applied at the global scale for food production, there is no land available for biomass production. Unfortunately, Wolf et al. (2003) did not address production sustainability and the long-term implications of potential soil degradation associated with corn biomass removal (soil carbon depletion), a major environmental concern (Grigal and Berguson, 1998; Mann et al., 2002; Wilhelm et al., 2004, 2007; Lemus and Lal, 2005; Johnson et al., 2006a,b, 2007; Hoskinson et al., 2007; Graham et al., 2007).

The basic challenge for soil conservation is not biofuel use, but the way in which biomass is produced (Plieninger and Bens, 2007). Plieninger and Bens (2007) stated, "Innovative land-use systems specifically designed for energy crops that have both high energy production per unit land area and support high structural and species diversity might offer a way to cope with this energy and environmental dilemma." Biofuels may be the renewable energy carrier with the highest relevance for biological conservation, but both conservation science and policy are just starting to understand the dimensions of the challenge (Larson, 1979). Life cycle assessments for biofuels are very complex and highly controversial; the system limits in terms of included environmental parameters and steps of the production process are often not standardized, and assessments can hardly keep pace with the rapid changes in the field. Significant

greenhouse gas emissions can be released in manufacture of nitrogen fertilizer inputs. Considering the choice of energy conversion, life-cycle assessments indicate that using biofuels for heat and electricity generation is generally superior to automotive fuels (Crutzen et al., 2007; Plieninger and Bens, 2007). In most cases energy crops are placed on fertile soils, where direct competition between food and fuel production arises. Currently there are few political and economic incentives for energy crops to encroach on marginal lands of high conservation value, although there is concern that seminatural grasslands might be converted to energy croplands with switchgrass (Liebig et al., 2005). Current energy cropping systems are largely derived from conventional intensive tillage agricultural and include monoculture crops and high inputs of synthetic nitrogen fertilizer. Crutzen et al. (2007) stated, "An estimated 83% of the global land area is already under direct human influence, and further extending the human footprint on land may be accompanied by negative ecological concomitants." Innovative land-use systems, specifically designed for energy crops, that both have high energy productivity per area and support a high structural and species diversity are needed. Potential strategies comprise the diversification of crop rotations, reductions in mineral fertilizer and herbicide use, the use of a broader diversity of crop species and varieties, the design of mixed cropping systems, longer harvest intervals, and increased physical landscape structure. Efficient conservation standards for bioenergy could help to integrate ecological knowledge and thus direct bioenergy into pathways that are compatible with landscape protection, biodiversity, soil conservation, and cultural issues. Future policies that provide better incentives for biofuels with a high energy efficiency and a high potential for greenhouse gas emission reduction must accommodate the long-term implications of potential soil degradation associated with corn biomass removal (soil carbon depletion) that is of major environmental concern (Mann et al., 2002; Reicosky et al., 2002; Wilhelm et al., 2004, 2007; Lemus and Lal, 2005; Johnson et al., 2006a,b, 2007; Graham et al., 2007).

Above- and belowground crop biomass provide the organic carbon input for building SOM (Johnson et al., 2006a). Soil organic matter is responsible for many of the characteristics associated with highly productive soils (Doran et al., 1998; Doran, 2002; Janzen et al., 1998). Soil organic matter improves soil aggregation and aggregate stability (Gollany et al., 1991; Tisdall and Oades, 1982; Tisdall, 1996; Six et al., 1998), which subsequently impacts soil infiltration, water-holding capacity (Gollany et al., 1992), aeration, bulk density (Gollany et al., 1992), penetration resistance, and soil tilth. Mann et al. (2002) reviewed existing literature to evaluate the major environmental impacts potentially associated with stover harvest from reduced tillage corn production sites. Mann et al. (2002) stated that "more information is needed on several topics to determine potential long-term effects of residue harvest including: (1) erosion and water quality, especially pesticides and nitrate; (2) rates of transformation of different forms of SOC; (3) effects on soil biota; and (4) SOC dynamics in the subsoil. Soil organic matter also impacts chemical properties including pH, nutrient availability and cycling, cation-exchange capacity and buffer capacity." This long list of soil carbon benefits from the plant biomass makes it difficult to understand the combination of intensive tillage and removal of biomass for bioenergy as a truly sustainable production system. Experts continue to debate the merits of growing biomass for bioenergy on farmland, but if society decides to proceed with biofuel crops, we will need to consider using no-till with diverse crop rotation to produce them sustainably (Reicosky et al., 2002; Moebius-Clune et al., 2008).

Primarily, organic C inputs to soil are from the unharvested aboveground, belowground biomass and rhizodeposition from cash crop plants and cover crops, and other organic inputs (e.g., animal manure). Studies have shown that manure application increased SOC and nutrient status in the soil (Webster and Goulding, 1989; Collins et al., 1992; Rochette and Gregorich, 1998) and labile carbon pools (Aoyama et al., 1999; Mikha and Rice, 2004). Total corn root-derived C (C in root biomass plus that in rhizodeposition) contributes from 1.5 times to more than 3 times more C to SOC than shoot-derived C (Balesdent and Balabane, 1996; Wanniarachchi et al., 1999; Allmaras et al., 2004; Wilts et al., 2004; Hooker et al., 2005). Hooker et al. (2005) attributed the difference to dissimilar C cycling rates of shoot and root material. Wilhelm et al. (2004) noted a critical caveat that even though a larger percentage of root C is incorporated into SOC, it does not negate the importance of shoot biomass in building and maintaining SOC. Despite the importance of roots to the formation of SOC, there is little information on total biomass (above and belowground) needed to maintain or build SOC. Most studies include only aboveground biomass (Reicosky et al., 2002; Johnson et al., 2006a; Wilhelm et al., 2007).

Johnson et al. (2006a) estimated the minimum amount of biomass necessary to prevent loss of SOC based on literature values from long-term field-studies. These estimates need to be improved to account for climatic and soil type effects. They are, however, the first published estimates of minimum biomass required to maintain productivity (i.e., SOC) and provide general guidelines to the cellulosic ethanol industry. These guidelines are stated as absolute amount of biomass input, not as a portion of the amount produced, as is more commonly stated, especially for preventing erosion. The soil C cycle works slowly. Even though SOC decomposition rates fluctuate with seasonal temperature and water condition, over time a mean rate of biomass input is required to replace C released from the soil system. The estimates by Johnson et al. (2006a; 2006b) are from limited data in the literature and based on long-term (mean) inputs. Crop yields fluctuate over seasons depending on the weather extremes.

Erosion prevention and C sequestration benefits associated with cover crop use make their use in conjunction with harvesting biomass appealing, provided that the added complexity to scheduling equipment and labor to accomplish additional tasks is not limiting. Preventing soil loss from erosion and increasing the influx of C to the soil by extending the photosynthetic season builds SOM and improves soil quality (Dabney et al., 2001). Cover crops in conjunction with conservation tillage practices sequester more C than conservation tillage alone (Causarano et al., 2006; Calegari et al., 2008). The C input from cover crops was linearly related to soil C concentration (Kuo and Jellum, 2002), consistent with others who reported linear increases in soil C with increased C inputs (Larson et al., 1972; Paustian et al., 1997; Follett et al., 2005). The question remains as to whether cover crops provide sufficient biomass to prevent erosion and loss of SOM in a system removing biomass for bioenergy (Moebius-Clune et al., 2008).

An important requirement for any biomass cropping is the sustainability of the production system (Volk et al., 2004). While the withdrawals of nutrients are probably small and the soil C contents will rather slightly increase than decrease under most scenarios (Lemus and Lal, 2005), soil fertility depletion can be an issue and has been the justification to fertilize bioenergy plantations with the ash from combustion residue (Park et al., 2005). Limited information on the impacts of tillage with residue removal suggests that tillage is more of a factor in carbon loss than the biomass return for soil carbon input (Karlen et al., 1994a, 1994b; Hooker et al., 2005; Moebius-Clune et al., 2008).

The minimum detectable difference in SOC was calculated as a function of variance and sample size for SOC changes after 5 yr under a herbaceous bioenergy crop (Garten and Wullschleger, 1999). The authors showed that the smallest difference that could be detected was about 1 t C ha^{-1}, and this could only be done using exceedingly large sample sizes. The minimum difference that could be detected with a reasonable sample size and a good statistical power (90% confidence) was 1 t C ha^{-1}. Most agricultural practices will not cause the soil to accumulate this during a 5-yr commitment period (Smith et al., 1998).

Loss of fertility is a major impact of soil erosion, especially in old and highly weathered soils in which SOC and plant nutrients are concentrated in the upper few centimeters of the soil profile. Loss of soil fertility is the principal cause of yield decline on eroded soils (Peterson, 1964). Nutrient losses are more severe on arable lands, where supplemental fertilizers applied can have a masking effect on crop yields (Cleveland, 1995). There are two mechanisms of fertility decline by erosion (Helvey et al., 1985). The greatest nutrient losses occurred with mass soil movements and the soil deposited in alluvial fans. High nutrient losses occurred in soils that did not receive chemical fertilizers. Controlled biomass burning may, in some cases, reduce risks of runoff and nutrient losses compared with uncontrolled burning. In Spain, Mangas et al. (1992) observed that nutrient loss in runoff after burning was between 8 and 35%

of that of the previous year, while the volume of runoff was only 3%, implying greater concentration of nutrients in runoff. Wallingford (1991) prepared a nutrient balance sheet for major U.S. crops and observed the N budget to be slightly positive and stable in the 1990s, the P budget was negative after being positive in the 1960s and 1970s, and the K budget was strongly negative. This budgeting exercise, however, did not take into consideration the losses due to erosion that in some cases may be substantial. Experiments conducted on Vertisols (fine, montmorillonitic, thermic, Udic Pellusterts) in east-central Texas showed that losses of plant nutrients in no-till and chisel till treatments were 3.8 and 8.1 kg ha^{-1} for N and 0.8 and 1.5 kg ha^{-1} for P, respectively (Chichester and Richardson, 1992). The combination of nutrient and soil loss with erosion is a double negative exacerbated by intensive tillage.

Summary and Conclusions

Humans require a secure and renewable natural-resource base to sustain their basic needs for future social and economic activity. However, while deriving natural resources from the terrestrial biosphere, humans also inadvertently modify their environment. The 20th century saw an expanding human population, increasing agricultural yields, and a decreasing land-resource base. To feed the world population, agriculture has expanded using intensive tillage, resulting in a greater impact on the environment, human health, and biodiversity. But, given our current knowledge of the planet's capacity, we now realize that producing sufficient food is not enough—it must also be done sustainably. Farmers need to generate adequate crop yields of high quality, conserve natural resources for future generations, make enough money to live on, and be fair and equitable to their workers and community. No-till farming is one system that has the potential to help realize this vision of a more sustainable agriculture. As with any new system, there are challenges and trade-offs. Nevertheless, growers in some parts of the world are increasingly abandoning their plows. Leaving crop residues on the soil surface provides soil protection and helps to increase water infiltration and limit runoff. Decreased runoff, in turn, can reduce pollution of nearby water sources with transported sediment, fertilizers, and pesticides. The residues also promote water conservation by reducing evaporation in drier areas. In instances where water availability limits crop production, greater water conservation can mean higher-yielding crops or new capabilities to grow alternative crops.

The balance between agricultural productivity and environmental quality relies on proper resource management. The sun, soil, water, and air are our primary resources for food security. The increasing global population requires improved management of these resources and challenges our human intellectual capacity to meet the food security needs of all society. As the global population continues to expand, our food security becomes a little more challenging when we recognize that we have a finite amount of land area for agricultural production. The increasing productivity required to meet this increasing demand for food must be done in an environmentally friendly way to maintain our quality of life. Improved soil management practices with emphasis on SOC to maintain soil physical, chemical, and biological properties to minimize soil degradation are urgently needed. Many of the environmental issues of intensive agriculture can be directly related to intensive tillage and its unintended consequences. Intensive tillage destroys the soil structural integrity, the natural soil fauna habitat, releases CO_2 and enhances soil mineralization and breakdown of SOM, causes tillage erosion on sloping lands, and sets the soil up for wind and water erosion that all contribute to degradation of soil, water, and air quality. There is compelling evidence that intensive tillage of our agricultural landscapes is responsible for environmental degradation in our agricultural ecosystems.

To conserve resources for future generations, we need alternatives to conventional farming practices. No-till systems simultaneously reduce the erosive force of runoff and increase the ability to hold soil in place, making these methods remarkably effective at curbing erosion. Although the effect of no-till on erosion rates depends on a number of site-specific factors, such as the soil type and crop, less intensive tillage can decrease soil erosion rates close to soil formation rates. There's a definite need for improved best management practices that lead to decreased tillage intensity and improved plant management techniques for capturing solar energy in the form of photosynthesis and for returning nutrients and carbon to the soil. While our society has a tremendous need for bioenergy from biomass, caution is suggested until the long-term implications of biomass and carbon removal for our food security are understood. More troubling, the environmental degradation caused by agriculture will likely worsen as the hungry human population grows to eight billion or ten billion in the coming decades. The need for critical research to develop the best management practices to maintain this delicate balance between agricultural productivity and environmental quality cries for our attention with more emphasis on reducing tillage intensity.

Acknowledgment

The authors gratefully acknowledge the assistance of Beth Burmeister for typing assistance, editorial comments, and crosschecking literature citations.

References

Allmaras, R.R., D.R. Linden, and C.E. Clapp. 2004. Corn residue transformations into root and soil carbon as related to nitrogen, tillage, and stover management. Soil Sci. Soc. Am. J. 68:1366–1375.

Alvarez, R. 2005. A review of nitrogen fertilizer and conservation tillage effects on soil organic carbon storage. Soil Use Manage. 21:38–52.

Angers, D.A., M.A. Bolinder, M.R. Carter, E.G. Gregorich, C.F. Drury, B.C. Liang, R.P. Voroney, R.R. Simard, R.G. Donald, R.P. Beyaert, and J. Martel. 1997. Impact of tillage practices on organic carbon and nitrogen storage in cool, humid soils of eastern Canada. Soil Tillage Res. 41:191–201.

Angers, D.A., and N.S. Eriksen-Hamel. 2008. Full-inversion tillage and organic carbon distribution in soil profiles: A meta-analysis. Soil Sci. Soc. Am. J. 72(5):1370–1374.

Angers, D.A., R.P. Voroney, and D. Côté. 1995. Dynamics of soil organic matter and corn residues as affected by tillage practices. Soil Sci. Soc. Am. J. 59:1311–1315.

Aoyama, M., D.A. Angers, A. N'Dayegamiye, and N. Bissonnette. 1999. Protected organic matter in water-stable aggregates as affected by mineral fertilizer and manure applications. Can. J. Soil Sci. 79:419–425.

Baker, J.M., T.E. Ochsner, R.T. Venterea, and T.J. Griffis. 2007. Tillage and soil carbon sequestration—What do we really know? Agric. Ecosyst. Environ. 118(1–4):1–5.

Balesdent, J., and M. Balabane. 1996. Major contribution of roots to soil carbon storage inferred from maize cultivated soils. Soil Biol. Biochem. 28:1261–1263.

Bauer, A., and A.L. Black. 1994. Quantification of the effect of soil organic matter content on soil productivity. Soil Sci. Soc. Am. J. 58:185–193.

Beinroth, F.H., H. Eswaran, P.F. Reich, and E. Van den Berg. 1994. Land related stresses in agroecosystems. p. 131–148. In S.M. Virmani et al. (ed.) Stressed ecosystems and sustainable agriculture. Oxford & IBH Publ. Co., New Delhi, India.

Blanco-Canqui, H., and R. Lal. 2004. Mechanisms of carbon sequestration in soil aggregates. Crit. Rev. Plant Sci. 23(6):481–505.

Blanco-Canqui, H., and R. Lal. 2008. No-tillage and soil-profile carbon sequestration: An on-farm assessment. Soil Sci. Soc. Am. J. 72:693–701.

Brown, L.R. 1994. Facing food security. p. 177–197. *In* L.R. Brown et al. (ed.) State of the world. Earthscan Publ., London.

Brown, L.R., and J.E. Young. 1990. Feeding the world in the nineties. p. 59–78. *In* L.R. Brown et al. (ed.) State of the world 1990: A Worldwatch Institute report on progress toward a sustainable society. Worldwatch Inst., Washington, DC.

Buringh, P. 1989. Availability of agricultural land for crops and livestock production. p. 69–83. *In* L.D. Pimentel and C.W. Hall (ed.) Food and natural resources. Academic Press, San Diego.

Calderón, F.J., L.E. Jackson, K.M. Scow, and D. Rolston. 2001. Short-term dynamics of nitrogen, microbial activity and phospholipid fatty acids after tillage. Soil Sci. Soc. Am. J. 65:118–126.

Calegari, A., W.L. Hargrove, D. Dos Santos Rheinheimer, R. Ralisch, D. Tessier, S. de Tourdonnet, and M. de Fatima Guimarães. 2008. Impact of long-term no-tillage and cropping system management on soil organic carbon in an Oxisol: A model for sustainability. Agron. J. 100:1013–1019.

Causarano, H.J., A.J. Franzluebbers, D.W. Reeves, and J.N. Shaw. 2006. Soil organic carbon sequestration in cotton production systems of the southeastern United States: A review. J. Environ. Qual. 35:1374–1383.

CCSP. 2008. The effects of climate change on agriculture, land resources, water resources, and biodiversity. A Report by the U.S. Climate Change Science Program and the Subcommittee on Global Change Research. P. Backlund, A. Janetos, D. Schimel, J. Hatfield, K. Boote, P. Fay, L. Hahn, C. Izaurralde, B.A. Kimball, T. Mader, J. Morgan, D. Ort, W. Polley, A. Thomson, D. Wolfe, M. Ryan, S. Archer, R. Birdsey, C. Dahm, L. Heath, J. Hicke, D. Hollinger, T. Huxman, G. Okin, R. Oren, J. Randerson, W. Schlesinger, D. Lettenmaier, D. Major, L. Poff, S. Running, L. Hansen, D. Inouye, B.P. Kelly, L Meyerson, B. Peterson, R. Shaw. USEPA, Washington, DC.

Chen, R.S. 1990. Global agriculture, environment, and hunger: Past, present and future links. Environ. Impact Assess. Rev. 10(4):335–358.

Chichester, F.W., and C.W. Richardson. 1992. Sentiment and nutrient loss from clay soils as affected by tillage. J. Environ. Qual. 21:587–590.

Christopher, S.F., R. Lal, and U. Mishra. 2009. Regional study on no-till effects on carbon sequestration in the Midwestern United States. Soil Sci. Soc. Am. J. 73:207–216.

Cleveland, C.J. 1995. Resource degradation, technical change, and productivity of energy use in U.S. agriculture. Ecol. Econ. (Amsterdam) 13:185–201.

Collins, H.P., P.E. Rasmussen, and C.L. Douglas. 1992. Crop rotation and residue management effects on soil carbon and microbial dynamics. Soil Sci. Soc. Am. J. 56:783–788.

Cornish, P.S., and J.E. Pratley (ed.) 1987. Tillage: New direction in Australian agriculture. Inkata Press, Sydney, Australia.

Crutzen, P.J., A.R. Mosier, K.A. Smith, and W. Winiwarter. 2007. N_2O release from agro-biofuel production negates global warming reduction by replacing fossil fuels. Atmos. Chem. Phys. Discuss. 7:1–15.

Dabney, S.M., J.A. Delgado, and D.W. Reeves. 2001. Using winter cover crops to improve soil and water quality. Commun. Soil Sci. Plant Anal. 32:1221–1250.

Dalal, R.C., W.M. Strong, E.J. Weston, J.E. Cooper, K.J. Lehane, A.J. King, and C.J. Chicken. 1995. Sustaining productivity of a Vertisol at Warra, Queensland with fertilisers, no-tillage, or legumes. 1. Organic matter status. Aust. J. Exp. Agric. 35:903–913.

Deen, W., and P.K. Kataki. 2003. Carbon sequestration in a long-term conventional versus conservation tillage experiment. Soil Tillage Res. 74:143–150.

Derpsch, R. 2001. Frontiers in conservation tillage and advances in conservation practice. p. 248–254. *In* D.E. Stout et al. (ed.) Sustaining the global farm. International Soil Conservation Organization in cooperation with the USDA and Purdue University, West Lafayette, IN.

Dick, W.A., and J.T. Durkalski. 1997. No-tillage production agriculture and carbon sequestration in a Typic Fragiudalf soil of northeastern Ohio. p. 59–71. *In* R. Lal et al. (ed.) Management of carbon sequestration in soil. CRC Press, Boca Raton, FL.

Doran, J.W. 1980. Microbial changes associated with residue management with reduced tillage. Soil Sci. Soc. Am. J. 44:518–524.

Doran, J.W. 1987. Microbial biomass and mineralizable nitrogen distributions in no-tillage and plowed soils. Biol. Fertil. Soils 5:68–75.

Doran, J.W. 2002. Soil health and global sustainability: Translating science into practice. Agric. Ecosyst. Environ. 88:119–127.

Doran, J.W., E.T. Elliott, and K. Paustian. 1998. Soil microbial activity, nitrogen cycling, and long-term changes in organic carbon pools as related to fallow tillage management. Soil Tillage Res. 49:3–18.

Doran, J.W., D.G. Fraser, M.N. Culik, and W.C. Leibhardt. 1987. Influence of alternative and conventional agricultural management on soil microbial processes and N availability. Am. J. Alternative Agric. 2:99–106.

Drinkwater, L.E., P. Wagoner, and M. Sarrantonio. 1998. Legume-based cropping systems have reduced carbon and nitrogen losses. Nature 396:262–265.

Du Pisani, J.A. 2006. Sustainable development-historical roots of the concept. J. Integrative Environ. Sci. 3(2):83–96.

Ehrlich, P.R., A.H. Ehrlich, and G.C. Daily. 1993. Food security, population, and the environment. Popul. Dev. Rev. 19:1–32.

Ellert, B.H., and H.H. Janzen. 1999. Short-term influence of tillage on CO_2 fluxes from a semi-arid soil on the Canadian Prairies. Soil Tillage Res. 50:21–32.

Eswaran, H., P.F. Reich, and F.H. Beinroth. 1997. Global distribution of soils with acidity. p. 159–164. *In* A.Z. Moniz et al. (ed.) Plant-Soil Interactions at Low pH: Sustainable Agriculture and Forestry Production. Proc. of the 4th International Symp. on Plant-Soil Interactions at Low pH, Belo Horizonte, Minas Gerais, Brazil. 17–24 Mar. 1996.

FAO. 2009. Conservation agriculture. Available at http://www.fao.org/ag/ca/ (verified 25 Aug. 2010).

FAOSTAT. 2009. FAOSTAT online statistical service. Available at http://faostat.fao.org/default.aspx (verified 25 Aug. 2010).

Fischer, G., and G.K. Heilig. 1998. Population momentum and the demand on land and water resources. p. 8–29. *In* D.J. Greenland et al. (ed.) Land resources: On the edge of the Malthusian precipice? CAB International, New York.

Follett, R.F., J.Z. Castellanos, and E.D. Buenger. 2005. Carbon dynamics and sequestration in an irrigated Vertisol in Central Mexico. Soil Tillage Res. 83:148–158.

Garcia, J.P., C.S. Wortmann, M. Mamo, R. Drijber, and D. Tarkalson. 2007. One-time tillage of no-till: Effects on nutrients, mycorrhizae, and phosphorus uptake. Agron. J. 99:1093–1103.

Gardner, G. 1996. Shrinking fields: Cropland lost in a world of 8 billion. World Watch Institute, Washington, DC.

Garten, C.T., and S.D. Wullschleger. 1999. Soil carbon inventories under a bioenergy crop (switchgrass): Measurement limitations. J. Environ. Qual. 28:1359–1365.

Gebregziabher, S., M.M. Mouaen, H. Van Brussel, H. Ramon, J. Nyssen, H. Verplancke, M. Behailu, J. Deckers, and J. De Baerdemaeker. 2006. Animal drawn tillage: The Ethiopian ard plough, *maresha*: A review. Soil Tillage Res. 89:129–143.

Giampietro, M., S. Ulgiati, and D. Pimentel. 1997. Feasibility of large-scale biofuel production. Bioscience 47(9):587–600.

Gilley, J.E., and J.W. Doran. 1997. Tillage effects on soil erosion potential and soil quality of a former conservation reserve program site. J. Soil Water Conserv. 52:184–188.

Gollany, H.T., T.E. Schumacher, P.D. Evenson, M.J. Lindstrom, and G.D. Lemme. 1991. Aggregate stability of an eroded and desurfaced Typic Argiustoll. Soil Sci. Soc. Am. J. 55:811–816.

Gollany, H.T., T.E. Schumacher, M.J. Lindstrom, P. Evenson, and G.D. Lemme. 1992. Topsoil thickness and desurfacing effects on properties and productivity of a typic Argiustoll. Soil Sci. Soc. Am. J. 56:220–225.

Govers, G., K. Vandaele, P.J. Desmet, J. Poesen, and K. Bunte. 1994. The role of tillage in soil redistribution on hillslopes. Eur. J. Soil Sci. 45:469–478.

Graham, R.L., R. Nelson, J. Sheehan, R.D. Perlack, and L.L. Wright. 2007. Current and potential U.S. corn stover supplies. Agron. J. 99:1–11.

Grandy, A.S., G.A. Porter, and M. Erich. 2002. Organic amendment and rotation crop effects on the recovery of soil organic matter and aggregation in potato cropping systems. Soil Sci. Soc. Am. J. 66:1311–1319.

Grandy, A.S., G.P. Robertson, and K.D. Thelen. 2006. Do productivity and environmental trade-offs justify periodically cultivating no-till cropping systems? Agron. J. 98:1377–1383.

Greenland, D.J., P.J. Gregory, and P.H. Nye (ed.) 1998. Land resources: On the edge of the Malthusian precipice? CAB International, New York.

Greenland, D.J., and I. Szabolc (ed.) 1994. Soil resilience and sustainable land use. CAB International, Wallingford, UK.

Grigal, D.F., and W.E. Berguson. 1998. Soil carbon changes associated with short-rotation systems. Biomass Bioenergy 14:371–377.

Hatfield, J., K. Boote, P. Fay, L. Hahn, C. Izaurralde, B.A. Kimball, T. Mader, J. Morgan, D. Ort, W. Polley, A. Thomson, and D. Wolfe. 2008. Agriculture. *In* The effects of climate change on agriculture, land resources, water resources, and biodiversity. A Report by the U.S. Climate Change Science Program and the Subcommittee on Global Change Research. Washington, DC.

Havlin, J.L., D.E. Kissel, L.D. Maddux, M.M. Classen, and J.H. Long. 1990. Crop rotation and tillage effects on soil organic carbon and nitrogen. Soil Sci. Soc. Am. J. 54:448–452.

Heimlich, R.E., and W.D. Anderson. 2001. Development at the urban fringe and beyond: Impacts on agriculture and rural land. Agric. Econ. Rep. 803. USDA-ERS, Washington, DC.

Heimsath, A.M., W.E. Dietrich, K. Nishiizumi, and R.C. Finkel. 1999. Cosmogenic nuclides, topography, and the spatial variation of soil depth. Geomorphology 27(1–2):151–172.

Helvey, J.D., A.R. Tiedmann, and T.D. Anderson. 1985. Plant nutrient losses by soil erosion and mass movement after wildfire. J. Soil Water Conserv. 40:168–173.

Hooker, B.A., T.F. Morris, R. Peters, and Z.G. Cardon. 2005. Long-term effects of tillage and corn stalk return on soil carbon dynamics. Soil Sci. Soc. Am. J. 69:188–196.

Hoskinson, R.L., D.L. Karlen, S.J. Birrell, C.W. Radtke, and W.W. Wilhelm. 2007. Engineering, nutrient removal, and feedstock conversion evaluations of four corn stover harvest scenarios. Biomass Bioenergy 31:126–136.

Huggins, D.R., R.R. Allmaras, C.E. Clapp, J.A. Lamb, and G.W. Randall. 2007. Corn–soybean sequence and tillage effects on soil carbon dynamics and storage. Soil Sci. Soc. Am. J. 71:145–154.

Huggins, D.R., and J.P. Reganold. 2008. No-till: The quiet revolution. Sci. Am. 7:70–77.

IPCC. 2007. Climate change 2007: The physical science basis, summary for policymakers. The 4th Assessment Report. Available at www.ipcc.ch (verified 25 Aug. 2010).

Ismail, I., R.L. Blevins, and W.W. Frye. 1994. Long-term no-tillage effects on soil properties and continuous corn yields. Soil Sci. Soc. Am. J. 58:193–198.

Izaurralde, R.C., N.J. Rosenberg, and R. Lal. 2001. Mitigation of climate change by soil carbon sequestration. Adv. Agron. 70:1–75.

Jackson, L.E., F.J. Calderón, K.L. Steenwerth, K.M. Scow, and D. Rolston. 2003. Responses of soil microbial processes and community structure to tillage events and implications for soil quality. Geoderma 114:305–317.

Janzen, H.H., C.A. Campbell, E.G. Gergorich, and B.H. Ellert. 1998. Soil carbon dynamics in Canadian agroecosystems. p. 57–80. *In* R. Lal et al. (ed.) Soil processes and the carbon cycle. CRC Press, Boca Raton, FL.

Jastrow, J.D., T.W. Boutton, and R.M. Miller. 1996. Carbon dynamics of aggregate-associated organic matter estimated by carbon-13 natural abundance. Soil Sci. Soc. Am. J. 60:801–807.

Jenkinson, D.S. 1991. The Rothamsted long-term experiments: Are they still of use? Agron. J. 83:2–10.

Johnson, J.M.F., R.R. Allmaras, and D.C. Reicosky. 2006a. estimating source carbon from crop residues, roots and rhizodeposits using the national grain-yield database. Agron. J. 98:622–636.

Johnson, J.M.-F., M.D. Coleman, R. Gesch, A. Jaradat, R. Mitchell, D. Reicosky, and W.W. Wilhelm. 2007. Biomass-bioenergy crops in the United States: A changing paradigm. The Americas. J. Plant Sci. Biotechnol. 1(1):1–28.

Johnson, J.M., D.C. Reicosky, R.R. Allmaras, D.W. Archer, and W.W. Wilhelm. 2006b. A matter of balance: Conservation and renewable energy. J. Soil Water Conserv. 61(4):120A–125A.

Kanchikerimath, M., and D. Singh. 2001. Soil organic matter and biological properties after 26 years of maize–wheat–cowpea cropping as affected by manure and fertilization in a Cambisol in semiarid region of India. Agric. Ecosyst. Environ. 86:155–162.

Karlen, D.L., N.C. Wollenhaupt, D.C. Erbach, E.C. Berry, J.B. Swan, N.S. Eash, and J.L. Jordahl. 1994a. Corn residue effects on soil quality following 10 years of no-till corn. Soil Tillage Res. 31:149–167.

Karlen, D.L., N.C. Wollenhaupt, D.C. Erbach, E.C. Berry, J.B. Swan, N.S. Eash, and J.L. Jordahl. 1994b. Long-term tillage effects on soil quality. Soil Tillage Res. 32:313–327.

Kern, J.S., and M.G. Johnson. 1993. Conservation tillage impacts on national soil and atmospheric carbon levels. Soil Sci. Soc. Am. J. 57:200–210.

Khan, S.A., R.L. Mulvaney, T.R. Ellsworth, and C.W. Boast. 2007. The myth of nitrogen fertilization for soil carbon sequestration. J. Environ. Qual. 36:1821–1832.

Kuo, S., and E.J. Jellum. 2002. Influence of winter cover crop and residue management on soil nitrogen availability and corn. Agron. J. 94:501–508.

Lafond, G.P., D.A. Derksen, H.A. Loeppky, and D. Struthers. 1994. An agronomic evaluation of conservation tillage systems and continuous cropping in East Central Saskatchewan. J. Soil Water Conserv. 49:387–393.

Lal, R. 1987. Effects of soil erosion on crop productivity. Crit. Rev. Plant Sci. 5(4):303–367.

Lal, R. 1989. Land degradation and its impact on food and other resources. p. 65–68. *In* D. Pimentel and C.W. Hall (ed.) Food and natural resources. Academic Press, San Diego, CA.

Lal, R. 1994. Sustainable land use systems and soil resilience. p. 41–68. *In* D.J. Greenland and I. Szaboles (ed.) Soil resilience and sustainable land use. CAB International, Wallingford, U.K.

Lal, R. 1995. Global soil erosion by water and carbon dynamics. p. 131–140. *In* R. Lal et al. (ed.) Soils and global change. Lewis Publ., Boca Raton, FL.

Lal, R. 1997. Degradation and resilience of soils. Phil. Trans. R. Soc. Land. B. 352:997–1010.

Lal, R. 1998. Soil erosion impact on agronomic productivity and environmental quality. Crit. Rev. Plant Sci. 17(4):319–464.

Lal, R. 1999a. Soil management and restoration for C sequestration to mitigate the accelerated greenhouse effect. Prog. Environ. Sci. 1:307–326.

Lal, R. 1999b. Global carbon pools and fluxes and the impact of agricultural intensification and judicious land use. p. 45–52. *In* Prevention of land degradation, enhancement of carbon sequestration and conservation of biodiversity through land use change and sustainable land management with a focus on Latin America and the Caribbean. World Soil Resources Rep. 86. FAO, Rome.

Lal, R. 2003. Soil erosion and the global carbon budget. Environ. Int. 29(4):437–450.

Lal, R. 2004. Soil carbon sequestration impacts on global climate change and food security. Science 304:1623–1627.

Lal, R., J.M. Kimble, R.F. Follet, and V. Cole. 1998. Potential of U.S. cropland for carbon sequestration and greenhouse effect mitigation. Ann Arbor Press, Chelsea, MI.

Lal, R., D.C. Reicosky, and J.D. Hanson. 2007. Evolution of the plow over 10,000 years and the rationale for no-till farming. Soil Tillage Res. 93:1–12.

Larson, W.E. 1979. Crop residues: Energy production or erosion control? J. Soil Water Conserv. 34:74–76.

Larson, W.E., C.E. Clapp, W.H. Pierre, and Y.B. Morachan. 1972. Effects of increasing amounts of organic residues on continuous corn. II. Organic carbon, nitrogen, phosphorus and sulfur. Agron. J. 64:204–208.

Larson, W.E., F.J. Pierce, and R.H. Dowdy. 1983. The threat of soil erosion to long-term crop production. Science 219:458–465.

Lemus, R., and R. Lal. 2005. Bioenergy crops and carbon sequestration. Crit. Rev. Plant Sci. 24:1–21.

Liebig, M.A., J.A. Morgan, J.D. Reeder, B.H. Ellert, H.T. Gollany, and G.E. Schuman. 2005. Greenhouse gas contributions and mitigation potential of agricultural practices in northwestern USA and western Canada. Soil Tillage Res. 83:25–52.

Lindstrom, M.J., W.W. Nelson, and T.E. Schumacher. 1992. Quantifying tillage erosion rates due to moldboard plowing. Soil Tillage Res. 44:243–255.

Lindstrom, M.J., J.A. Schumacher, and T.E. Schumacher. 2000. TEP: A tillage erosion prediction model to calculate soil translocation rates from tillage. J. Soil Water Conserv. 55(1):105–108.

Lobb, D., E. Huffman, and D. Reicosky. 2007. Importance of information on tillage practices in the modelling of environmental processes and in the use of environmental indicators. J. Environ. Manage. 82:377–387.

Lobb, D.A., R.G. Kachanoski, and M.H. Miller. 1995. Tillage translocation and tillage erosion on shoulder slope landscape positions measured using cesium-137 as a tracer. Can. J. Soil Sci. 75:211–218.

Lowdermilk, W.C. 1953. Conquest of the land through 7000 years. Bull. 99. USDA Soil Conservation Service, Washington, DC.

Mangas, V.J., J.R. Sanchez, and C. Oritz. 1992. Effects of a fire on runoff and erosion on the Mediterranean forest soils in SE Spain. Pirineos 140:37–51.

Mann, L., V. Tolbert, and J. Cushman. 2002. Potential environmental effects of corn (*Zea Mays* L.) stover removal with emphasis on soil organic matter and erosion. Agric. Ecosyst. Environ. 89:149–166.

Mikha, M.M., and C.W. Rice. 2004. Tillage and manure effects on soil and aggregate-associated carbon and nitrogen. Soil Sci. Soc. Am. J. 68:809–816.

Moebius-Clune, B.N., H.M. van Es, O.J. Idowu, R.R. Schindelbeck, D.J. Moebius-Clune, D.W. Wolfe, G.S. Abawi, J.E. Thies, B.K. Gugino, and R. Lucey. 2008. Long-term effects of harvesting maize stover and tillage on soil quality. Soil Sci. Soc. Am. J. 72(4):960–969.

Montgomery, D.R. 2007. Soil erosion and agricultural sustainability. Proc. Natl. Acad. Sci. 104(33):13268–13272.

Murage, E.W., and P. Voroney. 2008. Distribution of organic carbon in the stable soil humic fractions as affected by tillage management. Can. J. Soil Sci. 88(1):99–106.

Organization for Economic Co-operation and Development. 2001. Environmental indicators for Agriculture. Vol. 3. Methods and Results. Available at http://www.oecd.org/dataoecd/24/35/40680869.pdf (verified 25 Aug. 2010). Organization for Economic Co-operation and Development, Paris, France.

Olchin, G.P., S. Ogle, S.D. Frey, T.R. Filley, K. Paustian, and J. Six. 2008. Residue carbon stabilization in soil aggregates of no-till and tillage management of dryland cropping systems. Soil Sci. Soc. Am. J. 72:507–513.

Oldeman, L.R., R.T.A. Hakkeling, and W.G. Sombroek. 1991. World map of the status of human-induced soil degradation: An explanatory note. International Soil Information and Reference Centre, Wageningen, the Netherlands.

Park, B.B., R.D. Yania, J.M. Sahm, D.K. Lee, and L.P. Abrahamson. 2005. Wood ash effects on plant and soil in a Willow bioenergy plantations. Biomass Bioenergy 28:355–365.

Paustian, K., H.P. Collins, and E.A. Paul. 1997. Management controls on soil carbon. p. 15–49. *In* E.A. Paul et al. (ed.) Soil organic matter in temperate agroecosystems: Long-term experiments in North America. CRC Press, Boca Raton, FL.

Peterson, J.B. 1964. The relation of soil fertility to soil erosion. J. Soil Water Conserv. 9:15–18.

Phillips, R.E., and S.H. Phillips (ed.) 1984. No-tillage agriculture, principles and practices. Van Nostrand Reinhold Co., New York.

Phillips, S.H., and H.M. Young. 1973. No-tillage farming. Reiman Associates, Milwaukee, WI.

Pierce, F.C., M.-C. Fortin, and M.J. Staton. 1994. Periodic plowing effects on soil properties in a no-till farming system. Soil Sci. Soc. Am. J. 58:1782–1787.

Pimentel, D. 1993. World soil erosion and conservation. Cambridge Univ. Press, Cambridge, MA.

Pimentel, D., J. Allen, A. Beers, L. Guinand, R. Linder, P. McLaughlin, B. Meer, D. Musonda, D. Perdue, S. Poisson, S. Siebert, K. Stoner, R. Salazar, and A. Hawkins. 1987. World agriculture and soil erosion. Bioscience 37:277–283.

Pimentel, D., C. Harvey, P. Resosudarmo, K. Sinclair, D. Kurz, M. McNair, S. Crist, L. Sphpritz, L. Fitton, R. Saffouri, and R. Blair. 1995. Environmental and economic costs of soil erosion and conservation benefits. Science 267:1117–1123.

Plieninger, T., and O. Bens. 2007. How the emergence of biofuels challenges environmental conservation. Environ. Conserv. 34(4):273–275.

Poesen, J., B. van Wesemael, G. Govers, G.J. Martinez-Fernandez, P. Desmet, K. Vandaele, K.T. Quine, and G. Degraer. 1997. Patterns of rock fragment cover generated by tillage erosion. Geomorphology 18:183–197.

Postel, S. 1989. Water for agriculture: Facing the limit. Worldwatch Paper 93. Worldwatch Institute, Washington, DC.

Quincke, J.A., C.S. Wortmann, M. Mamo, T. Franti, R.A. Drijber, and J.P. Garcia. 2007. One-time tillage of no-till systems: Soil physical properties, phosphorus runoff, and crop yield. Agron. J. 99(4):1104–1110.

Ramankutty, N., J.A. Foley, and N.J. Olejniczak. 2002. People on the land: Changes in global population and croplands during the 20th century. Ambio 31(3):251–257.

Rasmussen, P.E., and C.R. Rohde. 1988. Long-term tillage and nitrogen fertilization affects on organic N and C in a semi-arid soil. Soil Sci. Soc. Am. J. 44:596–600.

Reeder, R., and D. Westermann. 2007. Soil management. p. 1–87. *In* M. Schnepf and C. Cox (ed.) Environmental benefits of conservation on cropland. Soil Water Conserv. Soc., Ankeny, IA.

Reicosky, D.C. 1997. Tillage-induced CO_2 emission from soil. Nutr. Cycl. Agroecosyst. 49:273–285.

Reicosky, D.C. 2001a. Conservation agriculture: Global environmental benefits of soil carbon management. p. 3–12. *In* L. Garcia-Torres et al. (ed.) Conservation agriculture: A worldwide challenge. XUL, Cordoba, Spain.

Reicosky, D.C. 2001b. Tillage-induced CO_2 emissions and carbon sequestration: Effect of secondary tillage and compaction. p. 265–274. *In* L. Garcia-Torres et al. (ed.) Conservation agriculture: A worldwide challenge. XUL, Cordoba, Spain.

Reicosky, D.C. 2002. Long-term effect of moldboard plowing on tillage-induced CO_2 loss. p. 87–97. *In* J.M. Kimble and R. Lal (ed.) Agricultural practices and policies for carbon sequestration in soil. CRC Press, Boca Raton, FL.

Reicosky, D.C. 2008. Carbon sequestration and environmental benefits from no-till systems. p. 43–58. *In* T. Goddard et al. (ed.) No-till farming systems. Spec. Publ. 3. World Assoc. Soil and Water Conserv., Bangkok, Thailand.

Reicosky, D.C., and R.R. Allmaras. 2003. Advances in tillage research in North American cropping systems. p. 75–125. *In* A. Shrestha (ed.) Cropping systems: Trends and advances. Haworth Press, New York.

Reicosky, D.C., S.D. Evans, C.A. Cambardella, R.R. Allmaras, A.R. Wilts, and D.R. Huggins. 2002. Continuous corn with moldboard tillage: Residue and fertility effects on soil carbon. J. Soil Water Conserv. 57:277–284.

Reicosky, D.C., R.W. Gesch, S.W. Wagner, R.A. Gilbert, C.D. Wente, and D.R. Morris. 2008. Tillage and wind effects on soil CO_2 concentrations in muck soils. Soil Tillage Res. 99:221–231.

Reicosky, D.C., J.L. Hatfield, and R.L. Sass. 2000. Agricultural contributions to greenhouse gas emissions. p. 37–55. *In* K.R. Reddy and H.F. Hodges (ed.) Climate change and global crop productivity. CABI Publishing, Wallingford, UK.

Reicosky, D.C., W.D. Kemper, G.W. Langdale, C.W. Douglas, Jr., and P.E. Rasmussen. 1995. Soil organic matter changes resulting from tillage and biomass production. J. Soil Water Conserv. 50:253–261.

Reicosky, D.C., and M.J. Lindstrom. 1993. Fall tillage method: Effect on short-term carbon dioxide flux from soil. Agron. J. 85:1237–1243.

Reicosky, D.C., and M.J. Lindstrom. 1995. The impact of fall tillage on short-term carbon dioxide flux. p. 177–187. *In* R. Lal et al. (ed.) Soils and global change. Lewis Publ., Chelsea, MI.

Reicosky, D.C., and K.E. Saxton. 2007a. The benefits of no-tillage. p. 11–20. *In* C.J. Baker et al. (ed.) No-tillage seeding in conservation agriculture. FAO and CAB International, Rome, Italy.

Reicosky, D.C., and K.E. Saxton. 2007b. Reduced environmental emissions and carbon sequestration. p. 257–267. *In* C.J. Baker et al. (ed.) No-tillage seeding in conservation agriculture. FAO and CAB International, Rome, Italy.

Robertson, G.P., E. Paul, and R.R. Harwood. 2000. Greenhouse gases in intensive agriculture: Contribution of individual gasses to the radiative forcing of the atmosphere. Science 289:1922–1925.

Robertson, G.P., and S.M. Swinton. 2005. Reconciling agricultural productivity and environmental integrity: A grand challenge for agriculture. Front. Ecol. Environ. 3(1):38–46.

Rochette, P., and E.G. Gregorich. 1998. Dynamics of soil microbial biomass C, soluble organic C, and CO_2 evolution after three years of manure application. Can. J. Soil Sci. 78:283–290.

Rosenzweig, C., and D. Hillel. 1995. Potential impacts of climate change on agriculture and food supply. Consequences 1(2). Available at http://www.gcrio.org/CONSEQUENCES/summer95/agriculture.html (verified 25 Aug. 2010).

Rosenzweig, C., M.L. Parry, G. Fischer, and K. Frohberg. 1993. Climate change and world food supply. Research Rep. 3. Oxford University, Oxford.

Russell, A.E., D.A. Laird, and A.P. Mallarino. 2006. Nitrogen fertilization and cropping system impacts on soil quality in Midwestern Mollisols. Soil Sci. Soc. Am. J. 70:249–255.

Sa, J.C.M., C.C. Cerri, W.A. Dick, R. Lal, S.P. Venske Fishlo, M.C. Piccolo, and B.E. Feil. 2001. Organic matter dynamics and carbon sequestration rates for a tillage chronosequence in a Brazilian oxisoil. Soil Sci. Soc. Am. J. 65:1486–1499.

Sanchez, J.E., R.R. Harwood, T.C. Willson, K. Kizilkaya, J. Smeenk, E. Parker, E.A. Paul, B.D. Knezek, and G.P. Robertson. 2004. Managing soil carbon and nitrogen for productivity and environmental quality. Agron. J. 96:769–775.

Sandor, J.A., and N.S. Eash. 1995. Ancient agricultural soils in the Andes of Southern Peru. Soil Sci. Soc. Am. J. 59:170–179.

Schumacher, T.E., M.J. Lindstrom, J.A. Schumacher, and G.D. Lemme. 1999. Modeling spatial variation and productivity due to tillage and water erosion. Soil Tillage Res. 51(3–4):331–339.

Shaver, T.M., G.A. Peterson, and L.A. Sherrod. 2003. Cropping intensification in dryland systems improves soil physical properties: Regression relations. Geoderma 116:149–164.

Shepherd, T.G., S. Saggar, R.H. Newman, C.W. Ross, and J.L. Dando. 2001. Tillage-induced changes to soil structure and organic carbon fractions in New Zealand soils. Aust. J. Soil Res. 39(3):465–467.

Six, J., E.T. Elliott, K. Paustian, and J.W. Doran. 1998. Aggregation and soil organic matter accumulation in cultivated and native grassland soils. Soil Sci. Soc. Am. J. 62:1367–1377.

Six, J., C. Feller, K. Denef, S.M. Ogle, and J.C.M. Sa. 2002. Soil organic matter, biota and aggregation in temperate and tropical soils—Effects of no-tillage. Agronomie (Paris) 22:755–775.

Six, J., S.M. Ogle, F.J. Breidt, R.T. Conant, A.R. Mosier, and K. Paustian. 2004. The potential to mitigate global warming with no-tillage management is only realized when practiced in the long-term. Glob. Change Biol. 10(2):155–160.

Smil, V. 1987. Energy, food, environment: Realities, myths, options. Oxford Univ. Press, New York.

Smith, P., C.M. Fang, J.J.C. Dawson, and J.B. Moncrieff. 2008. Impact of global warming on soil organic carbon. Adv. Agron. 97:1–43.

Smith, P., and D.S. Powlson. 2000. Considering manure and carbon sequestration. Science 287:428–429.

Smith, P., D.S. Powlson, M.J. Glendining, and J.U. Smith. 1998. Preliminary estimates of the potential for carbon mitigation in European soils through no-till farming. Glob. Change Biol. 4:679–685.

Smith, S.V., W.H. Renwick, R.W. Buddemeier, and C.J. Crossland. 2001. Budgets of soil erosion and deposition for sediments and sedimentary organic carbon across the conterminous United States. Global Biogeochem. Cycles 15:697–707.

Stockfisch, N., T. Forstreuter, and W. Ehlers. 1999. Ploughing effects on soil organic matter after 20 years of conservation tillage in Lower Saxony, Germany. Soil Tillage Res. 52:91–101.

Tiessen, H., and J.W.B. Stewart. 1983. Particle size fractions and their use in studies of soil organic matter: II.

Cultivation effects on organic matter composition in size fractions. Soil Sci. Soc. Am. J. 47:509–514.

Tinker, P.B. 1998. The environmental implications of intensified land use in developing countries. p. 163–173. *In* D.J. Greenland et al. (ed.) Land resources: On the edge of the Malthusian precipice? CAB International, New York.

Tisdall, J.M. 1996. Formation of soil aggregates and accumulation of soil organic matter. p. 57–96. *In* M.R. Carter and B.A. Stewart (ed.) Structure and organic matter storage in agricultural soils. CRC/Lewis Publ., Boca Raton, FL.

Tisdall, J.M., and J.M. Oades. 1982. Organic matter and water-stable aggregates in soils. J. Soil Sci. 33:141–163.

Titi, A.E. (ed.) 2003. Soil tillage in agroecosystems. CRC Press, New York.

Tracy, P.W., D.G. Westfall, E.T. Elliott, G.A. Peterson, and C.V. Cole. 1990. Carbon, nitrogen, phosphorus and sulfur mineralization in plow and no-till cultivation. Soil Sci. Soc. Am. J. 54:457–461.

Troeh, F.R., J.A. Hobbs, and R.L. Donahue. 1999. Soil and water conservation: Productivity and environmental protection. 3rd ed. Am. Acad. of Environ. Eng., Annapolis, MD.

Uhlen, G., and S. Tveitnes. 1995. Effects of long-term crop rotation, fertilizers, farm manure and straw on soil productivity. Norwegian J. Agric. Sci. 9:143–161.

U.S. Census Bureau. 2007. Total midyear population for the world: 1950–2050. Available at http://www.census.gov/ipc/www/worldpop.html (verified 25 Aug. 2010.)

VandenBygaart, A.J., and B. Kay. 2004. Persistence of soil organic carbon after plowing a long-term no till field in southern Ontario, Canada. Soil Sci. Soc. Am. J. 68:1394–1402.

Volk, T.A., T. Verwijst, P.J. Tharakan, L.P. Abrahamson, and E.H. White. 2004. Growing fuel: A sustainability assessment of Willow biomass crops. Front. Ecol. Environ. 2:411–418.

Wallingford, G.W. 1991. The U.S. nutrient budget in the red. Better Crops Plant Food 75:16–18.

Wanniarachchi, S.D., R.P. Voroney, T.J. Vyn, R.P. Beyaert, and A.F. MacKenzie. 1999. Tillage effects on the dynamics of total and corn-residue derived soil organic matter in two southern Ontario soils. Can. J. Soil Sci. 79:473–480.

Warkentin, B.P. 2001. The tillage effect in sustaining soil functions. J. Plant Nutr. Soil Sci. 164(4):345–350.

Warkentin, B.P. 2008. Soil structure: A history from tilth to habitat. Adv. Agron. 97:239–272.

Weast, R.C. 1968. CRC handbook of chemistry and physics. 48th ed. F-135. Chemical Rubber Co., Boca Raton, FL.

Weast, R.C. 1981. CRC handbook of chemistry and physics. 61st ed. F-202. Chemical Rubber Co., Boca Raton, FL

Webster, C.P., and K.W.T. Goulding. 1989. Influence of soil carbon content on denitrification from fallow land during autumn. J. Sci. Food Agric. 49:131–142.

West, T.O., and W.M. Post. 2002. Soil organic carbon sequestration rates by tillage and crop rotation: A global data analysis. Soil Sci. Soc. Am. J. 66:1930–1946.

Wiebe, K. 2003. Linking land quality, agricultural productivity, and food security. Agric. Econ. Rep. 823. Resource and Economics Division, USDA-ERS, Washington, DC.

Wilhelm, W.W., J.M.-F. Johnson, J.L. Hatfield, W.B. Voorhees, and D.R. Linden. 2004. Crop and soil productivity response to corn residue removal: A literature review. Agron. J. 96:1–17.

Wilhelm, W.W., J.M.F. Johnson, D.L. Karlen, and D.T. Lightle. 2007. Corn stover to sustain soil organic carbon further constrains biomass supply. Agron. J. 99:1665–1667.

Wilts, A.R., D.C. Reicosky, R.R. Allmaras, and C.E. Clapp. 2004. Long-term corn residue effects: Harvest alternatives, soil carbon turnover, and root-derived carbon. Soil Sci. Soc. Am. J. 68:1342–1351.

Wolf, J., P.S. Bindraban, J.C. Luijten, and L.M. Vleeshouwers. 2003. Exploratory study on the land area required for global food supply and the potential global production of bioenergy. Agric. Syst. 76:841–861.

Young, A. 1999. Is there really spare land? A critique of estimates of available cultivable land in developing countries. Environ. Dev. Sustain. 1:3–18.

3

Indices for Soil Management Decisions

Douglas L. Karlen, Brian J. Wienhold, Shujiang Kang,
Ted M. Zobeck, and Susan S. Andrews

Global efforts to identify and develop soil quality indices that can accurately and efficiently quantify effects of soil and crop management began to emerge around the world during the latter portion of the 20th century. This occurred as people became more aware that soil is a unique, nonrenewable resource that nurtures and sustains human civilizations (McNeill and Winiwater, 2004). These efforts have been further encouraged by a growing awareness of the multiple ecosystem services that soil resources provide to sustain food security, environmental quality, ecological functions, and most recently feedstock production for biofuels (Doran et al., 1996; Bouma, 2005; Lal, 2007). In addition to serving as assessment tools, soil quality indices also provide land managers with a better understanding of how their short-term, economically driven management decisions are affecting soil properties and processes over time.

Why Are Indices Needed?

Historically, human neglect of soil resources resulted in the demise of dominant societies and entire cultures (Lowdermilk, 1953; Hillel, 1991; Diamond, 2005). For example, soils of the Tikal rainforest never fully recovered from the Mayan occupation and abandonment that occurred more than 1000 years ago. In southern Mesopotamia, a once thriving land of lush fields is now largely desolate. What were once great cities are now barren mounds of clay rising out of the desert in mute testimony to the glory of a spent civilization.

In the United States, one of the most severe natural resource disasters occurred during the 1930s as a result of ignorance regarding the fragility of the Great Plains' soil resources, which just three decades earlier were described as "indestructible and immutable" in the 1909 Bureau of Soils Bulletin 55 (Whitney, 1909). Implementation of a wheat (*Triticum aestivum* L.)–fallow cropping system and use of intensive tillage throughout the Great Plains contributed to the Dust Bowl that fostered Hugh Hammond Bennett's 1933 indictment of Americans as "the great destroyers of land" (Baumhardt, 2003). Water erosion associated with cotton (*Gossypium hirsutum* L.) production in the southern United States and continuous oat (*Avena sativa* L.) and wheat in

---

D.L. Karlen, USDA-ARS, National Laboratory for Agriculture and the Environment (NLAE), 2110 University Blvd. Ames, IA 50011 (Doug.Karlen@ars.usda.gov); B.J. Wienhold, USDA-ARS, 305 Entomology Hall, University of Nebraska, East Campus, Lincoln, NE 68583; S. Kang and S.S. Andrews, USDA-NRCS, ENTSC, 200 E. Northwood, Ste. 410, Greensboro, NC 27401; T.M. Zobeck, USDA-ARS, Wind Erosion and Water Conservation Research Unit, 3810 4th St., Lubbock, TX 79415. The U.S. Department of Agriculture offers its programs to all eligible persons regardless of race, color, age, sex, or national origin, and is an equal opportunity employer.

doi:10.2136/2011.soilmanagement.c3

the Driftless Region (Major Land Resource Area [MLRA] 105) of the upper Midwest were also responsible for the destruction of fragile soil resources. By 1934, the U.S. government estimated that 1.4×10^7 ha (3.5×10^7 ac) of cultivated croplands had been "essentially destroyed" by soil erosion, while 4.0×10^7 ha (1×10^8 ac) had lost "all or most of the topsoil (USDA, 1934). Rapid and devastating loss of topsoil, and with it the homes and livelihoods of many Americans, led to the establishment of the Soil Erosion Service (now the Natural Resources Conservation Service [NRCS]) and the Coon Creek Watershed project to demonstrate how to best address erosion problems (Hart, 2009). With regard to indices, addressing soil erosion also led to the early development of tools, including the Universal Soil Loss Equation (USLE), Soil Loss Tolerance Standard (T), Revised Universal Soil Loss Equation (RUSLE), Water Erosion Prediction Project (WEPP), and Wind Erosion Equation (WEQ), in this region. During the past 75 years, these tools have helped land managers make much better management decisions and have significantly reduced erosion, but they do not address the full range of ecosystem services provided by soils (Soil and Water Conservation Society, 2008).

The Soil Quality Concept

The "soil quality" concept was introduced by Warkentin and Fletcher (1977) to guide the use and allocation of labor, fiscal resources, and other inputs to meet increasing demands being placed on agriculture. In subsequent decades, the soil quality concept has educated professionals, producers, and the general public about the critical functions soils perform. It has led to the development of assessment tools for comparing management practices and quantifying changes in dynamic soil properties through time. Among the factors that originally slowed acceptance of the concept were perceptions that soil quality assessment was simply an extension of productivity assessments or new soil suitability (interpretations of production capability) ratings as presented in soil surveys and not inclusive of other ecosystem functions or services. Several also argued that soil quality considerations can be traced back to ancient agricultural times when they were used for soil fertility

or productivity assessments (Krupenikov, 1981; Yaalon, 1997; Patzel et al., 2000).

Borggaard (2006) stated that although launching the soil quality concept definitely increased the focus on soils, the multifunctionality of the concept has been difficult to handle. For example, a highly fertilized soil may have high quality as a medium for agricultural crop production, but low quality with regard to protection of groundwater and surface water from nitrate pollution. The challenge is to develop the concept so it can integrate and operationally recognize the simultaneity of diverse and often conflicting soil functions. Others argue that the focus should simply be on "quality soil management" rather than on "soil quality" because of the impact that human decision-making and the management practices that are chosen have on highly variable and unique resources (Sojka and Upchurch, 1999; Sojka et al., 2003; Letey et al., 2003). In reality there is little difference between the two concepts—both focus on improved soil function, the latter attempting to offer assessment techniques to ensure quality soil management is working as intended. Nevertheless, this debate is consistent with that facing the entire soil science discipline. As soil science becomes more integrated with geosciences, environmental sciences, and engineering (Baveye, 2006; Lal, 2007), all are facing new demands that require many traditional disciplinary concepts and theories to be reexamined and perhaps even redefined in an interdisciplinary light.

The concepts of soil quality and land quality share many similar components, especially with regard to indexing land management and environmental issues (Carter, 2002; Bouma, 2002). Anderson and Magleby (1997) suggested that using soil quality to focus on soil functions would better meet the needs of environmentally sound land management. Herrick (2000) suggested that indexing soil quality under various landscapes would be an effective tool for land management. Such efforts could easily complement the land capability and suitability indices developed by the NRCS and thus provide a consistent approach for soil quality assessment (Lal, 1999; Bouma, 2004). Integrating land and soil quality indices could help solve environmental problems across spatial scales. Combining soil quality indexing with information regarding the specific capacity of soils to provide critical functions under different landscape features could help guide and improve land management, especially with regard to assessing impacts of various land use decisions. For example, in New Zealand, the national soil quality monitoring framework provided a major legislative basis for the Resource Management Act (Sparling and Schipper, 2002; Sparling et al., 2004). The European Union identified soil quality as a major focus for environmental assessment by adopting a Thematic Strategy on Soil Protection (Commission of the European Communities, 2006). Research and applications for soil quality assessment and indexing were also important topics at the 18th World Congress of Soil Sciences held in Philadelphia, PA in 2006.

Overall, we contend that both proponents and opponents of the soil quality concept want the same outcome—an improved public awareness of the importance of soil resources and a better understanding of how short-term economic decisions can affect long-term soil properties and processes. This is reflected in the USDA-NRCS strategic plan for 2005–2010 (USDA-NRCS, 2006) where understanding and promoting soil quality was identified as a foundation mission goal for ensuring that the United States continues to have productive lands and a healthy environment. Finally, the importance of focusing on soil quality and its assessment protocols was confirmed by the 2004 special section in *Science* (11 June 2004) that recognized soil as "The Final Frontier" to highlight the importance of this resource and to draw attention to our incomplete knowledge of soil properties, processes, and functions. The articles illustrated how processes occurring in the top few centimeters of Earth's surface are the basis of all life on dry land, but concluded that the opacity of soil has severely limited our understanding of how it functions (Sugden et al., 2004). Based

on the evolution of the concept during the past two decades, it seems likely that the soil quality concept, along with the theories, techniques, and logistics to support its assessment will continue to evolve with an ever-increasing understanding of soil resources and the changing needs associated with managing them for the benefit of humanity.

Soil Quality Assessment Methods

Soil quality scorecards were introduced during the 1990s as one of the first methods to assess soil quality (Harris et al., 1996; Romig et al., 1996; Shepherd, 2000; Shepherd et al., 2000). A scorecard and guidelines for tailoring them to local areas were among the first products developed by the NRCS-Soil Quality Institute (USDA-NRCS, 1999). The cards were developed and promoted primarily to build a basic awareness of soils and to help land managers document their efforts to improve them. Other assessment approaches include use of soil pits and the soil quality test kit (Fig. 3|1) developed by J.W. Doran, M. Sarrantonio, and others (Sarrantonio et al., 1996) to provide a hands-on understanding of how soil physical, chemical, and biological properties and processes change with time and from location to location. The kits, which emulate the "doctor's black bag," can be used to measure water

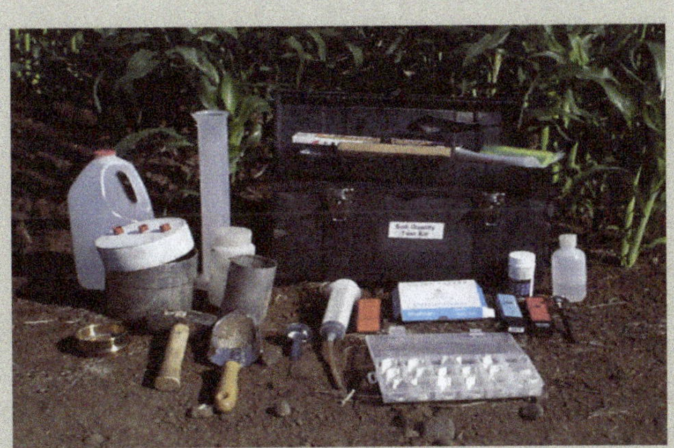

Fig. 3|1. A soil quality test kit, emulating the "doctor's black bag" was developed to demonstrate the importance of soil physical, chemical, and biological properties to the general practitioner and conservationist.

infiltration, bulk density, soil respiration at field capacity, soil stability, soil water content, water holding capacity, water-filled pore space, soil temperature, soil pH, electrical conductivity, and soil nitrate. When used with visual examination of soil profiles, the kit provides information that many conservationists, soil and crop consultants, and others have found useful for understanding spatial and temporal variability among soil resources (Doran et al., 1996; Liebig et al., 1996; USDA-NRCS, 1999).

More recently, the USDA-NRCS has recognized the importance of soil quality by incorporating the Soil Conditioning Index (SCI), a simple, linear predictive model to assess trends in soil organic carbon in crop management systems, into several policies and programs. The SCI was developed from data associated with a 12-yr field study (1948–1959) conducted near Renner, TX (Laws, 1961). The model was released initially for regional planning, and the NRCS Soil Quality Institute further calibrated the model before its national release and added a correction factor for soil texture to the original SCI. This improved the model's accuracy by requiring more biomass production to maintain the level of soil organic matter for coarse-textured soils (USDA-NRCS, 2003). The Institute then validated the SCI using data from long-term carbon studies around the United States. One evaluation, using nine long-term C studies, showed positive trends in soil C were reflected by positive trends in the SCI, while negative SCI trends were associated with negative soil C trends (Hubbs et al., 2002). Another study, using data from 52 western Texas sites, (Zobeck et al., 2007) showed that SCI values were not strongly correlated with total soil organic carbon. However, they were more strongly correlated with a specific soil C fraction known as particulate organic matter carbon, a more labile (changeable) form of C related to recent organic inputs such as animal or green manure, crop residues, or plant roots. A more recent study of different cropping systems on the same soil in Colorado (Zobeck et al., 2008) showed the SCI to be more highly correlated with total soil organic C. Obviously, this is an area of research that needs additional efforts for many different regions and cropping systems.

Following passage of the 2002 U.S. Farm Bill, the SCI was adopted nationally as one factor for determining eligibility for the USDA Conservation Security Program (CSP) and the Environmental Quality Incentives Program (EQIP). However, one limitation of the SCI is that it focuses only on potential changes in soil organic matter. This is justified because if only one indicator is to be used, soil organic matter is often agreed on to be the best choice because of the multitude of soil physical, chemical, and biological properties and processes it influences (USDA-NRCS, 2003). Another limitation is that, while it is well known that soil carbon change is asymptotic, the model does not predict where on the curve a particular system may be. It only provides positive or negative trend information, even when a system has reached a steady state for carbon.

The Soil Management Assessment Framework (SMAF), as described by Andrews et al. (2004), is a measurement-based approach for assessing soil quality. This tool evolved from studies applying principles of systems engineering (Karlen et al., 1994a,b), economics, and ecology (Andrews and Carroll, 2001) to interpret soil physical, chemical, and biological data collected from various soil management studies. The SMAF provides a consistent three-step approach or framework for evaluating all types of cropping systems and management goals by: (i) suggesting goal appropriate indicators, (ii) providing indicator interpretation within inherent soil and climatic context, and (iii) if desired, combining the ratings into an overall assessment of dynamic soil function (Andrews et al., 2002a,b, 2004). The SMAF has successfully distinguished between "dynamic soil properties" (or quality), which are responsive to current or recent management decisions on the human time scale, and "inherent soil properties," which are determined by basic soil forming factors and relatively unresponsive to recent management (Tugel et al., 2005).

A similar indexing approach has also been incorporated into the Agroecosystem Performance Assessment Tool (AEPAT). The AEPAT is a computer program designed to assess agronomic socioeconomic and environmental performance of soil and crop management practices (Liebig et al., 2004). Measured indicators are assigned by the user to various soil functions (e.g., food/feed production, nutrient cycling), as well as social and economic indicators such as net profit or quality of life indicators. The functions are weighted by the user, and individual function scores are combined into an index. The AEPAT was used to compare cropping system effects on soil quality using information from several long-term studies throughout the Great Plains (Wienhold et al., 2006), but it is designed primarily for soil scientists (most likely researchers) because indicators, their relationships to soil function, and weighting factors must all be defined by the user.

A simplified two-step version of SMAF with slightly different indicators is used in the Cornell Soil Health Assessment program (http://soilhealth.cals.cornell.edu/index.htm, verified 30 Aug. 2010), which was the first commercially available program to offer balanced assessments of soil physical, chemical, and biological quality (Gugino et al., 2007). This program was developed to facilitate education about soil health, guide farmers and land managers in their selection of soil management practices, provide monitoring for the NRCS, and indirectly increase land values by providing information regarding the soil's overall condition. Measured biological, chemical, and physical indicator values are interpreted using various nonlinear response curves, modified by soil texture. The tool has been found to be sensitive to soil and crop management practices (e.g., tillage, crop rotation, and animal manure) on hundreds of farms across New York and vicinity. Results are relevant to what has been defined as critical soil functions (Doran and Parkin, 1994), consistent and reproducible, easy to sample for, and economical for soil-testing laboratories to implement.

All three of these assessment tools (SMAF, AEPAT, and the Cornell Soil Health Test) focus on "dynamic soil quality," which describes the current soil condition created by recent soil management decisions, rather than "inherent soil quality," which reflects the basic soil forming factors of climate, parent material, time, topography, and vegetation (Seybold et al., 1998).

Development of Soil Quality Indices

Figures 3|2 and 3|3 illustrate two important points with regard to developing indices for soil quality assessment. Figure 3|2 illustrates inherent differences between soils and why meaningful comparisons can be made only by soil map unit component or phase (with similar surface texture and slope) for defined locations. The fluctuation about either soil (Fig. 3|2) shows there will be steady-state differences over time. The important interpretation that assessments must help identify is the trend in that fluctuation (Fig. 3|3). Are soil resources being improved, degraded, or at least maintained? With regard to the sometimes controversial issue of what baseline condition (e.g., native prairie, fencerow, cemetery, pasture, cultivated field) to use for indexing soil quality, we conclude that it does not matter. Since it is not possible to go back in time and many of the suggested reference conditions would not require the same soil functions as current land use, the most meaningful approach for examining long-term effects is to measure soil management effects every 3 to 5 yr using the same sampling and

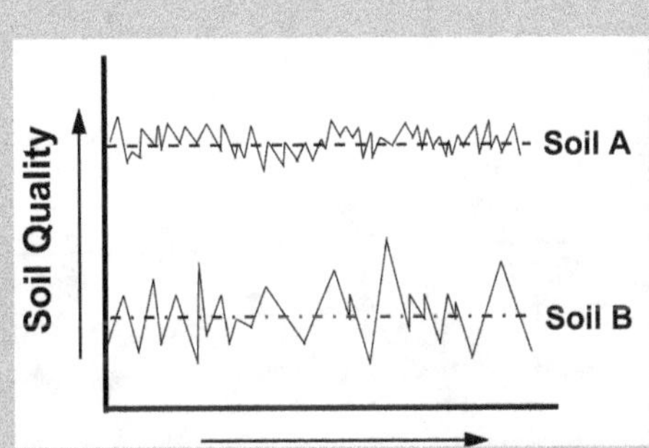

Fig. 3|2. Conceptualization of inherent soil quality differences between two soils. Adapted from Karlen et al. (2001).

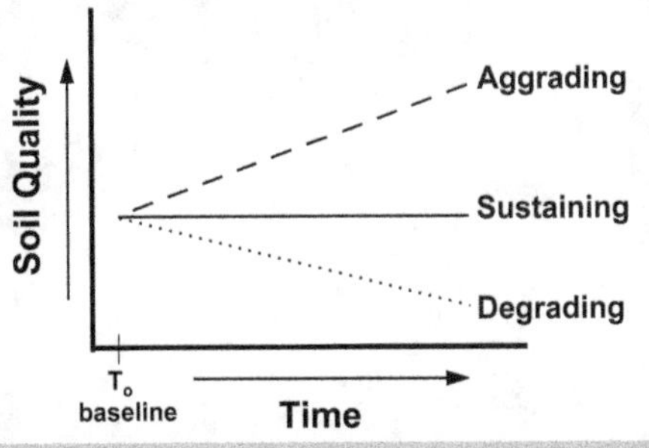

Fig. 3|3. Conceptualization of dynamic soil quality trends from time zero (T0). Adapted from Seybold et al. (1998).

indicator interpretations to quantify important trends. This approach provides the information needed to know if the practices being used are causing critical soil functions to improve, decline, or at least remain stable (Fig. 3|3). When monitoring over time is not possible, sampling similar soils under different steady-state management conditions allows an inventory of soil function related to management (Tugel et al., 2009).

Recent Soil Quality Assessment Studies

Examples of ongoing research to improve soil quality indices include a national effort to use the SMAF to quantify environmental benefits of public investment in conservation practices. Soil samples are being collected at 14 benchmark watershed sites associated with the ARS Conservation Effects Assessment Project (CEAP). The overall goal of CEAP is to quantify how agricultural management practices are influencing soil and water quality (NRC,

1993), thus providing an excellent dataset for validating the SMAF. Recognizing that high rates of soil erosion, loss of soil organic matter, imbalanced soil fertility, and chemical or heavy metal contamination continue to be critical soil quality issues (Larson and Pierce, 1991: Doran and Parkin, 1994; Karlen et al., 2001, 2003, 2006), the SMAF (Andrews et al., 2004) was chosen for this assessment because of its design to use biological, chemical, and physical indicators in an organized and consistent manner

A survey approach was chosen to identify the most limiting soil properties or processes within each benchmark watershed (Fig. 3|4). An initial assessment within the South Fork Watershed of the Iowa River (Karlen et al., 2008) provided the foundation for the overall CEAP soil quality program. Samples were collected from five to ten locations (as replicates) under three to five conservation practices within three to five soil map units in each watershed. At each sampling site, 20 soil cores were collected from the 0- to 5-cm depth using a soil probe with an inner diameter of 3.2

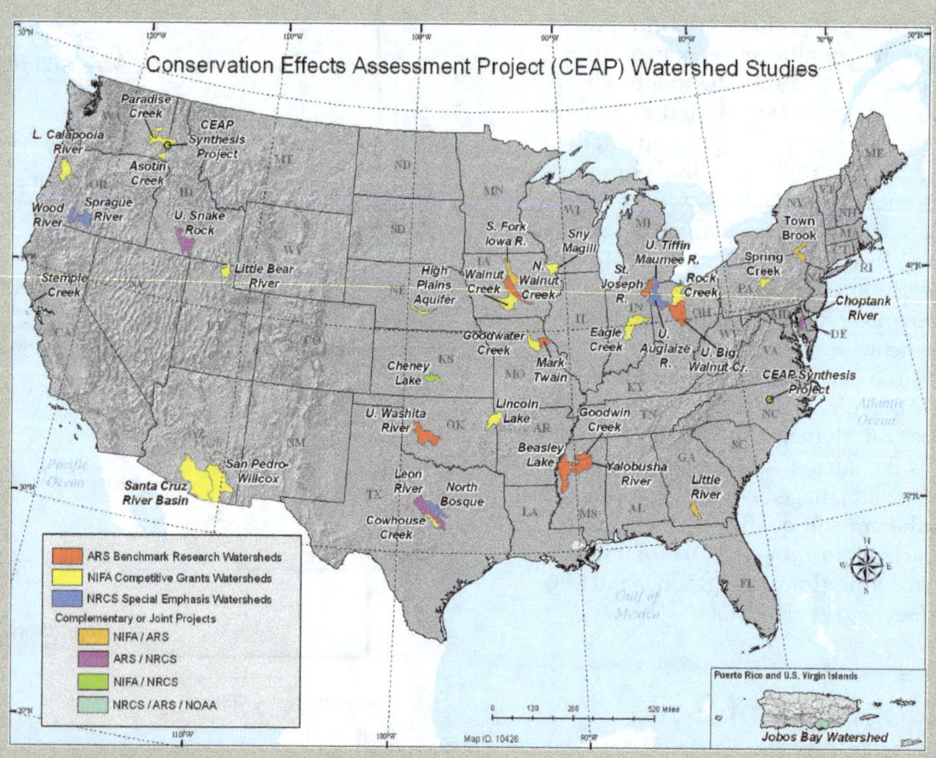

Fig. 3|4. Location of USDA-ARS Conservation Effects Assessment Project (CEAP) Benchmark Watersheds.

cm. Then, depending on the local research questions, either additional samples from lower depths were collected or more sites were sampled. All sampling sites were georeferenced, and soil map unit, landscape position, slope, and any evidence of wind, water, or tillage-induced soil erosion and periodic ponding or flooding was documented. Current and past management information from the land owner or operator was collected when possible. This included conservation practices, fertilizer and/or manure management histories, crop rotations, tillage practices, yields, and other pertinent information that is known to affect soil resources.

To date, 13 of 14 CEAP benchmark watersheds have been sampled, and soil analyses are nearly complete for five of them. A preliminary examination of the data shows that low SOM, especially on hilltops where water, wind, and tillage erosion (Schumacher et al., 2005) have decreased topsoil depth over time, is one of the most consistent findings. Areas receiving excess P through frequent animal manure applications often show increasing levels of soil-test P and an increased potential for surface water contamination through runoff that contains excessive levels of soluble P. This appears consistent with results from the initial South Fork watershed study (Karlen et al., 2008) that showed soil-test P ratings for upland soils were generally very high (>31 μg g^{-1}) (Mallarino et al., 2002) but not to the levels (e.g., >100 μg g^{-1}) at which severe environmental impact would be expected. Lower soil-test P ratings in the depression areas were consistent with the higher pH in those soils. Soil-test K in the initial South Fork study was generally in an optimum range (131–170 μg g^{-1}) for corn (*Zea mays* L.) and soybean [*Glycine max* (L.) Merr.] production, but some areas had surprisingly low K values. This could result in early season plant K deficiencies if no-tillage practices are used (Karlen and Kovar, 2005) to reduce soil erosion. Therefore, since reduced or no-tillage practices would be beneficial to increase soil C levels, close monitoring of K levels is recommended to prevent that essential plant nutrient from limiting crop yields.

A cropping systems study in Colorado (Zobeck et al., 2008), separate from the CEAP Watershed work, was used to compare the SCI and SMAF indices for their ability to detect management differences due to tillage, cropping sequence, and N fertilizer rate. Both indices differentiated among the three N rates with the SMAF index clearly identifying the plots that received very high N rates from those that received none. The intermediate N rate, however, was not significantly different from the two extremes. In contrast, the SCI identified distinct differences among all N rates, but the differences were the same as those found for crop yields and residue returned to the soil. The SMAF index was more sensitive and showed more distinct differences among crop management systems. The SMAF index values were reduced as tillage intensity increased and residue cover decreased.

In a Nebraska study, the SMAF was used to develop methods for conducting soil management assessments within spatially variable fields. Apparent electrical conductivity (EC$_a$) was intensively sampled for an entire field near Carleton, NE to evaluate spatial variability for several soil indicators. The predominant soil series at the site is a Muir silt loam (fine-silty, mixed, superactive mesic cumulic Haplustoll). The EC$_a$ survey was conducted using a Geonics EM-38 (Geonics Limited, Mississauga, ON, Canada) mounted on a nonmetallic sled pulled behind an all terrain vehicle.[1] All data was georeferenced as the survey was conducted with readings logged every 5 s. The survey consisted of 25 transects (20 m apart) and resulted in a total of 1958 EC$_a$ measurements. The survey data were processed using the ESAP software package (Lesch et al., 2000). This program uses spatial statistics to select sampling locations that reflect the observed spatial variability in EC$_a$ (Corwin and Lesch, 2003). Measured indicator data were also collected for 20 locations throughout the field. At each location a soil core was collected from the 0- to 90-cm depth and sectioned into 0- to 15-, 15- to 30-, 30- to 60-, and 60- to 90-cm increments, air-dried, and sieved. Soil bulk density, pH, electrical conductivity, organic matter content, and Bray-available P were determined.

The measured soil quality indicator data from the 20 points were used to calibrate the ESAP readings for those same points by calculating regression equations. Statistically

---

[1] Mention of trademark, proprietary product, or vendor is for information only and does not constitute a guarantee or warranty of the product by the USDA or imply its approval to the exclusion of other products or vendors that may also be suitable.

Table 3|1. Coefficient of determination (r^2) between Apparent Electrical Conductivity and select soil indicators for a Muir silt loam in southeastern Nebraska, USA.

Indicator	r^2	p value
Bulk density	0.47	0.014
Electrical conductivity	0.86	0.001
pH	0.63	0.002
Bray phosphorus	0.68	0.001
Soil organic matter	0.87	0.001

significant relationships were determined (Table 3|1) for five of the indicators currently being used by the SMAF to index soil quality. The calibration equations were then used to estimate indicator values at the other 1938 EC_a sample locations. The 1958 indicator values were then scored using the SMAF (Wienhold et al., 2008).

Values for EC_a ranged from 12 to 62 dS m^{-1} with high values observed in the northwest and southeast portions of the field and low values observed in the middle of the field. Salinity is not an issue at this site, so the variation in EC_a is most likely due to variation in clay content, soil organic matter content, and depth of topsoil (Johnson et al., 2001; Grigera et al., 2006). Values for Bray available P ranged from 3 to 45 mg kg^{-1} with high values on both ends and low values in the middle of the field. All values were below the threshold value (100 mg kg^{-1}) where the potential for environmental contamination is a concern and the SMAF scoring curve begins to lower the relative score (Andrews et al., 2004). Overall, this indexing approach was useful for identifying areas where additional P fertilizer would probably result in a positive yield response and where additional applications would not be beneficial.

These three studies and many others not reported here have demonstrated that indices can help quantify effects of agricultural management practices. Further assessments using the SMAF at field, farm, and watershed scales are needed, but preliminary results suggest this approach is appropriate and consistent with the goals stated in the publication, *Soil and Water Quality: An Agenda for Agriculture* (National Research Council, 1993).

On-Going Improvements for Soil Quality Indexing

The Soil and Water Conservation Society (2008) recently published results from an expert consultation that identified actions needed for more comprehensive soil assessment, management, and planning tools. That panel evaluated several soil management assessment tools, including the SMAF and the SCI. One recommendation was that the number of available scoring curves for interpreting measured soil indicators in the SMAF be increased. The original version of the SMAF (Andrews et al., 2004) had scoring curves for 10 soil attributes, but more than 60 other attributes were identified as having potential for being assessment indicators.

The approach being used to develop scoring curves for the SMAF involves a number of steps. The first is to identify a soil indicator that responds to management and affects a soil function of interest. Data sets containing indicator values and measures of soil function, preferably over a range of environmental conditions, must be identified or collected. These data sets are used to determine the shape of the curvilinear relationship between the indicator and the soil function and then to develop an algorithm describing that relationship. Abiotic factors that cause the relationship to change or the expected range to shift are identified to allow for appropriate interpretation of the indicator within its environmental context. Coefficients or logic statements modify each algorithm to mimic these environmental factors. The algorithm is then programmed into the SMAF and validated using additional data sets.

Recent efforts to develop additional scoring curves include Wienhold et al. (2009), who developed curves for a physical soil attribute (water-filled pore space), a chemical soil attribute (soil test K), and a biological soil attribute (β-glucosidase activity). Stott et al. (2010) also developed scoring curves for a suite of soil enzymes by using original data relating measured soil enzyme activity to management outcomes.

Further development of indices for soil quality assessment will be a continuous process, fostered by incremental improvements in our understanding of physical, chemical, and biological soil processes, as well as how they can be most effectively quantified. Assessing soil functions requires not only current soil science studies, but also information from associated disciplines such as geosciences, biology, hydrology, and engineering. Since soil quality depends on soil processes (Wagenet and Hutson, 1997), many are concerned about interpreting diverse soil functions with scores or indices based on one-time, snap-shot measurements of soil properties.

Directly quantifying capacities of soil functions and their associated processes is essential. The development of pedotransfer functions that emphasize soil processes and functions would also be useful to help calculate various soil capacities (Wösten, 1997). Applying the basic concepts and principles of soil–water–landscape dynamics being addressed by hydropedologists (Lin, 2003) will also enlighten the process-based indexing of soil quality.

Currently, the capacities for some soil functions such as soil resistance and resilience to change in function are not well understood, and it is therefore very difficult to develop reliable indices for quantifying such functions. One major difficulty with predicting each soil's resistance and resilience is that it will vary depending not only on inherent and dynamic (management induced) soil properties but also with type and intensity of soil disturbance and the specific functions of interest. In other words, one soil will have numerous resistances and resiliencies for different functions and disturbances. With a greater focus on soil quality and its relationship to environmental and ecological issues, we anticipate the soil functions influencing water quality and air quality will become components for holistic environmental quality assessments, with predictions of resistance and resilience a natural corollary of this work

Future Indexing Efforts

As indices to assess soil quality continue to evolve and improve, one of the future efforts to meet various resource management needs will be the development of soil quality information systems. An ideal soil quality information system would include a combination of soil quality databases, assessment tools, predictive models, and decision-making tools. To provide for a variety of users, this system would be expected to provide not only soil quality assessment scores and

indices over time but also compile data for determining soil capacities for diverse uses and the outcomes of those uses. The system would also provide inputs for environmental modeling and/or farm bill program evaluations. This soils information system would be open, allowing the introduction of new soil ecosystem functions and soil indicators or for renewing existing algorithms when an improved understanding of soil properties and processes necessitates change. Some current soil databases will be valuable to support the development of such systems. The STATSGO and SSURGO databases provide rich soil information for inherent soil quality assessment at different scales. The Natural Resources Inventory (NRI) program of the USDA-NRCS has performed nationwide resources monitoring every 4 yr since 1984 and now monitors a subset of points annually. The NRI datasets could provide valuable spatial and temporal land use and management information for plotting soil quality trends in this country. Combining CEAP with SMAF scoring (Potter et al., 2006) could provide contextual meaning to modeling outcomes that would be a valuable predictive tool. These databases could help develop national soil quality criteria and also improve soil quality tool development and validation. In the mean time, NRI and other soil databases could facilitate development of soil quality monitoring tools as suggested by Karlen et al. (2003). New, but not well quantified soil functions, such as resistance and resilience, urban soil quality assessment, and soil quality change patterns due to global warming, could also become major components of an extensive soils information system. Further improvements could be achieved as geographic information system (GIS), remote sensing, modeling, and data mining tools are developed and customized for indexing soil quality with regard to various needs and applications.

The soil quality information system should also include "prediction and uncertainty" guidelines to help interpret the indices and soil quality assessments. This is consistent with predictions by Tugel et al. (2005), who proposed a blueprint for quantifying soil changes through soil survey and decision-making processes. They also suggested that dynamic soil properties should be integrated into future soil databases. An ideal soil quality information system would not only assess the current states of soil quality but also provide trends and decision-making tools. Development of a soil quality information system such as this could be a pivotal bridge between soil science research and soil management practices.

Using Indices to Improve Soil Management

How might soil indices such as those described above be used to improve soil management decisions? One of the first applications would be to enhance routine soil test information that currently focuses almost exclusively on soil chemical or fertility parameters. Applying indices that account for physical, chemical, and biological properties and processes is the focus of the Cornell Soil Health Assessment program (Gugino et al., 2007). Current efforts to develop sustainable feedstock supplies for biofuel and other bioproducts offer another immediate application for indices such as the SMAF since initial estimates of feedstock supply were based solely on retaining sufficient surface cover to protect against wind and water erosion and not to sustain soil carbon (Wilhelm et al., 2007). Another application could be to help set land rental and purchase value based not only on potential productivity but also on the current physical, chemical, and biological status of the soil resource. This is an underlying reason for development of procedures for assessing soil change within soil survey and vegetation and ecological site inventories by the NRCS (Tugel et al., 2009). The critical point associated with these and other indexing applications is that soils are living, dynamic, and ever-changing bodies that are affected by our soil management decisions. Responses may be immediate, but more likely will be more insidious and hard to identify unless all aspects—physical, chemical, and biological—are monitored on a routine basis, perhaps every 3 to 5 yr. Incorporating such monitoring into long-range soil management plans will undoubtedly benefit not only the land owner and manager, but many others dependent on the ecological services that soil resources provide.

Summary

The importance and need for indices to guide improved soil management have become well established during the past three decades. As a result, efforts to develop soil quality assessment tools are underway and expected to go through continued development for several years. The soil quality assessment process is expected to be a holistic approach for examining multiple soil functions regarding productivity, environmental buffering, and ecosystem sustainability. Tools sensitive to soil biological, chemical, and physical indicators are needed to fully evaluate the impact of soil management decisions, such as when and where to harvest crop residues for biofuel feedstocks or when, where, and how to apply animal manures. The AEPAT, SCI, Cornell Soil Health Assessment, and SMAF are in various stages of development, release, refinement, or dormancy. The SCI has been incorporated into RUSLE2 software and is being used by the NRCS to assist with some program decisions. The Cornell Soil Health Assessment was successfully used on a trial basis in 2008 for several participatory research studies in New England. The SMAF has been evaluated at several scales and appears to be sensitive to various management scenarios. Scoring curves for three additional indicators (water-filled pore space, soil-test K, and the soil enzyme β-glucosidase were recently developed. Opportunities exist for adding many additional indices to the framework to make the tool even more robust and useful. Regardless of past perceptions of soil quality, we invite all readers to join in a concerted effort to move soil quality assessment beyond single factor analyses in a meaningful way so that soil management practices can be improved and everyone can benefit from our better understanding "The Final Frontier."

References

Anderson, M., and R. Magleby. 1997. Land and soil quality. p. 41–49. In Agricultural resources and environmental indicators. Agric. Handb. AH712. Available at http://www.ers.usda.gov/publications/arei/ah712/AH71213a.PDF (verified 30 Aug. 2010).

Andrews, S.S., and C.R. Carroll. 2001. Designing a decision tool for sustainable agroecosystem management: Soil quality assessment of a poultry litter management case study. Ecol. Appl. 11(6):1573–1585.

Andrews, S.S., D.L. Karlen, and C.A. Cambardella. 2004. The soil management assessment framework: A quantitative evaluation using case studies. Soil Sci. Soc. Am. J. 68:1945–1962.

Andrews, S.S., D.L. Karlen, and J.P. Mitchell. 2002a. A comparison of soil quality indexing methods for vegetable production systems in northern California. Agric. Ecosyst. Environ. 1760:1–21.

Andrews, S.S., J.P. Mitchell, R. Mancinelli, D.L. Karlen, T.K. Hartz, W.R. Horwath, G.S. Pettygrove, K.M. Scow, and D.S. Munk. 2002b. On-farm assessment of soil quality in California's Central Valley. Agron. J. 94:12–23.

Baumhardt, R.L. 2003. Dust Bowl era. p. 187–191. In B.A. Stewart and T.A. Howell (ed.) Encyclopedia of water science. Marcel-Dekker, New York.

Baveye, P. 2006. A future of soil science. J. Soil Water Conserv. 61:148–151.

Borggaard, O.K. 2006. Future of soil science. p. 19–21. In A.E. Hartemink (ed.) The future of soil science. International Union of Soil Sciences (IUSS), Wageningen, The Netherlands.

Bouma, J. 2002. Land quality indicators of sustainable land management across scales. Agric. Ecosyst. Environ. 88:129–136.

Bouma, J. 2004. Implementing soil quality knowledge in land-use planning. p. 283–296. In P. Schjønning et al. (ed.) Managing soil quality: Challenges in modern agriculture. CAB International, Wallingford, UK.

Bouma, J. 2005. Soil scientists in a changing world. Adv. Agron. 88:67–96.

Carter, M.R. 2002. Soil quality for sustainable land management: Organic matter and aggregation interactions that maintain soil functions. Agron. J. 94:38–47.

Commission of the European Communities. 2006. Questions and answers on the Thematic Strategy on soil protection. Available at http://europa.eu/rapid/pressReleasesAction.do?reference=MEMO/06/341&format=HTML&aged=0&language=EN&guiLanguage=en (verified 30 Aug. 2010).

Corwin, D.L., and S.M. Lesch. 2003. Application of soil electrical conductivity to precision agriculture: Theory, principles, and guidelines. Agron. J. 95:455–471.

Diamond, J. 2005. Collapse: How societies choose to fail or succeed. Viking Press, New York.

Doran, J.W., and T.B. Parkin. 1994. Defining and assessing soil quality. p. 3–21. In J.W. Doran, et al. (ed.) Defining soil quality for a sustainable environment. SSSA Spec. Publ. 35. SSSA and ASA, Madison, WI.

Doran, J.W., M. Sarrantonio, and M.A. Liebig. 1996. Soil health and sustainability. Adv. Agron. 56:1–55.

Grigera, M.S., R.A. Drijber, K.M. Eskridge, and B.J. Wienhold. 2006. Soil microbial biomass relationships with organic matter fractions in a Nebraska corn field mapped using apparent electrical conductivity. Soil Sci. Soc. Am. J. 70:1480–1488.

Gugino, B.K., O.J. Idowu, R.R. Schindelbeck, H.M. van Es, D.W. Wolfe, B.N. Moebius, J.E. Thies, and G.S. Abawi. 2007. Cornell Soil Health Assessment Training Manual. ed. 1.2.2. Available at http://soilhealth.cals.cornell.edu (verified 30 Aug. 2010). Cornell University, Geneva, NY.

Harris, R.F., D.L. Karlen, and D.J. Mulla. 1996. A conceptual framework for assessment and management of soil quality and health. p. 61–82. In J.W. Doran and A.J. Jones (ed.) Methods for assessing soil quality. SSSA Spec. Publ. 49. SSSA, Madison, WI.

Hart, J. 2009. Ground lost and gained in 75 years of conservation at Coon Creek. p. 102A–106A. In M. Anderson-Wilk (ed.) Relationship with the land. Soil Water Conserv. Soc., Ankeny, IA.

Herrick, J.E. 2000. Soil quality: An indicator of sustainable land management. Appl. Soil Ecol. 15:75–83.

Hillel, D. 1991. Out of the Earth: Civilization and the life of the soil. Univ. of California Press, Los Angeles.

Hubbs, M.D., M.L. Norfleet, and D.T. Lightle. 2002. Interpreting the soil conditioning index. p. 192–196. In E.V. Santen (ed.) Making conservation tillage conventional: Building a future on 25 years of research. Proc. 25th Annual Southern Conservation Tillage Conf. for Sustainable Agriculture. Spec. Rep. 1. Alabama Agric. Exp. Stn. and Auburn University, Auburn, AL.

Johnson, C.K., J.W. Doran, H.R. Duke, B.J. Wienhold, K.M. Eskridge, and J.F. Shanahan. 2001. Field-scale electrical conductivity mapping for delineating soil condition. Soil Sci. Soc. Am. J. 65:1829–1837.

Karlen, D.L., S.S. Andrews, and J.W. Doran. 2001. Soil quality: Current concepts and applications. Adv. Agron. 74:1–40.

Karlen, D.L., S.S. Andrews, B.J. Wienhold, and J.W. Doran. 2003. Soil quality: Humankind's foundation for survival. J. Soil Water Conserv. 58:171–179.

Karlen, D.L., E.G. Hurley, S.S. Andrews, C.A. Cambardella, D.W. Meek, M.D. Duffy, and A.P. Mallarino. 2006. Crop rotation effects on soil quality at three northern Corn/Soybean Belt locations. Agron. J. 98:484–495.

Karlen, D.L., and J.L. Kovar. 2005. Is K the Cinderella nutrient for reduced tillage systems? Fluid J. 13:8–11.

Karlen, D.L., M.D. Tomer, J. Neppel, and C.A. Cambardella. 2008. A preliminary watershed scale soil quality assessment in north central Iowa, USA. Soil Tillage Res. 99:291–299.

Karlen, D.L., N.C. Wollenhaupt, D.C. Erbach, E.C. Berry, J.B. Swan, N.S. Eash, and J.L. Jordahl. 1994a. Crop residue effects on soil quality following 10 years of no-till corn. Soil Tillage Res. 31:149–167.

Karlen, D.L., N.C. Wollenhaupt, D.C. Erbach, E.C. Berry, J.B. Swan, N.S. Eash, and J.L. Jordahl. 1994b. Long-term tillage effects on soil quality. Soil Tillage Res. 32:313–327.

Krupenikov, I.A. 1981. History of soil science. Nauka Publishers, Moscow, Russia.

Lal, R. 1999. Soil quality and food security: The global perspective. p. 3–16. In R. Lal (ed.) Soil quality and soil erosion. Soil Water Conserv. Soc., Inc., Ankeny, Iowa.

Lal, R. 2007. Soil science and the carbon civilization. Soil Sci. Soc. Am. J. 71:1425–1437.

Larson, W.E., and F.J. Pierce. 1991. Conservation and enhancement of soil quality. p. 175–203. In J. Dumanski et al. (ed.) Evaluation for sustainable land management in the developing world. Vol. 2. Technical papers. Proc. Int. Workshop, Chiang Rai, Thailand. 15–21 Sept. 1991. Int. Board for Soil Res. and Management, Bangkok, Thailand.

Laws, W.D. 1961. Farming systems for soil improvement in the Blacklands. Texas Research Foundation Bull. 10. Texas A&M University, College Stn., TX.

Lesch, S.M., J.D. Rhodes, and D.L. Corwin. 2000. The ESAP-95 version 2.01R user manual and tutorial guide. Research Rep. 146. USDA-ARS, George E. Brown, Jr., Salinity Laboratory, Riverside, CA.

Letey, J., R.E. Sojka, D.R. Upchurch, D.K. Cassel, K.R. Olson, W.A. Payne, S.E. Peterie, G.H. Price, R.J. Reginato, H.D. Scott, P.J. Smethurst, and G.B. Triplett. 2003. Deficiencies in the soil quality concept and its application. J. Soil Water Conserv. 58:180–187.

Liebig, M.A., J.W. Doran, and J.C. Gardner. 1996. Evaluation of a field test kit for measuring selected soil quality indicators. Agron. J. 88:683–686.

Liebig, M.A., M.E. Miller, G.E. Varvel, J.W. Doran, and J.D. Hanson. 2004. AEPAT: A computer program to assess agronomic and environmental performance of management practices in long-term agroecosystem experiments. Agron. J. 96:109–115.

Lin, H.S. 2003. Hydropedology: Bridging disciplines, scales, and data. Vadose Zone J. 5:317–340.

Lowdermilk, W.C. 1953. Conquest of the land through seven thousand years. Agric. Bull. 99. U.S. Gov. Print. Office, Washington, DC.

Mallarino, A.P., D.J. Wittry, and P.A. Barbagelata. 2002. Iowa soil-test field calibration research update:

Potassium and the Mehlich-3 ICP phosphorus test. ISU Dep. of Agronomy. Available at http://www.agronext. iastate.edu/soilfertility/ (verified 30 Aug. 2010).

McNeill, J.R., and V. Winiwater. 2004. Breaking the sod: Humankind, history, and soil. Science 304:1627–1629.

National Research Council. 1993. Soil and water quality: An agenda for agriculture. National Academy Press, Washington, DC.

Patzel, N., H. Sticher, and D.L. Karlen. 2000. Soil fertility—Phenomenon and concept. J. Plant Nutr. Soil Sci. 163:129–142.

Potter, S.R., S.S. Andrews, J.D. Atwood, R.L. Kellogg, J. Lemunyon, M.L. Norfleet, and D. Oman. 2006. Model simulation of soil loss, nutrient loss, and change in soil organic carbon associated with crop production. USDA-NRCS Conservation Effects Assessment Report. June 2006. Available at http://www.nrcs.usda. gov/technical/nri/ceap/croplandreport/ (verified 30 Aug. 2010).

Romig, D.E., M.J. Garlynd, and R.F. Harris. 1996. Farmer-based assessment of soil quality: A soil health scorecard. p. 39–60. *In* J.W. Doran and A.J. Jones (ed.) Methods for assessing soil quality. SSSA Spec. Publ. 49. SSSA, Madison, WI.

Sarrantonio, M., J.W. Doran, M.A. Liebig, and J.J. Halvorson. 1996. On-farm assessment of soil quality and health. p. 83–105. *In* J.W. Doran and A.J. Jones (ed.) Methods for assessing soil quality. SSSA Spec. Publ. 49. SSSA, Madison, WI.

Schumacher, J.A., T.C. Kaspar, J.C. Ritchie, T.E. Schumacher, D.L. Karlen, E.R. Venteris, G.M. McCarty, T.S. Colvin, D.B. Jaynes, M.J. Lindstrom, and T.E. Fenton. 2005. Identifying spatial patterns of erosion for use in precision conservation. Soil Tillage Res. 60:355–362.

Seybold, C.A., M.J. Mausbach, D.L. Karlen, and H.H. Rogers. 1998. Quantification of soil quality. p. 387–404. *In* R. Lal et al. (ed.) Soil processes and the carbon cycle. CRC Press, Boca Raton, FL.

Shepherd, T.G. 2000. Visual soil assessment. Vol. 1. Field guide for cropping and pastoral grazing on flat to rolling country. horizons.mw & Landcare Research, Palmerston North, New Zealand.

Shepherd, T.G., C.W. Ross, L.R. Basher, and S. Saggar. 2000. Visual soil assessment. Vol. 2. Soil management guidelines for cropping and pastoral grazing on flat to rolling country. horizons.mw & Landcare Research. Palmerston North, New Zealand.

Soil and Water Conservation Society. 2008. BEYOND T: Informing sustainable soil management. Available at http://www.swcs.org/en/publications/beyond_t/ (verified 30 Aug. 2010). Soil Water Conserv. Soc., Ankeny, IA.

Sojka, R.E., and D.R. Upchurch. 1999. Reservations regarding the soil quality concept. Soil Sci. Soc. Am. J. 63:1039–1054.

Sojka, R.E., D.R. Upchurch, and N.E. Borlaug. 2003. Quality soil management or soil quality management: Performance versus semantics. Adv. Agron. 79: 1–68.

Sparling, G.P., and L.A. Schipper. 2002. Soil quality at a national scale in New Zealand. J. Environ. Qual. 31:1848–1857.

Sparling, G.P., L.A. Schipper, W. Bettjeman, and R. Hill. 2004. Soil quality monitoring in New Zealand: Practical lessens from a 6-year trial. Agric. Ecosyst. Environ. 104:523–534.

Stott, D.E., S.S. Andrews, M.A. Liebig, B.J. Wienhold, and D.L. Karlen. 2010. Evaluation of β-glucosidase activity as a soil quality indicator. Soil Sci. Soc. Am. J. 74:107–119.

Sugden, A., R. Stone, and C. Ash. 2004. Ecology in the underworld. Science 304:1613.

Tugel, A.J., J.E. Herrick, J.R. Brown, M.J. Mausbach, W. Puckett, and K. Hipple. 2005. Soil change, soil survey, and natural resources decision making: A blueprint for action. Soil Sci. Soc. Am. J. 69:738–747.

Tugel, A.J., S.A. Wills, and J.E. Herrick. 2009. Soil change guide: Procedures for soil survey and resource inventory. Version 1.1. USDA-NRCS, National Soil Survey Center, Lincoln, NE.

USDA. 1934. Yearbook of agriculture. USDA, Washington, DC.

USDA-NRCS. 1999. Soil quality test kit guide. Available at http://soils.usda.gov/sqi/assessment/files/test_kit_complete.pdf (verified 30 Aug. 2010).

USDA-NRCS. 2003. Interpreting the soil conditioning index: A tool for measuring soil organic matter trends. Available at http://soils.usda.gov/SQI/management/files/sq_atn_16.pdf (verified 30 Aug. 2010).

USDA-NRCS. 2006. Productive lands & healthy environment: Natural Resources Conservation Service Strategic Plan 2005–2010. USDA-NRCS, Washington, DC.

Wagenet, R.J., and J.L. Hutson. 1997. Soil quality and its dependence on dynamic physical processes. J. Environ. Qual. 26:41–48.

Warkentin, B.P., and H.F. Fletcher. 1977. Soil quality for intensive agriculture. p. 594–598. Proceedings of the International Seminar on Soil Environment and Fertility Management in Intensive Agriculture. Society of Science of Soil and Manure, National Institute of Agricultural Science, Tokyo.

Whitney, M. 1909. Soils of the United States: Based upon the work of the Bureau of Soils to January 1, 1908. Bureau of Soils Bull. 55. USDA, Gov. Print. Office, Washington, DC.

Wienhold, B.J., S.S. Andrews, H. Kuykendall, and D.L. Karlen. 2008. Recent advances in soil quality assessment in the United States. J. Indian Soc. Soil Sci. 56:237–246.

Wienhold, B.J., D.L. Karlen, S.S. Andrews, and D.E. Stott. 2009. Protocol for Indicator Scoring in the Soil Management Assessment Framework (SMAF). Renewable Agric. Food Sys. 24(4):260–266.

Wienhold, B.J., J.L. Pikul, Jr., M.A. Liebig, M.M. Mikha, G.E. Varvel, J.W. Doran, and S.S. Andrews. 2006. Cropping system effects on soil quality in the Great Plains: Synthesis from a regional project. Renew. Agric. Food Sys. 21:49–59.

Wilhelm, W.W., J.M.-F. Johnson, D.L. Karlen, and D.T. Lightle. 2007. Corn stover to sustain soil organic carbon further constrains biomass supply. Agron. J. 99:1665–1667.

Wösten, J.H.M. 1997. Pedotransfer functions to evaluate soil quality. p. 221–245. *In* E.G. Gregorich and M.R. Carter (ed.) Soil quality for crop production and ecosystem health. Developments in Soil Science. Elsevier, Amsterdam, The Netherlands.

Yaalon, H.D. 1997. Introducing soils as an object of study. p. 1–67. *In* D.H. Yaalon and S. Berkowicz (ed.) History of soil science. Verlag GMBH, Reiskerchen, Germany.

Zobeck, T.M., J. Crownover, M. Dollar, R.S. Van Pelt, V. Acosta-Martinez, K.F. Bronson, and D. R. Upchurch. 2007. Investigation of soil conditioning index values for southern High Plains agroecosystems. J. Soil Water Conserv. 62:433–442.

Zobeck, T.M., A.D. Halvorson, B.J. Wienhold, V. Acosta-Martinez, and D.L. Karlen. 2008. Comparison of two soil quality indexes to evaluate cropping systems in northern Colorado. J. Soil Water Conserv. 63:329–338.

Section 2
Principles Underlying Management

Photo: Plow pan.
Photo courtesy of John A. Kelley, USDA-NRCS, http://SoilScience.info

4

Water Dynamics in Soils

M.B. Kirkham

To manage soil properly, we need to know the basic laws controlling water movement in the soil. This chapter will cover these laws for saturated and unsaturated soil. We first begin by defining terms.

Water can either be *static* or *dynamic* in the soil. Static water refers to the forces that retain water in soil. Water is retained in the soil largely by surface-tension forces (Kirkham, 1964). Textbooks describe these forces (Hillel, 2004, p. 28–29; Kirkham, 2005, p. 67–84; Jury and Horton, 2004, p. 37–73), and they will not be discussed in this chapter. *Dynamics* is the branch of physics that considers motion in terms of the forces that produce the motion (Shortley and Williams, 1971, p. 51). All of the vast body of knowledge called *classical mechanics*, including dynamics, has as its foundation the three principles of motion formulated by Sir Isaac Newton (Shortley and Williams, 1971, p. 51) in his monumental treatise *Philosophiae Naturalis Principia Mathematica*, or *Mathematical Principles of Natural Philosophy,* and published in 1687 (Sposito, 1976, p. 1). The three laws are described by Sposito (1976, p. 1–10) and will not be presented here. In this chapter, we shall consider the motion of water in soil and the principles underlying the motion using simple mathematics. Readers with advanced mathematical knowledge can learn about soil water dynamics from Warrick (2003), who uses exact, analytical equations.

Two important expressions used to describe the state of water in soil are *water content* and *water potential.* The water content is the measurement of the amount of water in the soil either by weight or volume and is defined as the water lost from the soil on drying to constant mass at 105°C (Soil Science Society of America, 1997). Water content tells us how wet a soil is, but it does not tell us the direction that the water is moving in the soil. The second expression utilizes the potential energy status of a small parcel of water (say a milligram) in the soil. If we know the potential energy of soil at two different points in the soil, we know the direction that the water is moving. Water moves in response to differences in water potential. The difference is called the *water potential difference.* The *water potential gradient* is the potential difference per unit distance of flow. Water moves from high potential energy to low potential energy (Kirkham, 2005, p. 60).

All water in the soil is subjected to force fields originating from four main factors: the presence of the solid phase (the matrix); the gravitational field; dissolved salts; and the action of external gas or water pressure (Kirkham, 2005, p. 55). Henceforth in this chapter, when we say "pressure," we shall be considering water pressure, not gas pressure. Other potentials can be defined, such as an overburden potential due to the weight of the soil itself, but they are generally neglected because they are usually small (Kirkham, 2005, p. 59–60).

M.B. Kirkham, Department of Agronomy, 2004 Throckmorton Plant Sciences Center, Kansas State University, Manhattan, KS 66506-5501 (mbk@ksu.edu).

doi:10.2136/2011.soilmanagement.c4

A force is a push or pull. Newton's second law of motion defines force. If the four main force fields in the soil are compared to a reference point, then they can be expressed on a potential energy basis, and each of the four factors can be assigned a separate potential energy value (Kirkham, 2005, p. 55). The sum of these four potential energy values is called the *water potential* of the soil, or the *total water potential* to emphasize that it is comprised of several factors. (For abbreviation, we often say "potential" instead of "potential energy" [Kirkham, 1961a, p. 40].) The reference point of these potential energies is taken as pure free water at some specified height or elevation (Kirkham, 2005, p. 56). Therefore, the total water potential is the sum of the *matric potential,* the *gravitational potential,* the *solute potential,* and the *pressure potential.* The matric potential used to be called the *capillary potential,* because over a large part of its range, the matric potential is due to capillary action. However, as the water content decreases in porous material, water that is held in pores due to capillarity becomes negligibly small compared to the water held directly on particle surfaces. The term *matric potential,* therefore, covers phenomena beyond those for which a capillary analogy is appropriate (Kirkham, 2005, p. 56).

If chemical (solute) forces are not considered, there are three principal forces that move water through soil. These are gravity, pressure, and capillarity (Kirkham, 1961a, p. 35). Under nonsaline, unsaturated conditions, the two most important potentials in the soil are the matric (capillary) potential and the gravitational potential. Under nonsaline, saturated conditions, the two most important potentials in the soil are the pressure potential and the gravitational potential.

We also can define our potentials in terms of heads. A *head* is a source of water kept at some height to supply, for example, a mill. Hence, it is a pressure. Instead of using potential terminology, we can express potential in terms of head (a length). Engineers prefer to work with lengths (heads) rather than potentials, because they are easier to measure and track. Under saturated conditions when there is no tension on the water in the soil and hence no matric head, the total head is the sum of the gravity head and the pressure head. When the

soil water is under tension, the total head is the sum of the tension (matric) head and the gravity head (Kirkham, 2005, p. 60–61). The total head is usually called the *hydraulic head.* The difference in the sum of the pressure or matric head and gravity head, called the *hydraulic head difference,* governs the soil water flow (Kirkham, 2005, p. 60).

To establish hydraulic heads, one needs a *hydraulic datum line* (H.D.L.). It is a reference level. It can be placed at any location, but it is usually placed either at the soil surface or at some arbitrary distance below the soil surface (Kirkham, 1961a, p. 37).

The symbol φ is commonly used for hydraulic head, especially by physicists. Physicists use φ for many potential energy problems. For water seepage, φ is a measure of the energy of 1 cm^3 of soil water at a point in the soil (Kirkham, 1961a, p. 40).

Now let us turn to water movement in saturated soil, where pressure and gravity are the two main forces that move water.

Saturated Soil
Darcy's Law

Understanding movement of water in saturated soil is important in drainage and groundwater studies. The French hydraulic engineer Henry Philibert Gaspard Darcy determined experimentally the law that governs the flow of water through saturated soil. It is called Darcy's law (Darcy, 1856). Figure 4|1 illustrates, for a vertical column of soil, Darcy's law. This law is basic in soil drainage and is, for a useful set of units (Kirkham, 1961a, p. 41)

$$V/t = k \left[(\varphi_1 - \varphi_2)/L \right](\pi R^2) \tag{1}$$

where V is volume of water (m^3), t is time for this volume V to discharge (d), k is hydraulic conductivity, sometimes called the permeability coefficient (m d^{-1}), φ_1 is hydraulic head, referred to an arbitrary hydraulic datum level (H.D.L.) for points in the surface of the top of the soil column (m). The subscript 1 in φ_1 refers to the first (top) surface that inflow water meets. φ_2 is hydraulic head, referred to the same H.D.L. for points in the bottom surface of the soil column (m). The subscript 2 in φ_2 refers to the second (bottom) surface that the water meets. L is length of soil column (m), and πR^2 is cross-sectional area of the soil column (m^2).

Suppose (Fig. 4|1) this inflow water is cut off. Then the water level above the soil will drop a small distance dh in a time dt, so that V in Eq. [1] is replaced by $\pi R^2 dh$ and t by dt, and for the condition that dh and dt (but not dh/dt) approach zero, the equation becomes

$$(dh\pi R^2)/dt = k[(\varphi_1 - \varphi_2)/L]\pi R^2 \qquad [2]$$

That is, one has

$$dh/dt = k[(\varphi_1 - \varphi_2)/L] \qquad [3]$$

which is also a form of Darcy's law. Now, dh/dt is the velocity of fall (m s^{-1}) of the surface water. Therefore, one may write the law as follows (do not confuse small v, velocity, with capital V, volume):

$$v = k[(\varphi_1 - \varphi_2)]/L \qquad [4]$$

a form in which no particular cross-sectional area of soil, as πR^2 in Fig. 4|1, is involved. The law is often further abbreviated by defining $(\varphi_1 - \varphi_2)/L$ as the hydraulic gradient i. That is,

$$i = (\varphi_1 - \varphi_2)/L \qquad [5]$$

Equations [4] and [5] then yield

$$v = ki \qquad [6]$$

which is the most compact form of Darcy's law.

In Darcy's law the hydraulic datum level does not appear: V or v cannot depend on the location of an arbitrary reference plane. Hydraulic gradients i in soil are ordinarily less than 1, except near drain tile (Kirkham, 1961a, p. 43). Some values of k are given in the section after the next.

Flow in a vertical soil column has been used to derive and illustrate Darcy's law. The law applies for flow in any direction and illustrations for other than vertical soil columns can be found (e.g., Kirkham and Powers, 1972, p. 47; see their Fig. 2.1). In each situation one has, by experi-

ment, $V/t = k[(\varphi_1 - \varphi_2)/L]$; A, $v = ki$, which are the results also of Eq. [1] and [6].

Interpretation of v and k

For Darcy's law, although it has just been seen that $v = dh/dt$ is a velocity of ponded surface water, one often uses v to refer to the fluid velocity *inside* the soil, without specifying that v is to be understood as an equivalent velocity of ponded surface water (Kirkham, 1961a, p. 43). Let f be the porosity (i.e., the fraction of bulk soil space not occupied by solid material) of the soil. Then the average velocity v_{pores} of the water in the soil pores is

$$v_{pores} = v/f \qquad [7]$$

which is a faster velocity than v. The velocity in the pores (v_{pores}) is called the *actual velocity*, and v is called the *Darcy velocity* (Kirkham, 2005, p. 86).

Equation [6] defines k as the velocity for *unit hydraulic gradient*. To understand unit hydraulic gradient and to obtain a more meaningful interpretation of k, we may write down Darcy's law with reference to a new figure (Fig. 4|2). The law, in view of Eq. [3] and Fig. 4|2, is

$$dh/dt = k[(L + h)/L] \qquad [8]$$

Therefore, as h approaches zero one has

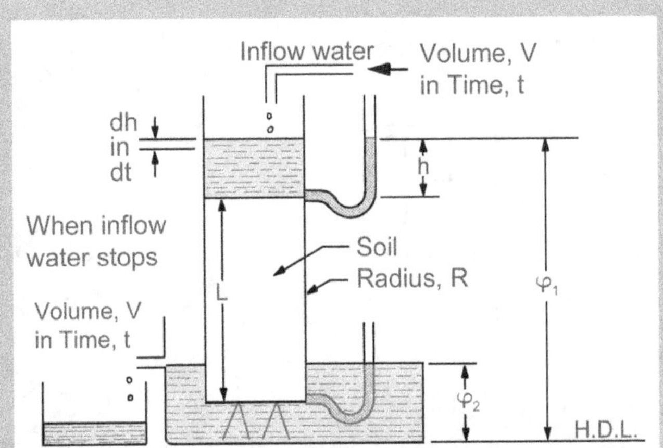

Fig. 4|1. Cylinder with soil and downward seeping water. The manometers are not needed to define φ_1 and φ_2. For abbreviations, see text. (Redrawn from Kirkham, 1961a, p. 41.)

$$dh/dt = k(L/L), (h \to 0) \qquad [9]$$

or has

$$dh/dt = k, (h \to 0) \qquad [10]$$

But, for $h = 0$ the pressure head at all points in the soil must be zero, so the only force causing movement of water in Fig. 4|2 when $h = 0$ is that of gravity. Thus, it is deduced using Eq. [10] that k is numerically equal to the rate of fall of a thin layer of ponded water into the soil, under the force of the Earth's gravitational pull. Equation [9] emphasizes also that k is the velocity under a unit hydraulic gradient (Kirkham, 1961a, p. 46).

Some Values of k

Hydraulic conductivity in natural field soil is governed by factors such as cracks, root holes, worm holes, and stability of soil crumbs (Kirkham, 1961a, p. 46). *Texture*, that is, the percentage of the primary particles of sand, silt, and clay, ordinarily has a minor role, except for disturbed soil materials. Table 4|1 gives hydraulic conductivities for some natural (undisturbed) field soils and for some disturbed soil materials (Kirkham, 1961a, p. 46; 1961b). The value for the Zuider Zee very fine sand in The Netherlands is very low because the silt and clay in this soil were apparently of just the size and amounts to almost clog the pores between the sand particles.

Hydraulic Conductivity and Infiltration Rate

The term *hydraulic conductivity* has been defined as the meters per day of water seeping into the soil under the pull of gravity or under a unit hydraulic gradient. The term should not be confused with *infiltration rate*. This latter term may be defined as the meters per day of water entering into the soil regardless of the types or values of forces or gradients (Kirkham, 1961a, p. 47). Infiltration rate need not refer to saturated conditions. If two raindrops of total volume 2 mm³ (= 0.000002 m³) fall per day on a square meter of soil, and are absorbed into the soil, the infiltration rate is 0.000002 m d⁻¹.

Is k in Darcy's Law Constant?

Ordinarily one considers k in $v = ki$ to be a constant. It is a constant if (i) the physical condition of the soil and of the water does not change in space or time as the water moves through the soil, and (ii) the type of flow is *laminar*, that is, not *turbulent* (Kirkham, 1961a, p. 47). In laminar flow, two particles of water seeping through the soil will describe paths (*streamlines*) that will never cross each other. In turbulent flow, eddies and whirls develop. The possibility of turbulent flow is considered in soil only if the soil is a coarse sand or gravel, and then only if the hydraulic gradients are large (larger than occur in most agricultural situations).

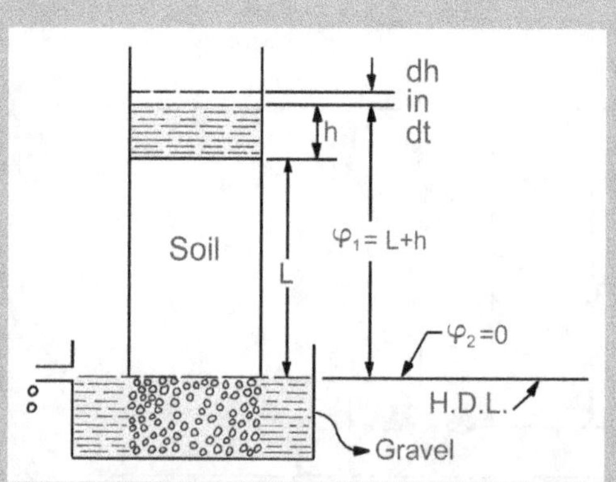

Fig. 4|2. Seepage situation for interpreting k of Darcy's law. For abbreviations, see text. (Redrawn from Kirkham, 1961a, p. 45.)

Table 4|1. Values of hydraulic conductivity for some natural soils in place and some disturbed soil materials (see Kirkham, 1961a,b).

Soils or soil material	Composition	Hydraulic conductivity
	% (w/w) silt + clay	m d⁻¹
Natural soils in place		
Webster silty clay loam	90	16–32
Marion silty clay loam	90	3
Luton clay	80	0.05–2
Zuider Zee very fine sand	20	0.05–0.15
Disturbed soil materials		
Silt and clay	25–71	0.02
Very fine and fine sand	6–28	0.7
Fine and medium sand	1–11	13
Coarse sand	0–8	82
Gravel	0–6	600

If molds and slimes develop as water seeps through soil, then the soil pores will clog and k will decrease with time. If the soil temperature falls, the viscosity of the water will increase and k will decrease. The value of k, for reclaimed soil, can be made higher or lower by soil management. Roots of crops (after decay) increase k; compaction of soil by animals or machinery decreases k, at least in the surface soil.

Simple Application of Darcy's Law

In Fig. 4|3 suppose the wetting front of water has penetrated to a depth x into the soil, and suppose that capillary forces can be neglected. Then a manometer established at the wetted front would read zero (gauge) pressure, which is the pressure in the soil air at the interface. [*Gauge pressure* is the amount by which a pressure exceeds that of the atmosphere (R.E. Allen, 1990).] With the hydraulic datum line as shown in Fig. 4|3, Darcy's law, then is, in accordance with Eq. [3] and Fig. 4|3:

$$dh/dt = k[(\varphi_1 - \varphi_2)/x] = k[(h + x)/x] \qquad [11]$$

Suppose one has $x >> h$ (x very much greater than h), as say $x = 2$ m and $h = 0.1$ m, then for practical purposes, h may be neglected compared to x, and one has from Eq. [11]:

$$dh/dt = k(x/x) = k \qquad (x >> h \text{ in Fig. 4|3}) \qquad [12]$$

Thus, for deep vertical drainage into uniform soil (no tight layers), one has $dh/dt = k$. That is, the number of meters of surface water (m³ m⁻²) that the soil will absorb per day is k. For a clay soil (Table 4|1) this might be 0.05 to 2 m d⁻¹, and for coarse sand, 80 m d⁻¹. It is clear that sand, unless underlain by clay, presents an irrigation water loss problem (Kirkham, 1961a, p. 49).

Laplace's Equation

To solve groundwater seepage and drainage problems, it is desirable to have a general differential equation. It is found that Laplace's equation applies, a familiar equation occurring in nearly all branches of applied mathematics (Kirkham, 1994, 2005, p. 88). Laplace's equation is derived from Darcy's law and the *equation of continuity*. The equation of continuity states mathematically that mass can be neither created nor destroyed. We can state the equation of continuity in words, as follows: For a volume element x times y times z, the change in velocity of water in the x direction plus the change in velocity of water in the y direction plus change in velocity of water in the z direction are equal to the total change in water content, θ, per unit time of the volume element under consideration. That is, inflow of water in the element minus outflow of water is equal to the water accumulated. Let us imagine a rectangular x, y, z system of coordinates that is established in a homogeneous porous medium of constant hydraulic conductivity, and let h be the hydraulic head referred to an arbitrary reference level for a point (x,y,z) and let time be t and v_x, v_y, and v_z be the velocity of water flowing in the x, y, and z directions, respectively; then, with θ being the volume of water per unit volume of bulk soil, from the equation of continuity,

$$-[(\partial v_x/\partial x) + (\partial v_y/\partial y) + (\partial v_z/\partial z)] = \partial\theta/\partial t \qquad [13]$$

and Darcy's law, one may, for incompressible steady state ($\partial\theta/\partial t = 0$) flow in a porous medium where k is constant, derive the expression

$$(\partial^2 h/\partial x^2) + (\partial^2 h/\partial y^2) + (\partial^2 h/\partial z^2) = 0 \qquad [14]$$

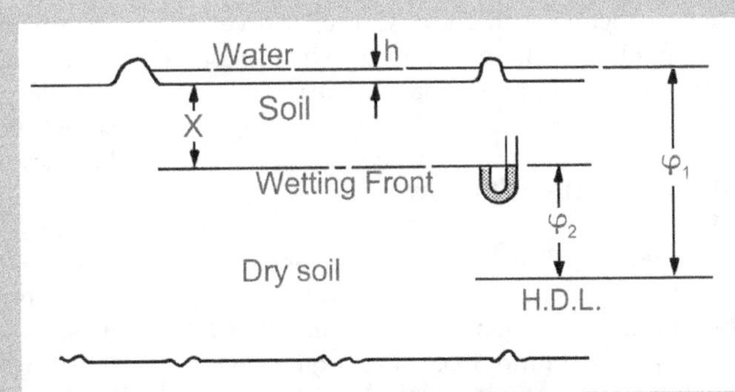

Fig. 4|3. Seepage of irrigation or leaching water into dry soil. For abbreviations, see text. (Redrawn from Kirkham, 1961a, p. 48.)

as the expression governing groundwater flow. Charles S. Slichter, a mathematician at the University of Wisconsin, was the first to show that Laplace's equation applies to the motion of groundwater (Slichter, 1899).

Unsaturated Soil

So far, we have been considering water flow in saturated soils. Although flow in saturated soils is important, soils are not generally water saturated (*saturated* meaning that the matric potential is equal to zero).

We now consider concepts of unsaturated flow (Kirkham, 1994). Using the equation of continuity and assuming that Darcy's law holds for unsaturated moisture flow, one may derive the following equation (Childs, 1957, p. 75; Kirkham, 1964):

$$\partial/\partial x)k(\partial h/\partial x) + (\partial/\partial y)k(\partial h/\partial y) + (\partial/\partial z)k(\partial h/\partial z) = \partial\theta/\partial t \tag{15}$$

where k is the hydraulic conductivity under unsaturated conditions, h is the hydraulic head, and θ is the fraction of the soil bulk volume occupied by water. If the soil is saturated, the equation reduces to Laplace's equation because k and θ become constants. For solutions of Eq. [15], the result is dependent on three factors:

1. We replace h, the hydraulic head, by

$$h = \psi + z \tag{16}$$

where ψ is the matric potential or capillary potential (rather than the pressure potential, because the soil is now not saturated) and z is the gravitational head. L.A. Richards, then a doctoral student at Cornell University, published the equation in which h is replaced by ψ and z (Richards, 1931):

$$(\partial/\partial x)k(\partial\psi/\partial x) + (\partial/\partial y)k(\partial\psi/\partial y) + (\partial/\partial z)k(\partial\psi/\partial z) + \partial k/\partial z = \partial\theta/\partial t \tag{17}$$

Equation [17] is known as the Richards equation, a nonlinear, partial differential equation. Such equations are difficult to solve. John R. Philip (1955) of Australia gave a solution.

2. We introduce a term called the *diffusivity* D, defined by

$$D = k(\partial\psi/\partial\theta) \tag{18}$$

The diffusivity D has units of millimeters squared per second. Ernest C. Childs (1936) in England noted that, under unsaturated conditions, water moves according to diffusion equations.

3. We now make use of a mathematical transformation called the *Boltzmann transformation*, which involves the time t to the one-half power. Arnold Klute (1952), as a doctoral student at Cornell University in New York, introduced this Boltzmann variable, λ:

$$\lambda = 1/2\, xt^{-1/2} \tag{19}$$

The Boltzmann variable λ has units of length divided by the square root of time. Klute (1952) wrote the flow equation in a diffusion form with water content, θ, as the dependent variable:

$$\partial\theta/\partial t = (\partial/\partial z)D(\partial\theta/\partial z) + (\partial k/\partial z) \tag{20}$$

He then restricted himself to the gravity-free case (horizontal flow into a soil column):

$$\partial\theta/\partial t = (\partial/\partial x)D(\partial\theta/\partial x) \tag{21}$$

Klute (1952) transformed the partial differential equation into an ordinary differential equation by using this Boltzmann variable.

Edgar Buckingham (1907) of the USDA Bureau of Soils in Washington, DC (Nimmo and Landa, 2005) already had published a bulletin entitled "Studies on the Movement of Soil Moisture," which established the mathematics of unsaturated soil water flow. He saw that the capillary conductivity (the unsaturated hydraulic conductivity) is a function of water content or matric (capillary) potential. The flow equation is sometimes known as the Darcy–Buckingham law, which honors both its discoverers in the saturated and unsaturated realms.

Thus, description of water flow in soil requires the functions $D(\theta)$ and $k(\theta)$. In general, it is difficult to measure $D(\theta)$, especially in situ. It is simpler to measure the total effect of the capillary attractiveness of soil, namely the *sorptivity* S (mm s$^{-1/2}$), a term defined by Philip (1969). The sorptivity is equal to the following:

$$S = \int_{\theta_n}^{\theta_s} \lambda_c d\theta \qquad [22]$$

and can be approximated by the following equation:

$$S_o^2 = \left[(\theta_o - \theta_n)/b\right]\int_{\theta_n}^{\theta_s} D d\theta \qquad [23]$$

where θ_n is the initial soil water content, θ_s is the saturated water content, λ_c is the macroscopic capillary length scale (mm) (Smettem and Clothier, 1989), θ_o is the water content to which the soil surface is wet, and $1/2 < b < \pi/4$, often $b \approx 0.55$. Thus, to interpret S_o in terms of D requires measurement of θ_n and θ_o, which can be easily determined (Smettem and Clothier, 1989). In theory, to describe flow in a uniform soil, one only requires instruments such as tension infiltrometers (also called disc permeameters) to measure the sorptivity and the conductivity function from saturation, θ_s, where $\theta_o = \theta_s$, which is at the free water condition of $\psi_o = 0$, where ψ_o is the supply potential, i.e., the tension with which the water is applied to the soil when one uses a tension infiltrometer, down to $\theta = \theta_n$ and $\psi = \psi_n$, where commonly $\psi_n \to -\infty$ (Smettem and Clothier, 1989).

Preferential Flow

The water flow equations have been derived using the assumption that the soil has a continuous solid matrix, which holds water in pores and films. Field soil, however, has a number of interconnected cracks, root holes, worm channels, and other voids, whose physical properties differ from the surrounding soil matrix (Kirkham, 1994). If filled, these continuing flow channels have the capacity to carry large amounts of water at velocities that greatly exceed those in the surrounding matrix. We first define and then consider the characteristics of these voids.

Microporosity and Macroporosity

Pores have been classified into different sizes, as follows (Clothier, 2008)

- Macropores: diameters ranging from >5000 to 75 µm;

- Mesopores: diameters ranging from 30 to 75 µm;

- Micropores: diameters ranging from 5 to 30 µm.

It is often more important to characterize soil pores in terms of their function, in particular with regard to their ability to store and conduct water, rather than their diameter (Kirkham, 1994). Transport of water with its dissolved chemicals through soil, as well as gaseous exchange, depends critically on soil pores, and especially on the continuous and connected macropores.

Functionally, we can distinguish between *macroporosity* and *matrix porosity* (Kirkham, 1994). Macroporosity refers to the interconnected pore space of voids, which causes preferential transport of both water and chemicals. When transport occurs through the macropores, there is limited exchange of water between the macropores and the pores of the matrix. Matrix porosity refers to those pores in which the flow through the body of the soil is slow enough so that there is extensive interpore mixing.

If we consider the unit soil pore to be a cylinder of radius r, then the Hagen-Poiseuille law can be used to describe the flow through the pore:

$$q = -(r^2/8\eta)\,(\Delta P/\Delta x) \qquad [24]$$

where q is the flux density (m/s), η is the viscosity of water (kg m^{-1} s^{-1}), and ΔP (Pa) is the pressure difference across the small pipe (pore) of length Δx (Kirkham, 1994). Because the flux density increases with the square of the radius, the macropores can have a great impact on soil transport processes. This is especially true because the macropores frequently form an interconnected network. However, they are fragile and can easily be disrupted, particularly at the soil surface, where they can be rendered ineffective by sealing. Much of the biologically created macroporosity drops off with depth as the populations of soil flora and fauna decline.

Macroporosity and Hydraulic Conductivity

Because of macroporosity, the hydraulic conductivity function $k(\psi)$ can vary dramatically, mainly as a result of changes in the characteristics of the larger surface pores that become filled only at high potentials near saturation (Kirkham, 1994). When the soil is wet, the role of macropores is paramount in determining $k(\psi)$. Here management (e.g., plowing) or natural events (e.g., crusting of the soil after rainfall) can rapidly modify the hydraulic conductivity by altering surface venting and subsurface connectedness. Tillage, especially, affects pore size. Pores are smaller in tilled soils because tillage pulverizes the soil. When the soil is not tilled, decaying roots and other organic matter create voids. Also, earthworms thrive on the organic matter, and their populations are greater in soil that has not been tilled. Earthworm holes and root channels are a prime reason for the difference in hydraulic conductivity between cultivated and no-till soil.

Mobile Water

Many water flow processes of interest, such as groundwater recharge, are concerned only with area-averaged water input (Kirkham, 2005, p. 161). Therefore, preferential flow of water through structural voids does not necessarily invalidate equations that assume homogeneous flow, like Darcy's law. However, preferential flow is of critical importance in solute transport, because it enhances chemical mobility and can increase pollution hazards. Many times we need to monitor chemical mobility along with hydraulic properties. We now shall see how we can determine mobility of chemicals.

Water and nutrients not taken up by roots move to depth and eventually to groundwater with deleterious consequences. With the tension infiltrometer (disc permeameter), we can measure hydraulic properties of soil that control infiltration and retention. (For a description of the tension infiltrometer or disc permeameter and how it works, see Kirkham, 2005, p. 151–155). In particular, we can distinguish *mobile* and *immobile* water with it.

The soil water content, θ, is made up of the mobile water content, θ_m, and the immobile water content, θ_{im}. Water is immobilized due to several factors: it can be in an occluded pore, it can be bound water, it can be in a dead-end pore, or it can be in the soil's microporosity and unable to move.

The portion θ_m is active in chemical transport, while the portion θ_{im} is not. If we know θ_m, we can develop management strategies to minimize leaching losses of chemicals.

The method of using the disc permeameter to obtain θ_m and θ_{im} relies on using it to supply a tracer solution to soil (Clothier et al., 1992). The tracer is not originally in the soil. If a tracer is added to a disc permeameter at a concentration c_m, then from the observed solution concentration c^* in soil samples extracted from underneath the disc, θ_m can be calculated from the dilution by the water of the immobile phase that must have remained in place during the passage past it of the invading solution of tracer. So we have (Kirkham, 2005, p. 163):

$$c^*\theta = c_m\theta_m + c_{im}\theta_{im} \qquad [25]$$

But if $c_{im} = 0$ (i.e., there was no tracer in the soil to begin with), then the last term on the right-hand side of Eq. [25] drops out and we have

$$\theta_m = \theta(c^*/c_m) \qquad [26]$$

In other words, the fraction θ_m/θ is directly proportional to the relative concentration of solute found in the soil, c^*, to that applied, c_m:

$$c^*/c_m = \theta_m/\theta \qquad [27]$$

We can determine the depth of penetration of the solute using the following one-dimensional equation:

$$Z_s^* = I/\theta_m \qquad [28]$$

where Z_s^* is depth of the solute front (mm) (the asterisk after the subscript "s" for solute indicates that we are dealing with a tracer), I is cumulative infiltration (mm), and θ_m is mobile water content ($m^3 \, m^{-3}$).

In the experiments of Clothier et al. (1992), the tracer was bromide (0.1 mol L^{-1} KBr). They found that the average mobile water content, θ_m, in the Manawatu fine sandy loam that they studied was 0.203 m^3 m^{-3}. The water content of the soil was 0.414 $m^3 \, m^{-3}$. Therefore, only about one-half of the water in the soil was mobile. The average I for their experiments was 15.7 mm. So $Z_s^* = $ 15.7 mm/(0.203 $m^3 \, m^{-3}$) = 77.3 mm. If all of the water were mobile, then $\theta_m = \theta$ and 15.7 mm/

$(0.414 \text{ m}^3 \text{ m}^{-3}) = 37.9$ mm. The observed front of the bromide determined experimentally at the end of the experiments showed that it was at about 77 mm, not at about 38 mm, so the experimental data supported the calculation. Because not all the water was mobile, the tracer penetrated to deeper depths than it would have had $\theta_m = \theta$.

The longitudinal mobility (the depth of penetration) is inversely related to θ_m. This is sometimes hard for students to understand when using the method initially because they think that the more mobile the water is, the deeper will be the penetration of a solute. This is not so. The larger the mobile volume fraction (i.e., θ_m), then the less longitudinally mobile the solute (the smaller Z_s^*) is. In other words, the solution carrying the dissolved solute travels through a larger volume fraction of the soil's wetted pore space, such that for a given amount of water infiltrated, I, the less the penetration into the soil. As θ_m becomes a smaller fraction of the wetted θ (total water content), then the smaller is the volume fraction of the soil's wetted pore space that transports the invading solution. Hence for an equivalent amount of I, the greater is the depth of penetration because Z_s^* is inversely related to θ_m.

Impacts of Soil Management Practices on Preferential Flow

Soil management practices can have important impacts on preferential flow, which results in the associated mobile and immobile soil water regions. As noted, the smaller the fraction of mobile water in a soil, the deeper is the penetration of water and solutes because they are moving through macropores instead of the matrix of the soil. Pore size and the percentage mobile water can be altered by soil management. The world's consumption of water is doubling every 20 years, and new sources of water are becoming scarcer (Green et al., 2008). The goal of irrigation is to use water parsimoniously. By knowing how to manage the water in soil with macropores, we can make wise use of the limited water available for agriculture.

If macropores are present at the soil surface, tillage can remove them so that water moves through the matrix of the soil. Tillage

pulverizes the soil, as shown by the work of Sauer et al. (1990). They compared the hydraulic properties of a Plainfield sand (Typic Udipsamment) in Wisconsin that had either been tilled for 5 yr by moldboard plow or not tilled. The crop was always corn (*Zea mays* L.). Under saturated conditions, the mean pore size (weighted in a way relevant to the flow of water through the soil surface) of the no-till soil was 1.34 mm, while in the plowed soil the mean pore size was only 0.19 mm. The large mean pore size for the no-till soil was presumed to be related to the macropore network associated with the decay of crop residues. The small mean pore size of the plowed soil reflected the annual pulverization of the soil surface by tillage. If we want water to move through the matrix of the soil and not macropores, then tillage might help ensure matrix flow. Conversely, placement of residue on the soil surface, such as through vertical mulching, should move water to depth. In vertical mulching, deep vertical channels are cut in the soil and filled with chopped organic matter. If the channels remain open to the surface, the large pores in the organic matter take in free water from rain or irrigation and transmit it deeply into the soil (Gardner, 1979).

The wetness of the soil affects mobile water content. Recall that the two main forces that pull water into soil under unsaturated conditions are capillarity and gravity. Capillarity dominates in a dry soil because the dry pores suck up the water. If we want to keep solutes (i.e., fertilizers) near the soil surface and in the root zone, so they do not escape to the groundwater (and pollute it), then we should maintain the soil on the dry side (Clothier and Green, 1994). Under dry conditions, water, with its associated solutes, is pulled out in all directions as it infiltrates into the soil. The water and solutes stay near the soil surface. If the soil is already wet, when water infiltrates into the soil it moves mainly by gravity because the pores are filled with water and the capillary force is reduced. Consequently, to retain water and solutes near the soil surface and in the root zone, we should irrigate in small, frequent amounts. The soil will dry between irrigations. When we irrigate again, water and solutes will be pulled out into the relatively dry soil surface. This means that fertilizer should be applied to dry soil, and subsequently washed into the soil with just a small amount of water. The applied chemicals are drawn by capillarity into more of the soil's microporosity, where they are rendered less likely to be leached by subsequent heavier irrigations or by rainfall (Clothier et al., 1994).

Measurements of mobile water show that, indeed, the percentage of mobile water content is greater in a dry soil than a wet soil. Clothier and Green (1994) found that when a Manawatu fine sandy loam was pre-wetted, the mobile water content was 40%. That is, less than one-half of the soil's water appeared to be involved in solute transport. Hence, in this instance, surface-applied chemicals are likely to be drawn to depth by any infiltrating irrigation water. When the soil was dry, the capillary force of the dry soil drew the invading solution directly into more of the soil's microporosity, and the mobile water content was 62%. Mobile water content also changes during a cropping season. Angulo-Jaramillo et al. (1997) found that on a sandy and rocky silt of fluvio-glacial origin near Grenoble, France, where corn was growing, the mobile water content rose from 13% at emergence to 18% just before harvest. The increase in the mobile water percentage was apparently due to an increase in the fine fraction of small pores between emergence and harvest, which sealed the preferential flow paths. They attributed the increase in the fine fraction to deposition of eroded small, easily transportable particles during the season.

Conversely, if we want to leach the soil of salts instead of retain them in the root zone, we should add water frequently. If water is added daily, much of it moves to depth by gravity, because capillary attraction at the surface is less (Kirkham and Kirkham, 1995). Much of the world's irrigated soil is affected by salinity. Under saline conditions, frequent irrigations are effective in leaching salts out of the soil (Miller et al., 1965). We cannot conserve water when we need to leach salts from salty soil.

The Future

We now turn to soil water as affected by elevated atmospheric carbon dioxide because increasing levels of atmospheric CO_2 will have major effects on agriculture, including soil water. The globally averaged atmospheric CO_2 concentration is constantly increasing. It has increased from 315 μmol mol^{-1} in 1958 (Taylor and MacCracken, 1990) to 382.7 μmol mol^{-1} in 2007 (Schnell, 2008). Much research has been done to consider the possibility of using of soil to sequester carbon as the concentration of CO_2 in the air increases (Lal et al., 1999). Less known is the direct effect of elevated atmospheric CO_2 on soil water content. To build a stable base for agriculture, it is necessary to know how to manage soil water as the CO_2 concentration in the atmosphere increases. L.H. Allen, Jr. (1990) said that the amount of CO_2 in the atmosphere might double within 100 years. Models predict that, in response to a doubling of the atmospheric CO_2 concentration, soil moisture will be reduced throughout an extensive mid-continental area of North America (Gleick, 1987; Manabe and Wetherald, 1987; Mitchell and Warrilow, 1987; Kellogg and Zhao, 1988), where temperate grasslands are important agriculturally since they provide natural pastures for grazing animals. The models are based on predicted increased temperatures and reduced rainfall.

Because apparently no measurements of the soil water content under augmented atmospheric CO_2 had been made, we measured the soil water content in a tall grass prairie in Kansas exposed to ambient (337 μmol mol^{-1}) and twice ambient (658 μmol mol^{-1}) levels of CO_2 during a 2-yr period (1989–1990) (Kirkham et al., 1993). The dominant species on the grassland is big bluestem (*Andropogon gerardii* Vitman), a warm-season grass. About 70% of the botanical composition of the grassland is big bluestem (Kirkham et al., 1991). The plants grew in closed-top chambers (1.5 m in diameter and 1.8 m tall). One-half of the plots (chambers) were kept at field capacity (high water level), and one-half of the plots were kept at one-half field capacity (low water level). The soil was a silty clay loam with a field capacity of 0.38 m^3 m^{-3}. Soil water content was measured with a neutron probe to the 2.0-m depth. Throughout both seasons, soil water content was greater in the plots with elevated CO_2 than in the plots with ambient CO_2. Figure 4|4 illustrates data obtained in June 1989 and June 1990. Under the low water level, in the total 2.0 m profile, soil exposed to elevated CO_2 had 40 mm (689–649 mm) and 21 mm (677–656 mm) more water in the profile in 1989 and 1990, respectively, than soil exposed to ambient CO_2. Under the high water level, these differences were 23 mm (725–702 mm) and 5 mm (730–725 mm) in 1989 and 1990, respectively.

The higher soil water content in the plots with elevated CO_2 was probably the result of a lower transpiration rate (Kirkham et al., 1991). Under the high and low water treatments, doubled CO_2 reduced transpiration of leaves by 25 and 35%, respectively, in 1989 and by 24 and 20%, respectively, in 1990. Transpiration was reduced with elevated CO_2 because stomatal resistance was increased (Kirkham et al., 1991). It has been known for a long time that CO_2 is an antitranspirant because it closes stomata (van Bavel, 1974). Data obtained after the 1989–1990 experiment have confirmed that soil water content in the mid-continental United States is greater under elevated CO_2 than under ambient CO_2 (Knapp et al., 1993; LeCain et al., 2003; Nelson et al., 2004). Climate modelers, who predict that soil will become drier during the summer in this area of North America in response to a doubling of atmospheric CO_2, must consider the effect of CO_2 on transpiration in their models.

The fact that there is more water in the soil with elevated CO_2 will not change the basic forces moving water in the soil. Water in the soil moves according to potential-energy gradients, not according to water-content gradients. Richards (1940) made this clear in his work (Kirkham, 2005, p. 63). Soil can be drier at a point in the soil compared to another, wetter point. But, moisture will not move from the wetter point to the drier point

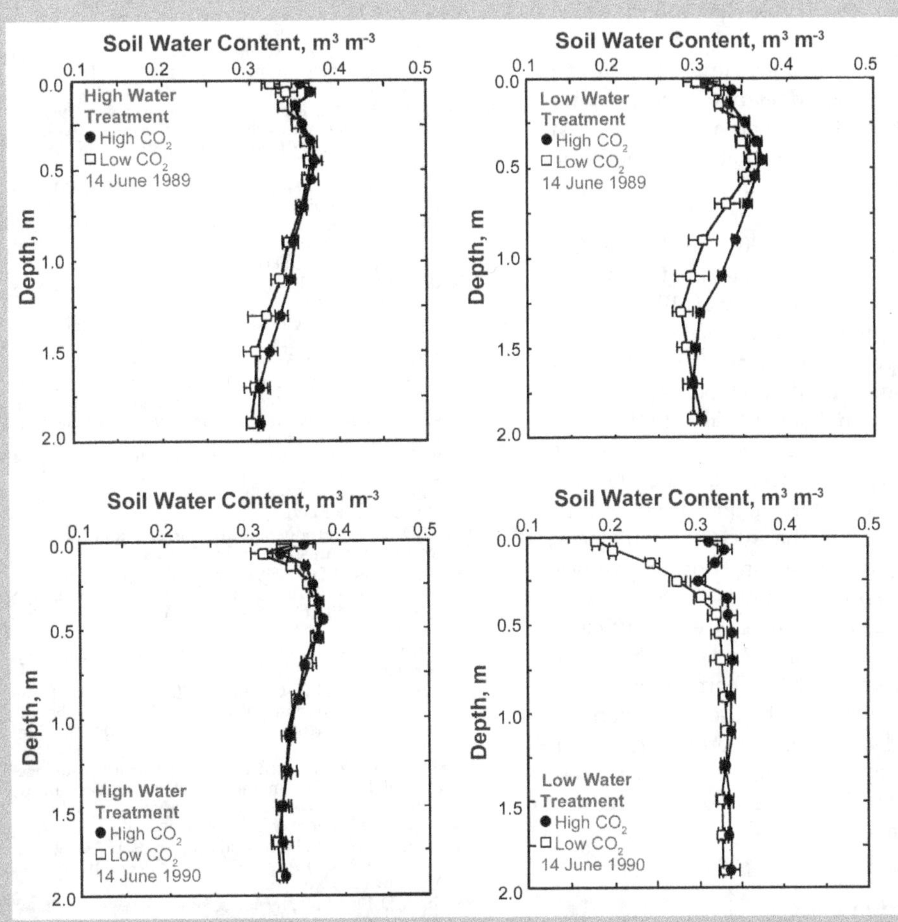

Fig. 4|4. Water content on 14 June 1989 (top) and 14 June 1990 (bottom) at different depths in a soil of a grassland with a high (twice ambient) and a low (ambient) atmospheric CO_2 concentration and a high (field capacity) and a low (half field capacity) water level. Horizontal bars = ± standard error (n = 4). (Redrawn from Kirkham et al., 1993.)

because the total hydraulic head includes both the matric potential energy (dryness of the soil) and the gravitational potential energy. The only way we can determine if water is moving up or down in a soil profile is to measure its total potential energy at different points in the profile. Water content measurements do not tell us the direction of movement of water in the soil. Because elevated atmospheric CO_2 affects an entire field, potential energy gradients in the soil probably will not be changed by elevated CO_2, even though the water content will be increased. Experiments need to be done in which the potential energy of the water in the soil is measured under elevated CO_2 (e.g., by using tensiometers).

In the very distant future, if we no longer can build a stable base for agriculture on Earth, humanity may attempt to live in space. There, it will be of vital importance to study the basic forces moving water in the soil, especially when gravity is no longer a force. Darcy's law does not work in space because it is dependent on gravitational forces. The little work that has been done on movement of water in porous media in space shows unexpected results (Jones and Or, 1999; Chau et al., 2005). In the absence of gravity, the arrangement of liquid in pore space is entirely defined by capillary forces, and liquid preferentially invades the smallest pore space. The effect of gravity is to force the liquid to fill large pores at the bottom of a domain that it would not otherwise occupy under a prescribed density and matric (capillary) potential (Chau et al., 2005; see their Fig. 3). Currently, plants are grown in space in special media, such as foam blocks, which keep the roots in constant contact with water so it does not need to move to the roots. If capillary forces are not sufficient to move water to plant roots, as they may not be (Kirkham, 2008), perhaps mechanical forces will need to be used. For example, springs or syringes could propel the water to roots for uptake. More research is needed to understand how water moves in the soil without gravity.

Acknowledgments

This is a contribution from the Kansas Agricultural Experiment Station, Manhattan, Kansas. I thank Brent E. Clothier for discussions about unsaturated flow.

References

Allen, L.H., Jr. 1990. Plant responses to rising carbon dioxide and potential interactions with air pollutants. J. Environ. Qual. 19:15–34.

Allen, R.E. (ed.) 1990. The concise Oxford dictionary of current English. Clarendon Press, Oxford Univ. Press, Oxford.

Angulo-Jaramillo, R., F. Moreno, B.E. Clothier, J.L. Thony, G. Vachaud, E. Fernandez-Boy, and J.A. Cayuela. 1997. Seasonal variation of hydraulic properties of soils measured using a tension disk infiltrometer. Soil Sci. Soc. Am. J. 61:27–32.

Buckingham, E. 1907. Studies on the movement of soil moisture. Bull. 38. USDA Bureau of Soils, Washington, DC.

Chau, J.F., D. Or, and M.C. Sukop. 2005. Simulation of gaseous diffusion in partially saturated porous media under variable gravity with lattice Boltzmann methods. Water Resour. Res. 41:W08410, doi:10.1029/2004WR003821.

Childs, E.C. 1936. The transport of water through heavy clay soils. I. J. Agric. Sci. 26:114–127.

Childs, E.C. 1957. The physics of land drainage, p. 1–78. In J.N. Luthin (ed.) Drainage of agricultural lands. ASA, Madison, WI.

Clothier, B.E. 2008. Soil pores. p. 693–699. In W. Chesworth (ed.) Encyclopedia of soil science. Springer, Dordrecht, The Netherlands.

Clothier, B.E., and S.R. Green. 1994. Rootzone processes and the efficient use of irrigation water. Agric. Water Manage. 25:1–12.

Clothier, B., S. Green, and G.N. Magesan. 1994. Soil and plant factors that determine efficient use of irrigation water and act to minimise leaching losses. Trans. 15th Int. Congr. Soil Sci. 2a:41–47.

Clothier, B.E., M.B. Kirkham, and J.E. McLean. 1992. In situ measurement of the effective transport volume for solute moving through soil. Soil Sci. Soc. Am. J. 56:733–736.

Darcy, H. 1856. Les Fontaines Publiques de la Ville de Dijon. (The public fountains of the city of Dijon). Victor Dalmont (ed.) (In French.) Librairé des Corps Impériaux des Ponts et Chaussées et des Mines, Quai des Augustins, 49, Paris.

Gardner, W.H. 1979. How water moves in the soil. Crops Soils 32(2):13–18.

Gleick, P.H. 1987. Regional hydrologic consequences of increases in atmospheric CO_2 and other trace gases. Clim. Change 10:137–161.

Green, S., B. Clothier, C. van den Dijssel, M. Deurer, and P. Davidson. 2008. Measurement and modeling the stress response of grapevines to soil-water deficits. p. 1–29. In L. Ahuja et al. (ed.) Response of crops to limited water: Understanding and modeling water stress effects on plant growth processes. Advances in Agricultural Systems Modeling Series 1. ASA, CSSA, and SSSA, Madison, WI.

Hillel, D. 2004. Introduction to environmental soil physics. Elsevier, Amsterdam, The Netherlands.

Jones, S.B., and D. Or. 1999. Microgravity effects on water flow and distribution in unsaturated porous media: Analyses of flight experiments. Water Resour. Res. 35:929–942.

Jury, W.A., and R. Horton. 2004. Soil physics. 6th ed. Wiley, Hoboken, NJ.

Kellogg, W.W., and Z.-C. Zhao. 1988. Sensitivity of soil moisture to doubling of carbon dioxide in climate model experiments. Part I. North America. J. Clim. 1:348–366.

Kirkham, D. 1961a. Lectures on agricultural drainage. Inst. of Land Reclamation, College of Agriculture, Alexandria Univ., Alexandria, Egypt. (Copy in the Iowa State University Library, Ames).

Kirkham, D. 1961b. Soil physical properties. p. 793–803. In C.B. Richey (ed.) Agricultural engineers' handbook. McGraw-Hill, New York.

Kirkham, D. 1964. Soil physics. p. 5-1–5-26. In Ven Te Chow (ed.) Handbook of applied hydrology. A compendium of water-resources technology. McGraw-Hill, New York.

Kirkham, D., and W.L. Powers. 1972. Advanced soil physics. John Wiley & Sons, New York.

Kirkham, M.B. 1994. Soil–water relationships. Vol. 4, p. 151–168. In C.J. Arntzen (ed.) Encyclopedia of agricultural science. Academic Press, San Diego.

Kirkham, M.B. 2005. Principles of soil and plant water relations. Elsevier, Amsterdam, The Netherlands.

Kirkham, M.B. 2008. Horizontal root growth: Water uptake and stomatal resistance under microgravity. Vadose Zone J. 7:1125–1131.

Kirkham, M.B., H. He, T.P. Bolger, D.J. Lawlor, and E.T. Kanemasu. 1991. Leaf photosynthesis and water use of big bluestem under elevated carbon dioxide. Crop Sci. 31:1589–1594.

Kirkham, M.B., and D. Kirkham. 1995. Chloride and water content in the root zone of barley grown under four salt-water irrigation regimes. p. 75–76. *In* D. Silva (ed.) Vadose zone hydrology: Cutting across disciplines. Int. Conf. Proc. Kearney Foundation of Soil Science and Hydrologic Science, Univ. of California, Davis.

Kirkham, M.B., D. Nie, H. He, and E.T. Kanemasu. 1993. Responses of plants to elevated levels of carbon dioxide. p. 130–161. *In* Proc. Symp. on Plant Growth and Environment, Suwon, Korea. October 1993. Korean Agricultural Chemical Society, Suwon, Korea.

Klute, A. 1952. A numerical method for solving the flow equation for water in unsaturated materials. Soil Sci. 73:105–116.

Knapp, A.K., E.P. Hamerlynck, and C.E. Owensby. 1993. Photosynthetic and water relations responses to elevated CO_2 in the C_4 grass *Andropogon gerardii*. Int. J. Plant Sci. 154:459–466.

Lal, R., J.M. Kimble, R.F. Follett, and C.V. Cole. 1999. The potential of U.S. cropland to sequester carbon and mitigate the greenhouse effect. Lewis Publ., Boca Raton, FL.

LeCain, D.R., J.A. Morgan, A.R. Mosier, and J.A. Nelson. 2003. Soil and plant water relations determine photosynthetic responses of C_3 and C_4 grasses in a semi-arid ecosystem under elevated CO_2. Ann. Bot. (Lond.) 92:41–52.

Manabe, S., and R.T. Wetherald. 1987. Large-scale changes of soil wetness induced by an increase in atmospheric carbon dioxide. J. Atmos. Sci. 44:1211–1235.

Miller, R.J., J.W. Biggar, and D.R. Nielsen. 1965. Chloride displacement in Panoche clay loam in relation to water movement and distribution. Water Resour. Res. 1:63–73.

Mitchell, J.F.B., and D.A. Warrilow. 1987. Summer dryness in northern mid-latitudes due to increased CO_2. Nature 330:238–240.

Nelson, J.A., J.A. Morgan, D.R. LeCain, A.R. Mosier, D.G. Milchunas, and B.A. Parton. 2004. Elevated CO_2 increases soil moisture and enhances plant water relations in a long-term field study in semi-arid short-grass steppe of Colorado. Plant Soil 259:169–179.

Nimmo, J.R., and E.R. Landa. 2005. The soil physics contributions of Edgar Buckingham. Soil Sci. Soc. Am. J. 69:328–342.

Philip, J.R. 1955. Numerical solution of equations of the diffusion type with diffusivity concentration-dependent. Trans. Faraday Soc. 51:885–892.

Philip, J.R. 1969. Theory of infiltration. Adv. Hydrosci. 5:215–296.

Richards, L.A. 1931. Capillary conduction of liquids through porous mediums. Physics 1:318–333.

Sauer, T.J., B.E. Clothier, and T.C. Daniel. 1990. Surface measurements of the hydraulic properties of a tilled and untilled soil. Soil Tillage Res. 15:359–369.

Richards, L.A. 1940. Hydraulics of water in unsaturated soil. Agric. Eng. 22:325–326.

Schnell, R.C. 2008. Carbon dioxide. Bull. Am. Meteorol. Soc. 89:S26–S27 (Special Supplement to July issue).

Shortley, G., and W. Williams. 1971. Elements of physics. 5th ed. Prentice Hall, Englewood Cliffs, NJ.

Slichter, C.S. 1899. Theoretical investigations of the motion of groundwater. USGS Annu. Rep. 19:295–384.

Smettem, K.R.J., and B.E. Clothier. 1989. Measuring unsaturated sorptivity and hydraulic conductivity using multiple disc permeameters. J. Soil Sci. 40:563–568.

Soil Science Society of America. 1997. Glossary of soil science terms. SSSA, Madison, WI.

Sposito, G. 1976. An introduction to classical dynamics. John Wiley & Sons, New York.

Taylor, K.E., and M.C. MacCracken. 1990. Projected effects of increasing concentrations of carbon dioxide and trace gases on climate. p. 1–17. *In* B.A. Kimball et al. (ed.) Impact of carbon dioxide, trace gases, and climate change on global agriculture. ASA, CSSA, and SSSA, Madison, WI.

van Bavel, C.H.M. 1974. Antitranspirant action of carbon dioxide on intact sorghum plants. Crop Sci. 14:208–212.

Warrick, A.W. 2003. Soil water dynamics. Oxford Univ. Press, New York.

5

Nutrient Cycling in Soils: Nitrogen

Robert W. Mullen

With the exception of water, nitrogen is typically the most limiting nutrient in non-legume crop production systems. Although higher plant species require adequate nitrogen fertilization to maximize agronomic productivity, excess nitrogen application for crop production can have negative environmental effects when it is transported away from production fields. Production agriculture strives to strike a balance between adequate nitrogen supplementation and minimizing nitrogen losses in a way that is both agronomically and economically sound. Thus it is important to have a general understanding of nitrogen cycling within the soil. The nitrogen cycle is a dynamic collection of complex processes that are affected by a number of controllable and uncontrollable factors that should be understood to improve nitrogen management (Fig. 5|1). The goal of this chapter is to provide a discussion of nitrogen transformations, losses, and additions within the soil system to provide some information as to the agronomic and environmental importance of nitrogen.

Nitrogen Forms in Soil

Nitrogen exists in two forms within the soil—organic and inorganic. From a plant's perspective, the inorganic form is the only form the plant can readily utilize for nutrition, inorganic nitrogen forms in the soil are ammonium-N (NH_4^+) and nitrate-N (NO_3^-). These are both forms that are available to higher plants that can be readily taken into the root system. Organic nitrogen, though not available directly to the plant, is critical to the supply of inorganic nitrogen and is by far the most abundant nitrogen form in the soil (>90%). Soil organic matter primarily consists of carbon, nitrogen, sulfur, and phosphorus in a ratio of 10:1:0.1:0.05. Assuming a bulk density of 1.47 g cm^{-3}, a soil with 1% organic matter in the top 15 cm will contain approximately 1100 kg of organic nitrogen per hectare. Thus, even a soil with relatively low organic matter has the potential of completely supplying the nitrogen requirement of most crop species, if environmental conditions are conducive to organic N release.

The natural first question when thinking about soil nitrogen is: How does nitrogen make its way into the soil? This is a good starting point for exploring the nitrogen cycle. The Earth's atmosphere is comprised of 78% diatomic nitrogen (N_2), which is an inert gas and not plant available. Before the advent of soil fertilization via organic waste or commercial fertilization, the bulk of soil nitrogen most likely came from direct fixation (breaking N_2 bond and combining N atoms with hydrogen) by soil microbes and plant fixation by legumes. Free-living (i.e., not associated with higher plants) microorganisms have the ability to reduce atmospheric N_2 to form amino-N

R.W. Mullen, 130 Williams, OARDC-Wooster, The Ohio State University, Wooster, OH 44691 (Mullen.91@osu.edu).

doi:10.2136/2011.soilmanagement.c5

and incorporate it into living cells. Notable examples of free-living soil microorganisms capable of N fixation include *Azotobacter*, *Clostridium*, and *Cyanobacteria*. In addition to free-living soil microorganisms, there are also bacteria that form symbiotic relationships with higher plants. *Rhizobium* bacteria are able to infect plant roots, where they fix atmospheric nitrogen for the host plant and extract plant sugars for their own energy needs. While it would be difficult to estimate how much total nitrogen is fixed by free-living soil microorganisms, leguminous plant species are estimated to fix 300 million metric tons of N annually (Socolow, 1999). Once atmospheric N is fixed directly in the soil by free-living microorganisms or by legumes, the organically bound N is deposited within the soil or on the soil surface as dead plant material or through an animal system in which the animal consumes the aboveground plant material and deposits it on the soil surface in the form of excrement.

Once organic N has found its way to the soil system what happens to it? Depending on climatic conditions and the nature of the material, heterotrophic microorganisms decay the plant material and animal waste, satisfying their carbon needs along the way. During this process, N is conserved in the soil system and carbon is lost to the atmosphere as carbon dioxide. Based on the carbon/nitrogen ratio of the plant material being returned to the soil, two general reactions occur—mineralization or immobilization.

Mineralization

Organic nitrogen is not available to plants; it must be rendered into an inorganic form by a process called *mineralization*. Mineralization is a biologically mediated reaction that takes organic nitrogen and ultimately results in the production of ammonium N.

$$\text{Organic N} \rightarrow \text{Amino N (R-NH}_2) + CO_2 + \text{energy} \qquad [1]$$

$$\text{Amino N (R-NH}_2) \rightarrow NH_3 + \text{energy} \qquad [2]$$

$$NH_3 + H_2O \rightarrow NH_4^+ + OH^- \qquad [3]$$

The first step in the process is known as the aminization of organic N into an amino N form that results in the oxidization of carbon and the release of energy to a biological system (Eq. [1]). The second step is the ammonification of amino N into inorganic free ammonia and the release of additional energy (Eq. [2]). The final step in the process is a simple chemical reaction between gaseous hydrophilic ammonia N and water that produces stable ammonium N (Eq. [3]).

To characterize and possibly predict N mineralization of organic matter two individual components have been proposed. The first component is the supplying potential of the soil known as the mineralization potential (N_0). Mineralization potential involves assessing mineralization under controlled laboratory conditions to measure how much cumulative N mineralization is able to occur. The second component is the amount of mineralization that is realized, known as the *rate factor* (*k*). This is affected by soil temperature, soil moisture content, soil microbiological activity, and their interactions. This is the most difficult component as it relates

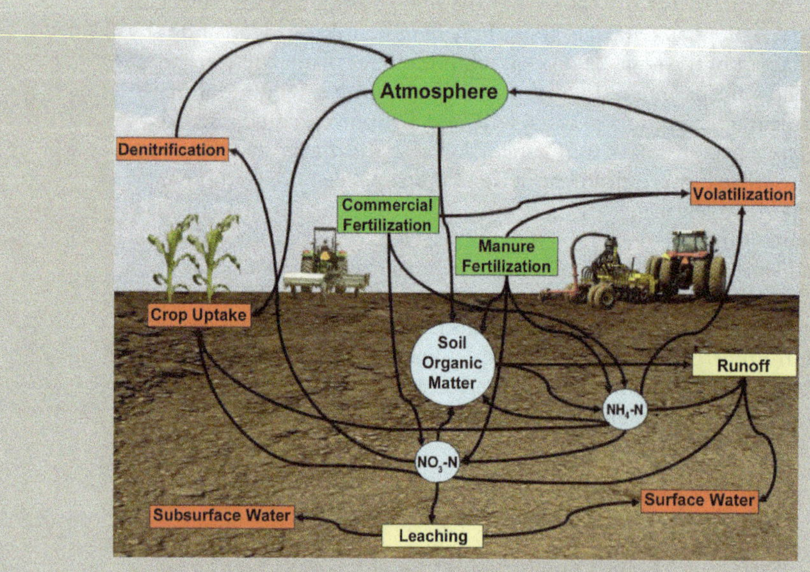

Fig. 5|1. Generalized illustration of the soil nitrogen cycle.

to translating mineralization potential (N_0) into N availability at the field level.

The nature of the organic matter is critical in determining how much mineralization will occur. Simply measuring organic matter as an estimate of potential mineralization has not been a reliable indicator of N mineralization (or N mineralization potential). Schmidt et al. (2002) reported that variable rate N recommendations based on soil organic matter alone did a poor job of indicating N mineralization within a given field. Their research indicates that soil organic matter alone is too simplistic to reflect what actually occurs at the field level. Measurement of more microbially active organic matter fractions may be a better estimator of net N mineralization potential than general organic matter level alone. Complex carbon compounds (such as lignin) are more stable and much less susceptible to breakdown and therefore are less likely to result in high levels of N mineralization when compared with a readily metabolized carbon source. Various organic matter fractions (light fraction and amino-sugar N), as well as various extractants (hot potassium chloride and phosphate borate), have been proposed as possible indicators of N mineralization potential (Curtin and Wen, 1999; Mulvaney et al., 2001). An inherent difficulty with N modeling is how these methods of quantifying potentially mineralizable N translate to actual field data (where a climate factor is necessary—the k factor). Some successes have been documented by some researchers (Mulvaney et al., 2001), but the results have been difficult to reproduce across multiple soils and regions (Laboski et al., 2008).

Soil organic matter is not just constituted by older residue additions, but is also made up of more recent additions. The nature of the recent residue additions can dramatically affect the mineralization rate of soil organic matter. A critical indication of mineralization potential is the carbon/nitrogen (C/N) ratio of the material being added to the soil. Carbon/nitrogen ratios are critical not only to plant residues being incorporated from a previous crop, but also for animal manure additions. The wider the gap between C and N, the lower the potential for net mineralization, but immobilization will likely occur. The narrower the gap between C and N of any residue incorporated into the soil, the greater the potential for N mineralization.

The general rule is that for C/N ratios greater than 30:1 net immobilization will occur. For C/N ratios between 20:1 and 30:1 mineralization will equal immobilization, and for C to N ratios less than 20:1 net mineralization will occur.

Cropping history and tillage practices also affect soil mineralization potential. Deng and Tabatabai (1999) evaluated N mineralization potential of different cropping systems and discovered that cropping systems sampled during meadow crops in a rotation showed greater N mineralization potential than row crops. Their work also concluded that more diverse crop rotations had higher cumulative N mineralization potential than two crop or monoculture cropping systems (Fig. 5|2).

Tillage has usually been thought to increase mineralization potential by introducing oxygen into the soil system, which results in increased microbial activity and subsequently increased release of inorganic N into soil solution. Recent research has revealed that tillage and the resultant disturbance of soil structure affects organic soil N mineralization. Kristensen et al. (2000) reported that tilling can destabilize protected organic-N pools (i.e., pools that are not active in short-term incubations), and these pools are less affected in no-till systems. As expected, they also found that soils from long-term no-till production systems (>20 yr) that were disturbed had initially greater organic-N release than soils that received annual tillage or no-till soils that remained undisturbed. Interestingly, Kristensen et al. (2000) also found that tilled soils (>20 yr) had similar mineralization potential to long-term no-till soils for the first 3 wk of their incubations, but during the last 2 wk of their 5-wk incubation undisturbed no-till soils actually had greater mineralization. This is likely due to the incorporation of high C/N ratio residue into the tilled treatments that began to decrease mineralization as that material was decomposed. More diverse cropping systems have greater organic-N mineralization potential, and disturbance of soil structure exposes protected pools of organic-N to mineralization.

The mechanics of the soil system are important to understand, but equally important are the environmental conditions that result in mineralization of organic N. Soil systems that are conducive to greater plant growth are also conducive to mineralization of organic N (i.e., warm, moist, oxygen-rich). Because mineralization is an environmentally driven phenomenon, it is difficult to model and predict given our inability to accurately forecast weather (at least at the level necessary to attempt to accurately model N mineralization). Several researchers have proposed various methods of measuring N mineralization in a lab setting under different environmental conditions (Stanford et al., 1973; Bonde and Lindberg, 1988; Cabrera and Kissel, 1988). The general conclusions are that warm, moist conditions are most conducive to N mineralization and that the length of the incubation is critical. Consistent moisture level from sampling time to incubation may not necessarily be the conditions most conducive to N mineralization. Cabrera (1993) reported that exposing a soil to drying and rewetting increased cumulative N mineralization compared to an undried soil. This is also more likely to be more indicative of what occurs at the field level. Conversely, Mikha et al. (2005) reported that constant soil water content resulted in increased cumulative N mineralization compared to a soil that was exposed to wetting and drying cycles. It should be noted that the Cabrera (1993) study did not expose the soil to periodic wetting and drying periods, but a single drying period followed by wetting to constant water content. Environmental conditions are critical to determining the extent of N mineralization under field conditions, and lab experiments have not been identified to most accurately represent conditions that will maximize N mineralization. Nor can meteorologists currently accurately predict weather conditions in the field. This is a challenge to modeling or predicting N mineralization at the field level.

Once ammonium N has been mineralized and released into soil solution, the molecule has several potential fates. First, as ammonium N in solution increases it can occupy exchange sites on clay edges and organic matter, being approximately the same size as a potassium (K^+) ion. Thus, the

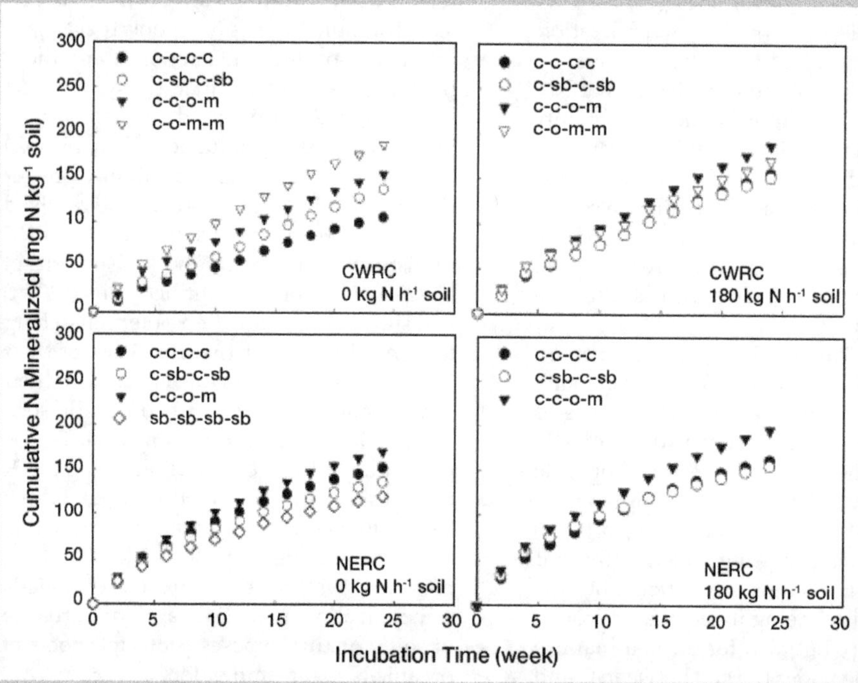

Fig. 5|2. Effect of different cropping systems on cumulative N mineralized from two experimental locations with different N rates. C–C–C–C is monoculture corn; C–SB–C–SB is corn–soybean rotation; C–C–O–M is corn–corn–oat (*Avena sativa* L.)–meadow; C–O–M–M is corn–oat–meadow–meadow; SB–SB–SB–SB is monoculture soybean (from Deng and Tabatabai, 1999).

same rules apply as that of any other monovalent cationic species in a temperate soil indicating that ammonium N is immobile. Second it can be taken up by higher plants. Third, it can be volatilized (as ammonia N [NH_3]) in high soil pH situations. Fourth, ammonium N can be immobilized and removed from soil solution, resulting in a decrease in N availability. Lastly, it can be oxidized to form nitrate-N.

Immobilization

Decay of plant residue can result in removal of inorganic N from soil solution if the C/N ratio of the residue being incorporated into the soil is greater than 30:1. As the microbes feed on the high C/N ratio residue they do not have adequate N from the material being consumed to produce stable organic matter. Thus, they will remove inorganic N from soil solution to achieve the correct C/N ratio. This removal of inorganic N from soil solution that results in decreased N availability for higher plants is called *immobilization*.

Plant residues have a wide range of C/N ratios based on when the crops are terminated. Crops terminated before reproductive stages have lower C/N ratios than crops allowed to reach full maturity (Table 5|1). In the Corn Belt of the United States the majority of the land grant universities have different N recommendations for corn production systems that follow a legume compared to corn production systems that follow a previous corn crop or other grass, like wheat (*Triticum* spp.). The primary reason being that a previous corn or wheat crop returns a tremendous amount of residue with a large C/N ratio. When fertilization occurs, the microbes compete for the inorganic- N supplied and will begin to immobilize it. Fertilization of a corn crop following soybean [*Glycine max* (L.) Merr.] does result in immobilization of fertilizer N, but the potential net immobilization is lower and the onset of mineralization is quicker because soybean residue has a lower C/N ratio and the total amount of residue being returned to the soil is lower (Green and Blackmer, 1995). This is why corn production systems that follow legume species have lower N recommendations than monoculture corn or corn after wheat production systems. Regardless of the amount of residue

Table 5|1. Carbon/nitrogen ratios of typical residue inputs.

Residue	C/N ratio
Wheat straw	80–100:1
Corn stover	60–80:1
Soybean stubble	20–40:1

returned, there is typically a net immobilization shortly after any N fertilization, but residue with narrower C/N ratios decreases the net amount of immobilization and the time needed to transition from immobilization to net mineralization (Green and Blackmer, 1995). Immobilization of N from soil solution is a critical step in the N cycle.

Nitrification

Ammonium N available in soil solution can be biologically transformed into nitrate-N in a two-step process known as nitrification. Nitrification of ammonium N requires similar environmental conditions to that of mineralization (being a microbial process this should not be a surprise), but it absolutely requires oxygen to proceed. The first reaction involves the bacteria *Nitrosomonas* and follows the reaction:

$$2\,NH_4^+ + O_2 - Nitrosomonas \rightarrow 2\,NO_2^-$$
$$+ 4\,H^+ + 2\,H_2O \qquad [4]$$

Notice that the product is nitrite N, not nitrate-N. This specific form of N does not accumulate in aerobic soils because the second step proceeds quickly, with some exceptions (Shen et al., 2003). This is fortunate considering that nitrite-N is toxic to plants at low concentrations (Oke, 1966). Adequate oxygen supply results in the second reaction:

$$2\,NO_2^- + O_2 - Nitrobacter \rightarrow 2\,NO_3^-$$
$$[5]$$

Thus, biological oxidation of ammonium N results in the release of hydrogen, and there may be concerns about soil acidification. Remember that the mineralization of organic matter resulted in the production of hydroxide, so the net effect would be that for every mole of nitrate-N produced from soil organic matter mineralization and nitrification only one mole of hydrogen (H^+) is produced. It could be further argued that as nitrate-N is assimilated by higher plants, the hydrogen is either neutralized or assimilated, so the use of organic N for higher plant N nutrition is not necessarily a soil acidifying process. Direct supplementation of ammonium N via commercial fertilization does result in soil acidification.

Nitrification is a relative quick process, and soils with temperatures adequate for microbial activity and abundant oxygen will convert ammonium N into nitrate-N in a matter of days or weeks (Table 5|2).

Nitrate in soil solution is also subject to different fates. First, it can be taken up by higher plants. Second, it can be immobilized just like ammonium N. Third, it can be leached from the root zone deeper into the soil profile. Finally, it can be denitrified to nitrous oxide or diatomic N and lost to the atmosphere. Since nitrate-N accumulation is the goal of crop production, we will ignore that aspect. We will instead focus on leaching and denitrification losses of nitrate-N.

Leaching

Leaching is the movement of plant nutrients (in this case nitrate-N) below the root zone where it can be stored in the subsoil, or lost to subsurface groundwater or subsurface drainage systems. Leaching has both economic and environmental impacts. As more leaching occurs in the crop production system, the more supplemental N is necessary to overcome a possible N deficiency. If nitrate-N finds its way to surface or subsurface waters it can have deleterious effects on water quality.

Nitrate leaching is more likely to occur in coarse-textured, sandy soils than fine-textured, clayey soils. The amount of nitrate-N leached from a soil is a function of the application rate and timing, soil texture, drainage, cropping system, and rainfall intensity. Loss mechanisms for N are a direct function of the amount of N supplemented to the system. Thus, as more N is put into the system, the greater the potential for loss. This is clearly true for nitrate-N and leaching. Jaynes et al. (2001) found that as N application rate increased for corn production the amount of nitrate-N found in tile drainage increased (Fig. 5|3). Note that even a low fertilizer N rate increased the nitrate-N concentration above 10 mg NO_3^--N L^{-1} level. It should be noted that for the 1996 corn production season the economic optimum N rate was between 67 and 135 kg N ha^{-1}, and thus the high N rate would be expected to be subjected to greater loss potential. In 1998, the economic optimum nitrogen rate was between 114 and 172 kg N ha^{-1}. This is clearly indicated in Fig. 5|3 since the total amount of nitrate-N lost was lower than in 1996. This particular tile drainage system was

Table 5|2. Influence of soil temperature on nitrification of ammonium N (adapted from Chandra, 1962).

Temperature sequence	Nitrification
	%
Continuous at 27°C for 24 d	100
12 d of 27°C, 12 d at 4°C	96
8 d at 27°C, 8 d at 16°C, 8 d at 4°C	74
12 d at 4°C, 12 d at 27°C	62
Continuous at 16°C for 24 d	59
8 d at 16°C, 8 d for 27°C, 8 d at 4°C	56
8 d at 4°C, 8 d at 16°C, 8 d at 27°C	45
Continuous at 4°C for 24 d	29

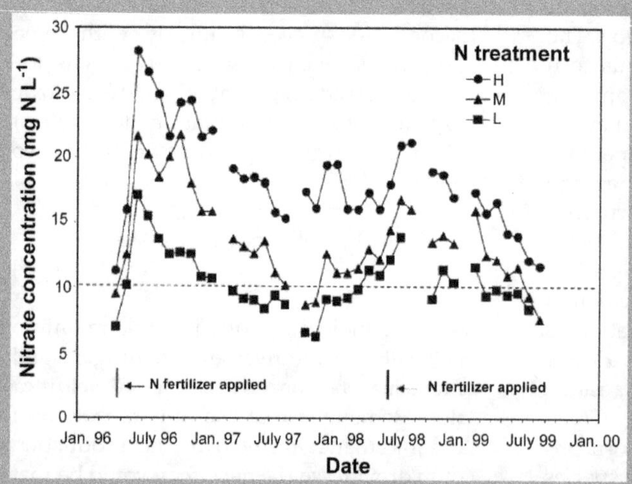

Fig. 5|3. Nitrate concentrations in tile drainage by month and N fertilizer rate versus time (from Jaynes et al., 2001).

installed to a depth of 1.45 m, and this study reveals that even at optimum N rates, some nitrate-N can be leached if rainfall patterns cause excess water to be drained from the soil profile and artificial drainage is present. A similar study conducted by Jemison and Fox (1994) determined that the amount of nitrate-N that leached (to a depth of 1.2 m) increased with the increasing N rate, and even the economic optimum N rate resulted in significant nitrate-N loss (Fig. 5|4). This is critical because to maximize economic return (not agronomic production) there is still significant loss of nitrate-N. Interestingly, in both of these studies at lower N rates (below economic optimum rates and including the zero-N check) there was still leaching of nitrate-N. This elucidates that agronomic production itself results in leaching of nitrate-N even when little or no fertilizer N is applied when artificial drainage is used. The study by Jemison and Fox (1994) happened to be a chisel-disk plow tillage system. One might think the increased leaching of nitrate-N was due to the tillage operation and alteration of organic C dynamics, but a study conducted by Zhu and Fox (2003) revealed that similar nitrate-N levels were leached from both a no-till and chisel plow soil. It should be noted that for each of the referenced studies all N fertilizer was applied before or near corn planting. Pre-plant and early season N applications made near planting are more subject to leaching losses because there is not a growing plant actively removing N from soil solution. This is especially true in the eastern portion of the Corn Belt, where these studies were conducted, because artificial drainage is often necessary. Pre-plant and early season N applications made in more arid environments are less likely to be lost by leaching.

Denitrification

Nitrate-N in solution may be subject to gaseous N losses if soil water contents approach field capacity through a process known as *denitrification*. Like most other N transformations in soil, denitrification is a biological reaction that converts nitrate-N to nitrous oxide

or diatomic N (N_2O, N_2) by the following reaction:

$$2\,NO_3^- - O_2 \rightarrow 2\,NO_2^- - O_2 \rightarrow$$
$$2\,NO^- - 1/2\,O_2 \rightarrow N_2O - 1/2\,O_2 \rightarrow N_2$$

[6]

Based on Eq. [6], this biological reaction occurs under anaerobic conditions, and this reaction requires an easily metabolized carbon source, available nitrate-N, and adequate temperature for biological activity. Denitrification occurs due to the absence of oxygen in a biological system. Oxygen, in most biological reactions, acts as the terminal electron acceptor, but when it is limiting or absent, some microbes have the ability to utilize other molecules as the terminal electron acceptor by stripping oxygen from other species. Most of these microbes are heterotrophic facultative anaerobes, meaning that they require the presence of an organic carbon source and that they can function in both an aerobic and anaerobic environment. The microbes that have been identified that can denitrify nitrate-N are *Pseudomonas*, *Bacillus*, *Paracoccus*, *Micrococcus*, *Achromabacter*, and *Thiobacillius denitrificans*.

Availability of nitrate-N is obviously an important component in determining the extent of potential denitrification. Denitrification increases as N application rates increase for crop production (MacKenzie et al., 1998). This is intuitive since the initial reactant in Eq. [6] is nitrate-N. Another important factor necessary for

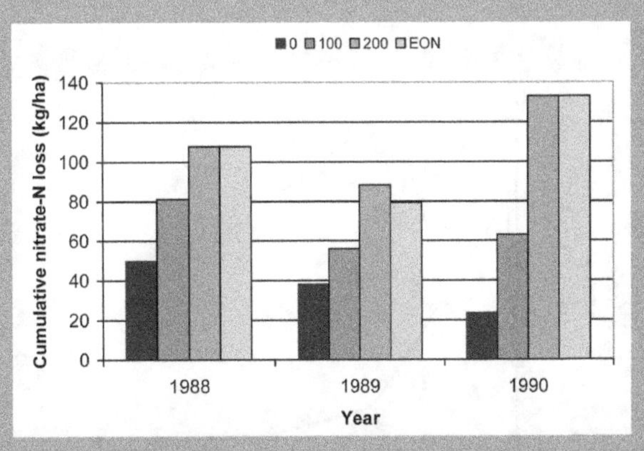

Fig. 5|4. Cumulative nitrate N loss as influenced by N rate (0, 100, and 200 kg N ha⁻¹, including the economic optimum N rate [EON]) of nonmanured corn fields in 1988–1990 (from Jemison and Fox, 1994).

denitrification is an easily mineralizable carbon source. Residues with lower C/N ratios would be considered more easily mineralizable, and when these residues are incorporated, denitrification does increase in an anaerobic environment. Aulakh et al. (1991) reported that incorporation of residue with a low C/N ratio (8:1) resulted in more denitrification of nitrate-N when water-filled pore space was 90% than with residues having higher C/N ratios (82:1). They also reported that denitrification occurs more rapidly with incorporated residues especially residues with low C/N ratios. Additional research has revealed that even though initial denitrification occurs more rapidly with incorporated residues in a conventional tillage system, established no-till soils have been observed to result in higher denitrification rates than conventionally tilled soils (MacKenzie et al., 1998). This is likely due to the increased biomass carbon, lower rates of air diffusion, and/or larger anaerobic soil aggregates in no-till soils (Lal, 1989; Aulakh et al., 1984). Availability of nitrate-N and an easily mineralizable carbon source can be critical rate limiters in denitrification. This is why denitrification of nitrate-N that has leached below the rooting zone is unlikely to occur.

Rainfall intensity and soil texture are also critical components to determining the extent of denitrification. Intense rainfall events can result in substantial denitrification losses of nitrate-N for most soils if nitrate-N, carbon source, and temperature are not limiting. Sexstone et al. (1985) found that intense rainfall events can result in short-term increases in denitrification. They also observed that coarser textured, sandy soils reach a maximum denitrification rate faster than fine textured soils, but coarser textured soils return to pre-rainfall conditions much faster than fine textured soils (Fig. 5|5). Thus, overall fine-textured soils are likely to lose more nitrate-N by denitrification than coarse-textured soils. Intense rainfall and resultant anaerobic conditions can result in substantial denitrification losses of nitrate-N especially in fine textured soils.

Surface Transport

Nitrogen can also move across the soil surface and be lost from the soil system through either erosion or through surface runoff. Erosion is the physical detachment and movement of soil across the landscape. Surface runoff is movement of water across the soil surface. Both erosion and surface runoff can transport particulate N, organic N, and solution inorganic N. If erosion or surface runoff ultimately connects to surface waters, the N transported can negatively affect water quality through eutrophication, especially in marine systems. Like the other loss mechanisms, the extent of N loss is a function of initial soil N content (organic matter level and inorganic N level), as well as rainfall intensity and initial soil water content. The form of nitrogen most frequently lost by erosion and surface transport is sediment-bound organic N (Schuman et al., 1973), but nitrate-N and ammonium N losses can be larger percentage-wise in production systems that decrease soil erosion (Udawatta et al., 2006). Minimum tillage production systems and permanently vegetated grass systems are two examples that would result in decreased total N transport, but increased soluble N transport as a percentage of total N surface flow.

Nitrogen Inputs for Crop Production

The soil N cycle is a dynamic combination of biological processes that affects availability and losses of an important plant nutrient, and up to this point in the discussion crop production inputs

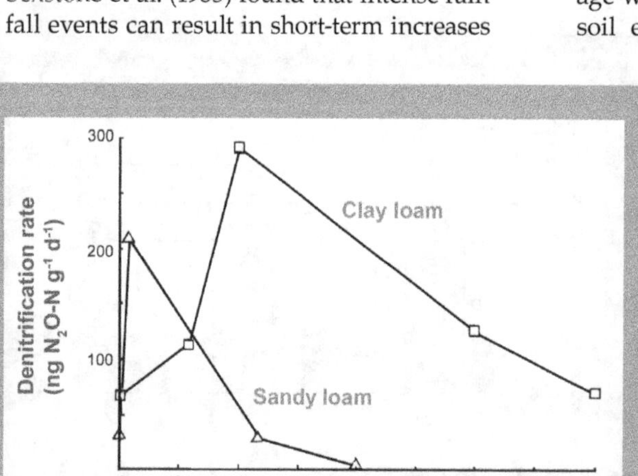

Fig. 5|5. Denitrification rate over time as a function of soil texture (from Sexstone et al., 1985).

have largely been unaddressed. It is critical to the understanding of the N cycle to address the issue of how N fertilizer (from commercial and animal manures) is supplemented to growing crops. Even more critical is how fertilizer management affects plant uptake thus decreasing it loss potential to the environment.

The initial focus will be on application rate decisions and how the dynamic nature of the N cycle can make determining how much N to supply in a given growing environment a difficult decision. As discussed in the section "Mineralization," the amount of inorganic N provided through natural reactions in the soil differs from field to field and from one season to the next. The nature of the organic matter, the nature of recent residue additions, and weather all play critical roles in determining how much N will be made available.

Evidence supporting the idea that mineralization changes from 1 yr to the next can be found in N response studies for corn conducted in Ohio between 2000 and 2005 (Fig. 5|6). The lowest N rates evaluated were 45 kg N ha^{-1} within this study, and the previous crop was always corn. Evaluation of yield level among different years at the 45 kg N ha^{-1} rate reveals that significant year to year variability exists, and this variability most likely exists as a result of N mineralization (subsequent leaching and denitrification also factors). This conclusion is arrived at based on the physiology of the corn plant. Assuming no loss of N by either denitrification or leaching and fertilizer N use is 100% efficient, the maximum attainable yield for corn fertilizer with only 45 kg N ha^{-1} would be near 3.5 Mg ha^{-1} (assuming a grain N content of 1.3%). In 5 of 6 yr, significantly higher yield levels were attained at the 45 kg N ha^{-1} rate, illustrating that some N must be made available somewhere in the soil system, and mineralization is the most likely process responsible. It is interesting to note that in at least 2 of the 6 yr 45 kg N ha^{-1} was enough for near maximum agronomic productivity.

Similar findings have been reported in long-term studies evaluating winter wheat (*Triticum aestivum* L.) response to

fertilizer N (Johnson and Raun, 2003). In this more than 30-yr N response study a true control (0-N) has been evaluated in a continuous wheat production system. The yield level of the control treatments varies tremendously from one year to the next because of variable mineralization rates. In some years, the amount of mineralization can be quite large and in others more modest, most likely the result of different weather conditions experienced throughout the growing season. Also noteworthy is that seasons with high N mineralization do not necessarily mean that agronomic productivity will be greatly improved with additional N inputs (year 2000 for the corn study and year 30 for the wheat study). It should also be noted that within these nitrogen response trials other loss factors are integrated into the identification of the optimum N rate. Thus, years with excessive leaching and denitrification would theoretically result in higher optimum N rates if yield levels are not suppressed by limiting factors. These results elucidate the need to identify not only mineralization potential, but also methods of directly measuring mineralization at the field level using innovative technologies or improving the accuracy and precision of weather forecasting. Additionally, it would be beneficial to identify the extent of leaching and denitrification losses during the growing season. The other challenge facing agriculture is that most N rate decisions are made before planting the crop, so for a real impact to be realized one of two things have

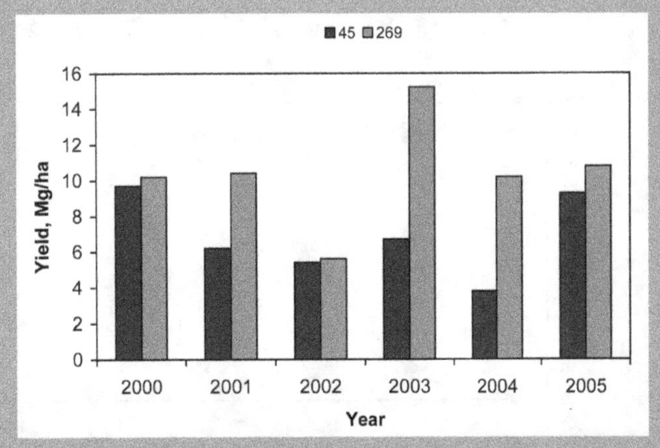

Fig. 5|6. Corn yield level as a function of N applied at the Northwest Research Station for corn following corn at two N rates (45 and 269 kg ha^{-1}) between 2000 and 2005.

to happen: (i) weather forecasting improves dramatically (at some unknown resolution) that allows for development of mineralization models, or (ii) cultural practices typically utilized for supplementing crop N changes to more in-season applications when more information is known about the crop growing environment and soil biological reactions.

There are other input factors that will affect how efficiently N input is utilized in the soil system. Surface application of urea-based fertilizers and animal manures can be subject to volatile losses of ammonia N that will obviously decrease N availability throughout the growing season. Urea is an organic compound that requires the presence of the enzyme urease to produce ammonia N by the following reactions:

$$CO(NH_2)_2 \text{ (urea)} + 2H^+ + 2\,H_2O - urease \rightarrow (NH_4)_2CO_3 \qquad [7]$$

$$(NH_4)_2CO_3 + H_2O \rightarrow 2\,NH_3 + 2\,H_2O + CO_2 \qquad [8]$$

$$2\,NH_3 + 2\,H_2O \leftrightarrow 2\,NH_4^+ + 2\,OH^- \qquad [9]$$

Once urea is applied to the soil surface it is hydrolyzed, and in the presence of the urease enzyme it converts to ammonia N (NH_3–N) (Eq. [7] and [8]). Notice that the reaction actually consumes hydrogen ions (H^+) from soil solution (Eq. [7]), which can result in an increase of soil pH if the system is poorly buffered (i.e., with coarse-textured soil). As soil solution pH increases, the amount of ammonia N present increases. At a pH of 9.3 an equal distribution of ammonia N and ammonium N exists (Fig. 5|7). This potentially translates into ammonia N volatilization. Similarly, higher pH soils can result in greater loss of ammonia N because the excess hydroxide in solution will force Eq. [9] to the left. Notice also that the reaction in Eq. [9] requires water to form stable ammonium N. Urea that has been hydrolyzed to ammonia N in moisture-limited situations can be subject to volatile ammonia N losses. Incorporation of urea fertilizers either mechanically or by rainfall or irrigation is by far the most efficient way of minimizing the risk of ammonia N volatilization, but urease inhibitors that deactivate the urease enzyme may also be considered. Additionally, applying the material in a concentrated band instead of a broadcast application can significantly decrease volatilization. This method of application is less efficient than incorporation, but it does decrease the surface area of urea exposed to soil and urease.

The presence of surface residue and temperature at the time of application affects how much ammonia volatilization will occur. Since urease is an active enzyme within the plant system, it is intuitive that surface deposition of plant residue will contain significant quantities of urease that can result in higher rates of ammonia N volatilization. Surface application of a urea-based N source in a conservational tillage operation designed to maintain significant amounts of plant residue does increase the risk of ammonia-N volatilization losses. Viswakumar et al. (2008) reported that liquid urea-ammonium nitrate broadcast applied in conservational tillage operations resulted in significant yield losses compared to split applications of N between starter and coulter injected urea-ammonium nitrate early in the growing season (Fig. 5|8). Conversely, more intensive tillage (disk/field cultivator) that had lower surface residue did not show a yield different between the two N application methodologies. This reveals that the presence of surface residue can increase the risk of volatilization of surface-applied urea-based fertilizers. Ammonia N volatilization

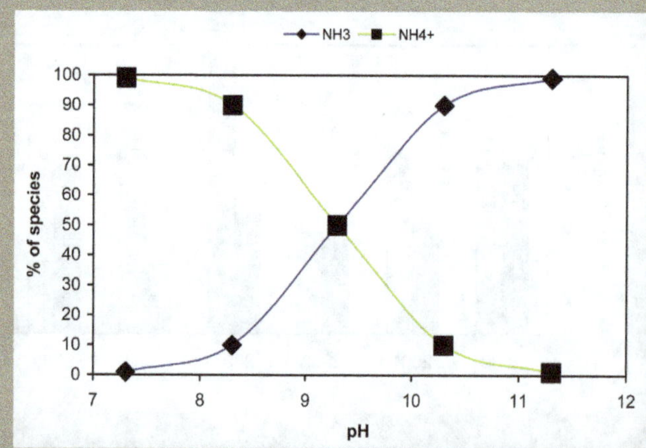

Fig. 5|7. Amount of ammonium N and ammonia N species present in solution as a function of pH.

rate increases as temperature increases; thus, applications in cooler environments are much less susceptible to volatile losses.

Surface-applied animal manures can also be subject to volatilization losses. The fraction of greatest concern is the ammoniacal (ammonia/ammonium N) portion of N within the manure because it is the fraction most readily available for loss. Typically, the amount of ammonia N volatilized is presented as a percentage of total ammoniacal N. Volatilization losses from surface application of animal manures are affected by several variables, including application rate, soil pH, soil buffer capacity, presence of residue, temperature, wind speed, and rainfall. As in all other N loss mechanisms (i.e., leaching, denitrification, and runoff), as the availability of the species capable of being lost increases, so too does the amount lost. Volatilization from surface applications of animal manure is no different. The higher the application rate, the more ammonia N volatilized. Similar to applications of urea-based fertilizers, higher soil pH level translates into more volatilization, and soils with lower buffer capacities are more likely to experience more volatilization. Presence of surface residue also increases the volatilization rate. Residue at the soil surface restricts contact between the waste material and soil, which increases the total surface area available for volatilization (Thompson et al., 1990a). Wind speed at the soil surface also affects ammonia volatilization of ammonia N from surface applications of animal waste. Higher wind speeds increase air exchange at the interface between the manure at the surface and the atmosphere (Thompson et al., 1990b). Finally, rainfall also influences ammonia/ammonium N infiltration into the soil. Rainfall events that effectively incorporated solution ammoniacal N decrease the risk of volatilization. Injection or rapid incorporation of animal waste materials dramatically decreases the risk of ammonia N volatilization.

Summary

The soil N cycle is a dynamic collection of complex biological processes that affects availability of an important nutrient for crop nutrition. Unfortunately, the system is capable of loss that can have negative environmental consequences. As pointed out here, the more N input into the system, the more N that is potentially lost. The goal of production agriculture at any scale should be to maximize agronomic and economic productivity and to minimize N overapplication. Understanding the natural N release mechanisms (mineralization) and loss mechanisms (leaching, denitrification, and surface runoff) is a starting point for making better N input decisions. These are lofty aspirations considering the dynamic nature of N and the influence of the climatic conditions, but it should be the goal nonetheless.

References

Aulakh, M.S., J.W. Doran, D.T. Walters, A.R. Mosier, and D.D. Francis. 1991. Crop residue type and placement effects on denitrification and mineralization. Soil Sci. Soc. Am. J. 55:1020–1025.

Aulakh, M.S., D.A. Rennie, and E.A. Paul. 1984. Gaseous nitrogen losses from soils under zero-till as compared with conventional-till management systems. J. Environ. Qual. 13–130–136.

Bonde, T.A., and T. Lindberg. 1988. Nitrogen mineralization kinetics in soil during long-term aerobic laboratory incubations: A case study. J. Environ. Qual. 17:414–417.

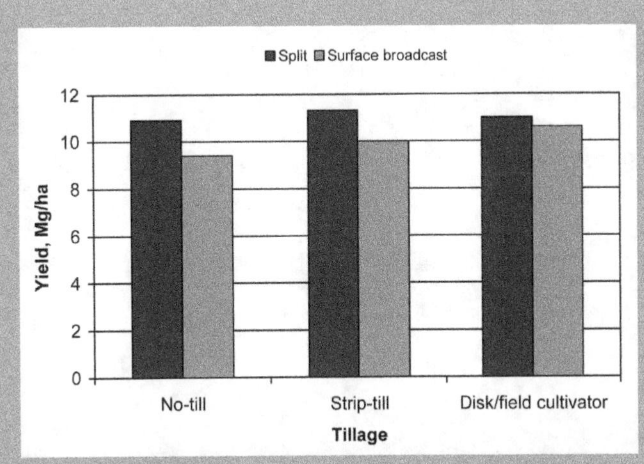

Fig. 5|8. Impact of tillage and N application methodology of liquid urea-ammonium nitrate on corn grain yield (from Viswakumar et al., 2008).

Cabrera, M.L. 1993. Modeling the flush of nitrogen mineralization caused by drying and rewetting soils. Soil Sci. Soc. Am. J. 57:63–66.

Cabrera, M.L., and D.E. Kissel. 1988. Potentially mineralizable nitrogen in disturbed and undisturbed soil samples. Soil Sci. Soc. Am. J. 52:1010–1015.

Chandra, P. 1962. Note on the effect of shifting temperatures on nitrification in a loam soil. Can. J. Soil Sci. 42:314–315.

Curtin, D., and G. Wen. 1999. Organic matter fractions contributing to soil nitrogen mineralization potential. Soil Sci. Soc. Am. J. 63:410–415.

Deng, S.P., and M.A. Tabatabai. 1999. Effect of cropping systems on nitrogen mineralization in soils. Biol. Fertil. Soils 31:211–218.

Green, C.J., and A.M. Blackmer. 1995. Residue decomposition effects on nitrogen availability to corn following corn or soybean. Soil Sci. Soc. Am. J. 59:1065–1070.

Jaynes, D.B., T.S. Colvin, D.L. Karlen, C.A. Cambardella, and D.W. Meek. 2001. Nitrate loss in subsurface drainage as affected by nitrogen fertilizer rate. J. Environ. Qual. 30:1305–1314.

Jemison, J.M., Jr., and R.H. Fox. 1994. Nitrate leaching from nitrogen-fertilized and manured corn measured with zero-tension pan lysimeters. J. Environ. Qual. 23:337–343.

Johnson, G.V., and W.R. Raun. 2003. Nitrogen response index as a guide to fertilizer management. J. Plant Nutr. 26:249–262.

Kristensen, H.L., G.W. McCarty, and J.J. Meisinger. 2000. Effects of soil structure disturbance on mineralization of organic soil nitrogen. Soil Sci. Soc. Am. J. 64:371–378.

Laboski, C.A.M., J.E. Sawyer, D.T. Walters, L.G. Bundy, R.G. Hoeft, G.W. Randall, and T.W. Andraski. 2008. Evaluation of the Illinois soil nitrogen test in the North Central Region of the USA. Agron. J. 100:1070–1076.

Lal, R. 1989. Conservation tillage for sustainable agriculture: Tropics versus temperate environments. Adv. Agron. 42:85–197.

MacKenzie, A.F., M.X. Fan, and F. Cadrin. 1998. Nitrous oxide emission in three years as affected by tillable, corn–soybean–alfalfa rotations, and nitrogen fertilization. J. Environ. Qual. 27:698–703.

Mikha, M.M., C.W. Rice, and G.A. Milliken. 2005. Carbon and nitrogen mineralization as affected by drying and wetting cycles. Soil Biol. Biochem. 37:339–347.

Mulvaney, R.L., S.A. Khan, R.G. Hoeft, and H.M. Brown. 2001. A soil organic nitrogen fraction that reduces the need for nitrogen fertilizer. Soil Sci. Soc. Am. J. 65:1164–1172.

Oke, O.L. 1966. Nitrite toxicity to plants. Nature 212:528.

Schmidt, J.P., A.J. DeJoia, R.B. Ferguson, R.K. Taylor, R.K. Youn, and J.L. Havlin. 2002. Corn yield response to nitrogen at multiple in-field locations. Agron. J. 94:798–806.

Schuman, G.E., R.E. Burwell, R.F. Piest, and R.G. Spomer. 1973. Nitrogen losses in surface runoff from agricultural watersheds on Missouri Valley loess. J. Environ. Qual. 2:299–302.

Sexstone, A.J., T.B. Parkin, and J.M. Tiedje. 1985. Temporal response of soil denitrification rates to rainfall and irrigation. Soil Sci. Soc. Am. J. 49:99–103.

Shen, Q.R., W. Ran, and Z.H. Cao. 2003. Mechanisms of nitrite accumulation occurring in soil nitrification. Chemosphere 50:747–753.

Socolow, R.H. 1999. Nitrogen management and the future of food: Lessons from the management of energy and carbon. Proc. Natl. Acad. Sci. USA 96:6001–6008.

Stanford, G., M.H. Frere, and D.H. Schwaninger. 1973. Soil nitrogen availability evaluations based on N mineralization potentials of soils and uptake of labeled and unlabeled nitrogen by plants. Plant Soil 39:113–124.

Thompson, R.B., B.F. Pain, and D.R. Lockyer. 1990a. Ammonia volatilization from cattle slurry following surface application to grassland: Influence of mechanical separation, changes in chemical composition during volatilization, and the presence of the grass sward. Plant Soil 125:109–117.

Thompson, R.B., B.F. Pain, and Y.J. Rees. 1990b. Ammonia volatilization from cattle slurry following surface application to grassland: Influence of application rate, wind speed, and applying slurry in narrow bands. Plant Soil 125:119–128.

Udawatta, R.P., P.P. Motavalli, H.E. Garrett, and J.J. Krstansky. 2006. Nitrogen losses in runoff from three adjacent agricultural watersheds with claypan soils. Agric. Ecosyst. Environ. 117:39–48.

Viswakumar, A., R.W. Mullen, A. Sundermeier, and C.E. Dygert. 2008. Tillage and nitrogen application methodology impacts on corn grain yield in a poorly drained soil. J. Plant Nutr. 31:1963–1974.

Zhu, Y., and R.H. Fox. 2003. Corn–soybean rotation effects on nitrate leaching. Agron. J. 95:1028–1033.

Potassium Cycling

Sylvie Brouder

Potassium is an essential nutrient required in relatively large quantities by all plants. After nitrogen and phosphorus, potassium is the third most likely nutrient to limit crop productivity. Chemically, K is known as an alkali metal (group IA of the periodic table: Li, Na, K, Rb, Cs, Fr); chemical compounds of K are highly soluble in water and the chemistry of K is predominantly ionic as it does not form strong bonds with oxygen, N, and carbon. In soil, K behavior and availability to crops is controlled by its size and charge in conjunction with the cation exchange capacity (CEC) of the soil and the weathering of primary and secondary minerals. Unlike N but similar to P, K does not form any gases that could subsequently be susceptible to loss. Also, while precipitation or irrigation can cause movement of K into water bodies via leaching or erosion, K is not thought to contribute to eutrophication of surface waters or generally considered toxic or a pollutant. Thus, in agriculture, K management decisions are primarily driven by a simple cost benefit analysis of the expected plant demand versus soil K supplying power.

Potassium is the seventh most abundant element in the Earth's crust. A commonly reported range for total K content of soil is 0.04 to 3.0% (0.4–30 g K kg^{-1} soil; Helmke, 2000). In the plow layer (0–0.2 m depth), a typical mineral soil can have 3,000 to 100,000 kg K ha^{-1}. However, 98% of this K is in a form not considered readily available to plants, while only about 2% occurs as solution or exchangeable K, the forms considered highly plant available (Sparks, 1987 and references cited therein). In an evaluation of 102 soils from the continental United States and Puerto Rico, total soil K ranged from 0.6 to 23.6 g kg^{-1}, with <1 to ~10% of this total K in plant-available form (Sharpley, 1989). Intensive agriculture regularly removes a portion of the plant-available soil K in harvested tissue, and K management must be intensified where removal rates are high.

High-yielding agricultural crops may contain K in quantities equivalent to or in excess of the N content. In plants, K is not actually incorporated into organic compounds but remains in ionic form (K$^+$) in cellular and vascular tissue solutions (e.g., cytoplasm, vacuole solution, xylem, and phloem). A primary plant K$^+$ function is the activation of more than 80 different enzymes for critical processes including sugar degradation, nitrate reduction, starch and protein synthesis, photosynthesis, and energy metabolism. Potassium is also a dominant driver in plant–water relations and turgor pressure-related phenomenon including stomatal movement, cell extension and plant growth, phloem transport and photonastic responses such as solar tracking (see Marschner, 1995 for extensive summary of K physiology). In cell solutions, required K concentrations range between 10 and 200 mM, which translates into dry weight K contents for gross plant tissue of approximately 1% (10 g K kg^{-1}). Reviews typically summarize the range in aerial K contents of most commercial crops as 100 to 300 kg K ha^{-1} (e.g., Haby et al., 1990). Variations in total crop K requirements and rates of demand are strongly linked to plant size and growth rate,

S. Brouder, Agronomy Department, Purdue University, 915 W. State St., West Lafayette, IN 47907-2054 (sbrouder@purdue.edu).

doi:10.2136/2011.soilmanagement.c6

respectively (Atkinson, 1990), and most agricultural soils must supply large quantities of K. Present and future higher yield systems can be expected to have significantly greater K demands, as crop K requirement is expected to scale with yield (Brouder and Volenec, 2008).

The general geographic distribution of plant-available, indigenous K can be easily linked to a combination of the distribution of parent material and its differential weathering, as well as to the subsequent association of released K with clay minerals. The main agent of weathering is water, and climate is often a major factor affecting mineral changes (Churchman, 2000). Thus, physical weathering is the dominant process in the formation of illites and chlorites in arid desert soils, while leaching and oxidation drive formation of kaolinite and iron oxides in hot (wet and dry) climates. Within the continental United States an inverse relationship has long been observed between geographic distribution of K and mean annual temperature and rainfall. Reitemeier (1957) observed that soil K content increased from east to west, and within the eastern United States from south to north. In an analysis of U.S. K geography, Bertsch and Thomas (1985) grouped soils first by stage of weathering, with subgroups reflecting differences in parent material. Lowest K soils were mapped to the South Atlantic region, where landscapes are old, rainfall is plentiful and frequently leaches soil profiles, and soils are the most weathered (Table 6|1). High K soils were mapped to regions where soils are younger, covered in deep loess, glacial till or alluvial material, and leaching has been minimized (e.g., west north-central region). Malavolta (1985) prepared a similar analysis for tropical and subtropical regions of the world. For agricultural purposes, surveys of results from public soil testing labs have been used to convert this general knowledge into maps of agronomic need for K management (e.g., Haby et al., 1990; Fixen, 2006). While information from U.S. soil testing labs is only for a subset of regional soils (mostly agricultural) and confounds indigenous K with fertilizer amendments, ranges in soil test levels have tended to reflect native fertility, and thus highlight regions where K fertilizer can be expected to be a key component of crop management.

Table 6|1. U.S. potassium availability by geographic region (derived from Bertsch and Thomas, 1885; Scott and Smith, 1987).

Region	States	Determinants of K status	Summary K status	Range in median soil test values†
New England & Mid Atlantic	ME, NH, VT, CT, MA, RI, NY, NJ, PA, MD, DE, WV	Parent material high and low in K. Glaciated. Younger soils. Moderate to wet summers → some seasonal leaching. Frozen ground in winter or freezing common. Fairly weathered.	fairly low	92–154
South Atlantic	VA, NC, SC, GA, FL	Unglaciated, low K parent material, most weathered. Rainfall > evapotranspiration by 30 to 58 cm yr⁻¹ → high leaching. Freezing uncommon.	low	43–134
East North Central	WI, MI, IL, IN, OH	East of Mississippi: Most of region glaciated, covered with loess or both. Relatively dry winter. West of Mississippi: Deep loess or glacial till, high in K, dry winters, not very leached or weathered.	moderate to relatively high	125–168
East South Central	KY, TN, MS, AL	Unglaciated, modest K (>Atlantic Coast); many soils have loess covering and are higher in K. Rainfall > evapotranspiration → high leaching. Freezing uncommon.	fairly low	88–132
West North & South Central	ND, SD, MN, NE, KS, IA, MO, OK, AR, LA, TX	Generally high K. Drier → leaching low except with irrigation. Frozen ground in winter or freezing common.	high (except in sandy soils)	110–364
Mountain & Pacific	MT, WY, ID, UT, CO, AZ, NM, WA, OR, CA	Agricultural soils formed in deep alluvial deposits. Very high K. Leaching high along coast only or with irrigation, otherwise dry. Freezing uncommon.	high	172–328

† Values are NH₄OAc-equivalent from Fixen (2006).

Phases and Forms of Potassium in Soil Systems

Potassium in soil exists in soluble ionic form in the soil water (soil solution) or held with varying degrees of strength on or within solid forms. From the perspective of plant availability and soil fertility, soil K is usually characterized in terms of the nature and quantity of solution K and K in three distinct solid-phase forms: mineral, nonexchangeable (or "fixed"), and exchangeable. Delineation of the solid-phase forms is usually presented in terms of the location of K^+ within the mineral or nonmineral phase and the bonding mechanism or strength. Typically, the relative order of pool quantity is structural >>> nonexchangeable ≥ exchangeable >> solution, while the relative order of general availability to plants and soil microbes is the reverse: solution > exchangeable >> nonexchangeable >>> structural (Fig. 6|1). In geologic time, metamorphasism can reform magma (new igneous rocks), and secondary minerals are continuously recycled. However, within timelines relevant to cropping cycles, K release from the crystalline structures of primary and many secondary minerals can be considered an irreversible process. In contrast, nonexchangeable, exchangeable, and solution K are considered interchangeable; the total and relative quantities of these forms and the readiness with which they interchange determine soil K fertility. The soil K forms have the following salient features:

Structural Potassium

Structural K is covalently bonded within the crystalline structure of primary and secondary minerals. As defined by Gaines et al. (1997), a mineral is an element or chemical compound that is normally crystalline with formation resulting from geological processes. Igneous and sedimentary rock constitute about 95 and 5%, respectively, of the Earth's crust, but sedimentary rock covers 75% of the land surface and therefore serves as parent material for much of the soil. For igneous rock, the stage of magmatic crystallization strongly influences the K content. Early formed igneous rocks such as basalt are generally low in alkali metals, whereas later-formed micas and K-feldspars are alkali metal enriched (Scott and Smith, 1987). Shales, the vast majority of an average sedimentary rock (Horstman, 1957), have high clay content and relatively large quantities of all alkali metals, while sandstones and limestones have alkali metals associated with residual particles of feldspars and mica as well as with clay impurities (Scott and Smith, 1987). As summarized by Malavolta (1985), within the igneous rock group, granites and syenites contain as much as 46 to 54 g K kg⁻¹, whereas basalts and peridotites contain approximately 7 and 2 g K kg⁻¹ respectively. Sedimentary materials also vary markedly in K content, with clayey shales containing approximately 30 g K kg⁻¹ and limestones averaging 6 g K kg⁻¹.

The feldspars and micas are considered the most important K-bearing primary minerals (Haby et al., 1990). Feldpars orthoclase, microcline, and sanidine have K contents of 110 to 150 g K kg⁻¹, while the K contents of micas muscovite, biotite, and phlogopite are approximately one-half as great. In both feldspars and micas, structural substitution of Si^{4+} by Al^{3+} results in a net negative charge to the crystalline structure that is compensated for by the insertion of another cation into the crystal lattice. The specific structural units and the degree of insertion of K versus

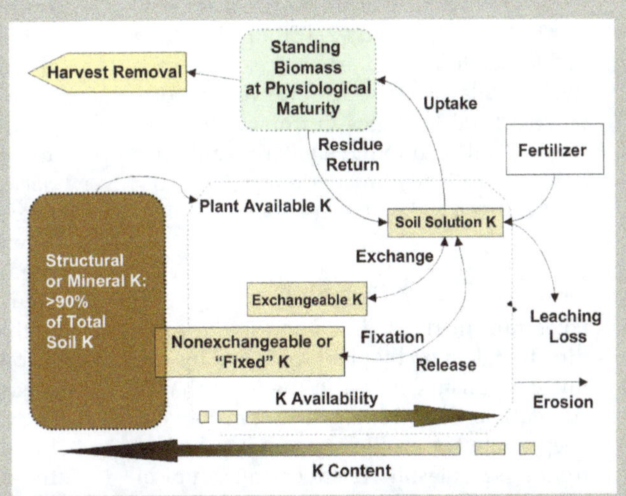

Fig. 6|1. Schematic of plant–soil K cycling in an agricultural system. See Table 6|2 for concentrations and quantities of K in different forms.

Na, Mg, and/or Fe distinguish among the various primary minerals. These primary minerals can be chemically and/or physically altered (weathered) into secondary minerals. The secondary minerals important for soil K fertility are mostly in the clay-sized fraction and include the structured aluminosilicate illites, vermiculites, kaolinites, and chlorites, and the amorphous allophone. The structured aluminosilicates are distinguished by the relative number of Si tetrahedral and Al octahedral sheets within their layers (2:1 or 1:1 Si:Al sheets), the extent to which water and other cations can penetrate mineral interlayers, and the overall particle charge density and surface area. See Olson et al. (2000) for details on structures of primary and secondary phylosillicates.

The combined action of chemical, physical, and biological forces very slowly (i.e., at the geologic time scale) releases K from the crystalline lattice into the surrounding soil water. For example, as mica is weathered to illite (hydrous mica) and eventually vermiculite or smectite, K content decreases from approximately 10 to <1% (Bertsch and Thomas, 1985). On a whole soil basis, Sharpley (1989) identified structural K contents of 0.03 to 21.05 g kg^{-1} for 102 U.S and Puerto Rican soils (Table 6|2). Kaolinitic soils dominated by 1:1 layer silicates in the clay fraction were markedly lower in structural K than those identified as smectitic and dominated by 2:1 layer silicates. Assuming an average bulk density of 1.3 g cm^{-3}, estimates for the total K content in a plow layer (0–0.2 m) are <100 to <21,000 kg ha^{-1} and >21,000 to >54,730 kg ha^{-1} for kaolinitic and smectitic soils, respectively. Malavolta (1985) reviewed reports on structural K ranges in tropical and subtropical soils and found similar magnitudes, although a slightly broader range (0.1–55.5 g kg^{-1}), across a wide array of soils.

Soil Solution Potassium

Potassium in the soil solution is the form directly taken up by plant roots and by soil microorganisms. It is also the form that is susceptible to leaching, which may move K deeper into the profile or cause it to be lost from the plant–soil system entirely. Typical soil solution K concentrations are low, and several studies of U.S. soils have reported similar ranges (Sharpley, 1989; Kovar and Barber, 1990, Brouder et al., 2003). Thus,

total solution K content in the plow layer is expected to be <100 kg ha^{-1}; in many soils, the solution K content will be far less than the amount required to grow a given crop for even one season (Table 6|2). However, exchangeable and nonexchangeable K can be interchanged with solution K$^+$ at a time-step relevant to the plant growth and nutrient uptake cycle. The rate of exchangeable K conversion to solution K$^+$ is expected to range from nearly instantaneous to several hours, while nonexchangeable K may be converted to solution K$^+$ over a period of several hours to several weeks (Haby et al., 1990). Because solution K$^+$ concentration can change rapidly with fluxes from other soil forms, plant uptake, fertilization, and/or leaching, values for a given soil can vary widely in space and time. Furthermore, the soil solution is polyionic and can be fairly concentrated, and both the thermodynamic activity and the concentration govern the ionic behavior and distribution. Sparks (1989) labeled this soil K form "enigmatic" not only because values tend to fluctuate, but because this pool is difficult to measure and the measurement difficult to interpret given the interactive effects of temperature with concentrations of multiple ions.

Exchangeable Potassium

Exchangeable K is electrostatically bound as an outer-sphere complex to the negative charge on the surfaces of clay minerals and humic substances, although it is not a structural component of the humus itself. As discussed above, isomorphic substitution of a cation of higher valence with one of lower valence within the crystal lattice leaves a net negative charge. This charge is termed *permanent charge* because it is independent of short- to medium-term changes due to environmental perturbation. Both minerals and organic matter can contribute an additional source of surface charge that is *pH dependent*, with positive and negative charge occurring at low and high pH, respectively. This charge is associated with functional groups such as hydroxyls (–OH) or carboxyls (–COOH). At high pH, the functional groups loose H$^+$ to the solution, imparting a negative charge to the surface, but at low pH, functional groups gain H$^+$ for an excess positive charge. Occasionally, isomorphic substitution can impart a net positive charge to minerals (e.g.,

Table 6|2. Example potassium phase or form concentrations and quantities within the K cycle of typical U.S. cropping systems.

Phase or form	Location	Concentration[†]	Total content to 0.2 m[‡]	Reference notes
		mg K kg^{-1}	kg K ha^{-1}	
Soil: structural	Tri- and dioctahedral mica, K-feldspar	Kaolinitic: 30–8,100[a] Smectitic: 8,160–21,050[a]	Kaolinitic: 78–21,060[a] Smectitic: 21,216–54,730[a]	[a] Grouping based on soil taxonomy and mineralogy of clay (<2 μm) fraction. Other soil studies do not group soils.
Soil: nonexchangeable	Vermiculite, hydrous mica, tri- and dioctohedral mica, others	Kaolinitic: 2–116[a] Smectitic: 38–1,548[a] Ungrouped: 32–1,082[b]	Kaolinitic: 5–302[a] Smectitic: 99–4,025[a] Ungrouped: 83–2,813[b]	[a,b] Estimated by difference between HNO$_3$-Ext ([a]) or NaTPB-Ext ([b]) and NH$_4$OAc-Ext values.
Soil: exchangeable	Colloidal exchange sites—clay and organic matter	Kaolinitic: 5–211[a] Smectitic: 19–874[a] Ungrouped: 58–516[b]; 21–651[c]	Kaolinitic: 13–549[a] Smectitic: 49–2,272[a] Ungrouped: 151–1,342[b]; 55–1,693[c]	[a] NH$_4$OAc-Ext K measured following extraction of water soluble K. [b,c] Estimated from difference between NH$_4$OAc-Ext and soil solution K.
Soil: water soluble	Soil solution	Kaolinitic: 1–12[a] Smectitic: 1–5[a] Ungrouped: 1–31[b]; 1–37[c]	Kaolinitic: 3–31[a] Smectitic: 3–39[a] Ungrouped: 3–81[b]; 3–96[c]	[a] Measured in 1:10 soil/water extract. [b,c] Measured in displaced solution at field capacity; converted to unit mass soil assuming θ = 0.25 ([b]) or using measures θ ([c]).
Crop: total uptake at physiological maturity	Aerial tissue	Corn: 10,400–14,400[d] Soybean: Alfalfa: 13,600–26,200[f] Miscanthus: 15,000–2,000[g]; 9,000–20,000[h]	Corn: 250–370[d] Soybean: 129–199[e] Alfalfa: 25–149[f] Miscanthus: 466[g]; 371–425[h]	[d] Range across years and populations. [f] Range from high-yielding plots across years and harvests (4 yr^{-1}) [g,h] Includes 59 ([g]) and 150 ([h]) kg K ha^{-1} in rhizomes; concentration varies with tissue and year.
Crop: residue	Aerial portion returned/incorporated	Corn: 4,800–19,500[i] Soybean: approx. 5,000[e] Alfalfa: Negligible Miscanthus: 11,300[g]	Corn: <40–>120[i] Soybean: 79–154[e] Alfalfa: Negligible Miscanthus: 113[g]; 52–166[h]	[i] Stover cut a ground level. [g] Feb. harvest. [h] Returns vary with year and time of harvest: Nov. vs. Feb.
Crop: removal	Grain, forage, or biomass	Corn: 3,900 ± 700[j] Soybean: 18,300 ± 1420[j]; 14,660–14,720[e] Alfalfa: 13,600–26,200[f] Miscanthus: 12,000[g]	Corn: 39–>60[j] Soybean: 60–>80[j]; 45–50[e] Alfalfa: 204–395[f] Miscanthus: 233[g]; 63–162[h]	[f] Removal values are sums of 4 harvests yr^{-1}; range of values from high yielding plots across years. [g] Feb. harvest. [h] Returns vary with year and time of harvest: Feb. vs. Nov.
Other: removal	Leachate	Silt loam: <3[k] Sand: <2–<3[l]	Silt loam: ~1[k] Sand: 13–19[l]	[k] Fallowed soil [l] Fallowed soil with and without 66 kg K ha^{-1}.

† Lowercase, superscripted letters denote the following reference sources: (a) Sharply, 1989; (b) Brouder et al., 2003; (c) Kovar and Barber, 1990; (d) Karlen et al., 1987 (e) Fernandez et al., 2008, 2009; (f) Lissbrant, 2008; (g) Beale and Long, 1997; (h) Himken et al., 1997; (i) Brouder, unpublished data, 2010; (j) Brouder and Volenec, 2008; (k) Stauffer, 1942; (l) Allison et al., 1959.
‡ Calculation assumes a bulk density of 1.3 g cm^{-3}.

Al^{3+} substituted for Mg^{2+} in brucite) and/or soils are sufficiently acid to have net positive pH-dependant charge, but for the most part, soil colloids have net negative charge, and, therefore, exchangeable ions are all cationic.

Exchangeable K is considered readily plant available as long as soil moisture itself is not restrictive because, as plant roots deplete solution K$^+$, exchangeable K is easily replaced on the exchange surface by other soil solution cations. At a given point in time, the quantity of exchangeable K is a function of the total number of exchange sites per unit mass of soil and the quantities of all exchangeable cations present. In addition to K, the major exchangeable cations in most agricultural soils are Na^+, Ca^{2+}, and Mg^{2+}. The most important factor governing selective retention of cations on surface charge is valence (individual ion charge) and thus the typical relative order of ion quantities is K $\cong$ Na < Mg < Ca (discussed further below). However, fertilization with ammoniated N sources can locally enrich soils with substantial quantities of NH_4^+; acid soils are relatively enriched in exchangeable Al^{3+} and hydrated H^+ or H_3O^+.

The exchangeable K pool is substantially larger than the solution pool. For a K deficient mineral soil from the eastern cornbelt, the proportion of K in solution may be only slightly over 5% of the quantity of exchangeable K (Fig. 6|2). When this same soil is maintained at an overall higher fertility level, the proportion of K in the soil solution is observed to increase slightly but remains below 10% of the quantity of K in the exchangeable form. Exchangeable K concentrations in typical Midwestern agricultural soils have been found to range from approximately 20 to > 650 mg K kg^{-1} soil, which translates into total exchangeable K contents of between 50 and 1700 kg K ha^{-1} in the plow layer (Table 6|2). Sharpley (1989) observed that overall exchangeable K contents in kaolinitic soils were lower than those of smetitic soils but that the proportion of K as solution versus exchangeable K was higher for the kaolinitic versus smectitic soils. In general, kaolinite has a relatively low charge density or CEC and surface area when compared with a smectite, such as montmorillinite, the most prominent member of the smectite group. The CEC and surface area of montmorillonite can easily be 100-fold that of kaolinite (Singer and Munns, 2002), and, thus, montmorillonite has a much greater capacity to retain K^+ on its colloidal surface.

Nonexchangeable Potassium

Nonexchangeable or fixed K is distinct from exchangeable K in that the binding sites on the colloids have specific preference for K that is beyond the simple electrostatic preferences related to ionic charge. However, it differs from structural K in that it is not covalently bonded within the crystalline structure of minerals. As described by Sparks (2000 and references cited therein), nonexchangeable K is held between adjacent layers of di- and trioctahedral micas, vermiculites, and the various transition minerals. When the binding forces between the clay surface and a solution K^+ ion are greater than the forces of hydration between individual solution K^+ ions, partial collapse of the crystalline structure can occur and physically trap K^+. Olson et al. (2000) remarked that these ions are not considered permanently fixed, but their release back to the soil solution will be relatively slow, especially if trapped deep within the interlayer of a vermiculite particle. Similarly, K in so called "wedge zones" is considered nonexchangeable because larger, hydrated ions such as Mg^{2+} and Ca^{2+} that are often in high concentrations in soil solutions cannot physically fit into these regions of the mineral. Wedge zones occur because expansion of interlayers during weathering is frequently not spatially continuous throughout an

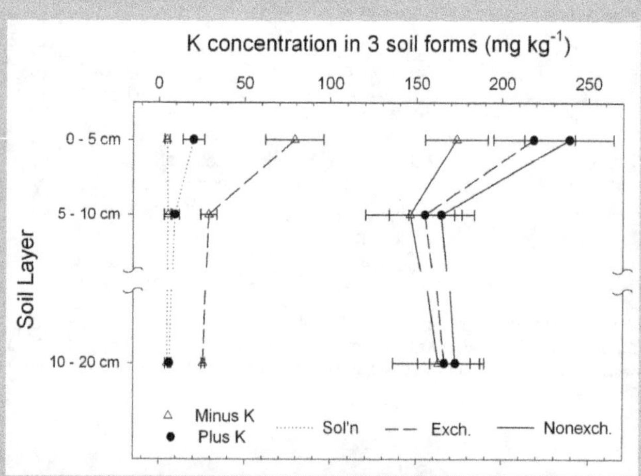

Fig. 6|2. Mean concentrations of solution, exchangeable and plant-available, nonexchangeable K in 0- to 5-, 5- to 10-, and 10- to 20-m soil layers of a no-tillage silt loam soil with a history of no or optimal K fertilization. Soybean yield on the unfertilized soil was K limited while yield on the fertilized soil was not K restricted (modified from Fernandez et al., 2008).

interlayer area of a mica particle (Huang, 2005). When interlayer expansion is not continuous, the internal termination point for the expansion creates a wedge-shaped configuration to the exchange site and only ions sized similarly to K such as NH_4^+ and H_3O^+ can physically fit into the wedge zones and exchange K from these sites.

Total nonexchangeable K contents in the tillable layer of soils can range from <100 to >4000 kg ha^{-1}. In many soils, nonexchangeable K content is substantially larger than the exchangeable K content (Table 6|2). While nonexchangeable K is generally termed *moderately to sparingly* available to plants (Mengel, 1985; Sparks and Huang, 1985; Sparks, 2000), the literature is replete with reports of crops acquiring significant proportions or even the majority of their K requirement from this pool without any apparent growth limitation associated with poor K fertility. For example, Meyer and Jungk (1993) observed that the change in exchangeable K in soil could only account for 32 and 43% of the K accumulated by 15-d-old wheat (*Triticum aestivum* L.) and sugar beet (*Beta vulgaris* L.), respectively, with the remainder attributed to nonexchangeable K. Likewise, Cox et al. (1999) used exhaustive cropping with wheat to demonstrate that the percentage of plant-available K in the initially nonexchangeable form ranged from negligible to 76% across 11 surface soils in varied textural classes from throughout the midwestern United States. Among all soils of agronomic importance, quantities of plant-available but nonexchangeable K have not been found to be easily predicted from either solution or exchangeable forms (e.g., Brouder et al., 2003), but linkages with mineralogy and charge density have been repeatedly demonstrated. For example, among soils with similar exchangeable K and clay contents, the quantity of plant-available but nonexchangeable K is greater in smecitic than in kaolinitic soils (Sharpley, 1989). This is consistent with expanding phylosilicates and 2:1-type clay minerals holding K more tightly than it is held by the external surface charge of 1:1-type clays.

It should be noted that K form distinctions are often intermixed with discussions of K extractability by the specific method or methods that have been experimentally linked to different chemical states and/or direct assays of plant uptake. Typically,

exchangeable K and other exchangeable cations are defined by their release from soil by extraction with salts from neutral solutions. In contrast, solution K is the quantity that can be removed by water alone. In practice, exchangeable K is the difference between neutral salt-extractable and water-extractible quantities of K. Theoretically, nonexchangeable K will not be released to the soil solution without reduction in the initial soil solution K$^+$ concentration. As reviewed by Helmke and Sparks (1996), most soils must have solution K reduced below 4 mg L^{-1} for nonexchangeable K release to occur with some mineral fractions requiring solution K levels <1 mg L^{-1} to trigger nonexchangeable K release. Hot acid extractions, continuous leaching, and other physical or chemical methods of reducing solution K concentrations are common approaches to indexing plant-available but nonexchangeable K. Not surprisingly, however, variations in protocols and extraction mechanisms or chemistries can result in large differences in the characterization of K distribution among forms for any given soil. The practical significance of such ambiguity is that any interpretation of results requires calibration across a sufficiently diverse array of soils for a method to be of any commercial use.

Disequilibria, Fluxes in Plant–Soil Systems, and Soil Potassium-Supplying Power

While the relative quantities of K forms in soil are often discussed in terms of equilibrium conditions, disequilibria should be considered the normal state. A primary driver of disequilibria and K flux among soil forms is plant uptake and subsequent return of plant-accumulated K to soil as leachate from senescing tissues on or above the soil surface or from incorporated residues and senescing roots and crowns. In intensively managed agricultural systems, large pulses of K are introduced as inorganic or organic fertilizers. Given that the quantity of K required by many agricultural crops is typically well in excess of total root-zone K in the soil solution, the K supplying power of a soil directly reflects the extent and rate of the transformation of exchangeable and/or nonexchangeable K into solution K$^+$. However, even without active plant removal, rapid and significant interchange among soil K forms can occur in response to several common physiochemical factors and processes. The following concepts and attributes are critical to understanding soil K supplying power in agricultural systems.

Fluxes to and from the Soil Matrix

Plants remove soil solution K^+ from throughout the rooting zone with the local intensity of removal reflecting the overall plant growth and associated K demand combined with the localized rooting intensity and soil K and moisture availability. In natural systems, K cycles between the standing biomass and the soils with very little off-site export, but in agricultural systems, some portion of the plant K is exported off the field in harvestable yields, while fallowed fields are particularly susceptible to K losses as leachate. Both the quantity and proportion of plant K exported offsite versus that locally returned to soil varies markedly among major crop species. Removal of K in grain crops where the straw is returned to the field tends to be relatively low. For example, corn (*Zea mays* L.) grain typically removes only 4 g K kg^{-1} dry matter in grain, with more than 80% of the K in aboveground tissues returned with the stover residue to the soil after harvest (Table 6|2). In contrast, the K removed by a high-yield alfalfa (*Medicago sativa* L.) stand can be greater than 26 g K kg^{-1} dry matter in shoots for total annual removals nearing 400 kg K ha^{-1}.

Other mechanisms of net soil K loss include leaching and erosion. The amount of K leached through the soil profile is a direct function of the amount of water percolating through the soil, which, in turn, is reflective of texture and the presence of vegetation. In most mature, natural systems with sufficient standing vegetation, leaching and erosion losses can be expected to be very low (<5 kg K ha^{-1} yr^{-1}), and these losses are easily replaced by weathering of primary and secondary minerals. Leaching losses in agricultural systems can be substantially greater. As reviewed by Bertsch and Thomas (1985), early leaching studies found 1 to 2 kg K ha^{-1} leached from fallowed silt loams, but, in general, losses from heavier-textured soils were negligible in the presence of crops. Documented leaching losses from fallowed sands ranged from 12 to >50 kg K ha^{-1} depending on fertilizer K additions but cropping could reduce these losses to less than 5 kg ha^{-1}.

Percolate quantity appears to integrate K leaching load losses across soil textures. Soil pH can be an additional modifying factor in K leaching losses. Several early studies clearly documented better soil K retention following lime additions to increase soil pH (Nolan and Pritchett, 1960; Shaw and Robinson, 1960). The mechanisms for better K retention with liming include the creation of additional pH-dependent negative charge to retain greater quantities of exchangeable K and potentially enhanced site selectivity for K relative to the prevalent, competing ions that dominate the soil solution at different pHs (Ca^{2+} at high pH versus Al^{3+} and or H_3O^+ at low pH; Richards et al., 1988). Finally, soil K can also be lost with erosion and runoff. This mechanism of K loss remains poorly characterized but, where erosion mitigation measures are needed but not employed, K losses can be expected to be large and scale with the magnitude of total soil losses.

Mass Flow and Diffusion

Even the densest root systems only occupy and directly contact a small portion of the total root zone volume of soil (e.g., <1%; Barber, 1995). In order for significant quantities of K to be interchanged among soil forms and acquired by roots for cycling through plants, K must physically move within the soil matrix. Plant availability of a nutrient is as much about the location of an ion relative to the root surface and the length of the pathway the nutrient must travel in soil to reach the root surface as it is about the soil chemical properties (reviewed by Jungk, 2002). Potassium movement occurs by mass flow and diffusion. Mass flow is the convective or bulk movement of solutes with the net movement of soil water. Gravity and excessive precipitation drive mass flow leaching losses of K in drainage water, while plant transpiration moves solution K^+ from the bulk soil to the root surface for plant uptake. For these two cases, the rate of K movement is approximated by the rate of infiltration and the mean rate of transpiration, respectively. The product of the season-long average for plant water use and soil solution K^+ concentration can be used to approximate the quantity of K that a plant can acquire via mass flow alone. Sample calculations show mass flow may only supply about 35% of the total K requirement to an annual crop (e.g., corn) grown in a fertile midwestern soil and using 2.5 to 3.0 million L H_2O ha^{-1} (Barber, 1995). Occasional studies have shown mass flow can deliver root surface K quantities equal to or in excess of plant demand, especially when transpiration is high and plant growth is poor (Hylander et al., 1999). More typically, however, mass flow contributions are far lower than plant demand and can be as low as 1 to 2% of total plant K accumulation (e.g., Strebel and Duynisveld, 1989). Certainly these calculations are overly simplistic, but the general magnitude of the results indicates that other mechanisms must be equally if not more important than mass flow for K movement in the soil.

The process of *diffusion*—movement in response to concentration gradients—occurs simultaneously with mass flow and is the major mechanism for K movement in the soil matrix toward absorbing roots. Fick's laws describe net ion movement in uniform media like water or air as a function of time, distance, concentration gradients, and the nature of the ion itself (Kemper, 1989). In soil, rates of ion diffusivity are modified by the soil water content, with other factors altering the path length an ion must travel between two locations and the soil-specific nature of the relationship between solid (exchangeable and nonexchangeable) and solution K$^+$ forms. Potassium does not volatilize and therefore can only move through a continuous film of water. In soils, the path between two points is lengthened by air in soil pore space and thus diffusion is slowed in dry soils. Ion diffusivity is also directly impeded by the average size of soil particles (texture) and by the extent to which particles are packed together (bulk density). In general, diffusion paths in soils are shortest at field capacity and a moderate bulk density (1.3 g cm^{-3}; Warncke and Barber, 1972). The rate of K diffusion in a plane of pure water (D) is 1.98×10^{-5} cm^2 s^{-1}. The average linear distance (D_{linear}) an ion can move in a given amount of time (t) is estimated as

$$D_{linear} = (2Dt)^{1/2} \qquad [1]$$

or 1.85 cm d^{-1} for K in water. Potassium movement in soil is on the order of 10^{-8} to 10^{-7} cm^2 s^{-1} or about 0.04 to 0.13 cm d^{-1} for linear travel (Barber, 1995).

Given K$^+$ diffusion is driven by concentration gradients, rates of K movement in soil are dramatically altered by K fertilizer additions and root uptake. Ching and Barber (1979) demonstrated that addition of 100 mg K kg^{-1}—equivalent to a fertilizer application of 260 kg K ha^{-1} incorporated to a depth of 0.2 m—could increase diffusivity 4.5-fold (from 1.5–6.9 cm^2 s^{-1}) in a Raub silt loam soil maintained at a low soil temperature. This increase corresponds to a more than doubling of the linear distance traveled (0.5–1.1 mm d^{-1}; Fig. 6|3). For the same soil maintained at a higher temperature, diffusivity increased 7- or 2.6-fold in linear distance traveled. Concomitantly, when plants are growing, root uptake of K strongly depletes concentrations immediately adjacent to the root surface, creating an even steeper concentration gradient to drive diffusion. The extent of depletion differs among crops and

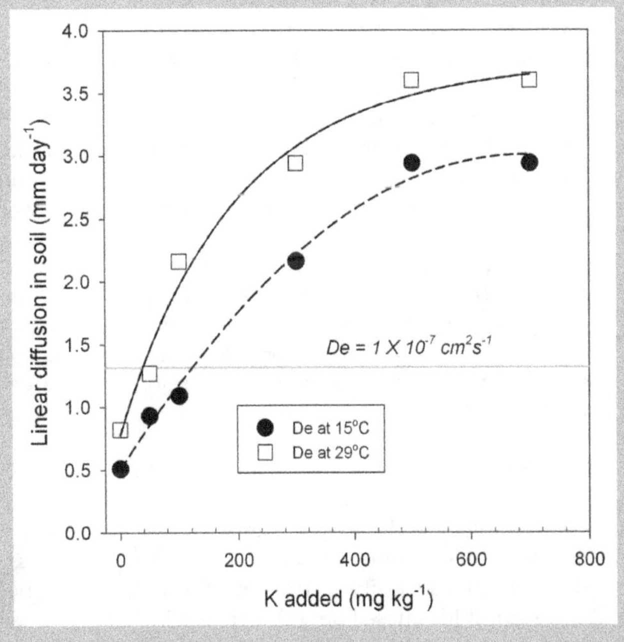

Fig. 6|3. Estimated linear diffusion rates in soil at 15 and 29°C calculated from K effective diffusion coefficients (D_e) as influenced by K fertilizer addition (estimated using Eq. [1] from D_e values in Ching and Barber, 1979). The reference line shows a typical D_e value for K in soils (Barber, 1995).

soils and is impacted by a host of transient environmental conditions. As reviewed by Barber (1995), studies of common crop species have found root-depleted soil solution K^+ concentration of <1 to about 5 μmol L^{-1}, which is 20-fold or more lower than typical solution K^+ levels in the absence of plant root activity. However, even at high soil fertility, maximum K^+ diffusivity is only on the order of a few millimeters per day (Barber, 1995). The generally short distance K can move by diffusion in combination with this mechanism's apparent importance relative to mass flow highlight why roots must proliferate throughout soil to obtain K.

Solution-Exchangeable Potassium Dynamics and Buffering

When the diffusion of nonadsorbed ions (e.g., NO_3^-) is considered, the volumetric water content and the length or tortuosity through the soil are the main factors reducing the rate of diffusivity in soil as compared with that observed in water. However, for cations, diffusivity is further reduced from pure water rates as cations interact with the negatively charged soil surface along the diffusion pathway. The release of exchangeable K into solution is *desorption*, and the association of solution K with colloidal exchange sites is termed *adsorption*. The concept of *buffer power* captures the dynamic nature of both processes and their importance in soil fertility as it characterizes the relation of the soil solution concentration to the concentration of ions in solid forms. Buffer power reflects the change in exchangeable or solid-phase K concentration that occurs in response to change in solution phase K^+ concentration. Nye and Tinker (1977) demonstrated that the effective rate of diffusion (D_e) of K^+ in soil could be estimated as

$$D_e = D_1 \theta f_1 (1/b) \qquad [2]$$

where D_1, θ, f_1, and b are the diffusion rate in water, the volumetric water content of the soil, the tortuosity of the path the ion is traveling (discussed above), and the buffer power, respectively. The process of ion exchange between solution and colloidal surfaces has long been identified as rapid, stochiometric (electronically neutral) and reversible. The stochiometric aspect of exchange dictates that one solution K^+ will replace one NH_4–X, Na-X or K-X, while two solution K^+ are required to replace one Ca-X or Mg-X ion (-X identifies the ion as held on an exchange surface). Soils, however, are not ideal exchange surfaces, and buffer power, most often calculated as the change in the amount of exchangeable K that occurs in conjunction with a unit change in the amount of solution K+, differs remarkably among soils, as well as within soils at different fertility levels. In sum, a soil's ability to maintain a given solution K^+ concentration against change is a key determinant of its K fertility.

In soils, the rate-limiting step for an exchange is typically not the exchange itself but the time required for an ion to diffuse through the soil water to the exchange site or the reverse. Numerous studies of exchange dynamics using pure clays have demonstrated that not only does the quantity of K held on the exchange complex vary with the type of clay mineral present, but the reaction rates and the binding energies of the exchange sites themselves are also mineral specific. Exchange rate differences reflect locations and geometries of exchange sites. Kaolinite and montmorillonite exchange K rapidly with soil solution, while exchange reactions in vermiculitic and micaceous minerals occur much more slowly (for a summary, see Sparks, 1987). In kaolinite, interlayer exchange sites are minimal, as strong H bonding holds adjacent clay sheets tightly together, and K exchange proceeds rapidly because most sites are on the easily accessible planar surfaces, with diffusion distances being short and not geometrically constrained. In contrast, montmorillonite layers are not tightly bound by H bonds and can swell with hydration. Thus, exchange is also relatively rapid because interlayer sites are made easily accessible to solution K^+ as long as soils are adequately hydrated. The relatively slower rates of K exchange that occur on vermiculitic and micaceous minerals are associated with restricted access to interlayer sites. In micas, interlayer space has been observed to be selective for certain types of cations (e.g., K^+, Cs^+), and the rate of K^+ exchange is determined by the relatively slow film diffusion process.

In addition to diffusivity, a soil's buffering capacity for K is impacted by the ease of K displacement on the exchange surface. In general, the energy in the bond holding K on exchange sites of 2:1 clays is greater than the bond energy of K on 1:1 clays, where K is more easily exchanged (Marshall, 1964). Potassium binding is weakest on pH-dependent exchange sites. Among all soil cations, the relative strength of a cation bond is a function of cation valence, hydration, and size. Soil cation exchanges are typically dominated by monovalent–divalent exchange among Ca^{2+}, Mg^{2+}, and K^+. Among biologically important monovalent cations in soils, Na is displaced more easily than K, reflecting the relatively larger size of hydrated Na. Because of their divalent charge, Mg and Ca are held more tightly on exchange sites than K, with exchangeable Mg more easily displaced than exchangeable Ca. Furthermore, at equilibrium, the relative distribution between exchangeable and solution forms of Mg, Ca, K, and other cations changes with dilution or concentration of the soil solution and/or changes in the ratios of monovalent to divalent cations in solution. The Schofield (1947) ratio law states that to maintain equilibrium between solid and solution forms, solution concentrations of divalent and trivalent cations will change by the square and cube, respectively, of the change ratio of a monovalent cation concentration. Thus, dilution or concentration of the soil solution affects concentrations of K^+ relatively less than Ca^{2+} and Mg^{2+}, and, in theory, high CEC soils will require relatively more exchangeable K than a lower CEC soil to maintain a given K/Ca and K/Mg ratio in the soil solution (McLean, 1976). From the soil management perspective, however, many studies have failed to substantiate this hypothesis of higher CEC soils needing greater levels of exchangeable K than lower CEC soils to provide similar quantities of plant-available K (e.g., Shaw et al., 1983; Favaretto et al., 2008) if Ca and Mg contents are not extreme.

In terms of plant-available K, Eq. [2] indicates that soils with lower buffering (small b values) will maintain greater rates of diffusivity in soil than soils with high b values. For a given soil, the relationship between solid and solution phase concentration is curvilinear. As K fertilizer is added to increase total soil K content, the amount of K remaining in the soil solution increases faster than the solid phase K as exchange sites become K saturated. Thus, soil-specific b values are lower when total soil content is greater, and K^+ diffusion proceeds at a faster rate. Faster movement in soil means solution phase K^+ concentration at the root surface will be more readily replenished by K from the bulk soil as the root accumulates K. Given that buffer capacity reflects interaction with the soil surface along the diffusion pathway, high CEC soils tend to have high b values and, therefore, lower K diffusivity when compared with low CEC soils. However, low diffusivity in a high CEC soil does not necessarily infer low soil fertility because the overall quantity of plant-available K in a high CEC soil is likely much higher than in a low CEC soil. Consequently, the soil volume required to supply a plant with a given quantity of K is smaller, and K diffusion distances shorter in a high CEC versus low CEC soil. This suggests that buffer power itself should be an important determinant of the minimum or critical soil solution K^+ concentrations required for unlimited plant growth. Barrow (1966) found 89% of variation in plant uptake could be accounted for if both solution K^+ concentration and soil buffering capacity were considered. Mengel and Busch (1982) worked with several soils varying widely in clay content and CEC to demonstrate a very close ($r^2 = 0.91$), hyperbolic relationship for solution K^+ critical concentrations for ryegrass (*Lolium multiflorum* Lam.) as a function of buffering with lower critical concentrations associated with higher b values. In sum, even though plants accumulate solution K, knowledge of the initial quantity of K in that form—occasionally termed *K intensity*—is not enough to determine soil K fertility. The dynamic, soil-specific buffering capacity or quantity–intensity (*Q/I*) relationships are widely considered the defining characteristic of soil K fertility.

Fixation–Release and Plant-Available, Nonexchangeable Potassium

The terms *fixation* and *release* describe solution K^+ movement into and out of nonexchangeable positions. The most simple and frequent characterization of soil b is as the slope of the extent of increase or decrease in exchangeable K plotted as a function of the associated change in solution K. This approach to b derivation emphasizes K quantities in exchangeable form, but in concept, buffering encompasses quantities of all solid forms and their rates of exchange, or fixation–release, with the soil solution. Using mechanistic simulation modeling, Brouder and Cassman (1994a) demonstrated that b values must encompass both nonexchangeable and exchangeable K dynamics in soils with potential to fix significant quantities of fertilizer K on nonexchangable sites. As discussed above, the rate at which added K moves out of solution to solid forms or returns to solution following solution depletion by root activity or leaching has long been considered indicative of the nature and geometry of the solid-phase binding site. Bolt et al. (1963) studied K exchange from illite and concluded rapid (1 h) and complete replacement by other cations indicated the K had been held on planar surfaces of the clay mineral, while rapid replacement but only by NH_4^+ indicated K occupied interlattice positions but near the edges of clay particles. Slower exchange and only by NH_4^+ as the exchanging ion indicated K occupied interlattice positions deep within layers. Early work with soils varying in hydrous mica content linked soil-specific rates of K release to significant differences in ryegrass yields (Grissinger and Jeffries, 1957). Across soils, the extent to which nonexchangeable K contributes to the buffer capacity and plant K uptake will be a function of the quantity and rate of release of nonexchangeable K relative to plant K requirements, including both quantity and rate of demand.

When K is added to soils with high charge density clays (e.g., weathered micas, montmorillonite, and vermiculite), nonhydrated K can fit into the holes on the interlayer surfaces of clay mineral silicate sheets (Barshad, 1951). The higher the clay charge density, the greater the extent of K fixation (Walker, 1957; Rich, 1968). With very high charge density, K fixation in adjacent clay sheets can stabilize the overall structure and can depress subsequent rates of K release (Barber, 1995). When soils are or have been acid and have iron and aluminum oxides precipitated on clay surfaces, K fixation can be lessened as the oxide precipitates prevent collapse of the crystalline structure that traps or fixes K. Early incubation studies with soils and clays (Rich and Obenshain, 1955; Horton, 1959; Rich and Black, 1964; Page and Ganje, 1964) confirmed that K fixation is greatest in soils with few oxides to prop open clay layers and also demonstrated fixation to be a relatively slow process, with the ongoing

K removal from solution and exchangeable pools occurring over many months.

Other noted modifiers of K fixation in soils include organic matter and NH_4^+. Decreased K fixation has been associated with increased soil organic matter in vermiculitic soils with high fixation capacity and low soil K fertility (Cassman et al., 1992; Olk and Cassman, 1993); Olk and Cassman (1995) demonstrated that young, N-rich fractions such as mobile humic acids can reduce fixation of added K and increase availability of K in K-deficient soils. Likewise, NH_4^+ addition may also increase plant-available K by limiting fixation. Bolt et al. (1963) observed that the rate of NH_4^+ diffusion to interlayer sites was higher than for K^+ and hypothesized penetration of interlayers by the slightly larger NH_4^+ ions may cause a slight opening of the lattice structure when compared with the effect of K^+. Lumbanraja and Evangelou (1992) examined K Q/I relationships in three soils and found increased labile or available K along with depressed potential buffer capacity in the presence of NH_4^+ as compared with its absence. Likewise, Brouder and Cassman (1994a) observed exchangeable K levels increased three- versus only twofold when a given amount of K was added with NH_4^+ versus without, respectively, to a K-fixing, vermiculitic soil, and cotton (*Gossypium hirsutum* L.) K uptake improved with soil NH_4^+ enrichment. Stehouwer and Johnson (1991) theorized preferential NH_4^+ fixation could decrease K fixation and increase exchangeable K following anhydrous injections. To date, however, it remains unclear what practical significance and lasting impact this effect has in K fixing soils receiving anhydrous applications.

Release of K from micas involves K exchange with hydrated cations to expand 2:1 layer silicates and/or dissolution followed by formation of weathering products; K release from feldspars involves hydrogen exchange and dissolution (Sparks and Huang, 1985; Sparks, 2000). As with fixation, release of nonexchangeable K is generally considered a relatively slow, diffusion-controlled process that is driven by reductions in exchangeable and solution K by plant removal, microbial activity, and/or excessive leaching. Given that K fixation can collapse lattice structures, rates of release can be markedly slower than fixation rates, and therefore fixation is generally viewed as reducing the effectiveness of fertilizer K. Cassman et al. (1989) demonstrated cotton K deficiency on vermiculitic soils could only be alleviated when fertilizer rates were high enough to at least partially saturate the fixation capacity of the soil, resulting in 50% greater fertilizer use efficiency. In contrast, other studies suggest that interlayer K can be released without a concomitant depletion of solution K^+ (e.g., Moritsuka et al., 2004) and that K release from interlayers of expandable 2:1 clay minerals (e.g., illite and vermiculite) occurs relatively easily, thereby providing significant percentages of K requirement to growing crops (Barber and Mathews, 1962; Richards et al., 1988; Mengel and Uhlenbecker, 1993; Rahmatullah et al., 1994; Cox et al., 1999; Srinivasarao et al., 2006). Potassium release with cropping has long been linked to particle size fractions (Doll et al., 1965). Cox and Joern (1997) studied rates of K release from a range of midwestern agricultural soils varying in texture and CEC and linked rapidly and slowly released nonexchangeable K to the fine clay and fine silt fractions, respectively, while the coarse clay contributed both rapid and slow release K. For wheat growing in soils of varying illite content, Cox et al. (1999) identified soil-specific critical concentrations for exchangeable K based on CEC and illite K concentration but also suggested such soil-specific critical levels could be more broadly derived from groupings of soils based on textural classes alone.

Acquiring significant quantities of nonexchangeable K is typically considered species and even cultivar specific. For example, studies comparing perennial ryegrass to various clovers (*Trifolium* spp.) repeatedly demonstrate higher nonexchangeable K uptake by the more densely rooted ryegrass (Graley, 1981; Mengel and Steffens, 1985). Mengel (1985) attributed the superior ability of monocots versus dicots to access nonexchangeable K to differences in root morphology. The collective works of Barber and colleagues with simulation modeling have repeatedly demonstrated the importance of root length, diameter, and surface area in K uptake (Barber, 1995 and references cited therein). Brouder and Cassman (1990, 1994b) demonstrated cotton cultivar differences in K uptake from fixing soils reflected root surface area differences

including mean root diameters and root system determinancy after peak bloom, root branching, and root/shoot dry matter allocation. Fergus and Martin (1974) linked crop-specific differences in exploitation of nonexchangeable K to root ability to lower soil solution K^+ concentration and Mengel and Steffens (1985) identified species differences in root hair length as a driver of nonexchangeable K uptake. Ultimately, biological activity of roots and microbes is a known but poorly understood driver of nonexchangeable K release. Certainly, root and microbial excretion of organic acids and root–microbe associations, including mycorrhizal infections, have documented potential to be important (Sparks, 2000 and references therein; Rygiewicz and Bledsoe, 1984) but remain little studied in the context of plant-available K.

Environmental Modifiers Moisture and Temperature

Variation in moisture and temperature can modify both exchangeable and nonexchangeable K dynamics. Net K^+ diffusion is driven by concentration gradients, but the molecular movement or Brownian motion is thermal. Therefore, in addition to the physical weathering effects of freezing and thawing that can release K (Walworth, 1992), all K dynamics including exchange and fixation–release reactions and rates of diffusion to the root surface change with above-freezing temperature fluctuations. Mortland (1958) found a linear relationship between rate of K release from biotite and increasing temperature. Subsequent studies with soil demonstrated rates of both K^+ adsorption and desorption increased with temperature (Sparks and Jardine, 1981), and, with higher temperatures, both buffer capacity and K^+ sorption relative to Ca^{2+} and Mg^{2+} sorption also decreased, resulting in higher available K (Sparks and Liebhardt, 1982). For low fertility soils, Ching and Barber (1979) found linear diffusion can double with a doubling of soil temperature, and, although the relative difference decreased with increasing K fertility, diffusivity remained substantially higher in warmer soils at all fertility levels (Fig. 6|3). Conversely, a given rate of linear diffusion can require twofold the K fertilizer in a cold soil as compared with a warm soil. Ching and Barber (1979) also showed increasing temperature from 15 to 29°C almost doubled solution K$^+$ concentrations at low fertility, and b values were reduced by 20 to 60% when soil was either unfertilized or received low K fertilizer rates ($\leq$100 mg K kg^{-1}). The interactive impacts of fertility level and temperature on nonexchangeable and exchangeable K dynamics are the theoretical underpinning for the practice of concentrating K in a band, especially near young, small root systems, to promote better nutrient uptake for early crop growth when cold soil conditions are expected to prevail (e.g., Bundy and Andraski, 1999), although practical benefits are not always realized (Kaiser et al., 2005; Rehm and Lamb, 2004).

Impacts of soil moisture on K dynamics can be even more dramatic than the effects of temperature. Not only is the effective rate of K diffusion in soil a direct function of soil water content (Eq. [2]) but wet–dry cycling concentrates/dilutes the soil solution, changing the soil solution's ionic strength and K$^+$ concentration and driving exchange and fixation–release reactions. For example, a soil that has a gravimetric moisture content of 0.28 mL H$_2$O g^{-1} soil at field capacity and a solution phase K concentration of 30 mg L^{-1} would have a K$^+$ concentration of 42 mg L^{-1} when dried to a soil solution volume of 0.20 mL g^{-1} soil provided no K adsorption on the solid phase occurred. The outcome of such concentrating of the soil solution will be a shift in the equilibrium condition among soil K forms, although the nature of the shift can be difficult to predict. Hanway and colleagues (Hanway et al., 1961, 1962) collected soils in 15-cm (6-in) depth increments to approximately 1 m (36 in) from throughout the midwestern United States and measured exchangeable K in samples at field moisture levels and following air drying. When moist soil were above ~110 mg K kg^{-1}, air drying had no effect on exchangeable K measurements, but when moist analyses were initially low, air drying increased exchangeable K (Fig. 6|4a). Furthermore, this effect was much more pronounced in soils collected from deep in the profile than in surface soils. For moist, subsurface soils testing below 50 mg kg^{-1}, drying increased exchangeable K levels as much as eightfold (Fig. 6|4b). Grava et al. (1961) also found greater K release with drying from unfertilized versus fertilized soils.

Many have theorized that drying can warp clay plates and/or peel back

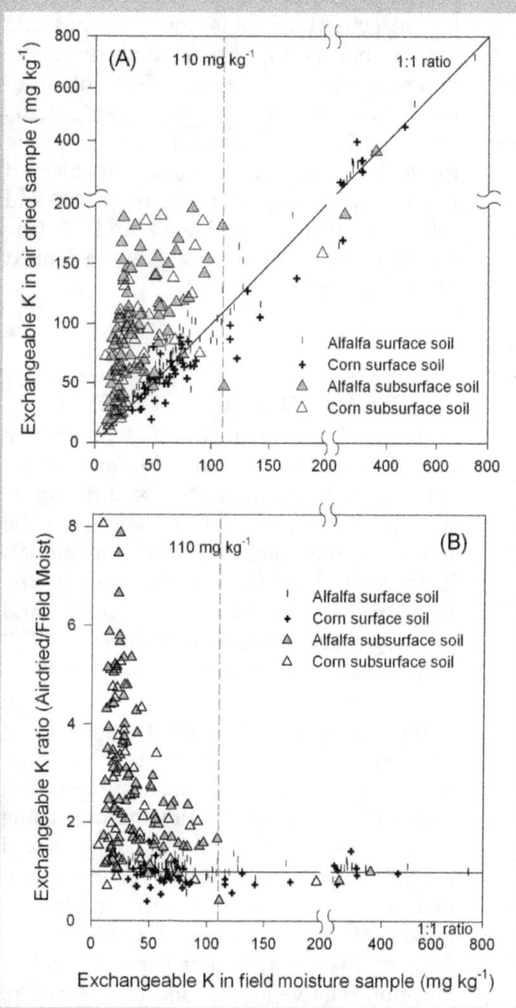

Fig. 6|4. Relationship between soil samples analyzed for exchangeable K at their field moisture content at time of collection and following air drying to constant weight. Data are from surface (|, +; 0–15 cm) and subsoil (▲, Δ; 60–75 cm) layers of a regional alfalfa (|,▲; Hanway et al., 1961) and corn (+, Δ; Hanway et al., 1962) experiments. Data are for (a) air-dry and (b) ratio of air-dry to moist NH$_4$OAc-extractable K concentration plotted as a function of moist NH$_4$OAc-extractable K concentration.

edges of layers, permitting easier access to water and exchanging ions on subsequent rewetting (Barber, 1995; others). Yet, in higher K soils, this effect may not be seen. Rich (1968) suggested that, in contrast to the K release observed in low K soils, fixation may occur with drying on high K soils. This effect can be readily observed following fertilization. For soils with a history of agricultural management, incubations of surface horizons with and without wet–dry cycles show relatively stable exchangeable K levels when no K is added irrespective of moisture regime (Fig. 6|5a). However, when K is added, wet–dry cycling drives a net K movement into fixed sites (Fig. 6|5d) with associated reductions in exchangeable K (Fig. 6|5c); this effect is most dramatic in the first few moisture cycles. In soils where the newly fixed K is tightly held, wet–dry

cycling can reduce fertilizer efficiency and curtail plant K uptake (Olk et al., 1995; Zeng and Brown, 2000).

Field soils experience temperature and moisture cycling in combination, and the combined effects of varying temperature and moisture regimes on plant K status are not simply additive. Temperature and moisture affect physiological factors that directly impact a plant's ability to acquire K from the soil. The Ching and Barber (1979) experiment also demonstrated that increasing soil temperature from 15 to 30°C increased water influx and root surface area by 70% across fertility levels in early vegetative corn. The principle driver of low K uptake on low K, low temperature soils was reduced root growth, but reduced K influx restricted K uptake on high K, low temperature soils. Mackay and Barber (1985) found that

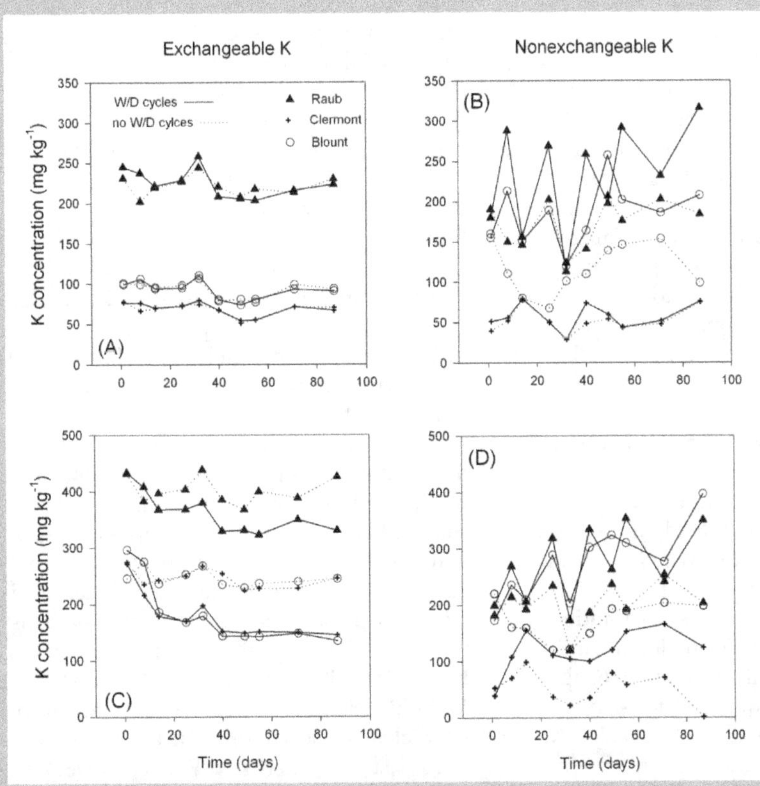

Fig. 6|5. (A,C) Exchangeable and (B,D) nonexchangeable K concentrations in three midwestern U.S. soils incubated for 87 d with and without 7-d wet–dry cycling and (A,B) without K fertilizer or (C,D) with 200 mg K kg^{-1} added before the first moisture cycle. Exchangeable K is the NH$_4$OAc-extractable amount including solution K$^+$; nonexchangeable K is the difference between 5-min tetraphenylboron extractable and NH$_4$OAc-extractable amounts (Brouder, unpublished data, 2010).

reducing volumetric water content by 20% reduced root growth and that root surface area was highly correlated with K uptake within but not across soil types. Barber and Mackay (1985) used simulation modeling to partition uptake into contributing factors and found the increased D_e associated with a 23% increase in soil water content accounted for 11 to 27% of the increase in corn K uptake, while the increased root growth could account for as much as 69% of additional K uptake. Certainly, integrating the temperature-impacted soil and plant processes with simulation modeling illustrates the complexity of the interactions that can occur within the timeframe of a few days and underscores the challenge of predicting the outcome of K dynamics for an entire growing season.

Assessment and Management Considerations

The cornerstone of K management for optimal crop productivity is soil testing. In the strictest sense the term *soil testing* refers to the rapid laboratory analysis of a soil sample for some fraction of its K content. However, as noted by Melsted and Peck (1973), the term has long been considered synonymous with the host of other factors that are encompassed in the development of a K fertilizer recommendation, including sampling depths, number of cores per sample, and the spatial distribution of single corings and composite samplings, sample handling before arrival at the laboratory and even the interpretation of the analytical results. The expectation for a well-calibrated soil test is twofold. First, it must indicate whether or not the available soil nutrient quantity is sufficient for a specific crop or crops in a rotation. Second, when soils are deficient, it must indicate the quantity of fertilizer to apply (Cope and Evans, 1985). For quick K tests, methods are expected to quantify "available" K, extracting within a few minutes an amount directly proportional to the season-long crop K requirement. In reality, this expectation is modified from the ideal by the inherent limits of the tests themselves, the strength of the science underpinning the recommendation, and economic considerations that may drive time and intensity of sampling.

Soil Testing—Principles and Practice

In the United States, the most broadly used soil testing protocols evaluate the quantity of the exchangeable K, as this form is considered highly plant available. In these "indexing" methods, the quantity of K extracted includes the solution phase, most of the exchangeable K, and, typically, small but varying proportions of the fixed or structural K (Haby et al., 1990). A decade or more ago, the one molar, neutral, ammonium (NH_4OAc) extraction method was probably the most commonly used method throughout the United States for indexing available soil K (McLean and Watson, 1985; Cope and Evans, 1985). The details of this method are given in numerous references, including Knudsen (1982). However, the 1M NH_4OAc protocol is not good for P; therefore, complete analyses of typical commercial samples require multiple extractions. The Bray P1 procedure (Bray and Kurtz, 1945) is effective for both P and K but fair to poor for Mg and Ca, respectively (Cope and Evans, 1985). Mehlich solutions can also serve as multi-element extractants. Recent studies report high correlation between NH_4OAc-extractable K and Mehlich 3-extractable K (e.g., Gartley et al., 2002). The economic benefits of a multi-element extractant coupled with equipment efficiencies in commercial labs are now shifting preference to the Mehlich 3 protocol, although only minimal new field correlation and calibration work has been done to directly validate interpretations.

It should be noted that Mehlich 3 is also an indexing method and, as such, does not overcome the known limitations of the original NH_4OAc test. From the earliest days of soil testing, static extraction methods that assess only exchangeable K have been criticized for giving anomalous results on soils where nonexchangeable K dynamics impact buffer capacity (e.g., Hoagland and Martin, 1935). Problems with the NH_4OAc test have been clearly shown for illitic (McLean, 1976; Portela, 1993; Eckert and Watson, 1996) and vermiculitic (Cassman et al., 1990) soils, but also among soils for which there is no significant correlation between K release and mineralogical composition (Binnie and Barber, 1964). Numerous extractants for indexing nonexchangeable K have also been investigated for their stand-alone ability to predict available K, but these tests are subject to the same limitations as the exchangeable K indexes—they do not capture the dynamic relationships among K forms that determine soil K-supplying power. Simple use of CEC and/or soil texture to aid interpretation of quick, indexing methods has not proved useful in practice. Many studies fail to demonstrate simple relationships between plant-available K and CEC (e.g., Shaw et al.,1983). Kovar and Barber (1990) examined 33 surface soils and found no relationship between CEC and changes K availability with fertilization. Similarly, while researchers have identified relationships between soil texture and K content, differences in parent material and weathering preclude a definitive relationship that is broadly applicable across all soils (Pratt, 1952; Scott and Smith, 1987).

In research, predictions of plant-available K from exchangeable K measures can be improved by integrating or correcting values with assessments of buffer power and/or K fixation–release potential (McLean et al., 1979, 1982; Nair et al., 1997; Cox et al., 1999). For routine use, one possibility is to develop correction factors or other necessary ancillary information for representative soils. Several protocols that mimic root function via precipitation or adsorption reactions to remove solution K$^+$ have shown promise for characterizing K release rates. Rapid tetraphenylboron extractions (5 min; Cox and Joern, 1997) to index easily released non-exchangeable K have proven to be even more sensitive than NH$_4$OAc extractions to laboratory (Brouder et al., 2003) and field (Fernandez et al., 2008) artifacts, but longer extractions show promise for determining soil-specific illite K buffering of exchangeable and solution K (Cox et al., 1999). Likewise, cation exchange resins have been used in research to quantify buffering and release of non-exchangeable K, and many studies have demonstrated a close correlation between K adsorbed on the resin and total plant K uptake (e.g.,

Rahmatullah and Mengel, 2000). Again, the presence of an ion exchanger as a sink for K$^+$ is theorized to mimic uptake by plant roots. Yet, the comprehensive work required to calibrate and validate any such ancillary tests for widespread commercial use is initially time-consuming and expensive and has yet to be undertaken for the two examples given. It is unknown whether subsequent improvements, if any, in soil test interpretation would prove cost-effective for commercial farming.

Soil Heterogeneity

Regardless of protocol selection, inherent soil heterogeneity at micro- to landscape scales must be considered in the soil test interpretation and fertilizer recommendation systems. Repeated analyses of laboratory reference soils that are theoretically uniform illustrate the confounded effects of run-to-run variability in protocol and equipment performance with sample-to-sample variation in exchangeable K. Within bucket standard deviations of 5 to >10 mg kg^{-1} (Fig. 6|6) and coefficients of variation <5 to >10% (Brouder et al., 2003) are common. Laboratory quality control plans often identify as "acceptable" measured values for reference standards that are within this general range (e.g., 15% and

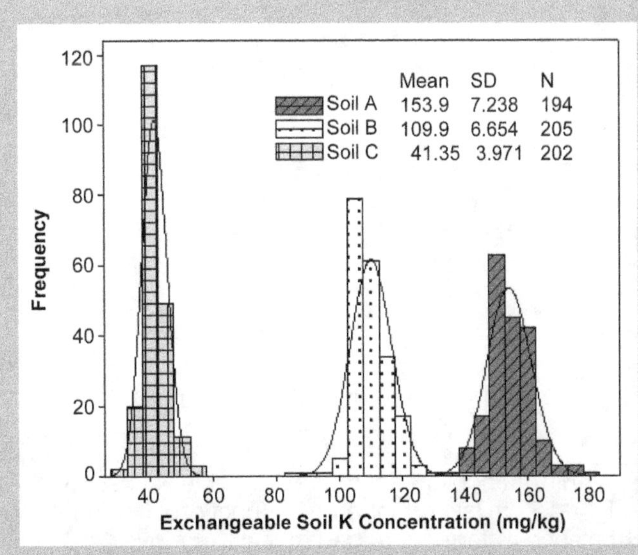

Fig. 6|6. Frequency distributions for individual NH$_4$OAc-extractable K values from approximately 200 analyses of three laboratory reference soils used as quality control samples in the Purdue University soil fertility program.

$\geq$15 mg kg^{-1} of the true value). Recommendations on numbers of cores to collect for a field composite sample extend this expected microvariability to the mesoscale to encompass variation that occurs over distances of a many meters—distances too short to be of management consideration. Sampling errors in the field are agreed to be at least as large as laboratory errors (Cope and Evans, 1985). For example, analysis of individual cores from research plots found values ranging from 150 to 195 (mean ± standard deviation:167 ± 13) and 150 to 270 (225 ± 36) mg kg^{-1} for Blount and Crosby silt loams, respectively (G. Van Scoyoc, personal communication, 1997). A simple power analysis demonstrates that a 90% probability (with 90% certainty) of attaining a soil test value within 15 mg kg^{-1} of the true mean value requires 7- and 50-core composites for soils, with standard deviations of 13 and 36 mg kg^{-1}, respectively. Reviews of numerous early studies support sampling intensities of 10 to > 30 cores per sample (Cope and Evans, 1985; Peck and Melsted, 1973). Such mesoscale variation can be expected to be higher in no-till systems, where interrows receive less residue and/

or where surface or shallow K banding has occurred (Varsa and Ebelhar, 2000).

The depth to which soils are sampled is another source of variation and is an increasingly important aspect of soil testing. Cycling by plants not only redistributes soil K horizontally across rows but vertically with depth in the profile. When soil pH and K fertility is managed within optimal ranges and soils are not excessively leached, K tends to become concentrated in surface soils relative to deeper soil layers. Because of slow soil diffusivity, K from aerial biomass returned to soil and from fertilizer tends to remain in the general vicinity where it is placed. Thus, many natural and agricultural soils have higher concentrations of plant-available K in surface horizons, with exchangeable and solution K decreasing markedly with depth in the profile. In the 1950s and 1960s, large K fertility studies of corn found exchangeable K levels in the 0- to 0.15-m layer to be threefold the levels observed at depths of 06 to 0.75 m (Fig. 6|4; Hanway et al., 1961, 1962). The long-term K fertility study of Li and Barber (1988) gives an excellent, focused example of the dynamic nature of this process with respect to the surface soil alone. Where K additions from fertilizer exceeded crop removal, exchangeable K content within the plow layer increased linearly with the net K additions to soil (Fig. 6|7), but, where plant removal equaled or exceeded the fertilizer K applications, the authors identified no relationship between surface soil exchangeable K and net K removal. However, reexamination of the data reveals that surface soil K content actually increased as a function of increasingly negative soil K balance. Subsurface soil K dynamics were not examined in this study, but these results suggest differences in subsurface and surface K cycling via plants accounted for the

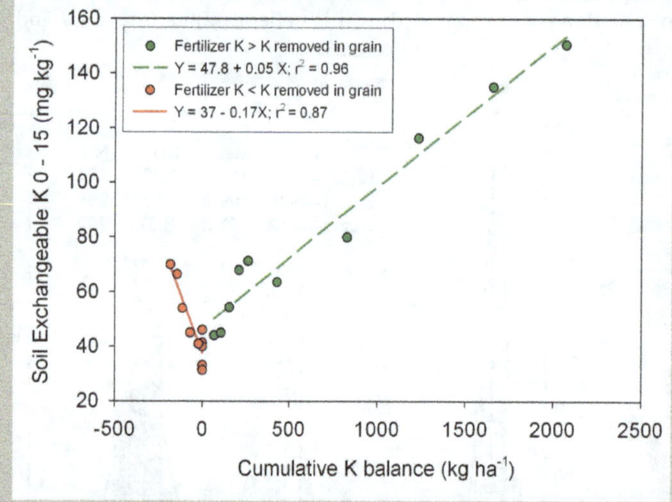

Fig. 6|7. Ammonium acetate extractable K in surface soil (0–15 cm layer) plotted as a function of cumulative K balance (K fertilizer added minus K removed in yield) following 20 years of cropping in a long-term K fertility experiment in West Lafayette, IN. Negative and positive cumulative K balance plots received 0 or 93 kg K ha^{-1} and 186 or 558 kg ha^{-1} every fourth year (adapted from Li and Barber, 1988).

difference in soil K dynamics between the positive and negative K balance treatments. Certainly there are soils where the K supplying power below the tillable layer is significantly different than is suggested by surface sample analysis (Woodruff and Parks, 1980), and soil test interpretation must encompass this layer's contribution to plant K status when the layer is not directly assessed.

The degree of stratification within the tillable layer itself (0–0.2 m depth) is strongly impacted by fertilizer and residue management. Typically, the bulk of K fertilizer is broadcast across the soil surface. When residue management entails moldboard plowing, K from fertilizer and from aerial and tillage-zone root residue is uniformly distributed throughout the depth of the tilled layer (Crozier et al., 1999; Cruse et al., 1983; Holanda et al., 1998). In contrast, reduced or no-tillage management often permits strong K stratification within the plow layer (Howard et al., 1999; Karathanasis and Wells, 1990). Across of a broad range of soils and soil K contents, no-till has been observed to enhance the exchangeable K content of the 0- to 0.05-m layer by 1.5- to twofold the content of the 0.05- to 0.10-m layer and by 2- to 2.5-fold the content of the 0.1- to 0.2-m layer (Buah et al., 2000; Hargrove, 1985). In a recent study of K balance in a long-term no-till soil, Fernandez et al. (2008) identified pronounced stratification within the tillable layer of all three forms of plant-available K—solution, exchangeable, and plant-available nonexchangeable—provided overall soil K levels from broadcast fertilization were sufficient to optimize yields (Fig. 6|2). However, when soybean yields were K limited, exchangeable K in the 0- to 0.05-m increment remained more than double the levels in the 0.05- to 0.10- and 0.10- to 0.20-m layers, but solution and nonexchangeable K levels were uniform throughout the tillage layer. Thus, for K fertility management to be optimized for reduced or no-tillage conditions, soil test recommendation and interpretation must accommodate varying K dynamics with depth in the soil profile.

Whether or not vertical nutrient stratification within the rooting zone is a benefit or limitation to plant K status is dependent on the distribution of water and roots. Where soil moisture and plant-available soil K are asynchronously distributed, root growth and K uptake efficiency will be reduced, and plant growth may be K limited (Brouder and Cassman, 1990). However, where vertical distribution of soil moisture and K are synchronous, stratification per se may be beneficial to soil K fertility (Fernandez et al., 2008). Studies of crop response to deep placement of K in stratified profiles are numerous, but results are inconsistent (Hairston et al., 1990; Borges and Mallarino, 2000; Buah et al., 2000; Rehm and Lamb, 2004). These studies have generally assumed soil moisture in surface soils limited K availability but did not monitor moisture dynamics. Determining whether or not deep placement will be beneficial requires the understanding of seasonal moisture fluctuation throughout the soil profile be coupled with an understanding of root growth, distribution, and both total plant K demand and K demand per unit surface area. For example, both corn (Mengel and Barber, 1974a,b) and soybean (Fernandez et al., 2009) have high K uptake per unit root length in the early season, suggesting K and water must be plentiful in surface soils near the seedling root system. As the growing season progresses, corn root uptake per unit root length steadily declines, reflecting the development of a dense root system and internal K translocation to meet relatively low grain K demands. In contrast, soybean root uptake declines in the late vegetative and flowering stages but increases sharply late in the growing season, reflecting a more generally sparse root system and a high late-season K demand during pod fill. Thus synchronous root zone water and K availability may be equally important to both crops early in the growing season but relatively more important to soybean later in the growing season.

Finally, most K management recommendations acknowledge that typical farm fields contain more than one soil type and that this variation is meaningful to K fertility. The historical approach to mapping and managing within-field variation in fertility has been to collect separate composites for discrete soil map units; large map units (e.g., >8 ha) and areas where significant variation was suspected were broken into one or more smaller areas (Peck and Melsted, 1973). However, apparently uniform fields can have a high degree of variability from a host of other factors, including nonuniform fertilizer and manure applications, other historic managements, erosion or land leveling, and uneven soil drainage. In the mid 1990s, the proliferation of georeferenced technologies spurred interest in managing smaller scales of fertility variation using more intensive point and gridded sampling strategies. Profitable precision management of a nutrient requires that yield-influencing variability exists, that it can be accurately identified, and that accurate nutrient application can be made. Phosphorus and K were an early focus because existing fertilizer recommendations could theoretically be applied to any scale, and it was perceived that these nutrients would have a high degree of spatial dependence and a low degree of temporal variation (Pierce and Nowak, 1999). However, much of the ensuing research failed to demonstrate profitable site- or soil-specific K management. Problems stemmed from a variety of factors including high, spatially independent K variation, short distances for K dependence, and variation in K levels that did not encompass low K tests (Franzen and Peck, 1995; Borges and Mallarino, 1997; Cambardella and Karlen, 1999; Chang et al., 1999; Mallarino et al., 1999). Mueller et al. (2001) examined the quality of K fertility maps generated by different sampling strategies and clearly demonstrated two key barriers to commercial implementation of soil-specific K management: (i) the 100-m grid unit commonly used in commercial samplings was grossly inadequate for identifying K variability, and (ii) the historical approach of composite sampling by soil map unit was just as efficient in predicting K need as various, more intensive strategies.

Future Directions

Any intensification of agriculture will likely increase K removal. In crop species that have been extensively improved for agriculture, nutrient concentrations, especially in grain, can be relatively constant when yields are not limited by other factors; therefore, a crop's nutrient removal can be expected to scale closely with yields (Brouder and Volenec, 2008). Realization of the bioenergy agenda may significantly increase K removal from agricultural fields if rotations are intensified (i.e., more crops per year), more plant tissues are harvested, and/or crop selections are shifted to favor more extractive crops. For the 74 million ha of irrigated rice (*Oryza sativa* L.) in Asia, Dobermann et al. (1996) estimated harvesting straw for fuel will increase crop K removal at least fivefold from 0.9 to 1.2 million t yr^{-1} to 5 to 9 million t yr^{-1}. If corn stover is used as a source of biomass for renewable fuel, recent studies suggest stover K removal will range 34 to 44 kg ha^{-1}, depending on the percentage of biomass removed (Hoskinson et al., 2007). Additionally, some of the perennial grasses advocated as candidates for cellulosic biomass require prodigious quantities of K for growth and may remove K at quantities greater than removed in the annual crops they replace. For example, *Miscanthus giganteus* J.M. Greef & Deuter ex Hodk. & Renvoize can have as much as 466 kg ha^{-1} in total plant K (shoots plus rhizomes) requiring new K uptake from soil >350 kg ha^{-1} yr^{-1} (Table 6|2; Himken et al., 1997; Beale and Long, 1997). Before senescing in the fall, aerial biomass can have >400 kg K ha^{-1}, with the actual quantity of K removed in harvested biomass varying significantly with the timing of biomass removal. Certainly our understanding of K management for less common crops including novel biomass species is fragmented and incomplete.

For most commonly grown crops, scientific understanding of plant uptake as a function of development is well characterized, in part because it is simple to do. The perception among producers and researchers is that quick tests for soil K availability can and should be improved for overall improved K management. Yet, both producers and researchers would be well served to temper any expectations for new tests with an understanding of the extent of innate variation that occurs across scales and within and among soils. In 1987, Scott and Smith remarked that the difficulties in determining plant-available soil K identified 35 years earlier by Reitemeier (1951) persisted, and now, another 20 years later, these difficulties remain. In no small part, these difficulties are a reflection of the fact that K content of soils is both greater and more variable than the contents of any other plant nutrients (Scott and Smith, 1987). Many of the factors that have been identified in empirical studies as determinants of plant-available K have not proved to be generalizable or distillable into practical and quantitative guides. Given the wealth of studies that have been conducted on almost every mineralogical, physiochemical, and biological aspect of soil K availability, future advances seem most like to occur by encompassing and not mastering soil K heterogeneity. More complex laboratory assays will likely be more expensive, deterring broad interest. Regardless, funding available for field correlation and validation research of new laboratory protocols remains sparse and unlikely to increase dramatically in the near term. Greatest advances may lie in integrated networks of commercial soil testing and crop, weather ,and soil data that are grounded in current knowledge but have the capacity to self assess, to iteratively evaluate local outcomes, and to "learn" and propose local management modifications for most probable improvements.

References

Allison, F.E., E.M. Roller, and J.E. Adams. 1959. Soil fertility studies in lysimeters. USDA Tech. Bull. 1199. U.S. Gov. Print. Office, Washington, DC.

Atkinson, D. 1990. Influence of root system morphology and development on the need for fertilizers and efficiency of use. p. 411–451. *In* V.C. Baligar and R.R. Duncan (ed.) Crops as enhancers of nutrient use. Academic Press, London.

Barber, S.A. 1995. Soil nutrient bioavailability. 2nd ed. John Wiley & Sons, New York.

Barber, S.A., and A.D. Mackay. 1985. Sensitivity analysis of the parameters of a mechanisitic mathematical model affected by changing soil moisture. Agron. J. 77:528–531.

Barber, T.E., and B.C. Mathews. 1962. Release of non-exchangeable soil potassium by resin-equilibration and its significance for crop growth. Can. J. Soil Sci. 42:266–272.

Barrow, N.J. 1966. Nutrient potential and capacity: II. Relationship between potassium potential and buffering capacity and the supply of potassium to plants. Aust. J. Agric. Res. 17:849–861.

Barshad, I. 1951. Cation exchange in soils: I. Ammonium fixation and its relation to potassium fixation and to determination of ammonium exchange capacity. Soil Sci. 77:463–472.

Beale, C.V., and S.P. Long. 1997. Seasonal dynamics of nutrient accumulation and partitioning in the perennial C_4–grasses *Miscanthus* × *giganteus* and *Spartina cynosuroides*. Biomass Bioenergy 12(6):419–428.

Bertsch, P.M., and G.W. Thomas. 1985. Potassium status of temperate region soils. p. 131– 162. *In* R.E. Munson (ed.) Potassium in agriculture. ASA, Madison, WI.

Binnie, R.R., and S.A. Barber. 1964. Contrasting release characteristics of potassium in alluvial and associated upland soils of Indiana. Soil Sci. Soc. Am. J. 28(3):387–390.

Bolt, G.H., M.E. Sumner, and A. Kamphorst. 1963. A study of the equilibria between three categories of potassium in an illitic soil. Soil Sci. Soc. Am. J. 27(3):294–299.

Borges, R., and A.P. Mallarino. 1997. Field-scale variability of phosphorus and potassium uptake by no-till corn and soybean. Soil Sci. Soc. Am. J. 61:846–853.

Borges, R., and A.P. Mallarino. 2000. Grain yield, early growth, and nutrient uptake of no-till soybean as affected by phosphorus and potassium placement. Agron. J. 92:380–388.

Bray, R.H., and L.T. Kurtz. 1945. Determination of total, organic, and available forms of phosphorus in soils. Soil Sci. 59:39–45.

Brouder, S.M., and K.G. Cassman. 1990. Root development of two cotton cultivars in relation to potassium uptake and plant growth in vermiculitic soil. Field Crops Res. 23:187–203.

Brouder, S.M., and K.G. Cassman. 1994a. Evaluation of a mechanistic model of potassium uptake by cotton in vermiculitic soil. Soil Sci. Soc. Am. J. 58:1174–1183.

Brouder, S.M., and K.G. Cassman. 1994b. Cotton root and shoot response to localized supply of nitrate, phosphate and potassium: Split-pot studies with nutrient solution and vermiculitic soil. Plant Soil 161:179–193.

Brouder, S.M., M. Thom, V.I. Adamchuck, and M.T. Morgan. 2003. Potential uses of ion-selective potassium electrodes in soil fertility management. Commun. Soil Sci. Plant Anal. 34(19&20):2699–2726.

Brouder, S.M., and J.J. Volenec. 2008. Impact of climate change on crop nutrient and water use efficiencies. Physiol. Plant. 133:705–724.

Buah, S.S.J., T.A. Polito, and R. Killorn. 2000. No-tillage soybean response to banded and broadcast and direct and residual fertilizer phosphorus and potassium applications. Agron. J. 92:657–662.

Bundy, L.G., and T.W. Andraski. 1999. Site-specific factors affecting corn response to starter fertilizer. J. Prod. Agric. 12(4):664–670.

Cambardella, C.A., and D.L. Karlen. 1999. Spatial analysis of soil fertility parameters. Precis. Agric. 1:15–25.

Cassman, K.G., D.C. Bryant, and B.A. Roberts. 1990. Comparison of soil test methods for predicting cotton reponse to soil and fertilizer potassium on potassium fixing soils. Commun. Soil Sci. Plant Anal. 21:1727–1743.

Cassman, K.G., B.A. Roberts, and D.C. Bryant. 1992. Cotton response to residual fertilizer potassium on vermiculitic soil: Organic matter and sodium effects. Soil Sci. Soc. Am. J. 56:823–830.

Cassman, K.G., B.A. Roberts, T.A. Kerby, D.C. Byant, and S.L. Higashi. 1989. Soil potassium balance and cumulative cotton repsonse to annual potassium additions on a vermiculitic soil. Soil Sci. Soc. Am. J. 53:805–812.

Chang, J., D.E. Clay, C.G. Carlson, D. Malo, and S.A. Clay. 1999. Precision farming protocols: Part 1. Grid distance and soil nutrient impact on the reproducibility of spatial variability measurements. Precis. Agric. 1:277–289.

Ching, P.C., and S.A. Barber. 1979. Evaluation of temperature effects on K uptake by corn. Agron. J. 71:1040–1044.

Churchman, G.J. 2000. The alteration and formation of soil minerals by weathering. p. F1– F76. *In* M.E. Sumner (ed.) Handbook of soil science. CRC Press, Boca Raton, FL.

Cope, J.T., and C.E. Evans. 1985. Soil testing. Adv. Soil Sci. 1:201–228.

Cox, A.E., and B.C. Joern. 1997. Release kinetics of non-exchangeable potassium in soils using sodium tetraphenylboron. Soil Sci. 162(8):588–596.

Cox, A.E., B.C. Joern, S.M. Brouder, and D. Gao. 1999. Plant-available potassium assessment with a modified sodium tetraphenylboron method. Soil Sci. Soc. Am. J. 63:902–911.

Crozier, C.R., G.C. Naderman, M.R. Tucker, and R.E. Sugg. 1999. Nutrient and pH stratification with conventional and no-till management. Commun. Soil Sci. Plant Anal. 30:65–74.

Cruse, R.M., G.A. Yakle, T.C. Colvin, D.R. Timmons, and A.L. Mussleman. 1983. Tillage effects on corn and soybean production in farmer-managed, university-monitored field plots. J. Soil Water Conserv. 38:512–514.

Dobermann, A., P.C. Sta. Cruz, and K.G. Cassman. 1996. Fertilizer inputs, nutrient balance, and soil nutrient-supply power in intensive, irrigated rice systems. I. Potassium uptake and K balance. Nutr. Cycl. Agro-ecosyst. 46:1–10.

Doll, E.C., M.M. Mortland, K. Lawton, and B.G. Ellis. 1965. Release of potassium from soil fractions during cropping. Soil Sci. Soc. Am. J. 29(6):699–702.

Eckert, D.J., and M.E. Watson. 1996. Integrating the Mehlich extractant into existing soil interpretation schemes. Commun. Soil Sci. Plant Anal. 27:1237–1249.

Favaretto, N.L., D. Norton, S.M. Brouder, and B.C. Joern. 2008. Gypsum amendment and exchangeable calcium and magnesium effects on plant nutrition under conditions of intensive nutrient extraction. Soil Sci. 173(2):108–118.

Fergus, I.F., and A.E. Martin. 1974. Studies in potassium. IV. Interspecific differences in uptake of non-exchangeable K. Aust. J. Soil Res. 12:147–158.

Fernandez, F.G., S.M. Brouder, C.A. Beyrouty, J. Volenec, and R. Hoyum. 2008. Assessment of plant-available potassium for no-till, rainfed soybean. Soil Sci. Soc. Am. J. 72(4):1085–1095.

Fernandez, F., S.M. Brouder, J. Volenec, C. Beyrouty, and R. Hoyum. 2009. Root and shoot growth, seed composition, and yield components of no-till rainfed soybean under variable potassium. Plant Soil 322:125–138.

Fixen, P.E. 2006. Soil test levels in North America. Better Crops 90(1):4–7.

Franzen, D.W., and T.R. Peck. 1995. Field soil sampling density for variable rate fertilization. J. Prod. Agric. 8(4):568–574.

Gaines, R.V., H.W. Skinner, E.F. Foord, B. Mason, and A. Rosenzweig. 1997. Dana's new mineralogy. John Wiley and Sons, New York.

Gartley, K.L., J.T. Sims, C.T. Olson, and P. Chu. 2002. Comparison of soil test extractants used in Mid-Atlantic United States. Commun. Soil Sci. Plant Anal. 33(5&6):873–895.

Graley, A.M. 1981. Assessing the availability of potassium in soils. Aust. J. Exp. Agric. Anim. Husb. 21:543–548.

Grava, J., G.E. Spalding, and A.C. Caldwell. 1961. Effect of drying upon amounts of easily extractable potassium and phosphorus in Nicollet clay loam. Agron. J. 53:219–221.

Grissinger, E., and C.D. Jeffries. 1957. Influence of continuous cropping on the fixation and release of potassium in three Pennsylvania soils. Soil Sci. Soc. Am. J. 21:409–412.

Haby, V.A., M.P. Russelle, and E.O. Skogley. 1990. Testing soils for potassium, calcium, and magnesium. p. 181–227. *In* R.L. Westerman (ed.) Soil testing and plant analysis. 3rd ed. SSSA Book Ser. 3. SSSSA, Madison, WI.

Hairston, J.E., W.F. Jones, P.K. McConnaughey, L.K. Marshall, and L.B. Gill. 1990. Tillage and fertilizer effects on soybean growth and yield on three Mississippi soils. J. Prod. Agric. 3:317–323.

Hanway, J.J., S.A. Barber, R.H. Bray, A.C. Caldwell, L.E. Engelbert, R.L. Fox, M. Fried, D. Hovland, J.W. Ketcheson, W.M. Laughlin, K. Lawton, R.C. Lipps, R.A. Olson, J.T. Pesek, K. Pretty, F.W. Smith, and E.M. Stickney. 1961. North central regional studies. I. Field studies with alfalfa. North Central Regional Publ. 124. Agric. and Home Econ. Exp. Stn., Iowa State University, Ames.

Hanway, J.J., S.A. Barber, R.H. Bray, A.C. Caldwell, M. Fried, L.T. Kurtz, K. Lawton, J.T. Pesek, K. Pretty, M. Reed, and F.W. Smith. 1962. North central regional potassium studies. III. Field studies with corn. North Central Regional Publ. 135. Res. Bull. 503. Agric. and Home Econ. Exp. Stn., Iowa State University, Ames.

Hargrove, W.L. 1985. Influence of tillage on nutrient uptake and yield of corn. Agron. J. 77:763–767.

Helmke, P.A. 2000. The chemical composition of soils. p. B3– B24. *In* M.E. Sumner (ed.) Handbook of soil science. CRC Press, Boca Raton, FL.

Helmke, P.A., and D.L. Sparks. 1996. Lithium, sodium, potassium, rubidium, and cesium. *In* D.L. Sparks (ed.) Methods of soil analysis. Part 3. SSSA Book Ser. 5. SSSA, Madison, WI. Pp 551– 574.

Himken, M., J. Lammel, D. Neukirchen, U. Czypionka-Krause, and H.-W. Olfs. 1997. Cultivation of *Miscanthus* under west European conditions: Seasonal changes in dry matter production, nutrient uptake and remobilization. Plant Soil 189(1):117–126.

Hoagland, D.R., and J.C. Martin. 1935. Absorption of potassium by plants and fixation by the soil in relation to certain methods for estimating available nutrients. Trans. 3rd Int. Congr. Soil Sci. 1:99–103.

Holanda, F.S.R., D.B. Mengel, M.B. Paula, J.G. Carvaho, and J.C. Bertoni. 1998. Influence of crop rotations and tillage systems on phosphorus and potassium stratification and root distribution in the soil profile. Commun. Soil Sci. Plant Anal. 29:2383–2394.

Horstman, E.L. 1957. The distribution of lithium, rubidium, and cesium in igneous and sedimentary rocks. Geochim. Cosmochim. Acta 12:1–28.

Horton, M.M. 1959. Influence of soil type on potassium fixation. M.S. thesis. Purdue Univ., West Lafayette, IN.

Hoskinson, R.L., D.L. Karlen, S.J. Birrell, C.W. Radtke, and W.W. Wilhelm. 2007. Engineering, nutrient removal, and feedstock conversion evaluations of four corn stover harvest scenarios. Biomass Bioenergy 31(2–3):126–136.

Howard, D.D., M.E. Essington, and D.D. Tyler. 1999. Vertical phosphorus and potassium stratification in no-till cotton soils. Agron. J. 91:266–269.

Huang, P.M. 2005. Chemistry of potassium in soils. p. 227–292. *In* M.A. Tabatabai and D.L. Sparks (ed.) Chemical processes in soils. SSSA Book Ser. 8. SSSA, Madison, WI.

Hylander, L., N. Ae, T. Hatta, and M. Sugiyama. 1999. Exploitation of K near roots of cotton, maize, upland rice, and soybean grown in an Ultisol. Plant Soil 208(1):33–41.

Jungk, A.O. 2002. Dynamics of nutrient movement at the soil-root interface. p. 587–616. *In* Y. Waisel et al. (ed.) Plant roots: The hidden half. 3rd ed. Marcel Dekker, New York.

Kaiser, D.E., A.P. Mallarino, and M. Bermudez. 2005. Corn grain yield, early growth, and early nutrient uptake as affected by broadcast and in-furrow starter fertilization. Agron. J. 97:620–626.

Karathanasis, A.D., and K.L. Wells. 1990. Conservation tillage effects on the potassium status of some Kentucky soils. Soil Sci. Soc. Am. J. 54:800–806.

Karlen, D.L., E.J. Sadler, and C.R. Camp. 1987. Dry matter, nitrogen, phosphorus, and potassium accumulation rates by corn on Norfolk loamy sand. Agron. J. 79(4):649–656.

Kemper, W.D. 1989. Solute diffusivity. p. 1007–1024. *In* A. Klute (ed.) Methods of soil analysis. Part 1. Agron. Monogr. 9. 2nd ed. ASA and SSSA, Madison, WI.

Knudsen, D., G.A. Peterson, and P.F. Pratt. 1982. Exchangeable and soluble potassium. p. 228–237. *In* A.L. Page et al. (ed.) Methods of soil analysis. Part 2. Agron. Monogr. 9. 2nd ed. ASA and SSSA, Madison, WI.

Kovar, J.L., and S.A. Barber. 1990. Potassium supply characteristics of thirty-three soils as influenced by seven rates of potassium. Soil Sci. Soc. Am. J. 54:1356–1361.

Li, R.-G., and S.A. Barber. 1988. Effect of phosphorus and potassium fertilizer on crop response and soil fertility in a long term experiment. Fert. Res. 15:123–136.

Lissbrant, I.S. 2008. Impact of long-term P and K fertilization on growth, persistence, and physiology of alfalfa (*Medicago sativa* L.). Ph.D. diss. Purdue Univ., West Lafayette, IN.

Lumbanraja, J., and V.P. Evangelou. 1992. Potassium quantity–intensity relationships in the presence and absence of NH_4 for three Kentucky soils. Soil Sci. 154:366–376.

Mackay, A.D., and S.A. Barber. 1985. Soil moisture effect on potassium uptake by corn. Agron. J. 77:524–527.

Malavolta, E. 1985. Potassium status of tropical and subtropical regions soils. p. 163–200. *In* R.E. Munson (ed.) Potassium in agriculture. ASA, Madison, WI.

Mallarino, A.P., E.S. Oyarzabal, and P.N. Hinz. 1999. Interpreting within-field relationships between crop yields and plant variables using factor analysis. Precis. Agric. 1(1):15–25.

Marschner, H. 1995. Mineral nutrition of higher plants. Academic Press, New York.

Marshall, C.E. 1964. The physical chemistry and mineralogy of soils. John Wiley, New York.

McLean, E.O. 1976. Exchangeable K levels for maximum crop yields on soils of different cation exchange capacities. Commun. Soil Sci. Plant Anal. 7:823–828.

McLean, E.O., J.L. Adams, and R.C. Hartwig. 1982. Improved corrective fertilizer recommendations based on a two-step alternative usage of soil tests: Recovery of soil-equilibrated potassium. Soil Sci. Soc. Am. J. 46:1189–1201.

McLean, E.O., T.O. Oloya, and J.L. Adams. 1979. Soil tests to inventory the initially available levels and to assess the fates of added P and K as bases for improved fertilizer recommendations. Commun. Soil Sci. Plant Anal. 10:623–630.

McLean, E.O., and M.E. Watson. 1985. Soil measurements of plant-available potassium. p. 277– 308. *In* R.D. Munson (ed.) Potassium in agriculture. ASA, CSSA, and SSSA, Madison, WI.

Melsted, S.W., and T.R. Peck. 1973. The principles of soil testing. *In* L.M. Walsh and J.D. Beaton (ed.) Soil testing and plant analysis. 2nd ed. SSSA, Madison, WI.

Mengel, D.B., and S.A. Barber. 1974a. Development and distribution of the corn root system under field conditions. Agron. J. 66(3):341–344.

Mengel, D.B., and S.A. Barber. 1974b. Rate of nutrient uptake per unit of corn root under field conditions. Agron. J. 66:399–402.

Mengel, K. 1985. Dynamics and availability of major nutrients in soils. Adv. Soil Sci. 2:65–131.

Mengel, K., and R. Busch. 1982. The importance of the potassium buffer power on the critical potassium level in soils. Soil Sci. 133(1):27–32.

Mengel, K., and D. Steffens. 1985. Potassium uptake of rye-grass (*Lolium perenne*) and red clover (*Trifolium pratense*) as related to root parameters. Biol. Fertil. Soils 1:53–58.

Mengel, K., and K. Uhlenbecker. 1993. Determination of available interlayer potassium and uptake by rye-grass. Soil Sci. Soc. Am. J. 57:561–566.

Meyer, D., and A. Jungk. 1993. A new approach to quantify the utilization of non-exchangeable soil potassium by plants. Plant Soil 149(2):235–243.

Moritsuka, N., J. Yanai, and T. Kosaki. 2004. Possible processes releasing nonexchangeable potassium from the rhizosphere of maize. Plant Soil 258(1):261–268.

Mortland, M.M. 1958. Kinetics of potassium release from biotite. Soil Sci. Soc. Am. Proc. 22:503–508.

Mueller, T.G., F.J. Pierce, O. Schabenberger, and D.D. Warncke. 2001. Map quality for site-specific fertility management. Soil Sci. Soc. Am. J. 65:1547–1558.

Nair, K.P.P., A.K. Sadanandan, S. Hamza, and J. Abraham. 1997. The importance of potassium buffer power in the growth and yield of cardamom. J. Plant Nutr. 20:987–997.

Nolan, C.W., and W.C. Pritchett. 1960. Certain factors affecting the leaching of potassium from sandy soils. Proc. Soil Crop Sci. Fla. 20:139– 145.

Nye, P.H., and P.B. Tinker. 1977. Solute movement in the soil–root system. Blackwell Scientific Publishers, Oxford, UK.

Olk, D.C., and K.G. Cassman. 1993. Reduction of potassium fixation by organic matter in vermiculitic soils. p. 307–315. *In* K. Mulongoy and R. Merckx (ed.) Soil organic matter dynamics and sustainability in tropical agriculture. Wiley-Sayce, Chichester, UK.

Olk, D.C., and K.G. Cassman. 1995. Reduction of potassium fixation by two humic acid fractions in vermiculitic soils. Soil Sci. Soc. Am. J. 59:1250–1258.

Olk, D.C., K.G. Cassman, and R.M. Carlson. 1995. Kinetics of potassium fixation in vermiculitic soils under different moisture regimes. Soil Sci. Soc. Am. J. 59(2):423–429.

Olson, C.G., M.L. Thompson, and M.A. Wilson. 2000. Phyllosilicates. p. F77– F123. *In* M.E. Sumner (ed.) Handbook of soil science. CRC Press, Boca Raton, FL.

Page, A.L., and T.J. Ganje. 1964. The Effect of pH on Potassium Fixed by an Irreversible Adsorption Process. Soil Sci. Soc. Am. J. 28(2):199–202.

Peck, T.R., and S.W. Melsted. 1973. Field sampling for soil testing. *In* L.M. Walsh and J.D Beaton (ed.) Soil testing and plant analysis. 2nd ed. SSSA, Madison, WI.

Pierce, F.J., and P. Nowak. 1999. Aspects of precision agriculture. Adv. Agron. 67:1–85.

Portela, E.A.C. 1993. Potassium supplying capacity of northeastern Portuguese soils. Plant Soil 154:13–20.

Pratt, P.F. 1952. Release of potassium from nonexchangeable forms from size fractions of several Iowa soils. Soil Sci. Soc. Am. J. 16:25–29.

Rahmatullah, B., and K. Mengel. 2000. Potassium release from mineral structures by H+ ion resin. Geoderma 96(4):291–305.

Rehm, G.W., and J.A. Lamb. 2004. Impact of banded potassium on crop yield and soil potassium in ridge-till planting. Soil Sci. Soc. Am. J. 68:629–636.

Reitemeier, R.F. 1951. Soil potassium. Adv. Agron. 3:113–164.

Reitemeier, R.F. 1957. Soil potassium and fertility. p. 101– 106. *In* A. Stefferud (ed.) Soil. 1957 Yearbook of Agriculture. U.S. Gov. Print. Office, Washington, DC.

Rich, C.I. 1968. Mineralogy of soil potassium. p. 79– 91. *In* V.J. Kilmer et al. (ed.) The role of potassium in agriculture. ASA, Madison, WI.

Rich, C.I., and W.R. Black. 1964. Potassium exchange as affected by cation size, pH, and mineral structure. Soil Sci. 97:384–390.

Rich, C.I., and S.S. Obenshain. 1955. Chemical and clay mineral properties of red-yellow podzolic soil derived from mica schist. Soil Sci. Soc. Am. Proc. 19:334–339.

Richards, J.E., T.E. Bates, and S.C. Sheppard. 1988. Studies on the potassium-supplying capacities of southern Ontario soils: I. Field and greenhouse experiments. Can. J. Soil Sci. 68:183–197.

Rygiewicz, P.T., and C.S. Bledsoe. 1984. Mycorrhizal effects of potassium fluxes in northwest coniferous seedlings. Plant Physiol. 76:918–923.

Schofield, R.K. 1947. A ratio law governing the equilibrium of cations in solution. Proc. 11th Int. Cong. Pure Appl. Chem., London 3:257–261.

Scott, A.D., and S.J. Smith. 1987. Sources, amounts, and forms of alkali elements in soils. Adv. Soil Sci. 6:101–147.

Sharpley, A.N. 1989. Relationship between soil potassium forms and mineralogy. Soil Sci. Soc. Am. J. 53(4):1023–1028.

Shaw, J.K., R.K. Stivers, and S.A. Barber. 1983. Evaluation of differences in potassium availability in soils of the same exchangeable potassium level. Commun. Soil Sci. Plant Anal. 14(11):1035–1049.

Shaw, W.M., and B. Robinson. 1960. Reaction efficiency of liming material as indicated by lysimeter leachate and composition. Soil Sci. 89:209–218.

Singer, M.J., and D.N. Munns. 2002. Soils: An introduction. 5th ed. Prentice Hall, Upper Saddle River, NJ.

Sparks, D.L. 1987. Potassium dynamics in soils. Advan. Soil Sci. 6:1–63.

Sparks, D.L. 1989. Kinetics of soil chemical processes. Academic Press, San Diego, CA.

Sparks, D.L. 2000. Bioavailability of soil potassium. p. D3–D53. *In* M.E. Sumner (ed.) Handbook of soil science. CRC Press, Boca Raton, FL.

Sparks, D.L., and P.M. Huang. 1985. Physical chemistry of soil potassium. p. 201–276. *In* R.E. Munson (ed.) Potassium in agriculture. ASA, Madison, WI.

Sparks, D.L., and P.M. Jardine. 1981. Thermodynamics of potassium exchange in soil using a kinetics approach. Soil Sci. Soc. Am. J. 45:1094–1099.

Sparks, D.L., and W.C. Liebhardt. 1982. Temperature effects on potassium exchange and selectivity in Delaware soils. Soil Sci. 133:10–17.

Srinivasarao, Ch., T.R. Rupa, A.S. Rao, G. Ramesh, and S.K. Bansal. 2006. Release kinetics of nonexchangeable potassium by different extractants from soils of varying mineralogy and depth. Commun. Soil Sci. Plant Anal. 37(3/4):473–491.

Stauffer, R.S. 1942. Runoff, percolate, and leaching losses from some Illinois soils. J. Am. Soc. Agron. 34:830–835.

Stehouwer, R.C., and J.W. Johnson. 1991. Soil adsorption interactions of band-injected anhydrous ammonia and potassium chloride fertilizers. Soil Sci. Soc. Am. J. 55:1374–1381.

Strebel, O., and W.H.M. Duynisveld. 1989. Nitrogen supply to cereals and sugar beet by mass flow and diffusion on a silty loam soil. Z. Pflanzenernaehr. Bodenkd. 152:135–141.

Varsa, E.C., and S.A. Ebelhar. 2000. Effect of potassium rate and placement on soil test variability across tillage systems. Commun. Soil Sci. Plant Anal. 31(11–14):2155–2161.

Walker, G.F. 1957. On the differentiation of vermiculites and smectites in clays. Clay Miner. Bull. 3:154–163.

Walworth, J.L. 1992. Soil drying and rewetting, or freezing and thawing, affects soil solution composition. Soil Sci. Soc. Am. J. 56:433–437.

Warncke, D.D., and S.A. Barber. 1972. Diffusion of zinc in soil. II. The influence of soil bulk density and its interaction with soil moisture. Soil Sci. Soc. Am. Proc. 36:42–46.

Woodruff, J.R., and C.L. Parks. 1980. Topsoil and subsoil potassium calibration with leaf potassium for fertility rating. Agron. J. 72:392–396.

Zeng, Q., and P. Brown. 2000. Soil potassium mobility and uptake by corn under differential soil moisture regimes. Plant Soil 221(2):121–134.

7

Nutrient Cycling in Soils: Sulfur

John L. Kovar and Cynthia A. Grant

Sulfur is an essential element required for normal plant growth, a fact that has been recognized since 1860 (Alway, 1940). It is considered a secondary macronutrient, following the primary macronutrients nitrogen, phosphorus, and potassium, but is needed by plants at levels comparable to P. Sulfur deficiency will impair basic plant metabolic functions, thus reducing both crop yield and quality. Deficiencies and responses to S amendments have been reported in crops worldwide (Tisdale et al., 1986; McGrath and Zhao, 1995; Scherer, 2001), and are becoming more common (Haneklaus et al., 2008). The likelihood of a response is determined by the balance between sulfur supply and crop demand. The main reasons for recent increases in documented S deficiencies include the reduction of SO_2 emissions from various industrial sources, mainly coal-fired power plants, an increase in the use of high-analysis fertilizers with little S, decreased use of S-containing pesticides, greater S removals with ever-increasing crop yields, and continued losses through leaching and erosion of topsoil. As pointed out by Haneklaus et al. (2008), in only a few years, the reputation of S has changed from that of an undesirable pollutant to a limiting factor in crop production.

In this chapter, we provide current information on the demand for S in various cropping systems, what we know about the soil supply of S, the best ways of assessing S status and managing S inputs, and how all of this information can be put together to optimize crop production. In each section, references will provide the reader with an opportunity to explore the topic in greater detail than can be given in these few pages.

Crop Demand for Sulfur

Substantial increases in the yields of major cereal and oilseed crops during the last four decades have greatly increased crop demand for S. With world population expected to rise to 9.2 billion by 2050, crop production and consequently S supply must increase as well.

The requirement for S or any other nutrient by a crop can be defined as the total amount of nutrient in the crop (kg ha^{-1}) or the concentration (g kg^{-1}) of the nutrient in the whole plant or specific plant part that is associated with optimum growth. Data on crop S contents are useful in calculating S removals from a field and for estimating S fertilizer needs. Critical concentrations of S in plant tissue are useful in diagnosing in-season S deficiencies. Numerous references provide critical plant tissue S concentrations for various crop species (Table 7|1) (Westerman, 1990; Bennett, 1993; Mills and Jones, 1996). Both public and commercial plant analysis laboratories

J.L. Kovar, USDA-ARS, National Laboratory for Agriculture and the Environment, 2110 University Boulevard, Ames, IA 50011-3120 (John.Kovar@ars.usda.gov); C.A. Grant, Agriculture and Agri-Food Canada, Brandon Research Centre, Brandon, Manitoba, Canada.

doi:10.2136/2011.soilmanagement.c7

Table 7|1. Critical sulfur concentrations in plant tissue of various crop species. Adapted from Mills and Jones (1996) and Dick et al. (2008).

Crop	Part sampled†	Time of sampling	Deficient	Low	Sufficient	High
					%	
Alfalfa	top 15 cm	early bud	<0.20	0.20–0.25	0.26–0.50	>0.50
Barley	whole top	heading			0.15–0.40	
Canola/rape	YMB	before flowering			0.35–0.47	
Cotton	YMB	early flowering			0.20–0.25	
Cowpea	YML	early bloom			0.17–0.22	
Maize	ear leaf	initial silk	<0.10	0.10–0.20	0.21–0.50	>0.50
Oats	top leaves	boot stage	<0.15	0.15–0.20	0.21–0.40	>0.40
Onion	whole top	half maturity			0.50–1.0	
Peanut	YML	pre-flowering			0.20–0.35	
Rice	whole top	max. tillering		0.10–0.20	0.20–0.30	>0.30
Ryegrass	young herbage	active growth			0.10–0.25	
Soybean	first trifoliate	early flower	<0.15	0.15–0.20	0.21–0.40	>0.40
Sugar cane	third leaf from tip	12–15 wk. after planting			0.14–0.20	
Sunflower	YML	mid-season			0.30–0.55	
White clover	young herbage	active growth			0.18–0.30	
Wheat	YEB/YMB	mid-late tillering			0.15–0.40	

† YEB, youngest emerged leaf blade; YMB, youngest mature leaf blade; YML, youngest mature leaf.

often provide critical values online. Sulfur concentration in most crop plants ranges between 0.1 and 1.5% S, although concentrations in excess of 3% have been reported for crops grown under saline conditions (Duke and Reisenauer, 1986). In general, S concentrations in grain are higher than in vegetative tissue.

Visual symptoms of S deficiency can be used as a diagnostic tool; however, symptoms will vary with crop species and the degree of deficiency (Duke and Reisenauer, 1986). Sulfur deficiency symptoms include reduced plant growth and chlorosis of the younger leaves, beginning with interveinal yellowing that gradually spreads over the entire leaf area. Unlike N, which can be readily remobilized in the plant, S is somewhat immobile, so that deficiency symptoms tend to occur first in younger leaves. With severe deficiencies, leaf cupping and a more erect leaf structure is often observed. This characteristic is common with canola (*Brassica napus* L. and *B. rapa* L.) (Franzen and Grant, 2008). Under mild to moderate S deficiency, however, visual symptoms may not always be a reliable indicator. Photos of S deficiency symptoms are available from many sources, including printed works (e.g., Bennett, 1993) and online sources (e.g., http://www.back-to-basics.net/nds/index.htm [verified 4 Feb. 2011]). Applications of soluble sulfate fertilizer often can correct a deficiency and increase crop yield and quality in the same growing season.

Responsive Crops

The S content of plants differs greatly among crop species, among cultivars within a species, and with developmental stage. Most species of the *Cruciferae* and *Liliaceae* families contain the largest amounts of S (Scherer, 2001). In general, the oilseed crops, such as oilseed rape (*Brassica napus* L.), canola, and sunflower (*Helianthus annuus* L.), and legumes, such as alfalfa (*Medicago sativa* L.) and soybean [*Glycine max* (L.) Merr.], have a much higher requirement for S than the small grains and maize (*Zea mays* L.) (Duke and Reisenauer, 1986). Whole plant S content is often higher during vegetative growth stages than at maturity. Malhi and Gill (2002) found that the demand for S by canola was greatest during flowering and seed set. Gregory et al. (1979) recorded a 50% decrease in plant S content of wheat during the period from anthesis to maturity, and speculated that efflux from roots into soil was the most likely pathway of loss. Plants also release measurable amounts of S into the atmosphere when S concentrations in foliage are high due to exposure to sulfur dioxide or excessive S uptake from soil (Janzen and Ellert, 1998).

Crop removal of S is a function of yield and S concentration in the harvested biomass (grain or dry matter). Sulfur removals by various crops as a function of yield are given in Table 7|2. Similar data are available from many sources (Spencer, 1975; Tabatabai, 1986; Jez, 2008). Currently, much of the S assimilated by the crop is retained in the system in plant residues returned to the soil. Intensification of cropping systems, however, has led to higher yields and accelerated crop S removal, which places greater demand on soil supply of S.

Sulfur Acquisition and Uptake

The majority of S required by a plant is absorbed from soil solution by roots in the form of the divalent sulfate anion, SO_4^{2-} (Barber, 1995). Similar to nitrate and phosphate, sulfate is taken up by specialized transporters in root cells and transported with the transpiration stream (Hawkesford and De Kok, 2006). Atmospheric SO_2 can be phytotoxic at high concentrations, but can also be captured and metabolized as a S source for plants when the S supply to roots is limiting (Westerman et al., 2000; Stuiver and De Kok, 2001). Sulfur that is captured directly from the atmosphere is eventually deposited in the soil as plant residue (Dick et al., 2008).

Because sulfate is an integral part of several metabolic pathways, an insufficient supply negatively affects plant metabolism. Sulfate taken up by roots must be reduced to sulfide (S^{2-}), before it is further metabolized. Reduction of sulfate to sulfide and its subsequent incorporation into cysteine (sulfate assimilation) occurs in the chloroplasts of the shoot (Droux, 2004). Cysteine is the precursor of methionine and most other organic sulfur compounds in plants, including thiols (glutathione), sulfolipids, and secondary

Table 7|2. Sulfur removals of various crops at the given yield levels. Adapted from Dick et al. (2008).

Crop	Plant component	Yield	S content
		Mg ha^{-1}	kg ha^{-1}
Alfalfa	biomass	13	34
Canola/rapeseed	grain	2.2	13
Cool-season grass	biomass	9.0	18
Cotton	lint	1.7	45
Grain sorghum	grain	9.4	25
	residue	–†	18
Maize	grain	11.5	15
	stover	6.9	10
Orange	fruit	60	31
Peanut	tuber	4.5	24
Potato	tuber	56	25
Rice	grain	7.8	13
Soybean	grain	4.0	13
	residue	–	15
Sugar Beet	tuber	67	50
Sunflower	seed	3.9	7
	residue	–	11
Tomato	fruit	67	46
Wheat	grain	5.4	8
	straw	–	17

† Sulfur removals in stover, straw, and crop residues are estimates based on typical values of a harvest index (i.e., the ratio of harvested grain to total plant biomass). In most cases, the crop residues are not harvested and the S would not be removed from the field.

metabolites (alliins, glucosinolates, and phytochelatins). These compounds are important for the physiology of plants and for resistance to environmental stresses and pests (Duke and Reisenauer, 1986). In addition to sulfate, S is moved within the plant in the reduced form as glutathione (Hawkesford and De Kok, 2006). Cysteine and methionine play a crucial role in the structure and function of plant proteins. Sulfur is involved in basic plant functions, such as photosynthesis and carbon and nitrogen metabolism (Droux, 2004). At present, however, the complex interactions between the shoot and roots that regulate S assimilation in relation to uptake and distribution are still poorly understood (Hawkesford and De Kok, 2006).

Sulfur compounds are important for crop quality. Haneklaus et al. (1992) found that insufficient S diminished the baking quality of wheat (*Triticum aestivum* L.) well before crop productivity decreased. Zhao et al. (1999) reported that a grain N/S ratio of 16:1 in wheat was the lower limit for optimum dough and bread-making properties. The S-containing amino acids in soybean are of particular nutritional importance in animal diets (Krishnan, 2008). Sulfur compounds in onion (*Allium cepa* L.), garlic (*Allium sativum* L.), and other *Allium* species determine the flavor profile of these crops (Boyhan, 2008). Defects in potato (*Solanum tuberosum* L.) tubers often result when S uptake is below optimum. Pavlista (2005) found that common scab and black scurf were reduced by early-season applications of elemental S, ammonium sulfate, or ammonium thiosulfate during a 6-yr study in the western United States. Haneklaus et al. (2008) concluded that a balanced nutrient supply, including S fertilization, for agricultural crops is the best guarantee for producing healthy foods.

Soil Supply of Sulfur

A general understanding of the basic processes involved in the soil S cycle is necessary to ensure proper S nutrition of crop plants. Total S in soils varies widely and depends on organic matter content, soil parent material, and the amount of S added via fertilizer amendments and atmospheric deposition (Scherer, 2009). Inorganic S is subject to adsorption, desorption, precipitation, and oxidation–reduction reactions, while organic S is subject to mineralization and immobilization (Fig. 7|1). Because soil S is continuously cycled between inorganic and organic forms, these processes determine the short- and long-term ability of a soil to supply available S. The soil S cycle has been reviewed extensively in the literature (Stevenson and Cole, 1999; Schoenau and Malhi, 2008; Scherer, 2009).

Inorganic Sulfur

As mentioned above, inorganic sulfate is the form of S absorbed by plant roots growing in soil. In general, less than 5% of total S in soil is the sulfate form. Sulfate can be present in soil solution, adsorbed on mineral surfaces, or coprecipitated with Ca and Mg. In well-drained surface soils with neutral to alkaline pH, sulfate exists mainly in the form of soluble salts of Ca, Mg, and Na. Solution sulfate concentrations of 3 to 5 mg L^{-1} are considered adequate for the growth of most crops, but concentrations change continuously depending on the balance

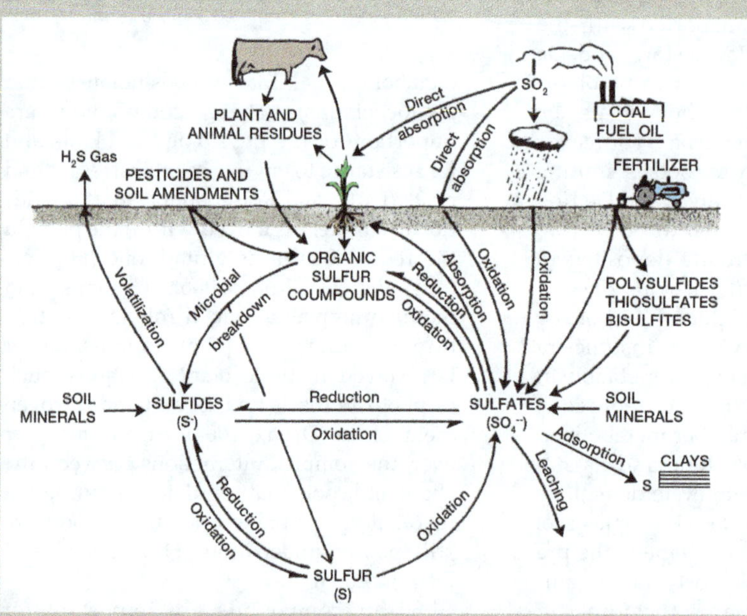

Fig. 7|1. Simplified version of the sulfur cycle in soils. Adapted from Stevenson and Cole (1999).

between plant uptake and mineralization–immobilization (Scherer, 2009).

Sulfate ions reach root surfaces via mass flow and diffusion. In soils with more than 5 mg L^{-1} soluble sulfate, essentially all S required by the crop is supplied by mass flow via the transpiration stream (Barber, 1995). Sulfate concentrations in soil solution are usually lowest in the winter and early spring because of leaching and slow mineralization rates (Castellano and Dick, 1990).

Adsorbed sulfate is in equilibrium with sulfate in soil solution. Adsorption is pH-dependent, and increases as pH decreases, reaching a maximum at pH 3 (Scherer, 2009). At pH levels greater than 6.5, adsorption is negligible, and the majority of soil sulfate is found in solution (Curtin and Syers, 1990). In acid soils, sulfate is often adsorbed on the surfaces of hydrous oxides of Fe and Al and edges of aluminosilicate clay minerals (Bohn et al., 1986). Adsorbed sulfate can significantly contribute to the S needs of plants growing in highly weathered, acidic soils because it is readily available. Sulfate adsorption is influenced by the presence of competing anions, such as phosphate, nitrate, and chloride (Tisdale et al., 1985). Adsorbed sulfate is held less strongly than ortho-phosphate (HPO_4^{2-}), so application of soluble P fertilizers will increase the availability of sulfate. Addition of lime also increases sulfate availability as a result of the competition of ortho-phosphate and hydroxyls (OH^-) with sulfate for adsorption sites on Fe and Al oxides (Scherer, 2009). Crops can utilize adsorbed sulfate in subsoils, but early season S deficiencies may occur until root development is sufficient. Deep-rooted crops are less likely to experience these early season deficiencies. Adsorption of sulfate can be a useful mechanism for retaining S in soils prone to leaching (Scherer, 2009).

Microbial oxidation of reduced inorganic S forms, such as elemental S, sulfides, and thiosulfates, to sulfates is an important process in soils (Stevenson and Cole, 1999). Microbial oxidation is performed by both autotrophic and heterotrophic microorganisms, such as *Thiobacillus*, *Pseudomonas*, and *Arthrobacter*. Reducing conditions found in flooded and waterlogged soils can result in sulfate conversion to sulfide. Sulfides are oxidized back to sulfates when the soil becomes aerobic again (Scherer, 2009).

Organic Sulfur

The organic S pool represents 95% or more of the total S in most noncalcareous surface soils. Organic S is present in plant and animal residues, microbial biomass and metabolites, and humus. Organic S is rendered plant available through the process of mineralization, while immobilization is the process by which sulfate is converted by soil biota into organic forms that cannot be taken up by roots. Hence, the amount of organic S in soil is highly correlated with organic C and total N. Unless S fertilizer is applied or atmospheric deposition is significant, mineralized S is the dominant input to the plant-available sulfate pool during a growing season (Schoenau and Malhi, 2008).

Soil organic S is a heterogeneous mixture of compounds, many with unknown chemical identity (Kertesz and Mirleau, 2004). Two main groups of S-containing compounds have been identified, namely ester sulfates (C–O–S) and carbon-bonded S (C–S), consisting of S-containing proteins and a variety of heterocyclic compounds. Delineation of these two groups is based on laboratory fractionation procedures in which ester sulfate is determined by hydriodic acid (HI) extraction, and C-bonded S is calculated from the difference between total S and ester sulfate (Tabatabai, 1996). McLaren et al. (1985) found that sulfate added to soil is quickly incorporated into the ester sulfate fraction, and that this pool, rather than the C-bonded S fraction, provides the majority of sulfate taken up by plants. Of the total organic S in soils, 30 to 70% is found in the organic sulfate fraction (Schoenau and Malhi, 2008). With time, ester sulfate S is converted to C-bonded S, indicating that C-bonded S is a more stable component of the soil organic S pool. Hence, the composition of the organic S pool in soil is an important determinant of the S-supplying capacity for crop plants.

Mineralization of ester sulfates in soil is accomplished by several sulfatase enzymes produced by soil microorganisms (Scherer, 2009). Ester sulfates in soil are hydrolyzed to release inorganic sulfate. Low levels of soil sulfate stimulate microbial production and release of sulfatases. Gupta et al. (1988) found that repeated application of S fertilizers resulted in a decline in sulfatase activity. Sulfur mineralization is greater when growing plants are present, presumably because of higher microbial populations in the rhizosphere, which increase sulfatase activity. There is some evidence that plant roots can produce and secrete sulfatase enzymes (Knauff et al., 2003), but further research is needed. Mineralization of C-bonded S occurs when soil microbes utilize the various compounds as a C source and release sulfate during the process (Scherer, 2009). However, mineralization of ester sulfates is much faster than that of C-bonded S compounds, so ester sulfates are more important contributors than C-bonded compounds for short-term S cycling.

The majority of organic S in crop residues is in the form of C-bonded S. Decomposition of residues results in conversion of these compounds into microbial biomass and humic products rich in organic sulfates. Microbial biomass S constitutes less than 3% of total soil S, but it is quite labile and considered a main factor controlling S turnover in soil (Yang et al., 2007). Greater amounts of biomass S often translate to greater amounts of S available for the crop. Factors controlling microbial activity and the release of plant-available S via mineralization include the C/S ratio in the residue being decomposed and environmental conditions (Pirela and Tabatabai, 1988). When the C/S ratio of organic residues is below 200, there is a net release of inorganic sulfate, while at C/S ratios greater than 400, there is a net loss of inorganic sulfate from the soil (Scherer, 2009). For C/S ratios between 200 and 400, sulfate can be either tied up or released from soil organic matter. Sulfur mineralization rates are greatest when soil water content is greater than 60% of field capacity and soil temperatures are in the range of 20 to 40°C (Stevenson and Cole, 1999). Under optimum soil temperature and moisture conditions, Tabatabai and Bremner (1972) showed that a significant amount of sulfate S will be mineralized in a short period of time.

Spatial (Landscape Scale) and Temporal Variability of Soil Sulfur

Sulfur availability is often associated with landscape position. As S distribution varies across a field, crop response to S fertilizer is also often strongly related to landscape position. Differential yield responses to landscape position have been documented (Haneklaus et al., 2006). Lower landscape positions tend to have higher soil S than upper landscape positions (Roberts and Bettany, 1985). Part of the difference between upper- and lower-slope positions is due to decreased organic matter and associated organic S in soils of the upper landscape surfaces. Higher water tables in lower landscapes result in higher subsoil sulfate (Haneklaus et al., 2008). Sulfur deficiencies are most often observed on hilltop and side-slope positions, especially on eroded, coarse-textured soils. However, an exception can occur where gypsum occurs near the surface on eroded knolls, provided that there is readily available sulfate for early growth of the crop. Sulfur deficiency is less common on foot-slope and toe-slope positions with medium- to heavy-textured soils high in organic matter. It is not unusual to find extremely high soil S concentrations and S deficiencies in the same field. The high variability in S concentration within a field poses challenges for soil testing (Bloem et al., 2001). If soil samples are composited, a sample with excessive S can elevate the results of the soil test and may lead to the conclusion that the field is well-supplied with S, when in fact the majority of the field is S deficient.

Assessing the Need for Sulfur

As the need to supplement S to achieve optimum crop production grows, greater attention will need to be paid to diagnostic tests that accurately predict responses. These tests must be reproducible and come at a reasonable cost. At present, soil tests that aim to extract some fraction of inorganic S and/or mineralizable organic S, and plant diagnostic tests that measure what the plant has captured at a specific stage of growth are available. Blanchar (1986), Jones (1986), Tabatabai (1996), and Dick et al. (2008) provide excellent reviews of testing methodology.

Soil Testing and Availability Indices

Although offered by many public and commercial laboratories, soil testing has not generally been very effective for predicting crop responses to available soil S (Dick et al., 2008). This is in part because a soil test cannot provide an estimate of the amount of atmospheric S that continuously supplies a crop with S. These atmospheric inputs vary with the time of year, amount of rainfall, and location of the field in relation to S sources. In general, the significant spatial and temporal variability in sulfate distribution creates problems in soil testing and subsequent recommendations for S fertilizers. Nevertheless, soil tests have been widely used for many years to predict crop requirements for S. Various extractants have been used, including water, acetates, carbonates, chlorides, phosphates, citrates, and oxalates (Jones, 1986; Kowalenko and Grimmett, 2008). The monocalcium phosphate or potassium phosphate extractant is commonly used in North America to predict S availability. There are many shortcomings to the procedure, which were identified early in its use (Hoeft et al., 1973). Blair et al. (1991) developed a method utilizing warm (40°C) potassium chloride solution for Australian soils, but this test has limited use elsewhere. Schoenau et al. (1993) reported good correlations between soil S measured with anion exchange resin membranes and plant S availability, but the method also has seen limited use.

The lack of a good correlation between soil tests and crop response has led to the consideration of N/S ratios in soils as an indication of sulfur supply (Janzen and Bettany, 1984). Total sulfur in a selected group of Canadian soils was highly correlated with organic carbon and total nitrogen (Bailey, 1985). It was suggested that soils with a high N/S ratio could be prone to sulfur deficiency.

Plant Analysis

Plant-tissue testing for S can also be used as an indication of S status of the crop (Jones, 1986; Mills and Jones, 1996). The plant growth stage and the plant part sampled are the most important variables to consider when using plant tissue testing to diagnose potential S problems. However, excess S can be taken up and stored as sulfate in the plant, which makes a plant diagnostic test more difficult. Although few studies have documented direct interaction of N and S fertility, plant N/S ratio has also been suggested as an indication of sulfur deficiency (Marschner, 1995). Bailey (1986) suggested that for maximum yield in Canadian soils, canola should have a total N/S ratio of 12 in the tissue at flowering, while barley (*Hordeum vulgare* L.) requires a ratio of 16 in the tissue at flag-leaf. Zhao et al. (1997), however, reported that sulfur-deficient rapeseed and that with sufficient S had similar N/S ratios. One of the problems with relying on plant analysis to diagnose sulfur problems is that if the problem is found, application of sulfur may come too late to benefit the crop during that growing season (Malhi et al., 2005).

Managing Sulfur Amendments

With a goal of maintaining or increasing crop production, any deficits in the S balance of the system are usually solved by the application of some form of S fertilizer. A wide range of inorganic and organic S fertilizers is available, several of which are listed in Table 7.3. Commercial S fertilizers tend to be inorganic materials that are directly manufactured or are produced as byproducts of other manufacturing processes. Animal manures, municipal biosolids, and composts are common soil amendments that often contain significant organic S. Detailed reviews of individual S fertilizer products, including their advantages and disadvantages in cropping systems, are available from various sources (Tisdale et al., 1985; Hagstrom, 1986; Boswell and Gregg, 1998; Scherer, 2001).

Inorganic Sulfur Sources

Sulfur-containing inorganic fertilizers can be divided into two main classes based on S form. Sulfate materials, such as ammonium sulfate $[(NH_4)_2SO_4]$ and gypsum $(CaSO_4)$, provide an immediate source of S the crop, but the sulfate can be highly susceptible to leaching (Curtin and Syers, 1990). For this reason, sulfate fertilizers should be managed similar to nitrate-N fertilizers. Elemental S materials provide a more gradual release of sulfate into soil because the S must first be oxidized to the sulfate form. This reduces the risk of leaching losses, but S availability to the crop is delayed and crop growth may not be improved (Janzen and Ellert, 1998). More importantly, oxidation of elemental S and other reduced fertilizer S forms produces acidity in the form of

Table 7|3. Examples of common inorganic and organic sulfur fertilizer sources. Adapted from Tisdale et al. (1985) and Dick et al. (2008).

Fertilizer sources	Nutrient concentrations	
	N–P–K	S
	—————%—————	
Inorganic sources		
Elemental S	0-0-0	88–98
Gypsum (calcium sulfate)	0-0-0	18
Ammonium sulfate	21-0-0	24
Ammonium thiosulfate	12-0-0	26
Magnesium sulfate	0-0-0	14
Potassium magnesium sulfate	0-0-18.2	22
Potassium sulfate	0-0-41.5	18
Aluminum sulfate	0-0-0	14
Ordinary superphosphate	0-9-0	11–12
Organic sources		
Municipal biosolids	–†	0.3–1.2
Cattle manure (liquid/solid)	–	0.15–0.8
Poultry litter	–	0.5
Sheep manure	–	0.35
Swine manure (liquid)	–	0.25
Composted biosolids	–	0.44
Composted dairy manure	–	0.22
Composted crop residues	–	0.10–0.22

† Nitrogen, phosphorus, and potassium levels in organic sources vary widely, so only typical sulfur concentrations for these materials are given.

sulfuric acid (H_2SO_4) as shown by following equation:

$$2\,S^0 + 3\,O_2 + 2\,H_2O \rightarrow 2\,H_2SO_4 \qquad [1]$$

In calcareous soils with high pH, this effect can be beneficial by improving the availability of phosphorus and most micronutrients. In some soils, however, soil acidification reduces populations of beneficial bacteria and fungi, which may affect cycling of S and other nutrients (Gupta et al., 1988).

Sulfate fertilizers can be further divided into sulfate and thiosulfate ($S_2O_3^{2-}$) forms. Gypsum is the most abundantly available sulfate material. In addition to being mined, gypsum is recovered from flue gases of coal-fired power plants, as well as from several industrial processes, such as production of phosphate fertilizers (Tisdale et al., 1985). Ammonium thiosulfate is the most common thiosulfate fertilizer; the clear liquid is widely used in the fluid industry. It can also be added to irrigation water. Following soil application, thiosulfate fertilizers break down to yield approximately equal parts of sulfate and elemental S (Hagstrom, 1986). The elemental S must undergo oxidation to sulfate before it can be captured by plant roots.

Elemental S fertilizers are the most S-dense materials (Table 7|3), but can vary greatly with respect to physical characteristics. Finer particle size allows more rapid conversion to sulfate. To avoid both the difficulties of handling finely divided S particles and the potential fire hazard of the dust, molten elemental S can be mixed with bentonite clay to produce a granular material that mixes well with other granular fertilizers on the market (Hagstrom, 1986). After application to soil, the bentonite clay absorbs water and swells, which then causes the granules to fracture and release the S. Because the S must be oxidized to sulfate, the effectiveness of the fertilizer can be inconsistent due to differences in both the fineness of the elemental S particles and soil properties, mainly aeration and temperature (Chapman, 1989). Particle fracturing and S dispersal is enhanced by soil wetting and drying cycles (Nuttall et al., 1993).

Research comparing sulfate sources with elemental S formulations indicates that in the initial year of application sulfate sources are more effective (Solberg et al., 2007). Cool, dry soils and the relatively short growing season that occurs in northern climates may restrict the oxidation of elemental S sources. However, research has shown that residual S from elemental S fertilizers will become available with time, thereby increasing yields in subsequent crops (Janzen and Ellert, 1998; Riley et al., 2000; Solberg et al., 2007). Between conversion to plant-available sulfate and S uptake by the crop, S from elemental S fertilizer is subject to leaching losses. Grant et al. (2004) found that the residual benefits of elemental S and ammonium sulfate were similar 3 yr after fertilizer application.

The combination of increasing S deficiency and strong demand for high analysis fertilizers that contain little or no S has lead the fertilizer industry to develop new

S-enhanced products. The S-enhanced materials are generally monoammonium phosphate (MAP) or diammonium phosphate (DAP) with microparticulate elemental S dispersed throughout the granules (Blair, 2009). Another material has one-half of the S in the sulfate form and the other half in the elemental S form that must be oxidized by soil bacteria to become available to plants. Lefroy et al. (1997) found that oxidation rates of elemental S are enhanced when S and P are mixed together in soil, possibly due to the P and S nutritional requirements of S-oxidizing microorganisms in the soil (Friesen, 1996). Recent research has shown that these new products may increase the agronomic efficiency of added S, which makes S inclusion in traditional P fertilizers an attractive option (Blair, 2009; Kovar and Karlen, 2010).

Organic Sulfur Sources

Sulfur-bearing organic amendments often contain significant amounts of sulfate and can be effective sources of plant-available S through mineralization. Organic S in these amendments, however, can vary considerably and appears to turn over relatively slowly (Eriksen et al., 1995). Tabatabai and Chae (1991) reported that there was a gradual linear increase in mineralized S with time in five soils amended with four types of animal manure, but that in some cases, S mineralization was slower in manure-amended soil than in unamended soil. These results and those of other studies (Eriksen et al., 1995) suggest that animal manures are not a good source of S in the short term. The type of feed and length of storage affect the plant availability of the S in the materials. In addition, animal manures, particularly liquid swine effluent, tend to be low in S relative to N, so that supplemental S fertilizer is needed to meet the needs of many crops (Schoenau and Davis, 2006).

The impact of municipal biosolids and compost applications on the S dynamics in agricultural soils depends on the C/N/S ratio in the material (Tabatabai and Chae, 1991). Sulfur mineralization can be significant in materials with a low (<<200) C/S ratio. Application of compost can also stimulate sulfatase activity in soil, as well as increase levels of microbial biomass S (Perucci, 1990). In some soils, S mineralization following biosolids application can be rapid and provide plant-available S within a few days (Tabatabai and Chae, 1991).

Timing, Placement, and Rate

The timing of S fertilizer applications, how the various forms of S should be applied to soil, and fertilizer rates for specific crops are all management decisions that require careful consideration. The growth and development of cereal grains, oilseed crops, and various legumes are quite different, so the demand for S varies considerably with growth stage. In general, research has shown that a sufficient S supply is needed during the early growth stages of cereal grains to ensure proper tiller development (Haneklaus et al., 2008). In contrast, insufficient S during the early part of the growing season may have little effect on canola yields if adequate S is available during flowering and seed set (Janzen and Bettany, 1984; Malhi and Gill, 2002; Franzen and Grant, 2008). Excellent reviews of S fertilizer management for specific crops are presented in Jez (2008).

The appropriate time of the year for S application also depends on the S form. Sulfate sources, such as ammonium or potassium sulfate, contain readily available S and should be applied at or near the time of planting to reduce S losses. In soils with low organic matter content, sandy texture, or rapid water movement through the profile, fall applications of sulfate materials should be avoided (Hagstrom, 1986). Soil or foliar applications of sulfate sources can also be used to correct S deficiencies during the growing season. To be effective, in-season soil (top-dress) applications depend on rainfall or irrigation to move the S into the root zone, although Kovar and Karlen (2010) found increased sulfate concentrations in the root zone approximately 4 wk after a surface application of liquid ammonium thiosulfate (Fig. 7|2). Elemental S sources must be applied early enough to allow oxidation of S to the sulfate form before the time of crop demand. Solberg et al. (2003) reported that fall application of elemental S allowed fertilizer granules to break down with freezing-thawing and wetting-drying cycles, thus aiding oxidation of elemental S during the growing season. However, Grant et al. (2004) found that even with fall application, conversion of elemental S to sulfate

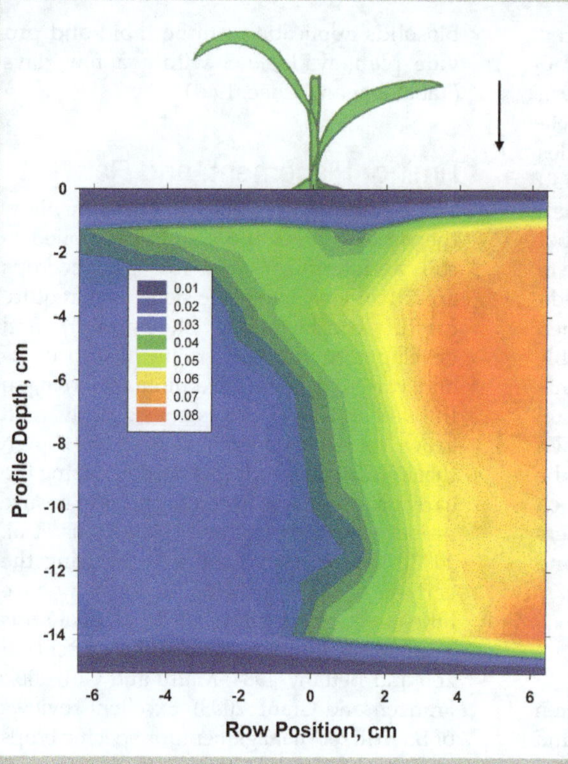

Fig. 7|2. Root zone distribution of bioavailable sulfur 26 d after application of ammonium thiosulfate (12–0–0–26 S) liquid fertilizer on the soil surface approximately 5 cm to the side of the maize row (arrow) in 2009 (Kovar and Karlen, 2010). Sulfur concentrations are micrograms sulfate (SO_4^{2-}) S cm^{-2} soil and were determined by extraction with bicarbonate-saturated exchange resin membranes.

may be too slow in the northern Great Plains to optimize yield of a spring crop.

The effectiveness of S fertilizer placement, as with the timing of S application, depends on the type of material applied and the soil to which the fertilizer is applied. Sulfate sources that are broadcast with or without incorporation at or near planting can provide readily available S to the crop (Malhi et al., 2005). In soils with adequate plant-available S in the subsoil, row or band application of sulfate sources at the time of planting can be quite effective (Hagstrom, 1986; Grant and Bailey, 1993). Care must be taken, however, to avoid seedling damage caused by excessive sulfate concentrations in contact with young roots. Elemental S sources generally should not be applied in bands, because this application practice reduces the contact of the S with oxidizing microorganisms in the soil (Nuttall et al., 1993). Broadcast application of elemental S should include tillage to mix the material with soil in the root zone. In flooded rice (*Oryza sativa* L.) systems, Blair and Lefroy (1998) suggest that S fertilizers should be placed on or near the soil surface to take advantage of the oxidized zone. Deep placement of sulfate sources decreases S availability as a result of reduction of sulfate to sulfide (Samosir et al., 1993).

The amount of S fertilizer needed for efficient production of a particular crop requires the integration of a significant amount of information. Nevertheless, general guidelines have been developed for important crops in specific regions. In the Midwest and northeastern United States, Hoeft and Fox (1986) found that an annual application of 28 kg S ha^{-1} was adequate for alfalfa production, and 17 kg S ha^{-1} were adequate for maize. Kamprath and Jones (1986) reported that S fertilization rates required for optimum maize yields in the southeastern United States ranged from 18 to 66 kg S ha^{-1}, with the higher amounts required on deeper, coarse-textured soils. For a soybean crop, 22 kg S ha^{-1} were adequate. When canola or other S-demanding crops are grown in the Great Plains of the United States and Canada, S fertilizer rates as high as 30 kg S ha^{-1} or more may be needed, depending on yield potential (Malhi et al., 2005). Blake-Kalff et al. (2000) found that oilseed rape grown in the UK requires more than 12 kg S ha⁻¹ for optimum yields, while a wheat crop requires less than 10 kg S ha⁻¹. Khurana et al. (2008) suggested that S fertilizer rates be increased for all crops grown in the Indo-Gangetic Plains of southern Asia. Application of 20 kg S ha⁻¹ is needed for raya (*Brassica juncea* L.) and lentil (*Lens culinaris* L.), while rice responds to applications up to 42 kg S ha⁻¹. Because of the widespread use of urea and S-free P and K fertilizers, rice production in Southeast Asia can benefit from S applications up to 60 kg S ha⁻¹ (Blair et al., 1979). The values listed here are broad averages based on reviews of available research. Results from a long-term fertility trial in Sweden (Kirchmann et al., 1996) indicated that when excessive amounts of S fertilizer are applied, leaching losses of S significantly increase. Therefore, S fertilizer recommendations, as those for other essential nutrients, must be site specific.

Challenges in Managing the Sulfur Fertility of Soils

Sulfur deficiencies will continue to be a growing problem due to ever-increasing crop yields, less atmospheric S deposition, less S applied as an impurity in fertilizers, and continued erosion of topsoil in which most mineralizable organic S is found. Decreased tillage affects the breakdown rate of residues and changes S release. Bioenergy feedstock production will result in greater S removals per unit of land area (Johnson et al., 2010), and increased drainage of agricultural lands will exacerbate leaching losses of S.

Sulfur deficiency not only impairs crop yield and quality, but also impacts environmental quality. Schnug (1991) found that for many European crops, N-use efficiency decreased when S was deficient, which led to significant increases in N losses through volatilization and leaching. Haneklaus et al. (2008) calculated that each kilogram of S deficit results in 15 kg of N loss to the environment.

Better methods for predicting crop S requirements are needed. Tissue tests provide information on plant capture of S from the soil and air, but are postmortem evaluations. Current soil tests more or less provide a snapshot of plant-available S. However, the balance between inputs and outputs from the available S pool during the growing season can have a significant impact on how much S is actually captured and utilized by the crop. This dynamic must be understood if accurate S fertilizer recommendations are to be made. The S balance of a crop production system on a local or regional scale will determine the external S requirements and the long-term stability of the system. If the S balance is negative, the system cannot be sustained.

Crop production systems are changing, but research addressing S nutrition lags. Many specialty fertilizers are coming onto the market. The agronomics and environmental impact of these materials are still uncertain. Little research addresses S-use efficiency of newer crop cultivars. Inter- and transdisciplinary efforts are necessary to unravel the interrelationships between S and other essential nutrients, and to understand their metabolic pathways within crop plants (Haneklaus et al., 2008). An understanding of the underlying mechanisms at the gene, cell, and whole-plant levels may allow us to grow crops with improved quality and resistance to stresses (Hawkesford and De Kok, 2006). This knowledge is also required if we hope to develop sophisticated nutrient management systems for future agricultural production.

References

Alway, F.J. 1940. A nutrient element slighted in agricultural research. J. Am. Soc. Agron. 32:913–921.

Bailey, L.D. 1985. The sulphur status of eastern Canadian prairie soils: The relationship of sulphur, nitrogen and organic carbon. Can. J. Soil Sci. 65:179–185.

Bailey, L.D. 1986. The sulphur status of eastern Canadian prairie soils: Sulphur response and requirements of alfalfa (*Medicago sativa* L.), rape (*Brassica napus* L.) and barley (*Hordeum vulgare* L.). Can. J. Soil Sci. 66:209–216.

Barber, S.A. 1995. Soil nutrient bioavailability—A mechanistic approach. 2nd ed. John Wiley and Sons, New York.

Bennett, W.F. 1993. Nutrient deficiencies and toxicities in crop plants. APS Press. St. Paul, MN.

Blair, G.J. 2009. Sulphur enhanced fertilizer (SEF). A new generation of fertilizers. Proc. Int. Plant Nutr. Colloq. XVI. Available at http://escholarship.org/uc/item/16h5b2dm (verified 7 Feb. 2011). Univ. of California, Davis.

Blair, G., and R. Lefroy. 1998. Sulfur and carbon research in rice production systems. Field Crops Res. 56:177–181.

Blair, G.J., E.O. Momuat, and C.P. Mamaril. 1979. Sulfur nutrition of rice. II. Effect of source and rate of S on growth and yield under flooded conditions. Agron. J. 71:477–480.

Blair, G.J., N. Chinoim, R.D.B. Lefroy, G.C. Anderson, and G.J. Crocker. 1991. A sulfur soil test for pastures and crops. Aust. J. Soil Res. 29:619–626.

Blake-Kalff, M.M.A., M.J. Hawkesford, F.J. Zhao, and S.P. McGrath. 2000. Diagnosing sulfur deficiency in field-grown oilseed rape (*Brassica napus* L.) and wheat (*Triticum aestivum* L.). Plant Soil 225:95–107.

Blanchar, R.W. 1986. Measurement of sulfur in soils and plants. p. 455–490. *In* M.A. Tabatabai (ed.) Sulfur in agriculture. Agron. Monogr. 27. ASA, CSSA, and SSSA, Madison, WI.

Bloem, E., S. Haneklaus, G. Sparovek, and E. Schnug. 2001. Spatial and temporal variability of sulphate concentration in soils. Commun. Soil Sci. Plant Anal. 32:1391–1403.

Bohn, H.L., N.J. Barrow, S.S.S. Rajan, and R.L. Parfitt. 1986. Reactions of inorganic sulfur in soils. p. 233–249. *In* M.A. Tabatabai (ed.) Sulfur in agriculture. Agron. Monogr. 27. ASA, CSSA, and SSSA, Madison, WI.

Boswell, C.C., and P.E.H. Gregg. 1998. Sulfur fertilizers for grazed pasture systems. p. 95–134. *In* D.G. Maynard (ed.) Sulfur in the environment. Marcel Dekker, New York.

Boyhan, G.E. 2008. Sulfur, its role in onion production and related alliums. p. 183–196. *In* J. Jez (ed.) Sulfur: A missing link between soils, crops, and nutrition. Agron. Monogr. 50. ASA, CSSA, and SSSA, Madison, WI.

Castellano, S.D., and R.P. Dick. 1990. Cropping and sulfur fertilization influence on sulfur transformations in soils. Soil Sci. Soc. Am. J. 54:114–121.

Chapman, S.J. 1989. Oxidation of micronized elemental sulphur in soil. Plant Soil 116:69–76.

Curtin, D., and J.K. Syers. 1990. Extractability and adsorption of sulphate in soils. J. Soil Sci. 41:295–304.

Dick, W.A., D. Kost, and L. Chen. 2008. Availability of sulfur to crops from soil and other sources. p. 59–82. *In* J. Jez (ed.) Sulfur: A missing link between soils, crops, and nutrition. Agron. Monogr. 50. ASA, CSSA, and SSSA, Madison, WI.

Droux, M. 2004. Sulfur assimilation and the role of sulfur in plant metabolism: A survey. Photosynth. Res. 79:331–348.

Duke, S.H., and H.M. Reisenauer. 1986. Roles and requirements of sulfur in plant nutrition. p. 123–168. *In* M.A. Tabatabai (ed.) Sulfur in agriculture. Agron. Monogr. 27. ASA, CSSA, and SSSA, Madison, WI.

Eriksen, J., J.V. Mortensen, V.K. Kjellerup, and O. Kristjansen. 1995. Forms and plant-availability of sulfur in cattle and pig slurry. Z. Pflanzenernaehr. Bodenkd. 158:113–116.

Franzen, D., and C.A. Grant. 2008. Sulfur response based on crop, source, and landscape position. p. 105–116. *In* J. Jez (ed.) Sulfur: A missing link between soils, crops, and nutrition. Agron. Monogr. 50. ASA, CSSA, and SSSA, Madison, WI.

Friesen, D.K. 1996. Influence of co-granulated nutrients and granule size on plant responses to elemental sulfur in compound fertilizers. Nutr. Cycling Agroecosyst. 46:81–90.

Grant, C.A., and L.D. Bailey. 1993. Fertility management in canola production. Can. J. Plant Sci. 73:651–670.

Grant, C.A., A.M. Johnston, and G.W. Clayton. 2004. Sulphur fertilizer and tillage management of canola and wheat in western Canada. Can. J. Plant Sci. 84:453–462.

Gregory, P.J., D.V. Crawford, and M. McGowan. 1979. Nutrient relations of winter wheat: 1. Accumulation and distribution of Na, K, Ca, Mg, P, S and N. J. Agric. Sci. 93:485–494.

Gupta, V.V.S.R., J.R. Lawrence, and J.J. Germida. 1988. Impact of elemental sulfur fertilization on agricultural soils. I. Effects on microbial biomass and enzyme activities. Can. J. Soil Sci. 68:463–473.

Hagstrom, G.R. 1986. Fertilizer sources of sulfur and their use. p. 567–581. *In* M.A. Tabatabai (ed.) Sulfur in agriculture. Agron. Monogr. 27. ASA, CSSA, and SSSA, Madison, WI.

Haneklaus, S., E. Bloem, and E. Schnug. 2006. Sulphur interactions in crop ecosystems. p. 17–58. *In* M.J. Hawkesford and L.J. De Kok (ed.) Sulfur in plants—An ecological perspective. Springer, Dordrecht, The Netherlands.

Haneklaus, S., E. Bloem, and E. Schnug. 2008. History of sulfur deficiency in crops. p. 45–58. *In* J. Jez (ed.) Sulfur: A missing link between soils, crops, and nutrition. Agron. Monogr. 50. ASA, CSSA, and SSSA, Madison, WI.

Haneklaus, S., E. Evans, and E. Schnug. 1992. Baking quality and sulphur content of wheat: II: Evaluation of the relative importance of genetics and environment including sulphur fertilization. Sulphur Agric. 16:35–38.

Hawkesford, M.J., and L.J. De Kok. 2006. Managing sulphur metabolism in plants. Plant Cell Environ. 29:382–395.

Hoeft, R.G., and R.H. Fox. 1986. Plant response to sulfur in the Midwest and northeastern United States. p. 345–356. *In* M.A. Tabatabai (ed.) Sulfur in agriculture. Agron. Monogr. 27. ASA, CSSA, and SSSA, Madison, WI.

Hoeft, R.G., L.M. Walsh, and D.R. Keeney. 1973. Evaluation of various extractants for available soil sulfur. Soil Sci. Soc. Am. Proc. 37:401–404.

Janzen, H.H., and J.R. Bettany. 1984. Sulfur nutrition of rapeseed. II. Effect of time of sulfur application. Soil Sci. Soc. Am. J. 48:107–112.

Janzen, H.H., and B.H. Ellert. 1998. Sulfur dynamics in cultivated, temperate agroecosystems. p. 11–44. *In* D.G. Maynard (ed.) Sulfur in the environment. Marcel Dekker, New York.

Jez, J. (ed.). 2008. Sulfur: A missing link between soils, crops, and nutrition. Agron. Monogr. 50. ASA, CSSA, and SSSA, Madison, WI.

Johnson, J.M., W.W. Wilhelm, D.L. Karlen, D.W. Archer, B.J. Wienhold, D.T. Lightle, D.A. Laird, J.M. Baker, T.E. Ochsner, J.M. Novak, A.D. Halvorson, F.J. Arriaga, and N.W. Barbour. 2010. Nutrient removal as a function of corn stover cutting height and cob harvest. BioEnergy Res. 3:342–352.

Jones, M.B. 1986. Sulfur availability indexes. p. 549–566. *In* M.A. Tabatabai (ed.) Sulfur in agriculture. Agron. Monogr. 27. ASA, CSSA, and SSSA, Madison, WI.

Kamprath, E.J., and U.S. Jones. 1986. Plant response to sulfur in the southeastern United States. p. 323–343. *In* M.A. Tabatabai (ed.) Sulfur in agriculture. Agron. Monogr. 27. ASA, CSSA, and SSSA, Madison, WI.

Kertesz, M.A., and P. Mirleau. 2004. The role of soil microbes in plant sulphur nutrition. J. Exp. Bot. 55:1939–1945.

Khurana, M.P.S., U.S. Sadana, and Bijay-Singh. 2008. Sulfur nutrition of crops in the Indo-Gangetic Plains of south Asia. p. 11–24. *In* J. Jez (ed.) Sulfur: A missing link between soils, crops, and nutrition. Agron. Monogr. 50. ASA, CSSA, and SSSA, Madison, WI.

Kirchmann, H., F. Pichlmayer, and M.H. Gerzabek. 1996. Sulfur balances and Sulfur-34 abundance in a long-term fertilizer experiment. Soil Sci. Soc. Am. J. 60:174–178.

Knauff, U., M. Schulz, and H.W. Scherer. 2003. Arylsulfatase activity in the rhizosphere and roots of different crop species. Eur. J. Agron. 19:215–223.

Kovar, J.L., and D.L. Karlen. 2010. Is sulfur limiting maize grown on eroded Midwestern U.S. soils? *In* R.J. Gilkes and N. Prakongkep (ed.) Proc. 19th World Congress Soil Sci., 1–6 Aug. 2010, Brisbane, Australia. [DVD.] IUSS, Brisbane, Australia.

Kowalenko, C.G., and M. Grimmett. 2008. Chemical characterization of soil sulfur. p. 251–263. *In* M.R. Carter and E.G. Gregorich (ed.) Soil sampling and methods of analysis. 2nd ed. CRC Press, Boca Raton, FL.

Krishnan, H.B. 2008. Improving the sulfur-containing amino acids of soybean to enhance its nutritional value in animal feed. p. 235–249. *In* J. Jez (ed.) Sulfur: A missing link between soils, crops, and nutrition. Agron. Monogr. 50. ASA, CSSA, and SSSA, Madison, WI.

Lefroy, R.D.B., Sholeh, and G. Blair. 1997. Influence of sulfur and phosphorus placement, and sulfur particle size, on elemental sulfur oxidation and the growth response of maize (*Zea mays*). Aust. J. Agric. Res. 48:485–495.

Malhi, S.S., and K.S. Gill. 2002. Effectiveness of sulphate-S fertilization at different growth stages for yield, seed quality and S uptake of canola. Can. J. Plant Sci. 82:665–674.

Malhi, S.S., J.J. Schoenau, and C.A. Grant. 2005. A review of sulphur fertilizer management for optimum yield and quality of canola in the Canadian Great Plains. Can. J. Plant Sci. 85:297–307.

Marschner, H. 1995. Mineral nutrition of higher plants. 2nd ed. Academic Press, London.

McGrath, S.P., and F.J. Zhao. 1995. A risk assessment of sulphur deficiency in cereals using soil and atmospheric deposition data. Soil Use Manage. 11:110–114.

McLaren, R.G., J.I. Keer, and R.S. Swift. 1985. Sulphur transformations in soils using sulphur-35 labelling. Soil Biol. Biochem. 17:73–79.

Mills, H.A., and J.B. Jones, Jr. 1996. Plant analysis handbook II. MicroMacro Publishing, Inc., Athens, GA.

Nuttall, W.F., C.C. Boswell, A.G. Sinclair, A.P. Moulin, L.J. Townley-Smith, and G.L. Galloway. 1993. The effect of time of application and placement of sulphur fertilizer sources on yield of wheat, canola, and barley. Commun. Soil Sci. Plant Anal. 24:2193–2202.

Pavlista, A.D. 2005. Early-season applications of sulfur fertilizers increase potato yield and reduce tuber defects. Agron. J. 97:599–603.

Perucci, P. 1990. Effect of the addition of municipal solid-waste compost on microbial biomass and enzyme activities in soil. Biol. Fertil. Soils 10:221–226.

Pirela, H.J., and M.A. Tabatabai. 1988. Sulphur mineralization rates and potentials of soils. Biol. Fertil. Soils 6:26–32.

Riley, N.G., F.J. Zhao, and S.P. McGrath. 2000. Availability of different forms of sulphur fertilizers to wheat and oilseed rape. Plant Soil 222:139–147.

Roberts, T.L., and J.R. Bettany. 1985. The influence of topography on the nature and distribution of soil sulfur across a narrow environmental gradient. Can. J. Soil Sci. 65:419–434.

Samosir, S.S.R., G.J. Blair, and R.D.B. Lefroy. 1993. Effects of placement of elemental S and sulfate on the growth of two rice varieties under flooded conditions. Aust. J. Agric. Res. 44:1775–1788.

Scherer, H.W. 2001. Sulphur in crop production—Invited paper. Eur. J. Agron. 14:81–111.

Scherer, H.W. 2009. Sulfur in soils. J. Plant Nutr. Soil Sci. 172:326–335.

Schnug, E. 1991. Sulphur nutritional status of European crops and consequences for agriculture. Sulphur Agric. 15:7–12.

Schoenau, J.J., and J.G. Davis. 2006. Optimizing soil and plant responses to land-applied manure nutrients in the Great Plains of North America. Can. J. Soil Sci. 86:587–595.

Schoenau, J.J., and S.S. Malhi. 2008. Sulfur forms and cycling processes in soil and their relationship to soil fertility. p. 1–10. *In* J. Jez (ed.) Sulfur: A missing link between soils, crops, and nutrition. Agron. Monogr. 50. ASA, CSSA, and SSSA, Madison, WI.

Schoenau, J.J., P. Qian, and W.Z. Huang. 1993. Assessing sulphur availability in soil using ion exchange resins. Sulphur Agric. 17:13–17.

Solberg, E.D., S.S. Malhi, M. Nyborg, and K.S. Gill. 2003. Fertilizer type, tillage, and application time effects on recovery of sulfate-S from elemental sulfur fertilizers in fallow field soils. Commun. Soil Sci. Plant Anal. 34:815–830.

Solberg, E.D., S.S. Malhi, M. Nyborg, B. Henriquez, and K.S. Gill. 2007. Crop response to elemental S and sulfate-S sources on S-deficient soils in the Parkland Region of Alberta and Saskatchewan. J. Plant Nutr. 30:321–333.

Spencer, K. 1975. Sulphur requirements of plants. p. 98–108. *In* K.D. McLachlan (ed.) Sulphur in Australasian agriculture. Sydney Univ. Press, Sydney.

Stevenson, F.J., and M.A. Cole. 1999. Cycles of soil—Carbon, nitrogen, phosphorus, sulfur, micronutrients. 2nd ed. John Wiley and Sons, New York.

Stuiver, C.E.E., and L.J. De Kok. 2001. Atmospheric H_2S as sulfur source for *Brassica oleracea*: Kinetics of H_2S uptake and activity of O-acetylserine (thiol)lyase as affected by sulfur nutrition. Environ. Exp. Bot. 46:29–36.

Tabatabai, M.A. (ed.) 1986. Sulfur in agriculture. Agron. Monogr. 27. ASA, CSSA, and SSSA, Madison, WI.

Tabatabai, M.A. 1996. Sulfur. p. 921–960. *In* D.L. Sparks (ed.) Methods of soil analysis. Part 3. Chemical methods. SSSA Book Series No. 5. ASA, CSSA, and SSSA, Madison, WI.

Tabatabai, M.A., and J.M. Bremner. 1972. Forms of sulfur, and carbon, nitrogen and sulfur relationships, in Iowa soils. Soil Sci. 114:380–386.

Tabatabai, M.A., and Y.M. Chae. 1991. Mineralization of sulfur in soils amended with organic wastes. J. Environ. Qual. 20:684–690.

Tisdale, S.L., W.L. Nelson, and J.D. Beaton. 1985. Soil fertility and fertilizers. 4th ed. Macmillan Publishing Co., New York.

Tisdale, S.L., R.B. Reneau, Jr., and J.S. Platou. 1986. Atlas of sulfur deficiencies. p. 295–322. *In* M.A. Tabatabai (ed.) Sulfur in agriculture. Agron. Monogr. 27. ASA, CSSA, and SSSA, Madison, WI.

Westerman, R.L. 1990. Soil testing and plant analysis. 3rd ed. SSSA Book Ser. 3. ASA, CSSA, and SSSA, Madison, WI.

Westerman, S., L.J. De Kok, C.E.E. Stuiver, and I. Stulen. 2000. Interaction between metabolism of atmospheric H_2S in the shoot and sulfate uptake by the roots of curly kale (*Brassica oleracea*). Physiol. Plant. 109:443–449.

Yang, Z., B.R. Singh, S. Hansen, Z. Hu, and H. Riley. 2007. Aggregate associated sulfur fractions in long-term (>80 years) fertilized soils. Soil Sci. Soc. Am. J. 71:163–170.

Zhao, F.J., P.E. Bilsborrow, E.J. Evans, and S.P. McGrath. 1997. Nitrogen to sulphur ratio in rapeseed and in rapeseed protein and its use in diagnosing sulphur deficiency. J. Plant Nutr. 20:549–558.

Zhao, F.J., M.J. Hawkesford, and S.P. McGrath. 1999. Sulphur assimilation and effects on yield and quality of wheat. J. Cereal Sci. 30:1–17.

8

Gas Exchange in Soils

Witold Stępniewski, Zofia Stępniewska, and Agnieszka Rożej

The soil medium is a place of intensive production, transformations, and consumption of a number of gases, which are transported to and from the atmosphere. The processes of gas exchange between soil and atmosphere affect not only the biomass production, but they are also very important from the environmental point of view.

The most important gas in soil air is oxygen, which is indispensable for respiration of plant roots and decisive for the direction of microbial metabolism as well as for numerous biochemical and chemical reactions in soil. It is transported from the atmosphere to the soil. The second important gas in soil is carbon dioxide, being an essential product of both oxic respiration of plant roots, of microorganisms and of meso- and macrofauna, as well as of anoxic microbial respiration and fermentation. Moreover it can be produced by some chemical reactions in soil. It is transported from the soil medium toward the atmospheric air. The most abundant component of the atmosphere and, in most cases, of soil air is nitrogen. Its physiological role is connected with nitrogen fixation by some groups of microbes. Nitrogen migration directed down the soil in the case of fixation and in the opposite direction in the case of its production due to denitrification, draws less attention because it is considered as nonlimiting because of its abundance.

Beside these macrocomponents of the soil air there are a number of gases occurring in trace amounts. The only exception and, to some extent intermediate component, is methane, which under dry land conditions occurs in trace amounts and under wetland conditions as well as under specific conditions such as landfill covers can occupy a substantial part of the soil air volume because of methane fermentation. Other microcomponents of soil air are nitrous oxide (N_2O), nitrogen oxide (NO), and dioxide (NO_2), ethylene, ammonia (NH_3), and hydrogen sulfide (H_2S).

Pathways of Gas Transport

Independently of the gas type and transport mechanism (diffusion or convection) there are two essential pathways of the gas exchange in the soil–plant–atmosphere continuum: (i) the soil pathway and (ii) the internal pathway through the plant tissues (Fig. 8|1). The former pathway is related to soils cultivated under dryland conditions. Under these conditions atmospheric oxygen and other gases consumed in the soil migrate from the atmosphere to the soil and particular roots through the soil bulk, while the gases generated in the roots and/or in the soil are transported in opposite direction through the soil medium.

W. Stępniewski (w.stepniewski@pollub.pl) and A. Rożej (a.rozej@wis.pol.lublin.pl), Lublin Univ. of Technology, Dep. of Land Surface Protection Engineering, Nadbystrzycka 40B, 20-618 Lublin, Poland; Z. Stępniewska, The John Paul II Catholic Univ. of Lublin, Kraśnicka 102, 20-551, Lublin, Poland (stepz@kul.pl)

doi:10.2136/2011.soilmanagement.c8

The second pathway is applicable to wetland conditions such as natural and constructed wetlands or paddy fields (e.g., Yan et al., 2000). In this situation the gas (or gasses) consumed in the soil and within the roots (mainly O_2) migrates from the atmosphere downward through the plant tissues, while the gases generated in soil and within the roots (most often CO_2 and CH_4) migrate to the atmosphere through the plant tissues. If the soil is deprived of plants the downward transport of gases practically ceases, while the gases generated in the soil (mainly CO_2 and CH_4) are released to the atmosphere by ebullition.

Between these two "model" pathways there are numerous intermediate situations with different relative contributions of soil- and plant-mediated transport of gases depending on the land use and tillage practices affecting air-filled porosity, gas diffusion coefficient, and air permeability.

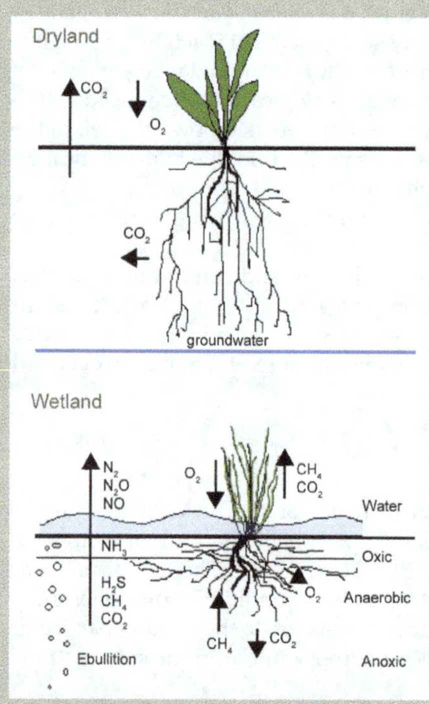

Fig. 8|1. Model of the two pathways of gas exchange in the atmosphere–plant–soil continuum. In dryland conditions oxygen is supplied to the plant roots via the soil. In wetland conditions oxygen is transported through the tissues of the plant itself.

Most of agricultural production is connected with dryland conditions, while wetland (mainly rice, *Oryza sativa* L.) cultivation delivers 20% of food calories (IRRI, 2005). The examples of intermediate systems are midseason drainage of paddy fields for preventing sulfide toxicity (Kanno et al., 1997) and combining lowland rice cultivation with dry season crops (So and Ringrose-Voase, 2000), as practiced in the Philippines and Indonesia.

Another example of this kind is the system of rice intensification (SRI), developed and practiced in Madagascar simultaneously with its ongoing evaluation in Asia (Dobermann, 2004). The most important feature of this system is the change from permanent flooding to intermittent irrigation. This improves the oxygen supply to rice roots, reducing formation of aerenchyma, and produces stronger and healthier root systems that favor nutrient uptake (Stoop et al., 2002). The SRI is considered to be an agroecologically sound method of rice cultivation, which allows the crop to reach its genetic yield potential. For this system rice yields in the range of 15 to 23 Mg ha^{-1} have been reported by Rafaralahy (2002).

An interesting system of water management in rice fields was proposed recently by Minamikawa and Sakai (2005, 2006). This system is based on direct control of soil redox potential to reduce methane emission from paddy field in Japan. As distinguished from traditional practice, this system depends on the desirable redox conditions.

Mechanism of Gas Exchange

There are two essential mechanisms of gas exchange in the soil medium: (i) mass flow, or advection (convection), and (ii) molecular diffusion. Both these mechanisms of gas movement can occur in soil pores as well as in plant tissues. The relative contribution of both mechanisms of gas exchange, similarly to the relative contribution of the two transport pathways (i.e., soil pathway and plant pathway) can vary in wide limits.

Mass Flow
Principles of Convection

Mass flow is forced by pressure gradients that may appear in soil due to atmospheric pressure fluctuation, diurnal temperature variation, groundwater depth changes, rainwater infiltration, and wind action. Of these drivers only the atmospheric pressure fluctuations may be of importance in the soils with very deep impervious layer (groundwater or permanent rock). If we assume that the diurnal pressure changes may be of order of 3%, then they will affect 3% of the soil profile. If the layer impervious for air is at a depth of 1 m, this will be only 3 cm of the topsoil. For the impervious layer at a depth of 10 m this will be a significant depth of 30 cm where a substantial part of the plant root system is situated. In many cases, such as in loess soils, the groundwater depth can be very deep (sometimes more than 100 m), and under such conditions the gas exchange due to atmospheric pressure fluctuation can be the main factor supplying oxygen to plant roots. According to Currie (1970), Bouyoucos and McCool (1924) placed a barometer in the soil at a depth of 3 m and found that changes of the pressure of the soil air followed those in the atmosphere without any visible delay.

The driving force of the advective gas movement in porous media is the pressure gradient. The flow can be of laminar or turbulent character. Under field conditions we usually find laminar flow when the Reynolds number is below 1 (Currie, 1970). Turbulent flow may, however, occur in landfill cover soils, where pressure gradients are usually higher.

Laminar flow of gases within a porous medium is described by Darcy's equation:

$$\frac{\mathrm{d}V}{\mathrm{d}t} = -A\frac{k}{\eta}\frac{\mathrm{d}p}{\mathrm{d}x} \qquad [1]$$

where $\mathrm{d}V/\mathrm{d}t$ is volumetric rate of gas flow ($m^3\ s^{-1}$), A is cross-sectional area of the porous medium (m^2), η is dynamic viscosity of the gas ($kg\ m^{-1}\ s^{-1}$), $\mathrm{d}p/\mathrm{d}x$ is the pressure gradient ($Pa\ m^{-1} = kg\ m^{-2}\ s^{-2}$), and k is gas permeability (m^2). The gas flow rate is proportional to the total pressure gradient and to the material constant k characterizing the conductive properties of the porous body.

It should be emphasized that advective gas flow is very dependent on the pore size. This is evident from the Hagen–Poisseuille law, which says that the gas flow through a single capillary is proportional to the fourth power of its diameter.

Air Permeability of Soil

Air permeability, k, depends in general terms on the quantity and quality of air-filled pores. The latter term relates to the pore-size distribution, continuity, and tortuosity, which in turn depend on the arrangement of soil particles, soil bulk density, and water content.

According to Gliński and Stepniewski (1985) the values of air permeability in soil are within the range of 0.01 to 500 × 10^{-12} m^2. The air permeability increases with soil water potential and with air-filled porosity of the soil, as illustrated in Fig. 8|2 and 8|3. The threshold value of pore water pressure at which the porous body becomes permeable to air is called the *water entry pressure*. This is connected with the appearance of continuous pores within the soil.

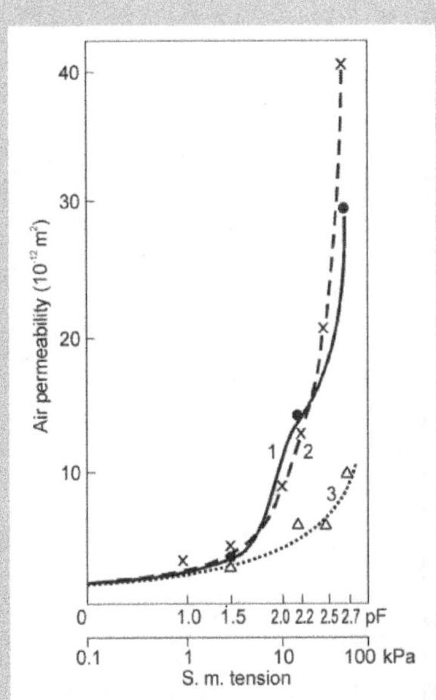

Fig. 8|2. Air permeability of a sandy loam Cambisol of different bulk densities (1–1.09, 2–1.21, and 3–1.41 Mg m⁻³) versus soil moisture tension (data of Turski et al., 1978).

There are several metabolic and morphological mechanisms of plant adaptation to flood conditions. Metabolic adaptations include the possibility to perform for a certain period of time, an anoxic fermentation instead of oxic respiration, lowering of the respiration rate and a reduction of the Pasteur effect, production of alternative product nontoxic for plant tissue (e.g., γ-amino butyric acid instead of ethanol), tolerance to elevated ethanol concentration, mechanism of ethanol removal by transpiration or by secretion through adventitious roots (Gliński and Stepniewski, 1985), and modification of the plant defense system toward higher tolerance to anoxia (Zakrzhevsky et al., 1995; Kalashnikov et al., 1999).

Morphological adaptations include development of a shallow root system, formation of adventitious roots characterized by increased porosity, formation of aerenchyma within roots (Gliński and Stepniewski, 1985), and formation of aerial roots or pneumatophores (e.g., Purnobasuki and Suzuki, 2005).

The plant's ability to transport gases through the pores system of its tissue can be a permanent feature or it may become apparent or enhanced under oxygen deficiency conditions, as presented in Fig. 8|4. There are two types of aerenchyma: schizogenous and lysigenous, with the occurrence of intermediate forms as well (Visser and Voesenek, 2004). *Lysigenous aerenchyma* is most abundant in roots and rhizomes and is induced by ethylene, a plant hormone produced from precursor cyclopropane-1-carboxylic acid (ACC) (Grichko and Glick, 2001; Maricle and Lee, 2002; Shiono et al., 2008). It is formed by the death of some cells leading to the appearance of voids filled with air. Lysigenous aerenchyma formation reduces the volume of respiring tissue, decreasing its oxygen demand.

Schizogenous aerenchyma is formed in an early stage of development through a separation of cells from each other. The presence of air between plant cells in aerenchyma improves the gas transport between the leaves and the tips of the roots (Maricle and Lee, 2002). It was found that aerobic rice genotypes were as flood tolerant as the irrigated lowland genotypes because of their similar abilities to enhance aerenchyma formation under flood conditions (Suralta and Yamauchi, 2008). Aerenchyma removes ethylene from plant roots in waterlogged soils (Shiono et al., 2008), preventing the accumulation of its root growth–inhibiting concentrations. Another type of morphological adaptation is *aerotropism*, the

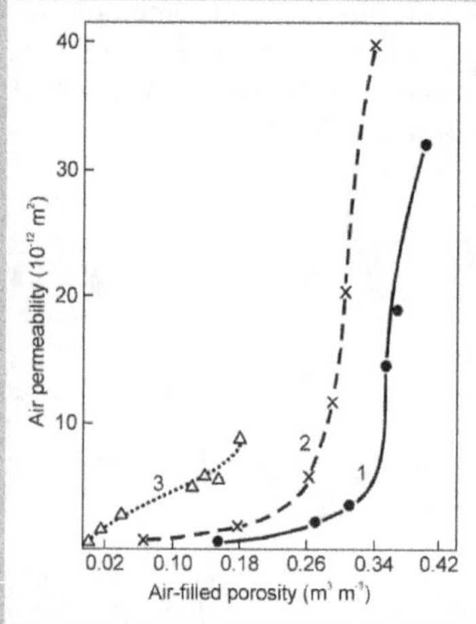

Fig. 8|3. Air permeability of a sandy loam Cambisol of different bulk densities (1–1.09, 2–1.21, and 3–1.41 Mg m^{-3}) versus air-filled porosity of the soil (data of Turski et al., 1978).

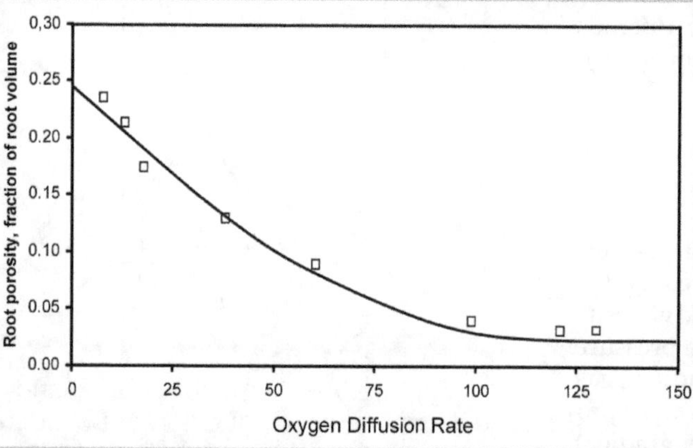

Fig. 8|4. The relationship between oxygen availability in the soil medium as characterized by oxygen diffusion rate (ODR) and the internal porosity of rice roots (modified from Ghildyal, 1982).

growth of roots in the direction of aerated areas of the soil even against geotropism, as observed by Wiersum (1967).

In many situations the internal transport of oxygen via plant roots into the soil is efficient enough to cover not only the oxygen demand of the roots themselves, but also to protect the plant against reducing substances, such as Fe^{2+} and H_2S, which may reach phytotoxic concentrations in waterlogged soils (Jayakumar et al., 2005). The oxidizing power of the root is the highest at the root tip. Flessa and Fischer (1992) observed an increase of redox potential in the rhizosphere of the primary rice root tips from −250 to 100 mV, while the redox potential at a distance of 1 mm from root surface was nearly constant.

There are literature data indicating that the maximum amount of oxygen transported via water, even in presence of wind induced turbulences, does not exceed 3 g O_2 m^{-2} d^{-1} (Kadlec and Knight, 1996), while the transport via plant roots is at least 5 g O_2 m^{-2} d^{-1} (McGechan et al., 2005).

The plant adaptation to waterlogging conditions by increase of internal porosity of root tissues increases both air permeability and the gas diffusion coefficient (see discussion below) of the tissues, but we are not aware of direct measurements of air permeability in plant roots.

It should be emphasized that according to Armstrong et al. (1996) a certain role in the gas exchange via plants is played by pressurized ventilation caused by humidity induced convection in leaves and Venturi convection caused by wind action.

It was reported that adventitious root porosity of dryland grass species ranged from 6 to 9% under drained conditions and increased to around 12% under waterlogged conditions (Smirnoff and Crawford, 1983), and from 1 to 6% under drained conditions to 2 to 18% under waterlogged conditions (Justin and Armstrong, 1987; Rubio et al., 1995).

Przywara and Stepniewski (2000) found that the increase in root porosity of wheat (*Triticum* spp.), peas (*Pisum* spp.), and maize (*Zea mays* L.) due to flooding did not exceed 2.5% of the root volume. Wheat plants at temperatures 15 and 20°C were not able to survive 12 days of flooding and died. The root penetration of the plants from an unflooded control treatment was several times deeper than in the flooded treatment.

Diffusion

The basic mechanism of gas exchange in the soil medium and within the plant tissues is the concentration diffusion, which is induced by appearance of concentration gradients not by external factors.

Usually during consideration of gas diffusion in soil medium a macro- and micro-diffusion are distinguished. Macrodiffusion relates to the entire soil profile disregarding heterogeneity in the micro scale. Microdiffusion, in turn, is related to the gas transfer within soil aggregates and to radial diffusion within the rhizosphere and within the plant roots.

Principles of Gas Diffusion in Porous Media

The diffusive flow f_x of a gas within a porous medium is described by the first Fick's law. It says that the uniaxial diffusion flow (f_x) of any diffusing agent through a unit cross section, in a unit of time, is proportional to the gradient of concentration (dC/dx) constituting the driving force of the flow, and to the diffusion coefficient D characterizing the mobility of the agent in a given medium. Thus, the diffusion coefficient relates to the properties of the diffusion medium. The dimensions of the diffusion coefficient are the square of length per unit of time.

$$f_x = -DdC/dx \qquad [2]$$

The concentration C of any gas in the Fick's equation refers to the unit volume of the soil air. The coefficient D is related to the soil as a whole (and not to the air-filled pores), and the flow f_x is related to a unit surface area of the soil.

When the concentration of the diffusing gas C in Eq. [2], is replaced by its content in a unit volume of the soil G, then we get

$$f_x = -D_G \frac{dG}{dx} \qquad [3]$$

where

$$G = C\left(E_g + \alpha_B \theta\right) \qquad [4]$$

and E_g and θ are the contents of air and water by volume, respectively, and α_B is Bunsen's solubility coefficient.

Coefficient D_G as defined by Eq. [3] is related to coefficient D defined by Eq. [2] as follows:

$$D_G = \frac{D}{E_g + \alpha_B \theta} \qquad [5]$$

In the case of a dry soil or of a soil fully saturated with water, we get

$$D_G = \frac{D}{E_o} \qquad [6]$$

where E_o is total soil porosity and D_G is the coefficient of diffusion in air-filled pores only (i.e., not in the bulk soil).

In the case of gases of low solubility (e.g., oxygen) diffusing through a soil partly saturated with water, we get a simplified approximated formula that can be applied (with the exception of when there is very high moisture content):

$$D_G \approx \frac{D}{E_g} = D^* \qquad [7]$$

where D^*, frequently called the *apparent diffusivity*, has the physical sense of the diffusion coefficient only within the pores filled with air.

Gas concentration in liquid phase C_θ is proportional to its solubility coefficient in water α_w and to its partial pressure P according to Henry's law:

$$C_\theta = \alpha_w P \qquad [8]$$

The gas concentration in air (C) is proportional to its partial pressure as follows:

$$C = \alpha_a P \qquad [9]$$

where α_a means "solubility" in the gaseous phase equal $1/RT$, where R is the universal gas constant and T is absolute temperature.

The first Fick's equation can be thus expressed in the form in which the driving force of the diffusion is the partial pressured gradient:

$$f_x = -K \frac{dP}{dx} \qquad [10]$$

where K is known as the Krogh's diffusion constant, which is related to the diffusion coefficient as follows:

$$K = \alpha_a D \frac{E_g + \alpha_B \theta}{E_o} \qquad [11]$$

The diffusion constant in air, K_o, is, therefore

$$K_o = \alpha_a D_o \qquad [12]$$

The diffusion constant in water, K_w, is equal to

$$K_w = \alpha_w D_w \qquad [13]$$

where D_o and D_w are the diffusion coefficients, and α_a and α_w are the coefficients of solubility of the gas under consideration, in air and water, respectively. Combining Eq. [12] and [13], we get

$$K_w = \frac{D_w \alpha_w}{D_o \alpha_o} K_o \qquad [14]$$

Having considered that

$$\frac{\alpha_w}{\alpha_a} = \alpha_B \qquad [15]$$

We obtain the formula:

$$K_w = \alpha_B \frac{D_w}{D_o} K_o \qquad [16]$$

As can be seen from Eq. [16], at the same gradient of partial pressure, the gas diffusion in water is $\alpha_B D_w / D_o$ times slower than in air. For oxygen it is approximately 300,000 times slower, and in the case of carbon dioxide it is about 10,000 times slower.

The diffusion coefficients depend on temperature and pressure as follows:

$$D_{T2P2} = D_{T1P1} \frac{P_1}{P_2} \left(\frac{T_2}{T_1} \right)^{1.5} \qquad [17]$$

where D_{T1P1} and D_{T2P2} are the diffusion coefficients at the temperature and pressure values indicated by the subscripts.

Applying the conservation law to gas movement in soil it is stated that the change in the quantity of the gas ∂G in a unit soil volume during time ∂t equals the change of the flow ∂f_x on the diffusion path ∂x minus sinks (e.g., in case of O_2) or plus the sources (e.g., in case of CO_2, CH_4, N_2O) of the gas in the soil q, according to the equation:

$$\frac{\partial G}{\partial t} = -\frac{\partial f_x}{\partial x} \pm q \qquad [18]$$

As already mentioned (see Eq. [4]) the amount of gas G in a unit soil volume is equal to the sum of its contents in the liquid and gaseous phases of the soil. Thus combining Eq. [2], [4], and [18], we get

$$\frac{\partial\left[C\left(E_g + \alpha_B\theta\right)\right]}{\partial t} = \frac{\partial\left(D\dfrac{\partial C}{\partial x}\right)}{\partial x} \pm q \qquad [19]$$

In the case of insoluble gases ($\alpha_B = 0$), Eq. [19] simplifies to

$$\frac{\partial\left(CE_g\right)}{\partial t} = \frac{\partial\left(D\dfrac{\partial C}{\partial x}\right)}{\partial x} \pm q \qquad [20]$$

In the soil profile all the parameters under consideration may vary with depth x and time t due to heterogeneity of the profile and changes (e.g., of water content and bulk density) occurring in the profile with time. Having considered that, we get a general diffusion equation as a starting point for consideration of particular situations:

$$\frac{\partial}{\partial t}\left\{C\left[E_g(\theta,t,x) + \alpha_B\theta(t,x)\right]\right\} =$$
$$\frac{\partial}{\partial x}\left[D(x,t,\theta)\frac{\partial C}{\partial x}\right] \pm q(x,t,\theta,C) \qquad [21]$$

If D, E_g, q, and θ are constant, it is possible to apply a simplified equation. If we assume that D (within the range of applicability of Fick's law) is independent on the concentration C, and if we neglect the solubility of the gas (which is possible, e.g., for oxygen, for which at 20°C the value of α_B is 0.033), this equation will be simplified to

$$\frac{\partial C}{\partial t} = \frac{D}{E_g}\frac{\partial^2 C}{\partial x^2} \pm \frac{q}{E_g} \qquad [22]$$

or

$$\frac{\partial C}{\partial t} = D^*\frac{\partial^2 C}{\partial x^2} \pm q^* \qquad [23]$$

where $D^* = D/E_g$ is apparent diffusivity, $q^* = q/E_g$ is apparent respiration. Equation [23] is known as the Fick's second law. For spatial diffusion of oxygen this equation, in any system of coordinates, has the form:

$$\frac{\partial C}{\partial t} = \frac{D}{E_g}\nabla^2 C \pm q^* \qquad [24]$$

where ∇^2 is div grad.

Some specific cases are considered in more detail by Gliński and Stepniewski

(1985) and Stepniewski and Stepniewska (2000).

For steady-state conditions Eq. [22] and [23] assume the form of Poisson's equation for uniaxial diffusion:

$$\frac{\partial^2 C}{\partial x^2} = \pm\frac{q}{D} = \frac{q^*}{D^*} \qquad [25]$$

In the absence of sources or sinks of the diffusing gas (i.e., $q = 0$), Eq. [22] is reduced to the formula:

$$\frac{\partial C}{\partial t} = \frac{D}{E_g}\frac{\partial^2 C}{\partial x^2} \qquad [26]$$

known as Fick's second law, and in the steady-state diffusion ($dC/dt = 0$), to an equation of the Laplace type:

$$\frac{\partial^2 C}{\partial x^2} = 0 \qquad [27]$$

Gas Diffusivity of Soil and of Plant Tissues

The coefficient of gas diffusion in soil depends on the kind of gas, its temperature and pressure, and also on the amount of air-filled pores, their continuity, and shape, which in turn depend on the spatial arrangement of soil particles and on distribution of water. The values of diffusion coefficients and other basic properties of soil gases are presented in Table 8|1.

Usually, the diffusive properties of a soil medium are characterized by means of the relative diffusion coefficient D/D_o, which is the ratio of gas diffusion coefficient in soil D to the diffusion coefficient D_o of the same gas in atmospheric air, under the same pressure and temperature conditions. This is convenient because the value of the relative diffusion coefficient does not depend either on the temperature or pressure nor the kind of diffusing gas.

The effect of soil bulk density and pore water tension on the relative gas diffusion coefficient is presented in Fig. 8|5. The value of D/D_o in soil is usually below 0.2, and it increases with the increase in the soil water tension, and decreases rapidly bulk density of the soil. Unlike the air permeability, the gas diffusion coefficient in a porous medium does not depend on the size of the pores, provided that the pore diameters

Table. 8|1. Densities, solubilities in water, and diffusion coefficients in the atmospheric air D_o of the basic gasses occurring in the soil (after Gliński and Stepniewski, 1985).

Gas	Density†	Solubility in water, α_B		Diffusion coefficient	
		0°C	20°C	In air, D_o†	In water, D_w‡
	kg m⁻³			10^{-5} m^2 s^{-1}	10^{-9} m^2 s^{-1}
N_2	1.251	0.0235		1.81	1.9
O_2	1.429	0.0489	0.0333	1.78	2.5
CO_2	1.977	1.713	0.942	1.39	1.96
CH_4	0.717	0.0556	0.0331	2.168§	
NH_3	0.771	1100.0		1.98	2.0
H_2S	1.539	4.67	2.582		
N_2O	1.978		0.629	1.43	
H_2	0.080	0.0215		6.34	5.85
H_2O vapor	0.768			2.82	
C_2H_4	1.261	0.226	0.122	1.37	
Ar	1.784				

† Normal conditions, pressure 101.3 kPa, temperature 0°C.
‡ Normal pressure, 101.3 kPa, temperature 25°C.
§ From Cowie and Watts (1971) 101.3 kPa, temp. 25°C.

are greater than the mean free path of the molecules of the gas under consideration. (Gliński and Stepniewski, 1985). This limit is reached at pore diameters of 0.10 μm. In the case of soil, pores of that size are emptied of water at soil moisture tension levels >3 MPa (pF > 4.5), that is, at moisture contents lower than the permanent wilting point. Thus, it can be concluded that gas diffusion within macropores (above 30 μm) as well in the mesopores (30–0.2 μm) usually containing plant-available water, does not depend on the pore size.

D/D_o usually shows a curvilinear relationship versus E_g, as shown in Fig. 8|6. It can be described by an empirical power model in the following form:

$$\frac{D}{D_o} = \gamma E_g^{\mu} \qquad [28]$$

where γ and μ are empirical coefficients characterizing the porous material.

We do not know of diffusion coefficient values from direct measurements in the living roots of agricultural plants. The literature data of Sorz and Hietz (2006) are related only to the pieces of wood tissues taken from trunks of forest trees, which may relate to some extent to the orchard tress. The conclusion of these studies, performed at different degrees of water contents in wood, is that oxygen is transported in tree stems mainly with transpiration stream, but in the case of zero sapflow the only source of oxygen for living trunk tissue is O_2 stored or diffusing radially through bark and xylem. Under studies performed by Sorz and Hietz (2006) the stem air-filled porosity of six tree

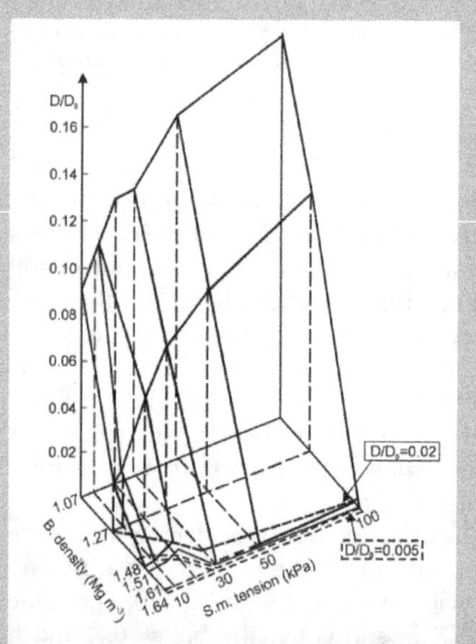

Fig. 8|5. Dependence of relative gas diffusion coefficient in a loamy textured Phaeosem (Kock, Poland) on soil moisture tension and bulk density (modified from Stepniewski, 1981).

species ranged from 0 to 60%. Their oxygen diffusion coefficients (D) in radial direction were between 10^{-11} and 10^{-7} m^2 s^{-1} (D/D_o in the range from 5×10^{-7} to 5×10^{-3}) and were strongly related to gas content in xylem. At axial direction diffusion was one to two orders of magnitude faster. The differences resulted from wood structure and were lowest in coniferous wood (*Picea abies* L.), highest in diffuse-porous (*Fagus silvatica* L. and *Carpinus betulus* L.) wood, and intermediate in ring-porous wood (*Quercus robur* L. and *Fraxinus excelsior* L.).

Diffusion within the Soil Profile

In the simplest case of a homogenous monolayer soil profile under steady-state conditions (i.e., q and D are constant with depth and time) solution of the diffusion equations for oxygen gives the following equation for oxygen distribution (C) with depth x:

$$C = C_o - \frac{q(2Lx - x^2)}{2D} \qquad [29]$$

where C_o represents oxygen concentration at the soil surface, and L is the depth of the biologically active layer (Gliński and Stepniewski, 1985).

In the situation when only a part of the biologically active layer is oxygenated, the depth of the oxic zone L_{ox} is given by the formula:

$$L_{ox} = \sqrt{\frac{2C_oD}{q}} \qquad [30]$$

The values of the oxygenation depth calculated from Eq. [30] for respiration activities normally occurring in soil (0.1–10 mg m^{-3} s^{-1}), and for the diffusion coefficient/constants likely to occur in soil, range from less than one millimeter (for compacted soils saturated with water, characterized by very high respiratory activities) to several meters (for dry loose soils of low respiratory activities) (Gliński and Stepniewski, 1985).

Three curves representing distribution of oxygen concentration in a uniform, monolayer soil profile calculated form Eq. [29] are presented in Fig. 8|7, while the measured distribution of oxygen and carbon dioxide in soil air is shown in Fig. 8|8.

Curve "a" presents a case in when oxygenation depth is higher than the thickness of the biologically active profile; that is, when the entire profile is oxygenated, and

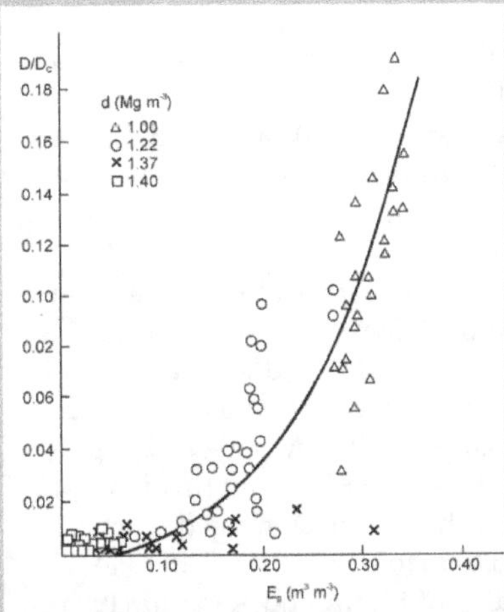

Fig. 8|6. Relationship of D/D_o to air-filled porosity E_g for a silt Fluvisol at different bulk densities (modified from Stepniewski, 1981).

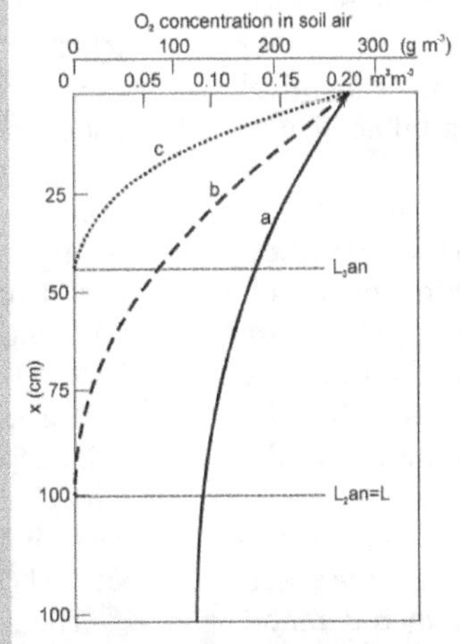

Fig. 8|7. Oxygen distribution in a homogenous soil profile of respiratory activity $q = 0.6$ mg m^{-3} s^{-1} at different oxygen diffusion coefficient D values: (a) 2×10^{-6} m^2 s^{-1}, (b) 1×10^{-6} m^2 s^{-1}, and (c) 0.2×10^{-6} m^2 s^{-1} (after Stepniewski, 1975).

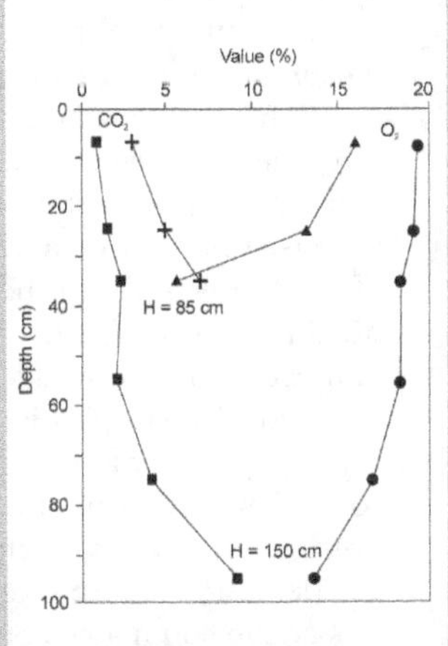

Fig. 8|8. Distribution of oxygen and carbon dioxide concentration in a gley meadow soil (Garbow, near Lublin, Poland) at two levels of groundwater on 19 June 1971 (groundwater level $H = 85$ cm) and the same point on 9 Apr. 1971 ($H = 150$ cm) (unpublished data, Stepniewski).

the oxygen concentration at the bottom of the biologically active zone is above zero.

Curve "b" relates to the situation when oxygen concentration at the bottom of the biologically active zone drops to zero. This means that the diffusion coefficient reaches its critical value.

Curve "c" shows a case when only the upper part of the biologically active soil profile is oxygenated because the oxygen diffusion coefficient is below its critical value.

It is necessary to emphasize that soil air composition in the intra-aggregate pores may differ from that in the interaggregate pores because of soil heterogeneity. The intra-aggregate pores contain less oxygen and more carbon dioxide than the inter-aggregate pores (e.g., Zausig et al., 1993; Højberg et al., 1994; Horn, 1994; Horn and Smucker, 2005).

Gas Fluxes between Soil and Atmosphere
Oxygen

Uptake

Soil respiration is influenced by environmental conditions, land use, and management practices. Oxygen is taken up by the soil due to respiration of microorganisms and plant roots. The respiration of mesofauna and the contribution of chemical reactions are considered negligible. Therefore, the term *soil oxygen demand* is used as synonymous with the soil respiration rate. Because under oxic conditions consumption of oxygen is equal to the volume of carbon dioxide produced, the soil respiration rate is most often measured by carbon dioxide evolution. The respiration of roots is usually higher by two orders of magnitude than the soil microbial respiration. Thus, the roots present themselves as threads of high oxygen demand on the background of the much less active soil matrix. Soil oxygen demand decreases with depth because both microbial respiration and root density (and thus root respiration) decrease with depth.

Respiration of soil microorganisms depends on the availability of oxygen, organic carbon, and nutrients, as well on temperature and water content. The temperature interval for microbial growth ranges from −12 to 110°C. Microorganisms with respect to their temperature requirements are divided into *psychrophilic* (maximum temperature for growth is <20°C), *mesophilic* (minimum temperature ~15°C, maximum <40°C), and *termophilic* (minimum temperature >40°C, maximum <70°C) organisms (Paul and Clark, 1998). A new term *hyper-termophiles* was introduced for extreme thermopiles growing in the temperature range from 60 to 110°C. Soil contains a mixture of different groups of microorganisms, and the overall microbial respiration increases two to three times with an increase of temperature by 10°C until the temperatures of break of metabolism (~60–70°C).

With respect to oxygen requirements, soil microorganisms are divided into *obligate oxic microorganisms* (traditionally called *aerobes*), *facultative anoxic organisms* (*facultative anaerobes*), and *obligate anoxic organisms* (*obligate anaerobes*). A special group of microorganisms is *microaerophiles*, which are characterized by oxygen optimum on the order of 2 to 10% by volume (Black, 1996). *Oxic microorganisms* utilize oxygen as a terminal electron acceptor from the cytochrome oxidase. The uptake of oxygen q by a single cell of an oxic microorganism is described by the Michaelis–Menten equation:

$$q = q_{max}C/(C + K_M) \qquad [31]$$

where C is the oxygen concentration, K_M is the Michaelis constant defined as the oxygen concentration when $q = 0.5q_{max}$, and q_{max} is the maximum respiration when the respiration process is not limited by the availability of oxygen. The first measurements of K_M performed for several strains of bacteria by Longsmuir (1954) gave values from 1.1 10^{-8} M to 3.57 × 10^{-6} M, and similar values (2.7 to 4.4 × 10^{-6} M) were obtained with four soil suspensions by Greenwood (1962). Also, Kroeckel and Stolp (1985) found that in mull rendzina soil the K_M for O_2 ranged from 3.0 ×10^{-6} M (2.55 hPa) to 16 × 10^{-6} M (13.3 hPa).

The studies of Sierra and Renault (1995) showed that Michaelis–Menten kinetics, including competitive inhibition by carbon dioxide, adequately describe the influence of oxygen and carbon dioxide on oxygen uptake in artificial aggregates. The K_M values for O_2 uptake were 1.69% O_2 (17 hPa) for artificial aggregates and 5.80% 0_2 (58hPa) for natural aggregates according to Sierra and Renault (1995).

Soil microbial activity depends both on oxygen and carbon dioxide concentration, independently. Carbon dioxide is a carbon source for autotrophic microflora. Dommergues and Mangenot (1970), quoted after Sierra and Renault (1995), reported 2 to 14% CO_2 as the optimal range for the growth of autotrophic bacteria. For other microorganisms, carbon dioxide as a respiration product is the inhibitor of respiration and growth and can completely inhibit oxygen consumption (Sierra and Renault, 1995).

The dependence of microbial respiration on water content shows a maximum corresponding to the range of optimal availability of oxygen and water. Initial increase of water content in soil is followed by an increase of oxygen uptake due to increase of the availability of water. Reaching the maximum respiratory activity a subsequent decrease of oxygen consumption occurs caused by limitation of oxygen transport within the soil medium. According to water requirements, soil microorganisms are divided (Gliński and Stepniewski, 1985) into *hydrophilic* (disappearance of activity at pore water pressure greater than –7.1 MPa), *mesophilic* (disappearance of activity within the pore water pressure interval from –7.1 to –30 MPa), and *xerophilic* (disappearance of activity at pore water pressures less than –30 MPa). The shape of the curve of oxygen uptake versus soil water content is related to the dependence of respiration on oxygen content or, more precisely, on its availability. Under field conditions the rate of soil microbial respiration is usually within the range 0.1 to 10 mg m^{-3} s^{-1} (Gliński and Stepniewski, 1985).

The rate of root respiration depends on internal factors of the plant itself, such as type of plant and development stage, root dimensions, and physiological age of the root tissue, as well as on external factors, such as temperature and the availability of soil water and nutrients. It should be emphasized that soil microbial activity increases in the presence of plant roots because root exudates are a source of easily available carbon for microorganisms. As already mentioned, respiration rate of the root tissues is about two orders of magnitude that of the soil itself and ranges from 10 to 50 mg m^{-3} s^{-1} (Gliński and Stepniewski, 1995).

Fluxes

Oxygen uptake by a cropped field is the sum of microbial and root respiration, and under field conditions is on the order of several tons of oxygen per hectare and year (Gliński and Stepniewski, 1985). Soil respiration under field conditions (Fig. 8|9) shows a maximum value in the afternoon and a minimum in the early morning. In the moderate climate zone the annual dynamics of oxygen uptake under field conditions are characterized by a summer maximum, as shown in Fig. 8|10. It should be emphasized that the presence of plant canopy may increase the uptake of oxygen by more than two times.

Carbon dioxide
Generation

As was mentioned above, the production of carbon dioxide under aerobic conditions is complementary to oxygen uptake, and all

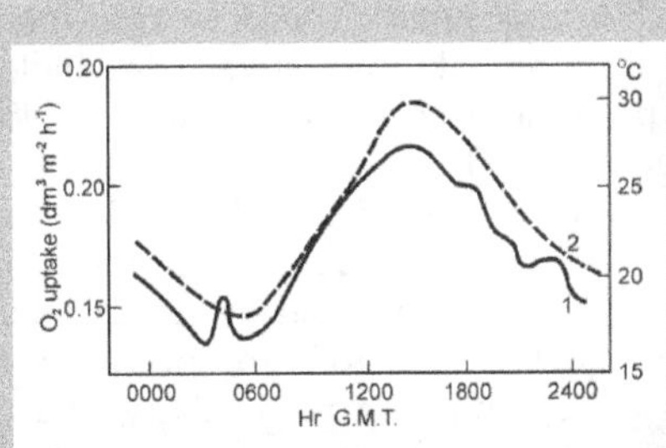

Fig. 8|9. Diurnal dynamics of soil respiration rate as measured by oxygen uptake rate (modified from Currie, 1975).

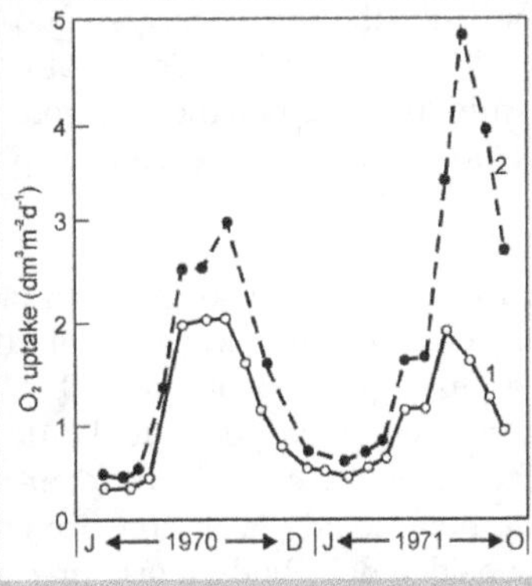

Fig. 8|10. Annual dynamics of soil respiration as measured by oxygen uptake rate in a uncropped (1) and cropped (2) soil under moderate climate conditions of England (modified from Currie, 1975).

the above listed factors affecting oxygen uptake also influence the emission of carbon dioxide. The generation of carbon dioxide of microbial origin is influenced directly by the availability of oxygen and organic carbon, water content, temperature, and the microbial population and its activity. The production of carbon dioxide by plant roots depends directly on root mass and its distribution and activity.

Due to the complementary character of oxygen uptake and carbon dioxide production the sum of oxygen and carbon dioxide concentrations in soil air is usually about 20% by volume. This also means that the oxygen and carbon dioxide fluxes between soil and the atmosphere are approximately equal. An example of changes in oxygen and carbon dioxide concentration in soil air with depth is shown in Fig. 8|8. As the groundwater table becomes shallower, an increase in the concentration gradients for both gases is observed.

The composition of soil air is a result of the dynamic equilibrium between the biological processes of generation and absorption of gases and the physical processes of their migration. It should be emphasized that the composition of soil air is a key factor for the microbial and plant metabolism, for oxidation status of numerous soil nutrients, for soil utility and crop productivity, and for the emission of greenhouse gases (Gliński and Stepniewski, 1985).

Under anoxic conditions carbon dioxide can be produced without oxygen consumption so the respiratory quotient increases above one. Then the sum of oxygen and carbon dioxide concentrations is above 20%, and the emission of carbon dioxide from the soil exceeds the volume of oxygen taken up.

Emission

The global CO_2 flux from soils is estimated at 64 to 72 Gt C yr^{-1} and accounts for 20 to 38% of annual global input of CO_2–C to the atmosphere (Baggs, 2006). It should be emphasized that the effect of different factors on carbon dioxide evolution from soil can be considered in both short time and long time periods. Evolution of carbon dioxide is directly interconnected with carbon sequestration in soil. Decrease of carbon content in soil in the long term denotes, beyond any doubt, elevated emission of carbon dioxide to the atmosphere.

It was found that the sequestration of carbon in agricultural ecosystems can be increased by reduced or no tillage (e.g., Alvaro-Fuentes et al., 2007; Chatskikh and Olesen, 2007), by enhancing crop rotation complexity, multiple cropping and agroforestry, use of cover crops year-round, selecting improved crops or cultivars, and planting deep-rooting crops or varieties (Lal and Kimble, 1997; West and Post, 2002).

The instantaneous emission of carbon dioxide from soil depends on numerous interrelated factors affecting directly or indirectly either its generation and/or its migration rate. Indirect effects originate from such factors as nutrient availability, climate, and physical modification of soil by tillage and land use practices.

The migration rate of carbon dioxide is affected directly by temperature, the air-filled pore volume, its tortuosity and continuity, and by gradients of its total and partial pressure. Indirect effects comprise all the factors modifying them, such as water content, land use, and tillage practice.

Soils in different land use are characterized by various annual CO_2 emissions but similar (as in the case of oxygen uptake) diurnal and seasonal patterns (Cao et al., 2004). The diurnal CO_2 efflux often fluctuates, as the result of fluctuations of soil temperature, between morning minimum at about 0500 h to noon/afternoon maximum occurring at about 1100 to 1400 h (Shi et al., 2006). The plant phenology is another important factor affecting carbon dioxide emissions. Shi et al. (2006) reported maximum CO_2 efflux when the leaf area index (LAI) and live root biomass (LRB)—but not temperature—were the highest.

Land Use

Land use is a key factor influencing the carbon cycle. The northern peatlands are a net sink for carbon dioxide (Vasander, 1996) and also the large area of organic carbon accumulation. On pristine mires the summer CO_2 efflux was estimated at between 100 and 500 mg m^{-2} h^{-1}.

In the northern peatlands the increased carbon dioxide emission occurs in fairly dry sites and when shrubs are present (Vasander, 1996). Depending on the vegetation structure and plant community the contribution of root respiration to total CO_2 efflux ranged from 10 to 40%.

Agriculture development and peat usage in numerous branches of industry have resulted in the drainage of peatland areas. After drainage the aerobic process of mineralization of the organic matter accumulated during long time periods starts, and CO_2 emission prevails. The enlargement of the oxic zone in peat additionally increases CO_2 emission due to the enhancement of microbial oxidation of methane generated in the deeper anaerobic zone. In Finland, 10% of the total peatland area (about 1 million hectares) has been drained for agricultural use, and 300,000 ha are still being cultivated (Maljanen et al., 2004). These drained organic soils constitute about 15% of the total arable land area in Finland, and they are responsible for 8% of the anthropogenic CO_2 emissions in this country (Kasimir-Klemendtsson et al. (1997) after Maljanen et al., 2004).

Conversion of native vegetation to cultivated cropland and conventional tillage practice reduces the soil organic carbon content and increases CO_2 efflux. Campos (2006) confirmed this trend measuring CO_2 emissions from soil surface in tropical cloud forest in Mexico and in areas where the native forest was transformed into arable plots and grassland.

Liu et al. (2008), during 2 years of continuous field experiments in South China (Fig. 8|11), found the dependence of CO_2 emissions on soil temperature and soil moisture. Annual mean CO_2 flux from soil of longan orchard was less than from soil of pine plantation (4.70 and 14.72 Mg CO_2–C ha^{-1} y^{-1}, respectively). Removal of surface litter reduced soil CO_2 fluxes by 17 to 25%.

Also, Frank et al. (2006) suggested that land use influences the CO_2 fluxes due to modification of respiration rate. They reported that higher CO_2 efflux from grassland (2.8 g CO_2–C m^{-2} d^{-1}) compared to continuous wheat (1.6 g CO_2–C m^{-2} d^{-1}) and wheat–fallow plots (1.9 g CO_2–C m^{-2} d^{-1}) was a consequence of higher root biomass, soil organic carbon, and microbial biomass carbon in these treatments.

They observed significant differences in the root biomass to the 0.3-m depth under undisturbed grassland (12.3 Mg ha^{-1}), continuous wheat (1.3 Mg ha^{-1}), and wheat–fallow (0.3 Mg ha^{-1}). Soil organic matter in grassland (84 Mg ha^{-1}) was 1.2 and 1.3 times greater than in continuous wheat and wheat–fallow rotation, respectively. Differentiation in microbial biomass carbon in these treatments was more significant, and the value for grassland (2.2 Mg ha^{-1}) was 3.6 times greater than for wheat and as much as 7.2 times greater for wheat–fallow plots.

Tillage

Tillage practice is an important factor influencing carbon turnover between soil and atmosphere. Tillage practice modifies microbial respiration by incorporating and mixing fresh plant residue with soil; by changing soil, moisture, temperature, and aeration; and by promoting macroaggregate carbon turnover by exposing soil organic matter to microorganisms (Alvaro-Fuentes et al., 2007). Jinbo et al. (2008) reported different short-term dynamics of carbon after tillage of intact wetlands and marsh soil cultivated for 10 yr in northeast China. Fifteen minutes after tillage, the intact wetlands released four times more CO_2 than the no-till control, while the cultivated soil emitted only 2.5 times more CO_2 than the control.

Gregorich et al. (2006) stated that differences in gas fluxes between tilled and no-till soils can be related to physical rather than biochemical aspects. Their laboratory and

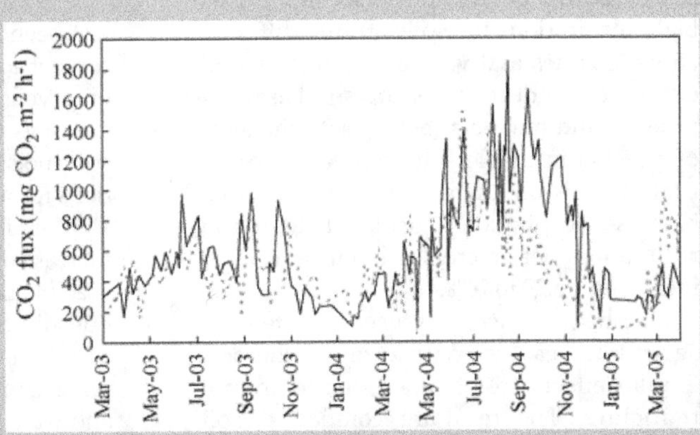

Fig. 8|11. Seasonal patterns of CO_2 fluxes measured at longan orchard in South China (Liu et al., 2008). The flux from soil without surface litter was marked by dashed line.

fields studies indicated the relative accumulation of CO_2 in soil profile in no-tilled soil and rapid gas diffusion through soil profile and efflux from soil surface in tilled soil. This explains immediate effect of high CO_2 emission after tillage.

Reduced tillage (disc harrow, chisel plow) alleviates the mechanical disturbance caused by the moldboard plow. These conservation tillage tools reduce the amount of carbon dioxide released from soil into the atmosphere immediately after tillage. Carbon dioxide efflux from soil 5 h after conservation tillage of a clay loam soil was 31% of the amount emitted from moldboard-plowed soil (Reicosky et al., 1997). Similarly Bauer et al. (2006) observed higher CO_2 flux rates from loamy sand of the southwestern USA in conventional tillage compared to a conservation tillage treatment.

Plowing depth influences the CO_2 release, especially immediately following tillage. Reicosky and Archer (2007) reported CO_2 emissions from soil with plowed depths 0.102, 0.152, 0.203, and 0.280 m being 3.8, 6.7, 8.2, and 10.3 times larger than from no-till soil, respectively. More evident differences in CO_2 emissions were reported by Gesch et al. (2007) from soil 24 h after different tillage practices. The deepest plow (300-mm depth) induced a 33-fold increase of CO_2 efflux, while intermediate tillage-disk harrow to the 78- to 145-mm depth resulted in only 2.3-fold more CO_2 emission than no-till treatment.

In the long term agricultural practices may reduce the CO_2 fluxes and change microbial activities. Ananyeva et al. (2008) compared the basal and specific soil microbial respiration in soils from different climatic zones and various treatments and observed the decrease of top soil basal respiration and microbial biomass (to the level of 52–74%) in arable soils compared to natural ones, while the specific soil microbial biomass respiration calculated as basal respiration per microbial biomass carbon increased by 29 to 74%.

The heavy machinery used in intensive agriculture causes compaction of the soil. Similar effects are obtained due to animal trampling. The resultant compacted soil layer reduces oxygen concentration in soil, retarding root growth, and consequently reduces crops yields. Huang et al. (2006) simulated the effect of a compacted layer on soil O_2 diffusion within and below it. They

reported that the compact layer is the obstacle to water infiltration down the soil profile during rainfall. Within the dense layer rainwater is stopped and forces the air out of soil pores, resulting in oxygen depletion. In the presence of a compacted layer the oxygen removal process by water is faster than in noncompacted soil.

The increase of soil bulk density results in CO_2 efflux reduction. Pengthamkeerati et al. (2005) reported 72% less CO_2 released from compacted soil in a laboratory study and 46% in a field study. The tillage practice can remediate compacted topsoil, while the improving of subsoil properties may be insufficient. The use of heavy wheel traffic, especially under wet conditions, increases the risk of irreversible damage of the subsoil structure (Lebert et al., 2007).

Other Factors

Frank et al. (2002) showed that soil temperature accounted for 65% of CO_2 flux variability from grassland, which is similar to findings of other studies—46% of CO_2 flux variability in tallgrass temperate prairie observed by Mielnick and Dugas (2000), 26 to 59% in subtropical red soil in China (Iqbal et al., 2008), and 40 to 75% of temporal variation of CO_2 emissions from grazing alpine meadow of Tibetan plateau (Lin et al., 2009). Lou et al. (2004) reported that the seasonal variability of CO_2 efflux followed the soil temperature by as much as 85.0 to 88.5%, while carbon dioxide variability followed dissolved organic carbon in 55.8 to 84.4% and rainfall only in 43.0 to 55.8%.

The variability of carbon dioxide emission has been found to be affected by soil water content to a lesser extent. Ruser et al. (2006) observed similar CO_2 emissions at different soil moisture contents, concluding that a wide soil moisture range provided proper conditions for heterotrophic activity in that soil.

Mielnick and Dugas (2000) observed the increase of annual CO_2 emissions with annual precipitation and estimated the quadratic relationship between CO_2 flux and soil moisture as 26% of the variability, but Iqbal et al. (2008) reported that interaction of soil temperature and water-filled pore space could account for 31 to 60% of soil CO_2 flux variations in subtropical red soils under various crops.

Respiration of grassland ecosystems is highly dependent on intensity of grazing (Frank et al., 2002; Cao et al., 2004). Cao et al. (2004) observed that heavy grazing (5.35 sheep ha^{-1}), leading to higher compaction of the soil, caused 25% decrease of CO_2 flux from grassland compared to light grazing (2.55 sheep ha^{-1}) site. The grazing animals affected the nutrient cycles and gas emissions from grassland ecosystems by excreta. Dung and urine patches are the "hot-spots" of the greenhouse gas emissions. Lin et al. (2009) estimated that cumulative CO_2 emission from yak dung patches on Tibetan plateau was about 36 to 50% higher than control plots during a 38- to 46-d experimental period.

Jassal et al. (2005) indicated that 75% of CO_2 efflux from a fir forest soil on Vancouver Island (Canada) originated in the top 20 cm of soil. Moncrieff and Fang (1999) confirmed that most CO_2 emitted from the soil was produced in the top 15 cm of soil. In their studies 53% of total carbon dioxide emitted from the soil surface originated from respiration of living roots and 47% from decomposition of organic matter. Table 8|2 presents the carbon dioxide emission data under different soil and climate conditions from various studies.

Table. 8|2. Carbon dioxide emissions from different arable soils.

Soil	CO_2 emission	Reference
Fine loam, southern Germany, potato plot, fertilized with 150kg N ha^{-1} (2 mo)		Ruser et al. (2006)
• ridge (D = 1.03 Mg m^{-3}, 40–90% water-filled pore space)	11.57–13.77 g CO_2–C m^{-2}	
• interrow area (1.24 Mg m^{-3}, 40–90% water-filled pore space)	10.78–8.29 g CO_2–C m^{-2}	
• tractor compacted interrow area (1.64 Mg m^{-3}, 60–98% water-filled pore space)	12.28–8.96 g CO_2–C m^{-2}	
Mexico silt loam, north-central Missouri, USA		Pengthamkeerati et al. (2005)
• unamended soil	2.51–7.42 mg CO_2–C kg^{-1} d^{-1}	
• litter-amended soil	5.42–12.50 mg CO_2–C kg^{-1} d^{-1}	
• uncompacted soil (1.27 Mg m^{-3}) field study	11.47 μmol CO_2–C m^{-2} s^{-1}	
• compacted soil (1.4 Mg m^{-3}) field study	6.12 μmol CO_2–C m^{-2} s^{-1}	
Mature slash pine plantation estimation	14 Mg C ha^{-1} yr^{-1}	Moncrieff and Fang (1999)
• site No. 1 hourly measurement	0.205 mg CO_2 m^{-2} s^{-1}	
• site No. 2 hourly measurement	0.095 mg CO_2 m^{-2} s^{-1}	
Oxisol developed on sandstone, South China		Liu et al. (2008)
• pine plantation (20 yr old)	4.70 Mg CO_2–C ha^{-1} yr^{-1}	
• longan orchard (12 yr old)	14.72 Mg CO_2–C ha^{-1} yr^{-1}	
Mat cry-gelic Cambisol, alpine meadow, Tibetan plateau, China, 3240 m above sea level		Cao et al. (2004)
• light grazing (2.55 sheep ha^{-1})	2040 g CO_2 m^{-2} yr^{-1}	
• heavy grazing (5.35 sheep ha^{-1})	1530 g CO_2 m^{-2} yr^{-1}	
Northern semiarid grassland, USA		Frank et al. (2002)
• dormant period (Oct.–Apr.) grazed prairie, 180 d	86 g CO_2–C m^{-2}	
• growing period (Apr.–Oct.) grazed prairie, 185 d	728 g CO_2–C m^{-2}	
Tallgrass prairie, Texas, USA, 1993–1998 clay	1.0–2.1 kg CO_2–C m^{-2} yr^{-1}	Mielnick and Dugas (2000)
Corn and winter wheat cropped Linwood muck, Ohio, USA		Elder and Lal (2008)
• moldboard/disking tilled	22.5 Mg CO_2–C ha^{-1} yr^{-1}	
• bare treated	20.8 Mg CO_2–C ha^{-1} yr^{-1}	
• no-till	18.9 Mg CO_2–C ha^{-1} yr^{-1}	
Arable peat soil, Kannus, western Finland		Maljanen et al. (2004)
• barley cropped	830 g CO_2–C m^{-2} yr^{-1}	
• grass cropped	395 g CO_2–C m^{-2} yr^{-1}	
Cropland, Tibetan Plateau, China	579 ± 13 g C m^{-2} yr^{-1}	Shi et al. (2006)
Subtropical arable red soil, China		Lou et al. (2004)
• sandy loam derived from tertiary red sandstone	2.13 kg CO_2 m^{-2} yr^{-1}	
• clay, derived from quaternary red clay	1.41 kg CO_2 m^{-2} yr^{-1}	
• sand, derived from granite	2.84 kg CO_2 m^{-2} yr^{-1}	
Subtropical arable red soil, China	(±SD)	Iqbal et al. (2008)
• paddy	901 ± 114 g CO_2 m^{-2} yr^{-1}	
• orchard	727 ± 55 g CO_2 m^{-2} yr^{-1}	
• upland	554 ± 22 g CO_2 m^{-2} yr^{-1}	
• woodland	533 ± 27 g CO_2 m^{-2} yr^{-1}	

Nitrous Oxide

Generation and Absorption

Under certain conditions in soil, gaseous dinitrogen as well as nitrogen oxides such as NO and N_2O are formed. Dinitrogen is paid less attention because it is ubiquitous and inert from the environmental point of view. Lower interest is also observed in case of nitric oxide (NO). The most attention is paid in the literature to N_2O being a strong greenhouse gas, which is produced due to nitrification as well as denitrification. Both nitrous oxide and nitric oxide can be generated or absorbed in soil depending on its redox conditions. Under oxic conditions they are oxidized by the soil, while under anoxic conditions they will be fixed due to reduction. It should be stressed that most N_2O emission at the global scale comes from soils (e.g., Beauchamp, 1997; Mosier et al., 1998; IPCC, 2001; Scott et al., 2002; Kavdir et al., 2008).

As was already mentioned, generation of N_2O proceeds due to both denitrification, taking place under anoxic conditions, and nitrification, occurring under oxic conditions (Gliński and Stępniewski, 1985). Bateman and Baggs (2005) found that at low and moderate soil water content (35–60% of water-filled pore space) N_2O was produced due to nitrification, while in soil containing more than 70% water-filled pore space N_2O is generated during denitrification.

Russow et al. (2008) reported that an average of 76 and 54% of N_2O was emitted due to denitrification from loess black-earth soils with total carbon content 2.2 and 1.8%, respectively. The emission of NO during denitrification contributed only an average of 12 to 17% of the total released NO. Russow et al. (2009) studied the origin of NO and N_2O under different aeration conditions in the black-earth soil using ^{15}N tracer techniques. They observed that at an oxygen concentration of 20%, 88% of the formed nitrite originated from nitrification and 12% was produced due to denitrification. The nitrite production via nitrification decreased quickly to zero due to oxygen partial pressure decline. Under strict anoxic conditions 100% of nitrite was generated by nitrate reduction. These authors stated that nitrite was almost always the direct precursor of NO formation under both aerobic and anoxic conditions. Under oxygen levels of 0 to 0.2% almost 100% of emitted N_2O originated from NO.

The soil water content is considered to be the key factor controlling N_2O emission (Choudhary et al., 2001) The content of water influences the oxygen supply and redox conditions, affecting the direction of N transformation. Weitz et al. (2001) agreed with this thesis, but they observed the highest N_2O emissions in wet soils shortly after fertilization and for up to 6 wk as "post fertilization flux," while fluxes from relatively dry soils increased only after rainfalls. During the periods of low N availability the N_2O emissions were weakly influenced by soil moisture.

Kroeckel and Stolp (1985) found that N_2O was produced due to denitrification in mull rendzina soil only at oxygen partial pressure, pO_2, below 40 hPa (47×10^{-6} M). They observed that the production of N_2O and consumption of O_2 occurred simultaneously with half maximum rates at pO_2 of 6.7 to 13.3 hPa. The highest denitrification and respiration rates obtained in remoistened, glucose- and nitrate-amended soil were 43 mL N_2O h^{-1} kg^{-1} soil and 130 mL O_2 h^{-1} kg^{-1} soil, respectively. Nitrite accumulated in soil below 40 hPa and increased with decreasing pO_2.

Between soil redox potential and the equilibrium N_2O concentration there is a tiny equilibrium (Fig. 8|12). As has been observed, N_2O can exist in equilibrium with soil only within a very narrow range of redox potentials (Yu et al., 2001; Wlodarczyk et al., 2005). Due to soil heterogeneity, N_2O can be produced within soil aggregates and simultaneously it can be absorbed and oxidized at the aggregate surfaces (Højberg et al., 1994). Variability of soil conditions can cause N_2O fluxes between particular soil layers. It can be produced at a certain soil depth and then can be reduced in the deeper horizons, or it can be oxidized in upper layers during its migration to the atmosphere. Closer to the soil surface it can be produced again due to nitrification. Thus, the net emission or absorption of nitrous oxide at the soil surface depends on the situation within the soil profile.

Denitrification is an important pathway of nitrogen loss from agricultural soils. Depending on N input and the farming system, the nitrogen loss via denitrification is 50 to more than 300 kg N ha^{-1} yr^{-1} (Aarts et al., 1999), quoted after van der Salm et al., 2007). Van der Salm et al. (2007) estimated that 25% of the N applied to the field as fertilizer was lost to the environment, and the denitrification can be responsible for 90% of this loss. The 75% of this denitrification process was placed in the upper 20 cm of the clay soil. In sandy soil the gaseous losses of nitrogen, applied as urea fertilizer, was calculated as 10% (Sánchez et al., 2001).

Emission

Wetlands are an important source of N_2O (Vasander, 1996). As was mentioned, 10% of the total peatland area in Finland (about 1 million hectares) has been drained for agricultural use and 300,000 ha are still being cultivated (Maljanen et al., 2004). These drained organic soils (~15% of the total arable land area in Finland) are responsible for 25% of the total anthropogenic N_2O emissions in this country (Kasimir-Klemendtsson et al., 1997, quoted after Maljanen et al., 2004).

Regina et al. (2004) measured emission of N_2O from a peat field in Northern and South Finland under different plant covers. The annual fluxes of N_2O in the north were 4.0, 4.4, and 13 kg N_2O–N ha^{-1} from the plots of grass, fallow, and barley (*Hordeum vulgare* L.), respectively. Emissions of N_2O from the south peat field were higher and reached up to 7.3, 15, 10, and 25 kg N_2O–N ha^{-1} yr^{-1} for grass, barley, potato (*Solanum tuberosum* L.), and fallow plots, respectively. Another study performed under the conditions of Finland by Maljanen et al. (2003) reported that the annual N_2O–N emissions from cultivated drained organic soil under barley and grass ranged from 8.3 to 11 kg N_2O–N ha^{-1} and were higher than those from the same soil kept bare (6.7–7.1 kg N_2O–N ha^{-1}). The same authors found that the emissions from the adjacent forest site with 22 yr birch (*Betula* spp.) was much lower and equaled 4.2 kg N_2O–N ha^{-1} yr^{-1}. Similar studies performed in the Netherlands by Langeveld et al. (1997) reported the emission of N_2O from a pasture on a drained peat soil in the range 14 to 61 kg N_2O–N ha^{-1} yr^{-1}.

According to Menneer et al. (2005) even a single treading event of 4.5 cows per 100 m^2 for 1.5 or 2.5 h caused a decrease of soil bulk density from 0.77 to 0.83 to 0.85 Mg m^{-3} and a significant short-term (21 d) increase in denitrification and N_2O emission from 2.3 to 52 g N_2O–N ha^{-1} d^{-1}.

Animal excreta deposited on grasslands from grazing animals are an important source of nitrogen returned into the soil and released into the atmosphere as N_2O. Lin et al. (2009) reported the cumulative N_2O emissions from both urine and dung patches in 2005 were 2.1 to 3.7 times greater, and in 2006 they were 1.8–3.3 times greater, than those for control plots during a 38- to 46-d experimental period. The temporal N_2O flux variations were the result of changes in both temperature and soil water-filled pore space (water-filled pore space), and the flux was 34% in urine patches, 48% in dung patches, and 56% in control plots (Lin et al., 2009). The values reported by Watanabe et al. (1997) for a grassland soil after surface application of cattle and swine excreta ranged from 0.1 to 1.0 mg N_2O–N m^{-2} h^{-1}.

Liu et al. (2008) measured the long-term tendencies of nitrous oxide emission from an Oxisol under two different land-use types in a subtropical region of South China

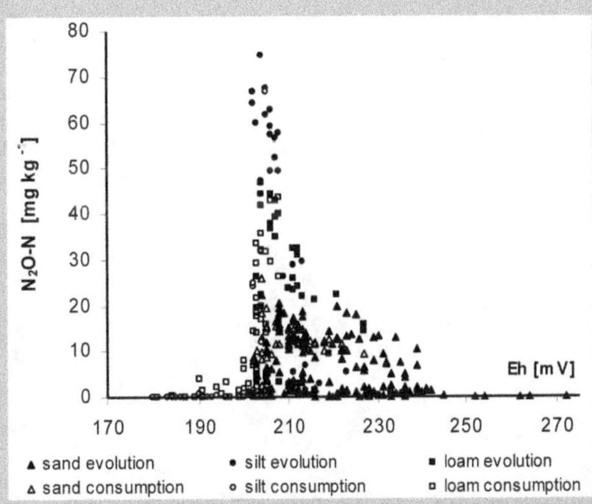

Fig. 8|12. Dependence of nitrous oxide concentration in the headspace air on redox potential in soil suspension. The results were obtained during incubation of 13 different soils (6 sandy, 3 silty, and 4 loamy). The right-hand side of the figure corresponds to the nitrous oxide production phase during first days of incubation and the left-hand side corresponds to the reduction phase (i.e., after maximum of N_2O concentration) (after Wlodarczyk et al., 2005).

(Fig. 8|13.). Annual mean N_2O flux from a 12-yr-old longan (*Domocarpus longan* Lour.) orchard was lower (3.03 kg N_2O–N ha^{-1} yr^{-1}) than from the same soil under a 20-yr-old pine (*Pinus massoniana* Lamb.) plantation (8.64 kg N_2O–N ha^{-1} yr^{-1}). The N_2O emission was influenced by soil temperature and moisture, but not so evidently as CO_2 emission. Removal of surface litter reduced soil N_2O fluxes by 31 to 34%.

Chen et al. (1997) measured emission of N_2O from fertilized fields (meadow brown soil) under cultivation of soybean [*Glycine max* (L.) Merr.], rice, and maize in northeast China. The emission from the rice field was 1.70 kg N_2O–N ha^{-1} yr^{-1}, and the emissions from soybean and maize fields were 3.12 kg N_2O–N ha^{-1} yr^{-1} and 7.10 kg N_2O–N ha^{-1} yr^{-1}, respectively. Authors of another study performed on paddy fields in China (Hua et al., 1997) reported that N_2O emission usually ranged from 0.005 to 0.5 mg N_2O–N m^{-2} h^{-1}. Hou et al. (2000) suggested that maintaining redox potential in soil in the range of –100 to 200 mV may be sufficient to minimize the emissions of nitrous oxide from paddy fields.

Long-term tillage and monoculture of crops with high demand for biogenic compounds influences the fertility of soil and microbial processes. Choudhary et al. (2001) observed less N_2O emissions from 34-yr maize-cropped soil than from similarly cropped soil for 17 yr and from permanent pasture and attributed this difference to organic carbon and total nitrogen depletion (Fig. 8|14).

The resultant emission of nitrous oxide depends on soil management practices through the modification of the distribution of sinks and sources of nitrous oxide, as well as by modification of gas movement. In their review of emission of greenhouse gases from soils under different land use and agricultural practices, Hatano and Lipiec (2004) reported that daily emissions range from <1 g to >80 g ha^{-1} but usually do not exceed 30 g ha^{-1}. Field measurements indicate that emissions of N_2O from grasslands are higher than those from cropped fields, forests, and woodlands, and that a large proportion of N_2O is emitted from agricultural soils in winter. According to estimation of Oenema et al.

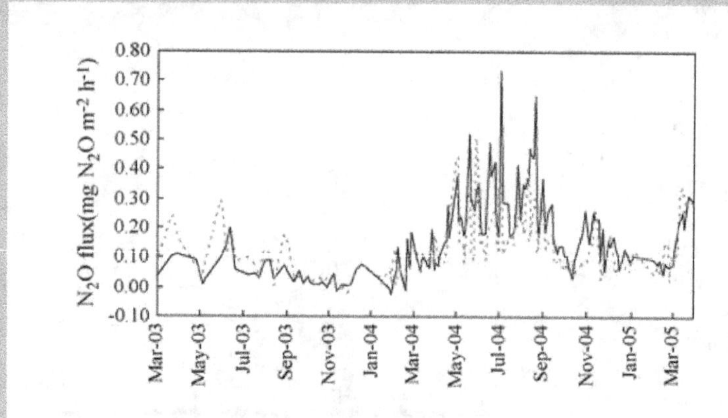

Fig. 8|13. Seasonal patterns of N_2O fluxes measured at longan orchard in South China (Liu et al., 2008). The flux from soil without surface litter was marked by dashed line.

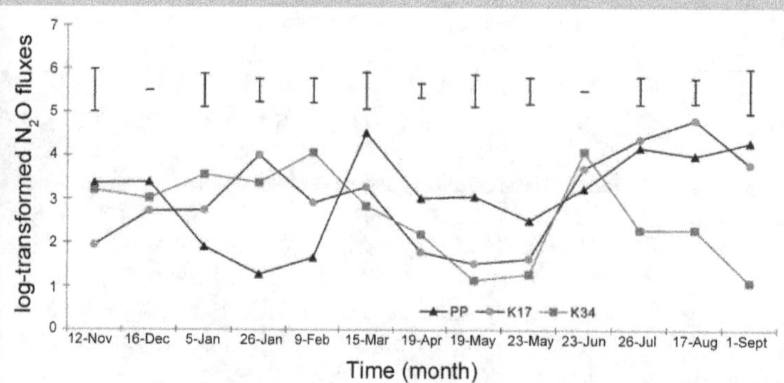

Fig. 8|14. The N_2O emission (log transformed) from permanent pasture (PP), 17-yr continuous cultivation (K17), and 34-yr continuous cultivation (K34) (Choudhary et al., 2001).

(1997), from 0.1 to 3.8% of urine N is emitted from grazed grassland to the atmosphere as N_2O, while for dung these values ranged from 0.1 to 0.7% of the applied nitrogen. A default urine N emission factor was identified as 2% (Mosier et al., 1998; IPCC, 2000).

It was found that the total N_2O emission from soil is proportional to the N fertilization level (Bouwman, 1996; Petersen et al., 2006) and is usually higher from no-tilled than from tilled soils (e.g., Lal et al., 1995; Jacinthe and Dick, 1997; Mummey et al., 1998; Hatano and Lipiec, 2004). It should be added that the opposite situations were also reported (Elder and Lal, 2008). The emissions of N_2O from agricultural fields ranged from 0.03 to 2.7% of the total nitrogen fertilizer applied (Franco, 2002; Gregorich et al., 2005; Hatano and Lipiec, 2004; Petersen et al., 2006).

Petersen et al. (2006) performed a study of 12-mo N_2O emissions (static chambers technique) in organic (N fertilizer only as manure) and conventional dairy crop rotations in five European countries (Austria, Denmark, Finland, Italy, and the United Kingdom), representing major cattle producing regions and main climatic zones. They found that the emissions of N_2O were usually lower from organic than from conventional crop rotations (with the exception of Austria). In their study 1.6% of total N applied was lost as N_2O, and the background emission was 1.4 kg N_2O N ha^{-1} yr^{-1}.

Special conditions in soil are formed by irrigation with wastewater. Nosalewicz and Stepniewska (2005) found that maximum N_2O emission (reaching the values up to 62 mg N_2O–N m^{-2} h^{-1}) was observed on the third day after irrigation of mixed grasses with pretreated municipal wastewater (i.e., wastewater after mechanical and biological treatment). In their studies up to 23% of N introduced with the wastewater (180 and 360 kg N_2O–N ha^{-1} were applied with the pretreated wastewater) was released in the form of N_2O during the first 10 d after irrigation, which is defined in the literature as a short time emission (Franco, 2002). It should be emphasized that these values are much higher than the values reported above for agricultural fields.

Tillage affects soil microbial activity by changing soil aeration conditions and enhances the processes of mineralization of organic matter in soil. No-till practice improves soil structure and increases soil water content. However, some reports have shown that no-till systems may increase ammonia and N_2O emissions (Skiba et al., 2002; Mkhabela et al., 2008). Mkhabela et al. (2008) observed two times higher emission of N_2O from no-tilled soil (Table 8|3), but more complete reduction of N_2O to N_2 was observed under no-till, when N_2O/N_2O+N_2 ratios were lower. Others authors reported the opposite (Passionato et al., 2003; Elder and Lal, 2008). Elder and Lal (2008) reported 3.5 times higher N_2O emissions from moldboard-tilled organic arable soil compared to no-till and bare treatments.

Soil compaction induces intense denitrification and N_2O emission. Bhandral et al. (2007) measured a sevenfold increase of N_2O flux from compacted soil compared to the uncompacted one. Sitaula et al. (2000) reported a 44% increase of N_2O emission from compacted, unfertilized soil compared to the uncompacted one and a 170% increase of emission of this gas in the treatment with NPK- fertilization.

The N cycling and emission of nitrogen oxides due to microbial processes is also influenced by plant community composition. Plants affect the soil diffusive parameters by root architecture and changes in water conditions. Root exudates influence the microbial activity by the organic carbon source, coenzymes, and other biologically active substances. Niklaus et al. (2006) reported that N_2O emissions decreased with plant community diversity and increased in the presence of legumes. The flux of N_2O depended on plant community composition and on the interaction of composition with soil type, as well as on soil nitrate.

Significant amounts of N_2O are emitted by agricultural soils at very low temperatures and under snow cover. Syväsalo et al. (2004) observed 60% of the annual fluxes from mineral agricultural soil in southern Finland outside the growing season, from October to April. High nitrous oxide emissions occurred during thawing–freezing cycles in winter and during thawing of soil in spring. The thawing of soil changes the structure of soil aggregates and enhances the denitrification by release of N and C from them and from microbial, animal, and plant biomass destroyed by freezing. Estimating the N_2O emissions the soil physiological properties and climatic processes

Table. 8|3. Nitrous oxide emissions from arable soils.

Soil	N_2O emission	Reference
Burkina Faso, west Africa		Brümmer et al. (2008)
• natural savanna,	$0.52–065$ kg N_2O–N ha^{-1} yr^{-1}	
• agricultural sites	$0.19–0.20$ kg N_2O–N ha^{-1}yr^{-1}	
Fine sandy loam, Palmerston North, New Zealand	(3 mo)	Bhandral et al. (2007)
• compacted	$2.62–61.74$ kg N_2O–N ha^{-1}	
• uncompacted	$1.12–4.37$ kg N_2O–N ha^{-1}	
Fine loam, southern Germany, potato plot, fertilized with 150kg N ha^{-1}	(2 mo)	Ruser et al. (2006)
• ridge (d = 1.03 g cm^{-3}, 90% water-filled pore space)	2.412 mg N_2O–N m^{-2} h^{-1}	
• interrow area (1.24 g cm^{-3}, 90% water-filled pore space)	1.159 mg N_2O–N m^{-2} h^{-1}	
• tractor compacted interrow area (1.64 g cm^{-3}, 98% water-filled pore space)	4.586 mg N_2O–N m^{-2} h^{-1}	
Oxisol developed on sandstone, south China		Liu et al. (2008)
• pine plantation (20 yr old)	3.03 Mg N_2O–N ha^{-1} yr^{-1}	
• longan orchard (12 yr old)	8.64 Mg N_2O–N ha^{-1} yr^{-1}	
Silt loam soil in mixed ryegrass–clover pasture, New Zealand		Menneer et al. (2005)
• 8 d after severe animal treading	52 g N_2O–N ha^{-1} d^{-1}	
• nil treading	2.3 g N_2O–N ha^{-1} d^{-1}	
Irrigated sandy soil, oat and maize cropped, central Spain	$0–1.2$ kg N ha^{-1} d^{-1}	Sanchez et al. (2001)
Silty clay loam, North Island, New Zealand	Average value	Choudhary et al. (2001)
• ryegrass and clover pasture	3.24 kg N_2O–N ha^{-1} yr^{-1}	
• 17 yr maize cropped soil	3.42 kg N_2O–N ha^{-1} yr^{-1}	
• 34 yr maize cropped soil	2.37 kg N_2O–N ha^{-1} yr^{-1}	
Nova Scotia, Canada: Gleyed Brunisolic Gray Luvisol or Gleyic Luvisol on sandy clay loam		Mkhabela et al. (2008)
• no-tillage	120 g N_2O–N ha^{-1} d^{-1}	
• conventional tillage	64 g N_2O–N ha^{-1} d^{-1}	
Nova Scotia, Canada: Humo-ferric Podzol and Gleyed Degraded Dystric Brunosol		Mkhabela et al. (2008)
• no-till	52 g N_2O–N ha^{-1} d^{-1}	
• conventional tillage	27 g N_2O–N ha^{-1} d^{-1}	
Southern Finland, plots of grass, barley and fallow		Syväsalo et al. (2004)
• clay	$3.7–7.8$ kg N_2O–N ha^{-1} yr^{-1}	
• loamy sand	$1.5–7.5$ kg N_2O–N ha^{-1} yr^{-1}	
Corn and winter wheat cropped Linwood muck, Ohio, USA		Elder and Lal (2008)
• moldboard/disking tilled	96.9 kg N_2O–N ha^{-1} yr^{-1}	
• bare treated	29.5 kg N_2O–N ha^{-1} yr^{-1}	
• no-till	35.9 kg N_2O–N ha^{-1} yr^{-1}	
Arable peat soil, Kannus, western Finland		Maljanen et al. (2004)
• barley cropped	848 mg N_2O–N m^{-2} yr^{-1}	
• grass cropped	275 mg N_2O–N m^{-2} yr^{-1}	
• bare soil	2350 mg N_2O–N m^{-2} yr^{-1}	
Tropical arable soil, annual cropped, Costa Rica (351-d measurement)		Weitz et al. (2001)
• loam, unfertilized	1.04 ± 0.72 ng N_2O–N cm^2 h^{-1}	
• loam, fertilized	3.54 ± 4.31 ng N_2O–N cm^2 h^{-1}	
• clay, unfertilized	1.43 ± 0.89 ng N_2O–N cm^2 h^{-1}	
• clay, fertilized	0.81 ± 0.79 ng N_2O–N cm^2 h^{-1}	

should be taken into account, as well as fertilizer application and soil N and C content.

Methane
Generation and Oxidation

Presence of methane (CH_4) in soil air and its absorption or emission rate depends on redox conditions. Production of CH_4 in soil takes place at redox potential values usually below -150 mV (e.g., Masscheleyn et al., 1993; Wang et al., 1993; Stepniewski and Stepniewska, 2000; Bennicelli et al., 2006; Yu et al., 2001) and is influenced by morphophysiological characteristics, for example, of rice (Das and Baruah, 2008).

As was already mentioned, soil can also act as a sink of CH_4, which can be absorbed due to oxidation at higher redox potential values (e.g., Bowman, 1990; Le Mer and Roger, 2001; Liu et al., 2008). Thus, generation of CH_4 takes place in the anoxic part of the soil profile. During its migration to the atmosphere it is oxidized in the oxic soil layer or in water stagnating on the surface of the flooded soil. The resultant net emission or absorption at the soil surface will depend on the relative contribution of the reduced and oxidized zones within the soil profile (e.g., Neue et al., 1996).

It should be emphasized that, due to soil heterogeneity, the production of CH_4 can take place within the centers of soil aggregates saturated with water, where redox potential is usually lower than in the exterior (e.g., Zausig et al., 1993), where CH_4 can be simultaneously oxidized by methanotrophs. The CH_4 concentration in the atmosphere of dryland soils is very low, but in rice fields under flood conditions it becomes the main component of the soil atmosphere together with carbon dioxide. Wetlands are an important source of CH_4, and its global emission is estimated at 92 to 232 Tg yr^{-1} (Wuebbles and Hayhoe, 2002). It contributes 76% of CH_4 emission from natural sources, while rice fields emit 25.6 Tg yr^{-1} of CH_4 (Yan et al., 2009).

The emission of CH_4 from rice fields depends on numerous factors such as temperature (optimum in the range from 37–45°C; Boone, 2000), cultivation method, water management, soil properties, rice variety, and fertilization.

According to Le Mer and Roger (2001) strategies aiming at lowering the emission of CH_4 from rice fields can be oriented toward decrease of its production, intensification of its oxidation, and reduction of its transport rate through the tissues of the plant itself.

Decrease of CH_4 production can be achieved by periodical dewatering of rice fields (40–55% reduction reported), selecting low CH_4–producing varieties (up to 60% reduction reported), introducing oxidants and other mineral fertilizers (possibility of 20–70% reduction), avoiding application of organic fertilization (responsible for elevating CH_4 emission by up to 50%; Wuebbles and Hayhoe, 2002), and managing water based on redox potential control (resulting in inhibition of CH_4 emission by 64%; Minamikawa and Sakai, 2006). It should be emphasized that there is a positive feedback between the temperature and the emission of CH_4 from rice fields and natural wetlands, as global warming increases production of CH_4, which in turn intensifies the global warming.

It should be noted that modifications of traditional water management systems in paddy fields allow us to lower CH_4 emission from the soil. According to Yagi et al. (1996) introduction of an intermittent rice irrigation system reduced emission of CH_4 to 45% compared to continuous a flooding treatment. As already mentioned, redox potential controlled water management system proposed by Minamikawa and Sakai (2006) deserves special attention. In their field experiment lasting 2 years, the drainage was performed whenever redox potential of the soil decreased below –150 mV, while flooding was performed when redox potential increased to 0 mV in the first year of the experiment and to 100 mV in the second year. The study showed that rice yield was not affected by the treatments, while methane emission in the redox potential control treatment was at a level of 36% of the emission from the continuous flooding treatment.

Methane oxidation is connected with occurrence of methanotrophic microorganisms (mainly bacteria) characterized by the possibility to oxidize CH_4 and to utilize it as the only source of carbon and energy (Conrad, 1996; Rozej et al., 1999; Stepniewski and Rozej, 2000). They are obligate aerobes, often with the optimum at subatmospheric concentrations of oxygen, and their activity depends on oxygenation status (Jones and Nedwell, 1993; King, 1994; Kightley et al., 1995).

The activity of methanotrophis microorganisms in soil depends on physical, chemical, biological, and physicochemical factors. According to Le Mer and Roger (2001) there two basic types of methanotrophs in soil. The first one is responsible for "high affinity oxidation" (i.e., of very low Michaelis constant value), which occurs at low CH_4 concentration (<12 ppm). This type of methanotrophic organism occurs in soils not exposed to high CH_4 concentrations. The methanotrophs of the second type, responsible for "low affinity oxidation" (i.e., with high Michaelis constant value), usually occur at higher CH_4 concentrations (>40 ppm). This type of oxidation is considered to be the proper methanotrophic activity. It occurs in such media as peatlands, rice fields, wetlands, and waste deposits. It should be emphasized that more than 90% of CH_4 produced in rice fields can be oxidized by the methanotrophs of the oxic zone. (Patrick and Jugsujinda, 1992; Le Mer and Roger, 2001). This creates the possibility of substantial reduction of CH_4 emissions from the rice fields.

Methanotrophs are generally neutrophiles (Dunfield et al., 1993), but they are capable of adapting to a wide pH range from 3.5 to 8.0 (Born et al., 1990). An important effect on methanotrophic activity of soil is exerted by its water content and temperature (Boeckx and Van Cleemput, 1996; Le Mer and Roger, 2001; Wuebbles and Hayhoe, 2002). The dependence of methanotrophic activity on temperature shows a maximum in the range of 25 to 31°C (Boeckx and Van Cleemput, 1996; Whalen et al., 1990). Also dependence on water content (Boeckx and Van Cleemput, 1996; Whalen et al., 1990) shows a maximum. Lowering the methanotrophic activity at low water contents is a result of limitation of availability of water for microorganisms, while its decrease at high water contents is caused by the limited availability of both substrates, i.e., of oxygen and CH_4 due to impairment of their transport within the soil (Omal'chenko et al., 1992; Mancinelli, 1995).

The influence of CH_4 concentration on methanotrophic activity was studied by Bender and Conrad (1995) and Pawlowska and Stepniewski (2004), among others. The latter authors observed a stabilization of methanotrophic activity in sandy materials at CH_4 concentrations above 6% (Fig. 8|15). Investigation of methanotrophic activity of different size fractions of sandy materials (Pawlowska et al., 2003) showed that the highest activity was measured for the 0.5- to 1.0-mm fraction.

Emission

Land use system and tillage influence the CH_4 emission to the atmosphere by alteration of gas diffusivity and modification of the distribution of redox potential within the soil profile, and thus, by the change of its generation and absorption in soil. It was found that atmospheric CH_4 oxidation potential of forest soils exceeded production potentials by 10 to 220 times (Bradford et al., 2001).

Two years of continuous measurements of CH_4 emissions from pine plantation and logan orchard soil in south China by Liu et al. (2008) showed that these soils were weak sinks of CH_4, and seasonal patterns of CH_4 fluxes were not clear without extremes, unlike CO_2 and N_2O.

The animal excreta patches in grazing ecosystems are the point-sources of CH_4 emissions in the generally CH_4 uptake area. The moderate grazing intensity when the dung patches cover small meadow area (0.3%) slightly weakened the CH_4 uptake potential of the whole area (Lin et al., 2009), but at the intensive grazing level the amount of CH_4 released from cowpats could exceed the CH_4 uptake potential from unmanured area in the same time (Flessa et al., 1996).

The CH_4 sink potential of agricultural soils depends on soil properties, land use, and agricultural practices and generally is reduced due to ammonium ions inhibition and tillage (Hütsch, 1998). Regina et al. (2007), studying CH_4 oxidation potential of arable soils in Finland, observed a wide range of annual CH_4 fluxes from uptake of –1.2 kg CH_4 ha^{-1} to emission of 40 kg CH_4 ha^{-1}. The higher content of macropores favored higher methane oxidation rate. The heaviest clay and poorly drained peat may be the place of methanogenesis on the water-saturated zone of this soil and decreased net CH_4 sink.

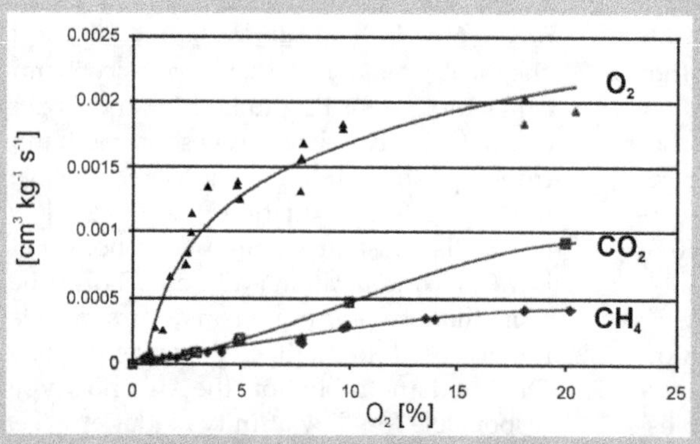

Fig. 8|15. The effect of oxygen concentration on methane oxidation rate as measured by oxygen and methane uptake, as well as by carbon dioxide production in a sandy material incubated at a room temperature in the atmosphere containing 8% of methane and different concentrations of oxygen (modified from Pawlowska and Stepniewski, 2004).

Hargreaves and Fowler (1998) found a negative linear correlation between water table depth in soil and methane flux to the atmosphere, confirmation that more and more CH_4 was oxidized with the increase of the oxic soil layer depth. The same authors found a positive linear correlation between the proportion of soil surface occupied by standing water and the CH_4 emission. The values of CH_4 emissions from different soil and climate conditions are shown in Table 8|4.

Ethylene

Ethylene (C_2H_4) is a plant growth regulator produced by the soil and plants under specific conditions. According to Nazli et al. (2003) methionine has been identified as a physiological precursor of C_2H_4 in microorganisms and higher plants, but two different biochemical pathways are involved here. The majority of microorganisms produce C_2H_4 from methionine via the 2-keto-4-methylthiobutyric acid (KMBA) pathway. Higher plants utilize the 1-aminocy-clo-propane-1-carboxylic acid (ACC) as a C_2H_4 precursor. Microbial production of ethylene is sensitive to antibiotics; the inhibition of biosynthesis was observed by Nazli et al. (2003) after glucose (as a C source) and NH_4NO_3 (N source) addition.

Ethylene-consuming microorganisms may play an important ecological role, preventing the influence of ethylene on plant growth. The threshold concentration in soil for plant response is below 0.1 μL C_2H_4 L^{-1} (Elsgaard, 2001).

Ethylene production is usually observed in anaerobic soils. Under oxic conditions C_2H_4 is degraded in top soil layer. Jackel et al. (2004) suggested the presence of two different ethylene-consuming groups of microorganisms exhibiting two different K_m values of 40 and 12,600 μL L^{-1}. Natural occurrence of ethylene in soil keeps up the populations of ethylene-degraders. Elsgaard (2001) showed that the arable, sandy soil without the indigenous C_2H_4 production required a few weeks of incubation to develop rapid C_2H_4 consumption compared to coniferous soil and litter.

Jackel et al. (2004) reported the inhibition of CH_4 oxidation by ethylene. Concentration of this gas in the range of 3, 6, 10 μL L^{-1} decreased methane uptake by 21, 63, and 98%, respectively. The earlier studies showed that ethylene can be co-oxidized as a nongrowth substrate by methanotrophic and nitrifying bacteria (Patel et al., 1982; Hommes et al., 1998). Rigler and Zechmeister-Boltenstern (1999) reported similar negative response of ethylene and methane oxidation kinetics

Table. 8|4. Methane emissions from arable soils.

Soil	CH_4 emission	Reference
Boreal arable soil in Finland		Regina et al. (2007)
• sandy soil	−0.8 to −1.2 kg CH_4 ha^{-1} yr^{-1}	
• clay soil	−0.1 to 0.6 kg CH_4 ha^{-1} yr^{-1}	
• peat soil (south)	−0.1 to −0.5 kg CH_4 ha^{-1} yr^{-1}	
• peat soil (north)	−0.2 to 40 kg CH_4 ha^{-1} yr^{-1}	
Oxisol developed on sandstone, south China		Liu et al. (2008)
• pine plantation (20 yr old)	−2.57 Mg CH_4–C ha^{-1} yr^{-1}	
• longan orchard (12 yr old)	−2.61 Mg CH_4–C ha^{-1} yr^{-1}	
Clay loam, alpine meadow, Tibet, China		Lin et al. (2009)
• meadow covered by 0.3% urine and 2.9% dung patches in 2005 (2006)	−0.30 (0.28) kg CH_4 ha^{-1} yr^{-1}	
• unmanured meadow	−0.33 (0.29) kg CH_4 ha^{-1} yr^{-1}	
Corn and winter wheat cropped Linwood muck, Ohio, USA		Elder and Lal (2008)
• moldboard/disking tilled	0.9 CH_4–C kg ha^{-1} yr^{-1}	
• bare treated	−4.5 CH_4–C kg ha^{-1} yr^{-1}	
• no-till	7.6 CH_4–C kg ha^{-1} yr^{-1}	
Arable peat soil, Kannus, western Finland		Maljanen et al. (2004)
• vegetated soil	−100 mg CH_4–C m^{-2} yr^{-1}	
• bare soil	−55 mg CH_4–C m^{-2} yr^{-1}	
Median values for different environments		Le Mer and Roger (2001), from different authors
• 5 upland soils temporarily submerged	3 g CH_4 ha^{-1} d^{-1}	
• 11 swamps	720 g CH_4 ha^{-1} d^{-1}	
• 4 peatlands	433 g CH_4 ha^{-1} d^{-1}	
• 23 rice fields	10^3 g CH_4 ha^{-1} d^{-1}	

to elevated concentrations of nitrate and ammonia N source and CO_2.

Ethylene seems to be a regulator of nodulation, impinging on both symbiotic nitrogen fixing Rhizobia and plant hosts. Tamimi and Timko (2003) observed the inhibition of nodules development in bean (*Phaseolus vugaris* L.) plants as well as of growth and proliferation of rhizobia by the ethylene-releasing compound 2-chloroethylphosphonic acid (ethephon). The ethylene synthesis inhibitors enhanced nodulation and did not inhibit growth of rhizobia.

References

Aarts, H.F.M., B. Habekotte, G.J. Hilhorst, G.J. Goskamp, F.C. van der Schans, and C.K. de Vries. 1999. Efficient resource management in dairy farming on sandy soil. Neth. J. Agric. Sci. 47:153–167.

Alvaro-Fuentes, J., C. Cantero-Martinez, M.V. Lopez, and J.L. Arrue. 2007. Soil carbon dioxide fluxes following tillage in semiarid Mediterranean agroecosystems. Soil Tillage Res. 96:331–341.

Ananyeva, N.D., E.A. Susyan, O.V. Chernova, and S. Wirth. 2008. Microbial respiration activities of soils from different climatic regions of European Russia. Eur. J. Soil Biol. 44:147–157.

Armstrong, J., W. Armstrong, P.M. Beckett, J.E. Halder, S. Lythe, R. Holt, and A. Sinclair. 1996. Pathways of aeration and the mechanisms and beneficial effects of humidity– and Venturi-induced convections in *Phragmites australis* (Cav.) Trin. ex Steud. Aquat. Bot. 54:177–197.

Baggs, E.M. 2006. Partitioning the components of soil respiration: A research challenge. Plant Soil 284:1–5.

Bateman, E.J., and E.M. Baggs. 2005. Contributions of nitrification and denitrification to N_2O emissions from soils at different water-filled pore space. Biol. Fertil. Soils 41:379–388.

Bauer, P.J., J.R. Frederick, J.M. Novak, and P.G. Hunt. 2006. Soil CO_2 flux from a norfolk loamy sand after 25 years of conventional and conservation tillage. Soil Tillage Res. 90:205–211.

Beauchamp, E.G. 1997. Nitrous oxide emission from agricultural soils. Can. J. Soil Sci. 77:113–123.

Bender, M., and R. Conrad. 1995. Effect of methane concentrations and soil conditions on the induction of methane oxidation activity. Soil Biol. Biochem. 27:1517–1527.

Bennicelli, R.P., A. Szafranek, and Z. Stepniewska. 2006. The influence of redox conditions on methane release from peat soils. p. 1114–1119. *In* Proc. 17th ISTRO Conf., Kiel, Germany.

Bhandral, R., S. Saggar, N.S. Bolan, and M.J. Hedley. 2007. Transformation of nitrogen and nitrous oxide emission from grassland soils as affected by compaction. Soil Tillage Res. 94:482–492.

Black, J.G. 1996. Microbiology. Principles and applications. 3rd ed. Prentice Hall, Upper Saddle River, NJ.

Boeckx, P., and O. Van Cleemput. 1996. Methane oxidation in a neutral landfill cover soil: Influence of moisture content, temperature, and nitrogen-turnover. J. Environ. Qual. 25:178–183.

Boone, D. 2000. Biological formation and consumption of methane. p. 42–62. *In* M. Khalil (ed.) Atmosheric methane: Its role in the global environment. Springer Verlag, New York.

Born, M., H. Dörr, and I. Levin. 1990. Methane consumption in aerated soils of the temperate zone. Tellus 42:2–8.

Bouyoucos, G.J., and M.M. McCool. 1924. The aeration of soils as influenced by air-barometric pressure changes. Soil Sci. 18:53–64.

Bouwman, A.F. 1996. Direct emission of nitrous oxide from agricultural soils. Nutr. Cycl. Agroecosyst. 46: 53–70.

Bowman, A.F. 1990. Soils and the greenhouse effect. John Wiley and Sons, New York.

Bradford, M.A., P. Ineson, P.A. Wookey, and H.M. Lappin-Scott. 2001. Role of CH_4 oxidation, production and transport in forest soil CH_4 flux. Soil Biol. Biochem. 33:1625–1631.

Brümmer, Ch., N. Brüggemann, K. Butterbach-Bahl, U. Falk, J. Szarzynski, K. Vielhauer, R. Wassmann, and H. Papen. 2008. Soil–atmosphere exchange of N_2O and NO in near-natural savanna and agricultural land in Burkina Faso (W. Africa). Ecosystems 11:582–600.

Campos, A.C. 2006. Response of soil surface CO_2–C flux to land use changes in a tropical cloud forest (Mexico). For. Ecol. Manage. 234:305–312.

Cao, G., Y. Tang, W. Mo, Y. v Wang, Y. Li, and X. Zhao. 2004. Grazing intensity alters soil respiration in an alpine meadow on the Tibetan plateau. Soil Biol. Biochem. 36:237–243.

Chatskikh, D., and J.E. Olesen. 2007. Soil tillage enhanced CO_2 and N_2O emissions from loamy sand soil under spring barley. Soil Tillage Res. 97:5–18.

Chen, G.X., G.H. Huang, B. Huang, K.W. Yu, J. Wu, and H. Xu. 1997. Nitrous oxide and methane emissions from soil–plant systems. Nutr. Cycl. Agroecosyst. 49:41–45.

Choudhary, M.A., A. Akramkhanov, and S. Saggar. 2001. Nitrous oxide emissions in soils cropped with maize under long-term tillage and under permanent pasture in New Zealand. Soil Tillage Res. 62:61–71.

Conrad, R. 1996. Soil microorganisms as controllers of atmospheric trace gases (H_2, CO, CH_4, OCS, N_2O, and NO). Microbiol. Rew. 60:609–640.

Cowie, M., and H. Watts. 1971. Diffusion of methane and chloromethanes in air. Can. J. Chem. 49:74–77.

Currie, J.A. 1975. Soil respiration. Tech. Bull. Min. Agric. Fish. Food. 29:461.

Currie, J.A. 1970. Movement of gases in soil respiration, in sorption and transport processes in soil. S.C.I. Monogr. 37. Soc. of Chemical Industry, London.

Das, K., and K.K. Baruah. 2008. A comparison of growth and photosynthetic characteristics of two improved rice cultivars on methane emission from rain fed agroecosystem of Northeast India. Agric. Ecosyst. Environ. 124:105–113.

Dobermann, A. 2004. A critical assessment of the system of rice intensification (SRI). Agric. Syst. 79:261–281.

Dommergues, Y., and F. Mangenot. 1970. Ecologie microbienne du sol. Mason, Paris.

Dunfield, P., R. Knowles, R. Dumont, and T.R. Moore. 1993. Methane production and consumption in temperate and subarctic peat soils: Response to temperature and pH. Soil Biol. Biochem. 25:321–326.

Elder, J.W., and R. Lal. 2008. Tillage effects on gaseous emissions from an intensively farmed organic soil in North Central Ohio. Soil Tillage Res. 98:45–55.

Elsgaard, L. 2001. Ethylene turn-over in soil, litter and sediment. Soil Biol. Biochem. 33:249–252.

Flessa, H., and W.R. Fischer. 1992. Plant induced changes in the redox potential of rice rhizospheres. Plant Soil 143:55–60.

Flessa, H., P. Dorsch, F. Beese, H. Konig, and A.F. Bouwman. 1996. Influence of cattle wastes on nitrous oxide and methane fluxes in pasture land. J. Environ. Qual. 25:1366–1370.

Franco, G.P. 2002. Inventory of California greenhouse gas emissions and sinks: 1990–1999. Staff Report, Gray Davis Governor.

Frank, A.B., M.A. Liebig, and J.D. Hanson. 2002. Soil carbon dioxide fluxes in northern semiarid grasslands. Soil Biol. Biochem. 34:1235–1241.

Frank, A.B., M.A. Liebig, and D.L. Tanaka. 2006. Management effects on soil CO_2 efflux in northern semiarid grassland and cropland. Soil Tillage Res. 89:78–85.

Gesch, R.W., D.C. Reicosky, R.A. Gilbert, and D.R. Morris. 2007. Influence of tillage and plant residue management on respiration of a Florida Everglades Histosol. Soil Tillage Res. 92:156–166.

Ghildyal, B.P. 1982. Nature, physical properties and management of submerged rice soils, Trans. 12th Int. Congr. Soil Sci. New Delhi, India. In Vertisols and Rice Soils of the Tropics, Symp. papers II, Shri S.N. Mehta, New Delhi, India.

Gliński, J., and W. Stepniewski. 1985. Soil aeration and its role for plants. CRC Press, Boca Raton, FL.

Greenwood, D.J. 1962. Nitrification and nitrate dissimilation in soil. II. Effect of oxygen concentration. Plant Soil 3:378–391.

Gregorich, E.G., P. Rochette, D.W. Hopkins, U.F. McKima, and P. St-Georges. 2006. Tillage-induced environmental conditions in soil and substrate limitation determine biogenic gas production. Soil Biol. Biochem. 38:2614–2628.

Gregorich, E.G., P. Rochette, B. Vanden, and D.A. Angers. 2005. Greenhouse gas contributions of agriculture soils and potential mitigation practices in Eastern Canada. Soil Tillage Res. 83:53–72.

Grichko, V.P., and B.R. Glick. 2001. Ethylene and flooding stress in plants. Plant Physiol. Biochem. 39:1–9.

Hargreaves, K.J., and D. Fowler. 1998. Quantifying the effects of water table and soil temperature on the emission of methane from peat wetland at the field scale. Atmos. Environ. 32:3275–3282.

Hatano, R., and J. Lipiec. 2004. Effects of land use and agricultural practices on greenhouse gas fluxes in soil. Monogr. Acta Agrophys. 109:50.

Højberg, O., N.P. Revsbech, and J.M. Tiedje. 1994. Denitrification in soil aggregates analyzed with microsensors for nitrous oxide and oxygen. Soil Sci. Soc. Am. J. 58:1691–1698.

Hommes, N.G., S.A. Russell, P.J. Bottomley, and D.J. Arp. 1998. Effects of soil on ammonia, ethylene, chloroethane, and 1,1,1-trichloroethane oxidation by Nitrosomonas europaea. Appl. Environ. Microbiol. 64:1372–1378.

Horn, R. 1994. Effect of aggregation of soils on water, gas and heat transport. p. 335–361. In E.E. Schulze (ed.) Flux control in biological systems.. Academic Press, New York.

Horn, R., and A. Smucker. 2005. Structure formation and its consequences for gas and water transport in unsaturated arable and forest soils. Soil Tillage Res. 82:5–14.

Hou, A.X., G.X. Chen, Z.P. Wang, O. Van Cleemput, and W.H. Patrick, Jr. 2000. Methane and nitrous oxide emissions from a rice field in relation to soil redox and microbiological processes. Soil Sci. Soc. Am. J. 64:2180–2186.

Hua, X., G.X. Xing, Z.C. Cai, and H. Tsuruta. 1997. Nitrous oxide emission fluxes from upland soils and paddy soils in China. Nutr. Cycl. Agroecosyst. 49:23–28.

Huang, C.H., Y. Ouyang, and J.-E. Zhang. 2006. Effects of a compact layer on soil O_2 diffusion. Geoderma 135:224–232.

Hütsch, B.W. 1998. Methane oxidation in arable soil as inhibited by ammonium, nitrite, and organic manure with respect to soil pH. Biol. Fertil. Soils 28:27–35.

IPCC. 2000. IPCC good practice guidance and uncertainty management in national greenhouse gas inventories. OECD/ODCE, Paris.

IPCC. 2001. Climate change 2001: The scientific basis. Cambridge Univ. Press, Cambridge, UK.

Iqbal, J., H. Ronggui, D. Lijun, L. Lan, L. Shan, C. Tao, and R. Leilei. 2008. Differences in soil CO_2 flux between different land use types in mid-subtropical China. Soil Biol. Biochem. 40:2324–2333.

IRRI. 2005. World rice statistics. Available at http://www.irri.org/science/ricestat (verified 16 Sept. 2010).

Jacinthe, P.A., and W.A. Dick. 1997. Soil management and nitrous oxide emissions from cultivated fields in southern Ohio. Soil Tillage Res. 41:221–235.

Jackel, U., S. Schnell, and R. Conrad. 2004. Microbial ethylene production and inhibition of methanotrophic activity in a deciduous forest soil. Soil Biol. Biochem. 36:835–840.

Jassal, R., A. Black, M. Novak, K. Morgenstern, Z. Nesic, and D. Gaumont-Guay. 2005. Relationship between soil CO_2 concentrations and forest-floor CO_2 effluxes. Agric. For. Meteorol. 130:176–192.

Jayakumar, B., C. Subathra, V. Velu, and S. Ramanathan. 2005. Effect of integrated crop management practices on rice (Oryza sativa L.) root volume and rhizosphere redox potential. J. Agron. 4(4):311–314.

Jinbo, Z., S. Changchun, and W. Shenmin. 2008. Short-term dynamics of carbon and nitrogen after tillage in a freshwater marsh of northeast China. Soil Tillage Res. 99:149–157.

Jones, H.A., and D.B. Nedwell. 1993. Methane emission and methane oxidation in land-fill cover soil. FEMS Microbiol. Ecol. 102:185–195.

Justin, S.H.F.W., and W. Armstrong. 1987. The anatomical characteristics of roots and plant response to soil flooding. New Phytol. 106:465–495.

Kadlec, R.H., and R.L. Knight. 1996. The inadequacy of first order treatment wetland models. Ecol. Eng. 15:105–119.

Kalashnikov, Y.E., T.I. Balakhnina, R.P. Bennicelli, W. Stepniewski, and Z. Stepniewska. 1999. Antioxidant activity and lipid peroxidation in wheat plants as related to cultivar tolerance to excessive soil moisture. Russian J. Plant Physiol. 46 (2):227–233. Translated from Fiziologiya Rastenii, 46(2):268–275.

Kasimir-Klemendtsson, Å., L. Klemedtsson, K. Berglund, P.J. Martikainen, J. Silvola, and O. Oenema. 1997. Greenhouse gas emissions from farmed organic soils: A review. Soil Use Manage. 13:245–250.

Kanno, T., Y. Miura, H. Tsuruta, and K. Minami. 1997. Methane emission form rice paddy fields in all of Japanese prefecture. Nutr. Cycl. Agroecosyst. 49:147–151.

Kavdir, Y., H.J. Hellebrand, and J. Kern. 2008. Seasonal variations of nitrous emission in relation to nitrogen fertilization and energy crop types in sandy soil. Soil Tillage Res. 98:175–186.

Kightley, D., D.B. Nedwell, and M. Cooper. 1995. Capacity for methane oxidation in landfill cover soils, measured in, laboratory-scale soil microcosms. Appl. Environ. Microbiol. 61:592–601.

King, G.M. 1994. Methanotrophic associations with the roots and rhizomes of aquatic vegetation. Appl. Environ. Microbiol. 60:3220–3227.

Kroeckel, L., and H. Stolp. 1985. Influence of oxygen on denitrification and aerobic respiration in soil. Biol. Fertil. Soils 1(4):189–193.

Lal, R., N.R. Fausey, and D.J. Eckert. 1995. Land use and soil management effects on emission of radiatively active gases in two Ohio soils. p. 41–59. In R. Lal et al. (ed.) Soil management and the greenhouse effect. Lewis/CRC Publ., Boca Raton, FL.

Lal, R., and J. Kimble. 1997. Conservation tillage and carbon sequestration. Nutr. Cycl. Agroecosyst. 49:243–253.

Langeveld, C.A., R. Segers, B.O.M. Dirks, A. van den Pol-van Dasselaar, G.L. Velthof, and A. Hensen. 1997. Emissions of CO_2, CH_4, and N_2O from pasture on drained peat soils in the Netherlands. Eur. J. Agron. 7:35–42.

Le Mer, J., and P. Roger. 2001. Production, oxidation, emission and consumption of methane by soils: A review. Eur. J. Soil Biol. 37:25–50.

Lebert, M., H. Boken, and F. Glante. 2007. Soil compaction—Indicators for the assessment of harmful changes to the soil in the context of the German Federal Soil Protection Act. J. Environ. Manage. 82:388–397.

Lin, X., S. Wang, X. Ma, G. Xu, C. Luo, Y. Li, G. Jiang, and Z. Xie. 2009. Fluxes of CO_2, CH_4, and N_2O in an alpine meadow affected by yak excreta on the Qinghai-Tibetan plateau during summer grazing periods. Soil Biol. Biochem. 41:718–725.

Liu, H., P. Zhao, P. Lu, Y.-S. Wang, Y.-B. Lin, and X.-Q. Rao. 2008. Greenhouse gas fluxes from soils of different land-use types in a hilly area of South China. Agric. Ecosyst. Environ. 124:125–135.

Longsmuir, I.S. 1954. Respiration rate of bacteria as a function of oxygen concentration. J. Biochem. 57:81.

Lou, Y., Z. Li, T. Zhang, and Y. Liang. 2004. CO_2 emissions from subtropical arable soils of China. Soil Biol. Biochem. 36:1835–1842.

Maljanen, M., V.-M. Komulainen, J. Hytönen, P.J. Martikainen, and J. Laine. 2004. Carbon dioxide, nitrous oxide and methane dynamics in boreal organic agricultural soils with different soil characteristics. Soil Biol. Biochem. 36:1801–1808.

Maljanen, M., A. Lijkanen, J. Silvola, and P.J. Martikainen. 2003. Nitrous oxide emission from boreal organic soils under different land use. Soil Biol. Biochem. 35:1–12.

Mancinelli, R.L. 1995. The regulation of methane oxidation in soil. Annu. Rev. Microbiol. 49:581–605.

Maricle, B.R., and R.W. Lee. 2002. Aerenchyma development and oxygen transport in the estuarine cordgrasses *Spartina alterniflora* and *S. anglica*. Aquat. Bot. 74:109–120.

Masscheleyn, P.H., R.D. De Laune, and W.H. Patrick, Jr. 1993. Methane and nitrous oxide emissions from laboratory measurements of rice soil suspension: Effect of soil oxidation–reduction status. Chemosphere 26:251–260.

McGechan, M.B., S.E. Moir, K. Castle, and I.P.J. Smit. 2005. Modelling oxygen transport in a reedbed-constructed wetland purification system for dilute effluents. Biosystems Eng. 91(2):191–200.

Menneer, J.C., S. Ledgard, C. McLay, and W. Silvester. 2005. Animal treading stimulates denitrification in soil under pasture. Soil Biol. Biochem. 37:1625–1629.

Mielnick, P.C., and W.A. Dugas. 2000. Soil CO_2 flux in a tallgrass prairie. Soil Biol. Biochem. 32:221–228.

Minamikawa, K., and N. Sakai. 2005. The effect of water management based on soil redox potential on methane emission from two kinds of paddy soils in Japan. Agric. Ecosyst. Environ. 107:397–407.

Minamikawa, K., and N. Sakai. 2006. The practical use of water management based on soil redox potential for decreasing methane emission from paddy field in Japan. Agric. Ecosyst. Environ. 116(3/4):181–188.

Mkhabela, M.S., A. Madani, R. Gordon, D. Burton, D. Cudmore, A. Elmi, and W. Hart. 2008. Gaseous and leaching nitrogen losses from no-tillage and conventional tillage systems following surface application of cattle manure. Soil Tillage Res. 98:187–199.

Moncrieff, J.B., and C. Fang. 1999. A model for soil CO_2 production and transport 2: Application to a Florida *Pinus elliotte* plantation. Agric. For. Meteorol. 95:237–256.

Mosier, A.R., C. Kroeze, C. Nevison, O. Oenema, S. Seitzinger, and O. Van Cleemput. 1998. Closing the global N2O budged: Nitrous oxide emissions through the agricultural cycle. Nutr. Cycl. Agroecosyst. 52:225–248.

Mummey, D.L., J.L. Smith, and G.R. Bluhm. 1998. Assessment of alternative soil management practices on N_2O emissions from U.S. agriculture. Agric. Ecosyst. Environ. 70:79–87.

Nazli, Z.H., M. Arshad, and A. Khalid. 2003. 2-Keto-4-methylthiobutyric acid-dependent biosynthesis of ethylene in soil. Biol. Fertil. Soils 37:130–135.

Neue, H.U., R. Wassmann, R.S. Lantin, M.C.R. Alberto, J.B. Aduna, and A.M. Javellana. 1996. Factors affecting methane emission from rice fields. Atmos. Environ. 30:1751–1754.

Niklaus, P.A., D.A. Wardle, and K.R. Tate. 2006. Effects of plant species diversity and composition on nitrogen cycling and the trace gas balance of soils. Plant Soil 282:83–98.

Nosalewicz, M., and Z. Stepniewska. 2005. N_2O emission from surface of Eutric Histosol irrigated with municipal wastewater (after second step of purification). Pol. J. Soil Sci. 37:111–118.

Oenema, O., G.L. Velthof, S. Yamulki, and S.C. Jarvis. 1997. Nitrous oxide emissions from grazed grassland. Soil Use Manage. 13:288–295.

Omal'chenko, L., N.D. Savel'eva, V. Vasil'eva, and G.A. Zavarzin. 1992. A psychrophilic methanotrophic community from tundra soil. Mikrobiologiya 61:1072–1076.

Passionato, C.C., T. Ahrens, B.J. Feigl, P.S. Stendler, J.A. do-Carmo, and J.M. Melillo. 2003. Emissions of CO_2, N_2O, and NO in conventional and no-till management practices in Rondonia. Brazil. Biol. Fertil. Soils 38:200–208.

Patel, R.N., C.T. Hou, A.I. Laskin, and A. Felix. 1982. Microbial oxidation of hydrocarbons: Properties of a soluble methane monooxygenase from a facultative methane-utilizing organism, *Methylobacterium* sp. strain CRL-26. Appl. Environ. Microbiol. 44:1130–1137.

Patrick, W.H., and A. Jugsujinda. 1992. Sequential reduction and oxidation of inorganic nitrogen, manganese, and iron in flooded soil. Soil Sci. Soc. Am. J. 56:1071–1073.

Paul, E.A., and F.E. Clark. 1998. Soil microbiology and biochemistry. Academic Press, San Diego.

Pawlowska, M., and W. Stepniewski. 2004. The effect of oxygen concentration on the activity of methanotrophs in sand material. Environ. Protect. Eng. 30(3):81–91.

Pawlowska, M., W. Stepniewski, and J. Czerwinski. 2003. The effect of texture on methane oxidation capacity in a sand layer—A model laboratory study. p. 339–354. *In* M. Pawlowski et al. (ed.) Environmental engineering studies: Polish research on the way to the EU. Kluwer Academic/Plenum Publishers, New York.

Pengthamkeerati, P., P.P. Motavalli, R.J. Kremer, and S.H. Anderson. 2005. Soil carbon dioxide efflux from a claypan soil affected by surface compaction and applications of poultry litter. Agric. Ecosyst. Environ. 109:75–86.

Petersen, S.O., K. Regina, A. Pollinger, E. Rigler, L. Valli, S. Yamulki, M. Esala, C. Fabbri, E. Syvasalo, and F.P. Vinther. 2006. Nitrous oxide emissions from organic and conventional crop rotations in five European countries. Agric. Ecosyst. Environ. 112:200–206.

Przywara, G., and W. Stepniewski. 2000. Influence of flooding and different temperatures of the soil on gas-filled porosity of pea, maize and winter wheat roots. Int. Agrophys. 14:401–410.

Purnobasuki, H., and M. Suzuki. 2005. Functional anatomy of air conducting network on the pneumatophores of a mangrove plant, *Avicennia marina* (Forsc.) Vierh. Asian J. Plant Sci. 4(4):334–347.

Rafaralahy, S. 2002. An NGO perspective on SRI and its origins in Madagascar. p. 17–22. *In* N. Uphoff et al. (ed.) Proc. Int. Conf. Assessment of the system for rice intensification (SRI), Sanya, China, 1–4 April. Cornell International Institute for Food, Agriculture and Development (CIIFAD), Ithaca, New York.

Regina, K., M. Pihlatie, M. Esala, and L. Alakukku. 2007. Methane fluxes on boreal arable soils. Agric. Ecosyst. Environ. 119:346–352.

Regina, K., E. Syväsalo, A. Honokaa, and M. Esala. 2004. Fluxes of N_2O from farmed peat soils in Finland. Eur. J. Soil Sci. 55:591.

Reicosky, D.C., and D.W. Archer. 2007. Moldboard plow tillage depth and short-term carbon dioxide release. Soil Tillage Res. 94:109–121.

Reicosky, D.C., W.A. Dugas, and H.A. Torbert. 1997. Tillage-induced soil carbon dioxide loss from different cropping systems. Soil Tillage Res. 41:105–118.

Rigler, E., and S. Zechmeister-Boltenstern. 1999. Oxidation of ethylene and methane in forest soils—Effect of CO and mineral nitrogen. Geoderma 90:147–159.

Rozej, A., W. Stepniewski, and W. Malek. 1999. Methanotrophic bacteria in ecosystems. (In Polish.) Postepy Mikrobiologii 38(4):295–313.

Rubio, G., R.S. Lavado, A. Rendina, M. Bargiela, C. Porcelli, and A.F. Delorio. 1995. Waterlogging effects on organic phosphorus fractions in a toposequence of soils. Wetlands 15:386–391.

Ruser, R., H. Flessa, R. Russow, G. Schmidt, F. Buegger, and J.C. Munch. 2006. Emission of N_2O, N_2 and CO_2 from soil fertilized with nitrate: Effect of compaction, soil moisture and rewetting. Soil Biol. Biochem. 38:263–274.

Russow, R., O. Spott, and C.F. Stange. 2008. Evaluation of nitrate and ammonium as sources of NO and N_2O emissions from black earth soils (Haplic Chernozem) based on ^{15}N field experiments. Soil Biol. Biochem. 40:380–391.

Russow, R., C.F. Stange, and H.-U. Neue. 2009. Role of nitrite and nitric oxide in the processes of nitrification and denitrification in soil: Results from ^{15}N tracer experiments. Soil Biol. Biochem. 41:785–795.

Sánchez, L., J.A. Díez, A. Vallejo, and M.C. Cartagena. 2001. Denitrification losses from irrigated crops in central Spain. Soil Biol. Biochem. 33:1201–1209.

Scott, M.J., R.D. Sands, N.J. Rosenberg, and R.C. Izaurralde. 2002. Future N_2O from U.S. agriculture: Projecting effects of changing land use, agricultural technology, and climate on N_2O emissions. Glob. Environ. Change 12:105–115.

Shi, P.L., X.Z. Zhang, Z.M. Zhong, and H. Ouyang. 2006. Diurnal and seasonal variability of soil CO_2 efflux in a cropland ecosystem on the Tibetan Plateau. Agric. For. Meteorol. 137:220–233.

Shiono, K., H. Takahashi, T.D. Colmer, and M. Nakazono. 2008. Role of ethylene in acclimations to promote oxygen transport in roots of plants in waterlogged soils. Plant Sci. 175:52–58.

Sierra, J., and P. Renault. 1995. Oxygen consumption by soil microorganisms as affected by oxygen and carbon dioxide levels. Appl. Soil Ecol. 2:175–184.

Sitaula B.K., S. Hansen, J.I.B. Sitaula, and L.R. Bakken. 2000. Effects of soil compaction on N_2O emission in agricultural soil. Chemosphere Global Change Sci. 2:367–371.

Skiba, U., S. van Dijk, and B.C. Ball. 2002. The influence of tillage on NO and N_2O fluxes under spring and winter barley. Soil Use Manage. 18:340–345.

Smirnoff, N., and R.M.M. Crawford. 1983. Variation in the structure and response to flooding of root aerenchyma in some wetland plants. Ann. Bot. (Lond.) 51:237–249.

So, H.B., and A.J. Ringrose-Voase. 2000. Management of clay soils for rain fed lowland rice-based cropping systems: An overview. Soil Tillage Res. 56:3–14.

Sorz, J., and P. Hietz. 2006. Gas diffusion through wood: Implications for oxygen supply. Trees (Berl.) 20:34–41.

Stepniewski, W., and A. Rozej. 2000. Methanotrophic bacteria and the impact of soil physical conditions on their activity. Int. Agrophys. 14:135–139.

Stepniewski, W. 1975. Influence of mineral fertilisation and watering of arable soil on its oxygenation. (In Polish.) Ph.D. diss. Agricultural University, Lublin, Poland.

Stepniewski, W. 1981. Oxygen diffusion and strength as related to soil compaction. II. Oxygen diffusion coefficient. Pol. J. Soil Sci. 14:3.

Stepniewski, W., and Z. Stepniewska. 2000. Oxygenology of treatment wetlands and its environmental effects. p. 671–678. *In* Proc. 7th Int. Conf. on Wetland Systems for Water Pollution Control. Vol. II. Lake Buena Vista, FL. 11–16 Nov.

Stoop, W.A., N. Uphoff, and A. Kassam. 2002. A review of agricultural research issues raised by the systems for resource-poor farmers. Agric. Syst. 71:249–272.

Suralta, R.R., and A. Yamauchi. 2008. Root growth, aerenchyma development, and oxygen transport in rice genotypes subjected to drought and waterlogging. Environ. Exp. Bot. 64:75–82.

Syväsalo, E., K. Regina, M. Pihlatie, and M. Esala. 2004. Emissions of nitrous oxide from boreal agricultural clay and loamy sand soils. Nutr. Cycl. Agroecosyst. 69:155–165.

Tamimi, S.M., and M.P. Timko. 2003. Effects of ethylene and inhibitors of ethylene synthesis and action on nodulation in common bean (*Phaseolus vulgaris* L.). Plant Soil 257:125–131.

Turski, R., H. Domzal, and A. Slowinska-Jurkiewicz. 1978. Air permeability as an index of the physical state of soil. (In Polish with English summary.) Roczn. Glebozn. 29:3.

Van der Salm, C., J. Dolfing, M. Heinen, and G.L. Velthof. 2007. Estimation of nitrogen losses via denitrification from a heavy clay soil under grass. Agric. Ecosyst. Environ. 119:311–319.

Vasander, H. (ed.) 1996. Peatlands in Finland. Finnish Peatland Society, Helsinki.

Visser, E.J.W., and L.A.C.J. Voesenek. 2004. Acclimation to soil flooding—Sensing and signal transduction. Plant Soil 254:197–214.

Wang, Z.P., R.D. De Laune, P.H. Masscheleyn, and W.H. Patrick, Jr. 1993. Soil redox and pH effects on methane production in a flooded rice soil. Soil Sci. Soc. Am. J. 57:382–385.

Watanabe, T., T. Osada, M. Yoh, and H. Tsuruta. 1997. N_2O and NO emissions from grassland soils after the application of cattle and swine excreta. Nutr. Cycling Agroecosyst. 49:35–39.

Weitz, A.M., E. Linder, S. Frolking, P.M. Crill, and M. Keller. 2001. N_2O emissions from humid tropical agricultural soils: Effects of soil moisture, texture and nitrogen availability. Soil Biol. Biochem. 33:1077–1093.

West, T., and W. Post. 2002. Soil organic carbon sequestration rates by tillage and crop rotation: A global data analysis. Soil Sci. Soc. Am. J. 66:1930–1946.

Whalen, S.C., W.S. Reeburgh, and K.A. Sandbeck. 1990. Rapid methane oxidations in a landfill cover soil. Appl. Environ. Microbiol. 56:3405–3411.

Wiersum, L.K. 1967. Presumed aerotrophic growth of roots of certain species. Naturwissenschaften 54:203.

Wlodarczyk, T., W. Stepniewski, and M. Brzezinska. 2005. Nitrous oxide production and consumption in Calcaric Regosols as related to soil redox and texture. Int. Agrophys. 19:263–271.

Wuebbles, D.J., and K. Hayhoe. 2002. Atmospheric methane and global change. Earth Sci. Rev. 57:177–210.

Yagi, K., H. Tsuruta, K. Kanda, and K. Minami. 1996. Effects of water management on methane emission form a Japanese rice paddy field: Automated methane monitoring. Global Biogeochem. Cycles 10:255–267.

Yan, X., H. Akiyama, K. Yagi, and H. Akimoto. 2009. Global estimations of the inventory and mitigation potential of methane emissions from rice cultivation conducted using the 2006 Intergovernmental Panel on Climate Change Guidelines. Global Biogeochem. Cycl. 23:GB2002, doi:10.1029/2008GB003299.

Yan, X., S. Shi, L. Du, and G. Xing. 2000. Pathways of N_2O emission from rice paddy soil. Soil Biol. Biochem. 32:437–440.

Yu, K.W., Z.P. Wang, A. Vermoesen, W.H. Patrick, Jr., and O. Van Cleemput. 2001. Nitrous oxide and methane emissions from different soil suspensions: Effect of soil redox status. Biol. Fertil. Soils 34:25–30.

Zakrzhevsky, D.A., T.I. Balakhnina, W. Stepniewski, Z. Stepniewska, R.P. Bennicelli, J. Lipiec. 1995. Oxidation and growth processes in roots and leaves of higher plants at different oxygen availability in soil. Russian J. Plant Physiol. 42(2):272–280 (Fiziol. Rast., 41:956–963, 1994, in Russian).

Zausig, J., W. Stepniewski, and R. Horn. 1993. Oxygen concentration and redox potential gradients in different model soil aggregates at a range of low moisture tensions. Soil Sci. Soc. Am. J. 57:906–916.

Soil Biology

Eileen J. Kladivko and M. Jill Clapperton

Soils are teeming with life. From the tiniest microorganisms to the large mammals that live in the soil, the variety and abundance of organisms in soils is truly amazing. These organisms interact in many complex ways and perform numerous functions in both natural and managed ecosystems. For agriculture to sustain productivity for the long term, better understanding and management of soil organisms is essential. This chapter focuses on soil fauna and their interactions with soil microorganisms.

How do soil management practices affect soil fauna? To understand the influence of the many different types of management practices existing today or devised in the future, it is helpful to consider the basic principles affecting soil fauna. How much physical disturbance of the soil are we doing? How much surface mulch or residue do we leave? How much food is there—of what quality and diversity, and over what time period? In general for soil fauna, the abundances and activity are usually greater in systems with less physical disturbances, more surface mulch, more quantities of food for longer periods during the year, and more diversity of food types. This means that systems including no-till, residue mulch, crop rotations, cover crops, perennials, and organic matter additions such as manure or biosolids, will usually have greater faunal activity. The degree of impact is different for different organisms, and thus generalizations should be understood for what they are, generalizations. First, we describe general ecological groups of soil fauna and the major impact of physical disturbance (tillage) on some of them. Next, we discuss individual faunal groups, their impacts on soils and plants, and their response to soil management practices. We then discuss in more detail the principles for increasing soil faunal abundances and activity, as well as the reasons we may want to do this. We end the chapter with some questions for future research.

Ecological Groupings of Soil Organisms

Soil fauna can be categorized in several different ways in addition to their genetic classification. We will adopt the functional guild classification described by Lavelle (1997, 2002) wherein organisms are classified by the type of relationships they have with microorganisms and the type of excrement or biogenic structures they produce. The first functional guild is the *microfood web*, which links the microorganisms to their predators. Protozoa, nematodes, and some microarthropods constitute this guild (Fig. 9|1). These microfauna significantly affect the populations and activity of bacteria and fungi and therefore affect nutrient release from microbial biomass. The importance of this "top-down" control by predators appears to be greater in bacterial-dominated

E.J. Kladivko, Agronomy Dep., Purdue University, West Lafayette, IN 47907 (kladivko@purdue.edu); M. Jill Clapperton, Earthspirit Consulting, 8337 Lamar Trail, Florence, MT 59833 (earthspiritconsulting@gmail.com).

doi:10.2136/2011.soilmanagement.c9

systems than in fungal-based systems (Wardle, 2002). Osler and Sommerkorn (2007) presented a framework for understanding the impacts of different soil faunal groups on nitrogen and carbon cycling and urged greater incorporation of these effects into soil foodweb models.

The second functional guild discussed by Lavelle (1997) is the *litter transformers*, comprising mesofauna and larger arthropods. These fauna physically fragment litter and ingest organic materials and typically do not ingest mineral soil nor dig the soil. Their fecal pellets often serve as an "external rumen" type of digestion (Lavelle, 1997) in which microbes in the fecal pellets degrade the complex organic substrates. The arthropods then reingest the pellets and absorb the released compounds. The fecal pellets have limited impact on soil structure both because they do not contain mineral soil and because they are often reingested by the same organisms.

The third functional guild is the *ecosystem engineers*, comprising earthworms, termites, and ants (Lavelle, 1997). As described by Jones et al. (1994) and Jones and Gutierrez (2007), ecosystem engineers modify or create habitats by "causing physical state changes in biotic or abiotic materials." In so doing, they change the availability of resources to other organisms, either directly or indirectly. Jones and Gutierrez (2007) contended

that an emphasis on *trophic* relations in ecology has resulted in a lack of recognition of the importance of the physical changes brought about by ecosystem engineers. Soil ecosystem engineers dig burrows and build nests, thereby changing soil pore structure and the flow of air, water, and nutrients. Whether the engineering is intentional or accidental may lead to differing impacts. Jouquet et al. (2006) suggested that "intentional" engineers such as termites modify and maintain the habitat for their own purposes, leading to isolated "hotspots" in soil. However, endogeic earthworms may be "accidental" engineers as they burrow to find food, and "accidentally" leave burrows and *casts* (fecal pellets) behind, leading to soil homogenization with time.

The ecosystem engineers often ingest both organic material and mineral soil. Digestion of complex organic substrates is accomplished through mutualistic relationships with the microbes within their guts. Their casts are a mixture of mineral soil and partially decomposed organic materials. These casts help improve soil structure and aggregation and may last for a much longer time than the organism that produced them. These "legacies," or the persistence of physical habitat changes, represent another feature of ecosystem engineering that goes beyond the trophic relationships.

Wardle (2002) suggested that mutualisms are a key part of soil food webs, particularly for the larger fauna. Although competition and predation are important in the microfood web, the meso- and macrofauna have more positive effects on the microorganisms through the "external rumen" (litter transformers) or within-gut (ecosystem engineers) mutualistic relationships. Wardle (2002, p. 43) contended: "classical food web theories… spectacularly fail to account for these sorts of relationships, or for habitat modification by larger organisms for smaller ones." Thus, particularly for the functional guilds of litter transformers and ecosystem engineers, their significance

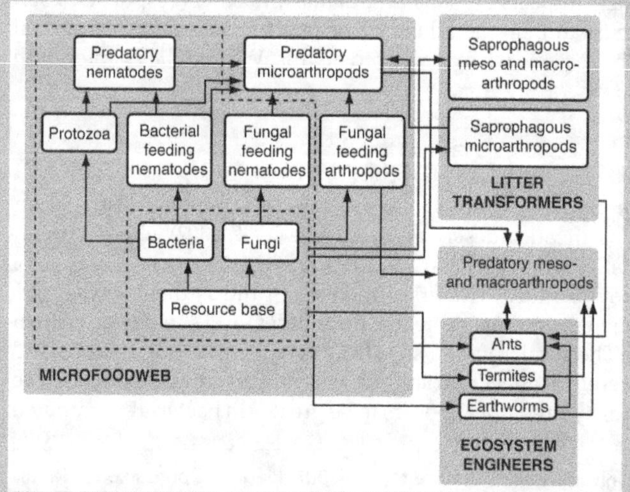

Fig. 9|1. Schematic of the soil food web including the three functional guilds of microfood web, litter transformers, and ecosystem engineers. Adapted from Wardle (2002).

in soil functioning goes well beyond their location within a soil food web structure.

With this background on the concept of functional guilds of soil fauna, we can discuss the first level of management impacts on these guilds of organisms before detailing the impacts on specific organism groups. One of the strongest impacts on soil fauna is one of the most common practices in agriculture, namely tillage. The physical disturbance of tillage, along with the burial of surface residues, greatly reduces the abundance and activities of the litter transformers and the ecosystem engineers, as discussed in reviews by Wardle (1995) and Kladivko (2001). In the case of full-width, full-inversion tillage, the litter-transformers and ecosystem engineers are usually greatly reduced in abundance and activity. While some of the functions of these two guilds are replaced by the action of the tillage implements—fragmenting litter, mixing organic materials into soil—the speed with which this occurs is obviously different with tillage vs. with soil fauna. In addition, some of the other functions of these faunal groups may not be accomplished by tillage, such as stabilization of continuous channels between topsoil and subsoil by deep-burrowing earthworms. It is generally hypothesized that for long-term sustainable agriculture, the soil biota should include organisms in all three functional guilds and not just the microfood webs. Having all three functional guilds present provides a more complex, diverse system, with many more points for potential manipulation and with more opportunity to retain nutrients that might otherwise be lost. Much more research is needed on how these groups can be manipulated to the best advantage for agricultural crops.

Soil Fauna and Rhizosphere Processes

The rhizosphere is undoubtedly the most biologically active region in soil. The plant roots and root exudates are a rich and complex food source for the bacteria and fungi, which in turn become a diverse food source for many of the soil fauna. Bacteria are grazed mostly by protozoa and bacterial-feeding nematodes (Bamforth, 2007), whereas the fungi are grazed primarily by fungal-feeding nematodes and microarthropods (Wardle, 2002). Soil texture and the stability of the soil pore network then interact with the soil biota and plant roots to modify and determine the characteristics of the soil as a habitat. Therefore, any soil or cropping practice that influences the plant, soil, or soil organisms will have direct and/or indirect consequences on rhizosphere processes. The effects on plant growth can range from beneficial (e.g., N_2 fixation and soil stabilization) to detrimental (e.g., disease infection and N immobilization) (Lynch, 1990). Soil fauna as active feeders, and in some cases burrowers, participate in rhizosphere dynamics and processes by affecting plant rooting, root architecture and growth, and soil structure. In this section we discuss microbial–soil fauna interactions in the rhizosphere as part of the detrital food web. We have also emphasized the groups of organisms that have been identified as bioindicators of soil health and ecosystem function. Blair et al. (1996) concluded that invertebrates were useful indicators of soil physical and/or chemical disturbance, and thus soil quality.

Microfauna

In a healthy plant rhizosphere, soil microfauna act primarily to regulate populations of bacteria and fungi by active grazing. The consumed nutrients are then, for example, synthesized into amino acids, nucleic acids, and proteins. The excess nitrogen and phosphorus are returned to the soil as fecal materials. These materials further stimulate biological activity, releasing valuable plant nutrients, and assisting with the continuing process of forming soil aggregates.

Protozoa

The majority of soil protozoa feed on bacteria and to a lesser extent on algae and fungi (Stout and Heal, 1967). This grazing activity can be specific, shifting species diversity in positive or negative directions (Coleman et al., 1984; Griffiths et al., 1993). Protozoa move easily in water films on and between the soil aggregates and organic matter and through water-filled pores. This means they are also sensitive to soil moisture gradients (Adl, 2007) and drought. The movement of many biocides is also linked to water films and moisture gradients; Foissner (1987, 1997) reviewed the effects of biocides on protozoa.

It has been suggested that protozoa are perhaps the best indicators of environmental disturbance (Foissner, 1987, 1997; Lousier and Bamforth, 1990; Ingham, 1994). Bamforth (2007) argued that rapid reproduction and close association with soil particles make protozoa excellent indicators of soil conditions. However, there is a lack of taxonomic expertise, even at the most rudimentary level (Foissner, 1997). In a study of the interaction between time in no-till and recovery of the populations and diversity of protozoa, Adl et al. (2006) showed few differences in abundance of the four most agriculturally relevant groups of protozoa. However, there was a gradual increase in diversity of protozoa with length of time in no-till. Foissner (1997) considered testate amoebae to be the most important indicator group in agroecosystems, although both flagellates and naked amoebae are more abundant in soil. Indeed, testate amoebae were also the most sensitive to tillage (Adl et al., 2006). It seems likely that, similar to their bacterial prey, protozoa are less affected by tillage than larger soil animals.

Protozoa, as tiny as they are, have a significant effect on nutrient cycling and plant uptake (Coleman et al., 1984; Ingham et al., 1985; Griffiths, 1990). They have higher turnover rates, and on a mass per mass basis, have a significantly greater effect on net N cycling (Bamforth, 2007) than most other animals in the rhizosphere. The populations and species diversity of these animals are strongly affected by bacterial populations, and it has been well argued that protozoa, bacteria, and plant rhizosphere interactions are positively correlated (Griffiths, 1990; Acosta-Mercado and Lynn, 2006). In the presence of naked amoebae, the root systems of rice (*Oryza sativa* L.) plants had more elongated lateral roots, indicative of more highly branched root systems. Moreover, the nutrients released from grazed bacteria by the protozoa combined with the changes in root architecture correlated to a 45% increase in N uptake by the plant (Kreuzer et al., 2006). However, the protozoa are less effective at stimulating plant growth and nutrient uptake when mycorrhiza are present in the root (Herdler et al., 2008).

In terms of the microfauna, protozoa and nematodes are both known to be important consumers of bacteria and fungi in the soil and rhizosphere. Griffiths (1990) concluded that the rhizosphere biomass of microbial-feeding nematodes was greater than that of protozoa and that nematodes were more important in terms of nutrient cycling than protozoa. More recently, a strong case has been made for using nematode diversity and functional groups as bioindicators of soil quality and health (Neher et al., 1995; Neher 2001; Yeates, 2003).

Nematodes

Nematodes are the most abundant multicellular animals on earth (Yeates, 2007a). They are also abundant and diverse in agroecosystems. Nematode abundance and species assemblages are strongly affected by soil texture and structure (Yeates, 2007a). Like protozoa, nematodes move on water films through water-filled pores and are strongly influenced by compaction (Yeates, 2007a). Although nematodes have long been studied primarily as plant pathogens (Overgaard Nielsen, 1967), nematodes have been increasingly shown to have positive influences on agroecosystem processes, with size and diversity of nematode populations likely contributing significantly to ecosystem productivity (Yeates, 2007a). Nematodes are known to be concentrated in the rhizosphere (Ingham et al., 1985), primarily grazing on bacteria and to a lesser extent on fungi. Some stylet-bearing nematodes apparently have no adverse effects on plant growth and have now been referred to as *root-associated nematodes* (Yeates et al., 1993; Yeates, 2007a). Slight root damage from root pathogenic nematodes, such as species of *Heterdera*, *Pratylenchus*, and *Meloidogyne*, can stimulate microbial activity in the rhizosphere due to the increased leakage of photosynthate (Denton et al., 1999; Yeates et al., 1999), although it is likely that this increased rhizodeposition may be a host plant–nematode specific reaction (Poll et al., 2007). Demonstrating a convoluted rhizosphere interaction, Atul-Nayyar et al. (2008) showed that in a mixed stand of Russian wild rye (*Psathostachys juncea* Fisch. Nevski) and alfalfa (*Medicago sativa* L.), the rye stimulated the populations of fungal-feeding and omnivorous nematodes, which under low P fertility levels conspired to reduce the essential mycorrhizal association for the alfalfa, leading to significant yield loss.

Nematode populations have been shown to increase in the rhizosphere in the presence of most crop species. However, significant population differences have been reported between specific crops, with peas (*Pisum sativum* L.) supporting greatest nematode biomass (Griffiths, 1990). It is well known that rhizosphere-associated microbial communities differ either subtly or dramatically depending on the plant species. Thus, it is not surprising that the species diversity and populations of bacterial-feeding nematodes can be affected by the species composition of the rhizosphere microbial community. In the presence of enhanced populations of bacterial feeding nematodes, there was a significant change in root architecture associated with increased content of indole acetic acid (IAA) and shifts in the structure of the microbial community (Mao et al., 2007). Selective grazing of specific groups of microbes by nematodes is known (Ruess et al., 2000; Venette and Ferris, 1998; Blanc et al., 2006) and clearly could affect rhizosphere community dynamics.

Nematode grazing has a strong influence on nutrient cycling in the rhizosphere. Griffiths (1990) argued that nematodes are more able to exploit low bacterial densities in the rhizosphere compared with protozoa, and that for a given biomass nematodes mineralize six times more N than protozoa. The net effect of nematodes on nutrient cycling in the rhizosphere and detrital food web is not easy to predict as plant species composition, organic matter quality, and soil texture strongly influence behavior, life cycles, species assemblages, and N output of nematodes (Yeates, 2007b).

Nematodes are also an important part of the detrital food web, being associated with mites and collembolans, feeding mostly in discreet resource patches (Griffiths and Caul, 1993; Chen and Ferris, 1999; Yeates et al., 1999). Indeed, results from long-term field studies have shown significant relationships between nematode functional groups and the processes regulating decomposition (Yeates et al., 1999). The C/N ratio of nematodes is usually higher than that of the bacteria they eat, leading to less than 40% of the N nematodes ingest being used for growth (Ferris et al., 1997). Thus, nematodes are net excreters of N

(Yeates, 2007b). Increases in fungal-feeding nematodes can result in some N immobilization (Ferris and Matute, 2003; Wang et al., 2004; Georgieva et al., 2005), and increased microbial turnover facilitated by nematodes, although mostly positive for N, can affect P dynamics (Forge et al., 2008).

Examining nematode abundance, species diversity, and associations has evolved into one of the most powerful bioindicators of soil quality and health, and ecosystem function (Porazinska et al., 1999; Neher, 2001; Ferris and Matute, 2003; Yeates, 2003). The abundance and assemblage of bacterial- and fungal-feeding nematodes appears to mirror other bacterial- and fungal-feeding soil fauna, reflecting the recent availability of their food, namely populations of bacteria and fungi (Chen and Ferris, 1999; Ferris et al., 2001; Neher, 2001; Ferris and Matute, 2003; Yeates, 2003).

Mesofauna

The mesofauna have a significant effect on nutrient cycling from residues and organic matter, and particularly enchytraeids have a significant effect on soil structural stability. These soil animals influence rhizosphere processes more indirectly compared with the mostly predator–prey relationships associated with the microfauna. Organisms in this group have been studied and identified as potential bioindicators of soil health. Oribatid mites in particular have been suggested as useful indicators of soil quality (Franchini and Rockett, 1996; Clapperton et al., 2007). However, the abundance and species diversity of oribatid mites can vary widely, making it difficult to decipher limiting factors (Paoletti et al., 2007; Osler et al., 2008). Collembola are used in standardized avoidance and toxicological testing because of their demonstrated sensitivity to environmental pollutants. Microarthropods and collembola tend to be less affected by soil physical disturbance such as tillage (Wardle, 1995; Miura et al., 2008; Osler et al., 2008). The enchytraeids, like their cousins the earthworms, are strongly favored by less soil disturbance (Topoliantz et al., 2000; Miura et al., 2008) and by agricultural practices that increase the quality of the soil organic matter.

Microarthropods

Mites

"More than any other habitat the soil litter stratum is the province of mites"
Walter and Proctor (1999).

The soil mites (Acari) are chelicerate arthropods related to spiders, and are perhaps the most abundant microarthropods in many soil types. They are primarily associated with the detrital food web, although there are a number of species that are predatory, including preying on other soil animals (Wallwork, 1976; Seastedt, 1984). Like most soil fauna, mites are concentrated where resources and food quality are greatest, the litter or residue layers and rhizosphere. In general, soil mites are considered to have a positive influence on plant health, as major predators of root pests, particularly nematodes (Walter and Proctor, 1999). Compared with the rhizosphere, interactions between mites and other members of the detrital food web are reasonably well understood, especially where predation is the primary role (Walter and Proctor, 1999). In the soil, mites contribute to regulating decay and cycling nutrients by chewing up organic matter into smaller pieces, preying on other decomposer fauna, feeding on bacteria and fungi, and transporting microbes to new resource patches (Moore et al., 1988). Transported microbes are mixed with fresh resources in fecal pellets that are distributed throughout the rhizosphere.

The more armored soil mites, such as the oribatid and astigmatid mites, feed by chewing fungal spores and hyphae and strands of algae and swallowing them. The larger, more armored oribatids will also eat dead vegetation and animal material (dung and carrion) that are colonized by the primary decomposers, bacteria and fungi. Fungal spores are often viable after gut passage and can establish easily in the fecal pellets. Thus, mites contribute directly to reducing the amount of *detritus*, or dead organic matter. The ability of soil mites to reduce detritus and microbial biomass to fecal pellets directly modifies the soil habitat, and the net results are various microhabitats for more soil microbes and fauna. The feeding and exploratory activities of soil mites create spaces in decaying litter. However, these same mites appear to have a limited ability to create soil pores or directly modify soil structure in the mineral soil, but will make spaces between soil aggregates, once again modifying the soil habitat (Norton, 1985). Additional information about mites of all kinds is presented by Walter and Proctor (1999).

Collembola.

Collembola are primitive Apterygote insects. They most often are called *springtails* because many species can jump by means of a lever or spring tail termed a *furcula* that is attached to the bottom of the abdomen. Collembola occur in most habitats, often reaching abundances of more than 100,000 m^{-2}. They tend to be distributed mostly in the upper 10 cm of the soil in patches, feeding mostly on fungi (Hunt et al., 1987). Plant–fungal interactions, including disease and symbiotic interactions, are likely strongly influenced by faunal grazing of the fungal hyphae and spores (Bonkowski et al., 2000; Gange, 2000). Collembola have been shown to prefer pathogenic (disease causing) fungi over saprophytic fungi (Lartey et al., 1994), and saprophytic fungi over mycorrhizal fungi (Klironomos and Kendrick, 1996). Collembola feeding on root pathogenic fungi is known to improve root health (Curl and Harper, 1990; Larsen et al., 2008) and alter competitive patterns between root pathogens (Sabatini et al., 2002). Thus, maintaining healthy populations of collembola in the rhizosphere can be beneficial to plant health and nutrient uptake.

Mycorrhizal fungi are known to form beneficial plant fungal associations with more than 85% of all land plants. It has been shown that grazing of mycorrhizal fungi by collembola can reduce the effectiveness of the mycorrhizal plant symbiosis (Schreiner and Bethlenfalvay, 2003; McGonigle and Fitter, 1988; Warnock et al., 1982). At first, this information leads to the conclusion that collembola are detrimental to plant health because grazing on mycorrhizal hyphae disrupts the plant benefits from the mycorrhiza. However, some hyphal grazing by collembola has been shown to stimulate nutrient uptake by plants, and generally collembola least prefer to feed on mycorrhizal fungi (Schreiner and Bethlenfalvay, 2003; Larsen et al., 2008). The essential lesson is that fungal–collembola interactions are species specific (Larsen et al., 2008), and collembola associations can influence the fungal activities (Tordoff et al., 2008). Similarly to mites, the collembola have been considered as possible bioindicators of soil health and quality. However, it was shown that both mites and collembola were not significantly affected by land use or type when evaluated for use in the index of soil biological quality (Parisi et al., 2005).

Enchytraeids

The enchytraeids belong to the phylum Annelida and class oligochaeta, like lumbricid earthworms. These animals range in size between 2 and 40 mm long and have colorless and segmented bodies. Unfortunately, they are often mistaken for root-feeding insect larvae, despite not having any visually obvious mouth-parts. Enchytraeids have some digging ability and may improve small-scale water and air infiltration in soil (Jänsch et al., 2005). The feeding activity of these animals gives soil a fine granular structure and increases aggregate stability (Topoliantz et al., 2000; Jänsch et al., 2005).

Enchytraeids tend to feed mostly in the upper 10 cm of the soil on slightly to strongly decomposed microbial and plant materials. Isotopic analysis using ^{14}C showed that many grassland species feed predominantly on organic material that is approximately 5 to 10 yr old (Briones and Ineson, 2002). Didden (1993) suggested that 80% of the enchytraeid diet was microbial remains and the other 20% organic matter. However, some species feed only on undecomposed organic matter (Didden, 1993). Thus, the feeding habits of enchytraeids suggest that these animals likely play a greater role in active nutrient cycling than their size might suggest (van Vilet et al., 2004; Marinissen and Didden, 1997).

Indeed, enchytraeids have the ability to significantly increase the decomposition of the particulate organic matter and stimulate N cycling in agroecosystems (Miura et al., 2008; van Vilet et al., 2004). They have been shown to use readily available C sources, specifically dissolved organic carbon decreasing the ratio of N immobilization to N mineralization in the decaying residues making more N available to the plants during the growing season (van Vilet et al., 2004). Microbial grazing by some species of enchytraeid can increase the release of P (Williams and Griffiths, 1989; Parfitt et al., 2005), which can influence N dynamics in pastures (Parfitt et al., 2005). These animals have a known role in soil nutrient cycling and structural stability, there is an availability of taxonomic expertise, and they have already been successfully used in ecological assessment studies. This supports the strong arguments made for considering both individual species and feeding groups of enchytraeids as indicators of soil biological health and soil quality, especially when they are considered within a suite of other bioindicators (Didden and Römbke, 2001; Jänsch et al., 2005).

Macrofauna

Ants, termites, millipedes, spiders, beetles, insect larvae, crustaceans, and earthworms often visually modify the soil habitat. They include the ecosystem engineers and some of the litter transformers as well as predators. They fragment and ingest plant and animal litter, mix it deeper into the soil profile, and release the nutrients that were previously unavailable. The macrofauna, much like the mesofauna, also carry microbial propagules and smaller animals on their bodies, and can excrete viable fungal spores and microbes into new microhabitats. The ecosystem engineers create new microhabitats and alter physical properties and processes in the soil, which can further modify activity of microorganisms and smaller soil fauna. The macrofauna are often operating outside of the rhizosphere, such as spiders and beetles in the detritusphere, and they may be the creator of their own functional domain, such as the *drilosphere* (earthworms), *termitosphere* (termites), or *myrmecosphere* (ants) (Beare et al., 1995; Lavelle, 2002).

The soil macrofauna are likely the main organisms connecting the aboveground (foliage-based) and belowground (detritus-based) food webs (Wardle, 1995, 2002). The predatory soil macrofauna (spiders and beetles) feed and are active mostly near the soil surface and litter layers (Wardle, 1995), while predators of larger soil fauna are often aboveground vertebrates (Wardle, 2002). The linkage between the above- and belowground biota is an active area of ecological research (Wardle et al., 2004).

Spiders (Araneidae) and the large litter-dwelling arthropods such as carabid beetles (Coleoptera: Carabidae) are generalist predators that can regulate insect populations (Reichert and Bishop, 1990; Thiele, 1977; Kromp, 1999). Their abundances are greatly reduced by tillage practices due to physical disturbance and abrasion from tillage as well as the reduction in residue cover (Wardle, 1995). No-till increases both populations and diversity of carabids (Hatten et al., 2007), by increasing the abundance of alternate prey choices and suitable microhabitats, especially in no-till systems that include intercropping or cover cropping (Cárcamo and Spence, 1994; Pavuk et al., 1997). Carabid beetles are known to reduce armyworm (Lepidoptera: Noctuidae), cabbage root fly (*Delia radicum* L.), and some aphids and aphid eggs (Clark et al., 1994; Finch, 1996). Spiders tend to be more sensitive to disturbance than carabids, with populations and diversity of spiders being adversely affected by both tillage and pesticides (Reichert and Lockley, 1984; Hummel et al., 2002). Spider abundance and diversity were found to increase when within-field heterogeneity increased, such as with reduced tillage, intercropping, or weedy patches (Samu et al., 1999). Carabids and spiders are critical for insect control, moving through the litter and consuming larvae and eggs. Overall, however, carabid beetles and spiders appear to have a lesser role in soil ecosystem function than the other fauna discussed in this section.

Ant nests are widespread in most habitats, and unlike termites, ants are mostly omnivorous, consuming seeds, plant tissue, insect carcasses, and dung in some cases (Whitford, 1996). The ants are known to take these food sources from aboveground and cache them in food storage chambers 10 to 30 cm belowground. These chambers are microbe and nutrient (particularly P) enriched, exerting a significant effect on the resident soil biota (Boulton et al., 2003). Dauber et al. (2008) reported greater root arbuscular mycorrhizal colonization of grasses due to ant activity. Amador and Görres (2007) found different microbial communities in ant mounds compared with earthworm casts or bulk soil, and speculated that at least some of the differences were due to the lower water-holding capacity of the primarily sand-sized particles in the ant mounds.

Termites in subtropical and tropical ecosystems are often the dominant soil fauna (Arshad et al., 1982). They are efficient at decomposing cellulose- and lignan-rich plant materials, and termite nests significantly alter soil properties (Arshad, 1982).

However, termites have not been shown to positively influence biota (Arshad et al., 1982). Indeed, it was difficult to find other soil fauna in the presence of subterranean termite mounds in pasture soils in Argentina (Clapperton, personal observation, 2004). Jiménez et al. (2008) measured higher concentration of C and NH_4 in termite mounds compared with ant nests or control soil, with the greatest difference occurring between the top center of the mound and the control soil. They also discussed the temporal and spatial variations that result from new material being added to the top center of an active mound and the implication this has for sampling and study of nutrient heterogeneity.

A general theme evolving from observations of the ecosystem engineers is one of heterogeneity of nutrients as a result of the engineering activity. The termite mounds and ant nests had different composition within different parts of the structure (Jiménez et al., 2008) and compared to bulk (control) soil. Earthworm casts and burrow walls are usually different than bulk soil (Edwards et al., 1995), with the degree and direction of difference depending on the timing of sampling compared with the deposition of the structure (Blair et al., 1995; Brown and Doube, 2004; Lavelle et al., 2004). Spatial heterogeneity of nutrients, carbon sources, and microbial communities as a result of the engineers suggests that many typical soil sampling protocols may not suffice to represent the potential impacts of the fauna on plant growth. The drilosphere, including casts, midden, and burrow wall lining, is a hotspot for microbial activity and carbon and nutrient cycling. Roots may proliferate in burrows and other structures left by the engineers. There is a great need to consider better ways of sampling the soil and characterizing the in situ structure of soil microhabitats and the ways that roots and soil biota interact with this complex, heterogeneous structure (Young and Crawford, 2004).

Earthworms

Earthworms are an important group of soil macrofauna in many soils. As ecosystem engineers (Lavelle et al., 2007) they have significant impacts on soil properties and subsequently on biological activity through their feeding, burrowing, and casting activities. Several excellent books provide detailed information about earthworm biology and ecology (Lee, 1985; Edwards and Bohlen, 1996). Two additional books contain reviews of earthworm ecology, their responses to agricultural management, and their effects on soil structure formation, microbial interaction, and carbon and nutrient cycling (Hendrix, 1995; Edwards, 2004).

Earthworms are often grouped into three major behavioral classes: *litter-dwellers* (*epigeic*), *shallow soil-dwellers* (*endogeic*), and *deep-burrowers* (*anecic*) (Bouché, 1977, as described by Lee, 1995). The litter-dwellers live in the litter layers of relatively undisturbed soils and are often absent from agricultural fields. The endogeics live primarily in the shallow soil layers, although they may move deeper during winter (hibernation) and summer (aestivation) resting phases. They ingest mineral soil and plant residues as they burrow throughout the topsoil and generally do not construct permanent burrows. The anecics dig large, deep, permanent burrows extending 1 to 2 m deep or more. *Lumbricus terrestris* L., the predominant anecic in northern temperate zones, pulls residues down into its burrow and constructs middens over the opening of the burrow. These middens, composed of casts, plant residues, and soil, not only serve as a food source and protection for the resident earthworm but also provide habitat for other earthworm species within the midden (Butt and Lowe, 2007).

Earthworm abundances are generally higher under no-till than conventional tillage systems (Kladivko, 2001). Residues left on the soil surface act as a mulch, buffering the changes in soil moisture and temperature and thereby providing longer time periods for activity. The surface residues are also available as a food source to the earthworms for a longer time than if they were mechanically incorporated with a tillage implement. The beneficial effects of no-till on earthworms is even greater for the deep-burrowers than the shallow soil-dwellers (Kladivko et al., 1997) because the deep-burrowers construct permanent burrows and feed primarily on surface residues.

Earthworms improve soil physical properties through their burrowing and casting activity. The deep, nearly vertical burrows of *L. terrestris* can increase water infiltration and root growth, while the shallow horizontal burrows of the endogeic species may increase porosity but not affect infiltration (Shipitalo and LeBayon, 2004). Soil aggregation is generally increased by earthworm casting and burrowing activity, although the specific effects vary with earthworm species, the food source, and many other soil and environmental factors (Tomlin et al., 1995). Earthworms also help mix surface organic materials and nutrients into the soil, which may be more important in agricultural systems with relatively little mechanical mixing, such as pastures (Lee, 1995) and no-till fields (House and Parmelee, 1985). Earthworms have also been found to be an important contributor to organic matter mixing and aggregate formation during soil development on open-cast coal mining spoil heaps following reclamation (Frouz et al., 2007).

The effects of earthworms on microorganisms, nutrient cycling, and organic matter degradation and stabilization are complex and depend on the time and spatial scales of observation. Earthworm casts usually contain higher concentrations of carbon, available nutrients, and microorganisms than the surrounding mineral soil because they contain a mixture of partially decomposed organic materials along with the mineral soil (Blair et al., 1995; Edwards et al., 1995). Overall microbial populations in freshly deposited casts are usually higher than in surrounding soil, although some species may be decreased during gut passage (Brown and Doube, 2004). Freshly deposited casts are hotspots of higher microbial activity and carbon and nutrient mineralization, providing a favorable environment for plant root growth (Brown and Doube, 2004; Lavelle et al., 2004; Bohlen et al., 2004). As the casts age, however, the organic carbon compounds may become physically protected in the stabilized casts, leading to lower mineralization rates. All these authors stressed that the effects of earthworms on the dynamics of soil organic matter stabilization or mineralization depend on the time scale considered, as well as on the soil type, food source, climate, and particular earthworm species. The transient effects of earthworms on soil processes has also been found with nitrous oxide emissions from tilled grassland soils, for example, in that emission rate were first higher and then lower in the presence of earthworms vs. their absence (Bertora et al., 2007).

Principles for Managing Soil Fauna

If we wish to manage the soil for the beneficial functions of soil fauna for production agriculture, we can begin with the idea of managing the soil as a habitat. This idea includes understanding the roles of the organisms themselves in creating and maintaining diverse soil habitats. Waid (1997) presented evidence for considering *metabiosis*, which describes relationships between organisms whereby the first organism or group of organisms must modify the habitat before other organisms dependent on the created habitat can move in. An example is the way plant root exudates influence the rhizosphere microbial community by actively recruiting beneficial organisms or antibiotic bacteria into the rhizosphere. The unique environment that defines the drilosphere is another example. Once the process of building the habitat has begun by limiting the amount of soil disturbance, and the soil structure develops a more continuous soil pore network (Young and Ritz, 2000), the functional groups of organisms will colonize the habitat.

Soil Biodiversity and Soil Function

Questions about the relative importance of soil biodiversity are receiving more attention in recent years (Hassall et al., 2006). On the one hand, there is the "enigma of soil diversity" (Wardle, 2002), whereby so many different species of organisms live in the soil, "seemingly in the same space" (Coleman, 2008). On the other hand, the question is raised about whether it is important to try to maintain or increase soil biodiversity, particularly within our highly managed agricultural systems. The diversity of soil organisms can be explained in part by the great diversity of microhabitats present in soil over time and space and the fact that many organisms are dormant for much of the time (Coleman, 2008). Many soil fauna are omnivorous and may feed at several trophic levels, and there appears to be a high degree of functional redundancy. Bardgett (2005) noted how much of the evidence suggests that the presence of specific species or functions or trophic levels seems to be more significant than the number of species per se. Thus, the presence of a key dominant species may be more important for soil functioning than a large number of different species within a group (Coleman, 2008). Soil biodiversity is often discussed in terms of species richness within an individual group (e.g., bacteria, protozoa) rather than in terms of the presence or absence of functional guilds, but in disturbed ecosystems, often the meso- and macrofauna are greatly reduced in abundance and diversity (Wardle, 1995), such that the guilds of litter transformers and ecosystem engineers may not contribute much to soil function. Therefore, an initial goal for agricultural management might be to promote the active presence of all three functional guilds, with a few key species from the litter transformers and ecosystem engineers, and secondarily to consider the species diversity within each of the groups. With earthworms, for example, one species from each of the three main functional groups (anecic, endogeic, epigeic) may be more valuable for soil function than five different endogeic species, since earthworms appear to have a high degree of redundancy (Lee, 1995). Thus, it appears that having representatives of organisms that perform different functions has an impact, whereas an increase in species number or richness per se may not (Bardgett, 2005).

The overall question of how much soil biodiversity is "enough" is far from resolved. As discussed by Wolters (2001), the identification of important species increases as more functions in the soil are considered. Functions discussed in this chapter include many that go far beyond nutrient cycling and have complex interactions with the soil, plant, and other organisms, including pest and disease suppression, organic matter stabilization or mineralization, and microhabitat formation. Even if the goal is to conserve only the important species, we would not know how to do it (Wolters, 2001). There is some evidence that ecosystem stability increases with increased soil biodiversity, as discussed by Brussaard et al. (2007) and Coleman (2008). Species that appear to be functionally redundant during "average" environmental conditions may respond differently to environmental stresses, buffering the ecosystem when conditions change (Cole et al., 2006). Using functional classifications of soil fauna may help discern the importance of functional redundancy in soil and the impact of this on ecosystem functioning (Decaëns et al., 2006). However, this is always limited by our current understanding of which functions are "important" from our anthropocentric view, as Lee (1995) cautioned.

Management Practices to Enhance Soil Fauna

Based on current knowledge, it seems to be a reasonable goal to try to increase soil faunal diversity and abundance in agroecosystems. As discussed in earlier sections of this chapter, many soil faunal groups have low abundance and diversity in row-crop agriculture, particularly in tilled systems. The goal would be to encourage the presence and activity of at least a few representatives of the ecosystem engineers and the litter transformers, and possibly to increase the diversity of the microfood web as well. How can this be accomplished? We have discussed this to some extent for individual groups throughout the chapter, but here we summarize and reiterate a few key concepts, in particular physical habitat, food quality and quantity, and the chemical environment.

Reducing tillage intensity is probably one of the most significant factors for increasing soil faunal activity (Kladivko, 2001). The increase in surface residue cover with no-till or reduced tillage provides a habitat and food source for anecic and epigeic worms, as well as many of the litter transformers. The surface mulch buffers soil moisture and temperature changes, leading to longer time periods for active functioning for many fauna. The reduced physical disturbance of the system promotes growth and activity of species highly sensitive to tillage. A greater variety of microhabitats can form and persist in these relatively undisturbed systems, which also fosters a diversity of fauna. Activity in the microfood web may also be increased by no-till, but the effects are generally smaller than for the meso- and macrofauna (Wardle, 1995).

Growing a greater diversity of crops or cover crops to improve food quality and increase quantity can be an important way to stimulate activity and diversity within the microfood web in addition to feeding the meso- and macrofauna. More or better food resources are often found to increase soil fauna populations, regardless of their trophic level, and may increase diversity as well. Although microorganisms may compete with plants for nutrients, their predators excrete excess nutrients and play an important role in nutrient cycling and release (Lavelle, 1997; Wardle, 2002; Osler and Sommerkorn, 2007). Providing regular sources of high quality organic matter can help keep the cycle going and provide benefits for plant nutrient uptake, although the synchrony of nutrient release and plant requirements is still poorly understood.

Practical ways to provide more and better food resources include adding and growing organic matter. Adding organic materials such as manure, composts, or biosolids can increase the food quality and quantity available in a typical row-cropping system. Such organic matter amendments are known to increase populations and diversity of soil fauna, especially beneficial nematodes (Sikora, 1992; Porazinska et al., 1999; Forge et al., 2005). Growing diverse cover crops or crop rotations, especially those with some hay crops in the rotation, provides a diversity of food resources and likely contributes to a greater diversity of soil fauna. The quantity of organic matter additions to the soil is greater with productive hay crops or pasture than in typical annual row-cropping systems due to a longer period of root growth and exudation, and the mulch protection and undisturbed soil provide additional benefits for the fauna. A diversity of plant types provides a more balanced C/N ratio of organic matter (Heal et al., 1997) and may help stimulate a wider variety of organisms in the microfood web within the rhizospheres of the different plants. Organic farming production methods often increase faunal abundance and diversity as well (Birkhofer et al., 2008).

Improved soil physical conditions can be both a cause and effect of increased faunal abundance and diversity. A well-structured soil has a wide variety of microhabitats, including a mixture of air- and water-filled pores of different sizes and shapes, different aggregates, and organic matter in different stages of decomposition. Soil microfauna (protozoa and nematodes) move through water films and water-filled pore networks to reach their prey and can be greatly affected by compaction and the resulting changes in pore continuity and size (Young and Ritz, 2000). Compaction can also cause anaerobic conditions that strongly inhibit protozoan and nematode metabolism. Many of the factors that are known to improve soil structure, such as organic matter additions, rotations with hay crops, and less intensive tillage, are also factors that stimulate soil faunal activity, which in turn further improves soil structure. This type of self-reinforcing system, or "self-organizing system" (Perry, 1995), can be used to encourage beneficial feedbacks and mutualisms for improvement of overall soil functioning. A limitation is that we have not been able to quantify all the economic and environmental benefits of soil faunal functions or the detriments of fauna-poor soils over different scales of time and space. Better quantification of soil fauna services, including ecosystem services, is needed before soil fauna can be value appropriately in the economic sphere (Huguenin et al., 2006).

The potential impacts of chemical disturbances from pesticide or fertilizer additions to the soil agroecosystem are the subject of much debate. Unfortunately, documenting the effects of pesticides on soil fauna is a chapter unto itself. There have been a number of reviews over the years on specific organisms, including those by Sikora (1992) (nematodes), Didden and Römbke (2001) (enchytraids), and Edwards and Bohlen (1992) (earthworms). However, a rather sweeping generalization based on review of the literature is that the effects of pesticides on soil fauna tend to follow the pattern: insecticides are more toxic than herbicides, herbicides tend to be mostly harmless, and fungicides have a more variable effect (Foissner, 1997; Hautier et al., 2005; Liphadzi et al., 2005; Pekár and Benes, 2008). The problem is separating the effects of pesticides on the target organisms from the effects on nontarget organisms (Joy et al., 2005). Broad-spectrum pesticides are just that and can have any number of nontarget effects on soil organisms (Didden and Römbke, 2001). Mixtures of pesticides can affect the dynamics of most soil animals, depending on the interactions between the pesticides, soil organic matter, and species sensitivities (Pereira et al., 2005). Pesticide-degrading microorganisms can interact with earthworms, affecting the distribution and survival of introduced bioremediation agents (Monard et al., 2008). Some pesticides, for example herbicides, have indirect effects on soil fauna by removing green canopy, which modifies the temperature and moisture and reduces the number of places to hide or wait for prey. This in turn can modify other predator–prey relationships and even diminish effectiveness of integrated pest management plans.

There continues to be much research into the short- and long-term effects of genetically modified and transgenic crops on soil biota. Again, generally speaking the conclusion from a number of varied research findings is that there are some small and transient negative and positive effects on soil fauna. Regardless, it has been concluded that other land management practices such as tillage, crop rotation, and pest management regime had significantly larger effects on the soil biology than the use of either insect- or herbicide-resistant crops (Griffiths et al., 2007a,b, 2008; Clark and Coats, 2006; Liphadzi et al., 2005). Addison et al. (2006) showed that spraying spores of *Bacillus thuringensis* for insect control also did not have any effect on nontarget soil microarthropods. The use of selective herbicides in herbicide-resistant crops can affect the weed community structure and N cycling dynamics that can indirectly influence soil fauna (Griffiths et al., 2007b; Bohan et al., 2005), as pointed out earlier in this section.

Conclusions and Areas for Future Research

Soil fauna are an important part of the soil ecosystem. Through their roles in nutrient cycling, organic matter degradation and stabilization, pest suppression, and soil physical property improvement, the fauna have significant impacts on soil ecosystem services. There is a great need to better understand the complex interactions among various groups of fauna, microorganisms, and plant roots to improve our management of agricultural soils. Interdisciplinary research at the soil community level should also consider the interactions between the fauna and the physical habitats within the soil, especially as modified by the soil ecosystem engineers.

Soil fauna are "patchy" in space and time, and this needs to be more fully considered in research and sampling design as well as in the implications for soil management for crop productivity or ecosystem services. Whether particular organisms increase or decrease carbon conservation or nutrient release to plants depends on the time scale considered, whether it is hours, days, months, or years. Patchiness in space can occur with organisms that concentrate in resource-rich patches (residues, animal dung), but it can also occur as a result of the activity of ecosystem engineers (e.g., in the drilosphere). Typical soil sampling protocols for soil management will not likely capture the full impact of soil fauna on the functioning of the soil ecosystem. New approaches and tools are needed to assess the in situ environment and the locations and functions of the organisms within that environment more appropriately over a continuum of time scales.

References

Acosta-Mercado, D., and D.H. Lynn. 2006. Contrasting soil ciliate species richness and abundance between two tropical plant species: A test of the plant effect. Microb. Ecol. 51:453–459.

Addison, J.A., I.S. Otvos, J.P. Battigelli, and N. Conder. 2006. Does aerial spraying of *Bacillus thuringensis* subsp. *Kurstaki* (BtK) pose a risk to nontarget soil microarthropods. Can. J. For. Res. 36:1610–1620.

Adl, S.M. 2007. Motility and migration rate of protozoa in soil columns. Soil Biol. Biochem. 39:700–703.

Adl, M.S., D.C. Coleman, and F. Read. 2006. Slow recovery of soil biodiversity in sandy loam soils of Georgia after 25 years of no-tillage management. Agric. Ecosyst. Environ. 114:323–334.

Amador, J.A., and J.H. Görres. 2007. Microbiological characterization of the structures built by earthworms and ants in an agricultural field. Soil Biol. Biochem. 39:2070–2077.

Arshad, M.A. 1982. Influence of the termite *Macrotermes michaelseni* (Sjost) on soil fertility and vegetation in a semi arid savanna ecosystem. Agro-Ecosystems 8:47–58.

Arshad, M.A., N.K. Mureria, and S.O. Keya. 1982. Effect of termite activities on soil microflora. Pedobiology 24:161–167.

Atul-Nayyar, C. Hamel, T.A. Forge, F. Selles, P.G. Jefferson, K. Hanson, and J. Germida. 2008. Arbuscular mycorrhizal fungi and nematodes are involved in negative feedback on a dual culture of alfalfa and Russian wild rye. Appl. Soil Ecol. 40:30–36.

Bamforth, S.S. 2007. Diversity of protozoa. p. 2005–2014. *In* G. Benckiser and S. Schnell (ed.) Biodiversity in agricultural production systems. CRC Press Taylor & Francis Group, Boca Raton, FL.

Bardgett, R.D. 2005. The biology of soil: A community and ecosystem approach. Oxford Univ. Press, Oxford, UK.

Beare, M.H., D.C. Coleman, D.A. Crossley, Jr., P.F. Hendrix, and E.P. Odum. 1995. A hierarchical approach to evaluating the significance of soil biodiversity to biogeochemical cycling. Plant Soil 170:5–22.

Bertora, C., P.C.J. van Vliet, E.W.J. Hummelink, and J.W. van Groenigen. 2007. Do earthworms increase N_2O emissions in ploughed grassland? Soil Biol. Biochem. 39:632–640.

Birkhofer, K., T.M. Bezemer, J. Bloem, M. Bonkowski, S. Christensen, D. Dubois, F. Ekelund, A. Fliessbach, L. Gunst, K. Hedlund, P. Maeder, J. Mikola, C. Robin, H. Setala, F. Tatin-Froux, W.H. Van der Putten, and S. Scheu. 2008. Long-term organic farming fosters below and aboveground biota: Implications for soil quality, biological control and productivity. Soil Biol. Biochem. 40:2297–2308.

Blair, J.M., P.J. Bohlen, and D.W. Freckman. 1996. Soil invertebrates as indicators of soil quality. p. 273–292. *In* J.W. Doran and A.J. Jones (ed.) Methods for assessing soil quality. SSSA Spec. Publ. 49. SSSA, Madison WI.

Blair, J.M., R.W. Parmelee, and P. Lavelle. 1995. Influences of earthworms on biogeochemistry. p. 127–158. *In* P.F. Hendrix (ed.) Earthworm ecology and biogeography in North America. Lewis Publ., Boca Raton, FL.

Blanc, C., M. Sy, D. Djigal, A. Brauman, P. Normand, and C. Villenave. 2006. Nutrition on bacteria by bacterial-feeding nematodes and consequences on structure of the soil bacterial community. Eur. J. Soil Biol. 42:70–78.

Bohan, D.A., C.W.H. Boffey, D.R. Brooks, S.J. Clark, A.M. Dewar, L.G. Firbank, A.J. Haughton, C. Hawes, M.S. Heard, M.J. May, J.L. Osborne, J.N. Perry, P. Rothery, D.B. Roy, R.J. Scott, G.R. Squire, I.P. Woiwod, and G.T. Champion. 2005. Effects of weed and invertebrate abundance and diversity of herbicide management in genetically modified herbicide-tolerant winter-sown oilseed rape. Proc. R. Soc. London Ser. B 272: 463–474.

Bohlen, P.J., R.W. Parmelee, and J.M. Blair. 2004. Integrating the effects of earthworms on nutrient cycling across spatial and temporal scales. p. 161–180. *In* C.A. Edwards (ed.) Earthworm ecology. 2nd ed. CRC Press, Boca Raton, FL.

Bonkowski, M., W. Cheng, B.S. Griffiths, J. Alphei, and S. Scheu. 2000. Microbial–faunal interactions in the rhizosphere and effects on plant growth. Eur. J. Soil Biol. 36:135–147.

Bouché, M.B. 1977. Stratégies lombriciennes. *In* U. Lohm and T. Persson (ed.) Soil organisms as components of ecosystems. Biol. Bull. 25 (Stockholm) 25:122–132..

Boulton, A.M., B.A. Jaffe, and K.M. Scow. 2003. Effects of a common harvester ant (*Messor Andrei*) on richness and abundance of soil biota. Appl. Soil Ecol. 23:257–265.

Briones, M.J.I., and P. Ineson. 2002. The use of ^{14}C carbon dating to determine the feeding behaviour of enchytraeids. Soil Biol. Biochem. 34:881–884.

Brown, G.G., and B.M. Doube. 2004. Functional interactions between earthworms, microorganisms, organic matter, and plants. p. 213–239. *In* C.A. Edwards (ed.) Earthworm ecology. 2nd ed. CRC Press, Boca Raton, FL.

Brussaard, L., P.C. de Ruiter, and G.G. Brown. 2007. Soil biodiversity for agricultural sustainability. Agric. Ecosyst. Environ. 121:233–244.

Butt, K.R., and C.N. Lowe. 2007. Presence of earthworm species within and beneath *Lumbricus terrestris* (L.) *middens*. Eur. J. Soil Biol. 43(SUPPL. 1):S57–S60.

Cárcamo, H.A., and J.R. Spence. 1994. Crop type effects on the activity and distribution of ground beetles (Coleoptera: Carabidae). Environ. Entomol. 23:684–692.

Chen, J., and H. Ferris. 1999. The effects of nematode grazing on nitrogen mineralization during fungal decomposition of organic matter. Soil Biol. Biochem. 31:1265–1279.

Clapperton, M.J., D.A. Kanashiro, and V.M. Behan-Pelletier. 2007. Changes in abundance and diversity of microarthropods associated with Fescue Prairie grazing regimes. Pedobiologia 46:496–511.

Clark, B.W., and J.R. Coats. 2006. Subacute effects of Cry1Ab Bt corn litter on the earthworm *Eisenia fetida* and the springtail *Folsomia candida*. Environ. Entomol. 35:1121–1129.

Clark, M.S., J.M. Luna, N.D. Stone, and R.R.Youngman. 1994. Generalist predator consumption of armyworm (Lepidoptera:Noctuidae) and effect of predator removal on damage in no-till corn. Environ. Entomol. 23:617–622.

Cole, L., M.A. Bradford, P.J.A. Shaw, and R.D. Bardgett. 2006. The abundance, richness and functional role of soil meso- and macrofauna in temperate grassland—A case study. Appl. Soil Ecol. 33:186–198.

Coleman, D.C. 2008. From peds to paradoxes: Linkages between soil biota and their influences on ecological processes. Soil Biol. Biochem. 40:271–289.

Coleman, D.C., R.V. Anderson, C.V. Cole, J.F. McClellan, L.E. Woods, J.A. Trofymow, and E.T. Elliot. 1984. Roles of protozoa and nematodes in nutrient cycling. p. 17–28. *In* R.L. Todd and J.E. Giddens (ed.) Microbial–plant interactions. SSSA, ASA, and CSSA, Madison WI.

Curl, E.A., and J.D. Harper. 1990. Fauna–microflora interactions. p. 367–387. *In* J.M. Lynch (ed.) The rhizosphere. Wiley, Chichester, UK.

Dauber, J., R. Niechoj, H. Baltruschat, and V. Wolters. 2008. Soil engineering ants increase grass root arbuscular mycorrhizal colonization. Biol. Fertil. Soils 44:791–796.

Decaëns, T., J.J. Jiménez, C. Gioia, G.J. Measey, and P. Lavelle. 2006. The values of soil animals for conservation biology. Eur. J. Soil Biol. 42(Suppl. 1):S23–S38.

Denton, C.S., R.D. Bardgett, R. Cook, and P.J. Hobbs. 1999. Low amounts of root herbivory positively influence the rhizosphere microbial community in a temperate grassland. Soil Biol. Biochem. 31:155–165.

Didden, W., and J. Römbke. 2001. Enchytraeids as indicator organisms for chemical stress in terrestrial ecosystems. Ecotoxicol. Environ. Saf. 50:25–43.

Didden, W.A.M. 1993. Ecology of terrestrial Enchytraeidae. Pedobiologia 37:21–29.

Edwards, C.A. (ed.) 2004. Earthworm ecology. 2nd ed. CRC Press, Boca Raton, FL.

Edwards, C.A., and P.J. Bohlen. 1992. The effects of toxic chemicals on earthworms. Rev. Environ. Contam. Toxicol. 125:23–99.

Edwards, C.A., and P.J. Bohlen. 1996. Biology and ecology of earthworms. Chapman and Hall, London.

Edwards, C.A., P.J. Bohlen, D.R. Linden, and S. Subler. 1995. Earthworms in agroecosystems. p. 185–213. *In* P.F. Hendrix (ed.) Earthworm ecology and biogeography in North America. Lewis Publishers, Boca Raton, FL.

Ferris, H., T. Bongers, and R.G.M. de Goede. 2001. A framework for soil food web diagnostics: Extension of the nematode faunal analysis concept. Appl. Soil Ecol. 18:13–29.

Ferris, H., and M.M. Matute. 2003. Structural and functional succession in the nematode fauna of a soil food web. Appl. Soil Ecol. 23:93–110.

Ferris, H., R.C. Venette, and S.S. Lau. 1997. Population energetics of bacterial-feeding nematodes: Carbon and nitrogen budgets. Soil Biol. Biochem. 29:1183–1194.

Finch, S. 1996. Effect of beetle size on cabbage root fly eggs by ground beetles. Entomol. Exp. Appl. 81:199–206.

Foissner, W. 1987. Soil protozoa: Fundamental problems, ecological significance, adaptations in ciliates and testaceans, bioindicators, and guide to the literature. Progress Protistol. 2:69–212.

Foissner, W. 1997. Protozoa as indicators in agroecosystems, with emphasis of farming practices, biocides and biodiversity. Agric. Ecosyst. Environ. 62:93–103.

Forge, T.A., S. Bittman, and C.G. Kowalenko. 2005. Responses of grassland soil nematodes and protozoa to multi-year and single-year applications of dairy manure slurry and fertilizer. Soil Biol. Biochem. 37:1751–1762.

Forge, T.A., E.J. Hogue, G. Neilsen, and D. Neilsen. 2008. Organic mulches alter nematode communities, root growth and fluxes of phosphorus in the root zone of apple. Appl. Soil Ecol. 38:15–22.

Franchini, P., and C.L. Rockett. 1996. Oribatid mites as "indicators" species for estimating the environmental impact of conventional and conservation tillage practices. Pedobiologia 40:217–225.

Frouz, J., V. Piž́l, and K. Tajovskỳı. 2007. The effect of earthworms and other saprophagous macrofauna on soil microstructure in reclaimed and un-reclaimed post-mining sites in central Europe. Eur. J. Soil Biol. 43(Suppl. 1):S184–S189.

Gange, A. 2000. Arbuscular mycorrhizal fungi, Collembola and plant growth. Trends Ecol. Evol. 15:369–372.

Georgieva, S., S. Christensen, and K. Stevnbak. 2005. Nematode succession and microfauna-microorganism interactions during root residue decomposition. Soil Biol. Biochem. 37:1763–1774.

Griffiths, B.S. 1990. A comparison of microbial-feeding nematodes and protozoa in the rhizosphere of different plants. Biol. Fertil. Soils 9:83–88.

Griffiths, B.S., and S. Caul. 1993. Migration of bacterial-feeding nematodes but not protozoa to decomposing grass residues. Biol. Fertil. Soils 3:207–210.

Griffiths, B.S., S. Caul, J. Thompson, A.N.E. Birch, J. Cortet, M.N. Andersen, and P.H. Krogh. 2007a. Microbial and microfaunal community structure in cropping systems with genetically modified plants. Pedobiologia 51:195–206.

Griffiths, B.S., F. Ekelund, R. Rønn, and S. Christiansen. 1993. Protozoa and nematodes on decomposing barley roots. Soil Biol. Biochem. 25:1293–1295.

Griffiths, B.S., L.H. Heckmann, S. Caul, J. Thompson, C. Scrigeour, and P.H. Krogh. 2007b. Varietal effects of eight paired lines of transgenic Bt maize and near-isogenic non-Bt maize on soil microbial and nematode community structure. Plant Biotechnol. J. 5:60–68.

Griffiths, B.S., S. Caul, J. Thompson, C.A. Hackett, J. Cortet, C. Pernin, and P.H. Krogh. 2008. Soil microbial and faunal responses to herbicide tolerant maize and herbicide in two soils. Plant Soil 308:93–103.

Hassall, M., S. Adl, M. Berg, B. Griffiths, and S. Scheu. 2006. Soil fauna-microbe interactions: Towards a conceptual framework for research. Eur. J. Soil Biol. 42(Suppl. 1):S54–S60.

Hatten, T.D., N.A. Bosque-Peréz, J. Johnson-Mynard, and S.D. Eigenbrode. 2007. Tillage differentially affects the capture rate of pitfall traps for three species of carabid beetles. Entomol. Exp. Appl. 124:177–187.

Hautier, L., J.P. Jansen, N. Mabon, and B. Schiffers. 2005. Selectivity lists of pesticides to beneficial arthropods for IPM programs in carrot—First results. Commun. Agric. Appl. Biol. Sci. 70:547–557.

Heal, O.W., J.M. Anderson, and M.J. Swift. 1997. Plant litter quality and decomposition: An historical overview. p. 3–30. *In* G. Cadish and K.E. Giller (ed.) Driven by nature: Plant litter quality and decomposition. CAB International, Wallingford, UK.

Hendrix, P.F. (ed.) 1995. Earthworm ecology and biogeography in North America. Lewis Publ., Boca Raton, FL.

Herdler, S., K. Kreuzer, S. Scheu, and M. Bonkowskia. 2008. Interactions between arbuscular mycorrhizal fungi (*Glomus intraradices*, Glomeromycota) and amoebae (*Acanthamoeba castellanii*, Protozoa) in the rhizosphere of rice (*Oryza sativa*). Soil Biol. Biochem. 40:660–668.

House, G.J., and R.W. Parmelee. 1985. Comparison of soil arthropods and earthworms from conventional and no-tillage agroecosystems. Soil Tillage Res. 5:351–360.

Huguenin, M.T., C.G. Leggett, and R.W. Paterson. 2006. Economic valuation of soil fauna. Eur. J. Soil Biol. 42(Suppl. 1):S16–S22.

Hummel, R.L., J.F. Walgenbach, G.D. Hoyt, and G.G. Kennedy. 2002. Effects of vegetable production systems on epigeal arthropod populations. Agric. Ecosyst. Environ. 93:177–188.

Hunt, H.W., D.C. Coleman, E.R. Ingham, R.E. Ingham, E.T. Elliot, J.C. Moore, S.L. Rose, C.P.P. Reid, and C.R. Morley. 1987. The detrital food web in a short grass prairie. Biol. Fertil. Soils 3:57–68.

Ingham, E.R. 1994. Protozoa, p. 491–515. *In* R.W. Weaver et al. (ed.) Methods of soil analysis. Part 2. SSSA Book Ser. 5. SSSA, Madison WI.

Ingham, R.E., J.A. Trofymow, E.R. Ingham, and D.C. Coleman. 1985. Interactions of bacteria, fungi and their nematode grazers: Effects on nutrient cycling and plant growth. Ecol. Monogr. 55:119–140.

Jänsch, S., J. Römbke, and W. Didden. 2005. The use of enchytraeids in ecological soil classification and assessment concepts. Ecotoxicol. Environ. Saf. 62:266–277.

Jiménez, J.J., T. Decaëns, and P. Lavelle. 2008. C and N concentration in biogenic structures of a soil-feeding termite and a fungus-growing ant in the Colombian savannas. Appl. Soil Ecol. 40:120–128.

Jouquet, P., J. Dauber, J. Lagerlof, P. Lavelle, and M. Lepage. 2006. Soil invertebrates as ecosystem engineers: Intended and accidental effects on soil and feedback loops. Appl. Soil Ecol. 32:153–164.

Jones, C.G., and J.L. Gutierrez. 2007. On the purpose, meaning, and usage of the physical ecosystem engineering concept. p. 3–24. *In* K. Cuddington et al. (ed.) Ecosystem engineers: Plants to protists. Elsevier, New York.

Jones, C.G., J.H. Lawton, and M. Shachak. 1994. Organisms as ecosystem engineers. Oikos 69:373–386.

Joy, V.C., R. Pramanik, and K. Sarkar. 2005. Biomonitoring insecticide pollution using non-target soil microarthropods. J. Environ. Biol. 26:571–577.

Kladivko, E.J. 2001. Tillage systems and soil ecology. Soil Tillage Res. 61:61–76.

Kladivko, E.J., N.M. Akhouri, and G. Weesies. 1997. Earthworm populations and species distributions under no-till and conventional tillage in Indiana and Illinois. Soil Biol. Biochem. 29:613–615.

Klironomos, J.N., and B. Kendrick. 1996. Palatibility of microfungi to soil arthropods in relation to the functioning of arbuscular mycorrhizae. Biol. Fertil. Soils 21:43–52.

Kreuzer, K., J. Adamczyk, M. Iijima, M. Wagner, S. Scheu, and M. Bonkowski. 2006. Grazing of a common species of soil protozoa (*Acanthamoeba castellanii*) affects rhizosphere bacterial community composition and root architecture of rice (*Oryza sativa* L.). Soil Biol. Biochem. 38:1665–1672.

Kromp, B. 1999. Carabid beetles in sustainable agriculture: A review on pest control efficacy, cultivation impacts and enhancement. Agric. Ecosyst. Environ. 74:187–228.

Larsen, J., A. Johansen, S.E. Larsen, L.H. Heckmann, I. Jakobsen, and P.H. Krogh. 2008. Population performance of collebolans feeding on soil fungi from different ecological niches. Soil Biol. Biochem. 40:360–369.

Lartey, R.T., E.A. Curl, and C.M. Peterson. 1994. Interactions of mycophagous collembola and biological control fungi in the suppression of *Rhizoctonia solani*. Soil Biol. Biochem. 26:81–88.

Lavelle, P. 1997. Faunal activities and soil processes: Adaptive strategies that determine ecosystem function. p. 93–132. *In* M. Begon and A.H. Fitter (ed.) Advances in ecological research. Academic Press, New York.

Lavelle, P. 2002. Functional domains in soils. Ecol. Res. 17:441–450.

Lavelle, P., S. Barot, M. Blouin, T. Decaëns, J.J. Jimenez, and P. Jouquet. 2007. Earthworms as key actors in self-organized soil systems. p. 77–106. *In* J.E. Cuddington et al. (ed.) Ecosystem engineers: Plants to protists. Elsevier, New York.

Lavelle, P., F. Charpentier, C. Villenave, J.-P. Rossi, L. Derouard, B. Pashanasi, J. Andre, J.-F. Ponge, and N. Bernier. 2004. Effects of earthworms on soil organic matter and nutrient dynamics at a landscape scale over decades. p. 145–160. *In* C.A. Edwards (ed.) Earthworm ecology. 2nd ed. CRC Press, Boca Raton, FL.

Lee, K.E. 1985. Earthworms: Their ecology and relationships with soils and land use. CSIRO, Sydney, Australia.

Lee, K.E. 1995. Earthworms and sustainable land use. p. 215–234. *In* P.F. Hendrix (ed.) Earthworm ecology and biogeography in North America. Lewis Publ., Boca Raton, FL.

Liphadzi, K.B., K. Al-Khatib, C.N. Bensch, P.W. Stahlman, J.A. Dillie, T. Todd, C. Rice, M.J. Horak, and G. Head. 2005. Soil microbial and nematode communities as affected by glyphosate and tillage practices in a glyphosate-resistant cropping system. Weed Sci. 53:536–545.

Lousier, J.D., and S.S. Bamforth. 1990. Soil protozoa. p. 97–136. *In* D. Dindal (ed.) Soil biology guide. John Wiley and Sons, New York.

Lynch, J.M. 1990. Introduction: Some consequences of microbial rhizosphere competence for pant and soil. p. 1–10. *In* J.M. Lynch (ed.) Rhizosphere. Wiley Interscience, Chichester UK.

Mao, X.-F., F. Hu, B.S. Griffiths, X.-Y. Chen, M.-Q. Liu, and H.-X. Li. 2007. Do bacterial-feeding nematodes stimulate root proliferation through hormonal effects? Soil Biol. Biochem. 39:1816–1819.

Marinissen, J.C.Y., and W.A.M. Didden. 1997. Influence of the Enchytraeid worm *Buchholzia appendiculata* on aggregate formation and organic matter decomposition. Soil Biol. Biochem. 29:387–390.

McGonigle, T.P., and A.H. Fitter. 1988. Growth and phosphorus inflows of *Trifolium-repens* L. with a range of indigenous vesicular-arbuscular mycorrhizal infection levels under field conditions. New Phytol. 108:59–65.

Miura, F., T. Nakamoto, S. Kaneda, S. Okano, M. Nakajima, and T. Murakami. 2008. Dynamic of soil biota at different depths under tow contrasting tillage practices. Soil Biol. Biochem. 40:406–414.

Monard, C., F. Binet, and P. Vandenkoornhuyse. 2008. Short-term response of soil bacteria to carbon enrichment in different soil microsites. Appl. Environ. Microbiol. 74:5589–5592.

Moore, J.C., D.E. Walter, and H.W. Hunt. 1988. Arthropod regulation of micro- and mesobiota in below-ground detrital food webs. Annu. Rev. Entomol. 33:419–439.

Neher, D.A. 2001. Role of nematodes in soil health and their use as indicators. J. Nematol. 33:161–168.

Neher, D.A., S.L. Peck, J.O. Rawlings, and C.L. Campbell. 1995. Measures of nematode community structure and sources of variability among and within agricultural fields. Plant Soil 170:167–181.

Norton, R.A. 1985. Aspects of the biology and systematics of soil arachnids, particularly saprophagous and mycophagous mites. Quaest. Entomol. 21:523–541.

Osler, G.H.R., L. Harrison, D.K. Kanashiro, and M.J. Clapperton. 2008. Soil microarthropod assemblages under different arable crop rotations in Alberta, Canada. Appl. Soil Ecol. 38:71–78.

Osler, G.H.R., and M. Sommerkorn. 2007. Toward a complete soil C and N cycle: Incorporating the soil fauna. Ecology 88:1611–1621.

Overgaard Nielsen, C. 1967. Nematoda. p. 197–211. *In* A. Burgess and F. Raw (ed.) Soil Biology. Academic Press, London.

Paoletti, M.G., G.H.R. Osler, A. Kinnear, D.G. Black, L.J. Thomson, A. Tsitsilas, D. Sharley, S. Judd, P. Neville, and A. D'Inca. 2007. Detritivores as indicators of landscape stress and soil degradation. Aust. J. Exp. Agric. 47:412–423.

Parfitt, R.L., G.W. Yeates, D.J. Ross, A.D. MacKay, and P.J. Budding. 2005. Relationships between soil biota, nitrogen and phosphorus availability, and pasture growth under organic and conventional management. Appl. Soil Ecol. 28:1–13.

Parisi, V., C. Menta, C. Gardi, C. Jacomini, and E. Mozzanica. 2005. Microarthropod communities as a tool to assess soil quality and biodiversity: A new approach. Agric. Ecosyst. Environ. 105:323–333.

Pavuk, D.M., F.F. Purrington, C.E. Williams, and B.R. Skinner. 1997. Ground beetle (Coleptera: Carabidae) activity density and community composition in vegetationally diverse agroecosytems. Am. Midl. Nat. 138:14–28.

Pereira, J.L., A.A. da Silva, M.C. Picanço, E.C. de Barros, and A. Jakelaitis. 2005. Effects of herbicide and insecticide interaction on soil entomofauna under maize crop. J. Environ. Sci. Health. Part B, Pestic. Food Contamin. Agric. Wastes 40:45–54.

Perry, D.A. 1995. Self-organizing systems across scales. Trends Res. Ecol. Evol. 10:241–244.

Pekár, S., and J. Benes. 2008. Aged pesticide residues are detrimental to agrobiont spiders (Araeae). J. Appl. Entomol. 132:614–622.

Poll, J., S. Marhan, S. Haase, J. Hallmann, E. Kandeler, and L. Ruess. 2007. Low around of herbivory by root-know nematodes affect microbial community dynamics and carbon allocation in the rhizosphere. FEMS Microbiol. Ecol. 62:268–279.

Porazinska, D.L., L.W. Duncan, R. McSorley, and J.H. Graham. 1999. Nematode communities as indicators of status and processes of a soil ecosystem influenced by agricultural management practices. Appl. Soil Ecol. 13:69–86.

Reichert, S.E., and L. Bishop. 1990. Prey controlled by an assemblage of generalist predators: Spider in garden test systems. Ecology 71:1441–1450.

Reichert, S.E., and T. Lockley. 1984. Spiders as biological control agents. Annu. Rev. Entomol. 29:299–320.

Ruess, L., E.J. Garcia Zapata, and J. Dighton. 2000. Food preferences of a fungal-feeding *Aphelenchoides* species. Nematology 2:223–230.

Sabatini, M.A., P. Grazioso, C. Altomare, and G. Innocenti. 2002. Interactions between *Onychiurus armatus* and *Trichoderma harianum* in take-al disease suppression in a simple experimental system. Eur. J. Soil Biol. 38:71–74.

Samu, F., K.D. Sunderland, and C. Szinetar. 1999. Scale-dependent dispersal and distribution patterns of spiders in agricultural systems: A review. J. Arachnol. 27:325–332.

Seastedt, T.R. 1984. The role of microarthropods in decomposition and mineralization processes. Annu. Rev. Entomol. 29:25–46.

Schreiner, R.P., and G.J. Bethlenfalvay. 2003. Crop residue and collembola interact to determine the growth of mycorrhizal pea plants. Biol. Fertil. Soils 39:1–8.

Shipitalo, M.J., and R.-C. Le Bayon. 2004. Quantifying the effects of earthworms on soil aggregation and porosity. p. 183–200. *In* C.A. Edwards (ed.) Earthworm ecology. 2nd ed. CRC Press, Boca Raton, FL.

Sikora, R.A. 1992. Management of the antagonistic potential in agricultural ecosystems for the biological control of plant parasitic nematodes. Annu. Rev. Phytopathol. 30:245–270.

Stout, J.D., and O.W. Heal. 1967. Protista. p. 149–195. *In* A. Burgess and F. Raw (ed.) Soil biology. Academic Press, London.

Thiele, H.U. 1977. Carabid beetles and their environments. Springer, Berlin, Germany.

Tomlin, A.D., M.J. Shipitalo, W.M. Edwards, and R. Protz. 1995. Earthworms and their influence on soil structure and infiltration. p. 159–183. *In* P.F. Hendrix (ed.) Earthworm ecology and biogeography in North America. Lewis Publishers, Boca Raton, FL.

Topoliantz, S., J.-F. Ponge, and P. Viaux. 2000. Earthworms and enchytraeid activity under different arable farming systems, as exemplified by biogenic structures. Plant Soil 225:39–51.

Tordoff, G.M., L. Boddy, and T.H. Jones. 2008. Species-specific impacts of collembola grazing on fungal foraging ecology. Soil Biol. Biochem. 40:434–442.

van Vilet, P.C.J., M.H. Beare, D.C. Coleman, and P.F. Hendrix. 2004. Effects of enchytraeids (Annelida: Oligochaeta) on soil carbon and nitrogen dynamics in laboratory incubations. Appl. Soil Ecol. 25:147–160.

Venette, R.C., and H. Ferris. 1998. Influence of bacterial type and density on population growth of bacterial-feeding nematodes. Soil Biol. Biochem. 30:949–960.

Waid, J.S. 1997. Metabiotic interactions in plant litter systems. p. 145–156. *In* G. Cadisch and K.E. Giller (ed.) Driven by nature: Plant litter quality and decomposition. CAB International, Wallingford, UK.

Walter, D.E., and H.C. Proctor. 1999. Mites: Ecology, evolution and behaviour. CABI Publishing, New York.

Wallwork, J.A. 1976. The distribution and diversity of soil fauna. Academic Press, London.

Wang, K.-H., R. McSorley, A.J. Marshall, and R.N. Gallaher. 2004. Nematode community changes associated with decomposition of *Crotalaria juncea* amendment in litterbags. Appl. Soil Ecol. 27:31–45.

Wardle, D.A. 1995. Impacts of disturbance on detritus food webs in agro-ecosystems of contrasting tillage and weed management practices. p. 105–185. *In* M. Begon and A.H. Fitter (ed.) Advances in ecological research. Vol. 26. Academic Press, New York.

Wardle, D.A. 2002. Communities and ecosystems: Linking the above-ground and belowground components. Princeton Univ. Press, Princeton, NJ.

Wardle, D.A., R.D. Bardgett, J.N. Klironomos, H. Setälä, W.H. Van der Putten, and D.H. Wall. 2004. Ecological linkages between aboveground and belowground biota. Science 304:1629–1633.

Warnock, A.J., A.H. Fitter, and M.B. Usher. 1982. The influence of a springtail *folsomia-candida* (insect, collembola) on the mycorrhizal association of leek *allium-porrum* and the vesicular-arbuscular mycorrhizal endophyte *glomus-fasciculatus*. New Phytol. 90:285–292.

Whitford, W.G. 1996. The importance of biodiversity of soil biota in arid ecosystems. Biodivers. Conserv. 5:185–195.

Williams, B.L., and B.S. Griffiths. 1989. Enhanced nutrient mineralization and leaching from decomposing sitka spruce litter by enchytraeid worms. Soil Biol. Biochem. 21:183–188.

Wolters, V. 2001. Biodiversity of soil animals and its function. Eur. J. Soil Biol. 37:221–227.

Yeates, G.W. 2003. Nematodes as soil indicators: Functional and biodiversity aspects. Biol. Fertil. Soils 37:199–210.

Yeates, G.W. 2007a. Diversity of nematodes. p. 205–236. *In* G. Benckiser and S. Schnell (ed.) Biodiversity in agricultural production systems. Taylor and Francis, Boca Raton, FL.

Yeates, G.W. 2007b. Abundance, diversity, and resilience of nematode assemblages in forest soils. Can. J. For. Res. 37:216–225

Yeates, G.W., T. Bongers, R.G.M. de Goede, D.W. Freckman, and S.S. Georgieva. 1993. Feeding habits in soil nematode families and genera—An outline for soil ecologists. J. Nematol. 25:313–331.

Yeates, G.W., S. Saggar, C.B. Hedley, and C.F. Mercer. 1999. Increasing the ^{14}C-carbon translocation to the soil biomass when five plant-parasitic nematodes infect roots of white clover. Nematology 1:295–300.

Young, I.M., and J.W. Crawford. 2004. Interactions and self-organization in the soil–microbe complex. Science 304:1634–1637.

Young, I.M., and K. Ritz. 2000. Tillage, habitat space and function of soil microbes. Soil Tillage Res. 53:201–213.

10

Soil Management for Increasing Water Use Efficiency in Field Crops under Changing Climates

Jerry L. Hatfield

Crop production throughout the world is dependent on soil water availability either directly through precipitation captured in the soil profile or indirectly as soil water recharge applied via irrigation. Increasing water use efficiency (WUE) is critical to ensuring that we continue to produce the food, feed, fuel, and fiber needed to sustain the world's increasing populations. Optimizing the factors that affect WUE will enhance the stability of crop production across a range of climates; however, the ever-increasing problem of climatic change increases the urgency with which we should view this issue and begin to understand the implications of the interactions between soil management factors and WUE. The increasing variability in both temperature and precipitation throughout the world raises the question of how to enhance WUE under current cropping systems. This goal has to be coupled with the sobering fact that the soils of the world continue to be degraded, and many of the critical properties that are linked to WUE of cropping systems are being negatively impacted. Increasing our ability to efficiently increase food and feed production given changes in climate and soil will require that we better understand the interactions between the soil and crop production. Wallace (2000) summarized the need to increase WUE by more effectively using water resources for plant production. The challenge for us and future generations will be to provide a stable and secure food supply and the efficient use of our natural resources—soil, water, and air.

Hatfield et al. (2001) reviewed the literature on WUE and soil management to highlight many of the options for increasing WUE through improvements in soil management. Among these options were soil management practices that affected water availability and nutrient management practices that increased the nutrient availability to the crop. They summarized the potential impacts as a relationship shown in Fig. 10|1. Soil management practices related to nutrients or water availability could change the WUE by ± 15 to 25% compared to the baseline. These changes in WUE offer potential for how we can cope with changing climate and will be explored in the remainder of this chapter. It is important to begin this discussion by first defining WUE and the principal variables that affect WUE. There have been several different forms of relationship used to characterize WUE, and these have been summarized by Tanner and Sinclair (1983). Water use efficiency is described in mathematical form as

J.L. Hatfield, Laboratory Director and Plant Physiologist, National Laboratory for Agriculture and the Environment, 2110 University Blvd., Ames, IA 50011 (jerry.hatfield@ars.usda.gov).

doi:10.2136/2011.soilmanagement.c10

$$WUE = Y/ET \qquad [1]$$

where Y is the harvestable yield of the crop, either biomass or grain, and ET the combination of evaporation of water from the soil surface and plant leaves and transpiration through the stomates to the atmosphere. deWit (1958) first proposed this relationship after he observed there was a linear relationship between plant yield and transpiration in crop production regions with high solar radiation (e.g., the western United States) and described this relationship as

$$Y/T = m/T_{max} \qquad [2]$$

where Y is total dry matter production, T is transpiration, m is an empirical coefficient, and T_{max} is daily free water evaporation, generally obtained from evaporation pans. Water use from the crop (ET in Eq. [1]) generally is based on total water use (ET) from the crop surface and includes evaporation from soil and plant components because of the difficulty in separating evaporation (E) from transpiration (T). Although there has been substantial progress in being able to separate E from T, this remains a challenge for most experiments; thus, the more common ET term is used.

Soil management impacts on WUE will occur through factors that affect the availability of soil water to influence ET in Eq. [1] or factors that affect Y that are not directly related to water but affect plant growth. Soil management practices can affect WUE

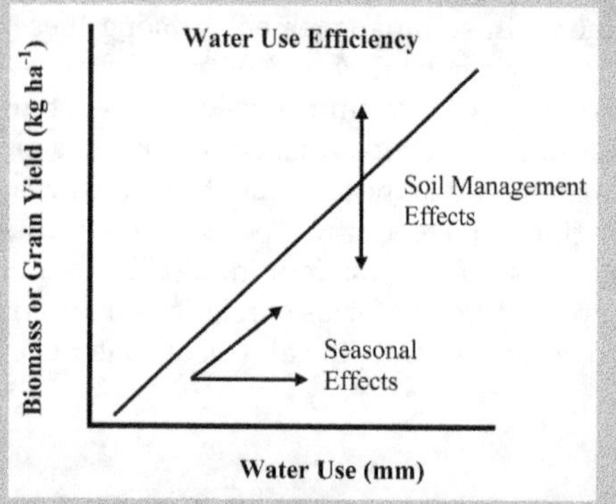

Fig. 10|1. Potential changes in water use efficiency as affected by seasonal and physical changes in soil and nutrient management (adapted from Hatfield et al., 2001).

through their direct effect on the surface energy balance:

$$ET = R_n - G - H - P \qquad [3]$$

where R_n is net radiation, G is soil heat flux, H is sensible heat flux, and P is photosynthetic flux. These terms are often expressed in a variety of different units (W m^{-2}, KJ m^{-2} s^{-1}). Changes in the energy exchanges (R_n, G, and H) and the plant photosynthetic (P) efficiency are the mechanisms by which WUE is changed because these components affect the soil water balance within and among growing seasons. The methods by which soil management practices modify the energy balance components and affect WUE will provide linkages among soil management practices and WUE discussed in this chapter.

Soil Management Practices
Modification of the Soil Surface

Soil management practices that influence WUE include manipulation of the soil surface, either by tillage system, residue management, or living mulches. The effectiveness of these practices in changing WUE varies among practices, climates, and cropping systems. All components—R_n, G, H, and P—of the energy balance (Eq. [3]) are affected by soil surface modifications. Water use efficiency has often been a concept that has been applied to either semiarid regions, where water is limited, or irrigated systems, where enhanced water management returns large dividends because of the positive impact of additional water on crop production. These areas are also those that may be the most affected by climate change impacts on precipitation patterns and amounts. Hatfield et al. (2001) summarized the range of WUE in different systems and provided an overview of the differences among soil management systems. Wallace (2000) described WUE of a crop as

$$WUE = \frac{e_w}{\left(1 + \dfrac{L + E_s + R + D}{E_t}\right)} \qquad [4]$$

where e_w is the ratio of carbon fixed per unit water transpired, L is the loss of irrigation water in storage and conveyance, E_s is the evaporation from the soil surface, R is the runoff, D is the drainage from the soil profile, and E_t is the transpiration from the crop. It is easy to see how the various factors that affect water impact WUE are linked to soil management practices.

Tillage

Tillage creates changes in the soil surface that breaks apart the surface soil layer, including soil crusts, which in turn leads to an initial increase in the rate of water infiltration into the soil and ultimately increases soil water storage. Disturbing the soil surface can also cause increased soil water evaporation compared to residue-covered surfaces or undisturbed surfaces because of the exposure of moist soil to the atmosphere. Lascano and Hatfield (1992) showed that soil water evaporation occurred from the soil surface until a very thin crust of dry soil was formed and eliminated the pathway for water exchange to the atmosphere. Conversely, removing the crust will increase evaporation. Burns et al. (1971) and Papendick et al. (1973) demonstrated that disturbing the soil surface with tillage increased soil water evaporation rates compared to untilled areas. Ritchie (1971) observed that soil water evaporation affected two surface features, surface soil water content and the amount of plant cover over the soil surface. Tillage moves moist soil up to the surface where drying losses are increased. Total soil water evaporation fluxes ranged from 10 to 12 mm for a three-day period following each cultivation operation in the spring in Iowa, while evaporation fluxes from no-till fields were less than 2 mm during this same time period were less than 2 mm (Hatfield and Prueger, unpublished data, 1999). Soil water availability in the seed zone could be reduced by as much as 20 to 30 mm with aggressive field cultivation operations in the spring. To replace this soil water lost from the seed zone it is necessary to have timely precipitation events to ensure germination and emergence of the crop. In semiarid areas, soil profile water contents that are near field capacity at the onset of the growing season are critical to crop production.

Water dynamics in soils are affected by tillage. In soils with no surface residue tillage has been found to increase the saturated hydraulic conductivity (i.e., the rate of water movement when the soil is saturated), while soil water content before tillage had no measureable effect (Cresswell et al., 1993). Tillage sequence affected unsaturated hydraulic conductivity (i.e., the rate of water movement at water contents that are less than field capacity), and excessive tillage created the lowest unsaturated hydraulic conductivities through the formation of more air-filled pores. Tillage is considered to have a positive impact on water infiltration, but excessive tillage may reduce infiltration because of the direct effect on hydraulic conductivity. Christensen et al. (1994) observed that soil water was conserved during fallow periods with no-tillage compared to clean-till, and his findings were opposite of those found by Cresswell et al. (1993). They found sorghum [*Sorghum bicolor* (L.) Moench] grain yields to be increased with adoption of no-tillage because water was conserved during the fallow periods accompanied with a deeper wetting of the soil profile in no-tillage systems.

There is not a strong relationship between tillage systems and WUE because it is not possible to discuss the tillage practices without considering the effect of mulch or crop residue management since residue management is closely linked with tillage practices. Pikul and Aase (1995) observed that infiltration rates were increased because residue protected the soil surface from the direct impact raindrop energy, which caused the infiltration rate over 3 h to be 52 mm under conventional tillage in a wheat fallow and 69 mm in the annual cropping system with no-tillage when these systems were compared in the northern Great Plains. Maintaining surface cover in no-tillage systems was advantageous compared to tillage systems because of the reduced soil crusting and erosion. Decreasing tillage intensity improved soil water availability because of reduced evaporation losses, which created a trend toward improved WUE (Aase and Pikul, 1995). Good and Smika (1978) found improved water storage with chemical fallow in wheat systems. In China, He et al. (2008) found that growing wheat (*Triticum aestivum* L.) on raised beds increased WUE under irrigation compared to the traditional tillage or zero-tillage because of the increased soil water and increased soil temperature in the root zone. They also found a reduced bulk density in the upper 30 cm of the soil profile in the raised beds of their study. They concluded that manipulation of the surface to create these raised beds increased the efficiency of irrigation water use in arid areas with limited irrigation water availability.

Water storage and evaporation losses are changed through tillage practices, but equally important is the maintenance of the soil profile. Tanaka (1990) observed soil loss in the northern Great Plains decreased WUE and dry matter production and noted that preservation of topsoil depth should be a priority outcome of soil management practices. The role of tillage on efficient water use and crop growth cannot be underestimated, and evaluation of tillage systems according to their impact on WUE provides a basis for being able to directly compare management systems.

Crop Residue Management
Soil Water Availability

Changes in WUE are a direct result of covering the soil surface with residue or mulch (Johnson and Davis, 1971). Modification of soil water evaporation by the addition of residues or mulches occurs through the reduction of soil temperature, impeding water vapor diffusion, absorption of water vapor onto mulch tissue, and decreasing the windspeed gradient at the soil surface–atmosphere interface (Greb, 1966). Sauer et al. (1996a) observed that surface residue decreased soil water evaporation by 34 to 50% and creating a 15-cm bare strip with tillage increased soil water evaporation only 7% compared to weathered residue cover. Deibert et al. (1986) stated that proper soil management could lead to both increases in precipitation storage efficiency and WUE; however, tillage effects on storage efficiency were minimal in their studies. They observed in the northern Great Plains that precipitation storage efficiency was similar among continuous wheat tillage systems but exhibited the largest variation among years and locations during the non-growing season. They defined precipitation storage efficiency as the soil water stored in the upper 1.2 m relative to the non-growing season precipitation. Differences among tillage systems were 56% with no-tillage and 47% with spring-sweep operations at Williston, ND, with no differences, from 59% with no-tillage compared to 57% with spring-sweep,

at Minot, ND. Variation among years was more noticeable for the tillage practices, and they found precipitation storage efficiencies ranged from 20 to 98%. This variation in storage efficiency was attributed to a combination of variation in annual precipitation and precipitation patterns. Yields under no-tillage were lower and were attributed to increased weed competition, foliar disease, and insect damage compared to spring-sweep or spring-plow operations, which resulted in a lower WUE with no-tillage (Deibert et al., 1986). In the Canadian prairies of British Columbia, Azooz and Arshad (1995) measured higher soil water contents under no-tillage plots compared to moldboard plow. Another study in eastern Canada, Ontario, found the corn (*Zea mays* L.) residue on the soil surface of no-tillage systems intercepted significant amounts of precipitation and reduced soil water evaporation (Zhai et al., 1990). An increase in available soil water in the upper meter of the soil profile was found in no-tillage versus other tillage practices in Wisconsin (Johnson et al., 1984). Reducing the tillage intensity in the upper Midwest and Canada generally increases soil water content. Reduction of tillage creates the potential for increased soil water content in the upper soil profile by increasing the physical barrier to soil water evaporation and reducing the disturbance of the soil surface that results in increased soil water evaporation.

In northern Great Plains cropping systems effective management of snow can have a significant impact on the soil water balance. Standing residue or stubble increases snow trapping and has been found to increase soil water content by 10 to 30 mm in spring (Aase and Siddoway, 1990). The effectiveness of standing residue vs. bare soil in increasing the soil water content was more evident with snow events than rainfall events. Energy exchange rates between the soil surface and the atmosphere affected by crop residue on the surface are albedo changes, altered aerodynamic coefficients, and diminished water vapor exchange rates (Eq. [3]). Sauer et al. (1996b) found the aerodynamic properties of corn stubble to change over the winter with roughness lengths and drag coefficients to be highest in the fall and lower in the spring because the residue had weathered and compacted beneath the snow layer. Increasing the roughness lengths and drag coefficients in the fall caused the water vapor exchange rates to increase. However, fresh residue on the soil surface in the fall has a larger amount of air-filled pore space, which offsets the increase caused by the altered aerodynamic properties. The addition of fresh residue on the soil surface creates the potential for rapid water loss, and the rate of water vapor movement through the stubble was the limiting factor. By the spring when the residue no long had snow cover, the aerodynamic properties were changed and the roughness lengths and drag coefficients were representative of a smoother surface and were the limiting factors to water vapor exchange.

Understanding the seasonality of changes in the aerodynamic properties of residue along with the properties of crop residue need to be evaluated to fully quantify how crop residue management can be altered to affect water management and potential water savings. Using wheat to protect young cotton (*Gossypium hirsutum* L.) plants from blowing sand in the southern High Plains offers the potential for effective management of soil water and decreasing the risk of blowing sand harming the plants. There was no observed difference in total seasonal evapotranspiration between conventional tillage practice (305 mm) and cotton planted into wheat residue with the growth terminated before maturity (304 mm) (Lascano et al., 1994). Wheat residue modified the microclimate by altering the partitioning of ET into the evaporation and transpiration components, increasing transpiration to 69% of the total ET compared to 50% for the conventional tillage practice. However, placing cotton into the wheat residue did not change cotton WUE. Hatfield (1990) observed that water vapor content increased and windspeed decreased within wheat residue, which resulted in a reduction of the water vapor gradient in wheat residue compared to bare soil. Increasing the water vapor content around the young cotton plants and decreasing the windspeed increased WUE in the early season by 25%, but this effect did not persist throughout the season because as the cotton grew above the wheat residue the effect of residue on water vapor and windspeed was no longer evident. Increasing the humidity and decreasing the windspeed around the young cotton seedling reduced the evaporation gradient,

which in turn created a favorable microclimate for the cotton plant. Observations from these types of studies show the potential for modifying WUE in cropping systems by altering residue management.

Sauer et al. (1998) observed large differences in the evaporation fluxes among days because the wetness of the corn residue layer had a large effect on the partitioning of available energy into evaporation and sensible heat. On radiation limited days (i.e., overcast), with a dry soil surface, the partitioning of net radiation into evaporation was observed to be between 50 and 75%, while on sunny days evaporation was less than 20% of the net radiation. On days when the soil surface was wet, there was no observable difference in partitioning of net radiation into evaporation fluxes (Sauer et al., 1998). An interesting observation in this study was the magnitude of the changes in the radiation components because albedo changed with age of the residue and the transmissivity of radiation through the residue increased with weathering (Sauer et al., 1997). Transmissivity of radiation is a measure of energy penetration onto the soil surface and is a function of the residue area index (the amount of residue covering the soil expressed as depth of residue, similar to leaf area index). Spatial variation of crop residue across a field is extremely dynamic because the wind rearranges the residue after harvest and before decomposition commences, which in turn affects the rate of decomposition. Changing the energy balance and the partitioning into evaporation by crop residue will affect the temporal and spatial dynamics of water storage and evaporation rates throughout the year and across a field or landscape.

Soil Temperature

Residue management affects soil temperatures, and soils with surface residue are generally cooler than tilled soils (Allmaras et al., 1964; Anderson and Russell, 1964; Greb, 1966). Cooler temperatures may cause slower early season crop growth and are the primary reason given to explain limited adoption of no-tillage in the upper Midwest. Observations by Hammel (1989) in northern Idaho revealed that reduced tillage and no-tillage increased soil impedance, and when combined with the increased cool, wet soil conditions in the spring, resulted in reduced root function and diminished crop growth potential. There is a tradeoff—the addition of crop residue on the surface can increase the soil water storage, but if there is a negative impact on crop growth caused by cooler temperatures, then there is little benefit from the additional soil water on WUE.

A solution to the negative impact of crop residue can be achieved by removing the corn residue from the seedbed; when this was done Kaspar et al. (1990) observed an increased rate of corn emergence caused by higher maximum soil temperatures in the seed zone, which affect germination and emergence. There is a difference among seasons on the effect of crop residue on soil temperatures. Hatfield and Prueger (1996) observed the greatest effect on soil temperature was in the fall when the residue was fresh compared to in the spring when the residue was weathered and minimal differences were observed. There is an additional complicating factor caused by the type of soil and its inherent thermal properties. In a Monona silt loam (fine-loamy, mixed, mesic, Typic Hapludoll) there was a 1 to 2°C cooler temperature than in a Nicollet loam (fine-loamy, mixed, mesic Aquic Hapludoll) caused by thermal conductivity differences and the effect of soil water on thermal properties even when the same amount of residue was added to both soils (Sauer et al., 1996a).

In a warmer climate, the High Plains of Texas, Unger (1988) observed that soil surface temperatures were affected more by season than by residue management practices. During the summer, the highest soil temperatures were found under the standing residue of dryland wheat, while during the winter, a no-tillage treatment with shredded residue had the highest temperatures. The effect of crop residue on soil temperatures is caused by changes in the soil water content and the interactions of water with soil thermal properties, and these interacting factors must be considered in evaluating the effectiveness of residue management.

Crop Growth and Yields

Increased soil water availability from the adoption of no-till systems or increasing or maintaining crop residue can have a positive effect on crop growth and yield. Adoption of no-till in western Kansas for wheat–row crop–fallow rotations increased corn yields by 31% (Norwood, 1999). The row crops in this study included corn, sorghum, sunflower (*Helianthus annuus* L.), and soybean [*Glycine max* (L.) Merr.], and the effect was not consistent among row crops—corn yields were increased in 3 years, sunflower and sorghum in 2 years, and soybean in only 1 year. In more arid climates conservation tillage has been coupled with irrigation. Unger (1994) found increased soil water use with conservation tillage, but these practices did not enhance grain yield of either wheat or grain sorghum. Sorghum is very efficient at using precipitation during the growing season; however, Jones and Popham (1997) did not find that continuous sorghum grain yields were improved by residue management compared to fallow systems on the southern High Plains. Unger (1991) found WUE varied among years, and for eight cultivars the highest yields were from cultivars with the highest water use amounts during the season.

An opposite result was found in Australia, where Gibson et al. (1992) observed that keeping sorghum stubble on the soil surface increased sorghum yield by 393 kg ha^{-1} because of increased WUE from the greater amount of water stored in and available to be used by the crop when extracted from the soil profile compared to conventional tillage. In this study, they found that decreasing tillage frequency increased soil water extraction, but no-tillage did not result in the optimum yield or WUE (Gibson et al., 1992). Water use efficiency can be enhanced by additional availability of soil water, and in the southern High Plains, the addition of soil water through irrigation increased WUE for wheat to 8 kg ha^{-1} mm^{-1} compared to 4 kg ha^{-1} mm^{-1} under dryland conditions (Musick et al., 1994). Increasing the soil water availability leads to increased WUE when there are no other limitations to crop yield. No standard set of recommendations exists on the effectiveness of different practices for WUE because the variation among years limits our ability to quantify the exact WUE response under a suite of management practices.

Additional management factors affect WUE. For example, in Saskatchewan Tompkins et al. (1991) observed that no-tillage winter wheat yields increased with seeding rate and decreased row spacing. Decreasing the row spacing from 36 to 9 cm and increasing the seeding rate from 35 to 140 kg ha^{-1} enhanced WUE. Using these changes in management caused grain yield to increase from 1.49 to 1.68 kg m^{-2} and WUE to increase from 9.4 to 10.3 kg ha^{-1} mm^{-1}. Although total water use increased with narrow row spacing and higher plant populations, the increased yield contributed the most to increased WUE (Tompkins et al., 1991). These results have been observed in other environments. For wheat in India, WUE was optimized at the 75 kg ha^{-1} seeding rate (Srivastava and Sidique, 1978). Jones and Johnson (1991) found for grain sorghum that WUE was not affected by plant density within the row but decreased with narrow rows in 1 out of 3 years. Variation in WUE among years attributable to the row width and plant density was 75%.

There are differences among crop response to tillage and residue management. Azooz and Arshad (1998) found barley (*Hordeum vulgare* L.) and canola (*Brassica campestris* L.) to vary among years. Comparing barley and canola water use and yield in no-tillage and a 75-mm strip till with conventional tillage in a silt loam and a sandy loam soil they found an increase in yield with no-tillage and strip till in dry years, but in wet years the highest yields were from the conventional tillage system. In the dry year, WUE was increased in barley by 21% with no-tillage and 18% with strip till in the silt loam soil and in the sandy loam, 19% with no-tillage and 10% with modified no-tillage compared to conventional tillage (Azooz and Arshad, 1998). Water use efficiency was highest with conventional tillage in the wet years in this study. There have been extensive studies on WUE response to crop management. For example, Liang et al. (1991) showed higher plant populations and higher fertilizer rates coupled with increased temperatures (heat units) and water inputs during the corn growing season increased yield and WUE. The implication from this study was that early season crop growth affected WUE because of the positive effects of increased heat units and water use on early season corn growth. A similar response for wheat was measured in the Mediterranean, where WUE was increased by agronomic factors that created high yields (Zhang and Qweis, 1999).

Differences in WUE are often observed among growing seasons. Chan and Heenan (1996) measured the water use in wheat–lupin (*Trifolium subterraneum* L.) rotation and observed that differences in crop water use among years were caused by early season growth of the wheat crop because the greater the early season growth, the greater the ability of the wheat crop to extract soil water. Lupin growth did not respond to differences in soil water among years. There are interactions between wheat growth and tillage practices; however, Dao and Nguyen (1989) concluded that in spite of these it was not necessary or feasible to develop cultivars for specific tillage methods. In their study at El Reno, OK, they found that no-tillage management under unfavorable growing conditions showed the greatest response in wheat growth and yield.

Evaluation of the impacts of soil management practices on WUE does not always yield definitive answers, and there is often variation among seasons that is not completely understood. In a study of WUE in sugarbeet (*Beta vulgaris* L.) and corn, Eck and Winter (1992) evaluated how modifying the soil profile affected water use and found that although water was extracted from deeper depths of the modified soil profile, this additional water did not lead to a consistent increase in yield. Water use efficiency was affected in only one year of the study, and Eck and Winter (1992) surmised that soil profile modification did not cause a consistent benefit because of the limited impact on yield. There is large variation among years on observed values for WUE, and when digitaria (*Digitaria eriantha* spp. *Eriantha*) was compared to lucerne (*Medicago sativa* L.) under sodic soils in New South Wales, WUE varied by 110% in the digitaria, 84% in the lucerne, and 72% for the mixture (Tow, 1993).

There were no observable differences among corn hybrids, and WUE values for grain yield and biomass were the same for short season and full season hybrids (Howell et al., 1998). There were; however, differences in the seasonal patterns of soil water extraction with hybrid maturity. There are also differences among soil types on water use patterns and corn yield (Tolk et al., 1998).

Soil Nutrient Status

The impacts of nutrients on WUE were first described by Viets (1962) when he observed a positive impact on WUE from the direct effect of nutrients on improving plant growth and yield. There have been some recent suggestions by Davis and Quick (1998) that cultivar selection for improved WUE could be based on quantifying the role of nutrient management on photosynthetic rate, yield, rooting characteristics, and transpiration. Optimization of WUE could be an outcome of enhanced cultivar selection and nutrient management practices (Davis and Quick, 1998). As a positive expression of these interactions, Payne (1997) reported a combination of N management, and increased plant population enhanced WUE of pearl millet [*Pennisetum glaucum* (L.) R. Br.] grown in the Sahel. The overall suggestion is that improved nutrient balance of the crop increases crop yields and should translate to improved WUE. A conclusion from these experimental results is that WUE improvements would be derived from a more in-depth knowledge of how nutrient management influences crop growth. Although there are some general conclusions, the current literature is not consistent in documenting the relationships between nutrient management and WUE. The following examples for specific nutrients provide evidence of nutrient management impacts on crop growth and yield and the potential WUE linkage.

Nitrogen

Soil type, tillage, N source (e.g., fertilizer, manure), crop rotation, and precipitation all affect N availability to a crop. Oberle and Keeney (1990) observed that for rainfed environments, preplant and early season precipitation amounts were important factors in explaining yield responses and were the factors that caused optimal N rates for maximum corn yield. In this study, N management caused variation in yield with no difference in amounts of water use. There are major differences among locations in N response. For example, in Alabama, Reeves et al. (1993) found maximum corn yields were obtained with N additions from 93 to 134 kg ha^{-1} in a legume-based conservation system, while in Minnesota, Jokela and Randall (1989) observed that grain and total dry matter yield of corn increased N additions up to 225 kg ha^{-1}. In both studies, large differences were observed across ies, large differences were observed across

the 3 yr of the study, and delayed N application did not influence dry matter or grain yield. Responses found for corn are different than those in wheat. Applying N at anthesis increased N use efficiency from 55 to 80% compared to N use efficiencies between 30 and 55% for preplant applications (Wuest and Cassman, 1992). Nitrogen management in wheat influences yield and grain quality, and protein content as a metric for grain quality is a critical parameter. Thus, the linkages between water and N will have to be addressed as components of the management system (Fowler et al., 1990). The findings of Jeuffroy and Bouchard (1999) demonstrated that N management in wheat influences grain number, and since grain number is a critical yield component, management practices need to be implemented that ensure the maximum number of grains per unit area are produced to obtain maximum yield. Improvements in wheat WUE can be made through N management because of its direct relationship to yield components like grain number per unit land area and grain size. Abbate et al. (1995) observed that N deficiency in wheat at anthesis affects grain number, and the number of grains per head is a function of the N content of the spikes. Strategies for improved N management to influence crop yield should consider the implications for WUE.

In addition to differences in soil and crop response, landscape position also affects N dynamics and availability to the crop. Across the landscape there are confounding interactions between water and N, Wood et al. (1991) showed slope position had little effect on plant N uptake or soil N dynamics, but aboveground biomass and plant residue production increased due to increased soil water availability from the top to the bottom of the landscape. Maskina et al. (1993) found that growth and N uptake by corn increased as residue amounts from previous crop production increased. This affect was more critical than tillage. Improvements in water availability and N increase the crop growth and potentially increase the amount of residue returned to the soil, and ultimately to the soil carbon (Halvorson et al., 1999). Increasing the cropping intensity in dryland regions, as suggested by Farahani et al. (1998), requires changes to N management practices since dryland soils have low N mineralization potential (Halvorson and Reule, 1994). The linkage between N management and water use rates is especially evident in dryland cropping systems. Changes in crop residue management used to increase WUE will have to be linked with N dynamics in the soil and across landscapes to achieve the maximum benefit of changing management practices.

There are direct effects on WUE from the addition of N fertilizer and incorporation of hairy vetch (*Vicia villosa* Roth.) residue into the soil (Corak et al., 1991). Increases in WUE from 6.1 to 8.5 kg ha^{-1} mm^{-1} in 1986 and from 9.1 to 16.6 kg ha^{-1} mm^{-1} in 1987 were found with the addition of 255 kg ha^{-1} N, with large variations between the 2 yr. Adding hairy vetch residue to the soil diminished the N fertilizer effect on WUE. There have been some general positive responses reported for N fertilizer effects on WUE for various crops, and these were attributed to the positive effect of increased biomass on WUE. Increases were found in grain sorghum (Varvel, 1995), native grasses (Smika et al., 1965), wheat (Campbell et al., 1992), and corn (Varvel, 1994).

Additional soil factors that link N management and WUE have been identified in poorly drained soils. In perennial ryegrass (*Lolium perenne* L.), Stout and Schnabel (1997) found that WUE decreased with poor drainage because denitrification reduced the available N, causing reduced plant growth. They observed reductions in WUE of 26% in the spring and 20% in the summer from the decrease in biomass production. There was an increase in WUE from 2.2 to 7.7 kg ha^{-1} mm^{-1} as N application increased from 0 to 126 kg ha^{-1} for these studies. There are differences in WUE among species, including observed values for orchardgrass (*Dactylis glomerta* L.) of 20.2 kg ha^{-1} mm^{-1} and 22.7 kg ha^{-1} mm^{-1} for tall fescue (*Festuca arundinacea* Schreb.) (Stout, 1992).

Phosphorus

The information on N effects on WUE is fairly abundant, but knowledge of the effect of phosphorus is much more limited.. Water use efficiency increased from 8.5 kg ha^{-1} mm^{-1} at 0 kg ha^{-1} of P to 12.2 kg ha^{-1} mm^{-1} at 100 kg ha^{-1} of P for chickpea (*Cicer arietinum* L.) because of the effect of additional P on improved yield, water use, and WUE (Singh and Bhushan, 1980). Improvements in WUE were due to increased soil water depletion with addition of P fertilizer and the accompanying increase in chickpea yield.

The effect of adding P is more pronounced in low-phosphorus soils; for example, addition of P fertilizer was found to increase both dry matter yield and WUE in pearl millet (Payne et al., 1992, 1995). Enhanced dry matter production in crops relative to water use rates and amounts from improved soil nutrient status will directly increase WUE.

Climate Change Impacts

There are differences in WUE among climates which are caused by the variations in the water use rate among crops. Zhang et al. (2000) observed that water use and WUE for chickpea and lentil (*Lens culinaris* Medikus) in northern Syria was dependent on the rainfall amounts and the patterns during the growing season. They found that yields increased during the wet seasons of this 12-season study and when supplemental irrigation was applied. The WUE for grain production was 3.8 kg ha^{-1} mm^{-1} for lentil and 3.2 kg ha^{-1} mm^{-1} for chickpea. They found that in the Mediterranean climate the lentil was better adapted to this climate. Sadras and Angus (2006) compared WUE for wheat in four environments, southeastern Australia, North American Great Plains, China Loess Plateau, and the Mediterranean Basin. In their study they compiled data from published studies from these sites and computed WUE. Based on this analysis the average WUE (kg ha^{-1} mm^{-1}) for grain production was 5.3 for the south-central Great Plains of North America, 7.6 for the Mediterranean Basin, 8.9 for the northern Great Plains of North America, 9.8 for the China Loess Plateau, and 9.9 for southeastern Australia. They observed that the variation in WUE was related to evapotranspiration around the time of flowering. The variation in yield was due to water availability during the critical time of flowering. Variation in rainfall among seasons was the primary factor creating differences in wheat yield and WUE.

Climate change impacts on agriculture have been compiled by Hatfield et al. (2008) in a summary of the potential effects on climatic factors, temperature, CO_2, and precipitation on crop growth and yield. Climate scenarios for the future were developed by Tebaldi et al. (2006) in which temperature and precipitation patterns across the United States for the next 50 yr show a warming trend for most of the United States of 1.5 to 2°C and a slight increase in precipitation over most of the United States. They also projected an increase in warm nights, defined as occurring when the minimum temperature is above the 90th percentile of the climatological distribution for the day. These changes are typical of other regions of the world and will impact WUE in several ways. The increase in temperature will increase the ET of the crop and increase the potential for water stress, thus lowering biomass and grain yield production. This will ultimately reduce the transpiration amounts and decrease production and lower WUE. Seasons with greater rainfall will benefit because of the potential positive impact on crop growth and development and increased WUE. These projected results are similar to the multisite comparison made by Sadras and Angus (2006), in which environments with seasonal water deficits at critical times would have reduced WUE. The increase in the nighttime temperatures and the negative impact on biomass and grain yield would reduce the WUE because the increased respiration at night would offset any gains during the day, and although soil management practices would increase the water availability to the crop, there may not be a positive gain from the increased water. Further analysis is required to determine the role that soil management could play in offsetting these impacts.

Increasing CO_2 has been linked with increasing WUE. Morison (1987) showed that for both C_3 and C_4 species, stomatal conductance was reduced about 40% with a doubling of CO_2 thereby increasing water conservation and reducing plant water deficits. A 12% reduction in seasonal transpiration and 51% increase in WUE was found for soybean crops in sunlit, controlled-environment chambers grown at ambient and doubled CO_2 (Jones et al., 1985). Doubling of CO_2 decreased transpiration in wheat by 8% (Andre and du Cloux, 1993). Also using environment chambers, Reddy et al. (2000) found transpiration was reduced by 8% in cotton canopies when CO_2 was doubled. Using lysimeters in Arizona for cotton experiments, Kimball and Idso (1983)

found a 4% reduction in seasonal water use of cotton at ambient versus 650 ppm CO_2. Reductions in ET reduction caused by increased CO_2 in soybean, ranged from 9 to 16% among seasons when grown at 550 compared to 375 ppm in Free-Air Carbon Exchange (FACE) experiments in Illinois (Bernacchi et al., 2007). There is an interaction of changing CO_2 and temperature on soybean ET, and Allen et al. (2003) detected a 9% reduction with doubling of CO_2 in sunlit, controlled-environment chambers for a 28/18°C treatment, but no reduction in ET when the plants were grown at 40/30°C. The conclusion is that the effect of CO_2 on reducing ET is temperature dependent and would also vary with species because of the variation in temperature responses among plants. A similar response was found by Horie et al. (2000) for rice, when a doubling of CO_2 caused a 15% reduction in ET at 26°C, but increased ET at 29.5°C. With doubled CO_2 and rice grown at 24 to 26°C, WUE increased by 50%, but as the temperature increased, the CO_2 enrichment effect diminished.

Interactions of CO_2 enrichment with climatic factors of water supply and evaporative demand are especially evident under water deficit conditions (Boote et al., 1997). Reductions in stomatal conductance with elevated CO_2 will potentially lead to improved soil water conservation and reduced plant water stress, especially for crops grown with periodic soil water deficit or under high evaporative demand. The changes in CO_2 will enhance the differences that have been reported by Sadras and Angus (2006) and increase the need for soil management practices that conserve soil water.

Climate changes in precipitation patterns will affect the amount and distribution of rainfall. These changes will render the impact of soil management practices even more important in the future. Stability of crop production will require the maintenance of crop yields and soil water availability will be one of the primary factors affecting WUE.

Challenges

Development of soil management practices that enhance WUE given the changes in climate over the next 50 years revolve around optimization of the availability of soil water to the crop. Even though increases in CO_2

have demonstrated their positive impact on plant growth and water use efficiency, these effects would be negated when soil water is limited due to lack of precipitation. Although there can be less positive impact on WUE from soil management practices when temperatures increase, there are still positive changes that can be made in WUE through improved nutrient management to enhance plant growth.

A challenge for the research community and producers will be to understand the interactions among the soil management factors, plant growth and yield, and the changing climate. There can be positive increases in WUE with improved management, as shown in Fig. 10I1. However, the response of these changes when plant growth is altered with the changed climate needs to be incorporated into our understanding. The challenge will be to extend these findings into areas where the research data are limited to be able to help producers understand the changes they can institute in their management decisions that will optimize WUE. Increasing WUE through management will help ensure a more stable food supply for future generations.

References

Aase, J.K., and J.L. Pikul, Jr. 1995. Crop and soil response to long-term no-tillage practices in the Northern Great Plains. Agron. J. 87:652–656.

Aase, J.K., and F.H. Siddoway. 1990. Stubble height effects on seasonal microclimate, water balance, and plant development of no-till winter wheat. Agric. Meteorol. 21:1–20.

Abbate, P.E., F.H. Andrale, and J.P. Culot. 1995. The effect of radiation and nitrogen on number of grains in wheat. J. Agric. Sci. 124:351–360.

Allmaras, R.R., W.C. Burrows, and W.E. Larson. 1964. Early growth of corn as affected by soil temperature. Soil Sci. Soc. Am. Proc. 28:271–275.

Allen, L.H., Jr., D. Pan, K.J. Boote, N.B. Pickering, and J.W. Jones. 2003. Carbon dioxide and temperature effects on evapotranspiration and water-use efficiency of soybean. Agron. J. 95:1071–1081.

Anderson, D.T., and G.C. Russell. 1964. Effects of various quantities of straw mulch on the growth and yield of spring and winter wheat. Can. J. Soil Sci. 44:109–118.

Andre, M., and H. du Cloux. 1993. Interaction of CO_2 enrichment and water limitations on photosynthesis and water use efficiency in wheat. Plant Physiol. Biochem. 31:103–112.

Azooz, R.H., and M.A. Arshad. 1995. Tillage effects on thermal conductivity of two soils in northern British Columbia. Soil Sci. Soc. Am. J. 59:1413–1423.

Azooz, R.H., and M.A. Arshad. 1998. Effect of tillage and residue management on barley and canola growth and water use efficiency. Can. J. Soil Sci. 78:649–656.

Bernacchi, C.J., B.A. Kimball, D.R. Quarles, S.P. Long, and D.R. Ort. 2007. Decreases in stomatal conductance of soybean under open-air elevation of CO_2 are closely coupled with decreases in ecosystem evapotranspiration. Plant Physiol. 143:134–144.

Boote, K.J., N.B. Pickering, and L.H. Allen, Jr. 1997. Plant modeling: Advances and gaps in our capability to project future crop growth and yield in response to global climate change. p. 179–228. In L. H. Allen, Jr. et al. (ed.) Advances in carbon dioxide effects research. ASA Spec. Publ. 61. ASA, CSSA, and SSSA, Madison, WI.

Burns, R.L., D.J. Cook, and R.E. Phillips. 1971. Influence of no tillage on soil moisture. Agron. J. 73:593–596.

Campbell, C.A., R.P. Zentner, B.G. McConkey, and F. Selles. 1992. Effect of nitrogen and snow management on efficiency of water use by spring wheat grown annually on zero-tillage. Can. J. Soil Sci. 72:271–279.

Chan, K.Y., and D.P. Heenan. 1996. Effect of tillage and stubble management on soil water storage, crop growth and yield in a wheat–lupin rotation in southern NSW. Aust. J. Agric. Res. 47:479–488.

Christensen, N.B., T.L. Jones, and G.J. Kauta. 1994. Infiltration characteristics under no-till and clean-till furrow irrigation. Soil Sci. Soc. Am. J. 58:1495–1500.

Corak, S.J., W.W. Frye, and M.S. Smith. 1991. Legume mulch and nitrogen fertilizer effects on soil water and corn production. Soil Sci. Soc. Am. J. 55:1395–1400.

Cresswell, H.P., D.J. Painter, and K.C. Cameron. 1993. Tillage and water content effects on surface soil hydraulic properties and shortwave albedo. Soil Sci. Soc. Am. J. 57:816–824.

Dao, T.H., and H.T. Nguyen. 1989. Growth response of cultivars to conservation tillage in a continuous wheat cropping system. Agron. J. 81:923–929.

Davis, J.G., and J.S. Quick. 1998. Nutrient management, cultivar development and selection strategies to optimize water use efficiency. J. Crop Prod. 1:221–240.

Deibert, E.J., E. French, and B. Hoag. 1986. Water storage and use by spring wheat under conventional tillage and no-till in continuous and alternate crop-fallow systems in the northern Great Plains. J. Soil Water Conserv. 41:53–58.

deWit, C.T. 1958. Transpiration and crop yields. Versl. Landouwk. Onderz. 64.6 Inst. of Biol. and Chem. Res. on Field Crops and Herbage, Wageningen, The Netherlands.

Eck, H.V., and S.R. Winter. 1992. Soil profile modification effects on corn and sugarbeet grown with limited water. Soil Sci. Soc. Am. J. 56:1298–1304.

Farahani, H.J., G.A. Peterson, D.G. Westfall, L.A. Sherrod, and L.R. Ahuja. 1998. Soil water storage in dryland cropping systems: The significance of cropping intensification. Soil Sci. Soc. Am. J. 62:984–991.

Fowler, D.B., J. Brydon, B.A. Darroch, M.H. Entz, and A.M. Johnston. 1990. Environment and genotype influence on grain protein concentration of wheat and rye. Agron. J. 82:655–664.

Gibson, G., B.J. Radford, and R.G.H. Nielsen. 1992. Fallow management, soil water, plant-available soil nitrogen and grain sorghum production in south west Queensland. Aust. J. Exp. Agric. 32:473–482.

Good, L.C., and D.E. Smika. 1978. Chemical fallow for soil and water conservation in the Great Plains. J. Soil Water Conserv. 33:89–90.

Greb, B.W. 1966. Effect of surface-applied wheat straw on soil water losses by solar distillation. Soil Sci. Soc. Am. Proc. 30:786–788.

Halvorson, A.D., and C.A. Reule. 1994. Nitrogen fertilizer requirements in an annual dryland cropping system. Agron. J. 86:315–318.

Halvorson, A.D., C.A. Reule, and R.F. Follett. 1999. Nitrogen fertilization effects on soil carbon and nitrogen in a dryland cropping system. Soil Sci. Soc. Am. J. 63:912–917.

Hammel, J.E. 1989. Long-term tillage and crop rotation effects on bulk density and soil impedance in Northern Idaho. Soil Sci. Soc. Am. J. 53:1515–1519.

Hatfield, J.L. 1990. Modification of the microclimate in rainfed agriculture by stubble mulch. p. 315–317. *In* P.W. Unger et al. (ed.) Challenges in dryland agriculture: A global perspective. Proc. Int. Conf. on Dryland Farming, Texas Agric. Exp. Stn., College Station, TX.

Hatfield, J.L., K.J. Boote, P. Fay, L. Hahn, C. Izaurralde, B.A. Kimball, T. Mader, J. Morgan, D. Ort, W. Polley, A. Thomson, and D. Wolfe. 2008. Agriculture. p. 21–74. *In* The effects of climate change on agriculture, land resources, water resources, and biodiversity in the United States. A report by the U.S. Climate Change Science Program and the Subcommittee on Global Change Research, Washington, DC.

Hatfield, J.L., and J.H. Prueger. 1996. Microclimate effects of crop residues on biological processes. Theor. Appl. Climatol. 54:47–59.

Hatfield, J.L., T.J. Sauer, and J.H. Prueger. 2001. Managing soils for greater water use efficiency: A Review. Agron. J. 93:271–280.

He, J., H. Li, A.D. McHugh, Z. Ma, X. Cao, Q. Wang, X. Zhang, and X. Zhang. 2008. Spring wheat performance and water use efficiency on permanent raised beds in arid northwest China. Aust. J. Soil Res. 46:659–666.

Horie, T., J.T. Baker, H. Nakagawa, T. Matsui, and H.Y. Kim. 2000. Crop ecosystem responses to climatic change: Rice. p. 81–106. *In* K.R. Reddy and H.F. Hodges (ed.) Climate change and global crop productivity. CAB International, New York.

Howell, T.A., J.A. Tolk, A.D. Schneider, and S.R. Evett. 1998. Evapotranspiration, yield, and water use efficiency of corn hybrids differing in maturity. Agron. J. 90:3–9.

Jeuffroy, M.H., and C. Bouchard. 1999. Intensity and duration of nitrogen deficiency on wheat grain number. Crop Sci. 39:1385–1393.

Johnson, W.C., and R.G. Davis. 1971. Research on stubble-mulch farming of winter wheat. USDA Conserv. Rep. 16. U.S. Gov. Print. Office, Washington, DC.

Johnson, M.D., B. Lowery, and T.C. Daniel. 1984. Soil moisture regimes of three conservation tillage systems. Trans. ASAE 27:1385–1390, 1395.

Jokela, W.E., and G.W. Randall. 1989. Corn yield and residual soil nitrate as affected by time and rate of nitrogen application. Agron. J. 81:720–726.

Jones, O.R., and G.L. Johnson. 1991. Row width and plant density effects on Texas High Plains sorghum. J. Prod. Agric. 4:613–621.

Jones, O.R., and T.W. Popham. 1997. Cropping and tillage systems for dryland grain production in the southern High Plains. Agron. J. 89:222–232.

Jones, P., J.W. Jones, and L.H. Allen, Jr. 1985. Seasonal carbon and water balances of soybeans grown under stress treatments in sunlit chambers. Trans. ASAE 28:2021–2028.

Kaspar, T.C., D.C. Erbach, and R.M. Cruse. 1990. Corn response to seed-row residue removal. Soil Sci. Soc. Am. J. 54:1112–1117.

Kimball, B.A., and S.B. Idso. 1983. Increasing atmospheric CO_2: Effects on crop yield, water use, and climate. Agric. Water Manage. 7:55–72.

Lascano, R.J., R.L. Baumhardt, S.K. Hicks, and J.L. Heilman. 1994. Soil and plant water evaporation from strip-tilled cotton: Measurement and simulation. Agron. J. 86:987–994.

Lascano, R.J., and J.L. Hatfield. 1992. Spatial variability of evaporation along two transects of a bare soil. Soil Sci. Soc. Am. J. 56:341–346.

Liang, B.C., A.F. MacKenzie, P.C. Kirby, and M. Remillard. 1991. Corn production in relation to water inputs and heat units. Agron. J. 83:794–799.

Maskina, M.S., J.F. Power, J.W. Doran, and W.W. Wilhelm. 1993. Residual effects of no-till crop residues on corn yield and nitrogen uptake. Soil Sci. Soc. Am. J. 57:1555–1560.

Morison, J.I.L. 1987. Intercellular CO_2 concentration and stomatal response to CO_2. p. 229–251. *In* E. Zeiger et al. (ed.) Stomatal function. Stanford Univ. Press, Stanford, CA.

Musick, J.T., O.R. Jones, B.A. Stewart, and D.A. Dusek. 1994. Water–yield relationships for irrigated and dryland wheat in the U.S. Southern Plains. Agron. J. 86:980–986.

Norwood, C.A. 1999. Water use and yield of dryland row crops as affected by tillage. Agron. J. 91:108–115.

Oberle, S.L., and D.R. Keeney. 1990. Soil type, precipitation, and fertilizer N effects on corn yields. J. Prod. Agric. 3:522–527.

Papendick, R.I., M.J. Lindstrom, and V.L. Cochran. 1973. Soil mulch effect on seedbed temperature and water during fallow in eastern Washington. Soil Sci. Soc. Am. Proc. 37:307–314.

Payne, W.A. 1997. Managing yield and water use of pearl millet in the Sahel. Agron. J. 89:481–490.

Payne, W.A., M.C. Drew, L.R. Hossner, R.J. Lascano, A.B. Onken, and C.W. Wendt. 1992. Soil phosphorus availability and pearl millet water-use efficiency. Crop Sci. 32:1010–1015.

Payne, W.A., L.R. Hossner, A.B. Onken, and C.W. Wendt. 1995. Nitrogen and phosphorus uptake in pearl millet and its relation to nutrient and transpiration efficiency. Agron. J. 87:425–431.

Pikul, J.L., Jr., and J.K. Aase. 1995. Infiltration and soil properties as affected by annual cropping in the northern Great Plains. Agron. J. 87:656–662.

Power, J.F. 1983. Soil management for efficient water use: Soil fertility. p. 461–470. *In* H.M. Taylor et al. (ed.) Limitations to efficient water use in crop production. ASA, Madison, WI.

Reddy, K.R., H.F. Hodges, and B.A. Kimball. 2000. Crop ecosystem responses to climatic change: Cotton. p. 161–187. *In* K.R. Reddy and H.F. Hodges (ed.) Climate change and global crop productivity. CAB International, New York.

Reeves, D.W., C.W. Wood, and J.T. Touchton. 1993. Timing nitrogen applications for corn in a winter legume conservation tillage system. Agron. J. 85:98–106.

Ritchie, J.T. 1971. Dryland evaporative flux in a subhumid climate. 1. Micrometeorological influences. Agron. J. 70:723–728.

Sadras, V.O., and J.F. Angus. 2006. Benchmarking water-use efficiency of rainfed wheat in dry environments. Aust. J. Agric. Res. 57:847–856.

Sauer, T.J., J.L. Hatfield, and J.H. Prueger. 1996a. Corn residue age and placement effects on evaporation and soil thermal regime. Soil Sci. Soc. Am. J. 60:1558–1564.

Sauer, T.J., J.L. Hatfield, and J.H. Prueger. 1996b. Aerodynamic characteristics of standing corn stubble. Agron. J. 88:733–739.

Sauer, T.J., J.L. Hatfield, and J.H. Prueger. 1997. Over-winter changes in radiant energy exchange of a corn residue-covered surface. Agric. For. Meteorol. 85:279–287.

Sauer, T.J., J.L. Hatfield, J.H. Prueger, and J.M. Norman. 1998. Surface energy balance of a corn-residue covered field. Agric. For. Meteorol. 89:155–168.

Singh, G., and L.S. Bhushan. 1980. Water use, water use efficiency and yield of dryland chickpea as influenced by P fertilization, stored soil water and crop season rainfall. Agric. Water Manage. 2:299–305.

Smika, D.E., H.J. Haas, and J.F. Power. 1965. Effect of moisture and nitrogen fertilizer on growth and water use by native grass. Agron. J. 57:483–486.

Srivastava, V.C., and M. Sidique. 1978. Moisture use efficiency of wheat as influenced by seeding rate and row spacing under limited moisture condition. J. Soil Water Conserv. India 28:42–49.

Stout, W.L. 1992. Water use efficiency of grasses as affected by soil, nitrogen, and temperature. Soil Sci. Soc. Am. J. 56:897–902.

Stout, W.L., and R.R. Schnabel. 1997. Water-use efficiency of perennial ryegrass as affected by soil drainage and nitrogen fertilization on two floodplain soils. J. Soil Water Conserv. 52:207–211.

Tanaka, D.L. 1990. Topsoil removal influences on spring wheat water-use efficiency and nutrient concentration and content. Trans. ASAE 33:1518–1524.

Tanner, C.B., and T.R. Sinclair. 1983. Efficient water use in crop production: Research or re-search? p. 1–27. *In* H.M. Taylor et al. (ed.) Limitations to efficient water use in crop production. ASA, Madison, WI.

Tebabaldi, C., K. Hayhoe, J.M. Arblaster, and G.E. Meehl. 2006. Going to the extremes: An intercomparison of model simulated historical and future changes in extreme events. Clim. Change 79:185–211.

Tolk, J.A., T.A. Howell, and S.R. Evett. 1998. Evapotranspiration and yield of corn grown on three High Plains soils. Agron. J. 90:447–454.

Tompkins, D.K., D.B. Fowler, and A.T. Wright. 1991. Water use by no-till winter wheat influence of seed rate and row spacing. Agron. J. 83:766–769.

Tow, P.G. 1993. Persistence and water use efficiency of a tropical grass and lucerne on a solodic soil on the far north-west slopes of New South Wales. Aust. J. Exp. Agric. 33:245–252.

Unger, P.W. 1988. Residue management effects on soil temperature. Soil Sci. Soc. Am. J. 52:1777–1782.

Unger, P.W. 1991. Ontogeny and water use of no-tillage sorghum cultivars on dryland. Agron. J. 83:961–968.

Unger, P.W. 1994. Residue management for winter wheat and grain sorghum production with limited irrigation. Soil Sci. Soc. Am. J. 58:537–542.

Varvel, G.E. 1994. Monoculture and rotation system effects on precipitation use efficiency of corn. Agron. J. 86:204–208.

Varvel, G.E. 1995. Precipitation use efficiency of soybean and grain sorghum in monoculture and rotation. Soil Sci. Soc. Am. J. 59:527–531.

Viets, F.G., Jr. 1962. Fertilizers and the efficient use of water. Adv. Agron. 14:223–264.

Wallace, J.S. 2000. Increasing water use efficiency to meet future food production. Agric. Ecosyst. Environ. 82:105–119.

Wood, C.W., G.A. Peterson, D.G. Westfall, C.V. Cole, and W.O. Willis. 1991. Nitrogen balance and biomass production of newly established no-till dryland agroecosystems. Agron. J. 83:519–526.

Wuest, S.B., and K.G. Cassman. 1992. Fertilizer-nitrogen use efficiency of irrigated wheat. I. Uptake efficiency of preplant versus late-season application. Agron. J. 84:682–688.

Zhai, R., R.G. Kachanoski, and R.P. Voroney. 1990. Tillage effects on the spatial and temporal variations of soil water. Soil Sci. Soc. Am. J. 54:186–192.

Zhang, H., M. Pala, T. Oweis, and H. Harris. 2000. Water use and water-use efficiency of chickpea and lentil in a Mediterranean environment. Aust. J. Agric. Res. 51:295–304.

Zhang, H., and T. Qweis. 1999. Water-yield relations and optimal irrigation scheduling of wheat in the Mediterranean region. Agric. Water Manage. 38:195–211.

11

Climatic Resources

Jerry L. Hatfield and John H. Prueger

Soil water and soil temperature patterns in the soil profile determine the overall biological response of plants, microbes, and other soil fauna. The impact of soil management practices on the soil microclimate depends primarily on how management practices affect the soil water and soil temperature patterns at the soil surface and within the soil profile throughout the day and across the year. As we begin to understand these interactions, the more opportunities we have to develop soil management practices that will have a positive impact on the soil. These impacts will improve plant production efficiency, decrease pressures from pests, and enhance the quality of the soil over time.

To understand how the soil microclimate is affected by soil management practices it is important to begin with an understanding of the physical processes that determine the temperature and water regimes in the soil profile. Manipulation of the soil surface by tillage, residue cover, cover crops, and the type of crop that is grown affects these dynamics of the energy balance, which defines the exchange of energy between the soil and the atmosphere. This process is relatively simple and is governed by the energy balance as shown in Eq. [1]:

$$R_n - G = H + \text{LE} \tag{1}$$

where R_n is the net radiation, G is the soil heat flux, H the sensible heat flux, and LE the latent heat flux with each parameter expressed in terms of watts per square meter (W m^{-2}).

Dissecting Eq. [1] into the components begins with the R_n component. This is the dominant parameter in the energy balance and is a function of the amount of sunlight and longwave radiation that impinges on the soil surface. Diagrammatically these components can be represented as shown in Fig. 11|1. Net radiation can be mathematically described as

$$R_n = (1 - \alpha)S_g + L_i - \varepsilon \sigma T_s^{\,4} \tag{2}$$

where α is the albedo of the surface which can be described as the reflectivity of the surface, S_g is the solar irradiance (W m^{-2}), L_i is the longwave irradiance from the sky, and ε is the emissivity of the soil surface, σ is the Stefan–Boltzman constant (5.67×10^{-8} W m^{-2} K^{-4}), and T_s is the surface temperature (K). Longwave radiation emitted from the atmosphere can be expressed in a similar form to the surface longwave in which the ε term is the emissivity of the atmosphere and the temperature term is expressed as the air temperature (T_a). Hatfield et al. (1983) compared a

---

J.L. Hatfield (jerry.hatfield@ars.usda.gov) and J.H. Prueger (john.prueger@ars.usda.gov), USDA-ARS National Laboratory for Agriculture and the Environment, 2110 University Blvd., Ames, IA 50011.

doi:10.2136/2011.soilmanagement.c11

number of different approaches to estimating ε from the atmosphere and the necessary precautions to be followed in applying these approaches. All of the methods use an empirical combination of air temperature and relative humidity and are often developed for specific locations.

The albedo of the surface represents the reflectivity, which can be thought of as amount of light that is reflected back to the atmosphere; therefore, the higher the albedo, the more light that is returned and the brighter the surface appears. For example, a dark soil that is wet has a low albedo, and as the soil dries the albedo increases. Similarly, a dark soil covered with fresh crop residue will have a higher albedo than a bare soil surface. The albedo of the surface is variable and depends on the soil type, organic matter content of the surface soil, amount of crop residue, age of residue, crusting, tillage, and surface wetness. Given all of these variables that affect albedo, it is difficult to assume a constant value throughout a growing season. An example is shown in Fig. 11|2,

which depicts the change in albedo over the course of a growing season. The presence of the residue material causes the albedo to be larger than the soil and during the season as the crop covers the soil surface the albedo increases with the presence of the crop. In a light-colored soil the growth of the crop actually decreases the albedo of the surface.

Solar irradiance is affected by a number of factors—the amount of sunlight that impinges on a surface depends on our location on the earth, the angle of the surface, and the time of year. Simply stated, the maximum solar energy is when the sun is directly overhead on a clear day, shining onto a level surface. The physics of this process are described in a number of textbooks (e.g., Monteith, 1973). There are physical equations that can be used to calculate the solar radiation impinging onto a surface on a clear day, and these are given in Ham (2005). There are actually two components of S_g, a direct and diffuse component. Direct sunlight is the direct beam of light from the sun, while diffuse is the amount of light that has been scattered by the atmosphere. The direct component is what causes a shadow, while the diffuse component allows us to have light in the shadow. On the soil surface, the direct component is a major energy source that impinges on the upper leaves of canopies or onto the surface, while the diffuse component is the energy that is present in the lower parts of the canopy or below the residue layer. The amount of direct and diffuse sunlight will vary throughout the year depending on the position on the earth, the slope, and cloudiness of the location.

Albedo and emissivity are dependent on parameters that are affected by soil management such as crop residue, surface drying, shape of the soil surface, or the soil organic matter content. As the albedo increases there is less energy that will be retained by the soil. If the emissivity increases, the amount of energy emitted from the soil surface will increase. There are large changes in the range of values induced by typical soil management practices, and albedo affects the energy available more than emissivity. The energy available in the solar radiation is larger than the longwave components during the day, while at night the longwave radiation is the only factor in the radiation

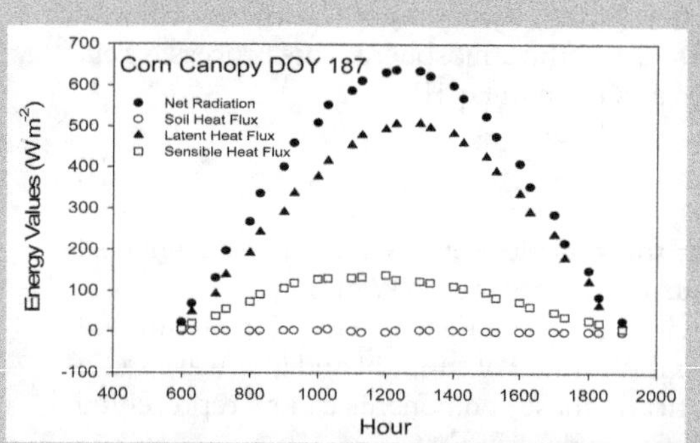

Fig. 11|1. Generalized description of the energy balance for a surface.

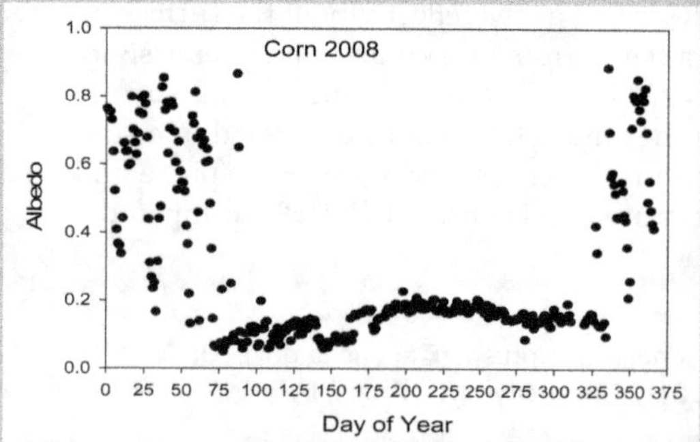

Fig. 11|2. Changes in albedo over a corn crop before planting until after harvest.

balance (Fig. 11|3). These illustrations from measurements over a cropped field demonstrate that the magnitude of these values change throughout the day. These patterns change throughout the year, and in summer the incoming shortwave is the dominant component in the radiation balance. This changes during the winter period when the outgoing longwave component is the largest (Fig. 11|4). This is to be expected since the cooling that occurs during the winter period is due to the loss of energy from the surface. These values are for the central United States and will change as we move with latitude around the Earth. For example, near the equator the exchange of radiation would be fairly consistent throughout the year, but as one moves to more northerly or southerly latitudes, then the patterns in the radiation components will change.

An important part of the balance of longwave and shortwave radiation is the shape of the surface. In soil management, there are changes of the surface due to tillage, and these changes will affect the angle of the surface relative to the angle of the sun. This change only affects the direct beam of incoming shortwave radiation and not the diffuse shortwave or the longwave components. If there is a ridge created by tillage then the south side of the ridge would warm more quickly because of greater exposure to direct beam radiation. This would cause this surface to dry and warm more quickly than the north side of the ridge. In the southern hemisphere the opposite effect would be seen, with the north side of the ridge being that warmer side. One way of considering the impact of a sloping surface is to consider that having a south-facing slope with a 10° angle would have the same exposure to the sun as being 10° further south in latitude. There is little effect on the ongoing longwave caused by the fact that warmer surfaces would emit more radiation. As we change the slope of the soil surface in northern latitudes these areas would tend to warm more quickly in the spring because their surface is oriented more directly toward the sun. The details of this process are described in many microclimate books (Monteith, 1973; Rosenberg et al., 1983).

The radiation balance is a large part of the overall energy balance for a surface in which

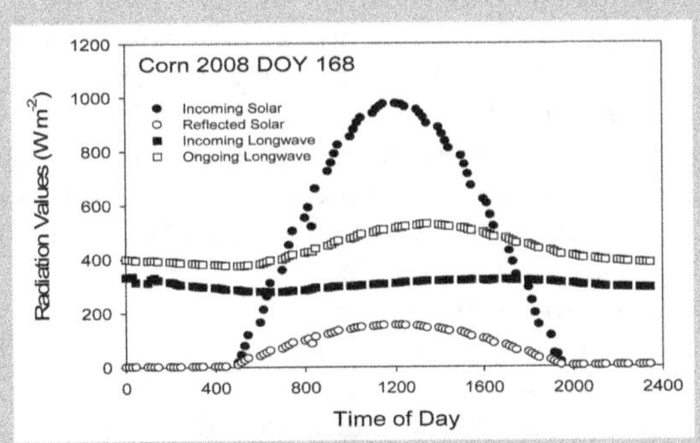

Fig. 11|3. Radiation balance for a typical summer day in central Iowa.

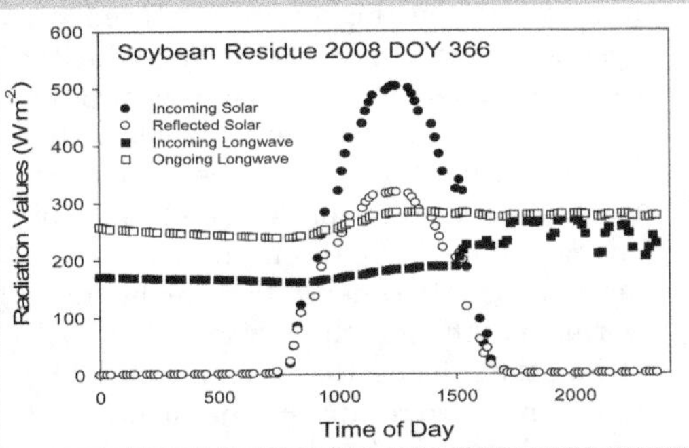

Fig. 11|4. Radiation balance for a typical winter day in central Iowa.

the available energy is partitioned into latent heat, sensible heat, and soil heat flux (Fig. 11|1). These data are for a typical day during the early spring in a northern latitude. There will be variation in these components caused by the cropping systems, tillage practice, locations, and time of year. To fully understand how soil management practices affect the energy balance it is important to briefly discuss each of the components.

Soil Heat Flux

Soil heat flux (G) is simply the amount of energy that is exchanged between the soil and the atmosphere and has recently been discussed in detail by Sauer and Horton (2005). This process proceeds primarily by conduction and is described as

$$G = -\lambda \frac{\delta T}{\delta z} \qquad [3]$$

where λ is the thermal conductivity of the soil, T is the temperature of the soil layer, and z the depth of the soil layer. The factor in Eq. [3] that is affected most by soil management is the thermal conductivity of the soil layers, which depends on mineral composition of the soil, particle size, amount of organic matter, soil bulk density, and water content.

Tillage loosens the soil, which reduces the bulk density of the soil, which in turn reduces the thermal conductivity of the upper layers of the soil. Azooz et al. (1997) showed that the soil heat flux was lower in a tilled soil than a non-tilled soil because of the impact of increased air spaces in the upper layers on reducing the thermal conductivity. Adding residue onto the soil surface creates a layer with a lower thermal conductivity because of all of the air spaces in the residue layer. The change in the thermal conductivity of this layer reduces the energy that can be transported into the soil; thus, crop residue will reduce the soil heat flux. Sauer et al. (1997) found that corn residue on the surface had an albedo higher than bare soil, presented a barrier to water vapor movement from the soil to the atmosphere, and reduced the amount of energy that could be partitioned into soil heat flux. Soils with a large amount of residue cover tend to be cooler, wetter, and have a smaller soil heat flux than soils without residue cover.

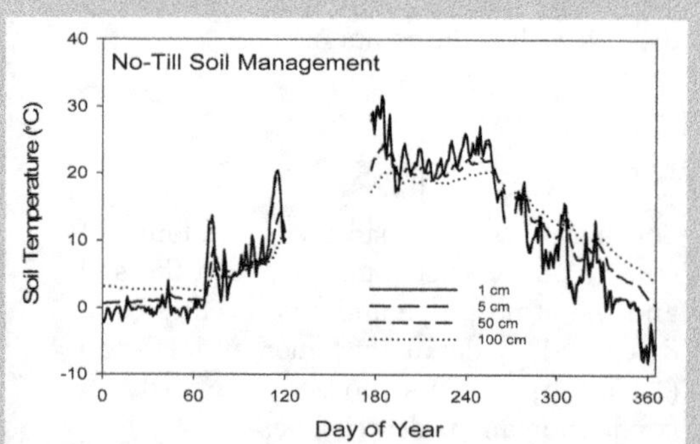

Fig. 11|5. Seasonal patterns in soil temperature throughout a year at multiple depths in the soil profile.

Soil Temperature

Soil heat flux provides the energy required to change the temperature of the soil. Soil temperature patterns are important for plant growth, biological activity, and water vapor exchange within the soil profile and between the soil surface and the atmosphere. Soil temperature is a soil parameter that is more often used to assess the impact of soil management practices because the question will be whether this change in practice will cause the soil temperatures to be either warmer or colder than what is optimum for plant growth and development. Soil temperatures are influenced by a number of factors, including meteorological conditions, soil surface conditions, type of crop, and growing season. Soil temperature patterns within the soil vary with time of day, time of year, and depth. Van Wijk and deVries (1966) were among the first to describe this process in detail and provided elaborate detail on the physics and mathematics of soil temperature patterns in soil. Soil temperatures within a field exhibit various patterns throughout the year, as shown in Fig. 11|5. The greatest variation over the year occurs in the upper layers of the soil profile and gradually diminishes with depth in the profile. At some depth, typically 2 m, there is no variation in soil temperature.

The effects of soil management on soil temperature have been extensively documented over the past 100 yr. For example, Burrows and Larson (1962) showed that corn residue reduced soil temperature and consequently corn growth. They found that plant height and plant biomass decreased as the amount of residue on the surface increases. Singh and Sandhu (1979) found a similar result in studies in India. Al-Darby and Lowery (1987) reported that soil temperatures at 5 cm were lower in no-till systems with undisturbed residue on the surface, and these lower temperatures affected emergence and growth of corn seedlings. Gupta et al. (1983) had previously reported that soil surface temperatures were lowest in no-till with surface residue and highest in no-till with the residue removed. Evaluation of the impact of residue on soil temperatures has to consider the annual changes in temperature. In comparing different tillage systems with and without corn residue, for example, fall plow, chisel-plow, and no-till, Benoit and van Sickle (1991) found that no-till with residue in Minnesota had the highest over-winter temperatures at the 5-, 10-, and 30-cm depths. They also found that the no-till with residue was the first soil to become frost free in the spring and had warmer temperatures until planting time. Earlier, Benoit et al. (1986) found that the reduced tillage systems with the residue increased the snow accumulation, which reduced the depth of the frost into the soil. Hatfield and Prueger (1996) compared continuous corn and corn–soybean rotations under no-till, chisel-low, and moldboard plow in central Iowa and found the largest effect of residue was in the fall after harvest when the no-till fields cooled more slowly than the tilled fields. They found that the diurnal temperature patterns were more affected by the presence of residue than the annual patterns. In a recent study, Dahiya et al. (2007) evaluated the effect of mulch on temperature patterns in a loess soil and found that tillage and mulch did not affect the soil thermal conductivity and changed the soil temperatures by less than 1.0°C.

The effect of crop residue and tillage of the surface on the soil temperature regimes with the soil profile is realized through changes in the soil thermal properties. Novak (2005) summarized the soil temperature regime as being affected by two major factors: those that affect the conduction of energy into the soil (Eq. [3]) and those that affect the volumetric heat capacity and soil thermal conductivity. The volumetric heat capacity of a soil is the sum of the individual heat capacities for the soil components weighted by their volumetric fraction. This can be expressed as a simple sum, as shown in Eq. [4]:

$$C = x_m M + x_{om} OM + x_w W + x_a A \qquad [4]$$

where C is the soil heat capacity and x_m, x_{om}, x_w, and x_a are the fraction of the soil volume comprised of minerals, organic matter, water, and air, respectively. The specific heat values for the individual components are shown in Table 11|1. As the composition of the soil changes there are large impacts on the heat capacity of the soil. Likewise, there are large differences in the thermal conductivity values for the different soil fractions (Table 11|1). As the fractions change within the soil there are large effects on how

Table 11|1. Thermal properties of the soil components.

Component	Specific heat capacity	Thermal conductivity
	J kg^{-1} K^{-1}	W m^{-2} K^{-1}
Mineral	755	2.9
Organic matter	1920	0.25
Water	4200	0.57
Air	1000	0.025

quickly the soil changes temperature. As tillage affects the amount of air space in the soil or the addition of organic materials into the soil volume, these factors can have large impacts on the soil temperature patterns within the soil profile.

Evapotranspiration

Evapotranspiration (ET) or latent heat of vaporization (LE in Eq. [1]) represents one of the largest components of the energy balance. For a crop with an adequate water supply in the middle of summer the total ET can be 6 to 7 and as high 10 mm d^{-1}. Soil management practices can have a large impact on ET, and in particular the evaporation of water from the soil surface. The presence of crop residue on the surface acts as a barrier to evaporation of water in the same way in which soil temperatures are affected by the residue. The presence of the residue acts as an entity in which the vapor diffusivity is quite low, slowing the transport of water vapor from the soil surface to the atmosphere.

There have been several different approaches to estimating ET from a surface, the most recognized of which is the Penman–Monteith equation (Penman, 1948; Monteith, 1964) given as

$$\mathrm{ET} = \frac{\Delta(R_n - G) + \dfrac{m \rho C_\rho \left[e*(z) - e(z)\right]}{r_a}}{\Delta + \dfrac{\gamma(r_a + r_c)}{r_a}} \qquad [5]$$

where Δ is the slope of saturation vapor pressure curve, γ is the psychrometric constant, λ is the latent heat of vaporization, m is the ratio of the molecular weight of water vapor to that of air (0.622), ρ the density of air, C_ρ is the volumetric heat capacity of air, e^* is the saturation vapor pressure of the air, e is the actual vapor pressure of the air both at some height z above the surface, r_a is the aerodynamic resistance to water vapor transfer, and r_c the canopy resistance to water vapor transfer. The major variables in Eq. [5] that are affected by soil management are the r_a term, r_c, R_n, G, and e. These are driving variables for ET that need to be examined. There are several forms of ET equations, but this form allows an examination of the factors that are affected by soil management. The resistance terms can be considered analogous to electrical resistors that affect the current flow. In the natural environment, the r_a term describes the rate of air movement from the surface to the atmosphere and is dependent on the wind speed, the roughness of the surface, and the impact of atmospheric stability that is affected by the temperature gradients in the lower atmosphere. The r_c term is the effect of the canopy on the release of water vapor from the leaf to the atmosphere. To place this in perspective, consider that a lush canopy with adequate water will have a minimal resistance, while a water-stressed or canopy with a large amount of senesced leaves will have a maximum r_c value.

Tillage disturbs the soil surface and also disrupts the soil crust, which in turn increases the rate of soil water evaporation from the surface. This is partially due to the exposure of wet soil to dry air in the atmosphere and the adsorption of energy into the surface, which evaporates water. Burns et al. (1971) and Papendick et al. (1973) showed that tillage disturbance of the soil surface increased soil water evaporation amounts compared to untilled areas. Ritchie (1971) found that soil water evaporation is affected by soil water content of the surface and degree of plant cover on the surface. Tillage moves moist soil up to the surface, where losses to drying may offset increased infiltration rates. Hatfield and Prueger (unpublished data, 1999) observed that total soil water evaporation fluxes were 10 to 12 mm for a three-day period following each cultivation operation in the spring in Iowa. Total evaporation fluxes from no-tillage fields were less than 2 mm during this same time period. Aggressive field cultivation operations in the spring could reduce soil water availability in the seed zone by as much as 20 to 30 mm. To replace this soil

water lost from the seed zone it is necessary to have timely precipitation events to ensure germination and emergence of the crop. In semiarid areas, soil profile water contents that are near field capacity at the onset of the growing season are critical to crop production. In a recent study in Kansas Klocke et al. (2009) found that surface residue reduced soil water evaporation. Soil water evaporation was reduced by nearly 50% compared to bare soil when either wheat or corn residue nearly covered the soil surface. When they changed the configuration of the surface residue so that there was only partial coverage then corn stover only had a slight impact on soil water evaporation rates. However, full surface coverage with residue reduced soil water evaporation by 50 to 65% compared to the bare soil surface. An interesting aspect of their study was that the suppression of soil water evaporation that led to greater soil water for the crop created an economic impact of $365.00 ha^{-1}. Manipulating the soil surface either with tillage or crop residue will affect the soil water evaporation. The presence of moist soil at the surface creates a more favorable microclimate for biological activity within the soil and the presence of the residue reduces the impact of raindrops onto the soil surface, thus reducing the potential for erosion by maintaining a larger infiltration rate into the soil.

Another form of a mulch on the surface is that of a dust mulch, in which a layer of dry soil occurs over a moist soil. The changes that occur in this layer serve to reduce the diffusivity of water vapor through the dust and which creates a situation in which the dust acts as barrier for evaporation. The presence of a layer of different diffusivity materials will alter the evaporation rate. In a similar fashion, adding residue to the surface also reduces the evaporation rate of water from the soil.

Soil Management Impacts on the Soil Microclimate

Soil management impacts can be detected in the soil through the effects on the factors that make up the soil microclimate, including the radiation balance, the thermal properties of the soil or crop residue, and the effect of the residue on heat or water vapor exchanges. The processes are governed by the available energy from solar radiation, which is dependent on the location and time of year. We can substantially alter the soil microclimate by how we shape the surface with tillage, remove or incorporate crop residue, or change crop cover during the season. All of these factors are interrelated. The challenge is to determine how to best manage the soil and crop system for a particular location to maximize crop production efficiency, minimize negative environmental impacts, and ensure that positive impacts on the soil increase with time. Evaporation from the surface is affected by different soil management practices. Tillage will temporarily increase soil water evaporation, dry the soil, and cause the soil temperature to rise more than if the soil had not been tilled. There would also be a change in the distribution of water content and soil temperature with depth in the soil profile between the tilled and un-tilled fields. Leaving residue on the surface will alter the radiation balance of the soil, thereby affecting the amount of energy available for heating the soil and evaporating water. Residue management on the soil surface can be effectively used as a method to alter soil water and soil temperature profiles.

Climatic Resources

Decisions about the proper management of the soil that are based on understanding and utilizing the soil microclimate require information about the general climatic conditions for a location. There are various sources of this information; these data often are available from meteorological agencies of a country (Leemans and Cramer, 1991; Lieth, 1972). However, there are some worldwide databases that are maintained by the Food and Agriculture Organization (FAO) that are available through FAOClim2-Net. This database covers monthly data for 28,100 stations and includes up to 14 observed and computed agroclimatic parameters. There are long-term averages for the period from 1961 through 1990 and time series for rainfall and temperature. These data can be retrieved by geographic area, time period, and parameter, and data can be downloaded in different formats for use with different analysis packages. The variables

available in this database include maximum air temperature, minimum air temperature, mean air temperature, mean nighttime air temperature, mean daytime air temperature, total daily rainfall, dew point temperature, relative humidity, actual vapor pressure, potential evapotranspiration using the Penman–Monteith equation, wind speed, global solar radiation, sunshine fraction, and sunshine hours. This would be a rich database for the assessment of the climate at any given location.

Challenges

There are many challenges in the assessment of soil management impacts on the soil microclimate, but the principles that affect these changes are relatively simple to understand, and the framework is contained in the energy balance for a given surface. Altering the surface with any soil management practices—tillage and residue management are the primary methods—changes the radiation balance through the albedo, the soil heat flux, and soil water evaporation rate. One challenge is to determine how these factors affect the development of the crop and the associated biological systems in the soil, including the microbes, weeds, pathogens, and insects. The primary challenge for those who manage the soil is to understand these dynamics and their impact on all of the biological systems so that soil management practices can be effectively used to enhance the growing conditions for the crop and diminish the negative impacts of pests on the economic crop.

References

Al-Darby, A.M., and B. Lowery. 1987. Seed zone soil temperature and early corn growth with three conservation tillage systems. Soil Sci. Soc. Am. J. 51:436–440.

Azooz, R.H., B. Lowery, T.C. Daniel, and M.A. Arshad. 1997. Impact of tillage and residue management on soil heat flux. Agric. For. Meteorol. 84:207–222.

Benoit, G.R., S. Mostaghimi, R.A. Young, and M.J. Linstrom. 1986. Tillage-residue effects on snow cover, soil water, temperature and frost. Trans. ASAE 29:473–479.

Benoit, G.R., and K.A. van Sickle. 1991. Overwinter soil temperature patterns under six tillage-residue combinations. Trans. ASAE 34:86–90.

Burns, R.L., D.J. Cook, and R.E. Phillips. 1971. Influence of no tillage on soil moisture. Agron. J. 73:593–596.

Burrows, W.C., and W.E. Larson. 1962. Effect of amount of mulch on soil temperature and early growth of corn. Agron. J. 54:19–23.

Dahiya, R., J. Ingwersen, and T. Streck. 2007. The effect of mulching and tillage on the water and temperature regimes of a loess soil: Experimental findings and modeling. Soil Tillage Res. 96:52–63.

Gupta, S.C., W.E. Larson, and D.R. Linden. 1983. Tillage and surface residue effects on soil upper boundary temperatures. Soil Sci. Soc. Am. J. 47:1212–1218.

Ham, J.M. 2005. Useful equations and tables in micrometeorology. p. 533–560. *In* J.L. Hatfield and J.L. Baker (ed.) Micrometeorology in agricultural systems. Agronomy Monogr. 47. ASA, CSSA, and SSSA, Madison, WI.

Hatfield, J.L., and J.H. Prueger. 1996. Microclimate effects of crop residues on biological processes. Theor. Appl. Climatol. 54:47–59.

Hatfield, J.L., R.J. Reginato, and S.B. Idso. 1983. Comparison of longwave radiation calculation methods over the United States. Water Resour. Res. 19(1):285–288.

Klocke, N.L., R.S. Currie, and R.M. Aiken. 2009. Soil water evaporation and crop residues. Trans. ASABE 52:103–110.

Leemans, R., and W. Cramer. 1991. The IIASA database for mean monthly values of temperature, precipitation and cloudiness on a global terrestrial grid. Res. Rep. RR-91-18. International Institute of Applied Systems Analyses, Laxenburg, Austria.

Lieth, H. 1972. Modelling the primary productivity of the earth. Nature and resources. UNESCO, VIII 2:5–10.

Monteith, J.L. 1964. Evaporation and environment. *In* State and movement of water in living organisms. 19th Symp. Soc. Exp. Biol. 205.

Monteith, J.L. 1973. Principles of environmental physics. Nottingham Press, London.

Novak, M.D. 2005. Soil temperature. p. 105–129. *In* J.L. Hatfield and J.L. Baker (ed.) Micrometeorology in agricultural systems. Agron. Monogr. 17. ASA, CSSA, SSSA, Madison, WI.

Papendick, R.I., M.J. Lindstrom, and V.L. Cochran. 1973. Soil mulch effect on seedbed temperature and water during fallow in eastern Washington. Soil Sci. Soc. Am. Proc. 37:307–314.

Penman, H.L. 1948. Evaporation from open water, bare soil, and grass. Proc. R. Soc. Lond. A 193:120–146.

Ritchie, J.T. 1971. Dryland evaporative flux in a subhumid climate. I. Micrometeorological influences. Agron. J. 70:723–728.

Rosenberg, N.J., B.L. Blad, and S.B. Verma. 1983. Microclimate: The biological environment. Wiley Interscience, New York.

Sauer, T.J., and R. Horton. 2005. Soil heat flux. p. 131–154. *In* J.L. Hatfield and J.L. Baker (ed.) Micrometeorology in agricultural systems. Agron. Monogr. 17. ASA, CSSA, SSSA, Madison, WI.

Sauer, T.J., J.L. Hatfield, and J.H. Prueger. 1997. Overwinter changes in radiant energy balance of a corn-residue-covered surface. Agric. For. Meteorol. 85:279–287.

Singh, B., and B.S. Sandhu. 1979. Effect of irrigation, mulch, and crop canopy on soil temperature in forage maize. J. Indian Soc. Soil Sci. 27:225–235.

Van Wijk, W.R., and D.A. de Vries. 1966. Periodic temperature variation in a homogeneous soil. p. 102–143. In van Wijk, W.R (ed.) Physics of plant environment. North Holland Publishing Co., Amsterdam, the Netherlands.

12

Pesticide Movement

Timothy J. Gish, Jared Williams, John H. Prueger, William Kustas, Lynn G. McKee, and Andy Russ

Pesticides generally include herbicides, insecticides, and fungicides and play an important role in maintaining worldwide food and fiber production by controlling weeds that compete for water and nutrients or by eliminating pests that reduce yields (Majewski and Capel, 1995). In the future, the role of pesticides and fertilizers in agriculture is likely to increase as marginal land is converted to agriculture to meet production needs (Helling, 1993). Furthermore, it has also been proposed that increasing food and fiber production by agriculture will be critical to maintaining political and social stability in many countries (Tilman et al., 2002). However, pesticides can be toxic to humans and other forms of life at low concentrations (Doull, 1989; Jin-Clark et al., 2002; Sparling et al., 2001; USEPA, 2008), so future research will have enhanced scientific, environmental, and regulatory significance (Jury and Flühler, 1992; Posner et al., 1995). To maintain productive and sustainable agricultural systems there is an immediate need to understand field-scale processes governing pesticide use and off-site movement.

Due to its size and diversity, agriculture is responsible for more than 75% of worldwide pesticide usage (Aspelin, 1994), resulting in more than 2.4 billion metric tonnes of pesticides applied to agricultural land during 2000 (USEPA Biological and Economic Analysis Division, 2004). Despite their global use, comprehensive pesticide field studies are rare, in part due to the complex interactions involving pesticide chemistry, soil properties, agricultural management practices, and local meteorological conditions (Table 12|1). Although detailed holistic studies on pesticides are rare, several national surveys have been conducted that shed light on pesticide fate and its potential human and environmental risks. One of the first national surveys was conducted by the USEPA (1990), which determined that more than 10% of community water system wells contained detectable amounts of at least one pesticide. From 1993 to 1995, the National Water-Quality Assessment program monitored 20 major basins in the United States and found pesticides in 54% of the 1034 sites sampled (Koplin et al., 1998). Furthermore, the U.S. Geological Survey observed that 97% of all streams sampled from agricultural and urban areas contain detectable concentrations of pesticides, while 65% of the streams in undeveloped areas contained observable levels of at least one pesticide (Gillion et al., 2006). In addition, Gillion et al. (2006)

T.J. Gish, USDA-ARS Hydrology and Remote Sensing Laboratory, Beltsville, MD 20705 (timothy.gish@ars.usda.gov); J. Williams, Department of Agribusiness, Plant and Animal Sciences, Brigham Young University-Idaho, Rexburg, ID; J.H. Prueger, USDA-ARS National Laboratory for Agriculture and the Environment, Ames, IA; W. Kustas, USDA-ARS Hydrology and Remote Sensing Laboratory, Beltsville, MD; L.G. McKee, USDA-ARS Hydrology and Remote Sensing Laboratory, Beltsville, MD; and A. Russ, Hydrology and Remote Sensing Laboratory, Beltsville, MD.

doi:10.2136/2011.soilmanagement.c12

Component	Property increasing loss
Pesticide characteristics	High vapor pressure High water solubility Large soil half-life $(T_{1/2})$ Low soil affinity $(K_{oc}$ or $K_d)$
Soil and landscape properties	High soil water contents for surface broadcast applied herbicides Low soil water contents for incorporated fungicides Low organic matter content $(K_{oc}$ or $K_d)$ Sandy soil texture pH Bulk density (affects soil air space) High temperature (especially if soil is moist)
Agricultural practice	Application method (foliar spray > surface broadcast > incorporated) High application rates Plant residue (amount and type) Pesticide formulation (wettable powder > emulsions > micro-encapsulated)
Local-meteorological	Precipitation (relative to time of application) High wind velocity and turbulent flow conditions High solar radiation (R_N) High relative humidity (RH)

reported that more than one-half the shallow groundwater aquifers also contained measurable levels of pesticides. Clearly, pesticide occurrence in streams, groundwater aquifers, and community wells are well documented, although concentrations are typically low.

Once applied to soil a portion of the pesticide will be adsorbed to soil particles, while the remaining pesticide mass will be associated with the liquid and/or vapor phases. The degree to which a pesticide is partitioned between adsorbed, liquid, and vapor phases strongly influences its subsequent dispersal into the environment (Cousins et al., 1999; Jury et al., 1983; Taylor and Spencer, 1990). Pesticide partitioning between these three phases is primarily a function of pesticide chemistry, pesticide formulation, soil properties, and local metrological conditions. For example, pesticides with high vapor pressure are more likely to favor the vapor phase while those with a high affinity for the soil matrix will favor adsorption to soil particles. Because pesticides exist in all three phases, there are several loss pathways that can simultaneously interact to influence water quality and the environment. As a result, methods for monitoring pesticide behavior at the field scale are expensive and difficult to interpret due to process interactions that are a function of scale, as well as soil properties that vary spatially and temporally (Fig. 12|1). Thus, although scientists have quantified the impact of various factors on pesticide movement in controlled environments, monitoring and interpreting field-scale pesticide behavior is more ambiguous. During application, pesticide emissions can occur as spray drift in concentrated droplets or as pesticide attached to dust particles (Symons, 1977; Majewski and Capel, 1995). Pesticide loss from drift and volatilization can be transported through the boundary layer of the atmosphere to forested areas and deposited directly into streams via tree wash-off (personal communication, 2010, Dr. Clifford Rice, USDA-ARS Beltsville, MD). Once on the soil surface, pesticides degrade in situ or move away from the targeted area by leaching into groundwater, runoff into adjacent streams, and/or volatilization into the atmosphere. The rate at which pesticides are lost from these three loss pathways is influenced by a number of small-scale factors like soil moisture content, organic matter content, soil hydraulic properties, as well as large-scale influences such as wind speed profiles, agricultural management practices, timing of rainfall events relative to application, and field slope. As a result, to reduce risks associated with pesticide use all major loss pathways must be evaluated so as to avoid simply shifting pesticides and their metabolites from one part of the environment to another.

Although field studies monitoring all possible loss pathways are essentially nonexistent, specific aspects of pesticide movement have been rigorously studied. Among the major loss pathways, pesticide runoff has been the most intensively studied. Results show that in general annual runoff losses are less than 3% of that applied, with the largest portion of this loss occurring near the time of application (Haith and Rossi, 2003). Enhanced pesticide runoff relative to leaching from tile-drained fields supports the hypothesis that pesticide runoff is more detrimental to the environment than pesticide leaching (LaFleur et al., 1975; Muir and Baker, 1976; Ng et al., 1995). Unfortunately, field-scale pesticide leaching losses in non-tile-drained fields is difficult to quantify due to soil heterogeneity, but it is thought that leaching losses due to preferential flow are probably <<5% of that applied. The third loss pathway, volatilization, is a major environmental concern, with as much as 90% of some pesticides applied being lost to the atmosphere (Taylor and Spencer, 1990). Unfortunately, efforts to determine the impact of field-scale variables like management and climate on pesticide volatilization are still in their infancy.

In this chapter, the primary principles and factors governing pesticide movement at the field scale will be discussed. Pesticide movement will be evaluated by examining how pesticides are partitioned between runoff, groundwater leaching, and volatilization processes.

Surface Runoff

Pesticide surface runoff is a concern in many watersheds since intensive agriculture may be adjacent to sensitive ecosystems (Capel et al., 2008). Although rainfall timing, intensity, and duration are the most critical factors governing pesticide runoff, the rate of application, pesticide formulation, management practice, and landscape features are also important (Caro, 1976; Wauchope, 1978; Hall et al., 1983; Felsot et al., 1990; Domagalski et al., 2008). Typically, pesticide losses from a single rainfall event are small—less than 1 to 2% of that applied. In rare situations, such as when a major rainfall event follows the application of a wettable powder formulation on a sloped field, as much as 15% of the pesticide applied can be lost through runoff (Baker, 1980; Haith and Rossi, 2003). Regardless of the pesticide mass lost from runoff, detrimental impacts decrease with increasing distance from the application site due to dilution from other runoff sites, streams, rivers, and lakes.

Pesticide in surface runoff occurs through two mechanisms that occur simultaneously: (i) erosion of pesticide adsorbed to sediment and (ii) dissolution of the pesticide into the runoff water. Pesticide *concentrations* in runoff sediment can be several times higher than that observed in the water phase; however, the greatest pesticide *mass* lost in runoff is from the water phase, as runoff water volumes are typically much greater than sediment mass losses (Wauchope, 1978).

Rainfall and Pesticide Runoff

The primary factors governing pesticide runoff losses are the intensity, duration, and timing of the rainfall events relative to application. To account for the interaction

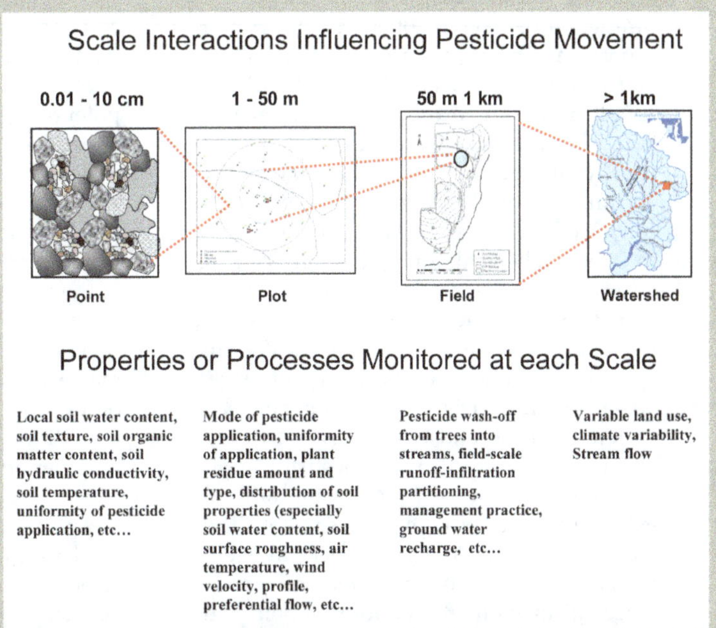

Fig. 12|1. Properties and processing influencing pesticide movement and their relative scales of observation.

between rainfall intensity, duration, and timing, three types of pesticide runoff events have been proposed: minor, critical, and catastrophic (Wauchope, 1978). Minor runoff events are a product of rain events that produce small amounts of runoff shortly after pesticide application (generally within 1–2 d). These minor events typically have high concentrations of pesticide in a relatively small amount of surface runoff and account for pesticide losses <1% of that applied (Wauchope, 1978). However, the high concentration of pesticides in these minor runoff events can significantly affect streams, rivers, and watershed ecosystems. The second type, a critical runoff event occurs within 2 wk of pesticide application and has at least 50% of the rainfall exiting the field through surface runoff. The amount of pesticide lost in a critical event is also dependent on pesticide soil persistence, adsorption affinity, formulation, and landscape features (Shipitalo et al., 2000; Ma et al., 2004). Pesticide runoff losses can be significant with a critical runoff event even if the pesticide has a high affinity for the soil matrix since sediment loss is common with such events. As a result, critical runoff events typically produce the bulk of the pesticide runoff loss from agricultural fields and account for 1 to 2% of the applied pesticide (Ma et al., 2004). The third type, catastrophic runoff events are rare and differ from a critical runoff event by the high intensity of the rainfall occurring shortly after pesticide application (Wauchope, 1978; Schulz et al., 1998). For example, catastrophic runoff events are typically caused by severe thunderstorms that produce large amounts of rainfall within 3 days of the pesticide application and can account for more than 5% of the pesticide applied. Although a *higher percentage* of the pesticide applied is lost in a catastrophic event, the *concentration* of pesticide in the water phase exiting the field is relatively low because of dilution. Occurrences of catastrophic runoff events are rare, because timing of a major storm and application of pesticide must coincide, and farmers typically try to avoid such situations. Although, rainfall is recognized as the principle factor governing pesticide runoff, pesticide soil persistence, formulation, method of application, and landscape features interact with rainfall to influence pesticide runoff losses.

The major factors influencing pesticide soil persistence include application method (Hall et al., 1983), adsorption affinity (Jenks et al., 1998; Spark and Swift, 2002), leaching potential (Webb et al., 2008), formulation and volatilization potential (Gish et al., 1994), and degradation processes (Kearney et al., 1969; Gan et al., 2005). Pesticide degradation is typically quantified as a half-life ($T_{1/2}$), which represents the time taken for one-half of the pesticide to degrade. Where pesticide persistence is high ($T_{1/2}$ > 2 wk) concentrations in the second and third runoff events may actually be higher than in the first runoff event (Caro et al., 1972; Gan et al., 2005). In addition, pesticides that persist in soil for long periods are more likely to be adsorbed to soil particles. A persistent pesticide poses an environmental threat to neighboring ecosystems through sediment loss, especially if critical or catastrophic runoff events occur. Pesticide application methods such as soil incorporation can reduce runoff, leaching, and volatilization losses even though soil persistence may increase (Caro et al., 1974). For example, soil incorporation can reduce pesticide runoff losses from 1/4 to 1/20 of the surface-applied pesticide loss (Baker and Laflen, 1979). Foliar applications should be avoided if possible because they tend to exhibit a short pesticide persistence and are easily washed off and transported off-site in runoff water before being adsorbed to the soil (Wauchope, 1978; Gevao et al., 2000). Applying pesticides as an encapsulated formulation can also increase soil persistence as this practice reduces pesticide susceptible to groundwater leaching by preferential flow (Gish et al., 1994). Multiple applications of a pesticide may result in metabolic pathways being established which increase biological degradation rates and reduces soil persistence (Kearney et al., 1969). Since persistent pesticides pose a greater environmental risk, most long-term persisting pesticides, such as DDT and other organochlorine pesticides, have been banned.

Pesticide formulation can also influence pesticide runoff losses. For example, wettable powder formulations have the highest runoff potential, with losses generally ranging from 2 to 5% of that applied (Wauchope, 1978). Since wettable powder formulations subjected to critical and catastrophic runoff events can result in pesticide runoff losses exceeding 5% of that applied, they should be avoided. Pesticides applied as an emulsion have the next highest potential for runoff, with typical losses of about 1% of that applied. Water-soluble pesticides have a much greater potential of being lost in water runoff, whereas, non-water-soluble pesticide often have an affinity for soil particles and will most likely be lost in the sediment portion of runoff. An example of the influence of formulation on pesticide runoff is observed with ester- and amine salt–based herbicides (Barnett et al., 1967; Caro, 1976). The amine salt, which is water soluble, rapidly dissolves in water and can be leached into the soil or moved by the water phase of a runoff event. Ester herbicides are relatively insoluble, but are readily adsorbed to soil particles and are primarily lost through the erosion of sediment. All other pesticide formulations (e.g., pelleted and micro-encapsulated) exhibit runoff losses less than 0.5% of that applied, except when a critical or catastrophic event occurs (Wauchope, 1978).

Landscape attributes interact with the type of runoff event, pesticide chemistry, and formulation to influence runoff losses. All other conditions being equal, pesticide runoff losses increase with increasing slope (Hall et al., 1983; Felsot et al., 1990). For example, pesticide runoff from a 3% slope can be as high as 2% of that applied, while slopes of 10 to 15% may result in pesticide runoff losses exceeding 5% of that applied (Wauchope, 1978). Furthermore, relative to other pesticide formulations, wettable powders will be more susceptible to runoff as slopes become steeper and as critical and catastrophic runoff events occur.

Soil properties that influence pesticide runoff losses include soil organic matter (Jenks et al., 1998; Spark and Swift, 2002), pH (Weber et al., 1972; Jenks et al., 1998), soil compaction (Baker and Laflen, 1979), soil moisture content (Spark and Swift, 2002), cation exchange capacity (Wauchope and Meyers, 1985), clay mineral content (Baskaran et al., 1996), and temperature (Caro, 1976). In general, soil properties influence pesticide runoff by affecting adsorption and desorption processes. High soil organic matter contents (>5%) will typically be the most important factor influencing pesticide absorption (Caro, 1976; Spark and Swift, 2002). In soils with a low soil organic matter content (<2%) clay mineral content may be the dominant factor because of the larger surface area of clay particles (Laird et al., 1992; Jenks et al., 1998). Soil pH has been shown to influence pesticide adsorption by altering the chemical composition of a pesticide, resulting in a net positive charge (Jenks et al., 1998). For example, some pesticides (e.g., atrazine) are more adsorptive in acidic soils because they react with H^+, making the pesticide cationic and chemically attracted to cation exchange sites (Bailey and White, 1964; Weber et al., 1972; Jenks et al., 1998). Although pesticides favoring adsorption are less susceptible to runoff from a minor runoff event, they are increasingly susceptible to critical or catastrophic runoff events where sediment erosion can be significant.

Soil moisture influences adsorption and desorption of pesticides because of water competition for adsorption sites on soil particles (Hamaker and Thompson, 1972; Cole et al., 1997). As soil moisture increases (through a rain or irrigation event), water is adsorbed to the soil matrix and pesticides desorb. Subsequently, the desorbed pesticide diffuses into the water phase, where it can be more readily transported off-site. Studies examining the application of pesticides to wet versus dry soils showed that runoff losses were significantly greater for wet soils because of the lower adsorption potential (Barnett et al., 1967; Baldwin et al., 1975; Asmussen et al., 1977).

Management Strategies to Reduce Pesticide Runoff

Pesticide runoff losses are primarily a function of the type of runoff event, with the bulk of the pesticide loss occurring early in the event and decreasing exponentially with time (Buttle, 1990; Reddy et al., 1994). Seasonal runoff losses are predominantly an accumulation of single-event losses, with minor loses in-between major storm or irrigation events (Haith and Rossi, 2003). As a result, best management practices for reducing pesticide runoff losses must consider reducing pesticide concentrations in both the water phase and the sediment. Erosion control practices such as minimum tillage and grass buffers may be effective in reducing erosion and runoff water, which greatly reduce pesticide loss via sediment erosion, but are ineffective in controlling losses in the water phase (Cole et al., 1997; Shipitalo et al., 2000). Practices for reducing pesticide losses in runoff include: (i) avoiding pesticide application during adverse weather conditions, such as when rain or high winds are anticipated within 48 h of application; (ii) determining the most appropriate pesticide type, rate, and persistence for weather and soil conditions; (iii) incorporating the pesticide if possible; (iv) using erosion control practices such as conservation tillage, contouring, and grass buffers around waterways to reduce runoff (Wauchope, 1978; Baker and Johnson, 1979; Fawcett et al., 1994; Cole et al., 1997; Shipitalo et al., 2000); and (v) avoiding pesticide wettable powder formulations where possible.

Groundwater Leaching

During the past four decades considerable effort has been made toward understanding the factors governing water and chemical transport through soil (Biggar and Nielsen, 1976; Addiscott and Wagenet, 1985; Jury and Flühler, 1992; Gish et al., 1998b; Kung et al., 2005). However, field-scale pesticide mass flux measurements of groundwater are rare. Typically, pesticide leaching studies focus on monitoring pesticide concentration profiles in soil as a function of time after application or effluent discharge from tile drains (Jury et al., 1984; Helling and Gish, 1986; Isensee et al., 1990; Kladivko et al., 1991; Harris and Catt, 1999; Novak et al., 2001). Unfortunately, uncertainty in soil water dynamics is extremely large, with soil hydraulic conductivity and soil water diffusivity parameters demonstrating coefficients of variation exceeding 200% (Libardi et al., 1980; Nielsen et al., 1973). The large uncertainty in soil water dynamics is most likely due to soil heterogeneities and preferential flow.

Preferential flow is likely the dominant flow mechanism governing pesticide movement to groundwater (Kladivko et al., 1991; Harris and Catt, 1999; Novak et al., 2001; Koplin et al., 1998; Jarvis, 2002). Preferential flow assumes that a small fraction of the total soil pore space is responsible for rapidly conducting solutes to groundwater (Germann and Beven, 1981). Preferential flow typically occurs through structureless soils by means of flow instabilities (Glass et al., 1988; Ghodrati and Jury 1992); flow through spatial voids resulting from decayed roots, shrinking clay minerals, sink holes, or created by soil fauna (Gish and Jury, 1983; Libra et al., 1984; Shipitalo et al., 1990; Ritsema and Dekker, 1995; Gish et al., 1998a; Williams et al., 2000); and/or flow along subsurface restricting layers (Kung, 1990). Unfortunately, quantifying preferential flow at the field scale is extremely difficult since there is uncertainty in identifying the location of subsurface flow pathways and when water is actually flowing in these pathways. As a result, pesticide mass fluxes are typically calculated, not monitored. Without mass flux procedures, it is impossible to quantify a total flux or the relevance of preferential flow at the field scale for various management, soil, and climatic scenarios.

Monitoring pesticide transport has been done for several decades, but with limited success. Early monitoring attempts focused on removal of soil cores and analyzing for pesticides as a function of time after application and profile depth. Unfortunately, soil field core data were found to be of limited usefulness. For example, Wyman et al. (1985, 1986) evaluated aldicarb transport through soil by collecting 48 soil cores (3.6 m long) four times during a growing season. Each soil core was segmented into 30-cm increments, homogenized, and subsampled. Because no aldicarb residues were detected below 2.4 m throughout a three-year period, it was concluded that aldicarb was completely degraded and did not pose a threat to groundwater quality. In contrast, Brasino (1986) conducted a coincident experiment which found that peak aldicarb concentrations in three groundwater monitoring wells (6-m depth) ranged from 27 to 76 ppb even though the same chemical application rate, irrigation scheme, and soil were used. As a result, early pesticide transport studies that focused on soil core data to evaluate pesticide leaching may be of limited value.

Recently, an innovative total flux monitoring procedure was evaluated in which tracers were surface applied in a narrow band (typically 3 by 25 m) parallel to a tile drain but offset by 0.3 m from the buried tile drain system (Kung et al., 2000a). Using this approach, a total mass flux for mobile tracers was monitored and the relevance of preferential and matrix flow processes were determined (Kung. et al., 2000a; Gish et al., 2004). Everts et al. (1989) and Kung et al.

(2000b) also used adsorbed tracers to simulate pesticide leaching behavior. Everts et al. (1989) observed that both the nonadsorbing tracer (bromide) and adsorbed tracer (rhodamine WT) simultaneously appeared in a tile drain shortly after the initiation of irrigation, while Kung et al. (2000b) observed that the nonadsorbed (bromide) and adsorbed (rhodamine WT) tracers arrived at the tile drains at the same time and that their breakthrough curves peaked at the same time. Thus, the soil water dynamics of preferential flow can dominate transport regardless of the chemical adsorption affinity for the soil. Unfortunately, these flux procedures have not yet been used to quantify the impact of tillage, formulation, or management practice on pesticide leaching.

Since pesticide flux monitoring through the soil profile is in its infancy, field observations discussed here will focus on field observations of pesticide concentrations. However, without flux measurements there is a greater likelihood that field experiments may generate contradictory results.

Field Observations of Pesticide Leaching

As is the case for surface runoff, the timing of a rainfall event relative to application influences pesticide leaching to groundwater. Pesticide transport is especially vulnerable to preferential transport shortly after application. For example, in a sandy soil in Beltsville, MD, surface-applied atrazine and metolachlor were rarely detected in a 3-m well for months before application (Fig. 12|2). Then pesticide concentrations peaked shortly after application when a 60-mm rain event had occurred and afterward declined exponentially with time. On the

other hand, once in the subsoil pesticides may be transported predominantly by the smaller pores of the soil matrix, which would increase the time it takes for a given pesticide to reach groundwater (Kung et al., 2000b; Delphin and Chapot, 2006). Thus, it appears that the first rainfall event after application has the highest risk of pesticide leaching to groundwater, but that modeling approaches will have to include interaction between matrix and preferential flow processes.

Controlled release formulations may reduce preferential transport of pesticides. In the laboratory and in field plots atrazine leachate concentrations were reduced by as much as 80% with starch encapsulation (Gish et al., 1991b; Schreiber et al., 1993). However, Brown et al. (1995) reported unexpectedly high leaching loses with encapsulation. Later, field evaluations

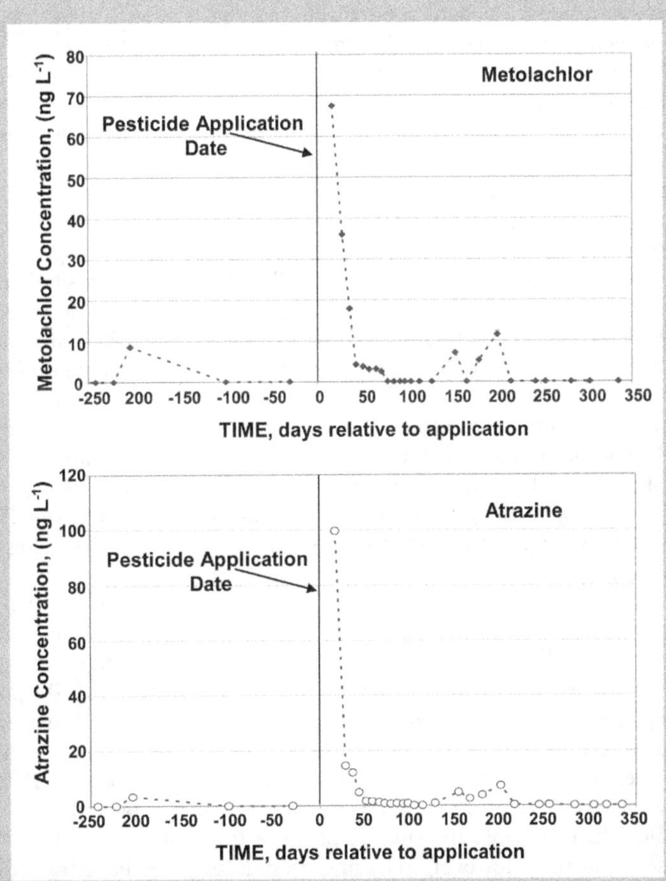

Fig. 12|2. Pesticide concentrations in a shallow 3-m groundwater well as a function of time. On the *x* axis, "0" represents the day of metolachlor application.

showed the starch encapsulated herbicides were more persistent in soil, which was attributed to being less available for leaching during rain events (Gish et al., 1994). However, pesticide chemistry must be considered when optimizing an encapsulation matrix to be effective in controlling the targeted pest and reducing environmental risks (Wienhold and Gish, 1994b).

There is some ambiguity regarding the benefit of incorporation of a pesticide relative to its leaching potential. Gish et al. (1991a) reported that soil incorporated carbofuran leached less than surface broadcast-applied atrazine, despite the much larger inherent mobility of carbofuran. However, Jones et al. (1995) suggested that soil incorporation of a pesticide after application had no impact on subsequent transport to tile drains. Obviously, more work needs to be done in this area so that the interactions among pesticide chemistry, pesticide formulation, soil type, rainfall intensity and duration, and pesticide leaching can be determined.

Preferential flow has been observed on all soils regardless of texture. However, the impact of preferential flow as a function of pesticide chemistry and soil texture has not yet been fully quantified. Initially, preferential leaching of pesticides was thought to occur on heavy or clayey textured soils via macropores or other spatial voids (Harris and Catt, 1999; Johnson et al., 1996). However, significant pesticide leaching has also been detected through loamy and silty textured soils (Kladivko et al., 1991; Brown et al., 1995; Zehe and Fluhler, 2001), as well as sandy soils (Ghodrati and Jury, 1992). On the other hand, there is some evidence that *total pesticide mass* losses in a clayey structured soil may be greater than from a sandy soil (Traub-Eberhard et al., 1995).

Management Strategies to Reduce Pesticide Leaching

Pesticide transport through soil is primarily a function of preferential flow, so for pesticides with a low solubility encapsulation will reduce leaching by increasing pesticide diffusion into the smaller pores of the soil matrix. However, reducing pesticide leaching through encapsulation may be countered by an enhanced runoff potential since encapsulation also increases soil persistence (Gish et al., 1994). Soil incorporation may also reduce leachability for pesticides that are insoluble and have low soil persistence. In addition, conservation tillage practices may temporarily enhance preferential pesticide transport through void root channels and bio-pores. In time, root channels and bio-pores will have clay and/or organic coatings where pesticide can be readily adsorbed and subsequently broken down by microbial degradation. After only a few years Gish et al. (1998b) observed higher pesticide metabolites under no-till relative to conventional tillage. This suggests that if the pesticide metabolites are less harmful than the parent compound, conservation tillage may be beneficial to groundwater quality. Farm managers should also avoid using pesticides with a long half-life or soil persistence as this will increases susceptibility to groundwater leaching.

Pesticide Volatilization

Pesticide volatilization can exceed 90% of that applied, but typical losses for many pesticides used in crop production range from 5 to 25% of that applied (Taylor and Spencer, 1990, Prueger et al., 1999; Glotfelty et al., 1989; Prueger et al., 2005). Once in the atmosphere, pesticides can be degraded or deposited in nontargeted areas via wet or dry deposition (Bidleman and Christensen, 1979; Bidleman, 1988; Burrows et al., 2002). Frequently, a portion of the applied pesticides volatilized into the atmosphere are transported and subsequently deposited in streams, rivers, and lakes (McConnell et al., 1998; Alegria and Shaw, 1999; Thurman and Cromwell, 2000; Kuang et al., 2003).

Pesticide volatilization occurs in two steps, evaporation of the pesticide from soil and plant residues followed by dispersion into the atmosphere by diffusion and turbulent mixing (Fig. 12|3) (Taylor, 1995; Prueger et al., 2005). From a thermodynamic perspective, the evaporative flux represents a phase change from liquid and/or solid phases to the vapor phase and can be described well as a diffusive process (Glotfelty and Schomburg, 1989). As a result, several methods have been developed to obtain estimates of pesticide volatilization at the field-scale. Parmele et al. (1972) developed an aerodynamic method based on gradients of wind speed, temperature, and pesticide concentrations collected over a uniform area. Denmead et al. (1977) developed an integrated horizontal flux approach that uses pesticide concentration and horizontal wind speeds profiles. For certain conditions a theoretical profile shape method may be useful (Wilson et al., 1982) which measures wind speed and pesticide concentration at a single height above the soil. Recently, wind, temperature, water, and pesticide profile data have been used to measure pesticide volatilization where turbulent flow conditions may exist (Prueger et al., 2005; Gish et al., 2009).

Pesticide Vapor Pressure

Perhaps the most crucial pesticide property influencing volatilization is the vapor pressure of the pesticide. As the vapor pressure increases, the pesticide increasingly favors the vapor phase and is more readily volatilized. Since fumigants have larger vapor pressures they are generally more volatile than herbicides (Yagi et al., 1995; Yates et al., 1996). In addition, in the field an "effective" pesticide vapor pressure is likely to be lower than the vapor pressure of the "pure" chemical due to interactions with the soil surface. For example, early studies detected a significant positive correlation between pesticide vapor pressure and pesticide volatilization (Farmer et al., 1972; Glotfelty et al., 1984). Later it was observed that dry soil conditions favored soil adsorption, which reduced the vapor pressure of the pesticide and lowered pesticide volatilization (Spencer and Cliath, 1974; Taylor and Spencer, 1990). In contrast, plant surfaces have a lower affinity for pesticides and as such exhibit pesticide vapor pressures that are closer to that of the "pure"

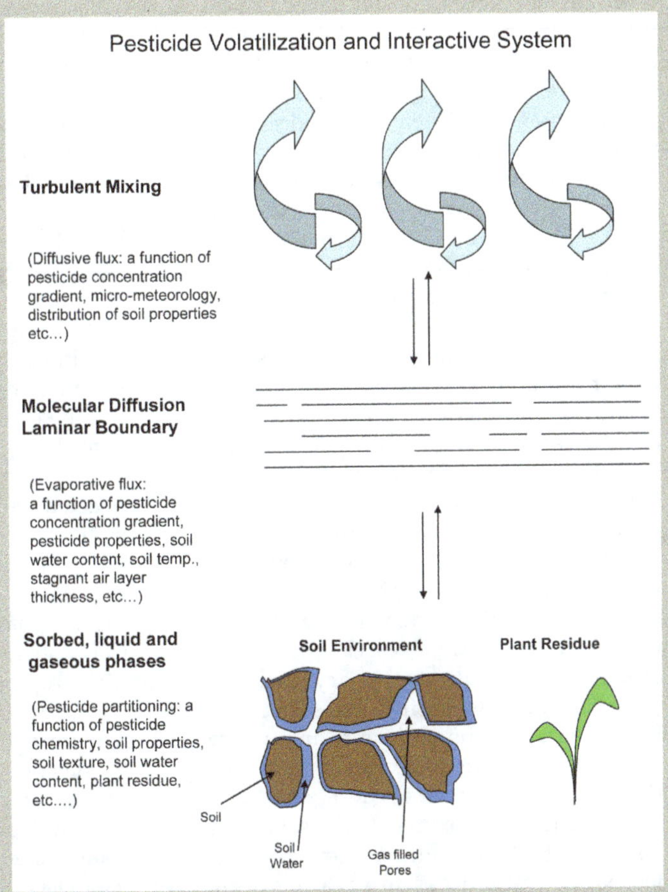

Fig. 12|3. Schematic of pesticide volatilization process. This process involves diffusive exchange of pesticide mass between the soil environment (adsorbed, gaseous pore space, liquid phase) and the atmosphere.

pesticide (Taylor and Spencer, 1990). For most pesticides, soil adsorption is primarily governed by the soil organic fraction (Rao and Davidson, 1980; Karickhoff, 1981). Thus, soil properties like organic matter, and to a lesser extent, texture (clay content), and soil pH, affect pesticide volatilization by increasing adsorption, thus reducing the liquid phase concentration and vapor pressure in the soil.

At the field scale, soil water content is perhaps the most critical soil property influencing herbicide volatilization. As the pesticide will be distributed among adsorbed, liquid, and vapor phases the amount of air space within a soil volume and the thickness of the water molecule layer adsorbed onto the soil particles will influence herbicide volatilization. Spencer and Cliath (1974) measured the pesticide vapor pressures in soil at various soil water contents and demonstrated greater volatilization losses from wet than dry soils. Glotfelty et al. (1984) demonstrated that pesticide vapor losses increased more with soil water content than organic matter content or soil temperature. Furthermore, in a 5-yr field investigation, Prueger et al. (2005) demonstrated that at high soil water contents, as much as 25% of surface-applied metolachlor could be lost through volatilization, compared to as little as 5% when soils were dry. Figure 12|4 shows the impact of soil moisture on field-scale herbicide volatilization (metolachlor) where soil type, climatic inputs, formulation, and management practices were identical. In 2004 and 2005 the wet location had soil water contents and metolachlor volatilization losses that were almost twice that of the dry location. Then during 2006, using the same locations, there was no difference in surface soil water contents due to drought conditions and so the previous "wet" and "dry" locations had approximately the same soil water content. In addition, several studies have shown peaks in early morning volatilization losses that were attributed to dew formation on the soil surface (Glotfelty et al., 1989; Taylor, 1995; Prueger et al., 2005). Increases in pesticide volatilization following a rain event are also common. These spikes can be relatively large if the soil had been dry before the rainfall event (Prueger et al., 2005). Although pesticide volatilization is influenced by soil bulk density, pH, and soil texture mineralogy, the dominant soil property appears to be soil moisture.

Although increased soil water contents enhance herbicide volatilization, this is not the case with fumigants. Because fumigants have a high vapor pressure, tarps are applied over the field and/or the fumigant is injected into the soil (Yagi et al., 1995; Yates et al., 1996). Higher soil water contents may reduce fumigant volatilization losses since the higher soil water contents will reduce the gas phase porosity. Recently, Yates et al. (2008) observed that the volatilization of 1,3 dichloropropene was significantly reduced when surface water was applied after fumigant injection.

Meteorological parameters are generally thought to be important only as they influence soil properties, such as surface soil water content. For example, relative humidity affects surface soil water content and can lead to enhanced pesticide volatilization if the fields are dry (Glotfelty et al., 1984; Taylor, 1995; Prueger et al., 2005). Additionally, increased solar radiation can increase air and soil temperatures, but their impact on pesticide volatility will likely depend on the surface soil water content. Rainfall can increase soil water contents, which, in turn, may increase pesticide volatilization in the short term, but depending on rainfall intensity and duration may actually move the pesticide deeper into the soil profile, where it is less likely to volatilize. Increasing wind velocity could increase pesticide volatilization if soils are wet,

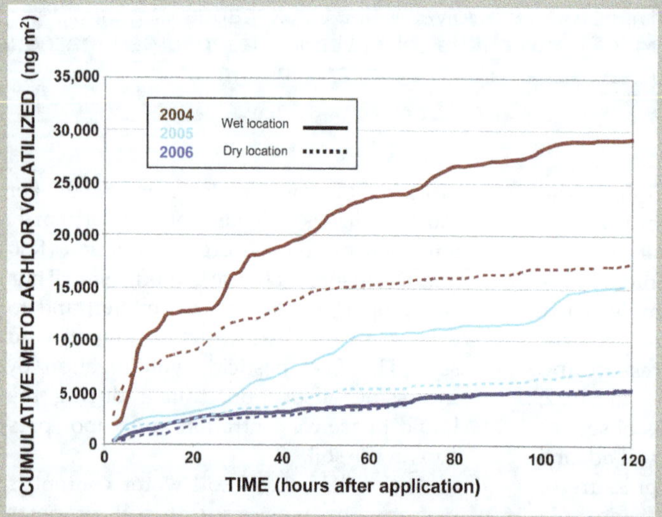

Fig. 12|4. Impact of surface oil moisture on field metolachlor volatilization. Soil water contents at the wet location were almost twice that of those around the dry location for 2004 and 2005. However, in 2006 soil water contents were not significantly different between the two locations.

but if the wind dries out the soil surface, pesticide volatilization could decrease.

Initially, soil temperature was thought to increase pesticide volatilization by increasing pesticide diffusion rates and the vapor pressure (Spencer et al., 1969). However, increased soil temperatures can also lead to a drying of the soil surface, leading to increased adsorption and a decrease in pesticide volatilization. The interaction between soil temperature and soil moisture on pesticide volatilization with identical solar radiation inputs is shown in Fig. 12|5. When the soil is "wet", increased temperatures result in a corresponding increase in pesticide volatilization (Fig. 12|5A). If, however, the soil is dry, no correlation between temperature and pesticide volatilization is observed (Fig. 12|5B). This interaction between soil temperature, soil water content, and pesticide volatilization supports the need for multi-year, field-scale investigations where soil surface and local meteorological interactions can be accurately monitored, modeled, and eventually predicted.

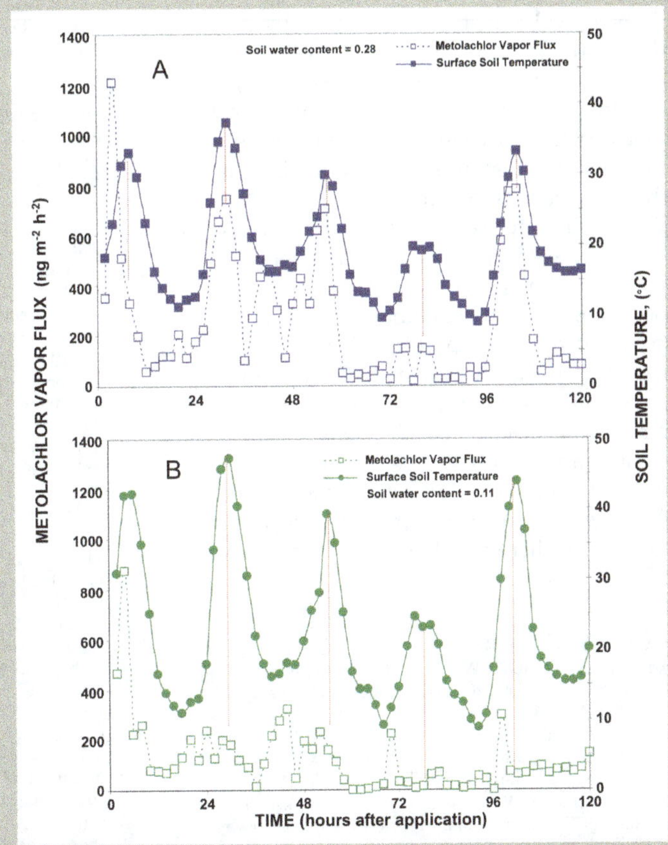

Fig. 12|5. Interaction of soil temperature and soil water content on metolachlor vitalization flux rates. Metolachlor fluxes were monitored simultaneously in A (wet location) and B (dry location) during 2005.

Agricultural Management

Agricultural management influences pesticide volatilization on several levels. First, soil incorporation of the pesticide decreases pesticide volatilization. Prueger et al. (1999) demonstrated that incorporating metolachlor in a band relative to a surface broadcast spray reduced pesticide volatilization losses from 22 to 6% of that applied. Although gas-phase diffusion is much greater than liquid-phase diffusion, only a small fraction of air voids is present in soil (Spencer and Cliath, 1970). Thus, by incorporating the pesticide the gas phase diffusion process is limited, and hence volatility is reduced. Second, increasing amounts of plant residue on the surface can increase pesticide volatilization since plants generally have a much lower affinity for pesticides than soil (Taylor and Spencer, 1990). Third, pesticide formulations such as control release or micro-encapsulated formulations can reduce pesticide volatility (Jackson and Lewis, 1978; Wienhold and Gish, 1994a; Gish et al., 1995). However, the effectiveness of the formulations is strongly dependent on pesticide chemistry. As pesticide solubility increases, the impact of formulation on reducing pesticide volatilization decreases (Wienhold and Gish, 1994b).

Reducing Pesticide Volatilization

Pesticide volatilization is governed by how the pesticide vapor pressure is influenced by interactions with soil properties, agricultural management practices, and local meteorology. In general, pesticides with a high vapor pressure should be avoided as they are more susceptible to volatilization. Since soil water content influences adsorption, field studies have shown that applying herbicides to a wet soil, or applying the pesticide when precipitation is anticipated (e.g., shortly after application) may be detrimental to the environment. On the other hand, rainfall or irrigation after fumigants have been injected will decrease fumigant volatilization. Best management practices for reducing pesticide volatilization also include the use of encapsulated formulations, and where possible, soil incorporation.

References

Addiscott, T.M., and R.J. Wagenet. 1985. Concepts of solute leaching in soils: A review of modeling approaches. J. Soil Sci. 36:411–424.

Alegria, H.A., and T.J. Shaw. 1999. Rain deposition of atrazine and trifluralin in coastal waters of the South Atlantic Bight. Environ. Sci. Technol. 33:850–856.

Asmussen, L.E., A.W. White, E.W. Hauser, and J.A. Sheridan. 1977. Movement of 2,4-D in a vegetated waterway. J. Environ. Qual. 6:159–162.

Aspelin, A.L. 1994. Pesticides industry and sales: 1992 and 1993 market estimates. 733-K-92-001. USEPA, Washington, DC.

Bailey, G.W., and J.L. White. 1964. Review of adsorption and desorption of organic pesticides by soil colloids, with implications concerning pesticide bioactivity. J. Agric. Food Chem. 12:324–332.

Baldwin, F.L., P.W. Santelmann, and J.M. Davidson. 1975. Movement of fluometuron across and trough the soil. J. Environ. Qual. 4:191–194.

Baker, J.L. 1980. Agricultural areas as nonpoint sources of pollution. p. 275–310. In M.R. Overcash and J.M. Davidson (ed.) Environmental impact of nonpoint source pollution. Ann Arbor Sci. Publ., Ann Arbor, MI.

Baker, J.L., and H.P. Johnson. 1979. The effect of tillage systems on pesticides in runoff from small watersheds. Trans. ASAE 22:554–559.

Baker, J.L., and J.M. Laflen. 1979. Runoff losses of surface applied herbicides as affected by wheel tracks and incorporation. J. Environ. Qual. 8:602–607.

Barnett, A.P., E.W. Hauser, A.W. White, and J.H. Holladay. 1967. Loss of 2,4-D in washoff of cultivated fallow lands. Weeds 15(2):133–137.

Baskaran, S., N.S. Bolan, A. Rahman, and R.W. Tillman. 1996. Pesticide sorption by allophonic and non-allophanic soils in New Zealand. N.Z. J. Agric. Res. 39:297–310.

Bidleman, T.F. 1988. Atmospheric process: Wet and dry deposition of organic compounds are controlled by their vapor-particle partitioning. Environ. Sci. Technol. 22:361–367.

Bidleman, T.F., and E.J. Christensen. 1979. Atmospheric removal processes for high molecular weight organochlorines. J. Geophys. Res. 84:7857–7862.

Biggar, J.W., and D.R. Nielsen. 1976. Spatial variability of the leaching characteristics of a field soil. Water Resour. Res. 12:78–84.

Brasino, J.S. 1986. A simple stochastic model predicting conservative mass transport through the unsaturated zone into groundwater. Ph.D. diss. Univ. of Wisconsin, Madison, WI.

Brown, C.D., R.A. Hodgkinson, D.A. Rose, J.K. Syers, and S.J. Wilcockson. 1995. Movement of pesticides to surface waters from a heavy clay soil. Pestic. Sci. 43:131–140.

Burrows, H.D., M. Canle, J.A. Santaballa, and S. Steenken. 2002. Reaction pathways and mechanisms of photodegradation of pesticides. J. Photochem. Photobiol. B 67:71–108.

Buttle, J.M. 1990. Metolachlor in surface runoff. J. Environ. Qual. 19:531–538.

Capel, P.D., K.A. McCarthy, and J.E. Barbash. 2008. National, holistic, watershed-scale approach to understand the sources, transport, and fate of agricultural chemicals. J. Environ. Qual. 37:983–993.

Caro, J. 1976. Pesticides in agricultural runoff. p. 91–119. In B.A. Stewart (ed.) Control of water pollution from cropland. Vol. II. USEPA 600275026b.

Caro, J.H., W.M. Edwards, B.L. Glass, and M.H. Frere. 1972. Dieldrin in runoff from treated watersheds. p. 141–160. In Proc. 1970 Symp. on Interdiscipl. Aspects of watershed management, Bozeman, MT. Am. Soc. Civil Eng., New York.

Caro, J.H., H.P. Freeman, and B.C. Turner. 1974. Persistence in soil losses in runoff of soil-incorporated carbaryl in a small watershed. J. Agric. Food Chem. 22:860–863.

Cole, J.T., J.H. Baird, N.T. Basta, R.L. Huhnke, D.E. Storm, G.V. Johnson, M.E. Payton, M.D. Smolen, D.L. Martin, and J.C. Cole. 1997. Influence of buffers on pesticides and nutrient runoff from bermudagrass turf. J. Environ. Qual. 26:1589–1598.

Cousins, I.T., A.J. Beck, and K.C. Jones. 1999. A review of the processes involved in the exchange of semi-volatile organic compounds (SVOC) across the air–sol interface. Sci. Total Environ. 228:5–24.

Delphin, J.E., and J.Y. Chapot. 2006. Leaching of atrazine, metolachlor and diuron in the field in relation to their injection depth into a silt loam soil. Chemosphere 64:1862–1869.

Denmead, O.T., J.R. Simpson, and J.R. Freney. 1977. A direct field measurement of ammonia emission after injection of anhydrous ammonia. Soil Sci. Soc. Am. J. 41:1001–1004.

Domagalski, J.L., S. Ator, R. Coupe, K. McCarthy, D. Lampe, M. Sandstrom, and N. Baker. 2008. Comparative study of transport processes of nitrogen, phosphorus, and herbicides to streams in five agricultural basins, USA. J. Environ. Qual. 37:1158–1169.

Doull, J. 1989. Pesticide carcinogenicity. p. 1–5. In N. Ragsdale and R.E. Menzer (ed.) Carcinogenicity and pesticides: Principles, issues and relationships. ACS, Washington, DC.

Everts, C.J., R.S. Kanwar, E.C. Alexander, Jr., and S.C. Alexander. 1989. Comparison of tracer mobilities under laboratory and field conditions. J. Environ. Qual. 18:491–498.

Farmer, W.J., J.P. Martin, W.F. Spencer, and K. Igue. 1972. Volatility of organochlorine insecticides from soil: I. Effect of concentration, temperature, air flow rate, and vapor pressure. Soil Sci. Soc. Am. Proc. 36:443–447.

Fawcett, R.S., D.P. Tierney, and B.R. Christensen. 1994. The impact of soil conservation on pesticide runoff into surface water: A review and analysis. J. Soil Water Conserv. 49:126–135.

Felsot, A.S., J.K. Mitchell, and A.L. Kenimer. 1990. Assessment of management practices for reducing pesticide runoff from sloping cropland in Illinois. J. Environ. Qual. 19:539–545.

Gan, J., S.J. Lee, W.P. Liu, D.L. Haver, and J.N. Kabashima. 2005. Distribution and persistence of pyrethroids in runoff sediments. J. Environ. Qual. 34:836–841.

Gevao, B., K.T. Semple, and K.C. Jones. 2000. Bound pesticide residue in soils: A review. Environ. Pollut. 108:3–14.

Germann, P.F., and K. Beven. 1981. Water flow in soil macropores: I. An experimental approach. J. Soil Sci. 18:363–368.

Ghodrati, M., and W.A. Jury. 1992. A field study of the effects of soil structure and irrigation method on preferential flow of pesticides in unsaturated soil. J. Contam. Hydrol. 11:101–125.

Gillion, R.J., J.E. Barbash, C.G. Crawford, P.A. Hamilton, J.D. Martin, N. Nakagaki, L.H. Nowell, J.C. Scott, P.E. Stackelberg, G.P. Thelin, and D.M. Wolock. 2006. Pesticides in the nation's streams and groundwater, 1992–2001. USGS Circular 1291. USGS, Reston, VA.

Gish, T.J., D. Gimenez, and W.J. Rawls. 1998a. Impact of roots on ground water quality. Plant Soil 200:47–54.

Gish, T.J., A.R. Isensee, R.G. Nash, and C.S. Helling. 1991a. Impact of pesticides on shallow groundwater quality. Trans. ASAE 34:1745–1753.

Gish, T. J., and W. A. Jury. 1983. Effect of plant roots and root channels on solute transport. Trans. ASAE 26:440–444, 451.

Gish, T.J., K.-J., S. Kung, D. Perry, J., Posner, G. Bubenzer, C.S. Helling, E.J., Kladivko, and T. S. Steenhuis. 2004. Impact of preferential flow at varying irrigation rates by quantifying mass fluxes. J. Environ. Qual. 33:1033–1040.

Gish, T.J., J.H. Prueger, W.P. Kustas, J.L. Hatfield, L.G. McKee, and A.L. Russ. 2009. Soil moisture and metolachlor volatilization observations over three years. J. Environ. Qual. 38:1785–1795.

Gish, T.J., A. Sadeghi, and B.J. Wienhold. 1995. Volatilization of alachlor and atrazine as influenced by surface litter. Chemosphere 31:2971–2982.

Gish, T.J., M.J. Schoppet, C.S. Helling, A. Shirmohammadi, M.M. Schreiber, and R.E. Wing. 1991b. Transport comparison of technical grade and starch-encapsulated atrazine. Trans. ASAE 34:1738–1744.

Gish, T.J., A. Shirmohammadi, C.S. Helling. K.-J.S. Kung, B.J. Wienhold, and W.J. Rawls. 1998b. Mechanisms of herbicide leaching and volatilization and innovative approaches for sampling, prediction and control. p. 107–134. In J.L. Hatfield et al. (ed.) Integrated weed and soil management. Ann Arbor Press, Chelsea, MI.

Gish, T.J., A. Shirmohammadi, and B.J. Wienhold. 1994. Field-scale mobility and persistence of commercial and starch-encapsulated atrazine and alachlor. J. Environ. Qual. 23:355–359.

Glass, R.J., T.S. Steenhuis, and J.Y. Parlange. 1988. Wetting front instability as a rapid and far reaching hydrologic process in the vadose zone. J. Contam. Hydrol. 3:207–226.

Glotfelty, D.E., M.M. Leech, J. Jersey, and A.W. Taylor. 1989. Volatilization and wind erosion of soil-surface applied atrazine, simazine, alachlor, and toxaphene. J. Agric. Food Chem. 37:546–555.

Glotfelty, D.E., A.W. Taylor, B.J. Turner, and W.H. Zoller. 1984. Volatilization of surface applied pesticides from a fallow soil. J. Agric. Food Chem. 32:638–643.

Glotfelty, D.E., and C.J. Schomburg. 1989. Volatilization of pesticides from soil. p. 181–207. In B.L. Sawhey and K. Brown (ed.) Reactions and movement of organic chemicals in soils. SSSA Spec. Publ. 22. SSSA, Madison, WI.

Hall, J.K., N.L. Hartwig, and L.D. Hoffman. 1983. Application mode and alternate cropping effects on atrazine losses from a hillside. J. Environ. Qual. 12:336–340.

Haith, D.A., and F.S. Rossi. 2003. Risk assessment of pesticides from turf. J. Environ. Qual. 32:447–455.

Hamaker, J.W., and J.M. Thompson. 1972. Adsorption. p. 49–143. In C.A. I. Goring and J.W. Hamaker (ed.)

Organic chemicals in soil environment, Vol. 1. Marcel Dekker, New York.

Harris, G.L., and J.A. Catt. 1999. Overview of the studies on the cracking clay soil at Brimstone Farm, UK. Soil Use Manage. 15:233–239.

Helling, C.S. 1993. Pesticides, agriculture, and water quality. p. 235–252. In Proc. Stockholm Water Symp. Stockholm, 1992. Available at http://www.worldwaterweek.org/documents/Resources/Synthesis/1992_Proceedings.pdf (verified 11 Feb. 2011). Stockholm Vatten, Sweden.

Helling, C.S., and T.J. Gish. 1986. Soil characteristics affecting pesticide movement into groundwaters. p. 14–38. In W.Y. Garner et al. (ed.) Evaluation of pesticides in ground water. ACS, Washington, DC.

Isensee, A.R., R.G. Nash, and C.S. Helling. 1990. Effects of no-tillage vs. conventional tillage corn production on the movement of several pesticides to ground water. J. Environ. Qual. 19:434–440.

Jackson, M.D., and R.G. Lewis. 1978. Volatilization of two methyl parathion formulations from treated fields. Bull. Environ. Contam. Toxicol. 20:793–796.

Jarvis, N.J. 2002. Macropore and preferential flow. p. 1005–1013. In J. Plimmer (ed.) The encyclopedia of agrochemicals. Vol. 3. John Wiley and Sons, Chelsea, MI.

Jenks, B.M., F.W. Roeth, A.R. Martin, and D.L. McCallister. 1998. Influence of surface and subsurface soil properties on atrazine sorption and degradation. Weed Sci. 46:132–138.

Jin-Clark, Y., M.J. Lydy, and K.Y. Zhu. 2002. Effects of atrazine and cyanazine on chlorpyrifos toxicity in Chironomus tentans (Diptera: Chironomidae). Environ. Toxicol. Chem. 21:598–603.

Johnson, A.C., A.H. Haria, C.L. Bhardwaj, R.J. Williams, and A. Walker. 1996. Preferential flow pathways and their capacity to transport isoproturon in a structured clay soil. Pestic. Sci. 48:225–237.

Jones, R.L., G.L. Harris, J.A. Catt, R.H. Bromilow, D.J. Mason, and D.J. Arnold. 1995. Management practices for reducing movement of pesticides to surface water in cracking clay soils. Weeds 2:489–498.

Jury, W.A., and H. Flühler. 1992. Transport of chemicals through soil: Movement, models, and field application. Adv. Agron. 47:141–202.

Jury, W.A., W.F. Spencer, and W.J. Farmer. 1983. Behavior assessment model for trace organics in soil: I. Model description. J. Environ. Qual. 12:558–564.

Jury, W.A., W.F. Spencer, and W.J. Farmer. 1984. Behavior assessment model for trace organics in soil: IV. Review of experimental evidence. J. Environ. Qual. 13:580–586.

Karickhoff, S.W. 1981. Semi-empirical estimation of sorption of hydrophobic pollutants on natural sediments. Chemisphere 10:833–846.

Kearney, P.C., R.G. Nash, and A.R. Isensee. 1969. Persistence of pesticide residues in soils. p. 54–67. In M.W. Miller and G.G. Berg (ed.) Chemical fallout. Current research on persistent pesticides. Charles C. Thomas Publ., Springfield, IL.

Kladivko, E.J., G.E. Van Scoyoc, E.J. Monke, K.M. Oates, and W. Pask. 1991. Pesticide and nutrient movement into subsurface tile drains on a silt loam soil in Indiana. J. Environ. Qual. 20:264–270.

Koplin, D.W., J.E. Barbash, and R.J. Gillion. 1998. Occurrence of pesticides in shallow groundwater of the United States: Initial results from the National Water-Quality Assessment Program. Environ. Sci. Technol. 32:558–566.

Kuang, Z., L.L. McConnell, A. Torrents, D. Meritt, and S. Tobash. 2003. Atmospheric deposition of pesticides to an agricultural watershed of the Chesapeake Bay. J. Environ. Qual. 32:1611–1622.

Kung, K.-J.S. 1990. Preferential flow in a sandy vadose zone: 1. Field observations. Geoderma 46:51–71.

Kung, K.-J.S., M. Hanke, C.C. Helling, E.J. Kladivko, T.J. Gish, T.S. Steenhuis, and D.B. Jaynes. 2005. Quantifying pore-size spectrum of macropore-type preferential pathways. Soil Sci. Soc. Am. J. 69:1196–1208.

Kung, K.-J.S., E.J. Kladivko, T.J. Gish, T.S. Steenhuis, G. Bubenzer, and C.S. Helling. 2000a. Quantifying preferential flow by breakthrough of sequentially applied tracers: Silt loam soil. Soil Sci. Soc. Am. J. 64:1296–1304.

Kung, K.-J.S., T.S. Steenhuis, E.J. Kladivko, T. Gish, G. Bubenzer, and C.S. Helling. 2000b. Impact of preferential flow on the transport of adsorbing and non-adsorbing tracers. Soil Sci. Soc. Am. J. 64:1290–1296.

LaFleur, K.S., W.R. McCaskil, and D.S. Adams. 1975. Movement of prometryne through Congaree soil into groundwater. J. Environ. Qual. 4:132–133.

Laird, D.A., E. Barriuso, R.H. Dowdy, and W.C. Koskinen. 1992. Adsorption of atrazine on smectites. Soil Sci. Soc. Am. J. 56:62–67.

Libra, R.D., G.R. Halberh, G.G. Resmeyer, and B.E. Hoyer. 1984. Groundwater quality and hydrogeology of Devonian-carbonate aquifers in Floyd and Mitchell Counties, Iowa. Iowa Geol. Surv. Open File Rep. 84-2. Part I. Iowa Geological Survey, Iowa City.

Libardi, P.L., K. Reichardt, D.R. Nielsen, and J.W. Biggar. 1980. Simple field methods for estimating soil hydraulic conductivity. Soil Sci. Soc. Am. J. 44:3–6.

Ma, Q.L., R.D. Wauchope, L. Ma, K.W. Rojas, R.W. Malone, and L.R. Ahuja. 2004. Test of the root zone water quality model (RZWQM) for predicting runoff of atrazine, alachor, and fenamiphos species from conventional-tillage corn mesoplots. Pest Manage. Sci. 60:267–276.

Majewski, M.S., and P.D. Capel. 1995. Pesticides in the atmosphere, distribution, trends, and governing factors. Ann Arbor Press, Inc., Chelsea, Michigan, USA.

McConnell, L.L., J.S. LeNoir, S. Datta, and J.N. Seiber. 1998. Wet deposition of current-use pesticides in the Sierra Nevada mountain range, California, USA. Environ. Toxicol. Chem. 17:1908–1916.

Muir, D.C., and B.E. Baker. 1976. Detection of triazine herbicide and their degradation products in tile-drain water from fields under intensive corn (maize) production. J. Agric. Food Chem. 24:122–125.

Ng, H.Y.F., J.D. Gaynor, C.S. Tan, and C.F. Drury. 1995. Dissipation and loss of atrazine and metolachlor in surface and subsurface drain water: A case study. Water Res. 29:2309–2317.

Nielsen, D.R., J.W. Biggar, and K.T. Erh. 1973. Spatial variability of field measured soil water properties. Hilgardia 42:215–259.

Novak, S.M., J.-M. Portal, and M. Schiavon. 2001. Effects of soil type on metolachlor losses in subsurface discharge. Chemosphere 42:235–244.

Parmele, L.H., E.R. Lemon, and A.W. Taylor. 1972. Micrometeorological measurement of pesticide vapor flux from bare soil and corn under field conditions. Water Air Soil Pollut. 1:433–451.

Posner, J.L., M.D. Casler, and J.O. Baldock. 1995. The Wisconsin integrated cropping system trial: Combining agro-ecology with production agronomy. Am. J. Alt. Agric. 10:98–107.

Prueger, J.H., T.J. Gish, L.L. McConnell, L.G. McKee, J.L. Hatfield, and W.P. Kustas. 2005. Solar radiation, relative humidity, and soil water effects on Metolachlor volatilization. Environ. Sci. Technol. 39:5219–5226.

Prueger, J.H., J.L. Hatfield, and T.J. Sauer. 1999. Field-scale metolachlor volatilization flux estimates from broadcast and banded application methods in central Iowa. J. Environ. Qual. 28:75–81.

Rao, P.S.C., and J.M. Davidson. 1980. Estimation of pesticide retention and transformation parameters required in nonpoint source pollution models. p. 23–67. In M.R. Overcash and J.M. Davidson (ed.) Environmental impact of nonpoint source pollution. Ann Arbor Sci. Publ., Ann Arbor, MI.

Reddy, K.N., M.A. Locke, and C.T. Bryson. 1994. Foliar washoff and runoff losses of lactofen, norflurazon, and fluometuron under simulated rainfall. J. Agric. Food Chem. 42:2338–2343.

Ritsema, C.J., and L.W. Dekker. 1995. Distribution flow: A general process in the top layer of water repellent soils. Water Resour. Res. 31:1187–1200.

Schreiber, M.M., M.V. Hickman, and G.D. Vail. 1993. Starch-encapsulated atrazine—Effects and transport. J. Environ. Qual. 22:443–453.

Shipitalo, M.J., W.A. Dick, and W.M. Edwards. 2000. Conservation tillage macropore factors that affect water movement and the fate of chemicals. Soil Tillage Res. 53:167–183.

Shipitalo, M.J., W.M. Edwards, W.A. Dick, and L.B. Owens. 1990. Initial storm effects on macropore transport surface-applied chemicals in no-till soil. Soil Sci. Soc. Am. J. 54:1530–1536.

Schulz, R., M. Hauschild, M. Ebeling, J. Nankodrees, J. Wogram, and M. Liess. 1998. A qualitative field method for monitoring pesticides in the edge-of-field runoff. Chemosphere 36:3071–3082.

Spark, K.M., and R.S. Swift. 2002. Effect of soil composition and dissolved organic matter on pesticide sorption. Sci. Total Environ. 298:147–161.

Sparling, D.W., G.M. Fellers, and L.L. McConnell. 2001. Pesticides and amphibian population declines in California, USA. Environ. Toxicol. Chem. 20:1591–1595.

Spencer, W.F., and M.M. Cliath. 1970. Desorption of Lindane from soil as related to vapor density. Soil Sci. Soc. Am. Proc. 34:574–578.

Spencer, W.F., and M.M. Cliath. 1974. Factors affecting vapor loss of trifuralin from soil. J. Agric. Food Chem. 20:987–991.

Spencer, W.F., M.M. Cliath, and W.J. Farmer. 1969. Vapor density of soil-applied dieldrin as related to soil-water content, temperature, and dieldrin concentration. Soil Sci. Soc. Am. Proc. 33:509–511.

Symons, P.E.K. 1977. Dispersal and toxicology of the insecticide fenitrothion; predicting hazards of forest spraying. Residue Rev. 68:1–36.

Taylor, A.W. 1995. The volatilization of pesticide residues. p. 257–306. In T.R. Roberts and P.C. Kearney (ed.) Environmental behavior of agrochemicals. Vol. 9. John Wiley and Sons, New York.

Taylor, A.W., and W.F. Spencer. 1990. Volatilization and vapor transport processes. p. 213–269. In H.H. Cheng (ed.) Pesticides in the soil environment: Processes, impacts, and modeling. SSSA Book Ser. 2. SSSA, Madison, WI.

Thurman, E.M., and A.E. Cromwell. 2000. Atmospheric transport, deposition, and fate of triazine herbicides and their metabolites in pristine areas at Isle Royale National Park. Environ. Sci. Technol. 34:3079–3085.

Tilman, D., K.G. Cassman, P.A. Matson, R. Naylor, and S. Polasky. 2002. Agricultural stability and intensive production practices. Nature 418:671–677.

Traub-Eberhard, U., K.-P. Henschel, W. Kordel, and W. Klein. 1995. Influence of different field sites on pesticide movement into subsurface drains. Pestic. Sci. 43:121–129.

USEPA. 1990. National pesticide survey: Project summary. EPA Rep. 570990NPSg.

USEPA. 2008. Drinking water contaminants. EPA 816-F-03-016. Available at http://www.epa.gov/safewater/contaminants/index.html (verified 29 Nov. 2010).

USEPA Biological and Economic Analysis Division. 2004. Pesticide industry sales and usage: 2000 and 2001 Market estimates. http://www.epa.gov/oppbead1/pestsales/.

Wauchope, R.D. 1978. The pesticide content of surface water draining from agricultural fields. A review. J. Environ. Qual. 7:459–472.

Wauchope, R.D., and R.S. Meyers. 1985. Adsorption-desorption kinetics of atrazine and linuron in freshwater sediment aqueous slurries. J. Environ. Qual. 14:132–136.

Webb, R.M.T., M.E. Wieczorek, BT. Nolan, T.C. Hancock, M.W. Sandstrom, J.E. Barbash, E. R. Bayless, R.W. Healy, and J. Linard. 2008. Variations in pesticide leaching related to land use, pesticide properties, and unsaturated zone thickness. J. Environ. Qual. 37:1145–1157.

Weber, J.B., S.B. Weed, and T.J. Sheets. 1972. Pesticides— How they move and react in the soil. Crops Soils 25(1):14–17.

Wienhold, B.J., and T.J. Gish. 1994a. Effect of Formulation and tillage practice on volatilization of atrazine and alachlor. J. Environ. Qual. 23:292–298.

Wienhold, B.J., and T.J. Gish. 1994b. Chemical properties influencing rate of release of starch encapsulated herbicides: Implications for modifying environmental fate. Chemosphere 28:1035–1046.

Williams, A.G., D. Scholefield, J.F. Dowd, N. Holden, and L. Deeks. 2000. Investigating preferential flow in a large intact soil block under pasture. Soil Use Manage. 16:264–269.

Wilson, J.D., G.W. Thurtell, G. Kidd, and E. Beauchamp. 1982. Estimation of the rate of gaseous mass transfer from a surface source plot to the atmosphere. Atmos. Environ. 16:1861–1867.

Wyman, J.A, J.O. Jensen, D. Curwen, R.L. Jones, and T.E. Marquardt. 1985. Effects of application procedures and irrigation on degradation and movement of aldicarb residues in soil. Environ. Toxicol. Chem. 4:641–651.

Wyman, J.A., J. Medina, D. Curwen, J.L. Hansen, and R.L. Jones. 1986. Movement of aldicarb and aldoxycarb residues in soil. Environ. Toxicol. Chem. 5:545–555.

Yagi, K., J. Williams, N.Y. Wang, and R.J. Cicerone. 1995. Atmospheric methyl bromide (CH_3Br) from agricultural soil fumigations. Science 267:1979–1981.

Yates, S.R., F.F. Ernst, J. Gan, F. Gao, and M.V. Yates. 1996. Methyl bromide emissions from a covered field. II. Volatilization. J. Environ. Qual. 25:192–202.

Yates, S.R., J. Knuteson, F.F. Ernst, W. Zheng, and Q. Wang. 2008. Effect of sequential surface irrigations on field-scale emissions of 1,3-dichloropropene. Environ. Sci. Technol. 42:8753–8758.

Zehe, E., and H. Fluhler. 2001. Slope scale variation of flow patterns in soil profiles. J. Hydrol. 247:116–132.

13

Subsurface Drainage Design and Management to Meet Agronomic and Environmental Goals

J.S. Strock, G.R. Sands, and M.J. Helmers

High commodity prices, the demand for food and energy security, and the expiration of Conservation Reserve Program (CRP) contracts are resulting in intensification of agricultural production on existing and previously retired cropland, and an increased interest in new management technologies. These developments are occurring despite current trends of increased energy costs, land values, and crop production input costs. One practice that has accompanied the intensification of agricultural production for millennia is the use of artificial subsurface drainage (a.k.a. "tile" drainage) to manage excess soil water for trafficability and crop production. Although still a common practice throughout the United States, practices such as subsurface drainage are being more closely scrutinized by society due to concerns about the impacts these practices have on natural resources and water quality. This chapter begins with a brief introduction to agricultural drainage, its extent in the United States, and some of the salient issues associated with the practice. A short section follows describing the different types of drainage, while the remainder of the chapter focuses on the agronomic and environmental benefits and challenges of subsurface drainage and how these may be addressed through drainage system design and management.

Brief History of Drainage in the United States

Water table management for agricultural production has been practiced for thousands of years. During the second century BCE, Marcus Porcius Cato, a Roman statesman, wrote *De Agri Cultura* (translated as "On Farming" or "On Agriculture"). In this early literary work on farm management, Cato very clearly described the practice of subsurface drainage designed to remove surplus water from the soil:

> If the land is wet, it should be drained with trough shaped ditches dug three feet wide at the surface and one foot at the bottom and four feet deep. Blind these ditches with rock. If you have no rock then fill them with green willow poles braced crosswise. If you have no poles, fill then with faggots (bundles of sticks). Then dig lateral trenches three feet deep and four feet wide in such way that the water will flowfrom the trenches into the ditches. (Cato, 1913)

J.S. Strock, University of Minnesota, Dep. Soil, Water and Climate and Southwest Research and Outreach Center, Lamberton, MN 56152 (jstrock@umn.edu); G.R. Sands, University of Minnesota, Dep. Bioproducts and Biosystems Engineering, St. Paul, MN 55108; and M.J. Helmers, Iowa State University, Dep. Agricultural and Biosystems Engineering, Ames, IA 50011.

doi:10.2136/2011.soilmanagement.c13

Drainage has been part of American farming and culture since colonial times. Drainage was encouraged to improve public highways, increase tonnage and decrease maintenance costs of transportation companies, reduce public health risks, promote increased crop yield and reduced yield variability, reduce surface runoff and erosion, and increase land value. The first cylindrical drain tiles were manufactured in England in approximately 1810 (Elliott, 1904), and the first documented subsurface tile installation in the United States occurred near Geneva, NY in 1835 by a wheat farmer named John Johnston (King and Lynes, 1946). Mr. Johnston reportedly installed more than 70 miles of drains on his 140-ha farm during his lifetime. Drainage practices flourished throughout the country and particularly in the upper Midwest during the late 1800s and 1900s. Readers interested in a detailed history of drainage and drainage methods will find a comprehensive treatment of the subject in Beauchamp (1987).

In many areas of the United States, profitable farming is still dependent on artificial drainage of the soil. Zucker and Brown (1998) summarized the 1985 USDA survey of agriculture drainage and concluded that more than 30% of the cropland in the humid north-central region, including the states of Illinois, Indiana, Iowa, Michigan, Minnesota, Missouri, Ohio, and Wisconsin, is drained for crop production. Irrigation and drainage are inseparably linked—approximately 25% of the irrigated cropland in the semiarid and arid western United States is artificially drained (Pavelis, 1987). In humid regions, drainage is needed to provide trafficable soil conditions for timely field operations and to protect growing crops from excess soil water conditions. In semiarid and arid regions subsurface drainage systems are commonly installed to prevent shallow water tables and for salinity control. Numerous published bulletins, articles, and textbooks have been dedicated to the theory, practice, and implications of artificial drainage (Pavelis, 1987; Ritzema, 1994; Skaggs and van Schilfgaarde, 1999)

Loss of nitrate-N through subsurface drainage systems in agricultural lands has been the focus of numerous research studies in the United States and abroad. Subsurface drainage has been identified as one of the primary sources of nitrate-N entering surface water (Randall and Goss, 2001) and has also been associated with the delivery of sediment, particulate, and dissolved phosphorus, pesticides, and coliform bacteria to surface waters (Carpenter et al., 1998; Scott et al., 1998; Gilliam et al., 1999; Kladivko et al., 1999; Addiscott et al., 2000; Randall and Goss, 2001; Jamieson et al., 2002; David et al., 2003). Hypoxia in the Gulf of Mexico has been strongly linked to nitrate-N losses from the upper Midwest and, thus, drained landscapes (Rabalais et al., 1996, 1999). Research has concentrated on factors including: precipitation; soil organic matter mineralization; cropping system; timing, source, and rate of nitrogen application; use of nitrification inhibitors and slow-release nitrogen fertilizers; tillage; and drainage system depth and spacing (Dinnes et al., 2002).

Environmental concerns in the United States are reflected in current policy discussions at state and federal levels. Both public and private entities desire increased accountability with regard to agriculture's impact on the environment, but do not want the cost of increased regulation. Thus, the adoption of more responsible, voluntary management practices is the preferred option of agricultural producers. Rethinking drainage system design and/or implementing drainage management practices to address both agronomic and environmental objectives are the latest goals of agricultural producers and agricultural drainage practitioners.

Types of Drainage

Land drainage can be accomplished by surface drainage or subsurface drainage and often incorporates features from both surface and subsurface systems. Surface drainage is accomplished by land forming, grading or smoothing, ditching, or surface inlets. The objective of these practices is to provide greater efficiency for surface water and runoff to leave the field. The orientation of the graded land and thus, direction of surface drainage is normally toward an adjacent open drain (ditch) or surface watercourse. If the receiving ditch or watercourse is sufficiently deep, it may also receive lateral seepage through the soil profile. A field is considered to have good surface drainage if it has a relatively small amount of depressional storage that must be satisfied before surface drainage occurs. Adequate surface drainage enhances overland runoff and thus, reduces the requirement for subsurface drainage compared to what would be needed if surface drainage were poor (Schwab et al., 1974; Skaggs, 1974). The primary environmental concerns regarding surface drainage include increased peak flow rates in receiving waters, loss of permanent and ephemeral wetlands, increased losses of sediment and sediment-bound nutrients such as phosphorus, and losses of other contaminants commonly found in surface runoff, such as pesticides and bacteria (Skaggs et al., 1994).

Subsurface drainage is accomplished with subsurface pipes or ditches designed to remediate saturated conditions in the root zone by removing excess pore water to a specific design depth, typically between 60 and 150 cm. Excess pore water is defined as that which exists at moisture contents between field capacity (1/3-bar tension) and saturation. When the water table rises above the drain depth, soil water is driven by hydrostatic forces from larger pores to subsurface drains. A subsurface drainage system may consist of parallel pipes or ditches covering an entire field or a more targeted system that drains isolated wet areas of a field. In many areas, open drains or ditches also function as outlets for field drainage systems.

From the early 1800s until the mid 1900s subsurface drainage pipes were manufactured from clay and cement and were installed in trenches, the earliest of which were hand-dug (Beauchamp, 1987). The term drain *tile* is a remnant of the use of these materials. The invention of polyethylene plastic in the early 1940s ultimately resulted in the use of flexible, corrugated, perforated plastic drain pipe or tubing for subsurface drainage systems in the late 1960s. Trenchless installation of the continuous plastic tubing with the "tile plow" greatly decreased the time and cost associated with subsurface drainage installation. The earliest known agricultural research using perforated plastic drainage pipe was conducted by G.O. Schwab beginning in 1947 (Schwab, 1955). Environmental concerns associated with subsurface drainage primarily include the transport of dissolved constituents such as nutrients (nitrate-N and phosphorus), pesticides, and liquid manure,

and the loss of permanent and ephemeral wetlands from the landscape.

Conventional Drainage System Design

The need for artificial drainage stems from the risks and damages on crop production associated with excess soil water. Drainage design, installation, and operation tend to be site-specific enterprises because these risks and damage to crop production are largely dependent on soil, climate, and crop. The need for drainage may also vary within a growing season because of seasonal climate patterns and changes in crop tolerance to excess water stress throughout the growing season, with some crops being more tolerant of poorly drained, wet conditions at different stages of growth than others.

The practice of subsurface drainage has evolved in the United States from hand-dug "random" drains that targeted isolated wet areas, to modern patterned drainage systems that cover entire fields. Changes in drainage materials (i.e., clay, concrete to plastic) and installation methods (hand digging to modern trenchers and drainage plows, laser guidance systems) have revolutionized the approach to drainage and drainage design. These modern materials and installation methods permit, if not encourage, drainage practitioners to implement artificial drainage over an entire agricultural field. Thus, the evolution of targeted drainage to patterned (whole-field) drainage has increased drainage *intensity* in agricultural landscapes over the years. In addition, these technologies have made economically feasible the drainage of soils that were once regarded as uneconomical to drain. The productivity, hydrologic, and water quality impacts of artificial drainage have thus been evolving since the introduction of drainage practices in the United States more than a century ago.

It may be instructive to discuss the typical approach to subsurface drainage design before developing a new approach that incorporates both environment and productivity goals. Sands (2003) described subsurface drainage design as a process of selecting a drainage intensity and accompanying drainage system layout so as to mitigate the damages associated with excess water, while balancing the costs associated with the drainage system. The drainage intensity that meets these criteria is termed the *design drainage rate*. Larger drainage intensities diminish the potential damage associated with excess water, but increase system costs. The optimum drainage design is ostensibly when the net returns on investment (benefits minus the costs of the system) are maximized over the life of the system. The costs associated with unwanted environmental effects of drainage have typically not been included in the traditional approach to design. We will return to this point later in the chapter.

The design drainage rate is a function of climate, site, and soil characteristics and thus, requires a site-specific approach. The interaction of these and other factors cannot be adequately described in the context of this chapter. However, generally, as soil texture becomes finer (i.e., silt and/or clay contents increase), the hydraulic conductivity decreases and narrower drain spacings are required to achieve the desired design drainage rate. Other factors also exhibit an influence on the design drainage rate. Higher value crops may dictate the need to further reduce potential risks of excess water by selecting narrower drain spacings than those used with more conventional crops. In some cases, the combination of site factors may result in drainage being uneconomical for the intended land use or cropping system.

A well-designed drainage system should be capable of reducing the depth of the water table to a suitable depth (typically 30 to 60 cm) within 24 to 48 h after a rainfall event.

Subsurface drains are usually installed at depths of 60 to 150 cm, depending on site, soil, and installation methods. When not constrained by shallow, restrictive soil layers or a lack of sufficient outlet depth, deeper drain depths that enable wider drain spacing are typically favored to reduce material costs. Drain spacing for parallel drains can be determined empirically for a given soil using the Hooghoudt or "drain spacing" equation (Eq. [1]), as described by van der Ploeg et al. (1999):

$$R = \frac{8K_2 dh}{L^2} + \frac{4K_1 h^2}{L^2} \qquad [1]$$

where L is the distance between parallel drains (*drain spacing*); R is the rainfall/ recharge rate, or *drainage intensity*; K_1 and K_2 are the hydraulic conductivities of the soil layers above and below the drain, respectively; h is the midpoint water table elevation above the drain; and d is the equivalent depth to the restrictive layer, as illustrated in Fig. 13|1. The Hooghoudt equation is an example of a steady-state drainage equation, where the water table is at an assumed, fixed depth and rainfall/recharge rate is equal to the drainage intensity. Non-steady-state, or transient, drainage equations can also be used to compute drain spacing wherein the drainage parameters and water table depth may vary throughout the drainage process.

The typical use of Eq. [1] is that a drain spacing L is computed for the site based on soil, crop, climate, and other considerations, with an assumed drainage intensity. Drainage intensities used in various regions of the world can be found in Framji et al. (1987) and typically range from 2 to 40 mm d^{-1}. Once the design drainage rate and drainage depth are selected, drain spacing can be determined solving Eq. [1] for L, the drain spacing. Examples using Eq. [1] and other drainage formulae can be found in drainage texts, such as Smedema et al. (2004).

Selection of the "best" design drainage rate poses a significant challenge when using Eq. [1] for drainage design because the interrelated effects of climate, soil, and other factors

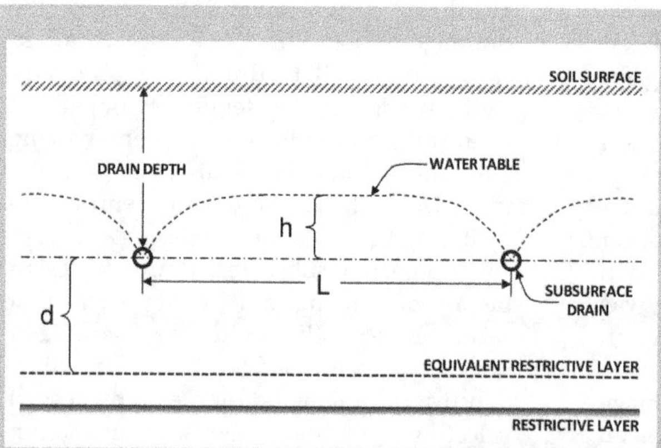

Fig. 13|1. Schematic of dimensions for describing subsurface drainage hydraulics.

are generally not well understood at the scale for which drainage design is conducted. Drainage design based simply on using empirical values of drainage intensities and/or soils information happens all too frequently, and runs a significant risk of under- or overdesign. In other cases, drainage design decisions may be based on the practitioner's experience of working in a certain region with certain soils. This type of experiential approach can lead to overdesign, however, because practitioners may naturally lean toward a higher drainage design rate to minimize potential for excess crop water stress and system underperformance. Both the empirical and experiential approaches are imprecise by nature, and offer little opportunity to fine-tune drainage design to accommodate both environment and productivity goals.

A more sophisticated approach to drainage design involves the application of a drainage simulation model, such as DRAINMOD (Skaggs, 1978), to estimate the optimum drainage design rate and drain spacing using a historical climatic record and measured soil and site characteristics for a given location. This approach has been illustrated by Skaggs et al. (2006). The modeling approach incorporates seasonal and annual climate variability into the decision process and by doing so, captures the frequency and magnitude of crop water stresses. Drainage design can thus be based on the occurrence of these stresses for the site of interest. The efficacy of the modeling approach depends on the quality of the soil and climate input data.

Skaggs et al. (2006) also noted that drainage system designs have evolved toward more intensive drainage to be on the "safe" side, to make up for uncertainties in the estimation of soil hydraulic conductivities and to provide protection from infrequent large storms or extended periods of wet weather. Skaggs et al. (2006) noted that the cost of such overdesign of drainage systems may be fairly low over the life of the investment, when unwanted environmental costs are not accounted for. Skaggs et al. (2006) used datasets from Indiana (Kladivko et al., 1999, 2004) and North Carolina (Gilliam and Skaggs, 1986) to illustrate the penalty paid in increased nitrate-N export from drainage systems designed with higher drainage intensities. The studies cited showed nitrate-N losses two to three times greater for drainage intensities of 2 cm d^{-1} compared to 0.5 cm d^{-1}. These studies, studies of controlled drainage (discussed below), and others, such as the study of shallow drainage by Sands et al. (2008) suggested that nitrate-N losses can be mitigated significantly by reducing drainage intensity, reducing drainage installation depths, and operating drainage systems on the wet end of the spectrum for crop production (Skaggs et al., 2006).

Designing for Productivity and the Environment

The conventional approach to subsurface drainage described above focuses on satisfying agricultural production goals. An alternative approach seeks to reduce the drainage intensities to mitigate or avoid unwanted environmental effects such as nitrate-N loss while simultaneously addressing production goals. Reducing drainage intensity can be accomplished by (i) reduced drain spacing and/or reduced drain depth and (ii) reduced drainage during times of the year when drainage is unnecessary (such as the off-season), unproductive, and perhaps damaging to crop production.

Drainage Intensity and Drain Depth

Opportunities and challenges for designing less intensive drainage systems may be significant. Skaggs et al. (2006) clearly illustrated through modeling of four soils that in some cases, drainage intensity may be reduced with only marginal consequences for net annual returns. Finding and taking advantage of opportunities to fine-tune drainage intensity should be a priority in designing all new drainage systems. The challenge of putting this knowledge into practice is that appropriate design drainage rates and soil hydraulic data are typically not well known, as described earlier, which leads to imprecise design, often overdesign. The use of generalized drain spacing and depth equations may be an important step toward more appropriate drainage design. Better still, would be a more detailed analysis using DRAINMOD, or another soil drainage model, to develop optimum design drainage rates and drain spacing for locations and soils throughout the United States.

Skaggs (2007), based on 50-year simulations of four soils and five locations throughout the eastern United States, developed the following equation for drainage design:

$$DDR = 0.004P - 0.0086D + 0.0095T - 0.396 \quad [2]$$

where DDR equals the design drainage rate for corn, P equals growing season precipitation, D equals drain depth, and T equals the transmissivity (the sum of saturated hydraulic conductivity $K_s \times$ depth for the soil profile).

Skaggs and Chescheir (2003) proposed and demonstrated through modeling, the use of reduced drain depths to conserve soil moisture and reduce the transport of nitrate-N losses. Their premise was that a reduced drainage depth would promote higher water tables and a greater saturated zone, which in turn, would facilitate denitrification and/or reduced drainage effluent volumes. In addition, Skaggs and Chescheir (2003) emphasized that the practice of shallow drainage could be applied anywhere new drains were being installed, rather than being limited by topography, as is the case for controlled drainage, which will be described in the next section. Sands et al. (2008) measured reductions in annual drainage volume and nitrate-N of approximately 20% over 6 years at a field site in south-central Minnesota when comparing a drainage depth of 90 cm to a more traditional depth of 120 cm. For their study, Sands et al. (2008) reduced the drain spacing of the 90-cm drain depth systems to maintain a drainage intensity equivalent to that of the 120-cm drain depth. Luo et al. (2010) subsequently conducted 90-year simulations with the same soil and climate conditions as the Sands et al. (2008) study and predicted a long-term average 20% reduction in drainage volume and nitrate-N. Helmers et al. (2010) measured reductions in annual drainage volume of approximately 40% when comparing a drain depth of 76 cm to a more traditional depth of 120 cm. For this study the drain spacing was 12.2 m for the shallow drain depth versus 18 m for the conventional drain depth. The spacing was narrower for the 76-cm drainage depth to maintain a drainage intensity equivalent to the 120-cm drain depth.

Reduced drain depth appears to have potential as a practice to mitigate nutrient losses from subsurface drainage systems. However, if narrower drainage spacings are used in conjunction with reduced drain depths, the increased cost of the system (greater quantity of drainage pipe) must be offset somehow to make the practice more attractive to landowners. In addition, reduced drain depths with decreased drain spacings may be viewed in a negative light by society, if naively regarded as "increased drainage" because of the reduced drain spacings.

Controlled Drainage or Managed Drainage

The practice known as *controlled drainage* (Fig. 13|2) or *managed drainage* was introduced in the early 1970s as a management technique to reduce the quantity of nitrate-N exported from artificially drained, irrigated, agricultural lands in California while maintaining a high level of agronomic productivity (Meek et al., 1970). Water control structures (Fig. 13|2), when installed in a ditch, main, sub-main, or lateral drainage line, allow the practitioner to set the drainage outlet at any level between the ground surface (undrained condition) and the drain depth (conventional drainage). The water table must rise above the elevation setting of the structure before drainage can occur. Thus, removal of excess soil water can be delayed and/or reduced, creating opportunities to both provide more optimum plant growth conditions for crop production while, at the same time reducing seasonal or

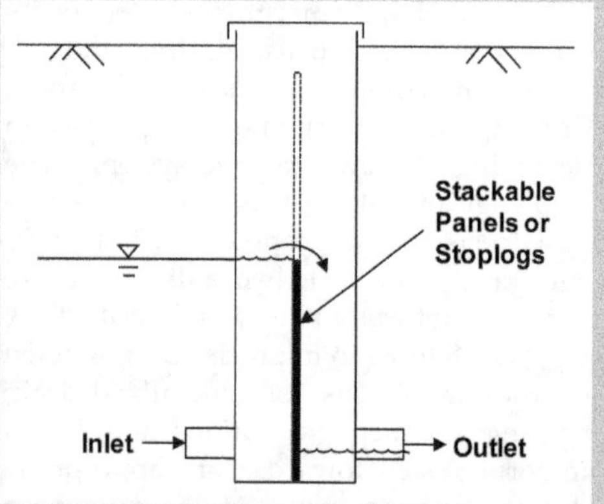

Fig. 13|2. Schematic of the controlled drainage concept, an example with an in-field water control structure.

annual drainage volumes. Unlike conventional drainage systems that remove excess water to the design drain depth whenever it occurs, controlled drainage conserves water by increasing the retention time of water in the soil profile, which, in turn, increases the opportunity for vertical or lateral seepage and for increased evapotranspiration. However, there may be a potential for increased surface runoff losses from controlled drainage systems (Riley et al., 2009) because of shallower water tables and reduced infiltration volumes.

Research shows that when drainage volumes are reduced, annual nitrate-N loads can be proportionately decreased. North Carolina has the longest record of continuous research on controlled drainage (Gilliam et al., 1979; Evans and Skaggs, 2004), where the practice received cost-sharing as part of North Carolina's effort to reduce nitrate-N loads in the Neuse River Basin. Field experiments with controlled drainage have concentrated on yield and water quality benefits. Research on controlled drainage has been conducted in the United States, Canada, and Europe (Gilliam et al., 1979; Bergstrom, 1987; Evans et al., 1995; Lalonde et al., 1996; Drury et al., 2001; Wesström et al., 2001). Although the practice has been in existence for nearly 40 years, its adoption and implementation have been rather localized, and have not been common in the U.S. Midwest.

When controlled drainage is practiced in a ditch or canal, the control structure can be retrofitted to an existing culvert to control the water level in the ditch. For in-field implementation of the practice, water control structures are placed at one or more points on the subsurface drainage system, and the area controlled by each structure may be termed a *water management zone*, as illustrated in Fig. 13|3.

Controlled drainage creates an opportunity to manage drainage systems according to seasonal needs, as illustrated in Fig. 13|4. In humid regions, the outlet elevation is raised after harvest to limit drainage outflow and reduce the mass of nitrate-N exported to surface waters during the non-growing season. In early spring before planting, the system is put into the conventional drainage mode by lowering the outlet elevation to the drain depth. This allows the soil to drain freely and permits timely field operations,

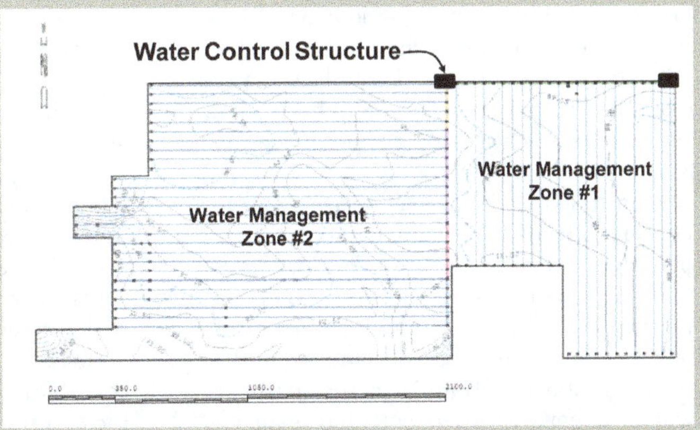

Fig. 13|3. An example of a controlled drainage system layout with two water control structures and two water management zones.

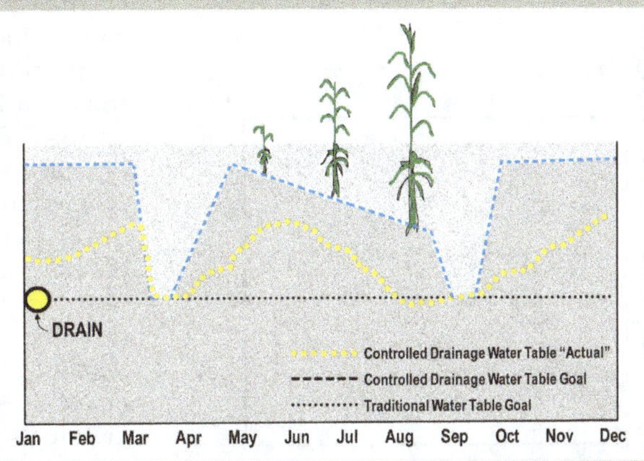

Fig. 13|4. Illustration of seasonal water table management goals for controlled drainage system. The blue and yellow dotted lines illustrate hypothetical "design" and "actual" water table depths, respectively. The shaded region above the drain elevation represents a shallower water table than the conventional drainage system.

such as tillage, planting, and spraying. The outlet elevation is raised again after spring field operations to reduce drainage outflow, conserve water for crop growth during the growing season, and reduce the losses of nitrate-N to surface waters. Before harvest the outlet elevation may be lowered to the drain depth again, if necessary, to allow for crop harvest. It is important to note that the practice of controlled drainage does not ensure that a water table will occur at the controlled outlet depth. Rather, the outlet depth setting (in the control structure) represents the *minimum* depth that the water table must achieve before drainage can begin, while the *actual* position of the water table depends on the occurrence of precipitation. Thus, there may be periods when it is impossible to establish a shallow water table with controlled drainage due to lack of precipitation, increased vertical and horizontal seepage, and/or high crop demand for water (Gilliam et al., 1979).

Controlled drainage has been shown to be very effective in reducing the outflow of water and nitrate-N from drainage systems. The reduction in nitrate-N has been primarily attributed to a reduction in outflow of water rather than a reduction in the concentration of nitrate-N in the drainage water. Other reported potential explanations for reductions in nitrate-N include denitrification, vertical (deep) seepage below the drainage system, or horizontal (lateral) seepage to natural water courses or open drains. An aggregation of results from 15 peer-reviewed controlled drainage studies (Fig. 13|5) shows that seasonal drainage volumes for conventional drainage systems exceeded controlled drainage volumes by an average of 40%. These studies represented a variety of observed and simulated environmental conditions in the United States, Canada, Belgium, Sweden, Italy, and Egypt (Amatya et al., 1998; Borin et al., 2001; Deal et al., 1986; Drury et al., 1996; Drury et al., 1997; El-Sadek et al., 2002; Fouss et al., 1987; Gilliam and Skaggs, 1986; Lalonde et al., 1996; Skaggs et al., 1995a,b; Tan et al., 1998; Wahba et al., 2001; Wesström et al., 2001, 2003).

For controlled drainage systems to be practical, certain conditions and criteria must be met: (i) artificial drainage of the land is necessary, and patterned systems are most suitable, rather than random drainage systems, and (ii) the land should have a field slope of no more than 0.5%. The field slope condition is based on both hydraulics and economics. Larger field slopes necessitate a greater number of control structures to achieve acceptable uniformity of water table control. The increased number of structures leads to increases in both initial costs and management costs of the controlled drainage system. The slope criterion can shift where cost-sharing incentives are available to offset the cost burden to the agricultural producer. In addition, larger field slopes may experience greater lateral seepage and pose greater difficulty for water table control. Moreover, if surface runoff is increased with the use of controlled drainage, this may be of greater concern in applications on larger field slopes.

An additional consideration for a controlled drainage design is the use of a narrower drain spacing than a conventional drainage system. Flow gradients are reduced in a controlled drainage system compared to a conventional system, when operating in the control mode. A narrower drain spacing may compensate for the reduced gradient, resulting in a faster water table drawdown following a rainfall event, reducing the risk of excess soil water conditions during the growing season.

Advantages and Disadvantages of Water Table Management

There are a number of potential benefits and concerns associated with the water table management practices of

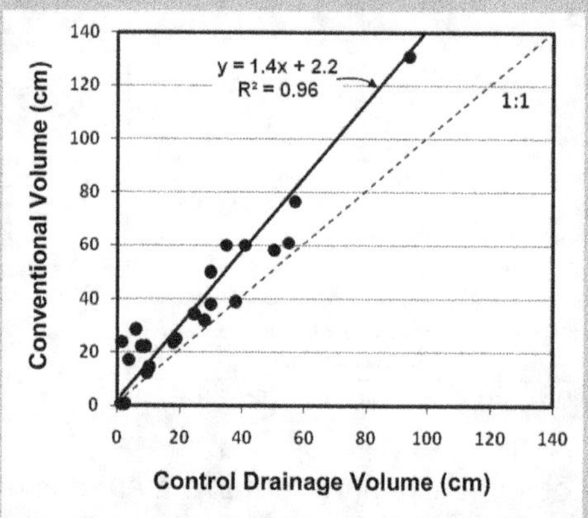

Fig. 13|5. Controlled drainage volume vs. conventional drainage volume as reported in 15 studies.

conventional and controlled drainage in agricultural landscapes:

1. Both practices are applicable on many relatively flat, poorly drained lands. Controlled drainage is best suited on landscapes with slopes of <0.5%, although installation of drains on topographic contours is possible up to 1%.

2. Water conservation is greater with the practice of controlled drainage than in conventional full-depth drainage. Controlled drainage has been reported to reduce drainage volume by as much as 40 to 50% compared to conventional drainage.

3. Controlled drainage has been reported to reduce nitrogen-N loss from drained lands by 40 to 50%. The loss of phosphorus to surface water has also been found to be reduced by 25 to 35%.

4. Crop yield response is known to vary spatially and temporally within a field, and with variations in climate and management practices. Studies from North Carolina indicate that, while increased yields with controlled drainage might be possible, the increases are strongly dependent on management. These studies indicated an average crop yield increase of about 5% when controlled drainage was effectively managed.

where conditions are favorable, controlled drainage offers a number of advantages over conventional drainage and can contribute to mitigation of water quality impairments due to excess nutrients and turbidity in the Total Maximum Daily Load (TMDL) process.

Precise management of agricultural drainage water using the practice of controlled drainage could represent a significant step forward for agriculture, with benefits that reach from conserving subsoil moisture on individual subsurface drained fields to reducing nutrient loading in lakes, rivers, and sensitive coastal estuaries. In the future, agricultural producers may view controlled and reduced drainage as soil and water management tools—rather than mitigation practices—designed to optimize agronomic production and minimize unwanted environmental quality impacts.

References

Addiscott, T.M., D. Brockie, J.A. Catt, D.G. Christian, N.N. Mirza, and K.R. Howse. 2000. Phosphate losses through field drains in a heavy cultivated soil. J. Environ. Qual. 29:522–532.

Amatya, D.M., J.W. Gilliam, R.W. Skaggs, M.E. Lebo, and R.G. Campbell. 1998. Effects of controlled drainage on forest water quality. J. Environ. Qual. 27:923–935.

Beauchamp, K.H. 1987. A history of drainage and drainage methods. p. 13–29. In G.A. Pavelis (ed.) Farm drainage in the United States. History, status, and prospects. Misc. Publ. 1455. USDA-ERS, Washington, DC

Bergstrom, L. 1987. Nitrate leaching and drainage from annual and perennial crops in tile-drained plots and lysimeters. J. Environ. Qual. 16:11–18.

Borin, M., G. Bonaiti, and L. Giardini. 2001. Controlled drainage and wetlands to reduce agricultural pollution: A lysimetric study. J. Environ. Qual. 4:1330–1340.

Carpenter, S.R., N.F. Caraco, D.L. Correll, R.W. Howarth, A.N. Sharpley, and V.H. Smith. 1998. Nonpoint pollution of surface waters with phosphorus and nitrogen. Ecol. Appl. 8:559–568.

Cato, M.P. 1913. Roman farm management. The Macmillan Co., New York.

David, M.B., L.E. Gentry, K.M. Starks, and R.A. Cooke. 2003. Stream transport of herbicides and metabolites in a tile-drained agricultural watershed. J. Environ. Qual. 32:1790–1801.

Deal, S.C., J.W. Gilliam, R.W. Skaggs, and K.D. Konyha. 1986. Prediction of nitrogen and phosphorus losses as related to agricultural drainage system design. Agric. Ecosyst. Environ. 18:37–51.

Dinnes, D.L., D.L. Karlen, D.B. Jaynes, T.C. Kaspar, J.L. Hatfield, T.S. Colvin, and C.A. Cambardella. 2002. Nitrogen management strategies to reduce nitrate leaching in tile drained Midwestern soils. Agron. J. 94:153–171.

Drury, C.F., C.S. Tan, J.D. Gaynor, T.O. Oloya, and T.W. Welacky. 1996. Influence of controlled drainage-subirrigation on surface and tile drainage nitrate loss. J. Environ. Qual. 25:317–324.

Drury, C.F., C.S. Tan, J.D. Gaynor, T.O. Oloya, I.J. Van Wesenbeeck, and D.J. McKenney. 1997. Optimizing corn production and reducing nitrate losses with water table control subirrigation. Soil Sci. Soc. Am. J. 61:889–895.

Drury, C.F., C.S. Tan, J.D. Gaynor, W.D. Reynolds, T.W. Welacky, and T.O. Oloya. 2001. Water table management reduces tile nitrate and loss in corn and in a soybean–corn rotation. p. 163–169. In Optimizing nitrogen management in food and energy production and environmental protection. Proceedings of the 2nd International Nitrogen Conference on Science and Policy. The Scientific World, Kirkkonummi, Finland.

Elliott, C.G. 1904. Drainage of farm lands. Farmers' Bull. 187. USDA, Washington, DC

El-Sadek, A., J. Feyen, R.W. Skaggs, and J. Berlamont. 2002. Economics of nitrate losses from drained agricultural land. J. Environ. Eng. 128:376–383.

Evans, R.O., R.W. Skaggs, and J.W. Gilliam. 1995. Controlled versus conventional drainage effects on water quality. J. Irrig. Drain. Eng. 121:271–276.

Evans, R.O., and R.W. Skaggs. 2004. Development of controlled drainage as a BMP in North Carolina. p. 1–15. In Drainage VIII: Proceedings of the 8th International Drainage Symposium. ASABE, St. Joseph, MI.

Fouss, J.L., R.W. Skaggs, and J.S. Rogers. 1987. Two-stage weir control for subsurface drainage in humid areas. Trans. ASAE 30:1713–1719.

Framji, K.K., B.C. Garg, and S.P. Kaushish (ed.) 1987. Design practices for covered drains in an agricultural land drainage system—A world-wide survey. International Commision on Irrigation and Drainage, Chanakyapuri, New Delhi, India.

Gilliam, J.W., J.L. Baker, and K.R. Reddy. 1999. Water quality effects of drainage in humid regions. p. 801–830. *In* R.W. Skaggs and J. van Schilfgaarde (ed.). Agricultural drainage. Agron. Monogr. 38. ASA, CSSA, and SSSA, Madison, WI.

Gilliam, J.W., and R.W. Skaggs. 1986. Controlled agricultural drainage to maintain water quality. J. Irrig. Drain. 112:254–263.

Gilliam, J.W., R.W. Skaggs, and S.B. Weed. 1979. Drainage control to diminish nitrate loss from agricultural fields. J. Environ. Qual. 8:137–142.

Helmers, M.J., R. Christianson, G. Brenneman, D. Lockett, and C. Pederson. 2010. Water table response to drainage water management in southeast Iowa. *In* Drainage IX: Proceedings of the 9th International Drainage Symposium, Quebec City, QC. ASABE, St. Joseph, MI.

Jamieson, R.C., R.J. Gordon, K.E. Sharples, G.W. Stratton, and A. Madani. 2002. Movement and persistence of fecal bacteria in agricultural soils and subsurface drainage water: A review. Can. Biosys. Eng. 44:1.1–1.9.

King, J.A., and W.S. Lynes. 1946. Tile drainage, 4th ed. Mason City Brick and Tile Co., Mason City, IA.

Kladivko, E.J., J.R. Frankenberger, D.B. Jaynes, D.W. Meek, B.J. Jenkinson, and N.R. Fausey. 2004. Nitrate leaching to subsurface drains as affected by drain spacing and changes in crop production system. J. Environ. Qual. 33:1803–1813.

Kladivko, E.J., J. Grochulska, R.R. Turco, G.E. van Scoyoc, and J.D. Eigel. 1999. Pesticide and nitrate transport into subsurface tile drains of different spacings. J. Environ. Qual. 28:997–1004.

Lalonde, V., C.A. Madramootoo, L. Trenholm, and R.S. Broughton. 1996. Effects of controlled drainage on nitrate concentrations in subsurface drainage discharge. Agric. Water Manage. 29:187–199.

Luo, W., G.R. Sands, M. Youssef, J.S. Strock, I. Song, and D. Canelon. 2010. Modeling the impact of alternative drainage practices in the northern Corn-belt with Drainmod-NII. Agric. Water Manage. 97:389–398.

Meek, B.D., L.B. Grass, L.S. Willardson, and A.J. MacKenzie. 1970. Applied nitrogen losses in relation to oxygen status of soils. Soil Sci. Soc. Am. Proc. 33:575–578.

Pavelis, G.A. (ed.) 1987. Farm drainage in the United States. History, status, and prospects. Misc. Publ. 1455. USDA, Washington, DC.

Rabalais, N.N., R.E. Turner, D. Justic, Q. Dortch, and J.W. Wiseman, Jr. 1999. Characterization of Hypoxia: Topic 1 Report for the Integrated Assessment on Hypoxia in the Gulf of Mexico. Decision Analysis Ser. 15. NOAA, Silver Spring, MD.

Rabalais, N.N., R.E. Turner, D. Justic, Q. Dortch, J.W. Wiseman, Jr., and B.K. Sen Gupta. 1996. Nutrient changes in the Mississippi River and system response on the adjacent continental shelf. Estuaries 19:385–407.

Randall, G.W., and M.J. Goss. 2001. Nitrate losses to surface water through subsurface, tile drainage. p. 95–122. *In* R.F. Follett and J.L. Hatfield (ed.). Nitrogen in the environment: Sources, problems, and management. Elsevier, Amsterdam.

Riley, K.D., M.J. Helmers, P.A. Lawlor, and R. Singh. 2009. Water balance investigation of controlled drainage in non-weighing lysimeters. Appl. Eng. Agric. 25:507–514.

Ritzema, H.P. (ed.) 1994. Drainage principles and applications. 2nd ed. Int. Inst. Land Reclam. Improv., Wageningen, the Netherlands.

Sands, G.R. 2003. Drainage coefficient. p. 118–120. *In* S.W. Trimble et al. (ed.) Encyclopedia of water science. Marcel Dekker, New York.

Sands, G.R., I. Song, L.M. Busman, and B. Hansen. 2008. The effects of subsurface drainage depth and intensity on nitrate load in a cold climate. Trans. ASABE 51:937–946.

Schwab, G.O. 1955. Plastic tubing for subsurface drainage. Agric. Eng. 36:86–89, 92.

Schwab, G.O., N.R. Fausey, and D.W. Michener. 1974. Comparison of drainage methods in a heavy textured soil. Trans. ASAE 17:424–425,428.

Scott, C.A., L.D. Geohring, and M.F. Walter. 1998. Water quality impacts of tile drains in shallow, sloping structured soils as affected by manure application. Appl. Eng. Agric. 14:599–603.

Skaggs, R.W. 1974. The effect of surface drainage on water table response to rainfall. Trans. ASAE 17:406–411.

Skaggs, R.W. 1978. A water management model for shallow water table soils. Tech. Rep. 134. Water Resources Research Inst., Univ. of North Carolina, Raleigh.

Skaggs, R.W. 2007. Criteria for calculating drain spacing and depth. Trans. ASABE 50:1657–1662.

Skaggs, R.W., M.A. Brevé, and J.W. Gilliam. 1994. Hydrologic and water quality impacts of agricultural drainage. Environ. Sci. Technol. 24:1–32.

Skaggs, R.W., M.A. Brevé, and J.W. Gilliam. 1995a. Predicting effects of water table management on loss of nitrogen from poorly drained soils. Eur. J. Agron. 4:441–451.

Skaggs, R.W., M.A. Brevé, A.T. Mohammad, J.E. Parsons, and J.W. Gilliam. 1995b. Simulation of drainage water quality with DRAINMOD. Irrig. Drain. Syst. 9:259–277.

Skaggs, R.W., and G.M. Chescheir, III. 2003. Effects of subsurface drain depth on nitrogen losses from drained lands. Trans. ASABE 46:237–244.

Skaggs, R.W., M.A. Mohamed, and G.M. Chescheir. 2006. Drainage design coefficients for eastern United States. Agric. Water Manage. 86:40–49.

Skaggs, R.W., and J. van Schilfgaarde (ed.) 1999. Agricultural drainage. Agron. Monogr. 38. ASA, CSSA, and SSSA, Madison, WI.

Smedema, S.K., L.K. Smedema, and W.F. Vlotman. 2004. Modern land drainage: Planning, design and management of agricultural drainage systems. Taylor and Francis Group, London.

Tan, C., C. Drury, M. Soultani, H. Ng, J. Gaynor, and T. Welacky. 1998. Effect of controlled drainage and tillage on soil structure and tile drainage nitrate loss at the field scale. Water Sci. Technol. 38:103–110.

van der Ploeg, R.R., M.B. Kirkham, and M. Marquardt. 1999. The Colding equation for soil drainage: Its origin, evolution, and use. Soil Sci. Soc. Am. J. 63:33–39.

Wahba, M.A.S., M. El-Ganainy, M.S. Abdel-Dayem, A. Gobran, and H. Kandil. 2001. Controlled drainage effects on water quality under semi-arid conditions in the western delta of Egypt. Irrig. Drain. 50:295–308.

Wesström, I., G. Ekbohm, H. Linner, and I. Messing. 2003. The effects of controlled drainage on subsurface outflow from level agricultural fields. Hydrol. Processes 17:1525–1538.

Wesström, I., I Messing, H. Linnér, and J. Lindström. 2001. Controlled drainage—Effects on drain outflow and water quality. Agric. Water Manage. 47:85–100.

Zucker, L.A., and L.C. Brown. 1998. Agricultural drainage: Water quality impacts and subsurface drainage studies in the Midwest. Ext. Bull. 871. Ohio State. Univ., Columbus.

14

Wind Erosion

Ted M. Zobeck and R. Scott Van Pelt

Wind erosion refers to the detachment, transport, and deposition of sediment by wind. It is a dynamic, physical process where loose, dry, bare soils are transported by strong winds. Geomorphologists and other earth scientists usually consider wind erosion as a specific sub-discipline of a more broad study of aeolian (also spelled eolian) processes. The term *aeolian* is derived from the Greek god Aeolus, the keeper of the winds, so aeolian processes refer to effects produced by the force of the wind interacting with surface features. Although aeolian research spans a wide range of topics, which may even include research on other planets, in this chapter we will limit our focus to erosion of soils by the wind on the Earth's surface, and more specifically on crop land and range land.

The movement of sediment by wind has been occurring for many eons, as demonstrated by aeolian cross-bedding seen in wind-blown sands of ancient sandstone bedrock. Loess deposits are ubiquitous accumulations of aeolian sediments of silt, and smaller amounts of clay and sand, derived from wind-blown glacial outwash deposits or from deserts or playa lakes. Large dune fields and sand seas around the world provide further evidence of current and past aeolian environments (Fig. 14|1). Fixed or stable dunes are no longer active in the current climate but were active sand seas or dune fields in the past.

Scientists have long been interested in the direct and indirect effects of wind erosion. The earliest publication relating to aeolian processes was written by a Flemish astronomer, Godefroy Wendelin, in 1646 (Stout et al., 2009). Wendelin's paper (Wendelin, 1646) described the purple rain of Brussels that we now recognize as wet deposition of African windblown dust. Charles Darwin collected dust over the Atlantic Ocean that had fallen during his voyages on the HMS Beagle (Darwin, 1845). Recent analysis of this dust indicated it originated from the western Sahara and molecular-microbiological methods demonstrated the presence of many viable microbes even today (Gorbushina et al., 2007).

Wind erosion is a soil degrading process that affects more than 500 million hectares of land worldwide and creates between 500 and 5000 Tg of fugitive dust annually (Grini et al., 2003). Perhaps the most memorable period of recent sustained wind erosion in the United States was the Dust Bowl era from about 1931 through 1939 (Baumhardt, 2003). During this period, wind erosion of rangeland and cropland reached an annual peak of 20 million hect-

T.M. Zobeck, USDA-ARS, Wind Erosion and Water Conservation Research Unit, 3810 4th St., Lubbock, TX 79415 (ted.zobeck@ars.usda.gov) and R.S. Van Pelt, USDA-ARS, Wind Erosion and Water Conservation Research Unit, 302 West I-20, Big Spring, TX 79720 (). The U.S. Department of Agriculture offers its programs to all eligible persons regardless of race, color, age, sex, or national origin, and is an equal opportunity employer.

doi:10.2136/2011.soilmanagement.c14

ares (Hurt, 1981), while the entire area affected encompassed almost 40 million hectares (Baumhardt, 2003).

Wind erosion has been estimated in the United States by the USDA-NRCS by means of a periodic Natural Resources Inventory (NRI). The NRI is a statistical survey of natural resource condition and trends on non-Federal U.S. land (USDA-NRCS, 2007). In 2003, the estimated erosion on cropland due to wind was 776 million tons per year. This represents a 7% reduction in erosion by wind estimated in a similar NRI compiled in 1997. In comparison, the amount of erosion on cropland due to water in 2003 was 971 million tons for the same year. These estimates are based on a longitudinal sample survey based on scientific sampling principles. Figure 14|2 shows the areas of total wind and water erosion. Although we are making progress in reducing wind erosion, it continues to be a national and international problem. In this chapter we will review the onsite and offsite effects of wind erosion, as well as details of the wind erosion process, prediction, and control measures.

Effects of Wind Erosion
Onsite Effects

The movement of large quantities of aeolian sediment as suspended windblown dust is clearly evident still today and produces dramatic on-site and off-site effects. Wind erosion winnows the finer, more chemically active components of the soil, especially nutrients affecting plant growth (Lyles, 1975; Sterk et al., 1996; Stetler et al., 1994; Van Pelt and Zobeck, 2007; Zobeck and Fryrear, 1986a,b). Other chemical species are lost in disproportionate amounts and unique chemical species such as anthropogenic radioisotopes may be used to estimate historic erosion rates in affected soils (Van Pelt et al., 2007). In addition to soil fertility degradation, the disproportionate loss of soil organic carbon (Van Pelt and Zobeck, 2007) and soil fines may affect soil water infiltration and holding capacity, further affecting soil productivity in semiarid regions.

In addition to soil loss from valuable agronomic systems and fragile natural ecosystems, wind erosion creates several other problems of great economic impact. In source fields, moving soil particles sandblast crop plants and can seriously damage a seedling stand (Armbrust, 1968; Fryrear and Downes, 1975; Skidmore, 1966). This damage often results in replanting decisions for producers (Fryrear, 1973). For example, cotton (*Gossypium hirsutum* L.) lint and kenaf (*Hibiscus cannibinus* L.) yields were reduced 40% and sorghum [*Sorghum bicolor* (L.) Moench] yields were reduced up to 58% in a study of a severely wind-eroded field in west Texas (Zobeck and Bilbro, 2001). In addition, for certain crops and certain growth stages, sandblast injury may result in increased rates of growth in surviving plants (Fig. 14|3; Baker, 2007). According to Farmer (1993), deposition of wind-blown soils on crops decreases their yield and hinders their processing.

Visibility reductions that may happen suddenly can result in a hazard to transportation and commerce on highways close to source fields. Dust storms often reduce visibility to less than 10 meters, causing numerous traffic accidents and deaths in developed countries. Numerous accidents have been attributed directly to wind-driven sand and dust (Skidmore, 1994). In one dust storm near Lubbock, TX in June 2006, 21 vehicles were involved in six different accidents sending 23 people to local hospitals, with one death reported (Blackburn, 2006).

Deposition of wind-driven sand along field margins, especially along weedy fence lines and in drainage ditches, results in costly, recurring maintenance tasks for landowners and government authorities (Fig. 14|4). Recent research indicates that most wind-eroded soil is deposited very close to the source field (Hagen et al., 2007). Wind-eroded soil that is not deposited along field margins enters the suspension mode and may be lofted tens to thousands of meters in altitude in the turbulent boundary layer (Gillette et al., 1997; Chen and Fryrear, 1996; Zobeck and Van Pelt, 2006). Dust that leaves the field and is transported significant distances is termed *fugitive dust*.

Offsite Effects

Fugitive dust impacts environmental, animal, and human health, as well as industry, transportation, and commerce for tens to hundreds of kilometers downwind. The

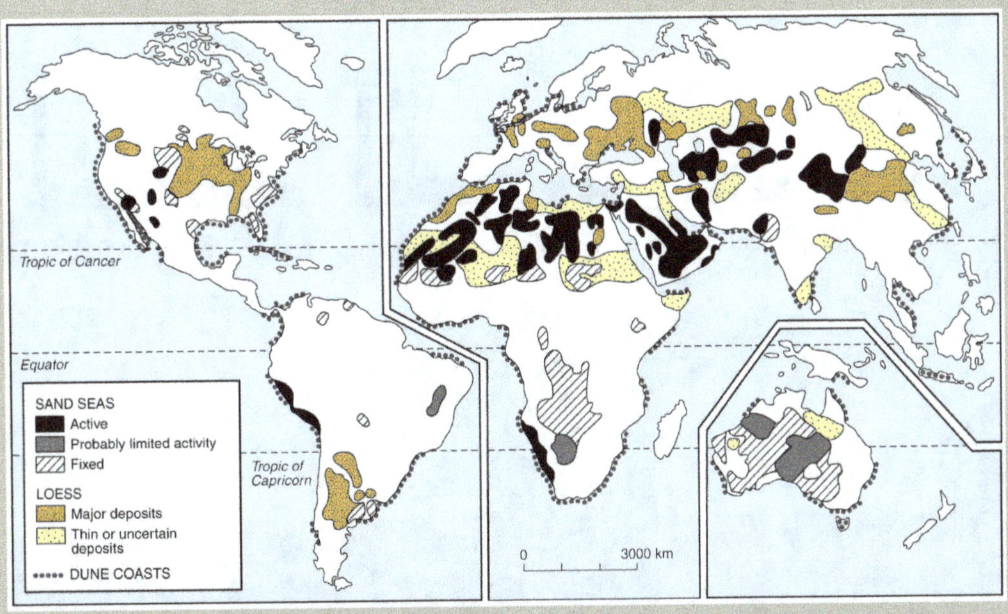

Fig. 14|1. Location of sand seas, loess, and dune coast deposits (with permission from Thomas and Wiggs, 2008).

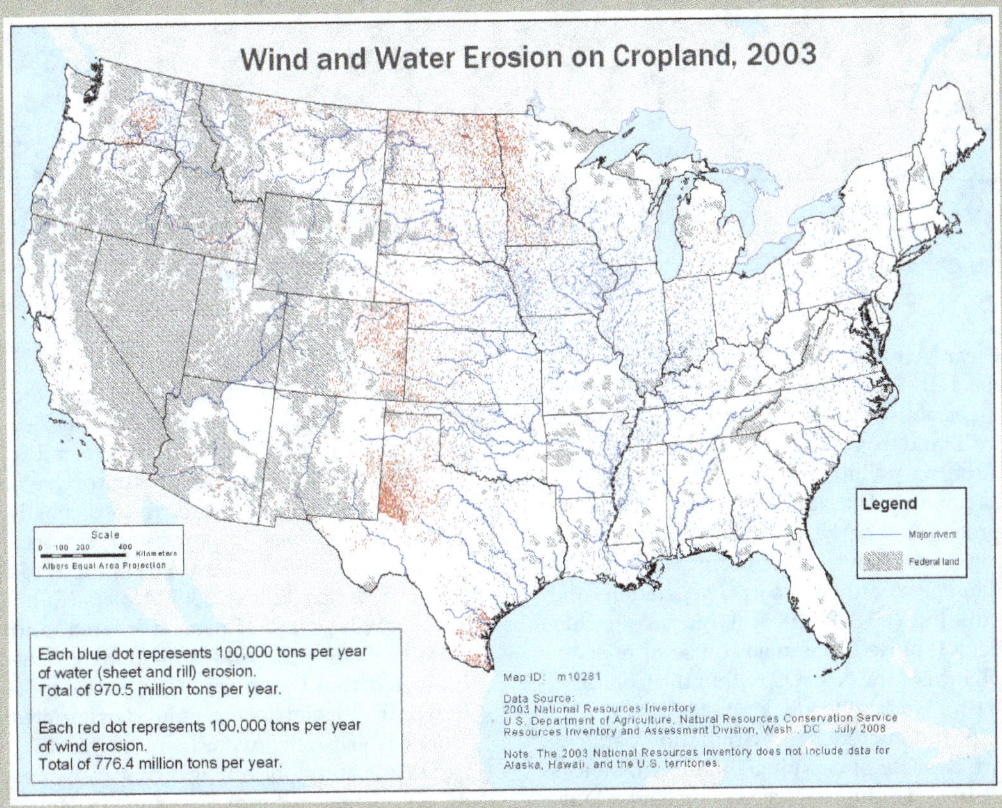

Fig. 14|2. Estimated average annual wind and water erosion on cropland in the United States estimated in the 2003 National Resources Inventory (USDA-NRCS, 2007).

Fig. 14|3. Examples of cotton plant damage after exposure to sand abrasion for 0, 5, 10, 20, 30, and 40 min, left to right (from Fig. 1 in Baker, 2007).

Fig. 14|4. Drainage ditch filled with sand following a severe dust storm in west Texas.

Clean Air Act, amended in 1990, required the U.S. Environmental Protection Agency to establish National Ambient Air Quality Standards (NAAQS) and set limits on airborne pollutants, including fine particulate matter. The standards were designed to protect public health and welfare, including protection against decreased visibility, damage to animals, corps, vegetation, and building (USEPA, 2008). Wind erosion has been reported as a major cause of noncompliance of the NAAQS within the Columbia Basin, Washington (Saxton, 1995).

Wind currents and circulation patterns are capable of carrying smaller diameters of fugitive dust between continents. Dust from the Saharan Desert in Africa has been documented to have fallen in Europe (Goudie, 1978), South America (Talbot et al., 1990), the Caribbean Sea (Delany et al., 1967), the North Atlantic Ocean (Prospero, 1996), and to the interior of North America, a distance of more than 9000 km from the source region (Gatz and Prospero, 1996). Similarly, dust from the deserts of northern China has been documented in Korea (Chung et al., 2003), Japan (Lee et al., 2003), North America (Shao, 2000), Alaska (Rahn et al., 1981), and Hawaii (Braaten and Cahill, 1986). Mineralogical analysis has indicated that the majority of dust deposited in the glaciers of Greenland originates from eastern Asia (Svensson et al., 2000).

Dust is an important agent for transporting soil parent material (Gile and Grossman, 1979; Reynolds et al., 2006), plant nutrients, trace metals (Van Pelt and Zobeck, 2007), soil biota (Delany et al., 1967), and toxic anthropogenics (Larney et al., 1999) between ecosystems and watersheds. The great loess deposits in various areas of the world (Fig. 14|1) are from aeolian deposition (Tsoar and Pye, 1987) and deposition of lesser amounts of aeolian sediments may affect the properties of soils weathered from bedrock or fluvial sediments (Rabenhorst et al., 1984). The mineralogy, chemical, and biotic characteristics of soil dust are determined by the surface from which it is entrained (Reheis and Kihl, 1995). Microbiological exudates such as fatty-acid methyl esters (Kennedy, 1998) or enzyme activities and arylsulfatase proteins (Acosta-Martinez and Zobeck, 2004) may be used to identify the probable source area of a given dust outbreak. Pathogenic microbes may also be transported on dust and affect distant ecosystems and human health (Leathers, 1981). Some agronomic ecosystems depend on the inputs from deposited dust (Sterk et al., 1996). Iron fertilization and resultant blooms of algae in the oceans has been documented and may result in increased carbon dioxide sink and photosynthetic production of oxygen (Mackie and Hunter, 2007). However, deposition of nutrient-rich dust in freshwater lakes and over terrestrial watersheds may result in undesirable algal blooms in freshwater bodies.

During transport, dust may enter into numerous chemical reactions and catalyze reactions of anthropogenic particulates in the atmosphere. Calcium carbonate is a common soil constituent in semiarid and arid regions and thus is a common constituent of soil dust (Gile and Grossman, 1979). Calcium carbonate originating from the Owens Lake dry lakebed is partially converted to calcium sulfate before it is deposited in the Los Angeles basin (Reheis and Kihl, 1995). Acid rain is a problem worldwide, but is partially ameliorated in regions where carbonate-rich dust interacts with the acid species in the clouds or in the soils of the affected watershed (Litaor, 1987; Trochkine et al., 2003). In regions distant from anthropogenic oxides of sulfur and nitrogen, the carbonates in dust may make normally mildly acidic rainwater alkaline (Zhang et al., 2003).

Calcium carbonate particles that have been modified by reactions with atmospheric acids are more hygroscopic and tend to form more effective condensation nuclei (Krueger et al., 2004). These wetted soil aerosols may attract and absorb gases and other aerosols from the adjacent atmosphere, allowing for rainout and effectively cleaning the atmosphere. Humic acid coatings on soil dust are highly attractive to hydrophobic organic species in the atmosphere (Chiou, 1989). The catalytic effect of humic acid coated soil dust in the atmosphere is enhanced at high relative humidities as the hydrophobic nature of the humic acid is overcome and the particles adsorb a thin layer of water (Brooks et al., 2004). Additionally, Miller et al. (1989) showed humus in the presence of sunlight to be an effective catalyst, creating highly reactive free radicals of oxygen that are instrumental in the oxidation of organic pollutants.

Wind Erosion Mechanics

Wind erosion occurs as the wind interacts with the soil surface to cause detachment (termed *entrainment*), transport, and finally deposition of soil particles. Detachment occurs as the wind exerts drag and lift forces to overcome the gravitational and cohesive forces that hold particles to the soil surface (Toy et al., 2002). Detachment also occurs as rolling or bouncing particles cause other particles to be released by impacts or abrasion (Fig. 14|5). The wind velocity at which sediment begins to move is termed the *threshold wind velocity*. Winds are considered erosive when they reach a speed of about 6 m s^{-1} (13 mph) at 0.3 m (1 ft) above the soil

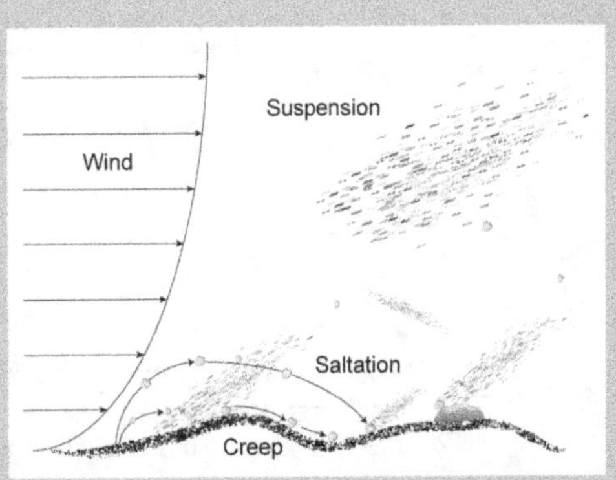

Fig. 14|5. Modes of particle transport due to the force of the wind on the surface. Length of wind arrows indicates relative strength of the wind.

surface or about 8 m s^{-1} (18 mph) at 9 m (30 ft) above the surface (USDA-NRCS, 2002).

After the wind exceeds the threshold wind velocity, soil particles or small stable aggregates begin to move in three primary ways, or modes, of transport: creep, saltation, and suspension (Fig. 14|5). Particles or soil aggregates in creep mode are about 0.5 to 1 mm in diameter and roll or scoot along the soil surface, propelled by the direct force of the wind or when bouncing (termed *saltating*) particles strike them. Individual saltating particles or soil aggregates are about 0.1 to 0.5 mm in diameter and move by bouncing along the soil surface, rarely exceeding heights of a few meters. These particles may be directly lifted off the soil surface by the force of the wind or be ejected from the soil surface as other saltating particles dislodge them on impact. As these saltating particles bounce along the soil surface they dislodge even more saltating particles, creating an avalanching effect. Saltating sediment may cause abrasion as particles bounce or collide with other sediment or the crusted soil surface, or they may become lodged in the soil surface.

Suspended sediment is generally less than 0.1 mm in diameter. Although some suspension-sized sediment is present in the soil, it is less susceptible to direct entrainment by the force of the wind. Suspended material is mainly created as saltating sediment abrades larger aggregates or strikes the soil surface in a process similar to sandblasting. Saltating particles collide with the surface with a force that is a function of their mass and velocity. However, recent studies in the Columbia Basin of Washington suggest direct emission of suspension is possible in some silty soils (Kjelgaard et al., 2004). Although particles less than 0.1 mm can be suspended, particles larger than about 0.02 mm diameter are unlikely to travel greater than 30 km from the source, settling back to the surface quite quickly when the turbulence associated with strong winds abates (Pye, 1987). The finer suspended sediments are carried up by turbulent eddies and, as mentioned above, may travel thousands of kilometers before settling back to the surface. In contrast, creep and saltating sediments are usually redeposited within or near the source field.

Wind

As the wind interacts with the Earth's surface, the surface exerts a drag on the wind, reducing the wind velocity nearer the surface (Fig. 14|5). During strong wind events, the boundary layer near the surface is usually statically neutral and the vertical profile of wind speed may be described by a well-known semilogarthmic equation of the form:

$$u(z) = \frac{u^*}{k} \ln\left(\frac{z}{z_0}\right) \tag{1}$$

where $u(z)$ is the wind speed at height z, u^* is the friction (or shear) velocity, k is the von Kármán constant (0.4), and z_0 is the aerodynamic roughness height. The friction velocity is a measure of the shear stress on the surface and has been used in predictive models as the driving force for wind erosion. It is indicative of the atmospheric turbulence and is proportional to the slope of the wind velocity profile when the height is represented on a logarithmic scale (Fig. 14|6). The aerodynamic roughness height refers to the theoretical height at which the wind speed near the surface falls to zero and depends on the characteristics of the surface. Numerous studies have found that z_0 is approximately equal to 1/30 the height of the roughness

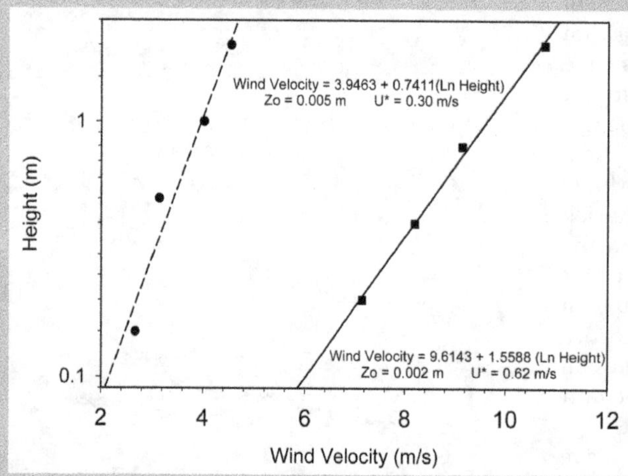

Fig. 14|6. Wind profiles above two soil surfaces. U* is the friction velocity and Zo is the aerodynamic roughness height.

elements. In vegetated surfaces, z_0 may vary with wind speed as the vegetation bends in the wind. In effect, the aerodynamic roughness is a representation of the capacity of the surface for absorbing momentum and is also an important quantity in wind erosion studies (Shao, 2000). In practice, if we plot wind speed on the y axis and the logarithm of the height on the x axis, we normally obtain a straight line with the slope $u*/k$ and the intercept $(u*/k)\ln(z_0)$.

Although the wind provides the energy to drive wind erosion, the characteristics and condition of the soil surface will ultimately control whether or not erosion occurs and its extent. In the next section we will explore how soil surface conditions affect wind erosion.

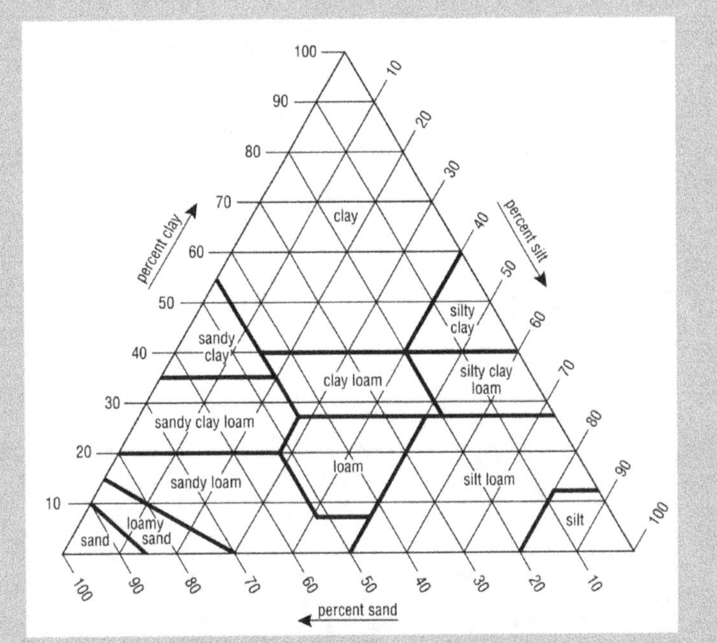

Fig. 14|7. Soil textural triangle.

Soil Surface Conditions

Soils have been described as having intrinsic or inherent soil properties that change very slowly and dynamic or temporal properties that vary through time. Dynamic soil properties may change very rapidly in response to weather factors, tillage, or other management and include properties such as bulk density and dry aggregate size distribution. Examples of inherent soil properties are soil texture, organic matter content, and mineralogy.

Soil Texture

The USDA-NRCS has established soil textural classes (Fig. 14|7) based on specific proportions of sand, silt, and clay contained in a sample (USDA-NRCS, 1993). Soil texture is one of the primary soil properties affecting soil susceptibility to wind erosion (also called wind erodibility). The USDA-NRCS has classified the wind erodibility of soils according to the soil texture and calcium carbonate content (Table 14|1). In general,

coarse soils such as sands are more erodible than finer-textured soils such as clay loam soils. Calcareous soils tend to have a higher erodibility than noncalcareous soils. Calcareous soils contain enough calcium carbonate to cause effervescence with the application of dilute acid. Soil texture and calcium carbonate content are inherent soil properties that change very slowly with time. Even so, this does not mean that it is impossible to change the texture. For example, soil scientists in west Texas have found that the surfaces of soils that have undergone wind erosion for a long period of time are now coarser than when originally mapped several decades ago. The relative increase in sand content over time has been caused by the winnowing of finer particles out of the soil by wind erosion.

Soil Moisture

Surface soil moisture content is an extremely important variable controlling both the entrainment (erodibility) and transport of sediment by wind (Nickling, 1994). Wind tunnel experiments have shown that soil moisture content clearly affects the wind threshold friction velocity at which

Table 14|1. Relation of soil texture and soil erodibility.†

Soil texture‡	Predominant soil texture class of surface layer	Wind erodibility group (WEG)	Soil erodibility index (I)§
			Mg ha⁻¹ yr⁻¹
C	Very fine sand, fine sand, sand, or coarse sand	1	694¶
			560
			493
			403
			358
C	Loamy very fine sand, loamy fine sand, loamy sand, loamy coarse sand, or sapric organic soil materials	2	300
C	Very fine sandy loam, fine sandy loam, sandy loam, or coarse sandy loam	3	193
F	Clay, silty clay, noncalcareous clay loam, or silty clay loam with more than 35% clay	4	193
M	Calcareous loam and silt loam or calcareous clay loam and silty clay loam	4L	193
M	Noncalcareous loam and silt loam with less than 20% clay, or sandy clay loam, sandy clay, and hemic organic soil materials	5	125
M	Noncalcareous loam and silt loam with more than 20% clay, or noncalcareous clay loam with less than 35% clay	6	108
M	Silt, noncalcareous silty clay loam with less than 35% clay, and fibric organic soil material	7	85
–	Soils not susceptible to wind erosion due to coarse surface fragments or wetness	8	–

† Adapted from the USDA National Agronomy Manual (USDA-NRCS, 2002).

‡ Soil texture: C, coarse; M, medium; F, fine.

§ The erodibility index is based on the relationship of dry soil aggregates greater than 0.84 mm to potential soil erosion.

¶ The I factors for WEG1 vary from 358 Mg ha⁻¹ yr⁻¹ for coarse sands to 694 Mg ha⁻¹ yr⁻¹ for very fine sands. For coarse sands gravel, use a low figure. For very fine sand without gravel, use a higher value. When unsure, use an I value of 493 Mg ha⁻¹ yr⁻¹.

particles begin to move (Belly, 1964; Bisal and Hsieh, 1966; Chepil, 1956). Early studies by Chepil (1956) suggested that soil erodibility by wind was about the same for soil that was oven-dried or air-dried when moisture content did not exceed one-third of the 15-atmosphere percentage (–1500 J kg⁻¹ matric potential). Beyond this range of moisture a distinct decrease in erodibility was observed. Above about 5% gravimetric moisture content, sand-sized material is inherently resistant to entrainment by most natural winds (Nickling, 1994). More recent studies have related the change in threshold friction velocity with soil water tension, derived from capillary force equations that consider the capillary forces developed at interparticle contacts surrounded by water (McKenna-Neuman and Nickling, 1989). The erodibility of soil by wind is so sensitive to the effects of moisture that even differences in relative air humidity modify the particle threshold wind velocity (McKenna-Neuman and Sanderson, 2008; Ravi et al. (2006a,b). A wind tunnel study of sand, sandy loam, and clay soils showed that the threshold friction velocity decreases with increasing values of relative humidity for values between 40 and 65%, while above and below this range the threshold fiction velocity increases with air humidity (Ravi et al., 2006b).

Surface Roughness

The aerodynamic roughness length is determined from the wind profile and is the height above the surface at which the mean wind speed becomes zero. This empirically derived value is related to roughness elements on the soil surface (e.g., clods, rocks, vegetation), as well as the surface microtopography or microrelief. The effects of vegetation on roughness will be discussed later in the chapter. Soil surface microrelief is a dynamic soil property that may change rapidly due to management or weather factors.

In tilled agricultural soils, tillage produces an oriented roughness (or ridges) parallel to the direction of tillage caused by pulling the tillage tool through the soil. In addition, a random roughness is produced by the random orientation of soil aggregates or clods on the surface. Research has shown that wind erosion is sensitive to the effects of both random and oriented roughness (Fryrear, 1984). The roughness caused by tillage will modify the wind profile to change z_0 and also protect the soil surface from the effects of abrading particles. In general, the aerodynamic roughness length increases as the size of the clods or ridges increases. The amount of change is related to the size and spacing of the ridges and clods. In addition, the effects of the roughness produced by tillage on wind erosion will depend on wind direction when ridges are present. When erosive winds blow perpendicular to a bare soil surface, ridges will act to physically protect a fraction of the soil surface, as illustrated in Fig. 14|8. The ridges have little effect when the wind blows in the direction parallel to tillage. A microrelief index called the cumulative shelter angle distribution (CSAD) is used to estimate the fraction of the tilled soil surface susceptible to abrasion by saltating grains (Potter and Zobeck, 1990). The CSAD has been shown to be sensitive to tillage tools, rainfall, and wind direction.

Aggregate Properties

Soil aggregates or peds are naturally occurring structural units composed of primary soil particles (USDA-NRCS, 1993). They are formed as a consequence of natural soil development. Cohesion within the aggregates is greater than the cohesion among

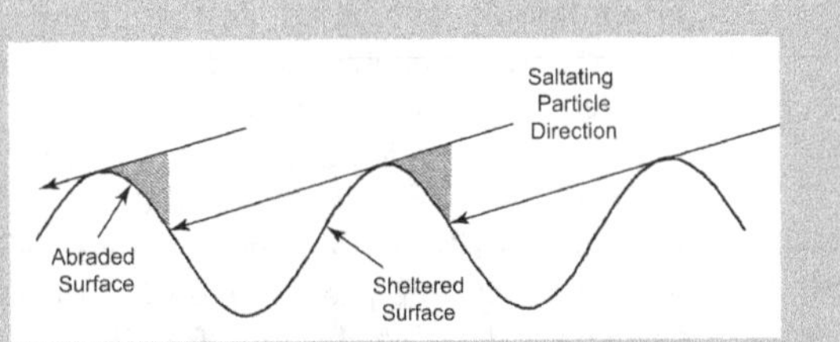

Fig. 14|8. Schematic representation of a ridged field. Part of the field is sheltered from abrasion by saltating grains.

adjacent aggregates. Thus, they are formed when stress causes the soil to rupture under predetermined planes of weakness. Soil clods are similar to aggregates, but soil forming processes have exerted very little or no control on the boundaries of clods. They are produced by tillage or other soil manipulations that cause the soil to rupture and break apart, and they may include pieces of aggregates. Following tillage, the soil surface typically contains clods and aggregates with a wide range of sizes.

Dry aggregate size distribution refers to the relative amounts of air-dry aggregates and clods, on a mass basis by size class, present on the soil surface (Zobeck, 1991b). A rotary sieve is used to determine the dry aggregate size distribution (Chepil, 1962). Wind erosion is related to the amount of aggregates >0.84 mm in diameter, called *nonerodible aggregates*. Table 14|2 lists the soil wind erodibility, also called the *I* value, as a function of the percentage of nonerodible aggregates. The dry aggregate size distribution

Table 14|2. Soil wind erodibility as determined by percentage of nonerodible soil (>0.84 mm diameter).†

| Tens‡ | Units | | | | | | | | | |
	0	1	2	3	4	5	6	7	8	9
					Mg ha⁻¹ yr⁻¹					
0	–	694	560	493	437	403	381	358	336	314
10	300	293	287	280	271	262	253	244	237	228
20	220	213	206	202	197	193	186	181	177	170
30	166	161	159	155	150	146	141	139	134	130
40	125	121	116	114	112	108	105	101	96	92
50	85	81	74	69	65	60	56	54	52	49
60	47	45	43	40	38	36	36	34	31	29
70	27	25	22	18	16	13	9	7	7	4
80	4	–	–	–	–	–	–	–	–	–

† Adapted from the USDA National Agronomy Manual (USDA-NRCS, 2002).

‡ Columns and rows represent the percentage of nonerodible aggregates. For example, to find 33% go to the nonerodible aggregates tens row at 30 and units column at 3 (30 + 3 = 33) to find 155 Mg ha⁻¹ yr⁻¹.

is commonly expressed as the geometric mean and geometric standard deviation derived from a lognormal distribution or as the shape and scale parameters of a Weibull distribution (Zobeck et al., 2003a). The Weibull distribution has been shown to be more accurate and precise in describing dry aggregate size distributions for tilled soils (Zobeck et al., 2003a).

Dry aggregate stability refers to the resistance of soil aggregates to breakdown from physical forces. It is a measure of the bonding strength of the bonding agents within aggregates (Skidmore and Powers, 1982). The physical forces causing aggregate breakdown may occur as a result of tillage but may also include the physical forces caused by the impact of saltating grains. The dry aggregate stability of bulk samples of tilled soils has been determined by repeated sieving using a rotary sieve (Chepil, 1958). In this case, the dry aggregate stability is calculated as the weight of the particles or aggregates greater than 0.84 mm in diameter after the second sieving divided by the weight of the particles or aggregates greater than 0.84 mm in diameter after the first sieving. The stability of individual soil aggregates is determined by measuring the force needed crush the aggregate to a known endpoint (Hagen et al., 1995). In this case, the energy needed to crush an aggregate approximately 15 mm in diameter is called the *crushing energy*. The stability of dry aggregates has been shown to be a dominant predictor of soil erosion from surface abrasion (Hagen, 1991a). Skidmore and Layton (1992) found that aggregate clay content and water content at the wilt point (-1500 J kg^{-1} matric potential) are good predictors of mean aggregate stability.

Surface Crusting

Surface crusting refers to a relatively thin consolidated soil surface layer or seal that is more compact and cohesive than the material immediately below it. When crusts are formed, particles are bound together and less susceptible to abrasion by blowing soils than the less stable material below the crust. Under natural conditions, crusts form from a variety of physical, chemical, and biological processes (Neave and Rayburg, 2007). Details of the interparticle forces contributing to the cohesion of crusts have been described by (Ishizuka et al., 2008). In rangelands, biological cryptogamic crusts may be particularly effective in stabilizing the soil by binding small particles into larger, nonerodible aggregates and protecting the soil from wind erosion (Eldridge and Greene, 1994; Leys and Eldridge, 1998).

In cropland soils, primary tillage acts to mix and loosen the surface. Rainfall is the primary agent that can create the crust or seal after tillage. The effects of rainfall on soil crust development have been studied for many years. Most studies have shown that the strength of the crust to withstand abrading particles, known as *crust stability*, is related to the soil properties and rainfall rate or energy used to create the crust (Zobeck, 1991b). The dislodgement of surface particles was shown to decrease with increasing crust strength in a laboratory study using artificially created crusts (Rice et al., 1996). In a wind tunnel study of 14 crusted soils that included a wide range of soils textures (from loamy sand to clay and one organic soil), loose, unconsolidated soil was on average about 40 times as erodible as crusts created using simulated rainfall at an intensity of 25 mm h^{-1} and 70 times as erodible as crusts created using simulated at an intensity of 64 mm h^{-1} (Zobeck, 1991a). In addition, the crust abrasion was positively correlated with the sand content and cation exchange capacity/clay ratio.

In sandy soils, loose, unconsolidated erodible material may be left exposed on the crust after rainfall. This loose, erodible material (LEM) is highly susceptible to wind erosion. If LEM is not present on a crusted surface, wind erosion generally will not occur. The LEM acts as projectiles or bullets as they bounce or saltate, abrading (sandblasting) the surface. The mass of LEM on crusted soils is affected by inherent soil properties, management, and climatic factors (Zobeck, 1991a). Rainfall simulation studies have shown that the logarithm of LEM was related to sand content, sampling location in relation to tillage ridges, and rainfall (Potter, 1990). Sandy soils tend to have much more LEM on the crust than finer textured soils. In Potter's (1990) rainfall simulation study of five soils ranging in texture from fine sandy loam to clay, the fine sandy loam soil had about 30 times the amount of LEM as the clay soil tested.

Stones, stable nonerodible aggregates, vegetation, and other nonerodible materials will physically protect or armor the soil surface from the direct force of the wind and from abrading sand particles. The effect of this protection is related to the amount of nonerodible material on the soil surface, as described in Fig. 14|9. The soil loss ratio described in Fig. 14|9 is the ratio of the erosion observed for the protected soil divided by the erosion on bare, unprotected soil. Covering the soil with 20% nonerodible material reduced soil loss by 57% and covering the soil with 50% nonerodible material reduced soil loss 95% (Fryrear, 1985).

Wind Erosion Models for Cropland

A wide variety of models have been developed to predict sediment entrainment and transport at scales ranging from small plots to global. A review of many of these models has been presented by Zobeck et al. (2003b). In the U.S., the Wind Erosion Equation (WEQ) (Woodruff and Siddoway, 1965) and the Wind Erosion Prediction System (WEPS) (Hagen, 1991b) have been the principle models used to predict wind erosion on cropland.

Wind Erosion Equation

The wind erosion equation (WEQ) was designed to predict long-term average annual soil erosion by wind based on a specific set of climatic and field conditions. It can also be used to predict erosion for specific time periods when using the appropriate factors in the equation. Details about WEQ are found in the USDA-NRCS *National Agronomy Manual* (USDA-NRCS, 2002). The WEQ is determined using the following equation:

$$E = f(IKCLV) \tag{2}$$

where E is the estimated annual soil loss (mass/unit area/time period), f indicates a nonlinear functional relationship among the variables, I is the soil erodibility index (mass/unit area), K is the soil surface roughness factor, C is the climatic factor, L is the unsheltered distance, and V is the vegetative cover factor.

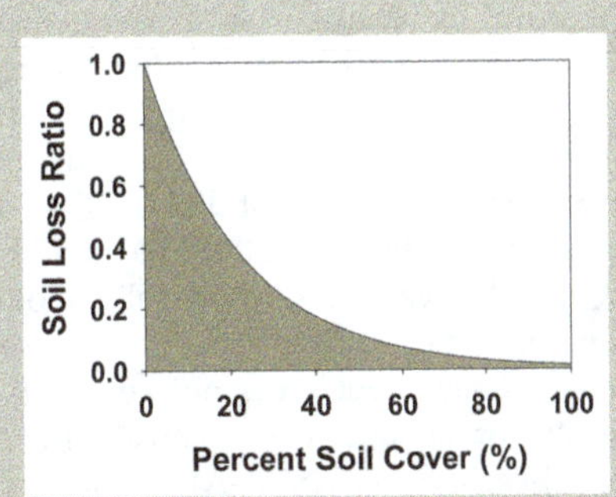

Fig. 14|9. Soil loss ratio as a function of soil cover by nonerodible elements.

In practice, the soil erodibility index (I) is determined by the surface soil texture and assigned a value as indicated in Table 14|1. The I value is the potential annual wind erosion for an isolated, unsheltered, level, wide, bare, smooth, loose, and noncrusted soil at a location where the climatic factor is equal to 100. The other individual factors are then multiplied, after their determination using measured values, nomographs, or other methods, as indicated in Eq. [2]. The K value adjusts the I value for tillage-induced random and oriented roughness. The C value adjusts after consideration of wind speed and potential evapotranspiration. This factor is expressed as a percentage of the C factor for Garden City, KS, which has a C factor of 100. The L factor considers the unprotected distance across the field along the direction of the prevailing wind direction. The V factor considers the kind, amount, and orientation of vegetation on the surface. The USDA-NRCS *National Agronomy Manual* (USDA-NRCS, 2002) has detailed instructions on the use of WEQ, and a spreadsheet version of WEQ for use in the United States was provided by (Sporcic et al., 1998). The wind erosion equation is currently used by USDA-NRCS for conservation planning and assessing soil erosion by wind for the USDA-NRCS National Resources Inventory.

Wind Erosion Prediction System

Although the WEQ has been used successfully for many years, it has several limitations. The WEQ predicts average annual erosion by summing erosion predicted for specific periods of time (i.e., 2-wk periods) but does not predict for daily events. The WEQ does not account for changes in many temporal surface features affecting wind erosion, such as changes in surface roughness and aggregation, and it does not account for the two-dimensional nature of fields. The climatic factor of WEQ is related to a standard location and does not stochastically model weather conditions. Many details in crop and soil management were not considered in the development of WEQ. In an attempt to resolve these issues, USDA-ARS embarked on the development of a new wind erosion prediction technology, the Wind Erosion Prediction System (WEPS).

The WEPS is a process-based, daily time-step, computer model that predicts soil erosion by wind by simulating the fundamental processes involved (Hagen, 1991b). The current version of WEPS (1.0) is designed to provide the user with a simple tool for inputting initial field and management conditions, calculating soil loss, and displaying simple or detailed outputs for designing erosion control systems. The WEPS is the computer implementation of a science model that simulates the processes involved with the wind erosion process. The science model is composed of the following major submodels (USDA-ARS, 1996):

- Weather—Uses historical statistical information of a wide variety of meteorological variables with stochastic techniques to determine the likelihood of various variables needed to drive processes in other submodels.

- Hydrology—Uses inputs from other submodels to compute water content in the various soil layers and at the soil–atmosphere interface throughout the simulation period.

- Management—Models the primary human-initiated actions that affect the susceptibility to wind erosion. These include all cultural practices applied to the field, such as tillage, planting, harvesting, and irrigation.

- Soil—Simulates soil temporal properties that affect the susceptibility to wind erosion on a daily basis in response to driving processes such as rainfall, tillage, and others.

- Crop—Calculates daily production of masses of roots, leaves, stems, reproductive organs, and leaf and stem areas.

- Decomposition—Simulates the decrease in crop residue biomass due to microbial activity and includes standing, surface, buried, and root biomass pools.

- Erosion—Uses parameters supplied by other submodels to simulate the process of soil movement. The submodel periodically updates any changes in the soil surface caused by soil movement and outputs estimates of soil loss or deposition from the simulation region.

In practice, users define a simulation region (field) and input management and soils information using a graphical user interface. The program includes a variety of databases from which information about barriers, soils, management, crop and decomposition, and climate is extracted. For example, after the location of the field is identified, the interface selects the appropriate weather station for which historical data are used to simulate weather parameters. The soils data are selected from a soils database supplied by the NRCS Soil Survey. WEPS provides a wide variety of output, including soil loss as saltation/creep, suspension, and particulate matter less than 10 μm (PM_{10}). By varying inputs, particularly those related to management, the user can easily evaluate various erosion control alternatives.

Initial tests of WEPS indicate that it produces reasonably good estimates of soil loss due to wind when compared with measured results. For example, soil loss measurements from 46 storms events from eroding fields in six states had reasonable agreement (R^2 = 0.71) with erosion simulated by WEPS (Hagen, 2004). A comparison of WEPS with measured soil loss in Germany showed excellent agreement ($R^2 > 0.9$) between measured and simulated soil loss (Funk et al., 2004). The USDA-NRCS is currently testing the WEPS for application in the United States as a replacement for WEQ. Detailed information about obtaining WEPS is available at http://www.weru.ksu.edu/weps/wepshome.html (verified 7 Dec. 2010).

Sensitivity analyses have shown that predictive models are very sensitive to soil surface conditions. The WEQ is particularly sensitive to soil texture, surface roughness, and residue. The sensitivity of WEQ to texture is so great that investigators have suggested using texture-adjusted I factors to calibrate the model for local soil conditions (Van Pelt and Zobeck, 2004). The WEPS has been shown to be sensitive to soil surface conditions, including soil surface wetness, dry aggregate stability, oriented roughness, and residue management (Hagen et al., 1999). In a separate evaluation of the WEPS for 46 wind events at six North American locations, Hagen (2004) noted that model inaccuracies may be due to average soil parameter values being used or time-dependent wind erosion-induced changes in soil erodibility after the date of soil sampling. In a very detailed sensitivity analysis, Feng and Sharratt (2005) reported that the WEPS was most sensitive to changes in biomass flat cover, near-surface soil water content, ridge height, and other management-related parameters, including crust cover and random roughness. The sensitivity of these models to management-related parameters is indicative of the profound effects that land management has on wind erosion.

Management Effects on Wind Erosion

Native Vegetation Communities

It is widely held that land management has a profound effect on erosion of the soil surface by wind. In humid and subhumid climates, the canopy of native vegetation communities is generally sufficient to prevent erosive wind energy from reaching the soil surface. For most forest ecosystems, native grasslands, and managed pastures, the literature is largely lacking reports of observed wind erosion. In semiarid and arid climates, however, native plant communities do not fully protect the soil surface from the erosive forces of wind. Semiarid ecosystems including grasslands, shrubland, savanna, woodland, and forests are all susceptible to wind erosion, especially when disturbed.

In a series of studies of several communities ranging from relatively undisturbed ponderosa pine (*Pinus ponderosa* P. Lawson & C. Lawson) forests to desert shrublands dominated by mesquite (*Prosopis glandulosa* Torr.). Breshears et al. (2008) measured sediment transport rates ranging from 0.17 to 27.4 g m^{-2} d^{-1}, respectively. These sites ranged from 75 to 0% woody canopy cover and from 98 to 38% herbaceous ground cover. For disturbed sites in the same or similar locations,

they measured sediment transport rates of 1.1 to 6002 g m^{-2} d^{-1}, with total canopy coverage of from 72 to 0%, respectively. They further concluded that sediment transport may be inherently greater for shrublands and that sparse shrublands have a greater influence on the wind profile by channeling the wind and increasing turbulence. Forests, dense shrublands, and grasslands protect the soil surface more evenly and tend to produce a "skimming flow" profile by the wind. Natural areas devoid of vegetation due to ephemeral flooding or water diversion are also susceptible to wind erosion (Pelletier, 2006). Ephemeral playa lakes, when dry, constitute locally and regionally important areas of dust generation (Prospero et al., 2002).

The effects of vegetation cover on wind erosion and soil loss have been investigated in a desert grassland of southern New Mexico (Li et al., 2007). The authors concluded that as lateral cover, a function of plant number density and vertical dimension, drops below 9%, wind erosion increases dramatically. Anthropogenic disturbance of desert grasslands by mechanical means or overgrazing often results in a sparse shrubland subclimax. The process begins by exposing the soil surface to wind (Sharifi et al., 1999; Liu and Wang, 2007) and becomes exacerbated by sandblasting of the remaining vegetation (Okin et al., 2001). The result is an anisotropic pattern of shrub vegetation (McGlynn and Okin, 2006), which leads to further degradation of the landscape by erosion of the soil surface in the bare alleys (Okin and Gillette, 2001) and deposition of the finer, more nutrient-rich samples in and under the canopies of the shrub patches. This redistribution of soil fines leads to heterogeneity of soil texture and infiltration rates that further influences the distribution of vegetation (Ravi et al., 2007). The recognition of the dependence of wind erosion on the distribution of vegetation in these disturbed communities has led to the development of a stochastically based model of wind erosion in sparsely vegetated communities (Okin, 2005, 2008).

Burning of native vegetation communities also increases the susceptibility of the soil to erosion by wind and subsequent degradation. Burning increases the erosion hazard primarily by removing the vegetation, which both slows the wind near the surface and prevents the re-entrainment of deposited sediments (Stout, 2006). The heat of the fire may also alter physical properties of the soil and affect the soil stability (Whicker et al., 2008). Fire has been credited with increasing the water repellency of soils and thus increasing their erosion susceptibility by maintaining a dry surface and modifying the surface soil threshold friction velocity (Ravi et al., 2006a). The effects of fire that increase the susceptibility of the soil to wind erosion may be short-lived, however, as vegetation grows back and protects the surface.

Cropped Ground

The development of land for production agriculture is often accomplished by total native vegetation removal and at least some smoothing of the land surface, leading to the increased susceptibility of the soil to wind erosion. Conventional cropland tillage practices that lead to the increased susceptibility of the surface to wind erosion and dust emissions include plowing, delaying primary tillage, leveling beds, planting, weeding, fertilizing, cutting and baling, spraying, and burning. Management methods that are used to control wind erosion on cropland include planting windbreaks to alter wind flow patterns, strip cropping, planting cover crops before or after low residue crops, cross-wind strips, vegetation barriers, retaining plant residue after harvest, stabilizing the surfaces using water or applied chemicals, and tilling the field to bury erodible particles, to increase roughness by increasing the percentage of nonerodible aggregates on the surface, and to create bed patterns perpendicular to the predominant winds (Nordstrom and Hotta, 2004; USDA-NRCS, 2009).

Windbreaks and shelterbelts have been used to decrease the erosive force of the wind in many local settings. They are typically rows of trees and shrubs planted along the margins of the field or farmstead they are intended to protect, but they may also be fences, rock walls, or earth berms. Such barriers effectively decrease the wind speed for a distance of about 10 to 15 times their height downwind and about three times their height upwind (Oke, 1987). Due to the limitations of tree growth in many regions, especially semiarid regions, this distance

rarely exceeds a few hundred meters downwind (Vigiak et al., 2003). Windbreaks and shelterbelts are not as common as they once were. Although there were approximately 65,000 km of them planted in the Great Plains of North America by the 1960s (Griffith, 1976), by the 1970s, many were dying or were being removed (Sorenson and Marotz, 1977).

Maintaining crop residues on the cropped ground is perhaps the most effective management solution for controlling wind erosion. The value of crop residues for controlling wind erosion has been recognized for at least six decades (Chepil, 1944). Residue protects the ground by offering elements that prevent saltating particles from cascading and by increasing the roughness height z_0 (Eq. [1]). The WEQ and other predictive models treat all crop residues, whether standing or flat on the ground as protection equivalent to flat small grain residues (Woodruff and Siddoway, 1965; Bilbro and Fryrear, 1985). Standing residues and growing crops provide greater protection than flat residues because they absorb much of the shear stress in the boundary layer (Skidmore, 1994). This displaces the effective roughness height, z_0, by a zero plane displacement height, d, which is a factor of the height, density, and stiffness of the vegetation (Oke, 1987). It may be approximated for a wide range of crops and trees by:

$$d = \frac{2}{3}h \qquad [3]$$

where h is the mean height of the standing crop or residue. The displacement height d is used to modify the logarithmic wind profile Eq. [1] over a crop under adiabatic atmospheric stability (Monteith, 1973):

$$u_z = \frac{u^*}{k}\ln\left(\frac{z-d}{z_0}\right) \qquad [4]$$

The effects of vegetation or other nonerodible material on the soil surface on soil loss is estimated using the soil loss ratio, an index calculated by dividing the amount of soil loss from a residue-protected soil surface by the loss from a similar bare surface (Fig. 14|9). The soil loss ratio decreases rapidly from 1.0 for a bare unprotected surface to a value of approximately 0.2, an inflection point characterized by 40% soil cover (Fryrear, 1985).

Another description of plant canopy or residue used by predictive models for standing vegetation is the plant silhouette through which the wind must pass. Bilbro and Fryrear (1985) observed a strong relationship between plant silhouette and the soil loss ratio. However, very sparse residue or other roughness element cover may actually increase soil loss by compressing airflow and creating localized super-critical wind velocities that exceed threshold (Sterk, 2000).

Tillage of cropped land tends to bury crop residues and thus diminish the protection provided. Tillage may also weaken aggregate stability by decreasing the soil organic carbon content (Fenton et al., 1999; Six et al., 1999) and thus increase the soil's intrinsic erodibility. Tillage is used to protect the soil surface by increasing the nonerodible fraction and the roughness of the surface (Fryrear, 1984). Oriented roughness elements, such as ridges created using tillage implements such as a lister, are very effective when oriented at angles greater than 13 degrees from the direction of the incident wind (Hagen and Armbrust, 1992). Nonoriented or random roughness is also used to lessen wind erosion by creating numerous nonerodible elements that provide shelter from abrading sand grains (Potter et al., 1990).

Best Management Practices for Controlling Wind Erosion

The best management practice (BMP) to prevent erosion is to prevent contact of wind with the soil surface by maintaining an effective cover of residue, such as a cover crop or carefully managed stubble. The emergence of no-till and conservation tillage practices has resulted in more effective post-harvest standing and flat residue over cropped ground. Advances in harvest equipment, such as finger or stripper headers on small grain combines, have also led to improvements in the post-harvest heights of standing residue. In semiarid regions that represent marginal dryland farming regions and with certain locally important crops such as cotton or sunflowers (*Helianthus annuus* L.), insufficient silhouette or flat residue may fail to protect

the soil. In addition, cultural practices in some crops, such as tillage for insect control in cotton, can contribute to loss of standing and flat residue.

In areas where rainfall is not limiting, such as the Red River Valley in Minnesota, sugar beets are protected with a low rate of spring barley (40 kg/ha [0.75 bu/ac]) before planting. The low rate of cover is killed with herbicides after the beets are established (M. Sporcic, personal communication, 2009).

On bare soils or soils with limited crop residues, tillage remains a common BMP to prevent erosion. Raising beds perpendicular to the prevailing wind direction increases the aerodynamic roughness and provides regularly spaced roughness elements offering shelter angle to prevent cascading saltation. By creating a surface dominated by nonerodible aggregates, a random roughness is formed that offers the same protective shelter angle to prevent cascading saltation. In fragile soils with low dry aggregate stability, erosion may start in localized areas of the field or at the downwind end of a long, frequently traveled, unpaved road. In such locations, it may be necessary to use a snow fence or other barrier to encourage deposition and discourage saltation. Intense rainfall on soils with low wet aggregate stability often results in a smooth crusted soil surface with loose sand-sized material on the surface. The use of crust breaking and clod forming tillage implements such as a rotary hoe or a sand fighter is often used after spring thunderstorms to create random roughness to the field surface. Once the crop is established and the canopy covers a significant portion of the soil, tillage is only used to control weeds.

Planting annual crop barriers is another BMP for soils in limited rainfall areas. For example, 102-cm (40-inch) strips of weeping love grass could be planted at 30- to 91-m (100–300 ft) intervals perpendicular to the erosive wind direction, depending on the soil properties. The interval can be determined using erosion models such as those described in this chapter (M. Sporcic, personal communication, 2009).

Wind erosion is a natural process that has formed and continues to form landscapes in anthropogenically disturbed and undisturbed locations. It is unlikely to think we can completely control wind erosion in every case. In cases where wind erosion is difficult to control, we need to evaluate all of the onsite and offsite costs and effects to determine whether or not we are pursuing the most prudent management. In most cases, control measures must be applied where economic losses and health or environmental effects from wind-driven soil movement are likely.

References

Acosta-Martinez, V., and T.M. Zobeck. 2004. Enzyme activities and arylsulfatase protein content of dust and the soil source: Biochemical fingerprints? J. Environ. Qual. 33:1653–1661.

Armbrust, D.V. 1968. Windblown soil abrasive injury to cotton plants. Agron. J. 60:622–625.

Baker, J.T. 2007. Cotton seedling abrasion and recovery from windblown sand. Agron. J. 99:556–561.

Baumhardt, R.L. 2003. Dust Bowl era, p. 187–191. In T. A. H. Howell and B. A. Stewart (ed.) Encyclopedia of water science. Marcel Dekker, New York.

Belly, P.-Y. 1964. Sand movement by wind. Tech. Memor. 1. U.S. ARMY Corps of Engineers, Coastal Engineering Research Center, Vicksburg, MS.

Bilbro, J.D., and D.W. Fryrear. 1985. Effectiveness of residues from six crops for reducing wind erosion in a semiarid region. J. Soil Water Conserv. 40:358–360.

Bisal, F., and J. Hsieh. 1966. Influence of moisture on erodibility of soil by wind. Soil Sci. Soc. Am. J. 102:143–146.

Blackburn, W. 2006. Unusual front leads to tragic scenario along area highway. In Lubbock Avalanche-Journal Online ed. Available at http://www.lubbockonline.com/stories/062406/loc_062406027.shtml (accessed 22 Feb. 2009, verified 7 Dec. 2010).

Braaten, D.A., and T.A. Cahill. 1986. Size and composition of dust transported to Hawaii. Atmos. Environ. 20:1105–1109.

Breshears, D.D., J.J. Whicker, C.B. Zou, J.P. Field, and C.D. Allen. 2008. A conceptual framework for dryland aeolian transport along the grassland–forest continuum: Effects of woody plant canopy cover and disturbance. Geomorphology 105:28–38.

Brooks, S.D., P.J. DeMott, and S.M. Dreidenweis. 2004. Water uptake by particles containing humic materials and mixtures of humic materials with ammonium sulfate. Atmos. Environ. 38:1859–1868.

Chen, W., and D.W. Fryrear. 1996. Grain-size distributions of wind-eroded material above a flat bare soil. Phys. Geogr. 17:554–584.

Chepil, W.S. 1944. Utilisation of crop residues for wind erosion control. Sci. Agr. 24:307–319.

Chepil, W.S. 1956. Influence of moisture on erodibility of soil by wind. Soil Sci. Soc. Am. Proc. 20:288–292.

Chepil, W.S. 1958. Soil conditions that influence wind erosion. USDA Tech. Bull. 1185. U.S. Gov. Print. Office, Washington, DC.

Chepil, W.S. 1962. A compact rotary sieve and the importance of dry sieving in physical soil analysis. Soil Sci. Soc. Am. J. 26:4–6.

Chiou, C.T. 1989. Theoretical considerations of the partition uptake of nonionic organic compounds by soil organic matter. p. 1–30. In B.L. Sawhney and K. Brown (ed.) Reaction and movement of organic chemicals in soils. SSSA Spec. Publ. 22. SSSA, Madison, WI.

Chung, Y.S., H.S. Kim, J. Dulam, and J. Harris. 2003. On heavy dustfall observed with explosive sandstorms in Chongwon-Chongju, Korea in 2002. Atmos. Environ. 37:3425–3433.

Darwin, C. 1845. Journal of researches into the natural history and geology of the countries visited during the voyage of H.M.S. Beagle round the world, under the command of Capt. Fitz Roy, R.N. John Murray, London.

Delany, A.C., A. Claire Delany, D.W. Parkin, J.J. Griffin, E.D. Goldberg, and B.E.F. Reimann. 1967. Airborne dust collected at Barbados. Geochim. Cosmochim. Acta 31:885–909.

Eldridge, D.J., and R.S.B. Greene. 1994. Microbiotic soil crusts: A review of their roles in soil and ecological processes in the rangelands of Australia. Aust. J. Soil Res. 32:389–415.

Farmer, A.M. 1993. The effects of dust on plants: A review. Environ. Pollut. 79:63–75.

Fenton, T.E., J.R. Brown, and M.J. Mausbach. 1999. Effects of long-term cropping on organic content of soils: Implications for soil quality. p. 95–124. In R. Lal (ed.) Soil quality and soil erosion. Soil and Water Conservation Society, Ankeny, IA; CRC Press, Boca Raton, FL.

Feng, G., and B. Sharratt. 2005. Sensitivity analysis of soil and PM_{10} loss in WEPS using the LHS-OAT method. Trans. ASAE 48:1409–1420.

Fryrear, D.W. 1973. Wind damage....should I replant? Am. Cotton Grower May, p. 12–25.

Fryrear, D.W. 1984. Soil ridges-clods and wind erosion. Trans. ASAE 27:445–448.

Fryrear, D.W. 1985. Soil cover and wind erosion. Trans. ASAE 28:781–784.

Fryrear, D.W., and J.D. Downes. 1975. Estimating seedling survival from wind erosion parameters. Trans. ASAE 18:888–891.

Funk, R., E.L. Skidmore, and L.J. Hagen. 2004. Comparison of wind erosion measurements in Germany with simulated soil losses by WEPS. Environ. Model. Softw. 19:177–183.

Gatz, D.F., and J.M. Prospero. 1996. A large silicon-aluminum aerosol plume in central Illinois: North African dust? Atmos. Environ. 30:3789–3799.

Gile, L.H., and R.B. Grossman. 1979. The desert project soil monograph: Soils and landscapes of desert region astride the Rio Grande Valley near Las Cruces, New Mexico. USDA-SCS, U.S. Gov. Print. Office, Washington, DC.

Gillette, D.A., D.W. Fryrear, T.E. Gill, T. Ley, T.A. Cahill, and E.A. Gearhart. 1997. Relation of vertical flux of particles smaller than 10 µm to total aeolian horizontal mass flux at Owens Lake. J. Geophys. Res. 102:26009–26015.

Gorbushina, A.A., R. Kort, A. Schulte, D. Lazarus, B. Schneteger, H.-J. Brumsack, W.J. Broughton, and J. Favet. 2007. Life in Darwin's dust: Intercontinental transport and survival of microbes in the nineteenth century. Environ. Microbiol. 9:2911–2922.

Goudie, A.S. 1978. Dust storms and the geomorphological implications. J. Arid Environ. 1:291–310.

Grini, A., G. Myhre, C.S. Zender, J.K. Sundet, and I.S.A. Isaksen. 2003. Model simulations of dust source and transport in the global troposphere: Effects of soil erodibility and wind speed variability. Institute Report Ser. 124. Dep. of Geosciences, Univ. of Oslo.

Griffith, P.W. 1976. Introduction to the problems. p. 3–7. In R. W. Tinus (ed.) Shelterbelts on the Great Plains. Publ. 78. Great Plains Agric. Council, Lincoln, NE.

Hagen, L.J. 1991a. Wind erosion mechanics: Abrasion of aggregated soil. Trans. ASAE 34:831–837.

Hagen, L.J. 1991b. A wind erosion prediction system to meet user needs. J. Soil Water Conserv. 46:106–111.

Hagen, L.J. 2004. Evaluation of the Wind Erosion Prediction System (WEPS) erosion submodel on cropland fields. Environ. Model. Softw. 19:171–176.

Hagen, L.J., and D.V. Armbrust. 1992. Aerodynamic roughness and saltation trapping efficiency of tillage ridges. Trans. ASAE 35:1179–1184.

Hagen, L.J., B. Schroeder, and E.L. Skidmore. 1995. A vertical soil crushing-energy meter. Trans. ASAE 38:711–715.

Hagen, L.J., S. Van Pelt, T.M. Zobeck, and A. Retta. 2007. Dust deposition near an eroding source field. Earth Surf. Process. Landforms 32:281–289.

Hagen, L.J., L.E. Wagner, and E.L. Skidmore. 1999. Analytical solutions and sensitivity analyses for sediment transport in WEPS. Trans. ASAE 42(42):1715–1721.

Hurt, R.D. 1981. The Dust Bowl, an agricultural and social history. Nelson-Hall, Chicago, IL.

Ishizuka, M., M. Mikami, J. Leys, Y. Yamada, S. Heidenriech, and Y. Shao. 2008. Effects of soil moisture and dried raindroplet crust on saltation and dust emission. J. Geophys. Res. 113:D24212.

Kennedy, A.C. 1998. Biological fingerprinting of dust aerosols. p. 49–50. *In* A. Busacca (ed.) Dust Aerosols, Loess Soils, and Global Change. Proc. Washington State Univ. Workshop October 11–14, 1998 Seattle, WA. Misc. Publ. MISC0190. College of Agriculture and Home Economics, Washington State Univ., Pullman.

Kjelgaard, J., B. Sharratt, I. Sundram, B. Lamb, C. Claiborn, K. Saxton, and D. Chandler. 2004. PM_{10} emission from agricultural soils on the Columbia Plateau: Comparison of dynamic and time-integrated field-scale measurements and entrainment mechanisms. Agric. For. Meteorol. 125:259–277.

Krueger, B.J., V.H. Grassian, J.P. Corwin, and A. Laskin. 2004. Heterogeneous chemistry of individual mineral dust particles from different dust source regions: The importance of particle mineralogy. Atmos. Environ. 38:6253–6261.

Larney, F.J., J.F. Leys, J.F. Muller, and G.H. McTainsh. 1999. Dust and endosulfan deposition in cotton-growing area of northern New South Wales, Australia. J. Environ. Qual. 28:692–701.

Leathers, L.H. 1981. Plant components of desert dust in Arizona and their significance for man. *In* T.L. Pewe (ed.) Desert dust: Origin, characteristics, and effect on man. Spec. Publ. 186. Geological Society of America, Boulder, CO.

Lee, H.N., T. Tanaka, M. Chiba, and Y. Igarashi. 2003. Long range transport of Asian dust from dust storms and its impact on Japan. Water Air Soil Pollut. Focus 3:231–243.

Leys, J.F., and D.J. Eldridge. 1998. Influence of cryptogamic crust disturbance to wind erosion on sand and loam rangeland soils. Earth Surf. Process. Landforms 23:963–974.

Li, J., G.S. Okin, L. Alvarez, and H. Epstein. 2007. Quantitative effects of vegetation cover on wind erosion and soil nutrient loss in a desert grassland of southern New Mexico. Biogeochemistry 85:317–322.

Litaor, M.I. 1987. The influence of eolian dust on the genesis of alpine soils in the Front Range, Colorado. Soil Sci. Soc. Am. J. 51:142–147.

Liu, S., and T. Wang. 2007. Aeolian desertification from the mid-1970s to 2005 in Otindag Sandy Land, Northern China. Environ. Geol. 51:1057–1064.

Lyles, L. 1975. Possible effects of wind erosion on soil productivity. J. Soil Water Conserv. 30:279–283.

Mackie, D.S., and K.A. Hunter. 2007. Australian dust: From uplift to uptake. Chem. N.Z. 71(3):82–83.

McGlynn, I.O., and G.S. Okin. 2006. Characterization of shrub distribution using high spatial resolution remote sensing: Ecosystem implications for a former Chihuahuan Desert Grassland. Remote Sens. Environ. 101:554–566.

McKenna-Neuman, C.M., and W.G. Nickling. 1989. A theoretical and wind tunnel investigation of the effect of capillary water on the entrainment of sediment by wind. Can. J. Soil Sci. 69:79–96.

McKenna-Neuman, C., and S. Sanderson. 2008. Humidity control of particle emissions in aeolian systems, J. Geophys. Res. 113:F02S14, doi:10.1029/2007JF000780.

Miller, G.C., V.R. Herbert, and W.W. Miller. 1989. Effect on sunlight on organic contaminants at the atmosphere-soil interface. p. 99–110. *In* B.L. Sawhney and K. Brown (ed.) Reaction and movement of organic chemicals in soils. SSSA Spec. Publ. 22. SSSA, Madison, WI.

Monteith, J.L. 1973. Principles of environmental physics. Edward Arnold, London.

Neave, M., and S. Rayburg. 2007. A field investigation into the effects of progressive rainfall-induced soil seal and crust development on runoff and erosion rates: The impact of surface cover. Geomorphology 87:378–390.

Nickling, W.G. 1994. Aeolian sediment transport and deposition, p. 293–350. *In* K. Pye (ed.) Sediment transport and depositional processes. Blackwell Scientific Publications, Oxford, UK.

Nordstrom, K.F., and S. Hotta. 2004. Wind erosion from cropland in the USA: A review of problems, solutions, and prospects. Geoderma 121:157–167.

Oke, T.R. 1987. Boundary layer climates. Methuen Ltd. London.

Okin, G.S. 2005. Dependence of wind erosion and dust emission on surface heterogeneity: Stochastic modeling. J. Geophys. Res. 110:D112208, doi:10.1029/2004JD005288.

Okin, G.S. 2008. A new model of wind erosion in the presence of vegetation. J. Geophys. Res. 113:F02S10, doi:10.1029/2007JF000758.

Okin, G.S., and D.A. Gillette. 2001. Distribution of vegetation in wind-dominated landscapes: Implications for wind erosion modeling and landscape processes. J. Geophys. Res. 106:9673–9683.

Okin, G.S., B. Murray, and W.H. Schlesinger. 2001. Degradation of sandy arid shrubland environments: Observations, process modelling, and management implications. J. Arid Environ. 47:123–144.

Pelletier, J.D. 2006. Sensitivity of playa windblown-dust emissions climatic and anthropogenic change. J. Arid Environ. 66:62–75.

Potter, K.N. 1990. Estimating wind-erodible materials on newly crusted soils. Soil Sci. Soc. Am. J. 150:771–776.

Potter, K.N., and T.M. Zobeck. 1990. Estimation of soil microrelief. Trans. ASAE 33:156–161.

Potter, K.N., T.M. Zobeck, and L.J. Hagen. 1990. A microrelief index to estimate soil erodibility by wind. Trans. ASAE 33:151–155.

Prospero, J.M. 1996. Saharan dust transport over the North Atlantic Ocean and Mediterranean: An overview. p. 131–151. *In* S. Guerzoni and R. Chester (ed.) The impact of desert dust across the Mediterranean. Kluwer Academic Press, the Netherlands.

Prospero, J.M., P. Ginoux, O. Torres, S.E. Nicholson, and T.E. Gill. 2002. Environmental characterization of global sources of atmospheric dust identified with the Nimbus 7 Total Ozone Mapping Spectrometer (TOMS) absorbing aerosol product. Rev. Geophys. 40:1002, doi:10.1029/2000RG000095.

Pye, K. 1987. Aeolian dust and dust deposits. Academic Press, Harcourt Brace Jovanovich, New York.

Rabenhorst, M.C., L.P. Wilding, and C.L. Girdner. 1984. Airborne dusts in the Edwards Plateau region of Texas. Soil Sci. Soc. Am. J. 48:621–627.

Rahn, K.A., R.D. Borys, and G.E. Shaw. 1981. Asian dust over Alaska: Anatomy of an Arctic haze episode. p. 37–70. *In* T.L. Pewe (ed.) Desert dust: Origin, characteristics, and effect on man. Spec. Publ. 186. Geological Society of America, Boulder, CO.

Ravi, S., P. D'Odorico, B. Herbert, T. Zobeck, and T.M. Over. 2006a. Enhancement of wind erosion by fire-induced water repellency. Water Resour. Res. 42:W11422, doi:10.1029/2006WR004895.

Ravi, S., P. D'Odorico, and G.S. Okin. 2007. Hydrologic and aeolian controls on vegetation patterns in arid landscapes. Geophys. Res. Lett. 34: L24S23, doi:10.1029/2007GL031023.

Ravi, S., T.M. Zobeck, T.M. Over, G.S. Okin, and P. D'Odorico. 2006b. On the effect of moisture bonding forces in air-dry soils on threshold friction velocity of wind erosion. Sedimentology 53:597–609.

Reheis, M.C., and R. Kihl. 1995. Dust deposition in southern Nevada and California, 1984–1989: Relations to climate, source area, and source lithology. J. Geophys. Res. 100:8893–8918.

Reynolds, R., J. Neff, M. Reheis, and P. Lamothe. 2006. Atmospheric dust in modern soil on aeolian sandstone, Colorado Plateau (USA): Variation with landscape position and contribution to potential plant nutrients. Geoderma 130:108–123.

Rice, M.A., B.B. Willetts, and I.K. McEwan. 1996. Wind erosion of crusted soil sediments. Earth Surf. Process. Landforms 43:21–32.

Saxton, K.E. 1995. Wind erosion and its impact on off-site air quality in the Columbia Plateau—An integrated research plan. Trans. ASAE 38:1030–1038.

Shao, Y. 2000. Physics and modelling of wind erosion. Kluwer Academic Publ., Boston.

Sharifi, M., A. Gibson, and P. Rundel. 1999. Phenological and physiological responses of heavily dusted creosote bush (*Larrea tridentata*) to summer irrigation in the Mojave desert. Flora 194:369–378.

Six, J., E.T. Elliot, and K. Paustian. 1999. Aggregate and soil organic matter dynamics under conventional and no-tillage systems. Soil Sci. Soc. Am. J. 63:1350–1358.

Skidmore, E.L. 1966. Wind and sandblast injury to seedling green beans. Agron. J. 58:311–315.

Skidmore, E.L. 1994. Wind erosion. p. 265–293. *In* R. Lal (ed.) Soil erosion research methods. 2nd ed. Soil and Water Conservation Society, Ankeny, IA.

Skidmore, E.L., and J.B. Layton. 1992. Dry-soil aggregate stability as influenced by selected soil properties. Soil Sci. Soc. Am. J. 56:557–561.

Skidmore, E.L., and D.H. Powers. 1982. Dry soil-aggregate stability: Energy-based index. Soil Sci. Soc. Am. J. 46:1274–1279.

Song, Z., J. Wang, and S. Wang. 2007. Quantitative classification of northeast Asian dust events. J. Geophys. Res. 112:D04211, doi:10.1029/2006DJ007048.

Sorenson, C.J., and G.A. Marotz. 1977. Changes in shelterbelt mileage statistics over four decades in Kansas. J. Soil Water Conserv. 32:276–281.

Sporcic, M., T. Keep, and L. Nelson. 1998. WEQ management period method wind erosion model worksheet. Available at http://www.nm.nrcs.usda.gov/technical/tech-notes/agro/ag55.xls. (verified 7 Dec. 2010).

Sterk, G. 2000. Flattened residue effects on wind speed and sediment transport. Soil Sci. Soc. Am. J. 61:616–632.

Sterk, G., L. Hermann, and A. Bationo. 1996. Wind-blown nutrient and soil productivity changes in southwest Niger. Land Degrad. Devel. 7:325–335.

Stetler, L.D., K.E. Saxton, and D.W. Fryrear. 1994. Wind erosion and PM_{10} measurements from agricultural fields in Texas and Washington. 87th. Paper 94-FA145.02. Air and Waste Management Assoc., Cincinnati, OH.

Stout, J.E. 2006. A field study of aeolian activity in burned semiarid grassland. *In* W.G. Nickling et al. (ed.) Abstracts of the Sixth Int. Conf. on Aeolian Research, 24–28 July 2006. Guelph, ON, Canada.

Stout, J.E., A. Warren, and T.E. Gill. 2009. Publication trends in aeolian research: An analysis of the Bibliography of Aeolian Research. Geomorphology 105:6–17.

Svensson, A., P. E. Biscaye, and F. E. Grousset. 2000. Characterization of late glacial continental dust in the Greenland Ice Core Project ice core. J. Geophys. Res. 105(D4):4637–4656.

Talbot, R.W., M.O. Andreae, H. Berresheim, P. Artaxo, M. Garstang, R.C. Harriss, K.M. Beecher, and S.M. Li. 1990. Aerosol chemistry during the wet season in central Amazonia: The influence of long-range transport. J. Geophys. Res. 95:16955–16969.

Thomas, D.S.G., and G.F.S. Wiggs. 2008. Aeolian system responses to global change: Challenges of scale, process and temporal integration. Earth Surf. Process. Landforms 33:1,396–1,418.

Toy, T.J., G.R. Foster, and K.G. Renard. 2002. Soil erosion: Processes, prediction, measurement, and control. John Wiley and Sons, New York.

Trochkine, D., Y. Iwasaka, A. Matsuke, M. Ymada, Y.S. Kim, D. Zhang, G.Y. Shi, Z. Shen, and G. Li. 2003. Comparison of the chemical composition of mineral particles collected in Dunhuang, China and those collected in the free troposphere over Japan: Possible chemical modification during long-range transport. Water Air Soil Pollut. Focus 3:161–172.

Tsoar, H., and K. Pye. 1987. Dust transport and the question of desert loess formation. Sedimentology 34:139–153.

USDA-ARS. 1996. WEPS technical documentation. Available at http://www.weru.ksu.edu/weps/docs/weps_tech.pdf (verified 7 Dec. 2010). USDA-ARS.

USDA-NRCS. 1993. Soil survey manual. Available at http://soils.usda.gov/technical/manual/ (verified 7 Dec. 2010). USDA-NRCS.

USDA-NRCS. 2002. National agronomy manual. 3rd ed. Available at http://www.nrcs.usda.gov/technical/agronomy.html (verified 7 Dec. 2010). USDA-NRCS.

USDA-NRCS. 2007. National resources inventory 2003. Available at http://www.nrcs.usda.gov/technical/NRI/2003/nri03eros-mrb.html (verified 7 Dec. 2010). USDA-NRCS.

USDA-NRCS. 2009. National conservation practice standards. Available at http://www.nrcs.usda.gov/technical/standards/nhcp.html (verified 7 Dec. 2010).

USEPA. 2008. Technology Transfer Network—National ambient air quality standards (NAAQS). Available at http://www.epa.gov/ttn/naaqs/ (verified 7 Dec. 2010).

Van Pelt, R.S., and T.M. Zobeck. 2004. Validation of the Wind Erosion Equation (WEQ) for discrete periods. Environ. Model. Softw. 19:199–203.

Van Pelt, R.S., and T.M. Zobeck. 2007. Chemical constituents of fugitive dust. Environ. Monit. Assess. 130:3–16.

Van Pelt, R.S., T.M. Zobeck, J.C. Ritchie, and T.E. Gill. 2007. Validating the use of ^{137}Cs measurements to estimate rates of soil redistribution by wind. Catena 70:455–464.

Vigiak, O., G. Sterk, A. Warren, and L.J. Hagen. 2003. Spatial modeling of wind speed around windbreaks. Catena 52:63–73.

Wendelin, G. 1646. Pluvia pupurea Bruxellensis. Apud Ludivicum de Heuqueville, Parisiis.

Whicker, J.J., J.E. Pinder III, and D.D. Breshears. 2008. Thinning semiarid forests amplifies wind erosion comparably to wildfire: Implications for restoration and soil stability. J. Arid Environ. 72:494–508.

Woodruff, N.P., and F.H. Siddoway. 1965. A wind erosion equation. Soil Sci. Soc. Am. Proc. 29:602–608.

Zhang, D.D., M. Peart, C.Y. Jim, Y.Q. He, B.S. Li, and J.A. Chen. 2003. Precipitation chemistry of Lhasa and other remote towns, Tibet. Atmos. Environ. 37:231–240.

Zobeck, T.M. 1991a. Abrasion of crusted soils: Influence of abrader flux and soil properties. Soil Sci. Soc. Am. J. 55:1091–1097.

Zobeck, T.M. 1991b. Soil properties affecting wind erosion. J. Soil Water Conserv. 46:112–118.

Zobeck, T.M., and J.D. Bilbro. 2001. Crop productivity and surface soil properties of a severely wind-eroded soil. p. 617–622. *In* D.E. Stott et al. (ed.) Sustaining the Global Farm. 10th International Soil Conservation Organization. 24–29 May, 1999. Purdue University and USDA-ARS National Soil Erosion Laboratory, West Lafayette, IN.

Zobeck, T.M., and D.W. Fryrear. 1986a. Chemical and physical characteristics of windblown sediment. I. Quantities and physical characteristics. Trans. ASAE 29:1032–1036.

Zobeck, T.M., and D.W. Fryrear. 1986b. Chemical and physical characteristics of windblown sediment. II. Chemical characteristics and total soil and nutrient discharge. Trans. ASAE 29:1037–1041.

Zobeck, T.M., T.W. Popham, E.L. Skidmore, J.A. Lamb, S.D. Merrill, M.J. Lindstrom, D.L. Mokma, and R.E. Yoder. 2003a. Aggregate-mean diameter and wind-erodible soil predictions using dry aggregate-size distributions. Soil Sci. Soc. Am. J. 67:425–436.

Zobeck, T.M., G. Sterk, R. Funk, J.L. Rajot, J.E. Stout, and R.S. Van Pelt. 2003b. Measurement and data analysis methods for field-scale wind erosion studies and model validation. Earth Surf. Process. Landforms 28:1163–1188.

Zobeck, T.M., and R.S. Van Pelt. 2006. Wind-induced dust generation and transport mechanics on a bare agricultural field. J. Hazard. Mater. 132:26–38.

15

Crusting

Guy J. Levy

Crust or seal (both terms are generally considered to be the same) formation at soil surfaces is a common phenomenon in many cultivated soils worldwide, and in particular in arid and semiarid regions (Shainberg and Letey, 1984). Soil crusts thus have been studied extensively over the years from a wide range of facets (Sumner and Stewart, 1992).

Surface crusts are thin (<2 mm) and are characterized by greater density, higher shear strength, finer pores, and lower saturated hydraulic conductivity than the underlying soil (McIntyre, 1958a; Bradford et al., 1987). Consequently, crusts have prominent effects on numerous soil phenomena; for example, when wet they decrease infiltration and increase runoff and erosion, and when dry they could slow soil–atmosphere gas exchange and interfere with seedling emergence.

Crusts are commonly classified according to the mode by which they are formed. There are *structural crusts*, which are formed by the impact energy of water drops whether from rain (Duley, 1939; McIntyre, 1958a), or from overhead sprinkler irrigation (Aarstad and Miller, 1973), and there are *depositional crusts*, which are formed by translocation and deposition of fine soil particles at a certain distance from their original location (Chen et al., 1980; Arshad and Mermut, 1988) as often occurs in furrow and basin irrigation (Kemper et al., 1985). Crusts have also been classified on the basis of their morphology (Valentin and Bresson, 1992).

The following discussion focuses on structural crusts, with attention given to the processes involved in the formation of crusts and the factors affecting it. Readers interested in the modeling aspects of the crusting phenomenon and flow processes in crusted soils are referred to Assouline (2004).

Processes Involved in Crust Formation

Crust formation is the result of two mechanisms: (i) physical disintegration of surface soil aggregates caused by wetting of the dry aggregates and/or the impact energy of the water drops, and subsequent compaction of the disintegrated aggregates by water drop impact, and (ii) physicochemical dispersion of soil clays, which then migrate into the soil with the infiltrating water, clog pores immediately beneath the surface, and form a layer of low permeability termed the *washed-in zone* (McIntyre, 1958a; Farres, 1978; Agassi et al., 1981). The two mechanisms are complementary. The physical disintegration of surface soil aggregates enhances clay susceptibility to dispersion under sodic conditions (Shainberg and Letey, 1984), while clay dispersion weakens

G.J. Levy, Institute of Soil, Water and Environmental Sciences, Agricultural Research Organization, The Volcani Center, Bet Dagan, Israel (vwguy@volcani.agri.gov.il).

doi:10.2136/2011.soilmanagement.c15

aggregate stability and renders them more sensitive to disintegration by wetting and/or the impact of water drops (Abu-Sharar et al., 1987; Mamedov et al., 2001).

Physical Mechanisms

The physical mechanisms that cause crust formation are determined by (i) the rate at which surface aggregates are wetted (Loch, 1994; Levy et al., 1997), (ii) drop kinetic energy (Betzalel et al., 1995; Mamedov et al., 2000), and (iii) soil aggregate stability (Moldenhauer and Kemper, 1969; Levy and Mamedov, 2002).

Disintegration of aggregates by wetting, termed *slaking*, is caused by differential swelling, entrapped air content, and the mechanical action of moving water (Quirk and Panabokke, 1962; Kay and Angers, 1999). The degree of aggregate disintegration by wetting depends on the rate at which they have been wetted—the faster the wetting rate, the greater the slaking of aggregates. Preventing aggregate slaking by slowly wetting the aggregates before exposing them to rain reduces the susceptibility of the soil to crust formation (Levy et al., 1997), especially in well-structured soils.

Of the various rain properties (e.g., intensity, duration, drop kinetic energy), raindrop kinetic energy is the most common property with which crusting is associated, as the kinetic energy of raindrops has a significant effect on crust formation and properties (e.g., Morin et al., 1981; Agassi et al., 1985; Thompson and James, 1985; Mohammed and Kohl, 1987; Valentin and Ruiz Figueroa, 1987; Betzalel et al., 1995; Mamedov et al., 2000). In soils exposed to drops with very low kinetic energy (< 0.01 kJ m^{-3}), no crust is formed. Conversely, when soils are exposed to drops with kinetic energy typical of rainstorms in Mediterranean climate (23 kJ m^{-3}), crusts of low permeability are formed (Agassi et al., 1985). The kinetic energy of rain, as well as of sprinkler drops, varies between these two extremes (Hudson, 1971).

In addition to its role in aggregate disintegration, drop kinetic energy also plays a dominant role in additional processes involved in surface crusting, such as compaction of the surface particles (Epstein and Grant, 1973) and rearrangement and realignment of the primary soil particles at the crusted surface (Eigel and Moore,

1983; Radcliffe et al., 1991). Concerning the latter, Morin et al. (1981) suggested the suction that develops at the crust–soil interface continuously adds clay particles that are in suspension at the soil surface to the crust and thus contributes to its densification and stabilization.

Physicochemical Mechanism

The role of the physicochemical clay dispersion in crust formation has been studied in detail (e.g., Shainberg and Levy, 1995, and references therein). Clay dispersion is sensitive to the electrolyte concentration and the sodium adsorption ratio (SAR) of the applied water (Quirk and Schofield, 1955). This is particularly true in the case where the soil surface is exposed to the mechanical action of falling water drops that lead to aggregate breakdown, and in turn, enhance clay susceptibility to chemical dispersion (Shainberg and Letey, 1984). Thus, not surprisingly, it has been shown that crust formation is very sensitive both to the level of soil sodicity (Kazman et al., 1983) and the electrolyte concentration and composition of the applied water (Agassi et al., 1981). High levels of soil sodicity and/ or low electrolyte concentration in the soil solution near the soil surface boost the formation of a more developed and less permeable crust.

Crust Morphology

The micromorphological characteristics of surface crusts have been studied extensively using microscopes with different powers of magnifications (e.g., petrographic, scanning electron microscope [SEM]) (see detailed review by West et al., 1992). Numerous micromorphological investigations of crust structure have shown that it consists of two distinct parts: (i) an upper skin crust, the washed-out zone (West et al., 1992), comprised of skeleton grains that have been stripped of the finer material, attributable to compaction by raindrop impact; and (ii) a zone of decreased porosity, termed the *washed-in zone*, attributed to the accumulation of fine material (e.g., McIntyre, 1958a,b; Tackett and Pearson, 1965; Gal et al., 1984; Luk et al., 1990; Wakindiki and Ben-Hur, 2002). However, some studies reported only of the presence of the skin layer (e.g., Chen et al., 1980; Gal et al., 1984; Tarchitzky et al., 1984; Bajracharya and Lal, 1999; Fan et al., 2008).

Gal et al. (1984), who tried to clarify this apparent discrepancy concluded, based on SEM micrograph studies of a sandy loam with exchangeable sodium percentage (ESP) of 1.0 and 11.6, respectively, exposed to distilled water rain, that the washed-in layer can be identified mainly in soils that are easily dispersed. At ESP 1.0, sand grains covered with a skin of clay were observed throughout the upper 2 mm, with no traces of a layer of accumulated clay. Crusting comprised a thin (0–0.1 mm thick) and compacted layer of broken aggregates. These observations indicate that only a small amount of clay dispersion or clay movement took place. Conversely, in the SEM micrographs of the sandy loam with ESP 11.6, a surface layer (0–0.25 mm thick) consisting of naked sand grains and a washed-in layer (0–0.75 mm thick) of accumulated clay was noted (Gal et al., 1984). Moreover, Onofiok and Singer (1984), who studied SEM micrographs from crusts of three different soil types, observed different morphological features in the crusts formed. Only in two of their soils did they observe the presence of a thin compacted surface layer and the washed-in zone. It was concluded that differences in the structure of crusts are determined by the physical, chemical, and mineralogical properties of the soil (Onofiok and Singer, 1984).

The fact that crust micromorphological studies are commonly based on small samples, as dictated by the use of microscopes, raises questions as to crust homogeneity. Levy et al. (1988a) observed that the structure of crusts is not homogeneous and consists of two distinct microtopographical features—(i) mounds that exhibit high permeability and a structure resembling an uncrusted soil and (ii) plains with low permeability and a crust that is clearly visible in SEM micrographs. The area occupied by the mounds and their permeability decreases with an increase in soil ESP (Levy et al., 1988a).

Characterization of surface crusts by their pore-size distribution and total porosity emerges as an interesting quantitative alternative to the traditional, qualitative studies of crust micrographs. Lee et al. (2008) used high-resolution X-ray computed tomography analysis to characterize the structure of soil surface crusts. Their results showed that after 60 min of rainfall the cumulative porosity of $\geq$15-μm equivalent diameter pores in a silt loam were reduced to 0.0156 mm mm^{-1} for a soil treated with a stabilizing agent (polyacrylamide) and 0.0004 mm mm^{-1} for an untreated soil. These results clearly indicate the distinct difference in the structure of the soil surface between the two treatments, and the formation of a dense crust in the case of the untreated soil (Lee et al., 2008).

However, it should be noted, as pointed out by Assouline (2004), that in all types of crust structure studies the samples studied require drying before analysis. The drying process may alter the crust's structure relative to that of a wet crust. Consequently, this may lead to some uncertainty when linking the micromorphological features of a dry crust to some of its other properties that are determined when wet (e.g., water transmission).

Factors Affecting Crust Formation
Measures of Crust Expression

Crust formation can be characterized using various approaches, whether direct, such as the already mentioned micromorphological approach, or indirect, such as raindrop splash detachment (McIntyre, 1958b; Bradford et al., 1986), surface strength (Morrison et al., 1985; Bradford et al., 1986; Govers and Poesen, 1986), and flow properties (e.g., Shainberg and Levy, 1995, and references therein). The apparent degree of crust formation may depend on the parameter chosen to characterize the crust (Remley and Bradford, 1989). Thus, ranking of soils based on their susceptibility to crusting could change depending on the parameter of choice (Bradford and Huang, 1992). The current discussion will focus on characterizing crust formation by changes in soil infiltration rate, which seems, based on the large volume of literature devoted to crusting and soil infiltration rate, to be the parameter most commonly used.

Infiltration rate

Infiltration is the term applied to the process of water entry into the soil, usually by downward flow through the soil surface. Hence, *infiltration rate* is defined as the volume flux of water flowing into the profile per unit of soil surface area under any set of circumstances. Under conditions where water is supplied to the soil without energy, infiltration rate is generally high during the early stages of infiltration, particularly when the soil is initially quite dry, but decreases monotonically to approach a constant rate asymptotically, due to a decrease in the matric suction gradient that occurs as infiltration proceeds (Baver et al., 1972; Hillel, 1980).

The decrease in infiltration rate from an initially high rate can also result from a gradual deterioration of soil structure and the formation of a dense surface crust. It is well known that crust formation at the soil surface predominates when there is a decrease of infiltration during rain (Duley, 1939; Epstein and Grant, 1973; Morin and Benyamini, 1977) or during irrigation with pressurized overhead sprinkler systems (Aarstad and Miller, 1973). The decrease in infiltration rate may result in a situation where the application rate of the water (i.e., rain/irrigation intensity) exceeds the infiltration rate of the soil. Consequently, water will start running off the field, and soil erosion will commence.

The infiltration rate values of crusted soils depend on the hydraulic conductivity of the crust formed at the soil surface; hence, some researchers tried to measure the hydraulic conductivity of the crust. McIntyre (1958a) found that the hydraulic conductivity of the upper and lower layers of the crust of a fine sandy loam soil were 2000 and 200 times lower than the hydraulic conductivity values recorded for the undisturbed soils. Bresler and Kemper (1970) measured a hydraulic conductivity value of 1.5×10^{-4} mm s^{-1} in the upper 2 to 3 mm of a crusted clay loam soil. However, since the crust is a very thin layer and it is very difficult to separate it from the rest of the soil profile, it is extremely difficult to determine its hydraulic conductivity. Consequently, infiltration rate is widely used to characterize water penetration into crusted soils.

Numerous equations, some entirely empirical and others theoretically based, have been proposed in attempts to express infiltration rate as a function of time or total volume of water that infiltrated into the soil (i.e., cumulative infiltration). The theoretical equations (Green and Ampt, 1911; Philip, 1957) arise out of mathematical solutions to physically based theories of infiltration. The empirical expressions (Horton, 1940) are not so restrictive as to the mode of water application, since they do not imply surface ponding from time zero on, as do the Green–Ampt and Philip equations (Hillel, 1980). An equation commonly used is a Horton-type one developed by Morin and Benyamini (1977), in which the infiltration process is described as a function of cumulative rain rather than cumulative time:

$$I_t = I_c + (I_i - I_c)e^{-\gamma pt} \qquad [1]$$

where I_t is the instantaneous infiltration rate (mm h^{-1}), I_c is the asymptotical final infiltration rate (mm h^{-1}), I_i is initial infiltration rate (mm h^{-1}), γ is a constant related to stability of soil surface aggregates (mm^{-1}), p is rain intensity (mm h^{-1}), and t is the time elapsed from the beginning of the storm (h).

Calculated infiltration rate values derived from this Eq. were in good agreement with values measured with laboratory and field rainfall simulators, such as those developed by Morin et al. (1967) and Miller (1987a), which are a common tool in seal formation studies.

Rain Properties
Rain Characterization

Rain is generally characterized by the following parameters: (i) rain intensity, (ii) raindrop median diameter, and (iii) final velocity of the median raindrop. The relationships among these parameters were examined (Laws, 1940; Wischmeier and Smith, 1951), and it was found that drops with a large diameter reach a high final velocity, and vice versa, and that the higher the rain intensity, the higher the percentage of large drops. The rate of crust formation was found to depend on median drop diameter and rain intensity (Ellison, 1947). On the other hand, Morin and Benyamini (1977) reported that rain intensity had no effect on the final infiltration rate of the crusts formed.

Raindrop Kinetic Energy

Of the various rain properties, raindrop kinetic energy has become the most common property associated with crust formation in recent years. The effect of drop kinetic energy on crust formation and the accompanying decrease in infiltration rate have been studied by many researchers (e.g., Agassi et al., 1985, 1988; Thompson and James, 1985; Mohammed and Kohl, 1987; Keren, 1990; Betzalel et al., 1995). Shevbs (1968) proposed that characterizing rain by kinetic energy per unit area and unit time is better than just using kinetic energy because it incorporates rainfall or sprinkler intensity. This concept was used by Mohammed and Kohl (1987) and Thompson and James (1985), who referred to it as droplet energy flux. These investigators showed that droplet energy flux can be used to determine the amount of water that can be applied to the field without obtaining runoff for different combinations of drop kinetic energy and application intensity. This concept, however, did not gain much popularity.

Agassi et al. (1988) showed that when soil is exposed to drops with kinetic energy below 0.01 J mm^{-1} m^{-2} (i.e., fog-type rain), no crust is formed, but when the kinetic energy is 23.0 J mm^{-1} m^{-2} (a kinetic energy typical of high-intensity rainstorms), a crust with a very low infiltration rate was formed. A detailed study on the combined effects of raindrop size and impact velocity on the infiltration rate of soils was performed by Betzalel et al. (1995). They applied rain of a constant intensity (40 mm h^{-1}) with 2.53- and 3.37-mm diameter drops falling from heights of 0.4, 1.0, 2.0, 6.0, and 10.0 m on two soils and measured the decay in infiltration rate. Aggregate disintegration and crust formation increased with an increase in the drops' impact energy; however, the two drop sizes studied had, in most cases, a similar effect on the final infiltration rate (Betzalel et al., 1995). In addition, Agassi et al. (1994) studied the combined effects of drop kinetic energy, soil sodicity, and water quality on infiltration rate of two kaolinitic soils. They observed that the final infiltration rate decreased exponentially as drop kinetic energy increased and that drop kinetic energy had an overriding effect on the reduction in final infiltration rate compared with soil sodicity and water quality.

The impact of rain kinetic energy on the rate of crust formation is not clear. Keren (1990) noted that the higher the kinetic energy of the drops, the steeper the drop in the infiltration rate of the soil. Conversely, Mohammed and Kohl (1987), who plotted the infiltration rate for four different levels of kinetic energy against cumulative kinetic energy observed that the various curves tend to converge in that section of the curve where the infiltration rate decreases sharply, thus suggesting that the rate of crust formation was independent of rain kinetic energy.

It is expected that the relation between crust formation and drops' kinetic energy will depend on the stability of the soil structure. Crust formation in soils with stable aggregate will take place only when the soils are exposed to high energy rain. Conversely, in soils with unstable structure, the crust may be formed already under low energy rain. This was verified by Shainberg and Singer (1988), who observed that sodic soils were more susceptible to crusting by rain of low impact drops than were Ca soils.

Electrolyte Concentration in the "Rain"

Clay dispersion is very sensitive to the chemistry (electrolyte concentration and cationic composition) of the applied water (Shainberg and Letey, 1984, and references therein). This is particularly true in the case where the soil surface is exposed to the mechanical action of the falling water drops, which enhances clay susceptibility to chemical dispersion. Using a laboratory rain simulator, Agassi et al. (1981) studied the effects of electrolyte concentration and soil sodicity on the infiltration rate of two loamy soils in Israel and noted its significant role in determining the infiltration rate (Fig. 15–1). Furthermore, the results of Agassi et al. (1981) revealed that, in comparison to soil hydraulic conductivity (Shainberg and Letey, 1984), the infiltration rate was by far more sensitive to the electrolyte concentration of the applied water. Similarly, Oster and Schroer (1979) found that cationic concentration greatly affected the infiltration rate of a Heimdal soil from North Dakota, even at low SAR values. They observed an increase in final infiltration rate from 2 to 28 mm h^{-1} as cation concentration in the applied water having SAR levels between 2 and 4.6 increased from 5 to 28 meq L^{-1}. The results

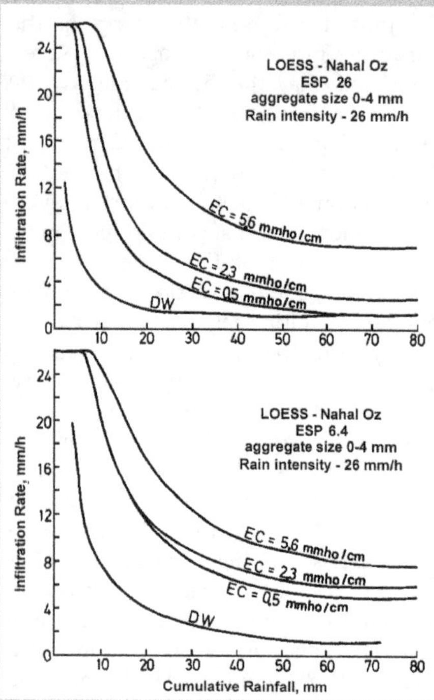

Fig. 15–1. Effect of electrolyte concentration in rain simulation experiments on the infiltration rate of loam (Agassi et al., 1981).

in Fig. 15–1 further show that salt concentration has an effect not only on the final infiltration rate but also on the rate at which the infiltration rate drops from its initial to the final value. The lower the concentration of the electrolyte solution, the faster the rate at which the infiltration rate decreases.

The degree of crust development thus depends on the potential of the soil clays to disperse, which increases with the increase in soil sodicity (ESP/SAR) and the decrease in the electrolyte concentration of the applied water. Consequently, crust formation by raindrops (i.e., water without electrolytes) is enhanced the greater the sodicity of the soil. It is possible to manipulate the conditions at the surface of soils exposed to rain by, for instance, adding a source of electrolytes such as gypsum to compensate for the lack of electrolytes in the rainwater and thus alleviate the hazard of clay dispersion and subsequent crust formation during rainstorms (Levy and Sumner, 1998, and references therein).

Soil Properties
Clay Content and Type

The tendency of soils to form crusts depends on the stability of their structure, which tends to increase with the increase in clay content. Increasing soil clay content leads to the increase in aggregate stability (Kemper and Koch, 1966; Levy and Mamedov, 2002). Clay particles act as cementing material, binding the particles together in the aggregates (Kemper and Koch, 1966). Thus, the stability of the aggregates against the impact action of the raindrops should also increase with an increase in clay content.

The effect of clay content on the infiltration rate of soils was studied by Ben-Hur et al. (1985) and later by Mamedov et al. (2001). It was noted that soils with 10 to 30% clay were most susceptible to crust formation and had the lowest infiltration rate. Apparently when clay content is >30%, soil structure is more stable, and a less developed and more permeable crust is formed. Conversely, in soils with lower clay contents (<10%), the amount of clay available to disperse and clog the soil pores is limited, therefore resulting in the formation of a less developed crust.

Most of the studies on crust formation have been conducted on soils in which the dominant clay minerals were smectites. These clay minerals are known to be more dispersive than illitic and kaolinitic clays (Frenkel et al., 1978; Goldberg and Glaubig, 1987; Singer 1994). However, a few kaolinitic and illitic soils from the southeastern United States (Miller, 1987b; Miller and Scifres, 1988) and Africa (Levy and van der Watt, 1988; Stern et al., 1991; Wakindiki and Ben-Hur, 2002) are known to form crusts and produce significant amounts of runoff. Stern et al. (1991) studied 19 kaolinitic and illitic soils (alfisols) from South Africa and divided them into stable and unstable (dispersive) soils on the basis of their susceptibility to crust formation when exposed to simulated rain. The final infiltration rate values for the stable soil were >14.5 mm h^{-1}, whereas for the dispersive soils (both smectitic and kaolinitic) the final infiltration rate values were <4.2 mm h^{-1}. Stern et al. (1991) attributed the susceptibility of the unstable kaolinitic soils to crust formation to the presence of smectite impurities in these soils. More recently, Wakindiki and Ben-Hur (2002) showed that the susceptibility of soils based on their clay mineralogy to aggregate slaking and crust formation was in the following decreasing order: smectitic $\geq$ illitic > kaolinitic.

Organic Matter Content

Organic matter (OM) is known to be a cementing agent in soils, contributing to the stabilization of soil aggregates (Kemper and Koch, 1966; Le Bissonnais and Arrouays, 1997). It is therefore expected that OM will enhance soil resistance to crust formation. Le Bissonnais and Arrouays (1997) found that reduction of the organic carbon content of a loamy soil below 1.5 to 2.0% decreased soil infiltration rate. Similarly, Fullan (1991) reported that a loamy sand with OM content <2% was prone to crusting and erosion. Conversely, Guerra (1994) found that an OM content of 3% was a threshold value, below which the aggregates of sandy loam soils were unstable and soil susceptibility to crust formation increased. Recently, Lado et al. (2004) observed for sandy loam soils from Spain that an increase in OM content from 2.3 to 3.5% had a considerable effect on mitigating crust development and consequently on maintaining a significantly higher final infiltration rate. Lado et al. (2004) suggested that in the low-OM soil there occurred a more extensive breakdown of surface aggregates, thus leading to a more continuous and less permeable crust compared with the high-OM soil.

Exchangeable Cations

As mentioned above, clay dispersion is one of the mechanisms responsible for crust formation and reduction in soil infiltration rate. The behavior of clay particles in aqueous solutions can be explained by the diffuse double layer theory, for which a description can be found in, for example, van Olphen (1977). According to the diffuse double layer theory, clay dispersion depends greatly on the valency of the exchangeable cations. For example, monovalent cations (Na, K) are considered dispersive, and bivalent cations (Ca, Mg) are considered nondispersive cations.

With respect to the effects of exchangeable sodium on the infiltration rate and crust formation, Kazman et al. (1983) noted that in a sandy loam a crust was formed and the infiltration rate dropped from an initial value of >100 mm h^{-1} to a final infiltration rate of 7.0 mm h^{-1} already at a very low ESP (1.0). Moreover, an ESP of 2.2 was enough to cause a further drop in the final infiltration rate of the sandy loam to 2.4 mm^{-1}. The depth of rain needed to approach the final infiltration rate was also affected by ESP; as the ESP of the soil increased, the depth of rain required to reach the final infiltration rate decreased (Kazman et al., 1983).

The high sensitivity of the soil surface to low ESP values is explained by three factors (Oster and Schroer, 1979; Kazman et al., 1983): (i) the mechanical impact of the raindrops, which enhances chemical dispersion; (ii) the absence of surrounding soil matrix (sand particles), which, when present, slows clay dispersion and movement; and (iii) the almost total absence of electrolytes in the applied distilled water. Indeed, the presence of some minerals (CaCO$_3$ and a few primary minerals) in the soil profile which readily release electrolytes to the percolating solution reduces the dispersive effect of sodicity and hence help maintain higher infiltration rate values (Shainberg and Letey, 1984).

The mechanical and chemical mechanisms responsible for crust formation are complementary. The mechanical impact of the raindrops has two effects: (i) breakdown of the soil aggregates and (ii) stirring of the soil particles. The latter mechanism enhances the rate of the chemical dispersion. Without the stirring process the rate of chemical dispersion is much slower. In hydraulic conductivity measurements, mechanical mixing of the soil is prevented. Thus, the infiltration rate is much more sensitive to ESP than is soil hydraulic conductivity (Shainberg and Letey, 1984).

Highly weathered soils in the southeastern United States also suffer from dispersion-related degradation of physical properties with additions of Na, either from fertilizers or wastewater sources. Miller and Scifres (1988) studied the effects of surface application of NaNO$_3$, at fertilizer N rates of 0.6 Mg ha^{-1}, on infiltration, runoff, and soil loss of a Greenville soil (clayey, kaolinitic, thermic, Rhodic Paleudult). Crust formation started only after 20 mm of rain, and the steady-state infiltration rate of the untreated soil stabilized at approximately 10 mm h^{-1}. Compared to other southeastern U.S. Piedmont soils (Miller, 1987b), such a steady-state infiltration rate is a reasonably high rate of water intake over the rainfall event, indicating a fair resistance to crusting by this soil. However, the NaNO$_3$ treatment resulted in a nearly immediate surface crusting, as noted from the rapid decline in infiltration reaching a steady-state infiltration rate of only 2 to 3 mm h^{-1}. These findings clearly demonstrate that, despite reports of the negligible impact of exchangeable Na on the hydraulic conductivity of kaolinitic soils (Frenkel et al., 1978; Chiang et al., 1987), once kaolinitic soils are exposed to mechanical stirring by the beating action of raindrops, aggregates from kaolinitic soils may disperse, in accordance with the level of exchangeable Na, resulting in a crust of low permeability.

Unlike the impact of exchangeable Na, the effect of exchangeable K on the water transmission properties is controversial (Chen et al., 1983, and references therein). Despite that, the effect of exchangeable K on crust formation and infiltration rate has received little attention. Levy and van der Watt (1990) compared the effect of exchangeable K on the infiltration rate of two South African soils to that of Ca and Na. Their results showed that, when K was the complementary cation to Ca in the soil exchange phase, an increase in exchangeable K led to a decrease in the infiltration rate. However, when compared to exchangeable Na, exchangeable K had a more favorable effect on soil infiltration rate than exchangeable Na. It was therefore concluded that exchangeable K has an intermediate effect, between that of Ca and Na, on infiltration rate of soils exposed to rain (Levy and van der Watt, 1990).

Between the two major divalent cations in the soil, exchangeable Mg has been known to cause deterioration in soil structure and develop a "magnesium solonetz" (Ellis and Caldwell, 1935). In addition, Mg enhances dispersion in montmorillonitic and illitic clays compared to Ca (Bakker and Emerson, 1973). Levy et al. (1988b) compared the effect of Mg to that of Ca as the complementary cation to Na on the infiltration rate of three South African soils. The results showed that exchangeable Mg had a similar effect to that of Ca on the infiltration rate. Keren (1990) compared the effect of Mg to that of Ca as the complementary cation to Na in Israeli soils under a rain kinetic energy range of 3.4 to 22.9 kJ m^{-3}. At the highest kinetic energy (similar to that used by Levy et al., 1988b), the effect of Mg on the infiltration rate was similar to that of Ca. For low to medium kinetic energy (8.0–12.5 kJ m^{-3}) the infiltration rate in the Mg–Na treated soil was lower than that in the Ca–Na soil. The adverse effect of exchangeable Mg on soil permeability seems to depend on the weight of the chemical dispersion in determining crust development. When the chemical clay dispersion is dominant in the crust formation process (e.g., raindrops with low to medium kinetic energy), the adverse effect of Mg on clay dispersion has a considerable impact on the infiltration rate of the crust; however, under rain with high kinetic energy, clay dispersion plays only a secondary role in determining soil crusting and permeability, and thus the impact of exchangeable Mg is comparable to that of Ca.

Slope Steepness

Reports in the literature indicate that, in general, increasing slope steepness leads to a more permeable crust. Poesen (1984) reported that only soils susceptible to crusting were affected by changing of slope, whereas in stable soils increasing slope steepness had only a negligible effect on the infiltration of the crusted soil. Warrington et al. (1989) observed for a smectitic sandy loam that for slopes ranging from 5 to 25% final infiltration rate increased linearly with the increase in slope steepness. Increasing slope steepness increased the final infiltration rate also in the case of kaolinitic soils (Norton et al., 1993). Conversely, Bradford and Huang (1992) reported that the relationship between slope steepness and crusting differs among soils, and that in some cases increasing the slope from 9 to 20% could lead to a lower final infiltration rate.

There are a number of possible reasons for the formation of more permeable crusts on steeper slopes (Bradford and Huang, 1992): (i) continuous erosion of the crusted layer caused by greater interrill erosion on the steeper slopes, (ii) an increase in rill density and depth with increase in slope, (iii) a decrease in the normal component of drop impact and number of drops impacts per unit surface area with the increase in slope steepness, and (iv) an increase in surface water depth on shallower slopes through which the compactive force of the impacting raindrops increases.

Conditions Prevailing in the Soil

The interest in the effects of prevailing soil conditions on susceptibility to crust formation stems from the realization that aggregate stability depends on conditions prevailing in the soil such as rate of aggregate wetting, antecedent moisture content in the aggregates, and aging duration (Panabokke and Quirk, 1957; Francis and Cruse, 1983; Kemper and Rosenau, 1984; Truman et al., 1990; Loch, 1994; Kjaergaard et al., 2004). Aggregate stability is closely related to crust formation, in addition to being closely associated with various soil properties (e.g., organic matter, clay percentage and oxides content). Aggregate disintegration by wetting, termed *slaking*, increases with the increase in their rate of wetting and is ascribed to the explosion of aggregates by entrapped air and differential swelling (Panabokke and Quirk, 1957; Quirk and Panabokke, 1962). Wetting and keeping the soil at a given moisture (i.e., aging) induces an increase in aggregate strength through the development of cohesive bonds between the soil particles (Blake and Gilman, 1970; Kemper and Rosenau, 1984; Attou et al., 1998) and the production of cementing agents by microbial activity (Martin et al., 1995). The effects of aging duration on aggregate stability, however, seem to depend on the antecedent moisture content of the soil (Gerard, 1965; Utomo and Dexter, 1981). The effects of these conditions on soil crusting, as well as their interactions with one another and with several soil and rain properties, were summarized in a review by Ben-Hur and Lado (2008).

Wetting Rate

Early studies of the effect of wetting rate on soil vulnerability to crusting clearly indicate that when slow wetting is used (i.e., eliminating aggregate slaking), surface aggregates are considerably less susceptible to disintegration by the impact of raindrops, and thus less developed and more permeable crusts are formed (Loch, 1994; Levy et al., 1997). However, results of various studies show that use of a slow wetting rate to prevent slaking and maintain high infiltration rate occurs mainly in fine-textured soils (Mamedov et al., 2001; Shainberg et al., 2003; Lado et al., 2004).

Mamedov et al. (2001) studied the combined effects of WR (2, 8, and 64 mm h^{-1}), soil sodicity (ESP levels between 0.9 and 20.4), and texture (8.8–68.3% clay) on infiltration rate and runoff from six Israeli soils. In soils with low clay content (8.8%), the effect of wetting rate on seal formation was negligible, but the effect of ESP was significant. Conversely, in the clay soils (>52.1%), wetting rate had a predominant effect on infiltration rate and runoff, while the effect of ESP was notable but secondary to that of wetting rate. The soils with intermediate clay content (22.5–40.2% clay) were the soils most susceptible to seal formation, with both wetting rate and ESP having moderate effects on seal formation.

Mamedov and Levy (2001) compared the importance of preventing aggregate slaking by using slow wetting on soil crusting (final infiltration rate and runoff) to that of preventing clay dispersion via the addition of 5 Mg ha^{-1} of phosphogypsum to the soil surface. Comparison of the two treatments (Table 15–1) indicated that their efficacy depended on soil clay content. In soils having low-to-medium clay content (<40%), prevention of clay dispersion via phosphogypsum application together with fast wetting of the soil was more effective than using a slow wetting rate in controlling crust formation. In clay soils (>40% clay), the opposite is true (Mamedov and Levy, 2001).

Shainberg et al. (2003) evaluated the relative importance of wetting rate to that of rain kinetic energy in the process of crust formation in four soils varying in texture. They found that in a loam (22.5% clay) rain kinetic energy played a significant role and wetting rate a minor one in crust formation. In clay soils (51.3–61.2% clay) wetting rate played a significant role and rain kinetic energy only a negligible one. In a sandy clay (38.1% clay) both played a significant role in the process of crusting (Shainberg et al., 2003).

Antecedent Moisture Content

The dependence of aggregate stability on antecedent moisture content has been linked to aggregate detachment, soil erosion, and seal formation (Farres, 1987; Bradford et al., 1992; Le Bissonnais, 1996; Levy and Mamedov, 2002). Furthermore, moisture content together with aging duration have been found to determine aggregate breakdown mechanism, the resulting particle-size distribution, and the evolution of the structure

Table 15–1. Final infiltration rate (FIR) and runoff data for the different treatments, including the fast and slow wetting rate (WR) and phosphogypsum (PG) (Mamedov and Levy, 2001).

Soil	Clay	Treatment		
		Fast WR	Fast WR+PG	Slow WR
	%			
FIR				
		$mm\ h^{-1}$		
Loamy sand	9.5	4.88 g†	10.00 e	5.38 fg
Loam	21.2	3.38 h	10.63 de	6.50 f
Sandy clay	38.6	3.50 h	11.75 d	6.00 fg
Clay Y	54.6	3.38 h	11.13 de	18.75 b
Clay E	64.0	3.63 h	13.25 c	36.00 a
Runoff				
	%	mm		
Loamy sand	9.5	39.42 cd†	18.60 i	35.05 e
Loam	21.2	46.30 a	30.40 g	36.66 de
Sandy clay	38.6	43.35 b	28.56 g	31.99 f
Clay Y	54.6	43.83 b	29.74 gh	4.09 j
Clay E	64.0	40.80 c	27.05 h	1.13 k

† For FIR and runoff, values followed by same letter are not significantly different at the 0.05 level.

of the seal (Kemper and Rosenau, 1984; Le Bissonnais, 1990). Results of studies that evaluated the impact of antecedent moisture content directly on seal formation and soil loss were inconsistent. Luk (1985) reported a fivefold increase in soil loss for loam and silt loam soils when increasing moisture content from wilting point to saturation. The author concluded that increasing antecedent moisture content may have led to shear strength reduction in the surface aggregates. Other studies also showed that soil susceptibility to detachment increased as soil antecedent moisture content increased (Cruse and Larson, 1977; Al-Durrah and Bradford, 1981; Francis and Cruse, 1983; Le Bissonnais et al., 1995; Froese et al., 1999). Wangemann et al. (2000) observed in three loam soils that infiltration rate decreased with the increase in antecedent moisture content in the range of 5 to 20%. In contrast, for loam and clayey soils (30 and 42% clay content) Le Bissonnais and Singer (1992) obtained significantly higher infiltration rate levels for wetted (24 h by a matric potential of 1.0 kPa) soil samples compared with air-dry ones. Truman and

Bradford (1990) noted that wetting of the soil (five soils with 11–53% clay) gradually for 48 h had no significant effect on the infiltration rate compared with that of an air-dried soil. Reichert and Norton (1994) also reported that wetting 12 dry mostly clay soils (10 clay, 2 loam) for 2 h at −0.54 kPa generally had little effect on the infiltration rate despite an observed increase in aggregate stability. Bradford and Huang (1992) reported that the effect of antecedent moisture content on seal formation depended on soil texture; in clay soils wetting increased the final infiltration rate compared with air-dried soils, whereas in loam soils the opposite was found. Recently, Mamedov et al. (2006) reported the existence of an optimal range of antecedent moisture content, generally between field capacity and wilting point, at which runoff levels were 30% lower than those obtained below or above this antecedent moisture content range (Fig. 15–2).

Aging

Levy et al. (1997) studied the affect of aging duration on the infiltration rate of a loam and a clay soil and noted the existence of an interaction between aging and wetting rate in their effect on the infiltration rate. Aging duration of 18 h was only effective in maintaining higher infiltration rate in cases where the wetting rate was not fast (<6 mm h^{-1}). When fast wetting rate was used, similar infiltration rate curves were obtained for aged- and non-aged samples (Levy et al., 1997). Ben-Hur et al. (1998) observed for a non-sodic and a sodic soil that 4 d of aging resulted in a more gradual decrease in the infiltration rate curve and a higher final infiltration rate compared to a non-aged soil.

Mamedov et al. (2006) studied the effect of 0 to 7 d of aging on crust formation and soil loss. They observed that the level of antecedent moisture content determined the beneficial effects of aging on improving the resistance of soil to crust formation. Increasing aging duration was most effective in improving soil infiltration rate at the optimal level of antecedent moisture content studied (antecedent moisture content between field capacity and wilting point). It was suggested that the combined favorable impact of aging and antecedent moisture content on improving soil stability is associated with water-filled pores that are of the size belonging to the clay fabric (pF

2.4–4.2). Thus, clay movement and reorientation have been considered key factors in the development of cohesive forces between and within soil particles during aging at optimal antecedent moisture content levels (Mamedov et al., 2006).

Crusting under Consecutive Rainstorms

Alternate applications of rain and saline water (by irrigation) predominate in many semiarid regions of the world. The stability and reversibility of the crusts formed were found to depend on the period of drying, water quality and degree of crust development. Hardy et al. (1983) studied the effect of subsequent saline water sprinkling (i.e., simulation of irrigation water) on the stability of crusts originally formed under rainwater. They found that application of saline water in the second storm caused an increase in the final infiltration rate of a loamy sand to 5.6 mm h^{-1} compared to 2 mm h^{-1} in the first storm with rainwater. They observed that in this soil the impact of the saline water drops caused a breakdown of the old crust and a formation of a new, more permeable one due to the use of saline water. Conversely, in a silt loam only a slight increase in the final infiltration rate was observed in the consecutive saline water storm. Hardy et al. (1983) concluded that in the silt loam the structure of the crust formed in the first storm was stable and did not break during the second storm. In a further study on crust stability, Levy et al. (1986) observed for a silt loam that, if the crust is sufficiently dry, applying a successive storm with rainwater leads to a final infiltration rate similar to that obtained in the first storm, suggesting that successive storms of similar water quality do not affect final permeability of the crust. Similarly, Le Bissonnais and Singer (1992) reported that drying of crusts between successive storms of the same water quality delays runoff initiation due to the formation of cracks in the dry crust but does not affect the final infiltration rate of the crust, which was similar in the three consecutive storms studied. However, use of saline water in successive storms increased the final infiltration rate of a dry crust in a silt loam to 6.4 mm h^{-1} compared to 2.5 mm h^{-1} at the end of the former rain storm (Levy et al., 1986). In the case where only partial drying occurred, changing from rainwater in the first storm to saline water in the consecutive storm did not increase the final infiltration rate (Levy et al., 1986). It can be concluded that sufficient drying of the crust enhances its sensitivity to slaking on rewetting and thus enhances its breakdown by the beating action of the raindrops in the subsequent storm. In such a case, the quality

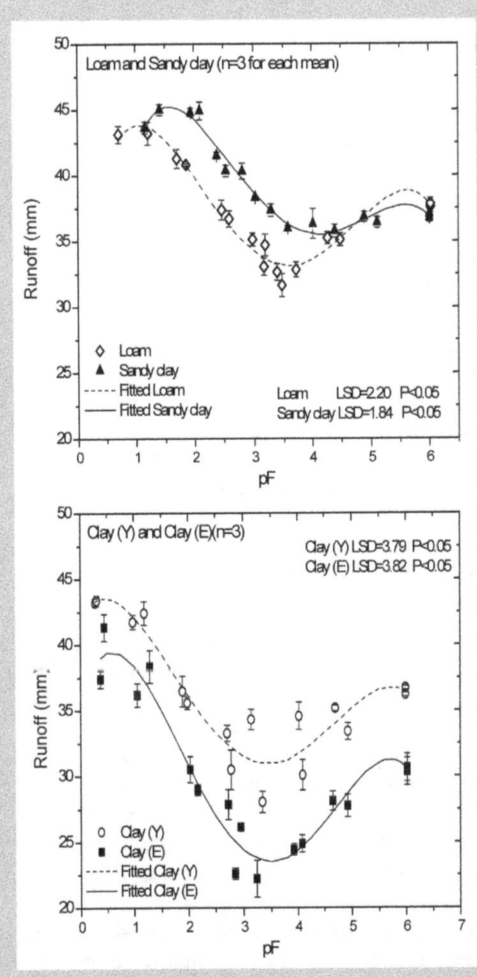

Fig. 15–2. Effect of antecedent moisture content (expressed as pF) on runoff with aging duration of 3 to 7 d. Runoff and soil loss data were fitted to the pF data by a polynom of the fourth order ($r^2 > 0.90$, $P < 0.01$ in all cases). Bars indicate the least significant differences at a probability level $P < 0.05$. Means are significantly different where bars do not overlap (adapted from Mamedov et al., 2006).

of the water in that storm will determine the characteristics of the newly formed crust.

Unlike fully dried crusts, the stability and reversibility of wet crusts was found to depend on whether the crust was fully or partially developed. Agassi et al. (1988) drew a distinction between fully developed crusts that are formed by a distilled water rain of high energy and sufficient duration and partially developed crusts that are formed by rain with low energy, or by high energy rain of short duration, or by saline rainwater of high energy and long duration. A fully developed wet crust was considered by Agassi et al. (1988) as stable, and its permeability may change only slightly when the soil is exposed to elevated electrolyte concentration in the applied water of the successive storm. Conversely, a partially developed crust is not stable, and therefore its permeability may be affected not only by the energy of the water drops but even by changes in the electrolyte concentration of the water applied (Agassi et al., 1988).

Concluding Comments

Cultivated soils are structurally unstable and form a crust at the soil surface when exposed to rain or overhead sprinkler irrigation. Crust formation is a destructive phenomenon that leads to a reduction in rain infiltration and subsequently to an increase in runoff volume and soil erosion. The susceptibility of soils to crusting depends on a wide array of factors, including rain properties (drop kinetic energy and electrolyte concentration in the applied water), soil properties (clay content and mineralogy, organic matter content and cationic composition of the exchange phase), soil slope, and conditions prevailing in the soil (rate of wetting, antecedent moisture content and aging). Crust formation may also depend on soil management (method of cultivation, type of crops).

There are currently no good predictive models that quantify, in terms of soil hydraulic properties, the effects of the aforementioned factors on crust formation (Assouline, 2004). Further research on the relation among those factors in their effects on soil crusting is needed to enhance our ability to quantitatively predicate with a sound degree of confidence the changes water flow processes undergo in crusted soils. The urgency to develop such quantitative predictive models to portray the crusting process is related not only to the importance of the crusting phenomenon and its derivatives from the agricultural perspective but also from environmental considerations. The phenomenon of soil crusting has bearing on the environment because runoff and soil loss derived from crusted soils may contaminate surface water bodies with nutrients, herbicides, pesticides, and other pollutants that may reside in the soil at the time of crust development.

References

Aarstad, J.S., and D.E. Miller. 1973. Soil management practices for reducing runoff under center-pivot sprinkler system. J. Soil Water Conserv. 28:171–173.

Abu-Sharar, T.M., F.T. Bingham, and J.D. Rhoades. 1987. Stability of soil aggregates as affected by electrolyte concentration and composition. Soil Sci. Soc. Am. J. 51:309–324.

Agassi, M., D. Bloem, and M. Ben-Hur. 1994. Effect of drop energy and soil and water chemistry on infiltration erosion. Water Resour. Res. 30:1187–1193.

Agassi, M., J. Morin, and I. Shainberg. 1985. Effect of raindrop impact energy and water salinity on infiltration rates of sodic soils. Soil Sci. Soc. Am. J. 49:186–190.

Agassi, M., I. Shainberg, and J. Morin. 1981. Effect of electrolyte concentration and soil sodicity on infiltration rate and crust formation. Soil Sci. Soc. Am. J. 45:848–851.

Agassi, M., I. Shainberg, and J. Morin. 1988. Effects on seal properties of changes in drops energy and water salinity during a continuous rainstorm. Aust. J. Soil Res. 26:651–659.

Al-Durrah, M.M., and J.M. Bradford. 1981. New methods of studying soil detachment due to water drop impact. Soil Sci. Soc. Am. J. 45:949–953.

Arshad, M.A., and A.R. Mermut. 1988. Micromorphological and physicochemical characteristics of soil crust types in Northwestern Alberta, Canada. Soil Sci. Soc. Am. J. 52:724–729.

Assouline, S. 2004. Rain-induced soil surface sealing: A critical review of observations, conceptual models, and solutions. Vadose Zone J. 3:570–591.

Attou, F., A. Bruand, and Y. Le Bissonnais. 1998. Effect of clay content and silt-clay fabric on stability of artificial aggregates. Eur. J. Soil Sci. 49:569–577.

Bajracharya, R.M., and R. Lal. 1999. Land use effects on soil crusting and hydraulic response of surface crusts on a tropical Alfisol. Hydrol. Processes 13:50–72.

Bakker, A.C., and W.W. Emerson. 1973. The comparative effect of exchangeable calcium, magnesium and sodium on some physical properties of red-brown earth subsoils. III: The permeability of Shepperton soil and comparison methods. Aust. J. Soil Res. 11:159–165.

Baver, L.D., W.H. Gardner, and W.R. Gardner. 1972. Soil physics. 4th ed. John Wiley and Sons, New York.

Ben-Hur, M., M. Agassi, R. Keren, and J. Zhang. 1998. Compaction, aging and raindrop-impact effects on hydraulic properties of saline and sodic vertisols. Soil Sci. Soc. Am. J. 62:1377–1383.

Ben-Hur, M., and M. Lado. 2008. Effect of soil wetting conditions on seal formation, runoff and soil loss in arid and semiarid soils—A review. Aust. J. Soil Res. 46:191–202.

Ben-Hur, M., I. Shainberg, D. Bakker, and R. Keren. 1985. Effect of soil texture and $CaCO_3$ content on water infiltration in crusted soil as related to water salinity. Irrig. Sci. 6:281–294.

Betzalel, I., J. Morin, Y. Benyamini, I. Agassi, and I. Shainberg. 1995. Water drop energy and soil seal properties. Soil Sci. 159: 13–22.

Blake, G.R., and R.D. Gilman. 1970. Thixotropic changes with aging of synthetic soil aggregates. Soil Sci. Soc. Am. Proc. 34:561–564.

Bradford, J.M., J.E. Ferris, and P.A. Ramley. 1987. Interrill soil erosion processes: I. Effect of surface sealing on infiltration. Soil Sci. Soc. Am. J. 51:1566–1571.

Bradford, J.M., and C. Huang. 1992. Physical components of crusting. p. 55–72. In M.E. Sumner and B.A. Stewart. (ed.). Soil crusting: Chemical and physical processes. Lewis Publ., Boca Raton, FL.

Bradford, J.M., P.A. Ramley, J.E. Ferris, and J.B. Santini. 1986. Effect of soil surface sealing on splash from a single waterdrop. Soil Sci. Soc. Am. J. 50:1547–1552.

Bradford, J.M., C.C. Truman, and C. Huang. 1992. Comparison of three measures of resistance of soil surface seals to raindrop splash. Soil Technol. 5:47–56.

Bresler, E., and W.D. Kemper. 1970. Soil water evaporation as affected by wetting methods and crust formation. Soil Sci. Soc. Am. Proc. 34:3–8.

Chen, Y., A. Banin, and A. Borochovitch. 1983. Effect of potassium on soil structure in relation to hydraulic conductivity. Geoderma 30:135–147.

Chen, Y., J. Tarchitzky, J. Brouwer, J. Morin, and A. Banin. 1980. Scanning electron microscope observation of soil crusts and their formation. Soil Sci. 130:49–55.

Chiang, S.D., D.E. Radcliffe, W.P. Miller, and K.D. Newman. 1987. Hydraulic conductivity of three southeastern soils as affected by sodium, electrolyte concentration, and pH. Soil Sci. Soc. Am. J. 51:1293–1299.

Cruse, R.M., and W.E. Larson. 1977. Effect of soil shear strength on soil detachment due to raindrop impact. Soil Sci. Soc. Am. J. 41:777–781.

Duley, F.L. 1939. Surface factors affecting the rate of intake of water by soils. Soil Sci. Soc. Am. Proc. 4:60–64.

Eigel, J.D., and I.D. Moore. 1983. Effect of rainfall energy on infiltration into bare soil. Proc. of the Nat. Conf. on Advances in Infiltration, Chicago, IL. 12–13 Dec. 1983. ASAE, St. Joseph, MI.

Ellis, J.H., and O.G. Caldwell. 1935. Magnesium clay solonetz. Trans. 3rd Int. Congr. Soil Sci. 348–350.

Ellison, W.D. 1947. Soil erosion studies. III: Some effects of erosion on infiltration and surface runoff. Agric. Eng. 28:245–248.

Epstein, E., and W.J. Grant. 1973. Soil crust formation as affected by raindrop impact. p. 195–201. In A Hadas et al. (ed.) Physical aspects of soil water and salts in ecosystems. Ecological Studies 4. Springer-Verlag, New York.

Fan, Y., T. Lei, I. Shainberg, and Q. Cai. 2008. Wetting rate and raindepth effects on crust strength and micromorphology. Soil Sci. Soc. Am. J. 72:1604–1610.

Farres, P. 1978. The role of time and aggregate size in the crusting process. Earth Surf Process. 3:243–254.

Farres, P.J. 1987. The dynamics of rain splash erosion and the role of soil aggregate stability. Soil Sci. Soc. Am. J. 14:119–130.

Francis, P.B., and R.M. Cruse. 1983. Soil water matrix potential effects on aggregate stability. Soil Sci. Soc. Am. J. 47:578–581.

Frenkel, H., J.O. Goertzen, and J.D. Rhoades. 1978. Effects of clay type and content, exchangeable sodium percentage, and electrolyte concentration on clay dispersion and soil hydraulic conductivity. Soil Sci. Soc. Am. J. 48:32–39.

Froese, J.C., R.M. Cruse, and M. Ghaffarzadeh. 1999. Erosion mechanics of soils with an impermeable subsurface layer. Soil Sci. Soc. Am. J. 63:1836–1841.

Fullan, M.A. 1991. Soil organic matter and erosion processes on arable loamy sand soils in the West Midlands of England. Soil Technol. 4:19–31.

Gal, M., L. Arcan, I. Shainberg, and R. Keren. 1984. The effect of exchangeable sodium and phosphogypsum on the structure of soil crusts. Soil Sci. Soc. Am. J. 48:872–878.

Gerard, C.J. 1965. The influence of soil moisture, soil texture, drying conditions and exchangeable cations on soil strength. Soil Sci. Soc. Am. Proc. 29:641–645.

Goldberg, S., and R.A. Glaubig. 1987. Effect of saturating cation, pH and aluminum and iron oxides on the flocculation of kaolinite and montmorillonite. Clays Clay Miner. 35:220–227.

Govers, G., and J. Poesen. 1986. A field scale study of surface sealing and compaction of loam and sandy loam soils. Part I. Spatial variability of surface sealing and crusting. p. 171–182. In F. Callebaut et al. (ed.) Assessment of soil surface sealing and crusting. State Univ. of Ghent, Ghent, Belgium.

Green, W.H., and G.A. Ampt. 1911. Studies on soil physics. I: Flow of air and water through soils. J. Agric. Sci. 4:1–24.

Guerra, A. 1994. The effect of organic matter content on soil erosion in simulated rainfall experiments in West Sussex, UK. Soil Use Manage. 10:60–64.

Hardy, N., I. Shainberg, M. Gal, and R. Keren. 1983. The effect of water quality and storm sequence upon infiltration rate and crust formation. J. Soil Sci. 34:665–676.

Hillel, D. 1980. Application of soil physics. Academic Press, New York.

Horton, R.E. 1940. An approach toward a physical interpretation of infiltration capacity. Soil Sci. Soc. Am. Proc. 5:399–417.

Hudson, N. 1971. Soil conservation. Cornell Univ. Press, Ithaca, NY.

Kay, B.D., and D.A. Angers. 1999. Soil structure. p. A229–A276. In M.E. Sumner (ed.) Handbook of soil science. CRC Press, Boca Raton, FL.

Kazman, Z., I. Shainberg, and M. Gal. 1983. Effect of low levels of exchangeable Na (and phosphogypsum) on the infiltration rate of various soils. Soil Sci. 135:184–192.

Kemper, W.D., and E.J. Koch. 1966. Aggregate stability of soils from western U.S. and Canada. USDA Tech. Bull. 1355. U.S. Gov. Print. Office, Washington, DC.

Kemper, W.D., and R.C. Rosenau. 1984. Soil cohesion as affected by time and water content. Soil Sci. Soc. Am. J. 48:1001–1006.

Kemper, W.D., T.J. Trout, M.J. Brown, and R.C. Rosenau. 1985. Furrow erosion and water and soil management. Trans. ASAE 28:1564–1572.

Keren, R. 1990. Water-drop kinetic energy effect on infiltration in sodium–calcium–magnesium soils. Soil Sci. Soc. Am. J. 54:983–987.

Kjaergaard, C., L.W. de Jonge, P. Moldrup, and P. Schjonning. 2004. Water dispersible colloids: Effects of measurement method, clay content, initial soil matric potential, and wetting rate. Vadose Zone J. 3:403–412.

Lado, M., A. Paz, and M. Ben-Hur. 2004. Organic matter and aggregate size interactions in infiltration, seal formation and soil loss. Soil Sci. Soc. Am. J. 68:935–942.

Laws, J.O. 1940. Recent studies in raindrops and erosion. Agric. Eng. 21:431–433.

Le Bissonnais, Y. 1990. Experimental study and modeling of soil surface crusting processes. Catena Suppl. 17:13–28.

Le Bissonnais, Y. 1996. Aggregate stability and assessment of soil crustability and erodibility. I. Theory and methodology. Eur. J. Soil Sci. 47:425–437.

Le Bissonnais, Y., and D. Arrouays. 1997. Aggregate stability and assessment of soil crustability and erodibility.

II. Application to humic loamy soils with various organic carbon contents. Eur. J. Soil Sci. 48:39–48.

Le Bissonnais, Y., B. Renaux, and H. Delouche. 1995. Interactions between soil properties and moisture content in crust formation, runoff and interrill erosion from tilled loess soils. Catena 25:33–46.

Le Bissonnais, Y., and M.J. Singer. 1992. Crusting, runoff, and erosion response to soil water content and successive rainfalls. Soil Sci. Soc. Am. J. 56:1898–1903.

Lee, S.S., C.J. Gantzer, A.L. Thompson, S.H. Anderson, and R.A. Ketcham. 2008. Using high-resolution computed tomography analysis to characterize soil-surface seals. Soil Sci. Soc. Am. J. 72:1478–1485.

Levy, G.J., and A.I. Mamedov. 2002. High energy moisture characteristics aggregate stability as a predictor for seal formation. Soil Sci. Soc. Am. J. 66:1603–1609.

Levy, G.J., and M.E. Sumner. 1998. Mined and by-product gypsum as soil amendments and conditioners. p. 445–462. *In* A. Wallace and E.R. Terry (ed.) Handbook of soil conditioners. Marcel Dekker, New York.

Levy, G.J., and H.v.H. van der Watt. 1988. Effects of clay mineralogy and soil sodicity on the infiltration rates of soils. S. Afr. J. Plant Soil 15:92–96.

Levy, G.J., and H.v.H. van der Watt. 1990. Effect of exchangeable potassium on the hydraulic conductivity and infiltration rate of some South African soils. Soil Sci. 149:69–77.

Levy, G.J., P.R. Berliner, H.M. du Plessis, and H.v.H. van der Watt. 1988a. Microtopographical characteristics of artificially formed crusts. Soil Sci. Soc. Am. J. 52:784–791.

Levy, G.J., J. Levin, and I. Shainberg. 1997. Prewetting rate and aging effects on seal formation and interrill soil erosion. Soil Sci. 162:131–139.

Levy, G.J., I. Shainberg, and J. Morin. 1986. Factors affecting the stability of soil crusts in subsequent storms. Soil Sci. Soc. Am. J. 50:196–201.

Levy, G. J., H.v.H. van der Watt, and H.M. du Plessis. 1988b. Effect of sodium-magnesium and sodium-calcium systems on soil hydraulic conductivity and infiltration. Soil Sci. 146:303–310.

Loch, R.J. 1994. Structure breakdown on wetting. p. 113–132. *In* H.B. So et al. (ed.) Sealing, crusting and hardsetting soils. Australian Soil Sci. Soc., Queensland Branch, Brisbane, Australia.

Luk, S.H. 1985. Effect of antecedent soil moisture content on rainwash erosion. Catena 12:129–139.

Luk, S.H., W.E. Dubbin, and A.R. Memut. 1990. Fabric analysis of surface crusts developed under simulated rainfall on loess soil, China. p. 29–40. *In* R.B. Bryan (ed.) Soil erosion–experiments and models. Catena Suppl. 17. Catena-Verlag, Cremlingen-Destedt, Germany.

Mamedov, A.I., C. Huang, and G.J. Levy. 2006. Antecedent moisture content and aging duration effects on seal formation and erosion in smectitic soils. Soil Sci. Soc. Am. J. 70:832–843.

Mamedov, A.I., and G.J. Levy. 2001. Clay dispersivity and aggregate stability effects on seal formation and erosion in effluent irrigated soils. Soil Sci. 66:631–639.

Mamedov, A.I., G.J. Levy, I. Shainberg, and J. Letey. 2001. Wetting rate and soil texture effect on infiltration rate and runoff. Aust. J. Soil Res. 30:1293–1305.

Mamedov, A.I., I. Shainberg, and G.J. Levy. 2000. Irrigation with effluent water: Effect of rainfall energy on soil infiltration. Soil Sci. Soc. Am. J. 64:732–737.

Martin, J.P., W.P. Martin, J.B. Page, W.A. Raney, and J.D. de Met. 1995. Soil aggregation. Adv. Agron. 7:1–37.

McIntyre, D.S. 1958a. Permeability measurements of soil crusts formed by raindrop impact. Soil Sci. 85:185–189.

McIntyre, D.S. 1958b. Soil splash and the formation of surface crusts by raindrop impact. Soil Sci. 85:185–189.

Miller, W.P. 1987a. A solenoid-operated variable intensity rainfall simulator. Soil Sci. Soc. Am. J. 51:832–834.

Miller, W.P. 1987b. Infiltration and soil loss of three gypsum-amended Ultisols under simulated rainfall. Soil Sci. Soc. Am. J. 51:1314–1320.

Miller, W.P., and J. Scifres. 1988. Effect of sodium nitrate and gypsum on infiltration and erosion of a highly weathered soil. Soil Sci. 148:304–309.

Mohammed, D., and R.A. Kohl. 1987. Infiltration response to kinetic energy. Trans. ASAE 30:108–111.

Moldenhauer, W.C., and W.D. Kemper. 1969. Interdependence of water drop energy and clod size on infiltration and clod stability. Soil Sci. Soc. Am. Proc. 33:297–301.

Morin, J., and Y. Benyamini. 1977. Rainfall infiltration into bare soils. Water Resour. Res. 13:813–817.

Morin, J., Y. Benyamini, and A. Michaeli. 1981. The effect of raindrop impact on the dynamics of soil surface crusting and water movement in the profile. J. Hydrol. 52:321–335.

Morin, J., S. Goldberg, and I. Seginer. 1967. A rainfall simulator with a rotating disk. Trans. ASAE 10:74–79.

Morrison, M.W., L. Prunty, and J.F. Giles. 1985. Characterizing strength of soil crusts formed by simulated rainfall. Soil Sci. Soc. Am. J. 49:427–431.

Norton, L.D., I. Shainberg, and K.W. King. 1993. Utilization of gypsiferous amendments to reduce surface sealing in some humid soils of the Eastern USA. Catena Suppl. 24:77–92.

Onofiok, O., and M.J. Singer. 1984. Scanning electron microscope studies of soil surface crusts formed by simulated rainfall. Soil Sci. Soc. Am. J. 48:1137–1143.

Oster, J.D., and F.W. Schroer. 1979. Infiltration as influenced by irrigation water quality. Soil Sci. Soc. Am. J. 43:444–447.

Panabokke, C.R., and J.P. Quirk. 1957. Effect of initial water content on stability of soil aggregates in water. Soil Sci. 83:185–195.

Philip, J.R. 1957. The theory of infiltration. 4: Sorptivity and algebraic infiltration equations. Soil Sci. 84:257–264.

Poesen, J. 1984. The influence of slope angle on infiltration rate and Hortonian overland flow volume. Z. Geom. 49(Suppl.):117–131.

Quirk, J.P., and C.R. Panabokke. 1962. Incipient failure of soil aggregates. J. Soil Sci. 13:60–69.

Quirk, J.P., and R.K. Schofield. 1955. The effect of electrolyte concentration on soil permeability. J. Soil Sci. 6:163–178.

Radcliffe, D.E., L.T. West, R.K. Hubbard, and L.E. Asmussen. 1991. Surface sealing in coastal plains loamy sands. Soil Sci. Soc. Am. J. 55:223–227.

Reichert, J.M., and L.D. Norton. 1994. Aggregate stability and rain-impacted sheet erosion of air-dried and prewetted clayey surface soils under intense rain. Soil Sci. 158:159–169.

Remley, P.A., and J.M. Bradford. 1989. Relationship of soil crust morphology to inter-rill erosion parameters. Soil Sci. Soc. Am. J. 53:1215–1221.

Shainberg, I., and J. Letey. 1984. Response of soils to sodic and saline conditions. Hilgardia 52:1–57.

Shainberg, I., and G.J. Levy. 1995. Infiltration and seal formation processes. p. 1–22. *In* M. Agassi (ed.) Soil erosion and rehabilitation. Marcel Dekker, New York.

Shainberg, I., and J.M. Singer. 1988. Drop impact energy–soil ESP interactions in seal formation. Soil Sci. Soc. Am. J. 52:1449–1452.

Shainberg, I., A.I. Mamedov, and G.J. Levy. 2003. Role of wetting rate and rain energy in seal formation and erosion. Soil Sci. 168:54–62.

Shevbs, G.I. 1968. Data on the erosive action of water drops. Sov. Soil Sci. 2:262–269.

Singer, A. 1994. Clay mineralogy as affecting dispersivity and crust formation in Aridisols. p. 37–46. *In* J.D. Etchevers (ed.) Transactions of the 15th World Congress of Soil Science. Acapulco, Mexico. Vol. 8a. Int. Soc. Soil Sci. and Mexican Soc. Soil Sci., Acapulco, Mexico.

Stern, R., M. Ben-Hur, and I. Shainberg. 1991. Clay mineralogy effect on rain infiltration, seal formation and soil losses. Soil Sci. 152:455–462.

Sumner, M.E., and B.A. Stewart. (ed.). 1992. Soil crusting: Chemical and physical processes. Lewis Publ., Boca Raton, FL.

Tackett, T.L., and R.W. Pearson. 1965. Some characteristics of soil crusts formed by simulated rainfall. Soil Sci. 99:407–413.

Tarchitzky, J., A. Banin, J. Morin, and Y. Chen. 1984. Nature, formation and effects of soil crusts formed by water drop impact. Geoderma 33:135–155.

Thompson, A.L., and L.G. James. 1985. Water droplet impact and its effect on infiltration. Trans. ASAE 28:1506–1510.

Truman, C.C., and J.M. Bradford. 1990. Effect of antecedent soil moisture on splash detachment under simulated rainfall. Soil Sci. 150:787–798.

Truman, C.C., J.M. Bradford, and J.E. Ferris. 1990. Antecedent water content and rainfall energy influence on soil aggregates breakdown. Soil Sci. Soc. Am. J. 54:1385–1392.

Utomo, W.H., and A.R. Dexter. 1981. Age hardening of agricultural top soils. J. Soil Sci. 32:335–350.

Valentin, C., and M. Bresson. 1992. Morphology, genesis and classification of surface crusts in loamy and sandy soils. Geoderma 55:225–245.

Valentin, C., and J.F. Ruiz Figueroa. 1987. Effects of kinetic energy and water application rate on the development of crusts in a fine sandy loam soil using sprinkling irrigation and rainfall simulation. p. 401–408. *In* N. Fedoroff et al. (ed.). Micromorphologie des Sols. Proc. of the 7th International working meeting on soil micromorphology, Pars, July 1985. Assoc. Francaise pour L'Etude du Sol, Olivet, France.

van Olphen, H. 1977. An introduction to clay colloid chemistry. 2nd ed. John Wiley and Sons, New York.

Wakindiki, I.I.C., and M. Ben-Hur. 2002. Soil mineralogy and texture effects on crust micromorphology, infiltration and erosion. Soil Sci. Soc. Am. J. 66:897–905.

Wangemann, S.G., R.A. Kohl, and P.A. Molumeli. 2000. Infiltration and percolation influenced by antecedent soil water content and air entrapment. Trans. ASAE 43:1517–1523.

Warrington, D., I. Shainberg, M. Agassi, and J. Morin. 1989. Slope and phosphogypsum effects on runoff and erosion. Soil Sci. Soc. Am. J. 53:1201–1208.

West, L.T., S.C. Chiang, and L.D. Norton. 1992. The morphology of surface crusts. p. 73–93. *In* M.E. Sumner and B.A. Stewart. (ed.) Soil crusting: Chemical and physical processes. Lewis Publ., Boca Raton, FL.

Wischmeier, W.H., and D.D. Smith. 1951. Rainfall energy and its relation to soil loss. Trans. Am. Geophys. Union 39:285–308.

Photo: Considering soil structure.
Photo courtesy of John A. Kelley, USDA-NRCS, http://SoilScience.info

16

Manure Management

Francis J. Larney, Xiying Hao, and Edward Topp

The word manure can be traced to *manus,* Latin for "hand," alluding to the work involved in applying manure to the soil by hand in ancient times. As such, manure is related to "manual," "manufacture," and the scientific "manuscript." Despite its noble etymological lineage, manure has fallen into disrepute and we often hear about it in terms of a "problem," an "issue," or something to be "disposed of." Increasingly, manure is either disguised as a generic "agricultural by-product" or simply dismissed as a "waste."

This chapter examines the pros and cons of manure and peruses some of the challenges facing its utilization in current farming practices. With a focus on soil management, we look at post-land application aspects of manure but also refer to pre-land application influences on manure properties such as storage and/or processing (e.g., composting).

The Manure Dichotomy

Manure is a dichotomy: a valuable resource if used judiciously as a soil amendment or an environmental polluter if mismanaged. Domestication of animals ~5000 years ago allowed humans to exploit their abilities to convert otherwise inaccessible resources into useful products and services (Mallin and Cahoon, 2003). Animal production was necessarily resource-limited and since it was tightly coupled to the productivity of the landscape, manure supply would seldom have exceeded the assimilation capacity of the landscape. In fact, from the early days of agriculture, livestock were seen as necessary to maintain soil quality (Janzen, 2001). The traditional (somewhat idealized) "mixed farm" showed diversity in crops (grains, pastures, legumes, root crops) and livestock (cattle, *Bos* spp.; swine, *Sus scrofa*; sheep, *Ovis aries*; horses, *Equus caballus*; and poultry).

However, the traditional link between livestock and crop production has been broken, especially in the developed world. Agricultural subsidies and the availability of inexpensive fertilizers have allowed a spatial decoupling of livestock and crop production (Watson et al., 2005). This has increased the flow of nutrients occurring between farms, often separated by great distances, compared with recycling of nutrients within individual farms. For example, the beef cattle industry in southern Alberta imports nutrients in the form of feed corn (*Zea mays* L.) from

F.J. Larney, Agriculture & Agri-Food Canada, 5403 1st Avenue South, Lethbridge, Alberta, Canada T1J 4B1 (francis.larney@agr.gc.ca); X. Hao, Agriculture & Agri-Food Canada, 5403 1st Avenue South, Lethbridge, Alberta, Canada T1J 4B1 (xiying.hao@agr.gc.ca); E. Topp, Agriculture & Agri-Food Canada, 1391 Sandford St., London, Ontario, Canada N5V 4T3 (ed.topp@agr.gc.ca). Lethbridge Research Centre contribution no. 38708042.

doi:10.2136/2011.soilmanagement.c16

the U.S. Midwest while timothy (*Phleum pratense* L.) grown in southern Alberta is exported to Japan and Korea as a source of fiber for dairy cows and horses.

Farms that are 100% cropping rely on inorganic fertilizers as crop nutrient sources unless they obtain livestock manure from neighbors. In contrast, intensive livestock operations, whereby large numbers of animals or birds are raised in confined conditions (e.g., feedlots or specialized housing), often exist on land bases that are too small to accommodate the volume of manure produced. This decoupling leads to an oversupply of manure in certain areas, causing it to be undervalued and hence overapplied at land application. It is the overapplication of manure nutrients that causes subsequent detrimental effects on soil, water, and air quality.

At the Agriculture and Agri-Food Canada Research Centre in Lethbridge, Alberta, experimental treatments with manure have been used in long-term rotations since 1910 (Dubetz, 1983). Up to the 1960s, practically all manure research examined its nutritional value for crop production, especially under irrigation. In these agronomic experiments, manure (usually from cattle) was generally applied at low rates and regular intervals in extended rotations. Most results pointed to the beneficial effects of manure on crop production. It was not until the early 1970s that Sommerfeldt et al. (1973) raised the issue of environmental effects of manure on soil and water quality. This work coincided with the intensification of the beef cattle feedlot industry in the region and consequently in 1973, a long-term manure rate study (annual rates of 0, 30, 60, and 90 Mg ha^{-1} wet wt. under dryland, and 0, 60, 120, and 180 Mg ha^{-1} under irrigation) was initiated at Lethbridge. This study, among one of the oldest continuous annual manure application studies in the world (37 yr), has generated a legacy of research findings on beef feedlot manure application rate effects on soil phosphorus (Chang et al., 2005; Dormaar and Chang, 1995; Whalen and Chang, 2001, 2002b; Hao et al., 2004b, 2008), soil nitrogen (Chang and Entz, 1996; Chang and Janzen, 1996), soil chemistry (Sommerfeldt and Chang, 1985; Sommerfeldt et al., 1988; Chang et al., 1990, 1991; Dormaar and Sommerfeldt, 1986; Hao and Chang, 2002, 2003), soil physical properties (Sommerfeldt and

Chang, 1987; Gao and Chang, 1996; Miller et al., 2002a,b; Whalen and Chang, 2002a), nitrous oxide emissions (Chang et al., 1998), trace elements (Benke et al., 2008), and crop response (Chang et al., 1993, 1994; Hao et al., 2004c).

On the Lethbridge long-term manure plots, soil chemical properties have changed over time in response to manure application. Hao and Chang (2002) found that after 25 annual applications, the solution concentrations of Na, K, Mg, and total cations increased significantly with manure rate (Table 16|1). In solution, Na increased by as much as 3.5 times and Mg by 2 times. In contrast, Ca in solution decreased significantly with manure rate under both dryland and irrigated conditions. While Ca was the dominant cation in the control treatment (no manure), K became the dominant cation with manure addition to this calcareous soil.

The Upside of Manure
Nutrient Source for Crop Production

Manure adds nutrients mainly in the form of N and P as well as K, in addition to a host of minor and trace elements. Manure improves soil physical properties such as water retention, aggregate stability, and infiltration (Miller et al., 2002a,b; Whalen and Chang, 2002a) mainly via the input of organic matter (OM). This is especially relevant on lighter-textured irrigated soils where OM may become depleted due to intensive cultivation and low amounts of crop residues returned to the soil (Grandy et al., 2002; Porter et al., 1999).

The N and P nutrient supply from livestock manure is directly linked to the handling system, the amount applied, their concentrations and availability, and subsequent transformation and transport in soil. Larney et al. (2006a) reported that feedlot manure-handling treatment (fresh, stockpile, compost) influenced the amounts and forms of nutrients hauled to the field (Table 16|2). Unlike mineral fertilizer, which provides available nutrients on application, nutrient availability from livestock manure varies considerably. Nutrients may be released or immobilized after application depending on manure characteristics (Gagnon and Simard, 1999). Sørensen et al. (2003) reported a negative relationship between manure carbon/nitrogen (C/N)

Table 16|1. Effect of 25 yr of annual cattle manure applications on soluble and exchangeable cations in soil†.

Manure rate	Soluble					Exchangeable				
	Na	K	Mg	Ca	Total	Na	K	Mg	Ca	Total
Mg ha⁻¹ yr⁻¹			mmol L⁻¹					cmol$_c$ kg⁻¹		
Dryland										
0	2.91d	1.45c	1.49c	3.15a	9.00d	0.18c	1.65c	2.22d	15.4a	19.48c
30	4.13c	6.34b	1.87bc	2.98a	15.32c	0.23c	3.95b	2.73c	13.7a	20.66bc
60	5.47bA‡	8.25abA	2.36abA	2.58abA	18.66bA	0.42bA	5.14aA	3.64bA	15.0aA	24.21abA
90	6.99a	10.07a	2.54a	2.26b	21.86a	0.65a	5.69a	4.23a	14.5a	25.10a
Irrigated										
0	2.12c	1.34d	1.74c	3.68a	8.88d	0.06d	1.49d	2.26d	15.8b	19.6d
60	5.01bA	6.56cB	2.71bA	2.99bA	17.27cA	0.31cA	3.56cB	3.72cA	16.1bA	23.7cA
120	5.51ab	8.51b	2.98ab	2.60b	19.6b	0.40b	4.27b	4.68b	19.0ab	28.4b
180	7.69a	10.84a	3.64a	3.07b	25.24a	0.75a	5.42a	6.00a	21.3a	33.5a

† Hao, X., and C. Chang. 2002. Effect of 25 annual cattle manure applications on soluble and exchangeable cations in soil. Soil Sci. 167:126–134.

‡ Values with different lowercase letters in the same column indicate a significant manure rate effect under dryland or irrigated conditions, and values with different uppercase letters indicate a significant irrigation effect for the 60 Mg ha⁻¹ yr⁻¹ manure rate at 0.05 probability levels.

Table 16|2. Effect of manure-handling treatment on component concentrations in a haulage scenario using "as is" (wet wt.) materials†.

Component	Handling treatment		
	Fresh	Stockpile	Compost
		kg Mg⁻¹ wet wt.	
Total mass	1000	1000	1000
Dry matter	349c‡	428b	640a
Total C	108a	106a	104a
Total N	5.6b	6.6b	9.0a
Available N§	1.3a	1.9a	0.5b
Total P	1.6c	2.3b	3.3a
Available P¶	1.1c	1.5b	2.1a

† Adapted from Larney et al. (2006a).
‡ Within rows, values followed by the same letter are not significantly different from each other ($P = 0.05$).
§ Available N = (NO_3–N + NH_4–N) determined colorimetrically after extraction with 2 N KCl.
¶ Modified Kelowna extract (Ashworth and Mrazek, 1995).

ratio and N mineralization while Calderón et al. (2004) found net mineralization when manure C/N ratio was <16. Thomsen and Olesen (2000) reported greater amounts of N mineralized from anaerobically stored vs. composted manure.

The rate of N and P release is affected by soil and environmental factors in addition to the manure properties. Soil moisture and temperature are the two most important environmental factors since mineralization is controlled by soil microbes. Microbial activity generally increases with soil temperature and availability of water (Agehara and Warncke, 2005) and is also affected by wetting and drying cycles (Watts et al., 2007). Other studies suggest that soil type and texture play a role in the mineralization or immobilization of applied manure (Honeycutt et al., 2005; Powell et al., 2006).

Reviews on the beneficial aspects of beef (Stewart et al., 2000), dairy (Meyer and Schwankl, 2000), swine (Mikkelsen, 2000), and poultry (Cabrera and Sims, 2000) manure are available.

Use in Soil Reclamation

It is well established that organic amendments have positive effects on the restoration of degraded or disturbed soils. Degraded soils are normally low in OM, due to removal of topsoil by erosion or industrial activity. Manure replenishes OM and stimulates biological activity. In southern Alberta, Larney and Janzen (1996, 1997) and Larney et al. (2000a, 2000c) found livestock manure to be an excellent amendment for restoring productivity to eroded soils since reclamation with inorganic fertilizer is not economical (Smith et al., 2000b). The main drawback is that eroded soils are often too far away from manure sources for economical haulage. The reclamation success of oil and gas well-sites in agricultural areas depends on their capacity to sustain levels of biomass production similar to those that existed before soil disturbance. There is often a reclamation problem on older wellsites where the original topsoil was removed before drilling. Larney et al. (2003a, 2005) examined the effect of compost application with various levels of topsoil replacement depths on three abandoned wellsites in south-central Alberta. In the absence of topsoil, compost at a rate of 40 Mg ha^{-1} (dry wt.) had a reclamation capacity of 99% (based on cumulative wheat biomass yields over 4 yr) of the check treatment (100% topsoil replacement depth with no amendment). For a 50% topsoil replacement depth, addition of compost achieved a reclamation capacity of 108% of the check treatment. These results showed that manure acted as a substitute for topsoil, at least in the short-term (4 yr), as the longevity of the effect was not evaluated.

Compost as Value-Added Manure

The public perception of compost is much more positive than manure. The content is similar: basically OM and nutrients, but minus the undesirable features, namely pathogens, odor, excess water. Recently, composting has been gaining increased attention as an alternative means of handling manure generated by the livestock industry (Larney et al., 2006b, Larney and Hao, 2007). Composting is not a new technology; it essentially applies controlled conditions (e.g., optimum recipe of organic materials, aeration, turning frequency, time) to enhance a natural decomposition process.

A major advantage of composting is reduced mass, volume, and water content compared with fresh manure, which in turn reduces transportation requirements (Larney et al., 2000b, Larney et al., 2006a). On an equivalent wet weight basis ("as is"), composting allowed haulage of 56% more N, 84% more P, 91% more Zn, and 76% more Cu than fresh manure, which is advantageous in terms of moving nutrients and trace elements from high- to low-loading areas (Larney et al., 2008a). Concomitant benefits include elimination of pathogens (Larney et al., 2003b), parasites (Van Herk et al., 2004), weed seeds (Larney and Blackshaw, 2003), antibiotic (Dolliver et al., 2008) and pesticide (Büyük-sönmez et al., 2000) residues, and malodors on land application (Rynk, 1992).

Nutrients are stabilized during composting, which slows their release once soil-applied (Helgason et al., 2005, 2007). Compost also enhances soil physical (Carter et al., 2004) and biological (Lalande et al., 2003) properties, and has a disease-suppression effect (Litterick et al., 2004; Noble and Coventry 2005). However, like all manure-handling systems, C and N losses (Larney et al., 2008a,b) and greenhouse gas (GHG) emissions (Hao et al., 2001, 2004a) are associated with composting. Where the supply of manure currently exceeds land availability for application, or in some future scenario, if producers need to comply with stricter manure application rate regulations, composting may be an option to encourage nutrient export from high-loading watersheds to soils that may benefit from nutrient and OM inputs. Composting may be seen as a means of maximizing the potential for recycling manure nutrients by soils and crops while protecting surface and groundwater resources from manure-related contamination. Larney et al. (2006b), Larney and Hao (2007), Keener et al. (2000), and Peigné and Girardin (2004) provide further discussion on manure composting.

The Downside of Manure

The phrase "too much of a good thing" aptly describes the manure dichotomy. Manure is not necessarily bad per se, as outlined above. However, when produced in large volumes from today's intensive farming operations, it is continuously in the spotlight and often linked to environmental issues related to N, P, pathogens, odors, greenhouse gases (nitrous oxide and methane), and other contaminants (antibiotics, hormones). The process of land application takes manure from storage facilities (e.g., lagoons) and spreads it out to a much wider environment (e.g., agricultural fields), essentially a "solution to pollution is dilution" approach. This approach probably had merit when livestock (and human) populations were lower and when manure was somewhat more benign, e.g., it did not contain high levels of metals (Cu, Zn, Co, Mn from dietary supplements), antibiotics, or pharmaceuticals. We currently expect a lot of our soil: we need it to act as a buffer, to absorb manure components, to hold them in the root zone, and to prevent movement to adjacent environments, e.g., downward transport to groundwater, lateral runoff to surface water, or upward movement (volatilization) to the airshed.

Nitrogen Losses

Nitrogen losses from manure (feces, urine) in the form of ammonia (NH_3) begin as soon as it is excreted. In southern Alberta, McGinn et al. (2003) reported that NH_3 concentrations in air were highest (~200–850 $\mu g\ m^{-3}$) adjacent to feedlots and declined with increasing distance downwind (Fig. 16|1). At a 200-m distance, the concentration was reduced to 65 to 82% of that adjacent to the feedlots. Hao et al. (2006) found that NH_3 was responsible for sorption of 2.6 to 3.2 kg N ha^{-1} wk^{-1} by soil near a feedlot source, declining to about 0.3 kg N ha^{-1} wk^{-1} 1700 m downwind.

After land application, ~19% of the original N excreted by livestock may be lost via NH_3 emissions (Oenema et al., 2007). Manure characteristics such as total N, ammonium N (NH_4–N), pH and percent dry matter (DM) play an important role in NH_3 volatilization following manure application to soils. When dietary protein levels were decreased from 20 to 12%, swine slurry pH decreased by 1.3 units and NH_3–N fell from 5.43 to 1.92 g N kg^{-1}, contributing to a 63% reduction in NH_3 emissions when pig slurry was surface applied to soil (Portejoie et al., 2004).

Manure application method also affects NH_3 losses, with the largest losses occurring from surface-applied manure that is not soil-incorporated. Rapid incorporation of solid manure reduced NH_3 emission by 15 to 87% (Sagoo et al., 2007; Mkhabela et al., 2008). Applying liquid manure by injection (>2-cm depth) or soil banding also decreased NH_3 emissions by 23 to 57% (Misselbrook et al., 2002; Smith et al., 2000a). Lau et al. (2003) described a manure applicator that banded swine manure slurry directly over a row of slots made by an aerator. The applicator reduced NH_3 losses by 46 to 48% relative to surface broadcasting (Bittman et al., 2005).

Environmental factors, such as temperature, precipitation, and wind speed govern NH_3 solubility and volatilization and affect NH_3 losses after manure application. Irrigating fields following manure application (or application before an expected rainfall) can improve NH_3 retention by allowing it to dissolve in soil water. This water can then percolate into the soil, providing an opportunity for NH_4 and NH_3 to be sorbed by soil constituents and hence reducing emissions. Soil conditions, including the water content, texture, cation exchange capacity, pH, and plant or residue cover, affect the retention and movement of soluble N in

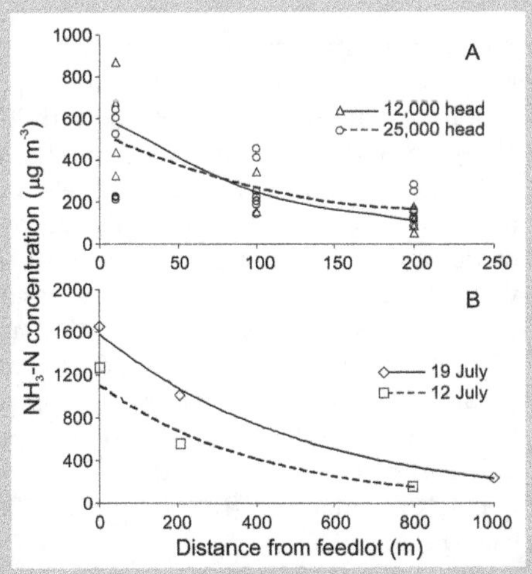

Fig. 16|1. Average ammonia concentration at distances downwind of (A) 12,000- and 25,000-head beef cattle feedlots in southern Alberta, measured over 2-d sampling durations in 1999 at fixed tower locations, and (B) dispersion measured on 12 July and 19 July 1999 with a portable analyzer using a 5-min sample duration (from McGinn et al., 2003).

soil and thus the amount of ammonia loss. As pH increases, the fraction in NH_3 form increases while that in NH_4^+ form decreases, thus increasing the likelihood of volatilization losses of NH_3. These losses may be reduced by lowering soil pH through the use of soil amendments.

McGinn et al. (2003) found a positive relationship between NH_3 concentration and odor intensity near a southern Alberta feedlot, showing that as well as contributing to N losses via volatilization, NH_3 was also a major player in odor. Before intensification of the livestock industry, odors emanating from manure were often trivialized as "the smell of money." However, with increased concentration of livestock numbers, odor has become a major air quality issue and, compared with other manure properties, it is often the one causing most public complaints. Millner and McConnell (2000) provide an excellent overview of odor issues associated with manure.

Nitrogen in manure is also lost as nitrous oxide (N_2O), a greenhouse gas. Nitrous oxide is produced naturally in soils through the microbial processes of nitrification and denitrification. Agricultural activities that increase mineral N availability in soils increase the amount available for nitrification and denitrification, and ultimately the magnitude of N_2O emissions. Land application of manure is among these activities, although addition of fertilizer, production of N-fixing crops and forages, and retention of crop residues also contribute. Nitrous oxide emissions from manure-amended soil vary considerably, being affected by manure type, application rate, and properties such as total and available C and N, and C/N ratio (Bhandral et al., 2007). Soil physical properties such as moisture and texture also play a role in manure decomposition and subsequent N_2O emissions (Rochette et al., 2006; Ciarlo et al., 2007).

Higher rates of N_2O emissions from manure-amended vs. fertilizer-amended soils have been reported (Stevens and Laughlin, 2001), in part due to an ample amount of biologically available C that stimulates or induces N_2O emission (Dittert et al., 2005). Similarly, manure with higher water-soluble C led to increased N_2O emissions from a clay loam soil in Ontario (Yang et al., 2008). However, Meng et al. (2005) reported an N_2O emission rate of 850 g N ha^{-1} yr^{-1} for soil receiving 300 kg N ha^{-1} yr^{-1}, regardless of whether the N source was organic manure or mineral fertilizer. Chantigny et al. (2007) reported similar rates of N_2O emission from soil receiving raw or biodigested pig manure.

Scaled-up estimates of N_2O emissions on a regional basis derived only from manure N application rates are inaccurate, as application technique has a considerable impact. Higher N_2O emissions were found after surface application of solid cattle manure to no-till vs. conventional-till soil (Mkhabela et al., 2008), attributed to higher SOM and water content under no-till. Moreover, application technique also affects the rate of N_2O emission and Perälä et al. (2006) advocated emission factors be developed to account for this variable.

The other main pathway for N loss is via NO_3–N leaching. Prerequisites for NO_3–N leaching are: (i) high soil NO_3–N content via overapplication of manure (or fertilizers) and (ii) capacity for downward water movement, e.g., heavy rainfall or irrigation. The timing of manure application and the type of crop also affect NO_3–N leaching. Van Es et al. (2006) showed that fall manure applications led to high NO_3–N leaching risks, especially on sandy soils, while manure applications to grass posed minimal concern compared with corn. Application of manure after fall harvest is commonplace, potentially increasing the risk for overwinter NO_3–N losses, since there is no crop present for N uptake. Simply changing the timing to a spring application may help. Ball-Coelho et al. (2004) found that overseeding cereal rye (*Secale cereale* L.) as a cover crop into standing corn minimized post-harvest NO_3–N leaching in southern Ontario.

Phosphorus Losses

Livestock manure applications that supply P in excess of crop requirements can increase soil P concentration and subsequently P loss to surface water (McDowell and Sharpley 2004; Volf et al., 2007). The application of livestock manure to soil increases both particulate P and soluble P loss in runoff (Smith et al., 2001), with the level of total P (TP) and dissolved reactive P (DRP) increasing with application rate (Miller et al., 2006).

Incorporating liquid swine manure into soil following application significantly reduced TP, DRP, and bioavailable P (BAP) concentrations in runoff (Allen and Mallarino 2008). Little et al. (2005) reported that P concentrations in runoff water were affected by tillage used for incorporation, with reductions in TP loads of 14% for double disk, 43% for cultivator, and 79% for moldboard plow compared with manure left on the soil surface. Applying manure immediately before rainfall also increases leaching losses. A 10- to 16-d rainfall delay reduced DRP, BAP and TP concentrations in runoff compared with when rainfall occurred within 24 h of application (Allen and Mallarino, 2008).

Miller et al. (2006) found that P concentrations were higher in runoff from fields receiving fresh vs. composted cattle manure. In contrast, McDowell and Sharpley (2004) found higher P leaching losses from soil amended with dairy manure compost (88 mg DRP and 148 mg TP) than dairy manure (26 DRP and 59 mg TP). They attributed greater P leaching from the compost-amended soil to blocking of P sorption sites. Phosphorus leaching losses were also affected by soil texture and application time. Van Es et al. (2004) reported 39 times higher P loss from preferential flow in clay loam (0.504 mg L^{-1}) than sandy loam soil (0.013 mg L^{-1}). Applying liquid dairy manure in early fall leads to greater P leaching than in spring. The overall potential environmental risks of P loss via runoff or leaching are generally higher with livestock manure P application than inorganic P fertilizer (Allen et al., 2006).

Contaminants in Manure

Protection of water resources adjacent to land receiving manure containing enteric microorganisms, veterinary pharmaceuticals, and hormones requires that these contaminants be killed, degraded, sequestered, or otherwise inactivated in the soil. This is best achieved through the application of manure to land at a judicious rate, and under suitable land, climate, and crop conditions to minimize the risk of offsite movement into adjacent surface or groundwater. The rate of microbial inactivation by soil at the point of application of manure in large part determines the risk for subsequent environmental problems, including contamination of food crops grown in manured fields (Islam et al., 2004). Understanding the behavior of manure contaminants following

application to soil is important in predicting exposure and managing risk to adjacent environments (Haas et al., 1999; Islam et al., 2004; Muirhead et al., 2005).

Pathogens

Zoonotic (i.e., transferable from animals to humans) pathogens represent a significant threat to water quality, as demonstrated through relationships between livestock production intensity, degradation of water quality, and the frequency of enteric illnesses in proximal populations (Michel et al., 1999; Cotruvo et al., 2004). The abundance and the identity of pathogenic microorganisms excreted by livestock can vary according to a variety of factors including the species of animal, herd or flock health, nutrition, and animal age (Inglis et al., 2004; Ojha and Kostrzynska, 2007). Many pathogens are characterized by episodic outbreaks, and consequently the seasonal detection and spatial distribution of pathogens from farm to farm can be highly variable (Hölzel and Bauer, 2008). The most commonly detected zoonotic pathogens for the major production systems include *Salmonella enterica*, *Yersinia* spp., *Campylobacter* spp. in pigs; *Cryptosporidium* spp., *Escherichia coli* O157, *Campylobacter jejuni* in cattle; and *Salmonella enterica* and *Campylobacter* spp. in poultry (Cotruvo et al., 2004). Overall, the maintenance of herd or flock health, sanitation, and welfare is the first crucial step in protecting both environmental and food quality from contamination with enteric pathogens (Gannon et al., 2004).

Various biological, chemical, and physical treatments can subject manures to conditions that promote faster loss of viability of viral, bacterial, and eukaryotic pathogens, and significantly reduce hazards (Gerba and Smith, 2005). Manure can be stored under varying redox conditions ranging from composts, aerated lagoons, anoxic lagoons, and biogas-producing anaerobic digesters (Larney et al., 2003b; Côté et al., 2006; Aitken et al., 2007). The combination of extremely high pH and exothermic heat production makes lime stabilization a potent chemical treatment option (Maguire et al., 2006). Overall, the microbial and chemical composition of manure will vary according to the length and method of storage, and treatment method to which it may be subjected (Bicudo and Goyal, 2003).

There is a wealth of information regarding the impact of climate, soil properties, and soil management factors on the persistence of enteric bacteria, notably *E. coli*. These include the effect of manure application timing and depth of placement, tillage, soil texture, soil temperature and moisture, presence of crop roots, protozoal grazing and the availability of invertebrate reservoirs, sunlight, and pH (Mawdsley et al., 1995; Stoddard et al., 1998; Maule, 2000; Bolton et al., 1999; Gagliardi and Karns 2002; Brown et al., 2002; Solomon et al., 2002; Natvig et al., 2002; Scott et al., 2006). It is generally unknown what innate factors contribute to variability of enteric bacterial persistence in soils, and whether persistence is associated in any way with attributes of public health significance, virulence potential or antibiotic resistance. The composition of *E. coli* populations changes profoundly in manure and in manured soils. Some *E. coli* clones can grow in soils following application of swine manure, apparently because of their ability to metabolize low molecular weight fatty acids (Topp et al., 2003; Scott et al., 2006). In swine manure, some genes that confer pathogenic potential are lost at high frequency, whereas antibiotic resistance is stably maintained (Duriez and Topp; 2007, Duriez et al., 2008). The infective resting structures of *Cryptosporidium* species (oocysts) and *Giardia* species (cysts) are notoriously persistent in environmental matrices (Mawdsley et al., 1996). In contrast, enteric viruses are rapidly inactivated.

Microorganisms are of sufficient size and inherent sorptivity that they are generally filtered out during the flow of water through the soil matrix. The key high risk exposure pathways for transport of microorganisms from manured soils to adjacent water are therefore surface runoff, and preferential flow via macropores to depth. Surface runoff is best managed through manure application methods and practices that physically incorporate the material with soil of good drainage class, that minimize soil saturation (in the case of slurry) by avoiding high antecedent soil moisture or significant post-application precipitation, that avoid sloping land, and that employ offset distances or buffer strips. Preferential flow is best minimized by employing manure application or tillage methods that incorporate material while disrupting macropores, and ensuring some significant depth to groundwater (Turpin et al., 2007). In the case of manure slurry, judicious rates of application and incorporation into the soil are important, particularly when land is systematically tiled for drainage (Ball Coelho et al., 2007; Lapen et al., 2008).

Veterinary Pharmaceuticals and Hormonal Substances

A variety of different classes (e.g., antibacterial, antiparasitic, antihistamine, steroidal and nonsteroidal anti-inflammatory) of pharmaceutical agents are approved for use in livestock and poultry production (Bloom, 2004). Antibacterials (antibiotics) are the pharmaceuticals of most concern because their widespread use as growth-promoting, prophylactic, and therapeutic agents is thought to contribute to the development of resistance to antibiotics that are of value in human medicine (Sarmah et al., 2006). Some antibiotics are persistent in treated animals, are excreted, and their residues detected on the farm or in the broader environment (Pope et al., 2008). Manure effluent may contaminate surface or groundwater with bacteria that have developed resistance to antibiotics (Sapkota et al., 2007). Humans and livestock may be exposed to antibiotic residues through consumption of crops that have taken up minute quantities of antibiotic residues from manured soils (Dolliver et al., 2007). Persistence in manure will vary according to the inherent recalcitrance of the antibiotic and the conditions under which it is held, in particular the redox conditions (e.g., composting, static anoxic, anaerobic digestion) and temperature (Pope et al., 2008). Antibiotics have been detected in soils associated with high livestock density e.g., floors of beef cattle feedlot pens (Aust et al., 2008).

Animals and poultry can naturally excrete significant amounts of reproductive hormones (Hanselman et al., 2003). In addition, in some production systems, synthetic hormonal agents designed to accelerate weight gain or synchronize reproductive cycles, are administered and subsequently excreted (Lorenzen et al., 2004). Certain legume-rich (e.g., soybean, [*Glycine max* (L.) Merr.]) feeds may contain elevated concentrations of phytoestrogenic chemicals, some of which are converted to more hormonally potent transformation products during anoxic manure storage (Burnison et al., 2003). Slurry can be more hormonally potent than manure stored as a solid, or deposited directly onto soil, because urinary excretion of conjugated hormones is a more important pathway that fecal excretion (Lorenzen et al., 2004; Hutchins et al., 2007).

Drugs that are of agricultural origin have been detected in surface and groundwater (Lissemore et al., 2006; Batt et al., 2006; Boxall et al., 2003). The dissipation characteristics and conditions that promote risk of field transport of veterinary pharmaceuticals are reviewed in Pope et al. (2008). The natural estrogenic hormones estradiol and estrone are rapidly and completely broken down in agricultural soils (Colucci and Topp, 2002; Hanselman et al., 2003). Testosterone is biodegraded rapidly in soil, with the transient accumulation of less hormonally potent steroidal transformation products (Lorenzen et al., 2005; Jacobsen et al., 2005). The synthetic anabolic steroids trenbolone-acetate and trendione are likewise biodegraded in soils with half-lives in the order of a few days (Khan et al., 2008).

Meeting the Manure Challenge
Best Management Practices

In agricultural settings, the term best management practice (BMP) generally refers to an effective, economical means of preventing or reducing negative impacts on soil, water, or air quality. Best management practices for manure minimize nutrient losses and optimize uptake by plants, thus improving both the economic value of manure as well as environmental quality (Alberta Cattle Feeders' Association and Alberta Agriculture, Food and Rural Development, 2002). Key BMPs for land application of manure include soil and manure analysis, nutrient management planning (e.g., matching nutrient application to projected crop demand), calibration of application equipment, soil injection or incorporation, critical area protection, vegetated waterways, riparian buffers or filter strips, use of cover crops, scavenger crops, and crop rotations, conservation or reduced tillage, ponds or retention structures, and rotational grazing or pasture management.

The scale of dealing with manure management issues has moved from the individual field scale to the watershed level in recent years, especially in the context of BMP adoption (Gitau et al., 2006).

While adoption of individual BMPs is helpful, their implementation as a suite is more effective. Inamdar et al. (2001) found BMPs (filter strips, nutrient management planning, no-till) reduced losses of some forms of nutrients, such as NH_4–N and particulate P, from the 1463 ha Nomini Creek watershed in Virginia, but concluded that additional BMPs were necessary to achieve significant reductions in all forms of N and P.

Nitrogen vs. Phosphorus-Based Land Application

Recommended manure application rates have typically been based on crop N requirements. However, most livestock manure has a narrower N/P ratio than crops need, so applying manure based on the desired N levels needed to meet crop growth supplies excess P, and ultimately leads to P accumulation in soil (Sharpley et al., 1984; Hao et al., 2008) and potential runoff. To control P runoff from manured fields and the risk of excess P to the environment and surface water quality, regulators have been switching from N-based to P-based manure application guidelines.

Sibbesen and Sharpley (1997) identified P-based manure application approaches: (i) the single P limit, (ii) P index, and (iii) percent P saturation. With the single P limit approach, P application restrictions take effect when soil test P reaches the critical level. Possible restrictions include no P application from any source, or at least application rates not exceeding crop removal, to reduce or eliminate P accumulation and erosion. The critical value varies from 75 to 200 mg P kg^{-1}, reflecting local soil, climate, and crop conditions. States using this approach include Arkansas, Delaware, Ohio, Oklahoma, Michigan, Texas, and Wisconsin.

Under USDA-USEPA joint strategies, the P index approach is based on assessment of various landforms and management practices to determine the risk of P export to the water system. This approach, first introduced by Lemunyon and Gilbert (1993), has undergone several modifications (Gbuerk et al., 2000). It uses information readily available to farmers and agricultural consultants to evaluate the potential for P in runoff from a specific field to enter a nearby stream and assigns a vulnerability rating to the site.

The percent P saturation approach is based on soil P levels and the potential for dissolved P to enter the water system. This approach is used in the Netherlands. Once calibrated for a specific soil, the P saturation approach could determine the capacity for a wide range of soils to release P. Unlike the P limit approach, the P saturation approach indicates how close a soil is to being an environmental risk for P loading. However, the approach has been criticized for relying solely on soil factors without taking other environmental and management practices into consideration. Regardless of the approach chosen by the regulators, P-based manure application rates reduce the amount of manure that can be applied to agricultural soil, which increases the amount of land needed to dispose of the same amount of manure (Ribaudo et al., 2003).

The Future for Manure
Role in Sustainable Agriculture

The traditional role of manure as a soil amendment has been diminished, exacerbated by industry changes toward larger production facilities and concentrations of livestock production in certain geographic regions (Petersen et al., 2007; Mallin and Cahoon, 2003). However, there is renewed interest in reintegrating crops and livestock because of concerns about long-term sustainability, natural resource degradation, profitability and stability of farm income, and increased regulation of livestock operations (Russelle et al., 2007). Integrated crop–livestock systems could foster diverse cropping systems and utilize animal manure, which enhances soil tilth, fertility, and C sequestration. Wilkins (2008) stressed the potential importance of mixed farming involving both crops and livestock, particularly when the systems incorporate biological N fixation and manure recycling. However, there was little evidence that the trend to farm-level specialization was declining in developed countries.

In much of the developing world, under small-scale crop production systems, the search for sustainable soil fertility replenishment techniques is an urgent need. A key resource in this respect is animal manure. The combined use of animal manure and mineral fertilizers has proved a promising alternative strategy for resource-poor small-scale farmers in Africa (Bayu et al., 2005; Rufino et al., 2007; Schlecht et al., 2006). Mkhabela (2005) pointed out that the use of manure, an old technology, is appropriate for small-scale farmers in South Africa, as most farmers practice mixed livestock and crop farming yet do not fully exploit available manure for replenishing the fertility of their soils.

In the larger context, Karmakar et al. (2007) identified a need to develop whole-farm decision-support systems that address all the major components of manure management systems, such as manure collection, storage, treatment, and land application. Grusenmeyer and Cramer (1997) recommended a total systems approach which expanded the focus further to include human and animal health, odor and fly control, nutrient import and export, diet manipulation and ration balancing, and the broader economic impacts of environmental regulation, compliance, and enforcement. Sandars et al. (2003) advised the use of a life cycle assessment to follow mass and energy balances of manure management practices from "cradle to grave" to ensure that improvements at one stage corresponded to an overall improvement and did not simply move problems up or down the chain

Recent research has aimed to manipulate dietary intake at the "front end" so as to change forms and concentrations of nutrients excreted at the "back end" of the animal. Diets can be formulated to increase animal N and P utilization, thereby reducing amounts excreted (Hao et al., 2005; Maguire et al., 2007; Yan et al., 2006). The use of dried distillers' grains with solubles (DDGS) in animal diets is increasing (Berger and Good, 2007) as the bioethanol industry expands. With the N and P concentrations in DDGS approximately three times those in unprocessed grain (Spiehs et al., 2002; Widyaratne and Zijlstra, 2007), adding DDGS to animal diets could potentially increase excretion of N and P nutrients. Livestock producers should consider the impact on manure when managing animal diet so as to become more environmentally sustainable.

The Carbon Connection

Manure, unlike fertilizer, contains C, which is often underappreciated compared with N and P in manure. Fellman et al. (2008) suggested that livestock manure represented a potential for long-term soil C gain. Their model simulations predicted that between 1980 and 1995, United States soils were responsible for the sequestration of 444 to 602 Tg C from livestock manure. During the same period, global soil C sequestration from livestock manure was 2810 to 4218 Tg C.

Pendell et al. (2006) reported that farm managers who land apply manure could benefit from selling C credits and collect larger per hectare payments than those using commercial N-fertilized systems because manure-fertilized systems have higher soil C-sequestration rates. This benefit may be even larger if land application of manure had fewer C-equivalent CH_4 emissions than other manure management methods and net soil C-sequestration rates were used as a basis for C credit payments.

Bioenergy Feedstock

The nutrients and OM in manure are beneficial to crop production by improving nutrient supply and soil quality when land applied. However, some regions have more manure than they can use without risking environmental damage. An alternative to land application is use of livestock manure as a source of energy. The increasing world demand for energy and the high cost of oil and natural gas has increased the interest in alternative and renewable energy sources, which is likely to grow in coming years. Cantrell et al. (2008) provide an extensive summary on opportunities for using livestock manure to generate bioenergy. Briefly, OM in the form of manure is converted into bioenergy either biochemically or thermochemcially. The biochemical processes include (i) anaerobic digestion and CH_4 biogas production, (ii) biohydrogen production, and (iii) biomethanol production. The thermochemical processes include (i) gasification, (ii) direct liquefaction, (iii) pyrolysis, and (iv) combustion. Anaerobic digestion and CH_4 biogas production is the most common process to convert livestock manure to bioenergy.

Anaerobic digestion breaks down complex organic material and produces biogas (CH_4), CO_2, and trace gases by a community of anaerobic microorganisms (Cantrell et al., 2008). The biogas yield depends on the temperature in the digester, the hydraulic retention time, and the specific load (Clemens et al., 2006). During fermentation, substrate properties, such as volatile DM content, NH_3 concentration, pH, and viscosity, may also affect biogas production.

Anaerobic digestion has the potential to mitigate GHG emissions from livestock manure and presents a means of biogas energy production without increasing GHG emissions during subsequent storage and field application. As an add-on component to bioenergy production, the organic solids by-product could be separated into solid and liquid components. The solids could be applied to cropland and used for local crop production (Loria et al., 2007). Alternatively, biodigested solids could be pelletized and shipped away from the source, representing nutrient export from a livestock facility. Additional advantages of bioenergy production include improved nutrient utilization, reduced odor from slurry storage, and reduced risk of leaching.

Rules and Regulations

A wide variety of strategies have been used to reduce environmental impacts of manure from intensive livestock operations (Sims et al., 2005). Initially, voluntary compliance and encouragement of BMPs were favored, but more recently "command and control" policies that rely more on regulations have predominated. Sims et al. (2005) remarked that despite decades of voluntary effort and increasing regulations, nutrient management from intensive animal agriculture remains an open question. They advocated that a fundamental guiding principle for manure management policies should be nutrient balance, at all spatial scales.

Typical manure application regulations include minimum setback distances from water bodies, protection of soil and water, soil testing, and record keeping (Alberta Agriculture and Rural Development, 2008). Manure regulations are becoming stricter, especially in Europe. Germany, Denmark, and the Netherlands have long-established regulations governing the rates and timing of manure spreading on agricultural land (Latacz-Lohmann and Hodge, 2003).

Manure regulations usually contain stocking rate limitations that effectively link all forms of livestock production to the land. More recently, a market approach has been adopted in the Netherlands, through levies on nutrient surpluses and opportunities for manure trading among farmers. However, neighbor complaints, odor, and spreading of weed seeds presented barriers to manure exchange in Michigan (Battel and Krueger, 2005).

O'Connor et al. (2005) pointed out that the traditional definition of land application, which emphasizes applying manure to land in a manner that protects human and animal health, safeguards soil and water resources, and maintains long-term ecosystem quality, is incomplete unless the earning of public trust in the practices is included.

Summary and Conclusions

In an idealized world, manure is produced by animals, applied to soils, and used by crops that are fed back to animals, hence closing the nutrient loop. However, modern farming practices, with high concentrations of animals on small land bases, often preclude closing the loop. Manure is a fairly unique agricultural commodity in that it has the ability to integrate numerous disciplines including animal nutrition, agricultural engineering, soil science, and agronomy. Fortunately, collaboration among these disciplines has increased in recent years in the face of continued pressure to reduce environmental impacts of manure. In an era of escalating energy prices, the future use of oil in agriculture has to be rationalized and made as efficient as possible. Recycling manure back to the land is hardly novel, yet innovate thinking is needed to facilitate and implement this "known technology."

With manure management, we must first recognize that nutrient losses occur and adopt practices that mitigate these losses. We should remember that most of the discussion on N and P losses from manure also applies to N and P fertilizers. However, because fertilizer is an expensive crop input and manure is essentially free (at least in the raw form), fertilizer rates are seldom exceeded whereas manure is often applied either with a "more is better"

philosophy or simply as a means of disposal. In some cases, legislation to restrict diffuse pollution may provide incentives for more diverse farming practices and reduce overapplication of manure. Alternatively, future price premia for produce from "environmentally friendly" farming systems, and subsidies for the provision of environmental services may provide economic incentives for the adoption of better manure management practices.

References

Agehara, S., and D.D. Warncke. 2005. Soil moisture and temperature effects on nitrogen release from organic nitrogen sources. Soil Sci. Soc. Am. J. 69:1844–1855.

Aitken, M.D., M.D. Sobsey, N.A. Van Abel, K.E. Blauth, D.R. Singleton, P.L. Crunk, C. Nichols, G.W. Walters, and M. Schneider. 2007. Inactivation of *Escherichia coli* O157:H7 during thermophilic anaerobic digestion of manure from dairy cattle. Water Res. 41:1659–1666.

Alberta Agriculture and Rural Development. 2008. Agricultural Operation Practices Act reference guide. Available at: http://www1.agric.gov. ab.ca/$department/deptdocs.nsf/all/epw5592 (verified 19 Oct. 2010).

Alberta Cattle Feeders' Association and Alberta Agriculture, Food and Rural Development. 2002. Beneficial management practices: Environmental manual for feedlot producers in Alberta. Alberta Cattle Feeders' Association, Calgary, AB.

Allen, B.L., and A.P. Mallarino. 2008. Effect of liquid swine manure rate, incorporation, and timing of rainfall on phosphorus loss with surface runoff. J. Environ. Qual. 37:125–137.

Allen, S.C., V.D. Nair, D.A. Graetz, S. Jose, and P.K.R. Nair. 2006. Phosphorus loss from organic versus inorganic fertilizers used in alley cropping on a Florida Utisoil. Agric. Ecosyst. Environ. 117:290–298.

Ashworth, J., and K. Mrazek. 1995. "Modified Kelowna" test for available phosphorus and potassium in soil. Commun. Soil Sci. Plant Anal. 26:731–739.

Aust, M.O., F. Godlinski, G.R. Travis, X. Hao, T.A. McAllister, P. Leinweber, and S. Thiele-Bruhn. 2008. Distribution of sulfamethazine, chlortetracycline and tylosin in manure and soil of Canadian feedlots after subtherapeutic use in cattle. Environ. Pollut. 156:1243–1251.

Ball-Coelho, B.R., R.C. Roy, and A.J. Bruin. 2004. Nitrate leaching as affected by liquid swine manure and cover cropping in sandy soil of southwestern Ontario. Can. J. Soil Sci. 84:187–197.

Ball Coelho, B.R., R.C. Roy, E. Topp, and D.R. Lapen. 2007. Tile water quality following liquid swine manure application into standing corn. J. Environ. Qual. 36:580–587.

Batt, A.L., D.D. Snow, and D.S. Aga. 2006. Occurrence of sulfonamide antimicrobials in private water wells in Washington County, Idaho, USA. Chemosphere 64:1963–1971.

Battel, R.D., and D.E. Krueger. 2005. Barriers to change: Farmers' willingness to adopt sustainable manure management practices. Available at http://www.joe. org/joe/2005august/a7p.shtml (verified 19 Oct. 2010). J. Extension 43(4).

Bayu, W., N.F.G. Rethman, and P.S. Hammes. 2005. The role of animal manure in sustainable soil fertility management in sub-Saharan Africa: A review. J. Sustain. Agric. 25:113–136.

Benke, M.B., S.P. Indraratne, X. Hao, C. Chang, and T.B. Goh. 2008. Trace element changes in soil after long-term cattle manure applications. J. Environ. Qual. 37:798–807.

Berger, L.L., and D.L. Good. 2007. Distillers dried grains plus solubles utilization by livestock and poultry. p. 97–111. *In* Corn-based ethanol in Illinois and the U.S: A report from the Department of Agricultural and Consumer Economics, University of Illinois. Available at http://www.farmdoc.uiuc.edu/policy/research_reports/ethanol_report/Ethanol%20Report%20-%20Ch%206.pdf (verified 19 Oct. 2010).

Bhandral, R., N.S. Bolan, S. Saggar, and M.J. Hedley. 2007. Nitrogen transformation and nitrous oxide emissions from various types of farm effluents. Nutr. Cycling Agroecosyst. 79:193–208.

Bicudo, J.R., and S.M. Goyal. 2003. Pathogens and manure management systems: A review. Environ. Technol. 24:115–130.

Bittman, S., L.J.P. van Vliet, C.G. Kowalenko, S. McGinn, D.E. Hunt, and F. Bounaix. 2005. Surface-banding liquid manure over aeration slots: A new low-disturbance method for reducing ammonia emissions and improving yield of perennial grasses. Agron. J. 97:1304–1313.

Bloom, R.A. 2004. Use of veterinary pharmaceuticals in the United States. p. 149–154. *In* K. Kümmerer (ed.) Pharmaceuticals in the environment: Sources, fate, effects, and risks. Springer, Berlin.

Bolton, D.J., C.M. Byrne, J.J. Sheridan, D.A. McDowell, and I.S. Blair. 1999. The survival characteristics of a non-toxigenic strain of *Escherichia coli* O157:H7. J. Appl. Microbiol. 86:407–411.

Boxall, A.B.A., D.W. Kolpin, B. Halling-Sørensen, and J. Tolls. 2003. Are veterinary medicines causing environmental risks? Environ. Sci. Technol. 37:286A–294A.

Brown, M.R.W., A.W. Smith, J. Barker, T.J. Humphrey, and B. Dixon. 2002. *E. coli* O157 persistence in the environment. Microbiology 148:1–2.

Burnison, B.K., A. Hartmann, A. Lister, M.R. Servos, T. Ternes, and G. Van der Kraak. 2003. A toxicity identification evaluation approach to studying estrogenic substances in hog manure and agricultural runoff. Environ. Toxicol. Chem. 22:2243–2250.

Büyüksönmez, F., R. Rynk, T.F. Hess, and E. Bechinski. 2000. Occurrence, degradation, and fate of pesticides during composting. Part II: Occurrence and fate of pesticides in compost and composting systems. Compost Sci. Util. 8:61–81.

Cabrera, M.L., and J.T. Sims. 2000. Beneficial uses of poultry by-products: Challenges and opportunities. p. 425–450. *In* J.F. Power and W.A. Dick (ed.) Land application of agricultural, industrial, and municipal by-products. SSSA, Madison, WI.

Calderón, F.J., G.W. McCarty, J.S. Van Kessel, and J.B. Reeves. 2004. Carbon and nitrogen dynamics during incubation of manured soil. Soil Sci. Soc. Am. J. 68:1592–1599.

Cantrell, K.B., T. Ducey, K.S. Ro, and P.G. Hunt. 2008. Livestock waste-to-bioenergy generation opportunities. Bioresour. Technol. 99:7941–7953.

Carter, M.R., J.B. Sanderson, and J.A. MacLeod. 2004. Influence of compost on the physical properties and organic matter fractions of a fine sandy loam throughout the cycle of a potato rotation. Can. J. Soil Sci. 84:211–218.

Chang, C., C. Cho, and H.H. Janzen. 1998. Nitrous oxide emission from long-term manured soils. Soil Sci. Soc. Am. J. 62:677–682.

Chang, C., and T. Entz. 1996. Nitrate leaching losses under repeated cattle feedlot manure applications in southern Alberta. J. Environ. Qual. 25:145–153.

Chang, C., and H.H. Janzen. 1996. Long-term fate of nitrogen from annual feedlot manure applications. J. Environ. Qual. 25:785–790.

Chang, C., H.H. Janzen, and T. Entz. 1994. Long-term manure application effects on the nutrient uptake of barley. Can. J. Plant Sci. 74:327–330.

Chang, C., T.G. Sommerfeldt, and T. Entz. 1990. Rates of soil chemical changes with eleven annual applications of cattle feedlot manure. Can. J. Soil Sci. 70:673–681.

Chang, C., T.G. Sommerfeldt, and T. Entz. 1991. Soil chemistry after eleven annual applications of cattle feedlot manure. J. Environ. Qual. 20:475–480.

Chang, C., T.G. Sommerfeldt, and T. Entz. 1993. Barley performance under heavy application of cattle feedlot manure. Agron. J. 85:1013–1018.

Chang, C., J.K. Whalen, and X. Hao. 2005. Increase in phosphorus concentration of a clay loam surface soil receiving repeated annual feedlot cattle manure applications in southern Alberta. Can. J. Soil Sci. 85:589–597.

Chantigny, M.H., D.A. Angers, P. Rochette, G. Bélanger, D. Massé, and D. Côté. 2007. Gaseous nitrogen emissions and forage nitrogen uptake on soils fertilized with raw and treated swine manure. J. Environ. Qual. 36:1864–1872.

Ciarlo, E., M. Conti, N. Bartoloni, and G. Rubio. 2007. The effect of moisture on nitrous oxide emissions from soil and the $N_2O/(N_2O+N_2)$ ratio under laboratory conditions. Biol. Fertil. Soils 43:675–681.

Clemens, J., M. Trimborn, P. Weiland, and B. Amon. 2006. Mitigation of greenhouse gas emissions by anaerobic digestion of cattle slurry. Agric. Ecosyst. Environ. 112:171–177.

Colucci, M., and E. Topp. 2002. Dissipation of part per trillion concentrations of estrogenic hormones from agricultural soils. Can. J. Soil Sci. 82:335–340.

Côté, C., D.I. Massé, and S. Quessy. 2006. Reduction of indicator and pathogenic microorganisms by psychrophilic anaerobic digestion in swine slurries. Bioresour. Technol. 97:686–691.

Cotruvo, J.A., A. Dufour, G. Rees, J. Bartram, R. Carr, D.O. Cliver, G.F. Craun, R. Fayer, and V.P.J. Gannon. 2004. Waterborne zoonoses: identification, causes and control. IWA Publishing, London.

Dittert, K., C. Lampe, R. Gasche, K. Butterbach-Bahl, M. Wachendorf, H. Papen, B. Sattelmacher, and F. Taube. 2005. Short-term effects of single or combined application of mineral N fertilizer and cattle slurry on the fluxes of radiatively active trace gases from grassland soil. Soil Biol. Biochem. 37:1665–1674.

Dolliver, H., S. Gupta, and S. Noll. 2008. Antibiotic degradation during manure composting. J. Environ. Qual. 37:1245–1253.

Dolliver, H., K. Kumar, and S. Gupta. 2007. Sulfamethazine uptake by plants from manure-amended soil. J. Environ. Qual. 36:1224–1230.

Dormaar, J.F., and C. Chang. 1995. Effect of twenty annual applications of excess feedlot manure on forms of soil phosphorus. Can. J. Soil Sci. 75:507–512.

Dormaar, J.F., and T.G. Sommerfeldt. 1986. Effect of excess feedlot manure on chemical constituents of soil under non-irrigated and irrigated management. Can. J. Soil Sci. 66:303–313.

Dubetz, S. 1983. Ten-year irrigated rotation U. 1911–1980. Contribution 1983–21E, Agriculture Canada, Research Branch, Lethbridge, AB.

Duriez, P., and E. Topp. 2007. Temporal dynamics and impact of manure storage on antibiotic resistance patterns and population structure of *Escherichia coli* isolates from a commercial swine farm. Appl. Environ. Microbiol. 73:5486–5493.

Duriez, P., Y. Zhang, Z. Lu, A. Scott, and E. Topp. 2008. Loss of virulence genes in *Escherichia coli* populations during manure storage on a commercial swine farm. Appl. Environ. Microbiol. 74:3935–3942.

Fellman, J.B., E.H. Franz, C.L. Crenshaw, and D. Elston. 2008. Global estimates of soil carbon sequestration via livestock waste: A STELLA simulation. Environ. Dev. Sustain. 11:871–885.

Gagliardi, J.V., and J.S. Karns. 2002. Persistence of *Escherichia coli* O157:H7 in soil and on plant roots. Environ. Microbiol. 4:89–96.

Gagnon, B., and R.R. Simard. 1999. Nitrogen and phosphorus release from on-farm and industrial composts. Can. J. Soil Sci. 79:481–489.

Gannon, V.P.J., C. Bolin, and C.L. Moe. 2004 Waterborne zoonoses: Emerging pathogens and emerging patterns of infection. p. 472–484. *In* J.A. Cotruvo et al. (ed.) Waterborne zoonoses: identification, causes and control. IWA Publishing, London.

Gao, G., and C. Chang. 1996. Changes in CEC and particle size distribution of soils associated with long-term annual applications of cattle feedlot manure. Soil Sci. 161:115–120.

Gbuerk, W.J., A.N. Sharpley, L. Heathwaite, and G.J. Folmar. 2000. Phosphorus management at the watershed scale: A modification to the phosphorus index. J. Environ. Qual. 29:130–144.

Gerba, C.P., and J.E. Smith. 2005. Sources of pathogenic microorganisms and their fate during land application of wastes. J. Environ. Qual. 34:42–48.

Gitau, M.W., T.L. Veith, W.J. Gburek, and A.R. Jarrett. 2006. Watershed level best management practice selection and placement in the Town Brook watershed, New York. J. Am. Water Resour. Assoc. 42:1561–1581.

Grandy, A.S., G.A. Porter, and M.S. Erich. 2002. Organic amendment and rotation crop effects on the recovery of soil organic matter and aggregation in potato cropping systems. Soil Sci. Soc. Am. J. 66:1311–1319.

Grusenmeyer, D.C., and T.N. Cramer. 1997. Manure management: A systems approach. J. Dairy Sci. 80:2651–2654.

Haas, C.N., J.B. Rose, and C.P. Gerba. 1999. Quantitative microbial risk assessment. John Wiley & Sons, New York.

Hanselman, T.A., D.A. Graetz, and A.C. Wilkie. 2003. Manure-borne estrogens as potential environmental contaminants: A review. Environ. Sci. Technol. 37:5471–5478.

Hao, X., and C. Chang. 2002. Effect of 25 annual cattle manure applications on soluble and exchangeable cations in soil. Soil Sci. 167:126–134.

Hao, X., and C. Chang. 2003. Does long-term heavy cattle manure application increase salinity of a clay loam soil in semi-arid southern Alberta? Agric. Ecosyst. Environ. 94:89–103.

Hao, X., C. Chang, H.H. Janzen, G. Clayton, and B.R. Hill. 2006. Sorption of atmospheric ammonia by soil and perennial grass downwind from two large cattle feedlots. J. Environ. Qual. 35:1960–1965.

Hao, X., C. Chang, and F.J. Larney. 2004a. Carbon, nitrogen balances and greenhouse gas emission during cattle feedlot manure composting. J. Environ. Qual. 33:37–44.

Hao, X., C. Chang, F.J. Larney, and G.R. Travis. 2001. Greenhouse gas emissions during cattle feedlot manure composting. J. Environ. Qual. 30:376–386.

Hao, X., C. Chang, and X. Li. 2004b. Long-term and residual effects of cattle manure application on distribution of P in soil aggregates. Soil Sci. 169:715–728.

Hao, X., C. Chang, and G.J. Travis. 2004c. Effect of long-term cattle manure application on relations between nitrogen and oil content in canola seed. J. Plant Nutr. Soil Sci. 167:214–215.

Hao, X., F. Godlinski, and C. Chang. 2008. Distribution of phosphorus forms in soil following long-term continuous and discontinuous cattle manure applications. Soil Sci. Soc. Am. J. 72:90–97.

Hao, X., P.S. Mir, M.A. Shah, and G.R. Travis. 2005. Influence of canola and sunflower diet amendments on cattle feedlot manure. J. Environ. Qual. 34:1439–1445.

Helgason, B.L., F.J. Larney, and H.H. Janzen. 2005. Estimating carbon retention in soils amended with composted beef cattle manure. Can. J. Soil Sci. 85:39–46.

Helgason, B.L., F.J. Larney, H.H. Janzen, and B.M. Olson. 2007. Nitrogen dynamics in soil amended with composted cattle manure. Can. J. Soil Sci. 87:43–50.

Hölzel, C., and J. Bauer. 2008. *Salmonella* spp. in Bavarian liquid pig manure: Occurrence and relevance for the distribution of antibiotic resistance. Zoonoses Public Health 55:133–138.

Honeycutt, C.W., T.S. Griffin, and Z. He. 2005. Manure nitrogen availability: Dairy manure in northeast and central U.S. soils. Biol. Agric. Hortic. 23:199–214.

Hutchins, S.R., M.V. White, F.M. Hudson, and D.D. Fine. 2007. Analysis of lagoon samples from different concentrated animal feeding operations for estrogens and estrogen conjugates. Environ. Sci. Technol. 41:738–744.

Inamdar, S.P., S. Mostaghimi, P.W. McClellan, and K.M. Brannan. 2001. BMP impacts on sediment and nutrient yields from an agricultural watershed in the Coastal Plain region. Trans. ASAE 44:1191–1200.

Inglis, G.D., L.D. Kalischuk, and H.W. Busz. 2004. Chronic shedding of *Campylobacter* species in beef cattle. J. Appl. Microbiol. 97:410–420.

Islam, M., J. Morgan, M.P. Doyle, and X.P. Jiang. 2004. Fate of *Escherichia coli* O157:H7 in manure compost-amended soil and on carrots and onions grown in an environmentally controlled growth chamber. J. Food Prot. 67:574–578.

Jacobsen, A.-M., A. Lorenzen, R. Chapman, and E. Topp. 2005. Persistence of testosterone and 17β-estradiol in soils receiving swine manure or municipal biosolids. J. Environ. Qual. 34:861–871.

Janzen, H.H. 2001. Soil science on the Canadian prairies—Peering into the future from a century ago. Can. J. Soil Sci. 81:489–503.

Karmakar, S., C. Laguë, J. Agnew, and H. Landry. 2007. Integrated decision support system (DSS) for manure management: A review and perspective. Comput. Electron. Agric. 57:190–201.

Keener, H.M., W.A. Dick, and H.A.J. Hoitink. 2000. Composting and beneficial utilization of composted by-product materials. p. 315–341. *In* J.F. Power and W.A. Dick (ed.) Land application of agricultural, industrial and municipal by-products. SSSA, Madison, WI.

Khan, B., L.S. Lee, and S.A. Sassman. 2008. Degradation of synthetic androgens 17α- and 17β-trenbolone and trendione in agricultural soils. Environ. Sci. Technol. 42:3570–3574.

Lalande, R., B. Gagnon, and R.R. Simard. 2003. Papermill biosolid and hog manure compost affect short-term biological activity and crop yield of a sandy soil. Can. J. Soil Sci. 83:353–362.

Lapen, D.R., E. Topp, M. Edwards, L. Sabourin, W. Curnoe, N. Gottschall, P. Bolton, S. Rahman, B. Ball-Coelho, M. Payne, S. Kleywegt, and N. McLaughlin. 2008. Effect of liquid municipal biosolid application method on tile and ground water quality. J. Environ. Qual. 37:925–936.

Larney, F.J., O.O. Akinremi, R.L. Lemke, V.E. Klaassen, and H.H. Janzen. 2003a. Crop response to topsoil replacement depth and organic amendment on abandoned natural gas wellsites. Can. J. Soil Sci. 83:415–423.

Larney, F.J., O.O. Akinremi, R.L. Lemke, V.E. Klaassen, and H.H. Janzen. 2005. Soil responses to topsoil replacement depth and organic amendments in wellsite reclamation. Can. J. Soil Sci. 85:307–317.

Larney, F.J., and R.E. Blackshaw. 2003. Weed seed viability in composted beef cattle feedlot manure. J. Environ. Qual. 32:1105–1113.

Larney, F.J., K.E. Buckley, X. Hao, and W.P. McCaughey. 2006a. Fresh, stockpiled and composted beef cattle feedlot manure: Nutrient levels and mass balance estimates in Alberta and Manitoba. J. Environ. Qual. 35:1844–1854.

Larney, F.J., and X. Hao. 2007. A review of composting as a manure management alternative for beef cattle feedlot manure in southern Alberta, Canada. Bioresour. Technol. 98:3221–3227.

Larney, F.J., and H.H. Janzen. 1996. Restoration of productivity to a desurfaced soil with livestock manure, crop residue and fertilizer amendments. Agron. J. 88:921–927.

Larney, F.J., and H.H. Janzen. 1997. A simulated erosion approach to assess rates of cattle manure and phosphorus fertilizer for restoring productivity to eroded soils. Agric. Ecosyst. Environ. 65:113–126.

Larney, F.J., H.H. Janzen, B.M. Olson, and C.W. Lindwall. 2000a. Soil quality and productivity responses to simulated erosion and restorative amendments. Can. J. Soil Sci. 80:515–522.

Larney, F.J., A.F. Olson, A.A. Carcamo, and C. Chang. 2000b. Physical changes during active and passive composting of beef feedlot manure in winter and summer. Bioresour. Technol. 75:139–148.

Larney, F.J., A.F. Olson, P.R. DeMaere, B.P. Handerek, and B.C. Tovell. 2008a. Nutrient and trace element changes during manure composting at four southern Alberta feedlots. Can. J. Soil Sci. 88:45–59.

Larney, F.J., B.M. Olson, H.H. Janzen, and C.W. Lindwall. 2000c. Early impact of simulated erosion and soil amendments on crop productivity. Agron. J. 92:948–956.

Larney, F.J., A.F. Olson, J.J. Miller, P.R. DeMaere, F. Zvomuya, and T.A. McAllister. 2008b. Physical and chemical changes during composting of wood chip-bedded and straw-bedded beef cattle feedlot manure. J. Environ. Qual. 37:725–735.

Larney, F.J., D.M. Sullivan, K.E. Buckley, and B. Eghball. 2006b. The role of composting in recycling manure nutrients. Can. J. Soil Sci. 86:597–611.

Larney, F.J., L.J. Yanke, J.J. Miller, and T.A. McAllister. 2003b. Fate of coliform bacteria in composted beef cattle feedlot manure. J. Environ. Qual. 32:1508–1515.

Latacz-Lohmann, U., and I. Hodge. 2003. European agri-environmental policy for the 21st century. Aust. J. Agric. Resour. Econ. 47:123–139.

Lau, A., S. Bittman, and G. Lemus. 2003. Odor measurements for manure spreading using a subsurface deposition applicator. J. Environ. Sci. Health, Part B 38:233–240.

Lemunyon, J.L., and R.G. Gilbert. 1993. The concept and need for a phosphorus assessment tool. J. Prod. Agric. 6:483–486.

Lissemore, L., C. Hao, P. Yang, P.K. Sibley, S. Mabury, and K.R. Solomon. 2006. An exposure assessment for selected pharmaceuticals within a watershed in southern Ontario. Chemosphere 64:717–729.

Litterick, A.M., L. Harrier, P. Wallace, C.A. Watson, and M. Wood. 2004. The role of uncomposted materials, composts, manures and compost extracts in reducing pest and disease incidence and severity in sustainable temperate agricultural and horticultural crop production—a review. Crit. Rev. Plant Sci. 23:453–479.

Little, J.L., D.R. Bennett, and J.J. Miller. 2005. Nutrient and sediment losses under simulated rainfall following manure incorporation by different methods. J. Environ. Qual. 34:1883–1895.

Lorenzen, A., R. Chapman, J.G. Hendel, and E. Topp. 2005. Persistence and pathways of testosterone dissipation in agricultural soil. J. Environ. Qual. 34:854–860.

Lorenzen, A., J.G. Hendel, K.L. Conn, S. Bittman, A.B. Kwabiah, G. Lazarovits, D. Massé, T.A. McAllister, and E. Topp. 2004. Survey of hormone activities in municipal biosolids and animal manures. Environ. Toxicol. 19:216–225.

Loria, E.R., J.E. Sawyer, D.W. Barker, J.P. Lundvall, and J.C. Lorimor. 2007. Use of anaerobically digested swine manure as a nitrogen source in corn production. Agron. J. 99:1119–1129.

Maguire, R.O., D.A. Crouse, and S.C. Hodges. 2007. Diet modification to reduce phosphorus surpluses: A mass balance approach. J. Environ. Qual. 36:1235–1240.

Maguire, R.O., D. Hesterberg, A. Gernat, K. Anderson, M. Wineland, and J. Grimes. 2006. Liming poultry manures to decrease soluble phosphorus and suppress the bacteria population. J. Environ. Qual. 35:849–857.

Mallin, M.A., and L.B. Cahoon. 2003. Industrialized animal production—A major source of nutrient and microbial pollution to aquatic ecosystems. Popul. Environ. 24:369–386.

Maule, A. 2000. Survival of verocytotoxigenic *Escherichia coli* O157 in soil, water and on surfaces. J. Appl. Microbiol. 88:71S–78S.

Mawdsley, J.L., R.D. Bardgett, R.J. Merry, B.F. Pain, and M.K. Theodorou. 1995. Pathogens in livestock waste, their potential for movement through soil and environmental pollution. Appl. Soil Ecol. 2:1–15.

Mawdsley, J.L., A.E. Brooks, R.J. Merry, and B.F. Pain. 1996. Use of a novel soil tilting apparatus to demonstrate the horizontal and vertical movement of the protozoan pathogen *Cryptosporidium parvum* in soil. Biol. Fertil. Soils 23:215–220.

McDowell, R.W., and A.N. Sharpley. 2004. Variation of phosphorus leached from Pennsylvanian soils amended with manures, composts or inorganic fertilizer. Agric. Ecosyst. Environ. 102:17–27.

McGinn, S.M., H.H. Janzen, and T. Coates. 2003. Atmospheric ammonia, volatile fatty acids, and other odorants near beef feedlots. J. Environ. Qual. 32:1173–1182.

Meng, L., W. Ding, and Z. Cai. 2005. Long-term application of organic manure and nitrogen fertilizer on N_2O emissions, soil quality and crop production in a sandy loam soil. Soil Biol. Biochem. 37:2037–2045.

Meyer, D., and L.J. Schwankl. 2000. Liquid dairy manure utilization in a cropping system: A case study. p. 409–423. *In* J.F. Power and W.A. Dick (ed.) Land application of agricultural, industrial, and municipal by-products. SSSA, Madison, WI.

Michel, P., J.B. Wilson, S.W. Martin, R.C. Clarke, S.A. McEwen, and C.L. Gyles. 1999. Temporal and geographical distributions of reported cases of *Escherichia coli* O157:H7 infection in Ontario. Epidemiol. Infect. 122:193–200.

Mikkelsen, R.L. 2000. Beneficial uses of swine by-products: Opportunities for the future. p. 451–480. *In* J.F. Power and W.A. Dick (ed.) Land application of agricultural, industrial, and municipal by-products. SSSA, Madison, WI.

Miller, J.J., E.C.S. Olson, D.S. Chanasyk, B.W. Beasley, F.J. Larney, and B.M. Olson. 2006. Phosphorus and nitrogen in rainfall simulation runoff after fresh and composted beef cattle manure application. J. Environ. Qual. 35:1279–1290.

Miller, J.J., N.J. Sweetland, and C. Chang. 2002a. Hydrological properties of a clay loam soil after long-term cattle manure application. J. Environ. Qual. 31:989–996.

Miller, J.J., N.J. Sweetland, and C. Chang. 2002b. Soil physical properties of a Chernozemic clay loam after 24 years of beef cattle manure application. Can. J. Soil Sci. 82:287–296.

Millner, P.D., and L.L. McConnell. 2000. Odor and other air quality issues associated with organic and inorganic by-products. p. 289–314. *In* J.F. Power and W.A. Dick (ed.) Land application of agricultural, industrial, and municipal by-products. SSSA, Madison, WI.

Misselbrook, T.H., K.A. Smith, R.A. Johnson, and B.F. Pain. 2002. Slurry application techniques to reduce ammonia emissions: Results of some UK field-scale experiments. Biosystems Eng. 81:313–321.

Mkhabela, T. 2005. A review of the use of manure in small-scale crop production systems in South Africa. J. Plant Nutr. 29:1157–1185.

Mkhabela, M.S., A. Madani, R. Gordon, D. Burton, D. Cudmore, A. Elmi, and W. Hart. 2008. Gaseous and leaching nitrogen losses from no-tillage and conventional tillage systems following surface application of cattle manure. Soil Tillage Res. 98:187–199.

Muirhead, R.W., R.P. Collins, and P.J. Bremer. 2005. Erosion and subsequent transport state of *Escherichia coli* from cowpats. Appl. Environ. Microbiol. 71:2875–2879.

Natvig, E.E., S.C. Ingham, B.H. Ingham, L.R. Cooperband, and T.R. Roper. 2002. *Salmonella enterica* serovar Typhimurium and *Escherichia coli* contamination of root and leaf vegetables grown in soils with incorporated bovine manure. Appl. Environ. Microbiol. 68:2737–2744.

Noble, R., and E. Coventry. 2005. Suppression of soilborne plant diseases with composts: A review. Biocontrol Sci. Technol. 15:3–20.

O'Connor, G.A., H.A. Elliott, N.T. Basta, R.K. Bastian, G.M. Pierzynski, R.C. Sims, and R.E. Smith. 2005. Sustainable land applications: An overview. J. Environ. Qual. 34:7–17.

Oenema, O., D. Oudendag, and G.L. Velthof. 2007. Nutrient losses from manure management in the European Union. Livest. Sci. 112:261–272.

Ojha, S., and M. Kostrzynska. 2007. Approaches for reducing *Salmonella* in pork production. J. Food Prot. 70:2676–2694.

Peigné, J., and P. Girardin. 2004. Environmental impacts of farm-scale composting practices. Water Air Soil Pollut. 153:45–68.

Pendell, D.L., J.R. Williams, C.W. Rice, R.G. Nelson, and S.B. Boyles. 2006. Economic feasibility of no-tillage and manure for soil carbon sequestration in corn production in northeastern Kansas. J. Environ. Qual. 35:1364–1373.

Perälä, P., P. Kapuinen, M. Esala, S. Tyynelä, and K. Regina. 2006. Influence of slurry and mineral fertiliser application techniques on N_2O and CH_4 fluxes from a barley field in southern Finland. Agric. Ecosyst. Environ. 117:71–78.

Petersen, S.O., S.G. Sommer, F. Béline, C. Burton, J. Dach, J.Y. Dourmad, A. Leip, T. Misselbrook, F. Nicholson, H.D. Poulsen, G. Provolo, P. Sørensen, B. Vinnerås, A. Weiske, M.P. Bernal, R. Böhm, C. Juhász, and R. Mihelic. 2007. Recycling of livestock manure in a whole-farm perspective. Livest. Sci. 112:180–191.

Pope, L., A.B.A. Boxall, C. Corsing, B. Halling-Sørensen, A. Tait, and E. Topp. 2008. Exposure assessment of veterinary medicines in terrestrial systems. p. 129–154. *In* M. Crane et al. (ed.) Veterinary medicines in the environment. CRC Press, Boca Raton, FL.

Portejoie, S., J.Y. Dourmad, J. Martinez, and Y. Lebreton. 2004. Effect of lowering dietary crude protein on nitrogen excretion, manure composition and ammonia emission from fattening pigs. Livest. Prod. Sci. 91:45–55.

Porter, G.A., G.B. Opena, W.B. Bradbury, J.C. McBurnie, and J.A. Sisson. 1999. Soil management and supplemental irrigation effects on potato: I. Soil properties, tuber yield, and quality. Agron. J. 91:416–425.

Powell, J.M., M.A. Wattiaux, G.A. Broderick, V.R. Moreira, and M.D. Casler. 2006. Dairy diet impacts on fecal chemical properties and nitrogen cycling in soils. Soil Sci. Soc. Am. J. 70:786–794.

Ribaudo, M.O., N.R. Gollehon, and J. Agapoff. 2003. Land application of manure by animal feeding operations: Is more land needed? J. Soil Water Conserv. 58:30–38.

Rochette, R., D.A. Angers, M.H. Chantigny, B. Gagnon, and N. Bertrand. 2006. In situ mineralization of dairy manures as determined using soil-surface carbon dioxide fluxes. Soil Sci. Soc. Am. J. 70:744–752.

Rufino, M.C., P. Tittonell, M.T. van Wijk, A. Castellanos-Navarrete, R.J. Delve, N. de Ridder, and K.E. Giller. 2007. Manure as a key resource within smallholder farming systems: Analysing farm-scale nutrient cycling efficiencies with the NUANCES framework. Livest. Sci. 112:273–287.

Russelle, M.P., M.H. Entz, and A.J. Franzluebbers. 2007. Reconsidering integrated crop-livestock systems in North America. Agron. J. 99:325–334.

Rynk, R. 1992. On-farm composting handbook. Publ. NRAES-54, Northeast Regional Agric. Engineering Serv., Coop. Ext. Serv., Ithaca, NY.

Sagoo, E., J.R. Williams, B.J. Chambers, L.O. Boyles, R. Matthews, and D.R. Chadwick. 2007. Integrated management practices to minimise losses and maximise the crop nitrogen value of broiler litter. Biosystems Eng. 97:512–519.

Sandars, D.L., E. Audsley, C. Cañete, T.R. Cumby, I.M. Scotford, and A.G. Williams. 2003. Environmental benefits of livestock manure management practices and technology by life cycle assessment. Biosystems Eng. 84:267–281.

Sapkota, A.R., F.C. Curriero, K.E. Gibson, and K.J. Schwab. 2007. Antibiotic-resistant enterococci and fecal indicators in surface water and groundwater impacted by a concentrated swine feeding operation. Environ. Health Perspect. 115:1040–1045.

Sarmah, A.J., M.T. Meyer, and A.B.A. Boxall. 2006. A global perspective on the use, sales, exposure pathways, occurrence, fate and effects of veterinary antibiotics (VAs) in the environment. Chemosphere 65:725–759.

Schlecht, E., A. Buerkert, E. Tielkes, and A. Bationo. 2006. A critical analysis of challenges and opportunities for soil fertility restoration in Sudano-Sahelian West Africa. Nutr. Cycling Agroecosyst. 76:109–136.

Scott, A., K. Conn, G. Lazarovits, and E. Topp. 2006. Dynamics of *Escherichia coli* in agricultural soils receiving swine manure slurry or liquid municipal biosolids. Can. J. Soil Sci. 86:841–849.

Sharpley, A.N., S.J. Smith, B.A. Stewart, and A.C. Mathers. 1984. Forms of phosphorus in soil receiving cattle feedlot waste. J. Environ. Qual. 13:211–215.

Sibbesen, E., and A.N. Sharpley. 1997. Setting and justifying upper critical limits for phosphorus in soils. p. 151–176. *In* H. Tunney et al. (ed.) Phosphorus loss from soil to water. CABI, Wallingford, UK.

Sims, J.T., L. Bergström, B.T. Bowman, and O. Oenema. 2005. Nutrient management for intensive animal agriculture: Policies and practices for sustainability. Soil Use Manage. 21:141–151.

Smith, K.A., D.R. Jackson, T.H. Misselbrook, B.F. Pain, and R.A. Johnson. 2000a. Reduction of ammonia emission by slurry application techniques. J. Agric. Eng. Res. 77:277–287.

Smith, K.A., D.R. Jackson, and P.J.A. Withers. 2001. Nutrient losses by surface run-off following the application of organic manures to arable land. 2. Phosphorus. Environ. Pollut. 112:53–60.

Smith, E.G., Y. Peng, M. Lerohl, and F.J. Larney. 2000b. Economics of N and P fertilization to restore wheat yields on three artificially eroded sites in southern Alberta. Can. J. Soil Sci. 80:165–169.

Solomon, E.B., S. Yaron, and K.R. Matthews. 2002. Transmission of *Escherichia coli* O157:H7 from contaminated manure and irrigation water to lettuce plant tissue and its subsequent internalization. Appl. Environ. Microbiol. 68:397–400.

Sommerfeldt, T.G., and C. Chang. 1985. Changes in soil properties under annual applications of feedlot manure and different tillage practices. Soil Sci. Soc. Am. J. 49:983–987.

Sommerfeldt, T.G., and C. Chang. 1987. Soil-water properties as affected by twelve annual applications of cattle feedlot manure. Soil Sci. Soc. Am. J. 51:7–9.

Sommerfeldt, T.G., C. Chang, and T. Entz. 1988. Long-term annual manure applications increase soil organic matter and nitrogen, and decrease carbon to nitrogen ratio. Soil Sci. Soc. Am. J. 52:1668–1672.

Sommerfeldt, T.G., U.J. Pittman, and R.A. Milne. 1973. Effect of feedlot manure on soil and water quality. J. Environ. Qual. 2:423–427.

Sørensen, P., M.R. Weisbjerg, and P. Lund. 2003. Dietary effects on the composition and plant utilization of nitrogen in dairy cattle manure. J. Agric. Sci. 141:79–91.

Spiehs, M.J., M.H. Whitney, and G.C. Shurson. 2002. Nutrient database for distiller's dried grains with solubles produced from new ethanol plants in Minnesota and South Dakota. J. Anim. Sci. 80:2639–2645.

Stevens, R.J., and R.J. Laughlin. 2001. Effect of liquid manure on the mole fraction of nitrous oxide evolved from soil containing nitrate. Chemosphere 42:105–111.

Stewart, B.A., C.A. Robinson, and D.B. Parker. 2000. Examples and case studies of beneficial reuse of beef cattle by-products. p. 387–407. In J.F. Power and W.A. Dick (ed.) Land application of agricultural, industrial, and municipal by-products. SSSA, Madison, WI.

Stoddard, C.S., M.S. Coyne, and J.H. Grove. 1998. Fecal bacteria survival and infiltration through a shallow agricultural soil: Timing and tillage effects. J. Environ. Qual. 27:1516–1523.

Thomsen, I.K., and J.E. Olesen. 2000. C and N mineralization of composted and anaerobically stored ruminant manure in differently textured soils. J. Agric. Sci. 135:151–159.

Topp, E., M. Welsh, Y.-C. Tien, A. Dang, G. Lazarovits, K. Conn, and H. Zhu. 2003. Strain-dependent variability in growth and survival of Escherichia coli in agricultural soil. FEMS Microbiol. Ecol. 44:303–308.

Turpin, K.M., D.R. Lapen, M.J.L. Robin, E. Topp, M. Edwards, W.E. Curnoe, G.C. Topp, N.B. McLaughlin, B. Ball Coelho, and M. Payne. 2007. Slurry-application implement tine modification of soil hydraulic properties under different soil water content conditions for silt–clay loam soils. Soil Tillage Res. 95:120–132.

Van Es, H.M., R.R. Schindelbeck, and W.E. Jokela. 2004. Effect of manure appliction timing, crop, and soil type on phosphorus leaching. J. Environ. Qual. 33:1070–1080.

Van Es, H.M., J.M. Sogbedji, and R.R. Schindelbeck. 2006. Effect of manure application timing, crop, and soil type on nitrate leaching. J. Environ. Qual. 35:670–679.

Van Herk, F.H., T.A. McAllister, C.L. Cockwill, N. Guselle, F.J. Larney, J.J. Miller, and M.E. Olson. 2004. Inactivation of Giardia cysts and Cryptosporidium oocysts in beef feedlot manure by thermophilic windrow composting. Compost Sci. Util. 12:235–241.

Volf, C.A., G.R. Ontkean, D.R. Bennett, D.S. Chanasyk, and J.J. Miller. 2007. Phosphorus losses in simulated rainfall runoff from manured soils of Alberta. J. Environ. Qual. 36:730–741.

Watson, C.A., I. Öborn, J. Eriksen, and A.C. Edwards. 2005. Perspectives on nutrient management in mixed farming systems. Soil Use Manage. 21:132–140.

Watts, D.B., H.A. Torbert, and S.A. Prior. 2007. Mineralization of nitrogen in soils amended with dairy manure as affected by wetting/drying cycles. Commun. Soil Sci. Plant Anal. 38:2103–2116.

Whalen, J.K., and C. Chang. 2001. Phosphorus accumulation in cultivated soils from long-term annual applications of cattle feedlot manure. J. Environ. Qual. 30:229–237.

Whalen, J.K., and C. Chang. 2002a. Macroaggregate characteristics in cultivated soils after 25 annual manure applications. Soil Sci. Soc. Am. J. 66:1637–1647.

Whalen, J.K., and C. Chang. 2002b. Phosphorus sorption capacities of calcareous soils receiving cattle manure applications for 25 years. Commun. Soil Sci. Plant Anal. 33:1011–1026.

Widyaratne, G.P., and R.T. Zijlstra. 2007. Nutritional value of wheat and corn distiller's dried grains with solubles: Digestibility and digestible contents of energy, amino acids and phosphorus, nutrient excretion and growth performance of grower-finisher pigs. Can. J. Anim. Sci. 87:103–114.

Wilkins, R.J. 2008. Eco-efficient approaches to land management: A case for increased integration of crop and animal production systems. Philos. Trans. Roy. Soc. B 363:517–525.

Yan, T., J.P. Frost, R.E. Agnew, R.C. Binnie, and C.S. Mayne. 2006. Relationships among manure nitrogen output and dietary and animal factors in lactating dairy cows. J. Dairy Sci. 89:3981–3991.

Yang, X.M., C.F. Drury, T.Q. Zhang, A. Ajakaiye, C.W. Forsberg, M.Z. Fan, and J.P. Philip. 2008. Short-term carbon dioxide emissions and denitrification losses from soils amended with low-P manure from genetically modified pigs. Nutr. Cycling Agroecosyst. 80:153–160.

17

Disease Management

Timothy C. Paulitz

Plants can be attacked by a variety of parasitic microorganisms, primarily fungi, bacteria, nematodes, and viruses. From 2001 to 2003, an average of 7% to 15% of the major world crops (wheat, rice, potato, maize, and soybean) were lost due to fungi and bacteria (Oerke, 2005). Along with weeds and insects, plant pathogens are the major biotic limitation to crop health and yield. Many of these pathogens are foliar and attack aboveground parts of plants, with inoculum spread by wind and rain. Examples are rust, powdery mildew, and foliar leaf pathogens such as the fungus *Septoria*. However, some of the most severe, intractable, and difficult to control pathogens are soil-borne pathogens, which live in the soil for part or all of their life cycle and interact with the soil biota and the edaphic environment. These pathogens can survive in the soil and infect the root systems of plants. Fungi, fungus-like Stramenopiles (Oomycetes) and nematodes are probably the most important of the soil-borne pathogens. Fungi are eukaryotic organisms that form threadlike filamentous hyphae that can spread through the soil and form resistant structures such as oospores or sclerotia. These structures allow the fungus to survive in the absence of the host, or during unfavorable environments such as heat, cold, or dry soils. When these resting structures encounter a seed or root in the soil, they are stimulated to germinate, chemotactically grow toward the root, and infect the epidermal cells. Some fungi can also destroy seedlings before they emerge from the soil. Once the root is infected, fungi can spread inside the root, rotting the root by producing enzymes and toxins. Fungi also destroy lateral roots, feeder roots, and root hairs. As a result, the plant loses its ability to absorb water and nutrients. Above ground, plants are stunted and show nutrient deficiencies, and yields are reduced. Some pathogens can also move up the roots to the base of the plant, girdling the base or infecting the lower stem. Finally, another group of fungal pathogens can induce wilt by colonizing the xylem system, restricting the conduction of water to the leaves.

Nematodes are microscopic invertebrate wormlike organisms. They can feed on the root system using a stylet, a hollow needlelike feeding structure that can penetrate root cells. Some can move to foliar plant parts, but most feed on root systems. They spend most of their life in the soil, feeding on the outside of the root (ectoparasites), while others can feed and move inside the root system (endoparasites). Root knot and cyst nematodes become sedentary in the root and the females will set up a feeding site by modifying plant cells and lay eggs. But all plant parasitic nematodes spend significant parts of their life cycle in the soil.

Monetary losses due to soil-borne diseases in the United States are estimated to exceed $4 billion per year (Lumsden et al., 1995), and losses due to parasitic nematodes exceed $100

T.C. Paulitz, USDA-ARS, Root Disease and Biological Control Research Unit, Pullman, WA 99164-6430 (paulitz@wsu.edu).

doi:10.2136/2011.soilmanagement.c17

billion per year worldwide (Bird and Kaloshian, 2003). Detailed studies on wheat crops in the Pacific Northwest (PNW) have documented losses of up to 36% due to the Oomycete *Pythium*, the fungi *Fusarium* and *Rhizoctonia*, and the nematode *Pratylenchus* (Cook et al., 1987, 2002; Smiley et al., 2005a,c). These genera, along with the Oomycete *Phytophthora*, are common in other major agronomic crops.

Another group of fungal pathogens spend part of their life in or on the soil, but infect the aboveground portions of the plant, and survive on crop residues resting on the soil surface. Examples would be apple scab [*Venturia inaequalis* (Cooke) G. Winter], Sclerotinia white rot of canola and beans [*Sclerotinia sclerotiorum* (Lib.) de Bary] and Fusarium head blight of wheat and barley [*Fusarium graminearum* Schwabe = *Gibberella zeae* (Schwein. : Fr.) Petch]. This latter disease has become epidemic in the upper Midwest since the mid 1990s, produces mycotoxins in the grain, and has caused billions of dollars of losses (McMullen et al., 1997). These pathogens can overwinter in the crop residue, and the following spring produce air-borne ascospores that can infect the wheat at flowering. Part of the reason for the increase in disease has been the move toward no-till. Under conventional tillage, the residue is plowed under and decomposes quickly, because of the action of saprophytic fungi and actinomycetes. This deprives the pathogen of a niche to survive. If the residue is left on the surface, the pathogen can overwinter and produce air-borne inoculum in the spring.

Unlike foliar pathogens, soil-borne pathogens are difficult to diagnosis and control. Foliar pathogens can be controlled with systemic or protectant fungicides applied as sprays to the leaves. But there are few chemicals that can be applied to soil and effectively control soil-borne pathogens. Soil fungicides can be leached away, bound to soil particles, or broken down by microorganisms. A soil-applied chemical must protect a constantly growing and expanding root system, but no chemicals move in the phloem down into the roots system. In addition, a large volume of soil must be treated, since the root systems of most plants can extend down more than 1 m. There are some nematicides that can be applied to the soil,

but only for high-value crops. Soil-applied fungicides and nematicides may also have environmental effects, destroying beneficial fungi and nematodes, or contaminating groundwater. Fumigation with compounds such as methyl bromide is effective, but is very expensive and is being phased out because this compound is destroying the ozone layer.

Genetic resistance has been the basis of sustainable management of many foliar pathogens. But there are few examples of genetic resistance to soil-borne root-rotting organisms, possibly because of the nature of these pathogens, which kill the host ahead of the advancing fungus and do not show a gene-for-gene coevolution like biotrophic pathogens. Because of the lack of chemical and genetic tools to manage these diseases, growers have relied on cultural methods to control soil-borne pathogens. Tillage is one cultural method that has the most profound effect on soil-borne pathogens. Tillage has been a part of agriculture since its inception almost 10,000 years ago. The domestication of cereals has selected for *r*-strategist plants, which can thrive in disturbed areas, grow quickly, and produce reproductive structures (seeds). Tillage has served several important purposes: preparation of the seed bed, reduction of compaction for soil aeration and better root growth, weed control, incorporation of fertilizer and organic amendments, and residue management (Gebhardt et al., 1985). But tillage has also been an important disease management tool for soil-borne pathogens.

Tillage can control pathogens by several mechanisms. Many pathogenic fungi (and beneficial fungi such as arbuscular mycorrhizal fungi) form networks of filamentous hyphae in the soil. Tillage can break apart these networks, which connect the fungi to food sources such as residues and plant roots (Deacon, 1996; Kabir, 2005). Tillage results in increases in microbial activity (Aon et al., 2001), as organic matter is exposed to microbial degradation, and the increase in microbial activity can suppress pathogen activity. Many pathogens produce inoculum and survive in infected crop residue. Tillage places this debris into a lower soil profile, which accelerates the decomposition of this material, depriving the pathogenic fungus of a niche. Pathogen inoculum is also reduced by the antagonistic action of soil microbes and microfauna. Examples are mycoparasites, organisms that degrade the cell walls of fungi, and fungal-feeding nematodes and invertebrates.

Conversely, some pathogens may be favored by tillage. Tillage reduces soil organic matter and microbial activity in the long term, which may reduce populations of antagonistic microbes. Tillage can produce compaction and tillage pans, decreasing soil infiltration and drainage. Root-rotting pathogens such as *Phytophthora* and *Pythium* require wet soils so motile spores (zoospores) can be formed and released, and swim to the root surface. Soils that are too wet can also become anaerobic, making the dying roots yet more susceptible to pathogens.

Tillage causes severe degradation of soil resources and soil erosion, as covered in other chapters of this book. One solution is reduced tillage or no-till (direct seeding). In direct seeding or no-till, the soil is left undisturbed from harvest to planting. The crop is seeded directly into the previous stubble, using a no-till drill. Planting is done in one operation, with no seedbed preparation, leaving most of the residue from the previous crop on the surface. No-till offers growers several economic and environmental advantages. With the residue of standing stubble covering the soil, soil loss from wind erosion is significantly reduced (Papendick, 2004). Soil particles less than 10 μm in diameter have been shown to have adverse effects on human health, and can contribute to pollution in downwind urban areas. Fuel, labor, and machine costs are reduced. In conventional tillage in eastern Washington with summer fallow, a grower may require seven or more tillage or cultivation operations in a single season, compared with a no-till grower who uses herbicides for primary weed control. The increasing cost of diesel fuel in the past few years has resulted in rising costs, and no-till can reduce diesel consumption by up to 80% (Baker et al., 1996). Organic matter increases with no-till (Dao, 1993; Douglas and Goss, 1982), improving soil structure and water infiltration. No-till can improve the soil water balance with crop residues that trap more snow, and by earthworm and root channels that allow water to penetrate without running off. Gradual reductions in tillage

from 1982 to 1997 have reduced soil loss from farmland by an estimated 1 billion tons per year (NRCS, 2000).

But reduced tillage can increase disease severity of many soil-borne pathogens. The lack of tillage increased Rhizoctonia diseases in Australia and U.S. (MacNish, 1985; Rovira, 1986; Pumphrey et al., 1987). Recent work in the PNW has shown that *R. solani* J.G. Kühn becomes a major problem during the third and fourth year of the transition from conventional tillage to no-till (Schroeder and Paulitz, 2006). Tillage may break up hyphal networks or induce microbes that are antagonistic to *Rhizoctonia*. Standing stubble on the soil will trap more snow than bare soil will, resulting in higher soil moisture levels in the spring. Residues also reflect solar radiation, so soils do not warm as quickly in the spring as black soil, which absorbs solar radiation. Reduced temperatures delay planting and slow plant growth, but increase the activity of pathogens that have low temperature optima. For example, *Rhizoctonia solani* AG-8 is highly pathogenic at 10°C (Ogoshi et al., 1990). *Pythium ultimum* Trow is active at 10°C, and *P. irregulare* Buisman is active at 5°C (Ingram and Cook, 1990). Residue can also act as a mulch, reducing evaporation from the soil surface, so soil remains wetter in the spring. As previously mentioned, pathogens such as *Pythium* are favored by wet soils

No-till can also influence soil nutrients and pH. Because the soil is not turned over, nutrients from fertilizers can accumulate in the top layer. For example, in the PNW, anhydrous ammonium is widely used, which results in acidification of soil over long use. Take-all, caused by *Gaeumannomyces graminis* (Sacc.) von Arx & Olivier var. *tritici* Walker, is reduced by lower pH (MacNish, 1988), but other pathogens, such as Fusarium wilts, clubroot of crucifers, and Cephalosporium stripe of wheat, are increased at lower pH (Myers and Campbell, 1985; Hoper and Alabouvette, 1996, Murray and Walter, 1991).

Residue can act as a mulch to reduce the freezing and thawing cycles in the soil. Standing stubble reduces the wind velocity at the surface. Freezing and thawing may reduce the survival of some pathogens, but others may be increased. For example, Cephalosporium stripe, a vascular pathogen of wheat, infects fall-planted wheat through wounds in the roots and crowns caused by freezing and frost heaving (Specht and Murray, 1990).

Probably the most profound effect of reduced tillage is the increased amount of crop residue left on the soil surface, which decomposes slower than if the residue were covered by soil. Classic examples of diseases in wheat in the Great Plains of the United States that are favored by no-till residue are Fusarium head blight, tan spot caused by *Pyrenophora tritici-repentis* (Died.) Drechsler, *Septoria tritici* Roberge (cause of leaf blotch) and *Stagonospora nodorum* (Berk.) Castell. & E.G. Germano (cause of Nodorum leaf and glume blotch). These produce fruiting bodies on the residue in the spring, and ascospores or conidia can be ejected into the air or splash up on the foliage by rain or irrigation. In the PNW, it is too dry for foliar pathogens, since there is little rainfall in the summer. But Fusarium crown rot, caused by *F. pseudograminearum* O'Donnell & T. Aoki and *F. culmorum* (Wm.G. Sm.) Sacc. survives in the crop residue, and can be more severe under no-till conditions (Bailey et al., 2001)

This rest of this chapter will focus on this dilemma—how do growers preserve and enhance soil structure and health using no-till without exacerbating root diseases? How do growers manage soil-borne pathogens without resorting to extensive tillage?

I will use examples from our research on management of root diseases of cereals in the PNW to address this question. Soil erosion, both wind and water, is a major problem in eastern Washington State because of the hilly topography with steep slopes. Despite the advantages of no-till, the adoption rates by cereal growers in the US are fairly low compared with other countries or cropping systems. In the United States, 54.7% of field crops, mainly soybean and corn, were produced using conservation tillage and 31.5% using no-till (CTIC, 2006). Canada, Argentina, Brazil, and Australia have high rates of adoption in cereal production. But in the PNW of the United States, less than 15% of cereal growers use no-till (Smiley et al., 2005b), and in Whitman County, the most productive wheat county in Washington, adoption is less than 10% (H. Kok, personal communication, 2008). One of the reasons for this reluctance to adopt no-till is the increased threat of root diseases.

Methods for Reducing the Impact of Pathogens in No-Till Systems
Genetic Resistance

The most economically and environmentally sustainable method of controlling plant diseases is genetic resistance. The grower does not need to change cultural practices or use fungicides. However, for many of the major soil-borne pathogens of cereals, such as *Rhizoctonia*, *Pythium* and *Gaeumannomyces graminis* var. *tritici*, genetic resistance has not been found (Smith et al., 2002a,b; Higginbotham et al., 2004; Tineline et al., 1989). But effective resistance has been found for Fusarium crown rot in Australian lines (Wildermuth and McNamara, 1994; Wallwork et al., 2004; Wildermuth and Morgan, 2004) and is being incorporated into PNW varieties. Single gene resistance to the cereal cyst nematode has been deployed in wheat in Australia and other countries (Cook and Noel, 2002). Several other pathogens are now minor problems in the PNW, because of resistant or tolerant cultivars, including Cephalosporium stripe (Bruehl et al. 1986), eyespot or strawbreaker foot rot (Allan and Roberts, 1991), and snow mold (Gaudet, 1994). For most soil-borne pathogens, resistance is fairly stable and durable, unlike with foliar pathogens such as rusts that have a race structure and mutate to overcome resistance genes.

Chemical Control

Because of the low economic value of cereal crops, the only fungicides that are economically feasible are seed treatments. Some of these seed treatments are systemic and control seed-borne and foliar pathogens such as bunts and smuts. These include tebuconazole, triticonazole, difenoconazole, and other demethylation inhibitors. Some chemicals only protect the seed and do not move into the seedling (e.g., metalaxyl, which protects against Pythium damping-off). Seed treatments can improve seedling health but rarely give yield increases, and no chemical moves systemically to protect the growing root system throughout the life of the plant.

Suppressive Soils

Many soils have the ability to suppress the growth and activity of soil-borne pathogens. Some of this is a general background or nonspecific suppression, caused by the total microbial biomass in the soil, sometimes referred to as biological buffering. This type of suppression can be increased by the addition of organic matter, compost, or green manure (Bonamomi et al., 2007; Zinati, 2005; Noble and Coventry, 2005; Bailey and Lazarovits, 2003). If the microbes are removed by pasteurization or chemical treatment, more disease will result from a given amount of inoculum. However, there is a more specific suppression in some soils, due to a specific group of microbes active against a specific pathogen. Cook and Baker (1983) define suppressive soils as "soils in which the pathogen does not establish or persist, establishes but causes little or no damage, or establishes and causes disease for a while but thereafter the disease is less important, although the pathogen may persist in the soil." The opposite of this would be a conducive soil. Specific suppressiveness can be eliminated by heating the soil to 60°C, and can be transferred to a conducive soil, indicating a biological basis. Specific suppression has been identified for a number of soil-borne fungi including *Fusarium oxysporum* Schltdl. : Fr., *Rhizoctonia solani*, actinomycetes such as *Streptomyces scabies* (Thaxt.) Waksman & Henrici, and nematodes such as *Heterodera schachtii* Schmidt (Alabouvette, 1999; Roget, 1995; Liu et al., 1996; Menzies, 1959; Westphal and Becker, 1999). A classic example is take-all decline in wheat, first discovered almost 60 yr ago. This phenomenon has been reported worldwide, and may result from a number of mechanisms (Weller et al., 2002). Monocropping of wheat will produce increasing take-all disease over the first few years, often peaking in years 3 to 4, but then the disease will start to decline with continued monoculture, reaching a low, economically unimportant level. In the PNW, take-all decline results from the buildup of *Pseudomonas fluorescens* that produce the antibiotic 2,4-diacetylphloroglucinol (Raaijmakers and Weller, 1998). These bacteria colonize the rhizosphere of plants infected with the take-all pathogen, *Gaeumannomyces graminis* var. *tritici*, especially the lesions caused by the pathogen. 2,4-diacetylphloroglucinol is produced in the rhizosphere (Raaijmakers et al., 1999) and the pathogen is sensitive to the antibiotic (Mazzola et al., 1995).

Cropping systems can be manipulated to favor and build up this specific suppressiveness. This suppressiveness, like genetic resistance, does not require any input from growers, and takes advantage of the natural microbial suppression.

Crop Rotation

One of the oldest and most effective methods of controlling disease is crop rotation. Some soil-borne pathogens have narrow host ranges, only attacking certain species of plants. If a resistant crop is planted, the inoculum in the soil cannot infect, and must remain dormant for a season, without a food source. Over the season without the host, inoculum can be reduced by the action of antagonistic soil microflora and microfauna, which feed on the inoculum. When a susceptible crop is planted the next year, inoculum is reduced below the economic threshold. For example, crop rotation with a broadleaf crop every third year effectively controls take-all caused by *Gaeumannomyces graminis* var. *tritici* (Hornby, 1998). However, many soil-borne pathogens such as *Rhizoctonia* and *Pythium*, have wide host ranges, and can attack the rotation crop, so crop rotation may not be as effective. In addition, pathogens that can survive in the soil for long periods of time cannot be controlled by rotation. For example, *F. culmorum*, cause of Fusarium crown rot, can survive for many years in the soil as chlamydospores (Inglis and Cook, 1986), and rotation is not effective.

Green Bridge Management

Many soil-borne pathogens can infect and survive in the roots of grassy weeds and crop volunteers. When these weeds are killed by herbicides before planting, the necrotrophic pathogens can colonize the dying plants and form more inoculum. Another reason is that herbicides such as glyphosate inhibit the biochemical pathway needed for plant defenses, and make the plant more susceptible to pathogens. This concept is known as "glyphosate synergy" (Lévesque and Rahe, 1992). If the crop is planted into these dying weeds, severe disease can result from this green bridge, where the pathogen moves from the dying plant to the young seedling (Smiley et al., 1992). On the other hand, the longer the inoculum is in the soil before planting the crop, the more inoculum activity can be reduced by antagonistic activity of soil microflora and microfauna. It is recommended that growers kill the green bridge at least 3 wk before planting to reduce this pathogen carryover (Fig. 17|1). Green bridge control has been shown to be effective against *Pythium*, *Gaeumannomyces graminis* var. *tritici* and *Rhizoctonia*.

Residue Management

Excessive residue can be removed by burning, but this has many environmental problems (smoke, health hazards, loss of organic matter from the soil, loss of nutrients from the straw). Growers can use harrows, light cultivators, or mowers to break up residue to allow for faster decomposition. Chaff spreaders can be used behind the combine to spread the straw and prevent it from being dumped in a row. Sickle bars in front of the combine reel can allow for the straw to be cut short. Straw can be baled and removed. However, for pathogens that survive in the root system and not in the crowns or straw, such as *Pythium* and *Rhizoctonia*, residue management may not be effective in reducing inoculum (Paulitz et al., 2010).

Fallow

Fallow is agriculture land that is left unseeded during the growing season, a soil without a crop. Growers in low-rainfall areas of the PNW practice summer fallow, and only plant winter wheat every

Fig. 17|1. Effect of timing of herbicide application on greenbridge carryover of *Rhizoctonia*. All plots of spring barley (cv. Baronesse) were planted at the same time, but glyphosate was applied 2 d, 1 wk, or 8 wk before planting. These plots had high levels of *Rhizoctonia solani* and *R. oryzae* in the soil, and crop volunteer (wheat) was allowed to overwinter on the plots before being killed out by glyphosate. Plots were 8 rows wide × 7.3 m long, planted at the USDA-ARS Palouse Conservation Farm, Pullman, WA.

other year. This allows the moisture to be stored in the soil for the following year. Growers can also use fallow for weed control, and weeds can be suppressed with chemical herbicides or mechanical cultivation. Since there is no plant present, soil-borne pathogens do not have a host to infect. During this time, inoculum will decline due to depletion of energy in resting structures and degradation by microflora and microfauna, similar to when a nonhost crop is grown. Can fallow reduce soil-borne pathogens? A period of 3 to 6 wk of fallow reduced *Rhizoctonia* in direct-seeded wheat in Australia (Roget et al., 1987). In a series of trials conducted in annual cropping areas in the PNW, chemical fallow was not effective in reducing *R. solani* or *R. oryzae* Ryker & Gooch, possibly because of the intact root systems that serve as a protected niche for the pathogen. But in a low-rainfall area, mechanical fallow with inverted sweeps and chemical fallow reduced the activity of *R. solani*, but not *R. oryzae*. The latter organism forms microsclerotia, which can probably survive 1 yr in the absence of a host. Fallow can also reduce take-all disease. Wheat after fallow had significantly less take-all disease in long-term trials in the southeastern US (Cunfer et al., 2006). Many species of *Pythium* were reduced to below detectable levels in fallow fields, when quantified with real-time PCR (Schroeder and Paulitz, unpublished data, 2008). However, two species, *P. abappressorium* Paulitz & M. Mazzola and *P. irregulare* Group IV, were not affected by fallow, so this effect may be species-dependent.

Application of Starter Fertilizer in the Seed Row

Because of the rotting and reduction in the root systems, feeder roots and root hairs of young seedlings infected by root pathogens, nutrient deficiencies may result, especially relatively immobile nutrients such as phosphorus. No-till growers can apply starter fertilizer below or to the side of the seed in the seed furrow. This fertilizer will be readily available to the young seedling, and may allow the plant to compensate for root loss. Nutrients do not directly affect plant resistance, but enhance seedling vigor to compensate for pathogen damage

(Patterson et al., 1998). Under Zn-deficient conditions, application of Zn decreased disease severity and patch area caused by *R. solani* AG-8 (Thongbai et al., 1993a,b; MacNish and Neate, 1996). However, additional Zn applications in patches in an experiment in the PNW did not provide a benefit (Cook et al., 2002). Application of starter fertilizer can be as effective as seed treatments in increasing yield of barley under high *Rhizoctonia* pressure (Paulitz and Reinertsen, 2005).

Altered Planting Date

Since disease can be increased by conditions that are more favorable for the pathogen or less favorable for the host, planting dates can be adjusted to avoid these periods. For pathogens such as *Rhizoctonia* and *Pythium* that are favored by wet, cool soil conditions, planting can be delayed in the spring until the soils warm and dry. However, this is often not possible with the short growing season of the PNW. Delays in planting past a certain date will severely reduce yield. With fall-planted wheat, early planting can increase Fusarium disease, since this disease is triggered by drought stress, and early-planted wheat would run out of water early in the spring.

Seed Quality

Although soil-borne pathogens are not generally seed-borne, use of quality seed that emerges quickly can reduce the impact of soil-borne pathogens such as *Pythium* spp. Older seed takes longer to germinate and emerge, increasing the time that the embryo and emerging seedling are susceptible to the pathogen (Hering et al., 1987). Deterioration of the seed coat in older seed also results in the release of more exudates that attract *Pythium*. Older seeds also produce more volatiles such as ethanol when germinating, especially under anoxic conditions (Rutzke et al., 2008), and ethanol is stimulatory to the growth of *Pythium* (Paulitz, 1991).

Conclusions

Soil-borne pathogens that cause root diseases spend most of their life cycle in or on the soil. Soil management decisions will influence the survival and growth of these pathogens, and the severity of disease. Many of the cultural methods that growers have relied on in the past to reduce the impact of these pathogens, such as tillage, can have negative effects on soil health and sustainability. But conversely, no-till practices, designed to improve sustainability, can exacerbate some diseases. Thus, growers must balance crop health vs. soil health in making management decisions. Innovative technologies, such as resistant cultivars and inducing and maintaining microbial suppressive soils, may solve this dilemma in the future.

References

Alabouvette, C. 1999. Fusarium wilt suppressive soils: An example of disease-suppressive soils. Australas. Plant Pathol. 28:57–64.

Allan, R.E. And D.E. Roberts. 1991. Inheritance of reaction to strawbreaker foot rot in two wheat populations. Crop Sci. 31:943–947.

Aon, M.A., M.N. Cabello, D.E. Sarena, A.C. Colaneri, M.G. Franco, J.L. Burgos, and S. Cortassa. 2001. 1. Spatiotemporal patterns of soil microbial and enzymatic activities in an agricultural soil. Appl. Soil Ecol. 18:239–254.

Bailey, K.L., B.D. Gossen, G.P. Lafond, P.R. Watson, and D.A. Derksen. 2001. Effect of tillage and crop rotation on root and foliar diseases of wheat and pea in Saskatchewan from 1991–1998: Univariate and multivariate analyses. Can. J. Plant Sci. 81:789–803.

Bailey, K.L., and G. Lazarovits. 2003. Suppressing soilborne diseases with residue management and organic amendments. Soil Tillage Res. 72:169–180.

Baker, C.J., K.E. Saxton, and W.R. Ritchie. 1996. No-tillage seeding: Science and practice. CABI, Wallingford, UK.

Bird, D.M., and I. Kaloshian. 2003. Are nematodes special? Nematodes have their say. Physiol. Mol. Plant Pathol. 62:115–123.

Bonamomi, G., V. Antignani, C. Pane, and E. Scala. 2007. Suppression of soilborne fungal diseases with organic amendments. J. Plant Pathol. 89:311–324.

Bruehl, G.W., T.D. Murray, and R.E. Allan. 1986. Resistance of winter wheats to Cephalosporiuim stripe in the field. Plant Dis. 70:314–316.

Conservation Technology Information Center. 2006. 2006 crop residue management survey. CTIC, West Layfayette, IN.

Cook, R.J., and K.F. Baker. 1983. The nature and practice of biological vontrol of plant pathogens. American Phytopathological Society Press, St. Paul, MN

Cook, R., and G.R. Noel. 2002. Cyst nematodes: Globodera and Heterodera species. p. 71–106 In J. L. Starr, R. Cook, and J. Bridge (ed.) Plant resistance to parasitic nematodes. CABI, Wallingford, U.K.

Cook, R.J., W.F. Schillinger, and N.W. Christensen. 2002. Rhizoctonia root rot and take-all of wheat in diverse direct-seed spring cropping systems. Can. J. Plant Pathol. 24:349–358.

Cook, R.J., J.W. Sitton, and W.A. Haglund. 1987. Influence of soil treatments on growth and yield of wheat and implications for control of Pythium root rot. Phytopathology 77:1172–1198.

Cunfer, B.M., G.D. Buntin, and D.V. Phillips. 2006. Effect of crop rotation on take-all of wheat in double-cropping systems. Plant Dis. 90:1161–1166.

Dao, T.H. 1993. Tillage and winter wheat residue management effects on water infiltration and storage. Soil Sci. Soc. Am. J. 57:1586–1595.

Deacon, J.W. 1996. Translocation and transfer in *Rhizoctonia:* Mechanisms and significance. p. 117–125 *In:* B. Sneh et al. (ed.). *Rhizoctonia* species: Taxonomy, molecular biology, ecology, pathology and disease control. Kluwer Academic Publishers, Dordrecht.

Douglas, J. T., and M.J. Goss. 1982. Stability and organic matter content of surface soil aggregates under different methods of cultivation and in grassland. Soil Tillage Res. 2:155–175.

Gaudet, D.A. 1994. Progress towards understanding interactions between cold hardiness and snow mold resistance and development of resistant cultivars. Can. J. Plant Pathol. 16:241–246.

Gebhardt, M.R., T.C. Daniel, E.E. Schweizer, and R.R. Allmaras. 1985. Conservation tillage. Science 230:625–630.

Hering, T.F., R.J. Cook, and W.-H. Tang. 1987. Infection of wheat embryos by *Pythium* species during seed germination and the influence of seed age and soil matric potential. Phytopathology 77:1104–1108.

Higginbotham, R.W., T.C. Paulitz, and K.K. Kidwell. 2004. Virulence of *Pythium* species isolated from wheat fields in eastern Washington. Plant Dis. 88:1021–1026.

Hoper, H., and C. Alabouvette. 1996. Importance of physical and chemical soil properties in the suppressiveness of soils to plant diseases. Eur. J. Soil Biol. 32:41–58.

Hornby, D. 1998. Take-all disease of cereals. CABI, Cambridge, UK.

Inglis, D.A., and R.J. Cook. 1986. Persistence of chlamydospores of *Fusarium culmorum* in wheat field soils of eastern Washington. Phytopathology 76:1205–1208.

Ingram, D.M., and R.J. Cook. 1990. Pathogenicity of four *Pythium* species to wheat, barley, peas, and lentils. Plant Pathol. (London) 39:110–117.

Kabir, Z. 2005. Tillage or no-tillage: Impact on mycorrhizae. Can. J. Plant Sci. 85:23–29.

Lévesque, C.A., and J. Rahe. 1992. Herbicide interactions with fungal root pathogens, with special reference to glyphosate. Annu. Rev. Phytopathol. 30:579–602.

Liu, D.Q., N.A. Anderson, and L.L. Kinkel. 1996. Selection and characterization of strains of *Streptomyces* suppressive to the potato scab pathogen. Can. J. Microbiol. 42:487–502.

Lumsden, R.D., J.A. Lewis, and D. Fravel. 1995. Formulation and delivery of biocontrol agents for use against soilborne plant pathogens. p. 166–182. *In* F. R. Hall and J. W. Barry (ed.) Biorational pest control agents, formulation and delivery. American Chemical Soc., Washington, DC.

MacNish, G.C. 1985. Methods of reducing rhizoctonia patch of cereals in Western Australia. Plant Pathol. 34:175–181.

MacNish, G.C. 1988. Changes in take-all (*Gaeumannomyces graminis* var. *tritici*), rhizoctonia root rot (*Rhizoctonia solani*) and soil pH in continuous wheat with annual applications of nitrogenous fertilizer in Western Australia. Aust. J. Exp. Agric. 28:333–341.

MacNish, G.C., and S.M. Neate. 1996. Rhizoctonia bare patch of cereals—An Australian perspective. Plant Dis. 80:965–971.

Mazzola, M., D.K. Fujimoto, L.S. Thomashow, and R. J. Cook. 1995. Variation in sensitivity of *Gaeumannomyces graminis* to antibiotics produced by fluorescent *Pseudomonas* spp. and effect on biological control of take-all of wheat. Appl. Environ. Microbiol. 61:2254–2259.

McMullen, M., R. Jones, and D. Gallenberg. 1997. Scab of wheat and barley: A re-emerging disease of devastating impact. Plant Dis. 81:1340–1348.

Menzies, J.D. 1959. Occurrence and transfer of a biological factor that suppresses potato scab. Phytopathology 49:648–652.

Murray, T.D., and C.C. Walter. 1991. Influence of pH and matric potential on sporulation of *Cephalosporium graminearum*. Phytopathology 81:79–84.

Myers, D.F., and R.N. Campbell. 1985. Lime and the control of clubroot of crucifers—Effects of pH, calcium, magnesium and their interactions. Phytopathology 75:670–673.

Noble, R., and E. Coventry. 2005. Suppression of soilborne plant diseases with composts: A review. Biocontrol Sci. Technol. 15:3–20.

NRCS. 2000. Summary Report 1997 National Resources Inventory. USDA-NRCS, Washington, DC.

Oerke, E.C. 2005. Crop losses to pests. J. Agric. Sci. 144:31–43.

Ogoshi, A., R.J. Cook, and E.N. Bassett. 1990. *Rhizoctonia* species and anastomosis groups causing root rot of wheat and barley in the Pacific Northwest. Phytopathology 80:784–788.

Papendick, R.I. 2004. Farming with the wind II: Wind erosion and air quality control on the Columbia Plateau and Columbia Basin, University Publishing, Washington State Univ., Pullman, WA.

Patterson, L.-M., R.W. Smiley, and S.M. Alderman. 1998. Effect of seed treatment fungicides and starter fertilizer on root diseases and yield of spring wheat. Fungic. Nematicide Tests 53:425.

Paulitz, T.C. 1991. Effect of *Pseudomonas putida* on the stimulation of *Pythium ultimum* by seed volatiles of pea and soybean. Phytopathology 81:1282–1287.

Paulitz, T.C., and S. Reinertsen. 2005. Effects of in-furrow nitrogen, tillage and seed treatment with Dynasty for control of Rhizoctonia root rot in spring barley, 2004. Fungic. Nematicide Tests 60:CF029.

Paulitz, T.C., K.L. Schroeder, and W.F. Schillinger. 2010. Soilborne pathogens of cereals in an irrigated cropping system: Effects of tillage, residue management, and crop rotation. Plant Dis. 94:61–66.

Pumphrey, F.V., D.E. Wilkins, D.C. Hane, and R.W. Smiley. 1987. Influence of tillage and nitrogen fertilizer on Rhizoctonia root rot (bare patch) of winter wheat. Plant Dis. 71:125–127.

Raaijmakers, J.M., R.F. Bonsall, and D.M. Weller. 1999. Effect of population density of *Pseudomonas fluorescens* on production of 2,4-diacetylphloroglucinol in the rhizosphere of wheat. Phytopathology 89:470–475.

Raaijmakers, J.M., and D.M. Weller. 1998. Natural plant protection by 2,4-diacetylphloroglucinol-producing *Pseudomonas* spp. in take-all decline soils. Mol. Plant-Microbe Interact. 11:144–152.

Roget, D.K. 1995. Decline in root rot (*Rhizoctonia solani* AG-8) in wheat in a tillage and rotation experiment at Avon, Sourth Australia. Aust. J. Exp. Agric. 35:1009–1013.

Roget, D.K., N.R. Venn, and A.D. Rovira. 1987. Reduction of Rhizoctonia root rot of direct-drilled wheat by short-term chemical fallow. Aust. J. Exp. Agric. 27:425–430.

Rovira, A.D. 1986. Influence of crop rotation and tillage on Rhizoctonia bare patch of wheat. Phytopathology 76:669–673.

Rutzke, D.J.J., A.G. Taylor, and R.L. Obendorf. 2008. Influence of aging, oxygen, and moisture on ethanol production from cabbage seeds. J. Am. Soc. Hortic. Sci. 133:158–164.

Schroeder, K.L., and T.C. Paulitz. 2006. Root diseases of wheat and barley during the transition from conventional tillage to direct seeding. Plant Dis. 90:1247–1253.

Smiley, R.W., J.A. Gourlie, S.A. Easley, L.M. Patterson, and R.G. Whittaker. 2005a. Crop damage estimates for crown rot of wheat and barley in the Pacific Northwest. Plant Dis. 89:595–604.

Smiley, R.W., A.G. Ogg, and R.J. Cook. 1992. Influence of glyphosate on Rhizoctonia root rot, growth, and yield of barley. Plant Dis. 76:937–942.

Smiley, R., M. Siemens, T. Gohlke, and J. Poore. 2005b. Small grain acreage and management trends for eastern Oregon and Washington. 2005 Dryland Agric. Res. Annu. Rep. Oregon State Univ. Spec. Rep. 1061. Corvallis, OR.

Smiley, R.W., R.G. Whittaker, J.A. Gourlie, and S.A. Easley. 2005c. Suppression of wheat growth and yield by *Pratylenchus neglectus* in the Pacific Northwest. Plant Dis. 89:958–968.

Smith, J.D., K.K. Kidwell, M.A. Evans, R.J. Cook, and R.W. Smiley. 2002a. Assessment of spring wheat genotypes for disease reaction to *Rhizoctonia solani* AG-8 in controlled environment and direct-seeded field evaluations. Crop Sci. 43:694–700.

Smith, J.D., K.K. Kidwell, M.A. Evans, R.J. Cook, and R.W. Smiley. 2002b. Evaluation of spring cereal grains and wild *Triticum* relatives for resistance to *Rhizoctonia solani* AG 8. Crop Sci. 43:701–709.

Specht, L.P., and T.D. Murray. 1990. Effects of root wounding and inoculum density on Cephalosporium stripe in winter wheat. Phytopathology 80:1108–1114.

Thongbai, P., R.D. Graham, S.M. Neate, and M.J. Webb. 1993a. Interaction between zinc nutritional status of cereals and Rhizoctonia root rot severity. 2. Effect of Zn on disease severity of wheat under controlled conditions. Plant Soil 153:215–222.

Thongbai, P., R.J. Hannam, R.D. Graham, and M.J. Webb. 1993b. Interaction between zinc nutritional status of cereals and Rhizoctonia root rot severity. 1. Field observations. Plant Soil 153:207–214.

Tineline, R.D., K.L. Bailey, and H. Harding. 1989. Role of plant breeding in controlling soil-borne diseases. Can. J. Plant Pathol. 11:158–165

Wallwork, H., M. Butt, J.P.E. Cheong, and K.J. Williams. 2004. Resistance to crown rot in wheat identified through an improved method for screening adult plants. Australas. Plant Pathol. 33:1–7.

Weller, D.M., J.M. Raaijmakers, B.B. McSpadden Gardener, and L.S. Thomashow. 2002. Microbial populations responsible for specific soil suppressiveness to plant pathogens. Annu. Rev. Phytopathol. 40:309–348.

Westphal, A., and J.O. Becker. 1999. Biological suppression and natural population decline of *Heterodera schachtii* in a California field. Phytopathology 89:434–440.

Wildermuth, G.B., and R.B. McNamara. 1994. Testing wheat seedlings for resistance to crown rot caused by *Fusarium graminearum* Group 1. Plant Dis. 78:949–953.

Wildermuth, G.B., and J.M. Morgan. 2004. Genotypic differences in partial resistance to crown rot caused by *Fusarium pseudograminearum* in relation to an osmoregulation gene in wheat. Australas. Plant Pathol. 33:121–123.

Zinati, G.M. 2005. Compost in the 20th century: A tool to control plant diseases in nursery and vegetable crops. HortTechnology 15:61–66.

18

Direct and Indirect Impacts of Weed Management Practices on Soil Quality

Richard G. Smith, Matthew R. Ryan, and Fabian D. Menalled

Weed management is an ever-present challenge to crop production. Weeds have the potential to usurp resources that would otherwise provide nourishment to growing crops or interfere with planting or harvesting operations. Because of these potential negative impacts, much research has been devoted to developing management strategies aimed at reducing weed populations, usually through mechanical disturbance or chemical applications (Zimdahl, 2004). Although research focused on understanding the factors that regulate crop productivity has demonstrated the linkage between soil quality, crop yield, and agricultural sustainability (Karlen et al., 2001), the impact that weed management strategies can have on soil quality should also be considered (Fig. 18|1).

The direct and indirect effects of weed management on soil quality can range from negative to positive. Direct effects of weed management practices on soil quality have been relatively well studied (Dick, 1992; Karlen et al., 1994; Lal et al., 1994; Reeves, 1997) and several aspects of the associated soil processes are fairly well understood (Sollins et al., 1996; Robertson et al., 1999; Six et al., 2002). In contrast, the indirect effects of weed management on soil quality—effects that occur through changes in weed abundance and species composition affecting the nature and quantity of organic matter inputs, as well as feedbacks between weeds and soil biota—are much less understood. While a number of variables can be used as indicators of soil quality, for the purpose of this discussion we focus mainly on soil organic carbon as a metric of soil health because of its impact on a suite of physical, chemical, and biological properties (Cannell and Hawes, 1994; Reeves, 1997).

The goal of this chapter is several-fold. First, we explore general principles regarding direct effects of common weed management practices including mechanical practices and herbicide applications on soil carbon dynamics. In this section we highlight a weed management–soil quality paradox: although effective weed management and enhancement of soil quality are necessary components of sustainable crop production, practices focused on enhancing one component could inhibit the other. We address the weed management–soil quality paradox in the second section of this chapter by examining strategies aimed at managing both weeds and soil quality simultaneously. In the third section of the chapter, we describe a recent greenhouse

R.G. Smith, Department of Natural Resources and the Environment, University of New Hampshire, Durham, NH 03824 (richard.smith@unh.edu); M.R. Ryan, Department of Crop and Soil Sciences, The Pennsylvania State University, University Park, PA 16802 (mrr203@psu.edu); and F.D. Menalled, Department of Land Resources and Environmental Sciences, Montana State University, Bozeman, MT 59717 (menalled@montana.edu).

doi:10.2136/2011.soilmanagement.c18

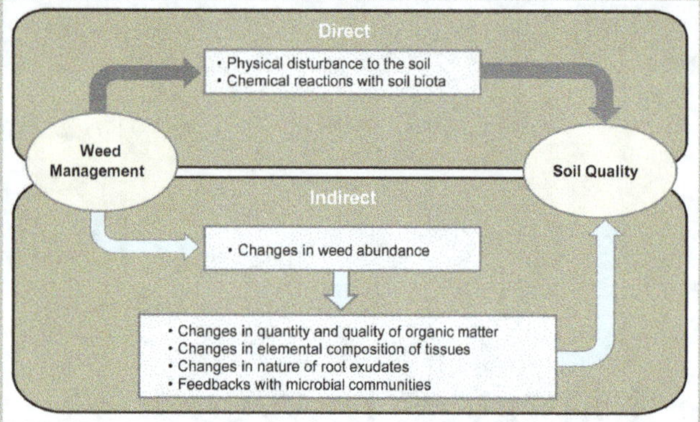

Fig. 18|1. Conceptual model of the direct and indirect effects of weed management on soil quality. Direct effects are primarily mediated through physical and chemical processes. Indirect effects are mediated through changes in weed abundance and species composition and their impacts on soil chemical and biological processes.

study aimed at evaluating potential indirect effects of weed management on soil quality and crop growth. Finally, we discuss recent applied agroecological research focused on developing weed-suppressive soils and agroecosystems, and offer suggestions for future research.

Direct Effects of Major Weed Management Practices on Soil Carbon

Tillage Systems and Soil Disturbance

Perhaps the most consequential change in soil quality associated with weed management occurs immediately after uncultivated land is put into production and mechanical management practices aimed at reducing weed abundance are implemented. For example, following initial cultivation, soil organic carbon can decrease by 30% relative to its precultivated condition (Davidson and Ackerman, 1993). The effects of subsequent cultivations on soil carbon dynamics generally depend on the nature of the implements, the severity of disturbance, and the condition of the soil at the time of implementation (Bowman, 1997; Franzluebbers et al., 1999; Mohler, 2001; Barberi, 2002; Koch and Stockfisch, 2006).

As a weed management strategy, tillage can be divided into three main categories: primary, secondary, and tertiary. Mohler (2001) provides a comprehensive overview of the different types and modes of action of various mechanical implements. The goal of primary tillage is to reduce the abundance, competitive ability, and reproductive potential of emerged weeds. Implements used for primary tillage vary in their intensity of soil disturbance and efficacy for weed control. Moldboard plow tillage is considered full-inversion tillage and results in the most physically intensive disturbance to the soil (Nichols and Reed, 1934). Inversion of the soil results in burial and redistribution of emerged weeds, weed seeds, and crop residues (Buhler and Daniel, 1988; Mohler, 1993; Buhler 1995; Clements et al., 1996). Less intensive forms of primary tillage, such as chisel plow tillage, result in reduced disturbance to the soil profile, but can also be less effective as a weed control practice. As a general rule, higher weed biomass is associated with chisel plow tillage compared with moldboard plow tillage (Fig. 18|2), demonstrating the tradeoff between weed suppression and reduced physical disturbance to the soil (Buhler and Oplinger, 1990; Ryan, unpublished data, 2002).

Secondary tillage is used to prepare the seedbed for planting. Implements used in secondary tillage include disks, rotovators, and harrows. As a weed management tool, secondary tillage can be important for reducing weed seeds that would otherwise germinate with the next crop. This practice, termed stale (or false) seedbed preparation, is often repeated several times to maximize the potential for reduction of the germinable weed seed bank (Mohler, 2001). Soil disturbances associated with secondary tillage are often confined to the upper 10 cm of the soil, and typically result in reduced impacts on soil organic carbon relative to practices that disturb the soil to a greater depth. For example, a loess soil maintained under a conservation tillage program in Germany where annual soil mixing was restricted to a depth of 10 cm contained 6% more soil organic carbon at the 0- to 46-cm depth than the same soil subjected to just three moldboard plow tillage events conducted over 3 yr (Koch and Stockfisch, 2006).

Tertiary tillage or cultivation involves direct interference with germinating or emerged weeds and it can be performed before or after crop emergence. Tertiary tillage implements range from scraping the soil surface to uproot recently emerged weed seedlings (tine weeder), to destroying weeds in the interrows (S-tine cultivator), to cutting the roots of weeds below the soil surface with minimal disturbance to the soil profile (lister or high residue cultivator set at a low angle with wide V-shaped sweeps), to transferring soil from the interrows to bury weed seedlings growing in the crop rows (concave discs and shovels). Tertiary tillage results in variable levels of soil disturbance and weed suppression depending on the implement. For example, when using tertiary cultivation equipment, weeds can recover and persist if soil clods are not adequately disturbed (Bond and Grundy, 2001).

and Düring, 1999; Wander and Bollero, 1999; Cambardella et al., 2004). Reduction or elimination of tillage has been associated with a shift in weed communities (Menalled et al., 2001) and increases in populations of agroecologically important soil macroinvertebrates, such as earthworms (Karlen et al., 1994; Smith et al., 2008a) and ground beetles (Menalled et al., 2007), which can have additional direct impacts on soil quality and weed population dynamics (Smith et al., 2005; Menalled et al., 2007).

While it is clear that reduced and no-tillage systems can have beneficial impacts on soil quality parameters, the heavy reliance on synthetic herbicides creates management challenges. Apart from the increased pressure to select for herbicide-resistant weed biotypes (Heap, 2010) and their nontarget impacts on water quality and human health (Liebman, 2001; Gilliom et al., 2006), herbicides can also have unintended effects on soil quality parameters (Roper and Gupta, 1995). These effects occur primarily through changes in the quality, quantity, and diversity of organic inputs into the soil from weed biomass (Hipps and Samuelson, 1991).

No-Tillage Cropping Systems, Herbicide, and Soil Quality

The wide availability of synthetic herbicides in the last 60 years has made it possible to reduce mechanical approaches to weed control and increased adoption of reduced and no-tillage crop production. The increased interest in reduced and no-tillage practices has also been driven by growing understanding of the detrimental effects of mechanically based weed control approaches on soil quality, particularly in regions susceptible to soil erosion (Gebhardt et al., 1985).

Reducing tillage intensity can have positive impacts on soil carbon dynamics, the formation rate of soil aggregates, water infiltration patterns, soil moisture retention, and the potential for soil crusting (Karlen et al., 1994; Six et al., 1999; Tebrügge

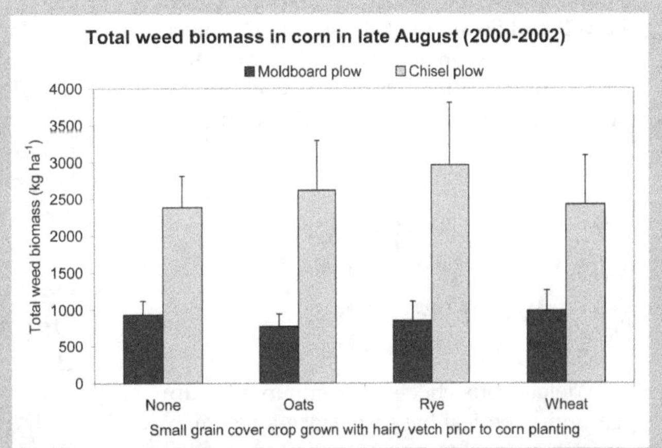

Fig. 18|2. Impact of seedbed preparation method and presence and type of cover crop on weed abundance from a study conducted at the Rodale Institute in Kutztown, Pennsylvania. Weed biomass was significantly higher in the chisel plow tillage plots compared with moldboard plow tillage plots ($P < 0.001$), whereas there was no effect from the cover crop combination ($P < 0.9423$), or the interaction between tillage and cover crop combinations ($P < 0.8437$). Data are pooled across years and bars represent the standard error of the mean.

Evidence for direct effects of repeated herbicide applications on soil microbial communities and their controls on soil quality have, to this point, been equivocal. In a recent review, Bünemann et al. (2006) reported few significant negative effects of herbicides on soil organisms. Similarly, Hartley et al. (1996) indicated that there were no effects of a residual herbicide (terbuthylazine) on in situ soil respiration or cellulose degradation when compared with several organic weed management practices. However, Santos et al. (2006) suggests that the effects of herbicides on soil microbial communities may depend on tillage systems. Furthermore, Zilli et al. (2008) found that while glyphosate and imazaquin-based herbicides had no effect on soil microbial biomass carbon, basal respiration, and metabolic quotient, there were changes in the bacterial profile in the rhizoplane, suggesting that herbicide applications may alter the structure of soil microbial communities. Similarly, when examining the impact of bromoxynil on soil-derived microbial communities, Baxter and Cummings (2006) found that the application of the herbicide exerted deterministic selection on the microbial community. Finally, in a critique of studies evaluating the effects of glyphosate on diseases associated with *Fusarium* species, Powell and Swanton (2008) indicate the need for further research aimed at determining if there are indeed direct negative effects on soil health from glyphosate applications.

Weed Management Coupled with Soil-Building Practices

Organic Matter Inputs

The deleterious effects of common mechanical and chemical-based weed management practices on soil organic carbon may be offset by soil-building practices such as applying manure or compost and growing cover crops. For example, in the long-term Rodale Institute Farming Systems Trial in Kutztown, Pennsylvania, two organic crop management systems were compared with a conventional system to assess the effects of management systems on various agroecosystem properties. The two organic systems received no applications of synthetic fertilizer or pesticides and included treatments with two different kinds of organic matter inputs: (i) organic animal-based (a diverse rotation including corn, soybeans, corn silage, wheat, and red clover–alfalfa hay, as well as a rye cover crop before corn silage and soybeans, with aged cattle manure applied as a supplemental nitrogen and organic matter source), and (ii) legume cover crop-based (a 5-yr rotation of hairy vetch [winter cover crop used as a green manure], corn, rye [winter cover crop], soybeans, and winter wheat). The conventional system (a 5-yr corn–corn–soybean–corn–soybean rotation) was managed with fertilizer and pesticide applications following Pennsylvania State University Cooperative Extension recommendations (Pimentel et al., 2005). Moldboard plow tillage and cultivation were performed four to eight times in any given cropping season in the organic systems and chisel plow tillage was used in the conventional system. Despite the intensive rate of soil disturbances applied in the organic systems, significantly higher soil microbial biomass, soil respiration, and arbuscular mycorrhizal fungi populations were reported in the organic systems than in the conventional system (Douds et al., 1993; Wander et al., 1994; Harris et al., 1994; Douds and Millner, 1999). Similarly, results from a 9-yr cropping systems trial comparing organic, no-tillage, and conventional tillage cropping systems, showed that the organic systems with cover crops and manure achieved higher levels of soil carbon, despite the use of more intensive tillage and cultivation (Teasdale et al., 2007).

Reduced Tillage in Low- and Organic-Input Systems

The practices of reduced and no-tillage have been restricted to systems in which high rates of herbicides can be applied. However, recent agricultural and technological developments may enable growers to manage weeds in reduced tillage systems with minimum or no herbicide inputs. Organic amendments and crop diversification through rotation and cover crops are important components of strategies aimed at improving soil quality while managing weeds in reduced- or no-herbicide systems (Liebman and Dyck, 1993; Liebman and Davis, 2000). For example, winter cereal cover crops can be effective components of integrated weed management systems that rely on lower herbicide rates in reduced tillage systems (Price et al., 2006). Cereal rye (*Secale cereale* L.) is a popular winter cereal cover crop in the northeast region, mainly because of its cold hardiness and ability to produce large amounts of biomass (~10,000 kg/ha), an attribute positively associated with weed suppression (Teasdale, 1996). Leguminous cover crops such as red clover (*Trifolium pratense* L.) and crimson clover (*Trifolium incarnatum* L.) can also suppress weeds in reduced-input and organic systems through competitive and allelopathic effects, provide habitat to seed-predating macroinvertebrates, and contribute to soil fertility and organic carbon (Davis and Liebman, 2003, Smith et al., 2008b).

Constraints to implementing reduced or no-tillage in low-input or organic systems include difficulty in terminating the cover crops before planting cash crops and managing weeds (Peigne et al., 2007). Development of new cover crop management systems, such as roller-crimpers may alleviate some of the constraints associated with cover crops, though these are still in their infancy. Depending on the design of the roller, crimped cover crop termination rates can range from 83% to 94% (Kornecki et al., 2006), and the residue provides a weed-suppressing layer of mulch that can reduce the need for cultivation.

An alternative approach to continuous no-tillage in reduced- and organic-input cropping systems could be to rotate tillage—necessary for weed management during some phases of the rotation—with no-tillage in other phases of the rotation when weed pressures are low (Peigne et al., 2007). This would allow the soil-building benefits of reduced tillage and organic matter inputs to be realized over some phases of the rotation, which might help offset the potential negative effects associated with more intensive tillage during particularly weed-sensitive phases of the rotation.

Indirect Effects of Weed Management on Soil Ecosystems

Understanding how plants alter the soil ecosystem is critical to the management of crop–weed interactions and the consequent impacts of weed management practices on soil quality. Weeds are often perceived as exerting only negative effects on crop growth through competition for limiting soil nutrients, water, and light, thus necessitating management practices aimed at reducing these effects. While there is ample evidence to support this perception (Knezevic et al., 2002; Zimdahl, 2004), there is also an increasing body of theory (reviewed in Tilman and Lehman, 2001; Hooper et al., 2005; Smith et al., 2010) and experimental evidence (Reynolds et al., 2003; Zak et al., 2003; Wardle et al., 2004; Jordan and Vatovec, 2004; Hooper et al., 2005; Pollnac et al., 2009) suggesting that the specific characteristics of the weed community (i.e., species composition and diversity of the weeds growing in a field) may affect the nature of crop–weed interactions, potentially offsetting or compounding the competitive effects of weeds on crops.

The diversity, composition, and abundance of weed communities could impact the competitive effects of weeds on crops and soil ecosystems through two potential pathways: (i) complementary resource use or facilitative interactions among co-occurring plant species (Hille Ris Lambers et al., 2004), and (ii) by influencing the composition and functioning of the soil microbial community, including the abundance of pathogenic and beneficial microorganisms (Bever et al., 1997; Wardle et al., 2003). The first of these two pathways implies that increased diversity could result in greater total resource acquisition in diverse than in simple (low species diversity) weed communities. In grassland ecosystems, many of the effects of plant diversity on ecosystem function are thought to be mediated by complementary resource use among species that differ in their requirements for and modes of acquiring limiting resources (Tilman et al., 1996; Hector et al., 1999; Tilman and Lehman, 2001). An important prediction of the resource-use (or niche) complementarity hypothesis is that plant communities containing a high diversity of plant species should have fewer resources left unexploited compared with communities containing fewer species (Elton, 1958; McNaughton, 1993; Naeem et al., 1994; Tilman et al., 1996; Tilman and Lehman, 2001). This suggests that in agricultural systems, for a given weed density, diverse weed communities might take up a greater quantity and diversity of soil resources compared with simple weed communities, affecting crop resource availability and the quantity and nature of soil organic inputs through root exudates and biomass turnover. To our knowledge, there have been few direct tests of this hypothesis (Tilman et al., 2001), especially within the context of weeds in managed agricultural systems. In a manipulative study assessing the joint impact of species density and species diversity on crop yield, Pollnac et al. (2009), determined that species richness had no effect on spring wheat biomass, yield, or relative growth rate, and that there was no evidence for reduced competitive effects on crop plants due to interspecific competition between neighbor weed species. However, weed species richness had a negative effect on the growth of individual weed species and competitive ability of dominant weed species.

The second pathway by which weed diversity may impact the soil ecosystem and weed–crop competition is by influencing the composition and functioning of the soil microbial community, including the abundance of pathogenic and beneficial microorganisms. Changes in the soil microbial community following plant invasion have been reported in a number of nonagricultural systems (Kourtev et al., 2003; Callaway et al., 2004; Hawkes et al., 2005; Wolfe and Klironomos, 2005). Changes in the microbial community mediated by shifts in plant community composition and diversity can then lead to additional feedbacks to the plant community (Bever, 1994; Bever et al., 1997; Klironomos, 2002; Reynolds et al., 2003; Clark et al., 2005; Mikola et al., 2005; Wolfe and Klironomos, 2005). These feedbacks are mediated through changes in decomposition rates (Hobbie, 1992; Knops et al., 2002), resource availability (Vinton and Burke, 1995; Ehrenfeld et al., 2001; Zak et al., 2003; Clark et al., 2005) and pathogenic and mutualistic interactions (Hamel et al., 2005; Wolfe and Klironomos, 2005).

In agricultural systems, it is known that continuously planting the same crop can lead to increased soil pathogen loads, which can severely decrease crop performance (Bruehl, 1987). Understanding this crop–pathogen link has lead to management practices such as crop rotation aimed at disrupting pathogen population build-up

(Bullock, 1992; Liebman and Dyck, 1993; Hamel et al., 2005). However, the potential for arable weeds to affect soil microbial communities in agroecosystems is less well understood (Li and Kremer, 2000). Changes in soil microbial communities driven by weed community composition and diversity could potentially lead to either positive or negative feedbacks on crop growth and competitive ability, depending on the weed-species specificity and the nature (positive or negative) of the feedback. Also unknown, is the degree to which changes in soil resource availability due to weed diversity are mediated through changes in the soil community (but see Zak et al., 2003). Here we present data from a recent greenhouse study aimed at elucidating the pathways mediating indirect effects of weeds on crops.

The experiment was conducted with field soils collected from an organic small grain–legume rotation near Bozeman, Montana that had been managed without synthetic inputs for 12 yr (Maxwell et al., 2007). The soil was collected from the field in fall 2005 and mixed into a blend that contained organically managed field soil, mineral soil, sand, and peat (1.0 part: 0.5 parts: 1.25 parts: 0.25 parts, respectively). The soil mixture was then subjected to two treatments: steam pasteurization at 70°C for 2 h to reduce the activity of the naturally occurring microbial community (hereafter SP), and no steam pasteurization (hereafter NSP). Each of the two soil mixtures were then used to fill large pots (25.4-cm diameter, 35.0-cm height) into which different weed community treatments were sown in a three-phase design, with each treatment replicated eight times. The goal of the first two phases was to prime soils with plant communities differing in plant species composition and diversity. The goal of the last phase was to assess the response of wheat to these primed soils. Species used in this study included five agriculturally important plant species, four weed species—wild oat (*Avena fatua* L.), green foxtail [*Setaria viridis* (L.) P. Beauv.], field pennycress (*Thlaspi arvense* L.), and redroot pigweed (*Amaranthus retroflexus* L.)—and one crop, spring wheat var. McNeal (*Triticum aestivum* L.). The reasoning supporting this study was that differences in wheat biomass in the third phase between soil pasteurization treatments and across plant communities would indicate the importance of each potential pathway (simple resource uptake or changes in soil communities) in mediating the observed responses (Fig. 18|3).

In the first two phases, each plant species was sown in monoculture, all possible three-species combinations, and in a mixture that contained all five species. Pots were watered with a drip irrigation system, as needed and plant communities were thinned to a density of 15 individuals following emergence (multispecies communities were thinned in such a way as to maintain an equivalent proportion of each species within the community). Plant communities were grown for ~40 d, after which time all aboveground biomass was removed, dried at 60°C, and weighed. This resulted in two "generations" of soil priming. Approximately 10 d after the second harvest, the third phase was initiated and wheat was planted into all pots to measure its response to the previous plant community treatments. Wheat populations in each pot were thinned to 15 individuals and the biomass was harvested after 40 d, dried, and weighed.

The three main results obtained in this study expand our understanding of the potential importance of interactions between weed and soil communities. First, the native soil community appeared to exhibit inhibitory control on the growth of each of the five plant species examined. By pasteurizing the soil, and reducing the activity of the native soil microorganisms, plant growth increased for each of the five species (Fig. 18|4). Second, while we observed a significant negative relationship between the diversity of the plant community in the first two phases of the study and wheat biomass

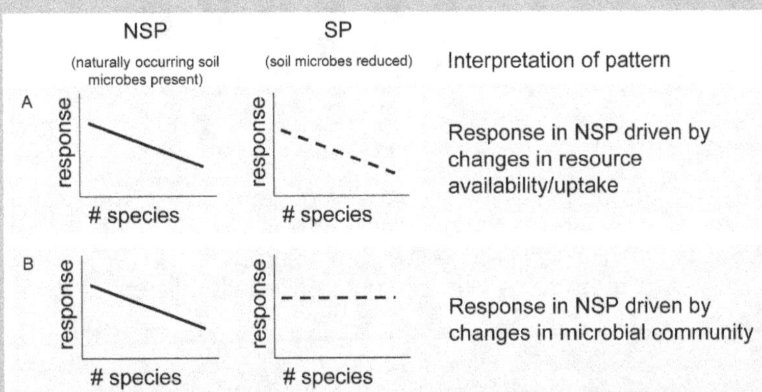

Fig. 18|3. Theoretical framework for interpreting plant response to manipulations of weed diversity. Responses observed in soil containing a naturally occurring microbial community (NSP) are compared with those observed in soil in which the naturally occurring microbial community has been minimized by means of steam pasteurization (SP).

Fig. 18|4. Growth of five agriculturally important plant species in soil containing a naturally occurring microbial community (left pot in each set) and soil in which the microbial community has been minimized by steam pasteurization (SP).

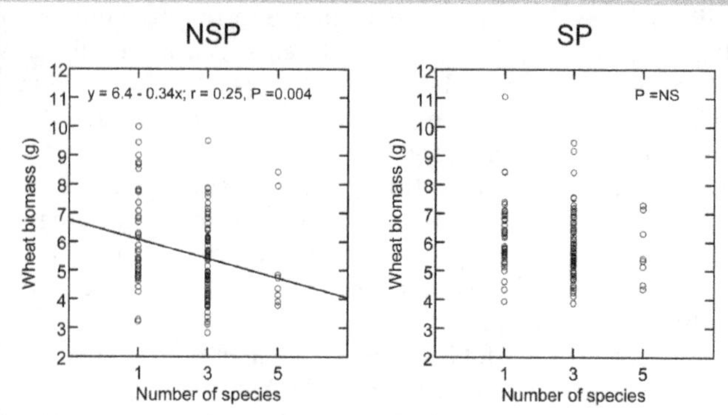

Fig. 18|5. Response of wheat to the number of plant species grown in a pot for two generations before the planting of wheat in soils containing a naturally occurring microbial community (NSP) and in which the naturally occurring microbial community was reduced by steam pasteurization (SP).

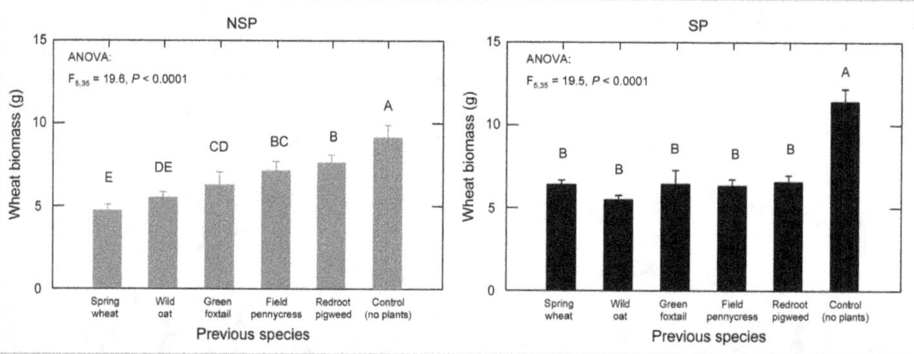

Fig. 18|6. Response of wheat to the identity of the previous plant grown for two generations in soils containing a naturally occurring microbial community (NSP) and in which the naturally occurring microbial community was reduced by steam pasteurization (SP). Data are means ± SE, N = 8. Bars sharing the same letter are not significantly different at P < 0.05 (Fisher's LSD test).

in the third phase of the study in the NSP treatment, no relationship between plant diversity and wheat biomass was observed in the SP treatment (Fig. 18|5). The fact that in the SP treatment we reduced microbial activity, but did not, in theory, alter resource availability, leads us to conclude that weed diversity reduced the growth of wheat primarily through impacts on soil communities rather than by simple nutrient uptake. Third, individual weed species differed in their effects on soil community-mediated feedbacks on wheat growth, and these effects depended on soil microbial activity (Fig. 18|6). In the NSP treatment, wheat biomass in the third phase of the study was lowest when wheat followed wheat, intermediate when it followed green foxtail, and highest when it followed redroot pigweed. In contrast, there was no relationship between the identity of the previous weed species and the growth of wheat in the SP treatment. Soil nitrate was measured before sowing wheat in the third phase of the study and indicated that differences in resource availability between the five species at the end of the second phase were not related to wheat biomass response in phase three, further suggesting that the observed response in wheat was driven by microbial-related feedbacks (data not shown).

Future Directions and Necessary Research

A better understanding of the complex interactions that can occur between weeds and soil communities may allow us to develop strategies for managing weeds while reducing direct and indirect negative impacts on soil quality (Kremer and Li, 2003; Smith et al., 2010). Managing soils in ways that make soils more resistant to weedy invasions or increase the abundance of weed-suppressive or antagonist microbes or other soil organisms is a promising avenue of research that has until now elicited only limited investigation (Kremer, 1993; Kremer and Li, 2003; Gallandt et al., 2004; Chee-Sanford et al., 2006; Davis et al., 2006). Although in a recent review, Vasquez et al. (2008) proposed that adequate nitrogen management can help create more invasion-resistant soils, more work is needed to understand how weedy invasions are mediated through the soil fungal and bacterial community and to identify the specific components of the soil communities that are important determinants of weed growth, fitness, and seed degradation (Davis et al., 2006).

It is also likely that greater progress will be made when research explicitly addresses strategies that integrate multiple tactics (Liebman and Gallandt, 1997). For example, combines can be modified to capture weed seeds during crop harvest to reduce weed populations in the soil seed bank (Davis et al., 2005). This tactic combined with other cultural practices, such as planting competitive cultivars or enhancement of weed seed predator populations through the establishment of noncrop habitat around crop fields (Landis et al., 2000), may result in enough cumulative effects on weed populations to reduce the necessity for intensive soil disturbances. Unfortunately, the majority of crop breeding programs have been performed to maximize growth response to fertilizer application, and in the absence of weed competition. Breeding efforts aimed at improving crop competitiveness in the presence of weeds could reduce the need for employing potentially soil-damaging management practices (Christensen, 1995; Pester et al., 1999; Lemerle et al., 2001; Olofsdotter et al., 2002). To date, there has been surprisingly little work in this arena, and opportunities to improve integrated weed management through competitive cultivar selection abound.

Lastly, because of the negative impact of weeds on crop yield and quality, soil-conditioning benefits of weedy plants are seldom discussed (Sturz et al., 2001). However, recent research suggests that crop plants may benefit from noncrop plants growing in their presence either at low densities or at limited times throughout the growing season (Sturz et al., 2001; Jordan and Vatovec, 2004). Potential soil quality benefits provided by weedy plants include: (i) sequestering carbon, (ii) cycling nutrients, (iii) hosting mycorrhizal fungi populations, and (iv) supporting plant-growth-promoting rhizobacteria. Other ecosystem services enhanced from increasing benign noncrop vegetation in arable landscapes include insect pest suppression, buffering of microclimate and hydrologic processes, providing cover and habitat, and detoxification of noxious chemicals (Altieri, 1999; Gerowitt et al., 2003; Marshall et al., 2003). Research focused on developing a better understanding of the potential benefits that weeds may provide to cropping systems could illuminate opportunities to enhance their usefulness, reduce their negative impacts on crop yields, and improve strategies to manage their populations in ways that are beneficial to soil quality.

Conclusions

A better understanding of the direct and indirect relationships between weed management practices and soil quality is a necessary step to sustain agricultural productivity in the face of a growing global population and a finite abundance of natural resources. Problems of erosion of the world's agricultural soils provide strong evidence of the tight coupling between agricultural management practices and soil quality (Montgomery, 2007). As public awareness grows regarding where and how food is produced, so too does the need for agricultural scientists to provide growers and land managers the tools they need to grow that food profitably and with minimal impact on the environment (Robertson and Swinton, 2005). An ecologically based approach to agroecosystem management relies on a keen understanding of processes that affect the weed populations and community dynamics and increase the relative competitive ability of crops. As the research reviewed here suggests, there may be opportunities to tip the balance between weeds and crops in ways that favor the crop, and that do not necessarily come at the expense of soil quality.

References

Altieri, M.A. 1999. The ecological role of biodiversity in agroecosystems. Agric. Ecosyst. Environ. 74:19–31.

Barberi, P. 2002. Weed management in organic agriculture: Are we addressing the right issues? Weed Res. 42:177–193.

Baxter, J., and S.P. Cummings. 2006. The application of the herbicide bromoxynil to a model soil-derived bacterial community: Impact on degradation and community structure. Lett. Appl. Microbiol. 43:659–665.

Bever, J.D. 1994. Feedback between plants and their soil communities in an old field community. Ecology 75:1965–1977.

Bever, J.D., K.M. Westover, and J. Antonovics. 1997. Incorporating the soil community into plant population dynamics: The utility of the feedback approach. J. Ecol. 85:561–573.

Bond, W., and A.C. Grundy. 2001. Non-chemical weed management in organic farming systems. Weed Res. 41:383–405.

Bowman, G.B. 1997. Steel in the Field: A farmer's guide to weed management tools. Sustainable Agriculture Network, USDA, National Agricultural Library, Beltsville, MD.

Bruehl, G.W. 1987. Soilborne plant pathogens. Macmillan Publishing Company, New York.

Buhler, D.D. 1995. Influence of tillage systems on weed population dynamics and management in corn and soybean in the Central USA. Crop Sci. 35:1247–1258.

Buhler, D.D., and T.C. Daniel. 1988. Influence of tillage systems on giant foxtail, *Setaria faberi*, and velvetleaf, *Abutilon theophrasti*, density and control in corn, *Zea mays*. Weed Sci. 36:642–647.

Buhler, D.D., and E.S. Oplinger. 1990. Influence of tillage system on annual weed densities and control in solid-seeded soybean (*Glycine max*). Weed Sci. 38:158–165.

Bullock, D.G. 1992. Crop rotation. Crit. Rev. Plant Sci. 11:309–326.

Bünemann, E.K., G.D. Schwenke, and L. Van Zwieten. 2006. Impact of agricultural inputs on soil organisms—a review. Aust. J. Soil Res. 44:379–406.

Callaway, R.M., G.C. Thelen, A. Rodriguez, and W.E. Holben. 2004. Soil biota and exotic plant invasion. Nature 427:731–733.

Cambardella, C.A., T.B. Moorman, S.S. Andrews, and D.L. Karlen. 2004. Watershed-scale assessment of soil quality in the loess hills of southwest Iowa. Soil Tillage Res. 78:237–247.

Cannell, R.Q., and J.D. Hawes. 1994. Trends in tillage practices in relation to sustainable crop production with special reference to temperate climates. Soil Tillage Res. 30:245–282.

Chee-Sanford, J.C., M.M. Williams, A.S. Davis, and G.K. Sims. 2006. Do microorganisms influence seed-bank dynamics? Weed Sci. 54:575–587.

Christensen, S. 1995. Weed suppression ability of spring barley varieties. Weed Res. 35:241–247.

Clark, B.R., S.E. Hartley, K.N. Suding, and C. de Mazancourt. 2005. The effect of recycling on plant competitive hierarchies. Am. Nat. 165:609–622.

Clements, D.R., D.L. Benoit, S.D. Murphy, and C.J. Swanton. 1996. Tillage effects on weed seed return and seedbank composition. Weed Sci. 44:314–322.

Davidson, E.A., and I.L. Ackerman. 1993. Changes in soil carbon inventories following cultivation of previously untilled soils. Biogeochemistry 20:161–193.

Davis, A.S., and M. Liebman. 2003. Cropping system effects on giant foxtail (*Setaria faberi*) demography: I. Green manure and tillage timing. Weed Sci. 51:919–929.

Davis, A., K. Renner, C. Sprague, L. Dyer, and D. Mutch. 2005. Integrated weed management: One year's seeding… Extension bulletin E-2931. Michigan State Univ. Ext., East Lansing, MI.

Davis, A.S., A.I. Kathleen, S.G. Hallett, and K.A. Renner. 2006. Weed seed mortality in soils with contrasting agricultural management histories. Weed Sci. 54:291–297.

Dick, R.P. 1992. A review: long-term effects of agricultural systems on soil biochemical and microbial parameters. Agric. Ecosyst. Environ. 40:25–36.

Douds, D., R. Janke, and S. Peters. 1993. VAM fungus spore populations and colonization of roots of maize and soybean under conventional and low-input sustainable agriculture. Agric. Ecosyst. Environ. 43:325–335.

Douds, D., and P. Millner. 1999. Biodiversity of arbuscular mycorrhizal fungi in agroecosystems. Agric. Ecosyst. Environ. 74:77–93.

Ehrenfeld, J.G., P. Kourtev, and W.Z. Huang. 2001. Changes in soil functions following invasions of exotic understory plants in deciduous forests. Ecol. Appl. 11:1287–1300.

Elton, C.S. 1958. The ecology of invasions by animals and plants. Chapman and Hall, London.

Franzluebbers, A.J., G.W. Langdale, and H.H. Schomberg. 1999. Soil carbon, nitrogen, and aggregation in response to type and frequency of tillage. Soil Sci. Soc. Am. J. 63:349–355.

Gallandt, E.R., E.P. Fuerst, and A.C. Kennedy. 2004. Effect of tillage, fungicide seed treatment, and soil fumigation on seed bank dynamics of wild oat (*Avena fatua*). Weed Sci. 52:597–604.

Gebhardt, M.R., T.C. Daniel, E.E. Schweizer, and R.R. Allmaras. 1985. Conservation tillage. Science 230:625–630.

Gerowitt, B., E. Bertke, S.K. Hespelt, and C. Tute. 2003. Towards multifunctional agriculture—weeds as ecological goods? Weed Res. 43:227–235.

Gilliom, R.J., J.E. Barbash, C.G. Crawford, P.A. Hamilton, J.D. Martin, N. Nakagaki, L.H. Nowell, J.C. Scott, P.E. Stackelberg, G.P. Thelin, and D.M. Wolock. 2006. The quality of our nation's waters– pesticides in the nation's streams and ground water, 1992–2001. U.S. Geological Survey Circular1291. USGS, Reston, VA.

Hamel, C., V. Vujanovic, R. Jeannotte, A. Nakano-Hylander, and M. St-Arnaud. 2005. Negative feedback on a perennial crop: *Fusarium* crown and root rot of asparagus is related to changes in soil microbial community structure. Plant Soil 286:1123–1127.

Harris, G.H., O.B. Hesterman, E.A. Paul, S.E. Peters, and R.R. Janke. 1994. Fate of legume fertilizer nitrogen-15 in a long-term cropping systems experiment. Agron. J. 86:910–915.

Hartley, M.J., J.B. Reid, A. Rahman, and J.A. Springett. 1996. Effect of organic mulches and a residual herbicide on soil bioactivity in an apple orchard. N.Z. J. Crop Hortic. Sci. 24:183–190.

Hawkes, C.V., I.F. Wren, D.J. Herman, and M.K. Firestone. 2005. Plant invasion alters nitrogen cycling by modifying the soil nitrifying community. Ecol. Lett. 8:976–985.

Heap, I. 2010. The international survey of herbicide resistant weeds. Available at www.weedscience.com (accessed 19 Feb. 2010; verified 24 Oct. 2010).

Hector, A., B. Schmid, C. Beierkuhnlein, M.C. Caldeira, M. Diemer, P.G. Dimitrakopoulos, J.A. Finn, H. Freitas, P.S. Giller, J. Good, R. Harris, P. Hogberg, K. Huss-Danell, J. Joshi, A. Jumpponen, C. Korner, P.W. Leadley, M. Loreau, A. Minns, C.P.H. Mulder, G. O'Donovan, S.J. Otway, J.S. Pereira, A. Prinz, D.J. Read, M. Scherer-Lorenzen, E.D. Schulze, A.S.D. Siamantziouras, E.M. Spehn, A.C. Terry, A.Y. Troumbis, F.I. Woodward, S. Yachi, and J.H. Lawton. 1999. Plant diversity and productivity experiments in European grasslands. Science 286:1123–1127.

Hille Ris Lambers, J., W.S. Harpole, D. Tilman, J. Knops, and P.B. Reich. 2004. Mechanisms responsible for the positive diversity—productivity relationship in Minnesota grasslands. Ecol. Lett. 7:661–668.

Hipps, N.A., and T.J. Samuelson. 1991. Effects of long-term herbicide use, irrigation and nitrogen-fertilizer on soil fertility in an apple orchard. J. Sci. Food Agric. 55:377–387.

Hobbie, S.E. 1992. Effects of plant species on nutrient cycling. Trends Ecol. Evol. 7:336–339.

Hooper, D.U., F.S. Chapin, J.J. Ewel, A. Hector, P. Inchausti, S. Lavorel, J.H. Lawton, D.M. Lodge, M. Loreau, S. Naeem, B. Schmid, H. Setala, A.J. Symstad, J. Vandermeer, and D.A. Wardle. 2005. Effects of biodiversity on ecosystem functioning: A consensus of current knowledge. Ecol. Monogr. 75:3–35.

Jordan, N., and C. Vatovec. 2004. Agroecological benefits from weeds. p. 137–158. *In* Inderjit (ed.) Weed biology and management. Kluwer Academic Publishers, Netherlands.

Karlen, D.L., S.S. Andrews, and J.W. Doran. 2001. Soil quality: Current concepts and applications. Adv. Agron. 74:1–40.

Karlen, D.L., N.C. Wollenhaupt, D.C. Erbach, E.C. Berry, J.B. Swan, N.S. Eash, and J.L. Jordahl. 1994. Long-term tillage effects on soil quality. Soil Tillage Res. 32:313–327.

Klironomos, J.N. 2002. Feedback with soil biota contributes to plant rarity and invasiveness in communities. Nature 417:67–70.

Knezevic, S.Z., S.P. Evans, E.E. Blankenship, R.C. Van Acker, and J.L. Lindquist. 2002. Critical period for weed control: The concept and data analysis. Weed Sci. 50:773–786.

Knops, J.M.H., K.L. Bradley, and D.A. Wedin. 2002. Mechanisms of plant species impacts on ecosystem nitrogen cycling. Ecol. Lett. 5:454–466.

Koch, H.J., and N. Stockfisch. 2006. Loss of soil organic matter upon ploughing under a loess soil after several years of conservation tillage. Soil Tillage Res. 86:73–83.

Kornecki, T.S., A.J. Price, and R.L. Raper. 2006. Performance of different roller designs in terminating rye cover crop and reducing vibration. Appl. Eng. Agric. 22:633–641.

Kourtev, P.S., J.G. Ehrenfeld, and M. Haggblom. 2003. Experimental analysis of the effect of exotic and native plant species on the structure and function of soil microbial communities. Soil Biol. Biochem. 35:895–905.

Kremer, R.J. 1993. Management of weed seed banks with microorganisms. Ecol. Appl. 3:42–52.

Kremer, R.J., and J.M. Li. 2003. Developing weed-suppressive soils through improved soil quality management. Soil Tillage Res. 72:193–202.

Lal, R., A.A. Mahboubi, and N.R. Fausey. 1994. Long-term tillage and rotation effects on properties of a central Ohio soil. Soil Sci. Soc. Am. J. 58:517–522.

Landis, D.A., S.D. Wratten, and G.M. Gurr. 2000. Habitat management to conserve natural enemies of arthropod pests in agriculture. Annu. Rev. Entomol. 45:175–201.

Lemerle, D., B. Verbeek, and B. Orchard. 2001. Ranking the ability of wheat varieties to compete with Lolium rigidum. Weed Res. 41:197–209.

Li, J.M., and R.J. Kremer. 2000. Rhizobacteria associated with weed seedlings in different cropping systems. Weed Sci. 48:734–741.

Liebman, M. 2001. Weed management: A need for ecological approaches. p. 1–39. *In* M. Liebman, C.L. Mohler, and C.P. Staver (ed.) Ecological management of agricultural weeds. Cambridge Univ. Press, Cambridge, UK.

Liebman, M., and A.S. Davis. 2000. Integration of soil, crop and weed management in low-external-input farming systems. Weed Res. 40:27–47.

Liebman, M., and E. Dyck. 1993. Crop rotation and intercropping strategies for weed management. Ecol. Appl. 3:92–122.

Liebman, M., and E.R. Gallandt. 1997. Many little hammers: Ecological management of crop-weed interactions. p. 291–343. *In* L.E. Jackson (ed.) Ecology in Agriculture. Academic Press, San Diego, CA.

Marshall, E.J.P., V.K. Brown, N.D. Boatman, P.J.W. Lutman, G.R. Squire, and L.K. Ward. 2003. The role of weeds in supporting biological diversity within crop fields. Weed Res. 43:77–89.

Maxwell, B.D., R.G. Smith, and M. Brelsford. 2007. Wild oat (*Avena fatua*) seed bank dynamics in transition to organic wheat production systems. Weed Sci. 55:212–217.

McNaughton, S.J. 1993. Biodiversity and function of grazing ecosystems. p. 361–363. *In* E.D. Schulze and H.A. Mooney (ed.) Biodiversity and ecosystem function. Springer-Verlag, Berlin, Germany.

Menalled, F.D., K.L. Gross, and M. Hammond. 2001. Weed aboveground and seedbank community responses to agricultural management systems. Ecol. Appl. 11:1586–1601.

Menalled, F.D., R.G. Smith, J.T. Dauer, and T.B. Fox. 2007. Impact of agricultural management on carabid communities and weed seed predation. Agric. Ecosyst. Environ. 118:49–54.

Mikola, J., K. Ilmarinen, M. Nieminen, and M. Vestberg. 2005. Long-term soil feedback on plant N allocation in defoliated grassland miniecosystems. Soil Biol. Biochem. 37:899–904.

Mohler, C.L. 1993. A model of the effects of tillage on emergence of weed seedlings. Ecol. Appl. 3:53–73.

Mohler, C.L. 2001. Mechanical management of weeds. p. 139–209. *In* M. Liebman, C.L. Mohler, and C.P. Staver (ed.) Ecological management of agricultural weeds. Cambridge Univ. Press, Cambridge, UK.

Montgomery, D.R. 2007. Soil erosion and agricultural sustainability. Proc. Natl. Acad. Sci. USA 104:13268–13272.

Naeem, S., L.J. Thompson, S.P. Lawler, J.H. Lawton, and R.M. Woodfin. 1994. Declining biodiversity can alter the performance of ecosystems. Nature 368:734–737.

Nichols, M.L., and I.F. Reed. 1934. Soil dynamics. VI. Physical reactions of soil to moldboard surfaces. Agric. Eng. 15:187–190.

Olofsdotter, M., L.B. Jensen, and B. Courtois. 2002. Improving crop competitive ability using allelopathy—an example from rice. Plant Breed. 121:1–9.

Peigne, J., B.C. Ball, J. Roger-Estrade, and C. David. 2007. Is conservation tillage suitable for organic farming? A review. Soil Use Manage. 23:129–144.

Pester, T.A., O.C. Burnside, and J.H. Orf. 1999. Increasing crop competitiveness to weeds through crop breeding. p. 59–76. *In* D. Buhler (ed.) Expanding the context of weed management. Haworth Press, New York.

Pimentel, D., P. Hepperly, J. Hanson, D. Douds, and R. Seidel. 2005. Environmental, energetic, and economic comparisons of organic and conventional farming systems. Bioscience 55:573–582.

Pollnac, F.W., B.D. Maxwell, and F.D. Menalled. 2009. Weed community characteristics and crop performance: a neighborhood approach. Weed Res. 49:3

Powell, J.R., and C.J. Swanton. 2008. A critique of studies evaluating glyphosate effects on diseases associated with *Fusarium* spp. Weed Res. 48:307–318.

Price, A.J., D.W. Reeves, and M.G. Patterson. 2006. Evaluation of weed control provided by three winter cereals in conservation-tillage soybean. Renew. Agric. Food Syst. 21:159–164.

Reeves, D.W. 1997. The role of soil organic matter in maintaining soil quality in continuous cropping systems. Soil Tillage Res. 43:131–167.

Reynolds, H.L., A. Packer, J.D. Bever, and K. Clay. 2003. Grassroots ecology: Plant-microbe-soil interactions as drivers of plant community structure and dynamics. Ecology 84:2281–2291.

Robertson, G.P., D.C. Coleman, C.S. Bledsoe, and P. Sollins. 1999. Standard soil methods for long-term ecological research. Oxford Univ. Press, New York.

Robertson, G.P., and S.M. Swinton. 2005. Reconciling agricultural productivity and environmental integrity: A grand challenge for agriculture. Front. Ecol. Environ. 3:38–46.

Roper, M.M., and V.V.S.R. Gupta. 1995. Management practices and soil biota. Aust. J. Soil Res. 33:321–339.

Santos, J.B., A. Jakelaitis, A.A. Silva, M.D. Costa, A. Manabe, and M.C.S. Silva. 2006. Action of two herbicides on the microbial activity of soil cultivated with common bean (*Phaseolus vulgaris*) in conventional-till and no-till systems. Weed Res. 46:284–289.

Six, J., E.T. Elliott, and K. Paustian. 1999. Aggregate and soil organic matter dynamics under conventional and no-tillage systems. Soil Sci. Soc. Am. J. 63:1350–1358.

Six, J., R.T. Conant, E.A. Paul, and K. Paustian. 2002. Stabilization mechanisms of soil organic matter: Implications for C-saturation of soils. Plant Soil 241:155–176.

Smith, R.G., K.L. Gross, and S. Januchowski. 2005. Earthworms and weed seed distribution in annual crops. Agric. Ecosyst. Environ. 108:363–367.

Smith, R.G., C.P. McSwiney, A.S. Grandy, P. Suwanwaree, R.M. Snider, and G.P. Robertson. 2008a. Diversity and abundance of earthworms across an agricultural land-use intensity gradient. Soil Till. Res. 100:83–88.

Smith, R.G., K.L. Gross, and G.P. Robertson. 2008b. Effects of crop diversity on agroecosystem function: Crop yield response. Ecosystems 11:355–366.

Smith, R.G., D.A. Mortensen, and M.R. Ryan. 2010. A new hypothesis for the functional role of diversity in mediating resource pools and weed-crop competition in agroecosystems. Weed Res. 50:37–48.

Sollins, P., P. Homann, and B.A. Caldwell. 1996. Stabilization and destabilization of soil organic matter: Mechanisms and controls. Geoderma 74:65–105.

Sturz, A.V., B.G. Matheson, W. Arsenault, J. Kimpinski, and B.R. Christie. 2001. Weeds as a source of plant growth promoting rhizobacteria in agricultural soils. Can. J. Microbiol. 47:1013–1024.

Teasdale, J.R. 1996. Contribution of cover crops to weed management in sustainable agricultural systems. J. Prod. Agric. 9:475–479.

Teasdale, J.R., C.B. Coffman, and R.W. Mangum. 2007. Potential long-term benefits of no-tillage and organic cropping systems for grain production and soil improvement. Agron. J. 99:1297–1305.

Tebrügge, F., and R.A. Düring. 1999. Reducing tillage intensity—a review of results from a long-term study in Germany. Soil Tillage Res. 53:15–28.

Tilman, D., and C. Lehman. 2001. Biodiversity, composition, and ecosystem processes: Theory and concepts. p. 9–41. *In* A.P. Kinzig, S.W. Pacala, and D. Tilman (ed.) The functional consequences of biodiversity: empirical progress and theoretical extensions. Princeton Univ. Press, New Jersey.

Tilman, D., D. Wedin, and J. Knops. 1996. Productivity and sustainability influenced by biodiversity in grassland ecosystems. Nature 379:718–720.

Tilman, D., J. Knops, D. Wedin, and P. Reich. 2001. Experimental and observational studies of diversity, productivity, and stability. p. 42–70. *In* A.P. Kinzig, S.W. Pacala, and D. Tilman (ed.) The functional consequences of biodiversity: empirical progress and theoretical extensions. Princeton Univ. Press, New Jersey.

Vasquez, E., R. Sheley, and T. Svejcar. 2008. Creating invasion resistant soils via nitrogen management. Invasive Plant Sci. Manage. 1:304–314.

Vinton, M.A., and I.C. Burke. 1995. Interactions between individual plant species and soil nutrient status in shortgrass steppe. Ecology 76:1116–1133.

Wander, M.M., and G.A. Bollero. 1999. Soil quality assessment of tillage impacts in Illinois. Soil Sci. Soc. Am. J. 63:961–971.

Wander, M.M., S.J. Traina, B.R. Stinner, and S.E. Peters. 1994. Organic and conventional management effects on biologically active soil organic matter pools. Soil Sci. Soc. Am. J. 58:1130–1139.

Wardle, D.A., G.W. Yeates, W. Williamson, and K.I. Bonner. 2003. The response of a three trophic level soil food web to the identity and diversity of plant species and functional groups. Oikos 102:45–56.

Wardle, D.A., R.D. Bardgett, J.N. Klironomos, H. Setala, W.H. van der Putten, and D.H. Wall. 2004. Ecological linkages between aboveground and belowground biota. Science 304:1629–1633.

Wolfe, B.E., and J.N. Klironomos. 2005. Breaking new ground: Soil communities and exotic plant invasion. Bioscience 55:477–487.

Zak, D.R., W.E. Holmes, D.C. White, A.D. Peacock, and D. Tilman. 2003. Plant diversity, soil microbial communities, and ecosystem function: Are there any links? Ecology 84:2042–2050.

Zilli, J.E., G.R. Botelho, M.C.P. Neves, and N.G. Rumjanek. 2008. Effect of glyphosate and imazaquin on the soybean rhizoplane bacterial community and microbiological soil characteristics. Rev. Bras. Cienc. Do Solo 32:633–642.

Zimdahl, R.L. 2004. Weed–crop competition: A review. Blackwell Publishing, Oxford, UK.

19

Fallow Effects on Soil

David C. Nielsen and Francisco J. Calderón

Fallow has been defined as a farming practice wherein no crop is grown and all plant growth is controlled by cultivation or chemicals during a season when a crop might normally be grown. (Haas et al., 1974). Fallow as a practice, associated with crop rotation, had its origins in Mediterranean agriculture (Karlen et al., 1994) and continues to be used throughout the semiarid and arid regions of West Asia and North Africa (Ryan et al., 2008). Additionally, summer fallow has been practiced widely across the 15 western states of the United States and the farmed areas of the prairie provinces of Canada in response to widely varying precipitation from year to year. For example, precipitation in any given year for a specific site in the central Great Plains region of the United States may range from double to less than half of the long-term average (Greb et al., 1974).

The primary reason for summer fallow is to stabilize crop production by forfeiting production in one season in anticipation that there will be at least partial compensation by increased crop production the next season. Summer fallow was almost universally adopted in the semiarid U.S. Great Plains in response to the 1930s dust bowl, higher wartime prices, and much improved tractor power systems and implements needed to control weeds during fallow (Greb, 1979). Other objectives of fallowing are to maximize soil water storage through improved water intake, snow trapping, and decreased evaporation; maximize plant nutrient availability; minimize soil erosion hazards; and minimize energy and economic inputs (Greb, 1979). Soil texture determines water holding capacity, thereby influencing how well fallow can buffer the influence of variable growing season precipitation on crop yield.

Fallow systems in semiarid regions can vary in fallow frequency (one crop in 2 yr, two crops in 3 yr, three crops in 4 yr, etc.), with the more frequently cropped systems generally producing more surface crop residues (Cantero-Martinez et al., 2006). Crop residue produced degrades over the fallow period at varying rates depending on the weed control methods used and climatic conditions. Fallow systems vary in intensity of tillage needed to control weeds during the noncrop period, and include maximum tillage (plowing and harrowing), conventional bare fallow (shallow disking and rod-weeding), stubble mulch (undercutting), minimum tillage (combinations of residual and contact herbicides with subsequent tillage), and no-tillage (use of only herbicides to control fallow weed growth). Tillage knocks down, cuts up, and incorporates standing and flat crop residue, facilitating organic matter mineralization by bringing together substrates, microbes, water, and oxygen. The various combinations of fallow frequency, tillage, and chemical weed control have effects on surface soil residue quantity, orientation, and duration, which subsequently affect surface soil organic matter content, soil physical structure, precipitation storage, nutrient availability, microorganisms,

D.C. Nielsen, USDA-ARS, Central Great Plains Research Station, 40335 County Road GG, Akron, CO 80720-0400 (david.nielsen@ars.usda.gov); F.J. Calderón, USDA-ARS, Central Great Plains Research Station, 40335 County Road GG, Akron, CO 80720-0400 (francisco.calderon@ars.usda.gov).

doi:10.2136/2011.soilmanagement.c19

erosion potential, and ultimately, crop production. This chapter reviews results of some of the past research conducted to determine the effects of varying methods of fallow on these parameters.

Effects on Organic Matter

Soil organic matter (OM) is a key indicator of soil quality as it influences biological activity, serves as a nutrient reservoir, and impacts soil aggregation (Doran and Parkin, 1996; Wienhold et al., 2006; Ryan et al., 2008). Rasmussen and Collins (1991) stated that soil OM assumes a pivotal role in semiarid rainfed areas because of its disproportionate influence on water and nutrient availability and on yield stability. Cropping systems that employ fallow generally have negative impacts on soil OM due both to decreased crop residue production and due to tillage used for weed control during the fallow period (Biederbeck et al., 1984; Campbell and Souster, 1982; Campbell et al., 2000; Mikha et al., 2006; Rasmussen and Collins, 1991; Rasmussen and Parton, 1994; Peterson et al., 1998; Ryan and Pala, 2007; Williams, 2004). When sod was broken and placed into a wheat (*Triticum aestivum* L.)–fallow rotation at Sidney, Nebraska, OM declined by 20% for no-till, 25% for mulch till, and 37% for plow till in just 16 yr (Follett and Schimel, 1989). Golchin and Asgari (2008) concluded that frequent tillage and use of a summer fallow–small grain system in northeast Iran caused soil quality to deteriorate through decreased soil organic C and increased erosion as structural stability declined. However, Williams (2008) showed that the typical decline in soil organic C in a tilled winter wheat–fallow system in north-central Oregon did not occur with applications of manure containing 145 kg N ha^{-1} per crop.

Annual cropping produces more crop residues than systems employing fallow and can gradually restore some of the organic matter lost in fallow systems. Bowman et al. (1999) and Ortega et al. (2002) reported increases in total soil organic C and N in the 0- to 5-cm depth, as fallow frequency decreased for dryland cropping systems in northeast Colorado. Similar increases in total organic C and N were reported by Campbell et al. (1999) for the 0- to 7.5-cm layer due to increased cropping intensity after 12 yr when comparing spring wheat–fallow to continuous spring wheat in southwestern Saskatchewan. Additionally, Bowman et al. (1999) found that particulate OM-C doubled and particulate OM-N and soluble organic C increased by one-third when a cropping system moved from wheat–fallow to continuous cropping. Pikul and Aase (1995) also reported significantly greater soil organic C in the 0- to 12-cm layer in an annual spring wheat system compared with a wheat–fallow system in Montana, and attributed the difference to 40% less crop residues returned annually to the soil in the wheat–fallow system. Mikha et al. (2006) found that tillage reduction and elimination of fallow increased total soil organic C in the 0- to 7.5-cm depth at four Great Plains locations extending from Bushland, Texas to Swift Current, Saskatchewan. Wood et al. (1991) reported a similar increase in soil organic C in response to a reduction in tillage and elimination of fallow. Ryan et al. (2008) found that soil OM levels in Syria were increased when fallow in a wheat–fallow cropping system was replaced with a crop (particularly a legume) and when N fertilizer was added, presumably in response to increased amounts of biomass produced.

Reducing tillage can likewise result in higher OM levels (Havlin et al., 1990). Halvorson et al. (1997) found 9% higher total C and 19% higher total N in the 0- to 2.5-cm soil layer of a no-till wheat–fallow system in northeastern Colorado compared with a conventional till wheat–fallow system 22 yr after implementation of the two systems. In a wheat–sorghum [*Sorghum bicolor* (L.) Moench]–fallow study in the Texas Panhandle, Unger (1991) reported a nonsignificant tendency for the no-till system to have higher OM levels than the stubble mulch tillage system 6 to 8 yr after the systems were established. Doran et al. (1998) reported that fallow tillage management practices greatly influenced the content and distribution of OM in the soil, with both organic C and N declining by 35 to 40% in the 0- to 7.6-cm layer following plowing of a native sod site in western Nebraska. They attributed the changes to (i) redistribution, mixing, and dilution with depth due to tillage; (ii) biological oxidation of soil OM (Doran and Smith, 1987); (iii) reductions in C and N inputs to soil due to changes in plant inputs due to crop and fallow management; and (iv) erosion loss from and deposition on surface soil. Losses of surface soil C and N are generally greatest during the first 8 to 10 yr after sod is cultivated, although losses continue to occur at a slower rate thereafter (Peterson and Vetter, 1971; Campbell et al., 1976). The generally moist, warm late spring and early summer periods in the central Great Plains create favorable conditions to decompose soil OM (Bowman et al., 1999), and those favorable conditions are enhanced by tillage operations (Doran et al., 1998).

Effects on Soil Physical Structure

Tillage for weed control during the fallow period is reported to have varying effects on the soil bulk density in the surface layer. The lack of a consistent response of bulk density to tillage in fallow systems may be attributable to amount of time and precipitation between tillage and sampling, type of tillage (disk vs. sweep vs. plow), depth of tillage, and soil type (Mikha et al., 2006). Results also vary with sampling depth. Generally, tillage will decrease bulk density in the tilled soil layer. However, tilled soil may eventually reconsolidate due to gravity, precipitation, and traffic. Mielke and Wilhelm (1998) found a bulk density of 1.19 Mg m^{-3} in the 0- to 7.6-cm layer of a silt loam soil in western Nebraska that had been plowed during the fallow period of a wheat–fallow system compared with a bulk density of 1.27 Mg m^{-3} for the system under no-till management. Halvorson et al. (1997) reported greater bulk densities in no-till vs. conventional till wheat–fallow after a 15-yr study comparing the two systems in northeastern Colorado. Those differences were seen at all sampling depths from 0 to 20 cm, with the greatest difference (17% higher bulk density in no-till) occurring in the 2.5- to 5.0-cm depth interval. Mielke et al. (1986) reported no difference in bulk density in the 0- to 7.5-cm layer of an Alliance silt loam (Aridic Argiustolls) in western Nebraska due to tillage in a wheat–fallow system 8 mo after the tillage occurred. On the other hand, they reported lower bulk density under no-till for a Duroc loam (Pachic Haplustolls). Mielke et al. (1984) and Unger (1991) reported no effect of tillage treatments on bulk density, while

Unger and Fulton (1990) found greater bulk density under conventional stubble mulch tillage than under no-till in the 4- to 7-cm depth. Pikul and Aase (1995) analyzed the combined effects of tillage and cropping intensity on bulk density in northeastern Montana. They found that after 9 yr of cropping, the spring wheat–fallow conventional till system had higher bulk density in the surface to 12-cm layer than in either the annual spring wheat no-till system or the annual spring wheat system with fall and spring tillage. Pikul et al. (1997) found no difference in bulk density in the 0- to 0.08-cm, 0.08- to 0.15-cm, and 0.15- to 0.30-cm surface soil layers of a Williams loam (fine-loamy, mixed Typic Agriboroll) in eastern Montana when comparing wheat–fallow no-till and conventional till systems and continuously cropped systems over a 5-yr period.

Pikul et al. (2006) reported greater water-filled pore space for systems employing fallow compared with continuously cropped systems at two northern Great Plains locations, but mixed results regarding the effects of tillage on water-filled pore space. The mixed results are likely to be a result of variations that occur with time of sampling and the large seasonal fluctuations that occur in water-filled pore space during different rotational phases. Mielke et al. (1986) found greater water-filled pore space in both the Duroc loam and Alliance silt loam mentioned above under no-till management, and lower air permeability and hydraulic conductivity under no-till.

Tillage for weed control during fallow periods can also create soil conditions that can restrict root growth and development. Results of a study conducted at Tribune, Kansas, where a sweep plow was used in a conventional till wheat–fallow system, showed increased bulk densities (compared with sod and no-till) at 30 to 40 cm (McVay et al., 2006). Many tillage operations at a consistent depth can lead to destruction of plant roots, and without plant roots to reinforce the soil, machine-induced compaction can occur (Ess et al., 1998). Pikul and Aase (1995) also identified a bulk density maximum occurring at about 10 cm in a wheat–fallow conventional tillage system, coinciding with the depth of the shallow sweep tillage operation conducted during the fallow periods.

Changes to the soil physical condition by reducing the frequency of fallow in a cropping system may take many years. After 15 yr of no-till management in dryland wheat systems varying in fallow frequency (which can conversely be seen as cropping intensity) from wheat–fallow to continuously cropped wheat–corn (*Zea mays* L.)–millet (*Panicum miliaceum* L.), Benjamin et al. (2007) found no effects of fallow frequency on bulk density, pore size distribution, water holding capacity, or saturated hydraulic conductivity in northeastern Colorado.

No-till fallow can affect soil aggregation through decreased tillage frequency and soil disruption. Unger and Fulton (1990) reported lower mean weight diameter of water-stable aggregates and lower porosity in the 4- to 7-cm layer under conventional stubble mulch tillage than under no-till. Pikul et al. (2006) showed that mean weight diameter was greater in a continuously cropped winter wheat system at Bushland, Texas than in a wheat–sorghum–fallow system leading to improved soil structure and greater resistance to erosion. Similarly, Blair et al. (2006) reported greater mean weight diameter under continuously cropped wheat compared with wheat–fallow for two Vertisol soils in northeastern New South Wales, Australia. They suggested that the poorer structural stability as well as less cover in fallow systems could result in increased erosion risk.

Effects on Soil Water

As stated earlier, one of the primary driving factors for the implementation of cropping systems that employ fallow periods is to store water in the soil profile to mitigate the effects of widely varying precipitation amounts in semiarid environments. Studies conducted over many years and at many locations confirm the increases in stored soil water at planting that occur in systems that employ fallow compared with continuously cropped systems.

Differences in amount of water stored in the soil over the fallow period vary due to differences in precipitation storage efficiency that occur with varying methods of weed control used during the fallow period and the variability in timing and amount of fallow period precipitation. As tillage intensity decreases through the use of minimum disturbance tillage implements such as rod-weeders and sweep plows and the use of herbicides, more crop residues remain on the soil surface for longer periods of time during the fallow period. Those increased residues are responsible for decreased runoff, decreased evaporation, and increased infiltration, resulting in greater precipitation storage efficiency (Unger, 1978).

Several studies have shown that crop residue management during fallow is important to reduce runoff and evaporation. Russel (1939) quantified the effects of winter wheat residues on runoff and soil water evaporation during the second summer of fallow in a winter wheat–fallow system in eastern Nebraska. He reported runoff decreasing from 60 to 0 mm as quantity of wheat residue on the soil surface went from 0 to 9 Mg ha^{-1}, and evaporation declining from 255 mm to 182 mm, respectively. Results from a study conducted on clay soils in southeastern Queensland, Australia, showed 22 to 35% greater runoff during the fallow period of a winter wheat–fallow system managed as bare fallow (5% soil surface cover by residue) compared with no-till management (64% surface cover) (Freebairn and Wockner, 1986). In another study from southeastern Queensland, Tullberg et al. (2001) investigated both tillage and controlled traffic effects on runoff in a dryland cropping system with periods of fallow between production of corn, sorghum, and wheat. They reported mean runoff over a 3-yr period to be 63 mm per year greater in a wheel-tracked area vs. a controlled traffic area and 38 mm per year greater under stubble mulch tillage vs. no-till. Decreased runoff in response to greater surface crop residues has been attributed to reduced soil crust formation and slowed flow rate across the surface due to greater flow path tortuosity and greater resistance to flow (Steiner, 1994).

In northern Mississippi where fallow is not a common production practice, Wilson et al. (2008) found shredded corn residues on the soil surface and disk-incorporated in the fallow year following corn production resulted in very little reduction in runoff compared with a bare soil surface, but soil losses were 54% lower with corn residues on the soil surface compared with bare soil. A laboratory study by Gilley et al. (1986) with corn residues also showed that residues were less effective for reducing runoff than for reducing soil loss. In the semiarid area of north-central Oregon, runoff was reduced during fallow periods of a conventional till winter wheat–fallow system receiving 145 kg N ha^{-1} per crop as manure for the previous 67 yr compared with the same system receiving 90 kg N ha^{-1} per crop as commercial fertilizer (Williams, 2004). Williams (2004) concluded that using manure amendments and not burning residue from the previous crop maintained soil organic C levels that reduced or retarded runoff. That same data set provided evidence of reduced soil erosion with the use of manure and maintenance of wheat residue on the soil surface (Williams, 2008).

Small grain harvest methods can greatly influence residue amount and orientation, and subsequently soil water evaporation during the fallow period. McMaster et al. (2000) showed that soil water evaporation could be reduced by 20 to 50% as wheat harvest cutting height increased from 0.1 m to 0.5 m, with the amount of evaporation reduction during the fallow period being dependent on standing residue stem population. Under very low stem population conditions (a result of poor seedling emergence and/or poor growing season rainfall), the use of a stripper-header (Fig. 19|1) (Henry et al., 2008) was advised to increase standing residue mass and height to minimize soil water evaporation over the fallow period.

The combined effects of decreased runoff and evaporation through reduced tillage and increased surface residues during the fallow period lead to increased infiltration. Baumhardt and Lascano (1996) measured cumulative infiltration in winter wheat residue increasing from 29 mm with 0 Mg ha^{-1} of residue to 47 mm with 2.5 Mg ha^{-1} of residue in response to 65 mm of simulated rainfall applied over an hour. They noted the ability of surface residues to absorb raindrop impact and retard runoff. In a study on a sandy clay loam in southwestern Queensland, Thomas et al. (2008) reported increased surface residue resulted in increases in time to runoff, final infiltration rate, and cumulative infiltration following 100 mm of simulated rainfall

to wheat residues at the end of a 6-mo fallow period. Pikul and Aase (1995) reported greater infiltration in an annual wheat no-till system than in a wheat–fallow conventional till system during the first hour of the first day in which measurements were taken, but that this difference disappeared over the course of the infiltration run. They concluded that the sandy loam soil of the experimental area settled firmly following rainfall, with textural size components that effectively filled the available void spaces of the soil with solids causing surface sealing. Pikul et al. (2006) found no significant cropping system effects on infiltration for locations that had the same tillage system but differing cropping intensity or crop species in the cropping system. However, where no tillage was compared with tillage, infiltration was greater following tillage and declined over time in tilled systems. They cautioned that conclusions regarding cropping system effects on infiltration should be made carefully due to the significant temporal variation in infiltration rate measurements.

No-tillage, however, does not always result in the most infiltration from a given precipitation event (Unger, 1992; Jones and Popham, 1997). Infiltration may be greater into a tillage-loosened than a no-tillage soil when precipitation amounts do not exceed the temporary storage capacity of the loosened soil layer. Also, infiltration into a tillage-loosened soil may be greater when

Fig. 19|1. Fallow wheat stubble following harvest with a stripper-header.

the water content of no-tillage soil is relatively high following precipitation, thereby resulting in slow infiltration and limited opportunity for additional water storage, which was the case on a Pullman clay loam (fine, mixed, thermic Torrertic Paleustoll) at Bushland, Texas (Jones and Popham, 1997).

An important fraction of the precipitation in parts of the central and northern Great Plains falls as snow (Fig. 19|2). Standing crop residues are more effective at reducing wind speed near the soil surface than flat residues (Siddoway et al., 1965; Bilbro and Fryrear, 1994) and therefore trap more snow during the winter period. Nielsen (1998) measured about 20 cm more stored soil water after winter in standing sunflower (*Helianthus annuus* L.) residue with a silhouette area index (residue height × diameter × population) of 0.07 m^2 m^{-2} than where the sunflower stalks were lying flat on the soil surface. This was in response to greater snow catch by the standing sunflower stalks.

The final effect of decreased runoff, decreased evaporation, and increased infiltration in response to more surface residues during the fallow period is greater precipitation storage efficiency and increased available soil water at the end of the fallow period. Peterson et al. (1996) summarized data from several Great Plains studies that showed the inefficiency of precipitation storage in no-till fallow systems during the late portion of the summer fallow period, a time when the soil profile is at or near field capacity, daily weather conditions are hot and dry, and little surface residue remains. During the 11-mo fallow period between winter wheat harvest and grain sorghum planting in the Texas Panhandle, Baumhardt et al. (1985) measured precipitation storage efficiency increasing from 22% under disk tillage to 31% under no-till management, which they attributed to evaporation suppression due to greater surface crop residues under no-till. Similar results have been reported at other Great Plains locations (Peterson et al., 1996; Smika, 1990; Smika and Wicks, 1968; Tanaka and Aase, 1987; Unger and Wiese, 1979; Lyon et al., 1998; Nielsen and Vigil, 2010). Available water at sorghum planting, following an 11-mo fallow period following after wheat harvest in Texas, was nearly 50 mm greater with 4 Mg ha^{-1} of wheat residues on the surface compared with 0 Mg ha^{-1} of surface residues (Unger, 1978). The 9-yr average available soil water at wheat planting at Akron, Colorado was 227 mm for no-till wheat–fallow systems and 156 mm for conventional till wheat–fallow systems (Nielsen et al., 2002). These results clearly indicate the effects of surface residue destruction and soil stirring on enhancing evaporation and decreasing precipitation storage efficiency when sweep plow tillage was used for weed control during the fallow period.

The effect of the fallow period frequency in a cropping system on available soil water was quantified by Nielsen et al. (2002) for winter wheat systems in northeastern Colorado. The 9-yr average available water content at wheat planting was 227 mm for a wheat–fallow system, but only 108 mm for the continuously cropped wheat–corn–millet system. Jones and Popham (1997) similarly reported a 10-yr average plant available soil water content at wheat planting at Bushland, Texas of 212 mm for wheat–fallow, 205 mm for wheat–sorghum–fallow, and 156 mm for continuous wheat.

The increased precipitation storage efficiency from the use of reduced and no-till systems for weed control during the fallow period of wheat–fallow systems in the central Great Plains has led to the implementation of more intensive cropping systems that reduce the frequency of fallow (Peterson et al., 1993; Anderson et al., 1999; Nielsen et al., 2002). If the increased soil water is not used by more intensive cropping systems that reduce the frequency of fallow,

Fig. 19|2. Snow deposition in fallow sunflower stalks.

water can move below the active root zone, taking with it N and potentially affecting groundwater quality. O'Connell et al. (2003) measured increased drainage below the root zone in a fallow system (fallow–wheat–pea [*Pisum sativum* L.]) compared with a system without fallow (mustard [*Brassica juncea* (L.) Czern.]–wheat–pea) in southeastern Australia. Introduction of fallow production methods in the semiarid areas of the Great Plains has sometimes led to the formation of saline seeps (Halvorson and Black, 1974) as water percolated below the root zone. This problem can be alleviated with the use of flexible crop rotations involving small grains, grasses, deep-rooted crops, and a minimum amount of summer fallow as crops are grown when sufficient soil water is present at planting to indicate likely successful crop production.

Effects on Nutrient Availability

Fallow enhances accumulation of nitrate through mineralization of organic matter (Smika, 1983a; Campbell et al., 1990). Cochran et al. (2006) stated that during the early years of crop production in the northern Great Plains, relatively high levels of organic matter supplied adequate nutrition as N mineralization was enhanced by aeration with tillage in conjunction with high soil moisture content during the fallow period. Prolonged cropping of these prairie soils depleted soil N such that fertilizer N is now required. Unger (1991) reported a nonsignificant tendency for a no-till wheat–sorghum–fallow system in Texas to have higher levels of N, NO_3–N, P, and K at the soil surface than the stubble mulch tillage system 6 to 8 yr after the systems were established. Mikha et al. (2006) found total soil N was significantly increased in the 0- to 7.5-cm depth by decreasing tillage and fallow frequency at several central and northern Great Plains locations. The effect of fallow tillage intensity on total N in the 0- to 7.6-cm depth in a wheat–fallow system in Texas was reported by Unger (1968). After 24 yr of management, 36% higher total N was found in the system where six to ten sweep tillage operations were delayed until the spring and summer following wheat harvest compared with the system where up to ten fallow-period one-way disk operations were used to control weeds throughout the entire fallow period.

Fallow no-till systems can also increase P and other micronutrients in the upper layers of soil. Unger (1991) also reported approximately 60% higher extractable P in the 0- to 4-cm soil layer from a wheat–sorghum–fallow system under no-till management compared with stubble mulch management in Texas. Follett and Peterson (1988) showed tillage intensity effects on several nutrients from a loam soil in western Nebraska that had been in wheat–fallow production for 16 yr. They found that total P, organic P, K, Zn, and Fe in the 0- to 5-cm layer declined with increasing tillage intensity (no-till > stubble mulch > moldboard plow). They attributed these results mainly to cycling of nutrients to the soil surface in plant parts and subsequent residue that was then mixed and diluted with soil from lower depths as tillage intensity increased.

Bowman and Halvorson (1997) conducted a detailed study of the effect of fallow frequency on P in the 0- to 5-cm soil layer in northeastern Colorado. As fallow frequency was reduced from one crop in 2 yr to two crops in 3 yr to three crops in 4 yr to continuous cropping, water-soluble P, anion-exchange resin P, total soil organic P, phosphatase activity, soil bicarbonate-extractable organic P, and total P all increased. They attributed the increase primarily to greater residue production in systems with less fallow. The P uptake from deeper in the soil profile was deposited at the soil surface through greater residue and litter production and subsequent leaching of P from the residue and decomposition of the residue in contact with soil. Additionally, there was probably enhanced P protection from wind erosion as summer fallow was eliminated. Decreasing fallow frequency similarly increased levels of Zn, Mn, and Fe in the 0- to 5-cm layer but did not affect Cu or SO_4–S levels (Bowman and Vigil, 2000).

Because crop water use generally exceeds precipitation in the semiarid regions that employ fallow systems, leaching of N beyond the root zone through downward water movement is rarely a loss mechanism for N (Ryan and Monem, 1998). In comparing fallow tillage systems in Queensland, Australia, Standley et al. (1990) found greater losses of N, P, S, and K in the surface 10 cm of a vertisol after 7 yr of conventional tillage compared with no-till. They attributed the N losses to greater losses of NO_3^- leached below the 0.6-m sampling zone and greater denitrification with conventional tillage.

Effects on Soil Microbial Activity and Diversity

Soil microbes are important for the functioning of agroecosystems because they serve as catalysts of essential nutrient cycling conversions and are an important part of the labile C and N pools in soil (Paul, 2007). Because of this, reducing plant residue inputs through the use of fallow causes a decline in soil OM that starves soil microbes, which in turn limits the metabolic capacity and biological function of the soil. Steenwerth et al. (2002) showed that soil microbial biomass, measured by phospholipid fatty acid content, declined sharply in soils under conventional till fallow for 2 yr compared with adjacent cropped and grassland soils in California coastal valleys. Experiments at the Ultuna Long Term Soil Organic Matter Experiment in Uppsala, Sweden (clay loam soils, Typic Eutrochrept) showed that important soil microbial variables such as potential denitrification and basal soil respiration rate were reduced in conventional till fallow soils relative to cropped soils (Enwall et al., 2005). The conventional till fallow had less total soil C and N content compared with the cropped soils, and the lowest microbial biomass as measured by substrate induced respiration (Enwall et al., 2007). Enwall et al. (2007) found that bare fallow had low potential ammonia oxidation and a different ammonia oxidizer diversity relative to the cropped soils, possibly due to the reduced N mineralization and lack of N fertilizer in the fallow soil, showing that bare fallow can also affect soil microbial activity indirectly through changes in soil properties such as N availability and pH, brought about by the presence or absence of N fertilizer.

Liebig et al. (2006) summarized the findings of several Great Plains studies by stating that cropping systems with intensive crop sequences (reduced fallow frequency) and/or reduced tillage possess more soil microbial biomass, potentially mineralizable N, and total glomalin. Glomalin is a gel-like substance produced by mycorrihzal fungi

that has been linked to higher soil C and better soil structure (Wright et al., 1999). Those trends were attributed to greater crop residue, root mass, and soil OM accumulation in the soil surface of these systems. They also stated that in the Great Plains, no-till management during fallow compared with conventional tillage resulted in increased fungal abundance, higher populations of denitrifying bacteria, and greater ester- and phospholipid-linked fatty acid methyl esters, resulting in conditions favoring growth and activity of soil microorganisms that improve soil structure, but also increasing gaseous N loss by denitrification.

Residue management can have long-lasting effects on soil microbiology. Biederbeck et al. (2005) conducted a 6-yr study comparing the effects of green fallow–wheat rotations, bare fallow–wheat, and continuous wheat in a Canadian silt loam (Aridic Haploboroll). Bare fallow–wheat had 34% less residue dry matter inputs and 48% less residue N inputs than the continuous wheat system, resulting in a lasting negative effect of the bare fallow on soil microbes. Soil bacterial counts, microbial biomass, mineralizable C, and the soil enzymes dehydrogenase, phosphatase, arylsulfatase, and urease were lowest in the bare fallow–wheat relative to the rest of the treatments, even though the soils were sampled after the wheat phase of the rotation. In contrast, the green-fallow systems, in which the legumes were grown to full bloom and then incorporated, had a positive effect on microbial counts, microbial biomass, and enzymes relative to bare fallow–wheat and continuous wheat.

Agricultural practices can affect the soil microbiological abundance and diversity indirectly by altering physical and chemical properties, which in turn can alter the immediate environment of soil microbes and cause shifts in microbial community composition (Paul, 2007). Steenwerth et al. (2002) studied soil microbial community composition on loam soils in California under irrigated and dryland management, annual and perennial grassland, and bare fallow. The fallow sites were previously annual grasslands that were tilled and then kept bare for 2 yr using herbicides. Total soil C and N (0- to 6-cm depth) suffered losses of 20 and 34% respectively due to the lack of C inputs and mineralization of soil OM during fallow. Phospholipid fatty acid analysis indicated that microbial community structure diverged from that of the adjacent grassland sites, partly due to increased abundance of Gram-negative bacterial markers.

Exceptions do occur, showing that not all microbial functions are equally sensitive to fallow. Enwall et al. (2005) showed that bare-fallow soils had similar genetic fingerprints as unfertilized cropped plots, indicating that fertilizer addition and pH may be more important drivers of microbial diversity than the presence or absence of plants in the soils studied.

The reduced C inputs in bare fallow can have negative effects on soil function through the reduction of microbial enzymes. Pankhurst et al. (2005) compared three crop break treatments—pasture, alternate crops, and bare fallow—as alternatives to continuous sugarcane in Queensland, Australia. The microbial biomass C decreased up to 43% and total free-living nematodes decreased up to 14% under bare fallow relative to continuous sugarcane. Interestingly, the soil microbial community of the bare-fallow break had a reduced capacity to utilize different C substrates than the soil microbial community under sugarcane, indicating a clear effect on soil microbial function in the absence of plant cover.

Weigand et al. (1995) investigated microbial C, catalase activity, and earthworm abundance in soils from different sites in Bavaria maintained under bare fallow for 6 yr. Microbial C was strongly correlated with soil organic C and with soil catalase (an important enzyme for aerobic metabolism). The soils under bare fallow, with no rhizodeposition or residue input, had less microbial biomass and lowered efficiency of microbial C metabolism as measured by substrate induced respiration.

Acosta-Martinez et al. (2007) showed that cropping intensity affected soil microbial composition and enzyme activity in long-term plots in northeastern Colorado. The study aimed to find more efficient alternatives to the traditional wheat–fallow common in the region. Increased fallow frequency coupled with conventional tillage was associated with reduced soil microbial biomass and soil enzymatic activity. Fatty acid methyl ester analysis indicated that the plots under the more intense rotations also had a different soil microbial community

structure than the wheat–fallow plots, as well as reduced fungal abundance in the wheat–fallow treatment.

Bare fallow can have a direct effect on obligately symbiotic organisms such as arbuscular mycorrhizal fungi that depend entirely on certain plant roots for their energy. The long-fallow disorder occurs when mycorrhizal fungal inoculum declines during a bare-fallow period because of the absence of active crop roots (Thompson, 1987; Hulugalle et al., 1998; Pankhurst et al., 2005). Mycorrhizae can benefit row crops by increasing P and water uptake, so the bare fallow can potentially result in reduced performance in subsequent crops. However, P and Zn fertilizer help alleviate the long-fallow effect by reducing nutrient deficiency and promoting mycorrhiza formation. Oliveira and Sanders (1999) observed the long-fallow effect on low-P soils in Leeds, England. As expected, fallow soils had lower mycorrhizal infectivity than recently cropped soils. Interestingly, prior cropping with a strongly mycorrhizal plant (*Zea mays* L.) increased infectivity relative to a wheat pre-crop, suggesting a strategy to minimize the negative effect of the fallow on mycorrhizal infection.

Bare fallow can also affect soil pathogenic microorganisms by removing the host crops, reducing crop residues, and changing the soil physical environment in such a way that the pathogen life cycles are disrupted. Hulugalle et al. (1998) compared long-fallow cotton (*Gossypium* L.) with continuous cotton in New South Wales, Australia. The long-fallow cotton affected the soil physical properties by reducing soil strength and plastic limit, and black root rot was lower during the cotton phase after long fallow. *Fusarium* fungi can cause important economic losses in crops because *Fusaria* can infect the xylem of the plant and cause head blight, root rot, crown rot, and seedling blight in several crop species. Contaminated crop residues are an important source of *Fusarium* propagules (Dill-Macky and Jones, 2000), so crop rotations and residue management are important options to control disease. Sturz and Johnston (1985) found that pathogenic *Fusaria* are found on stubble in soil, and tend to be in higher amounts in soils under continuous cropping than in bare-fallow soils due to the differences in crop residue. Because of this, they suggest fallowing as a strategy to control *Fusarium* in barley (*Hordeum vulgare* L.) and wheat in Canada.

Rhizoctonia fungi are root pathogens that can cause important economic losses due to root infection and seedling damping-off. Like *Fusarium*, *Rhizoctonia* normally survives saprophytically in plant debris and plant tissues before it infects a crop. However, Bell and Sumner (1987) demonstrated that *Rhizoctonia* sclerotia can survive for more than 40 wk in fallow soils without plant cover or fresh residues, and remain viable to infect corn afterward. The authors suggest that a bare fallow alone will not be effective in controlling *Rhizoctonia*, and that tillage or mulching with polyethylene will be required in addition to fallow for effective control. It is not surprising, then, that chemical fallow periods of up to 66 d were not effective in the control of *Rhizoctonia* in wheat grown in Western Australia (MacNish and Fang, 1987). In a study about carrot (*Daucus carota* L.) damage in the Central Valley of California, populations of *Rhizoctonia* and *Pythium* were similar following fallow, onion (*Allium cepa* L.), or carrots (Davis and Nunez, 1999). Like *Rhizoctonia*, other fungal pathogens such as *Sclerotium rolfsii* Sacc. produce resistant sclerotia, giving them the capacity to survive for prolonged periods of time in soil (Coley-Smith and Cooke, 1971), and it is possible that a fallow period will not be completely effective for their control.

Rotations with resistant crops and residue removal can be better than fallow for the control of root pathogens under some conditions. Johnson et al. (1997) studied the effectiveness of bare fallow and sod to suppress nematodes, *Pythium*, and *Rhizoctonia* on vegetable crops in a Tifton loamy sand (fine-loamy, siliceous, thermic Plinthic Paleudults). The plant parasitic nematode, *Rhizoctonia*, and *Pythium* numbers as well as *Fusarium* damage, were lower on vegetable crops preceded by sod than on plantings after fallow. Smiley et al. (1996) studied wheat diseases in wheat–fallow, wheat–pea, and continuous wheat in a Walla Walla silt loam (coarse-silty, mixed, mesic Typic Haploxeroll) near Pendleton, Oregon. Rhizoctonia root rot, Pythium root rot, eyespot, and take-all were suppressed by burning crop residues following harvest. These diseases can survive on infected crop residues, so removing the residue via burning lowers the incidence of disease in subsequent crops. Furthermore, crown rot increased with surface residue. Eyespot and crown rot were exacerbated by soil N and declined with soil pH. In this case, rotating crops was a good disease management strategy. In general, diseases were less damaging on the wheat–pea rotation relative to the continuous wheat and the wheat–fallow. Replacing chemical fertilizers with green manure and animal manure was also beneficial in suppressing soil-borne diseases.

Effects on Erosion

As stated earlier, standing residue is more effective than flat residue for decreasing wind speed near the soil surface, and consequently in controlling wind erosion. The standing residue absorbs more of the wind's energy and raises the zero-velocity point above the soil, thereby preventing much of the normal avalanching of soil material downwind (Woodruff et al., 1972; van de Ven et al., 1989). Smika (1983b) measured a 74% reduction in wind speed at the soil surface when standing wheat residue height was increased from 30 to 61 cm. Consequently, there may be significant differences in soil erosion during fallow periods depending on residue amount, type, quality, and orientation.

Nielsen and Aiken (1998) showed that a silhouette area index of 0.035 to 0.045 $m^2 m^{-2}$, achievable through leaving sunflower stalks standing during the fallow period following harvest (Fig. 19|3), can reduce saltation discharge to less than 5% of that predicted for bare surfaces. A silhouette area index of 0.04 $m^2 m^{-2}$ would be achieved with a stem population of 40,000 stems ha^{-1}, stem diameter of 1.75 cm, and stem height of 57 cm. Typical values of silhouette area index for standing wheat (0.54 $m^2 m^{-2}$; McMaster et al., 2000) and proso millet (0.08 $m^2 m^{-2}$; Henry et al., 2008) residues following harvest are well above the 0.04 $m^2 m^{-2}$ value, indicating that those small grain residues left standing will reduce erosion potential to nearly zero as they remain standing during the fallow period. The use of a stripper-header for the harvest of small grains leaves significantly more standing crop residues that will be retained longer than residues following grain harvest conventionally cut with a sickle-bar header (Henry et al., 2008). The result is more protection of the soil from wind erosion and greater precipitation storage efficiency during fallow periods.

Sharratt et al. (2007) quantified soil loss from conventionally tilled silt loam fallow fields (winter wheat–fallow system) in the Columbia Plateau of eastern Washington. Six high wind events occurred over a 2-yr period resulting in soil loss ranging from 43 to 2320 kg ha^{-1} per wind event and PM10 loss ranging from 5 to 210 kg ha^{-1} per wind event. The PM10 loss comprised 9 to 12 per cent of the total soil loss. They concluded that alternative tillage practices or cropping systems were needed for minimizing PM10 emissions and improving air quality in that region. Similar magnitudes of soil losses through wind erosion have been reported for silt loam soils in Colorado (Van Donk and Skidmore, 2003) and in Washington (Zobeck et al., 2001), but much higher losses (12,000 to 56,000 kg ha^{-1}) have been reported for sand and sandy loam soils in other locations (Zobeck et al., 2001; Larney et al., 1995). The higher soil losses may also be related to differences in wind event speed and duration, surface roughness, or surface cover. Using the wind erosion prediction system (WEPS; Hagen, 1991), Feng and Sharratt (2007) estimated an annual soil loss of 14,250 kg ha^{-1} from summer fallow fields in eastern Washington.

Soil loss under fallow management due to water erosion can also be significant. Boellstorff and Benito (2005) described the increase in bare (unseeded) fallow area in Europe that occurred following the adoption of the 1992 MacSharry reforms to the European Union's Common Agricultural Policy that included a set-aside program requiring farmers to take certain percentages of arable land out of production. In central Spain, even areas with sufficient precipitation to support seeded fallow with a cover crop were being put into traditional unseeded fallow with tillage. A study involving the use of the revised universal soil loss equation (RUSLE; Renard et al., 1991) indicated the use of seeded fallow in central Spain would cut the area estimated to have greater than 6 t ha^{-1} soil loss to one-third the area under that risk when in unseeded fallow (Boellstorff and Benito, 2005). In central Croatia, Basic et al. (2004) measured a 5-yr average soil loss of 87 t ha^{-1} from standard bare fal-

Fig. 19|3. Anemometers in fallow sunflower stalks.

low USLE protocol plots (Wischmeier and Smith, 1978) on a 9% slope.

Summary

Fallow production systems continue to be used throughout various regions of the world, but particularly in semiarid regions where precipitation is highly variable in timing and amount. Systems that reduce or limit fallow frequency and tillage intensity generally result in greater amounts of surface crop residues remaining during fallow periods. Those residue increases generally produce positive effects on soil quality for crop production, including increases in soil OM, nutrients, physical structure, water content, and microorganisms, as well as reductions in soil loss by wind and water erosion.

References

Acosta-Martinez, V., M.M. Mikha, and M.F. Vigil. 2007. Microbial communities and enzyme activities in soils under alternative crop rotations compared to wheat-fallow for the Central Great Plains. Appl. Soil Ecol. 37:41–52.

Anderson, R.L., R.A. Bowman, D.C. Nielsen, M.F. Vigil, R.M. Aiken, and J.G. Benjamin. 1999. Alternative crop rotations for the central Great Plains. J. Prod. Agric. 12:95–99.

Basic, F., I. Kisic, M. Mesic, O. Nestroy, and A. Butorac. 2004. Tillage and crop management effects on soil erosion in central Croatia. Soil Tillage Res. 78:197–206.

Baumhardt, R.L., and R.J. Lascano. 1996. Rain infiltration as affected by wheat residue amount and distribution in ridged tillage. Soil Sci. Soc. Am. J. 60:1908–1913.

Baumhardt, R.L., R.W. Zartman, and P.W. Unger. 1985. Grain sorghum response to tillage method used during fallow and to limited irrigation. Agron. J. 77:643–646.

Bell, D.K., and D.R. Sumner. 1987. Survival of Rhizoctonia solani and other soilborne basidiomycetes in fallow soil. Plant Dis. 71:911–915.

Benjamin, J.G., M.M. Mikha, D.C. Nielsen, M.F. Vigil, F. Calderón, and W.B. Henry. 2007. Cropping intensity effects on physical properties of a no-till silt loam. Soil Sci. Soc. Am. J. 71:1160–1165.

Biederbeck, V.O., C.A. Campbell, and R.P. Zentner. 1984. Effect of crop rotation and fertilization on some biological properties of a loam in southwestern Saskatchewan. Can. J. Soil Sci. 64:355–367.

Biederbeck, V.O., R.P. Zentner, and C.A. Campbell. 2005. Soil microbial populations and activities as influenced by legume green fallow in a semiarid climate. Soil Biol. Biochem. 37:1775–1784.

Bilbro, J.D., and D.W. Fryrear. 1994. Wind erosion losses as related to plant silhouette and soil cover. Agron. J. 86:550–553.

Blair, N., R.D. Faulkner, A.R. Till, and G.J. Crocker. 2006. Long-term management impacts on soil C, N and physical fertility. Part III: Tamworth crop rotation experiment. Soil Tillage Res. 91:48–56.

Boellstorff, D., and B. Benito. 2005. Impacts of set-aside policy on the risk of soil erosion in central Spain. Agric. Ecosyst. Environ. 107:231–243.

Bowman, R.A., and A.D. Halvorson. 1997. Crop rotation and tillage effects on phosphorus distribution in the central Great Plains. Soil Sci. Soc. Am. J. 61:1418–1422.

Bowman, R.A., and M.F. Vigil. 2000. Sulfur and micronutrient changes as a function of dryland cropping intensity. In Proc. Great Plains Soil Fertility Conf. Denver, CO., 7–8 Mar. 2000.

Bowman, R.A., M.F. Vigil, D.C. Nielsen, and R.L. Anderson. 1999. Soil organic matter changes in intensively cropped dryland systems. Soil Sci. Soc. Am. J. 63:186–191.

Campbell, C.A., V.O. Biederbeck, B.G. McConkey, D. Curtin, and R.P. Zentner. 1999. Soil quality—Effect of tillage and fallow frequency. Soil organic matter quality as influenced by tillage and fallow frequency in a silt loam in southwestern Saskatchewan. Soil Biol. Biochem. 31:1–7.

Campbell, C.A., E.A. Paul, and W.B. McGill. 1976. Effect of cultivation and cropping on the amounts and forms of soil N. p. 9–101. In W.A. Rice (ed.) Proc. Western Canada Nitrogen Symp., Calgary, AB, 21–21 Jan. Alberta Agric. Edmonton, AB.

Campbell, C.A., and W. Souster. 1982. Loss of organic matter and potentially mineralizable nitrogen from Saskatchewan soil due to cropping. Can. J. Soil Sci. 62:651–656.

Campbell, C.A., R.P. Zentner, J.J. Janzen, and K.E. Bowren. 1990. Crop rotation studies on the Canadian Prairies. Publ. No. 184/E. Canadian Gov. Publ. Centre, Supply Services Canada, Hull, QC.

Campbell, C.A., R.P. Zentner, B.C. Liang, G. Roloff, E.C. Gregorich, and B. Blomert. 2000. Organic C accumulation in soil over 30 years in semiarid southwester Saskatchewan: Effect of crop rotations and fertilizers. Can. J. Soil Sci. 80:179–192.

Cantero-Martinez, C., D.G. Westfall, L.A. Sherrod, and G.A. Peterson. 2006. Long-term crop residue dynamics in no-till cropping systems under semi-arid conditions. J. Soil Water Conserv. 61:84–95.

Cochran, V., J. Danielson, R. Kolberg, and P. Miller. 2006. Dryland cropping in the Canadian prairies and the U.S. northern Great Plains. p. 293–339. In G.A. Peterson, P.W. Unger, and W.A. Payne (ed.) Dry-

land Agriculture. 2nd ed. Agron. Monogr. 23. ASA, Madison, WI.

Coley-Smith, J.R., and R.C. Cooke. 1971. Survival and Germination of Fungal Sclerotia. Ann. Rev. Phytopathol. 9:65–92.

Davis, R.M., and J.J. Nunez. 1999. Influence of crop rotation on the incidence of Pythium- and Rhizoctonia-induced carrot root dieback. Plant Dis. 83:146–148.

Dill-Macky, R., and R.K. Jones. 2000. The effect of previous crop residues and tillage on Fusarium head blight of wheat. Plant Dis. 84:71–76.

Doran, J.W., E.T. Elliott, and K. Paustian. 1998. Soil microbial activity, nitrogen cycling, and long-term changes in organic carbon pools as related to fallow tillage management. Soil Tillage Res. 49:3–18.

Doran, J.W., and T.B. Parkin. 1996. Quantitative indicators of soil quality: A minimum data set. p. 25–37 In J.W. Doran and A.J. Jones (ed.) Methods of assessing soil quality. SSSA Spec. Publ. No. 49. SSSA, Madison, WI.

Doran, J.W., and M.S. Smith. 1987. Organic matter management and utilization of soil and fertilizer nutrients. p. 53–72. In R.F. Follett, J.W.B. Stewart, and C.V. Cole (ed.) Soil fertility and organic matter as critical components of production systems. Spec. Publ. No. 19. ASA–CSSA–SSSA. Madison, WI.

Enwall, K., K. Nyberg, S. Bertilsson, H. Cederlund, J. Stenström, and S. Hallin. 2007. Long-term impact of fertilization on activity and composition of bacterial communities and metabolic guilds in agricultural soil. Soil Biol. Biochem. 39:106–115.

Enwall, K., L. Philippot, and S. Hallin. 2005. Activity and composition of the denitrifying bacterial community respond differently to long-term fertilization. Appl. Environ. Microbiol. 71:8335–8343.

Ess, D.R., D.H. Vaughan, and J.V. Perumpral. 1998. Crop residue and root effects on soil compaction. Trans. ASAE 41:1271–1275.

Feng, G., and B. Sharratt. 2007. Scaling from field to region for wing erosion prediction using the Wind Erosion Prediction System and geographical information systems. J. Soil Water Conserv. 62:321–328.

Follett, R.F., and G.A. Peterson. 1988. Surface soil nutrient distribution as affected by wheat-fallow tillage systems. Soil Sci. Soc. Am. J. 52:141–147.

Follett, R.F., and D.S. Schimel. 1989. Effect of tillage practices on microbial biomass dynamics. Soil Sci. Soc. Am. J. 53:1091–1096.

Freebairn, D.M., and G.H. Wockner. 1986. A study of soil erosion on vertisols of the eastern Darling Downs, Queensland. I. The effect of surface conditions on soil movement within contour bay catchments. Aust. J. Soil Res. 24:135–158.

Gilley, J.E., S.C. Finkner, R.G. Spomer, and L.N. Mielke. 1986. Runoff and erosion as affected by corn residue. I. Total losses. Trans. ASAE 29:157–160.

Golchin, A., and H. Asgari. 2008. Land use effects on soil quality indicators in north-eastern Iran. Aust. J. Soil Res. 46:27–36.

Greb, B.W. 1979. Reducing drought effects on croplands in the west-central Great Plains. USDA Agric. Inf. Bull. No. 420. U.S. Gov. Print. Office, Washington, DC.

Greb, B.W., D.E. Smika, N.P. Woodruff, and C.J. Whitfield. 1974. Summer fallow in the central Great Plains. p. 51–85. In H.J. Haas, W.O. Willis, and J.J. Bond (ed.) Summer fallow in the western United States. USDA-ARS Conserv. Res. Rep. No. 17. U.S. Gov. Print. Office, Washington, DC.

Haas, H.J., W.O. Willis, and J.J. Bond (ed.) 1974. Summer fallow in the western United States. USDA-ARS Conserv. Res. Rep. No. 17. U.S. Gov. Print. Office, Washington, DC.

Hagen, L.J. 1991. A wind erosion prediction system to meet user needs. J. Soil Water Conserv. 46:106–111.

Halvorson, A.D., and A.L. Black. 1974. Saline-seep development in dryland soils in northeastern Montana. J. Soil Water Conserv. 29:77–81.

Halvorson, A.D., M.F. Vigil, G.A. Peterson, and E.T. Elliott. 1997. Long-term tillage and crop residue management study at Akron, Colorado. p. 361–370. *In* E.A. Paul et al. (ed.) Soil organic matter in temperate agroecosystems, long-term experiments in North America. CRC Press, New York.

Havlin, J.L., D.E. Kissel, L.E. Maddux, M.M. Claassen, and J.H. Long. 1990. Crop rotation and tillage effects on soil organic carbon and nitrogen. Soil Sci. Soc. Am. J. 54:448–452.

Henry, W.B., D.C. Nielsen, M.F. Vigil, F.J. Calderón, and M.S. West. 2008. Proso millet yield and residue mass following direct harvest with a stripper-header. Agron. J. 100:580–584.

Hulugalle, N.R., P. Entwistle, J.L. Cooper, S.J. Allen, and D.B. Nehl. 1998. Effect of long-fallow on soil quality and cotton lint in an irrigated, self-mulching, grey Vertosol in the central-west of New South Wales. Aust. J. Soil Res. 36:621–640.

Jones, O.R., and T.W. Popham. 1997. Cropping and tillage practices for dryland grain production in the Southern High Plains. Agron. J. 89:222–232.

Johnson, A.W., G.W. Burton, D.R. Sumner, and Z. Handoo. 1997. Coastal bermudagrass rotation and fallow for management of nematodes and soilborne fungi on vegetable crops. J. Nematol. 29:710–716.

Karlen, D.L., G.E. Varvel, D.G. Bullock, and R.H. Cruse. 1994. Crop rotations for the 21st century. Adv. Agron. 53:1–45.

Larney, F., M. Bullock, S. McGinn, and D. Fryrear. 1995. Quantifying wind erosion on summer fallow in southern Alberta. J. Soil Water Conserv. 50:91–95.

Liebig, M., L. Carpenter-Boggs, J.M.F. Johnson, S. Wright, and N. Barbour. 2006. Cropping system effects on soil biological characteristics in the Great Plains. Renew. Agric. Food Syst. 21:36–48.

Lyon, D.J., W.W. Stroup, and R.E. Brown. 1998. Crop production and soil water storage in long-term winter wheat-fallow tillage experiments. Soil Tillage Res. 49:19–27.

MacNish, G.C., and C.S. Fang. 1987. Effect of short chemical fallow on Rhizoctonia bare patch and root rot of wheat at Esperance, Western Australia. Aust. J. Exp. Agric. 27:671–677.

McMaster, G.S., R.M. Aiken, and D.C. Nielsen. 2000. Optimizing wheat harvest cutting height for harvest efficiency and soil and water conservation. Agron. J. 92:1104–1108.

McVay, K.A., J.A. Budde, K. Fabrizzia, M.M. Mikha, C.W. Rice, A.J. Schlegel, D.E. Peterson, D.W. Sweeney, and C. Thompson. 2006. Management effects on soil physical properties in long-term tillage studies in Kansas. Soil Sci. Soc. Am. J. 70:434–438.

Mielke, L.M., J.W. Doran, and K.A. Richards. 1986. Physical environment near the surface of plowed and no-tilled soils. Soil Tillage Res. 7:355–366.

Mielke, L.N., and W.W. Wilhelm. 1998. Comparison of soil physical characteristics in long-term tillage winter wheat-fallow tillage experiments. Soil Tillage Res. 49:29–35.

Mielke, L.N., W.W. Wilhelm, K.A. Richards, and C.R. Fenster. 1984. Soil physical characteristics of reduced tillage in a wheat-fallow system. Trans. ASAE 27:1724–1728.

Mikha, M.M., F. Vigil, M.A. Liebig, R.A. Bowman, B. McConkey, E.J. Deibert, and J.L. Pikul, Jr. 2006. Cropping system influences on soil chemical properties and soil quality in the Great Plains. Renew. Agric. Food Syst. 21:26–35.

Nielsen, D.C. 1998. Snow catch and soil water recharge in standing sunflower residue. J. Prod. Agric. 11:476–480.

Nielsen, D.C., and R.M. Aiken. 1998. Wind speed above and within sunflower stalks varying in height and population. J. Soil Water Conserv. 53:347–352.

Nielsen, D.C., and M.F. Vigil. 2010. Precipitation storage efficiency during fallow in wheat-fallow systems. Agron. J. 102:537–543.

Nielsen, D.C., M.F. Vigil, R.L. Anderson, R.A. Bowman, J.G. Benjamin, and A.D. Halvorson. 2002. Cropping system influence on planting water content and yield of winter wheat. Agron. J. 94:962–967.

O'Connell, M.G., G.J. O'Leary, and D.J. Connor. 2003. Drainage and change in soil water storage below the root-zone under long fallow and continuous cropping sequences in the Victorian Mallee. Aust. J. Agric. Res. 54:663–675.

Oliveira, A.A., and F.E. Sanders. 1999. Effect of management practices on mycorrhizal infection, growth and dry matter partitioning in field-grown bean. Pesquisa Agropecu. Bras. 34:1247–1254.

Ortega, R.A., G.A. Peterson, and D.G. Westfall. 2002. Residue accumulation and changes in soil organic matter as affected by cropping intensity in no-till dryland agroecosystems. Agron. J. 94:944–954.

Pankhurst, C.E., G.R. Stirling, R.C. Magarey, B.L. Blair, J.A. Holt, M.J. Bell, and A.L. Garside. 2005. Quantification of the effects of rotation breaks on soil biological properties and their impact on yield decline in sugarcane. Soil Biol. Biochem. 37:1121–1130.

Paul, E.A. 2007. Soil microbiology, ecology, and biochemistry. Academic Press, Burlington, MA.

Peterson, G.A., A.D. Halvorson, J.L. Havlin, O.R. Jones, D.J. Lyon, and D.L. Tanaka. 1998. Reduced tillage and increasing cropping intensity in the Great Plains conserves soil C. Soil Tillage Res. 47:207–218.

Peterson, G.A., A.J. Schlegel, D.L. Tanaka, and O.R. Jones. 1996. Precipitation use efficiency as affected by cropping and tillage systems. J. Prod. Agric. 8:180–186.

Peterson, G.A., and D.J. Vetter. 1971. Soil nitrogen budget of wheat-fallow area. Nebraska Farm, Ranch, and Home Quarterly. 12:23–25.

Peterson, G.A., D.G. Westfall, and C.V. Cole. 1993. Agroecosystem approach to soil and crop management research. Soil Sci. Soc. Am. J. 57:1354–1360.

Pikul, J.L., and J.K. Aase. 1995. Infiltration and soil properties as affected by annual cropping in the northern Great Plains. Agron. J. 87:656–662.

Pikul, J.L., Jr., J.K. Asae, and V.L. Cochran. 1997. Lentil green manure as fallow replacement in the semiarid Northern Great Plains. Agron. J. 89:867–874.

Pikul, J.L., R.C. Schwartz, J.G. Benjamin, R.L. Baumhardt, and S. Merrill. 2006. Cropping system influences on soil physical properties in the Great Plains. Renew. Agric. Food Syst. 21:15–25.

Rasmussen, P.E., and H.P. Collins. 1991. Long-term impacts of tillage, fertilizer, and crop residue on soil organic matter in temperate semiarid regions. Adv. Agron. 45:93–134.

Rasmussen, P.E., and W.J. Parton. 1994. Long-term effects of residue management in wheat-fallow: I. Inputs, yield, and soil organic matter. Soil Sci. Soc. Am. J. 58:523–530.

Renard, K.G., G.R. Foster, G.A. Weesies, and G.A. Porter. 1991. RUSLE: Revised universal soil loss equation. J. Soil Water Conserv. 46:30–33.

Russel, J.C. 1939. The effect of surface cover on soil moisture losses by evaporation. Soil Sci. Soc. Am. Proc. 4:65–70.

Ryan, J., and M.A. Monem. 1998. Soil fertility management for sustained production in the West Asia-North Africa region: Need for long-term research. p. 155–174. *In* R. Lal (ed.) Soil quality and agricultural sustainability. Ann Arbor Press, Chelsea, MI.

Ryan, J., and M. Pala. 2007. Syria's long-term rotation and tillage trials: Potential relevance to carbon sequestration in Central Asia. p. 223–234. *In* R. Lal (ed.) Carbon sequestration in Central Asia. Taylor and Francis, Leiden, the Netherlands.

Ryan, J., M. Singh, and M. Pala. 2008. Long-term cereal-based rotation trials in the Mediterranean Region: Implications for cropping sustainability. Adv. Agron. 97:273–319.

Sharratt, B., G. Feng, and L. Wendling. 2007. Loss of soil and PM10 from agricultural fields associated with high winds on the Columbia Plateau. Earth Surf. Process. Landf. 32:621–630.

Siddoway, F.H., W.S. Chepil, and D.V. Armbrust. 1965. Effect of kind, amount and placement of residue on wind erosion control. Trans. ASAE 8:327–331.

Smika, D.E. 1983a. Cropping practices: Introduction. p. 293–295. *In* H.E. Dregne and W.O. Willis (ed.) Dryland Agriculture. ASA–CSSA–SSSA, Madison, WI.

Smika, D.E. 1983b. Soil water changes as related to position of wheat straw mulch on the soil surface. Soil Sci. Soc. Am. J. 47:988–991.

Smika, D.E. 1990. Fallow management practices for wheat production in the central Great Plains. Agron. J. 82:319–323.

Smika, D.E., and G.A. Wicks. 1968. Soil water storage during fallow in the central Great Plains as influenced by tillage and herbicide treatments. Soil Sci. Soc. Am. Proc. 32:591–595.

Smiley, R.W., H.P. Collins, and P.E. Rasmussen. 1996. Diseases of wheat in long-term agronomic experiments at Pendleton, Oregon. Plant Dis. 80:813–820.

Standley, J., H.M. Hunter, G.A. Thomas, G.W. Blight, and A.A. Webb. 1990. Tillage and crop residue management affect Vertisol properties and grain sorghum growth over seven years in the semi-arid sub-tropics. II. Changes in soil properties. Soil Tillage Res. 18:367–388.

Steenwerth, K.L., L.E. Jackson, F.J. Calderón, M.R. Stromberg, and K.M. Scow. 2002. Soil microbial community composition and land use history in cultivated and grassland ecosystems of coastal California. Soil Biol. Biochem. 34:1599–1611.

Steiner, J.L. 1994. Crop residue effects on water conservation. p. 41–76. *In* P.W. Unger (ed.) Managing agricultural residues. Lewis Publ., Chelsea, MI.

Sturz, A.V., and H.W. Johnston. 1985. Characterization of Fusarium colonization of spring barley and wheat produced on stubble or fallow. Can. J. Plant Pathol. 7:270–276.

Tanaka, D.L., and J.K. Aase. 1987. Fallow method influences soil water and precipitation storage efficiency. Soil Tillage Res. 9:307–316.

Thomas, G.A., D.N. Orange, and A.J. King. 2008. Effects of crop and pasture rotations and surface cover on rainfall infiltration on a Kandosol in south-west Queensland. Aust. J. Soil Res. 46:203–209.

Thompson, J.P. 1987. Decline of vesicular-arbuscular mycorrhizae in long fallow disorder of field crops and its expression in phosphorus deficiency of sunflower. Aust. J. Agric. Res. 38:847–867.

Tullberg, J.N., P.J. Ziebarth, and Y. Li. 2001. Tillage and traffic effects on runoff. Aust. J. Soil Res. 39:249–257.

Unger, P.W. 1968. Soil organic matter and nitrogen changes during 24 years of dryland wheat tillage and cropping practices. Soil Sci. Soc. Am. J. 32:427–429.

Unger, P.W. 1978. Straw mulch rate effect on soil water storage and sorghum yield. Soil Sci. Soc. Am. J. 42:486–491.

Unger, P.W. 1991. Organic matter, nutrient, and pH distribution in no- and conventional-tillage semiarid soils. Agron. J. 83:186–189.

Unger, P.W. 1992. Infiltration of simulated rainfall: Tillage system and crop residue effects. Soil Sci. Soc. Am. J. 56:283–289.

Unger, P.W., and L.J. Fulton. 1990. Conventional- and no-tillage effects on upper root zone soil conditions. Soil Tillage Res. 16:337–344.

Unger, P.W., and A.F. Wiese. 1979. Managing irrigated winter wheat residues for water storage and subsequent dryland grain sorghum production. Soil Sci. Soc. Am. J. 42:582–588.

van de Ven, T.A.M., D.W. Fryrear, and W.P. Spaan. 1989. Vegetation characteristics and soil loss by wind. J. Soil Water Conserv. 44:347–349.

Van Donk, S.J., and E.L. Skidmore. 2003. Measurement and simulation of wind erosion, roughness degradation and residue decomposition on an agricultural field. Earth Surf. Process. Landf. 28:1243–1258.

Weigand, S., K. Auerswald, and T. Beck. 1995. Microbial biomass in agricultural topsoils after 6 years of bare fallow. Biol. Fertil. Soils 19:129–134.

Wienhold, B.J., J.L. Pikul, Jr., M.A. Liebig, M.M. Mikha, G.E. Varvel, J.W. Doran, and S.S. Andrews. 2006. Cropping system effects on soil quality in the Great Plains: Synthesis from a regional project. Renew. Agric. Food Syst. 21:49–59.

Williams, J.D. 2004. Effects of long-term winter wheat, summer fallow residue and nutrient management on field hydrology for a silt loam in north-central Oregon. Soil Tillage Res. 75:109–119.

Williams, J.D. 2008. Soil erosion from dryland winter wheat-fallow in a long-term residue and nutrient management experiment in north-central Oregon. J. Soil Water Conserv. 63:53–59.

Wilson, G.V., K.C. McGregor, and D. Boykin. 2008. Residue impacts on runoff and soil erosion for different corn plant populations. Soil Tillage Res. 99:300–307.

Wischmeier, W.H., and D.D. Smith. 1978. A universal soil-loss equation to guide conservation farm planning. Int. Congr. Soil Sci. Trans. 7:418–425.

Wood, C.W., D.G. Westfall, and G.A. Peterson. 1991. Soil carbon and nitrogen changes on initiation of no-till cropping systems. Soil Sci. Soc. Am. J. 55:470–476.

Woodruff, N.P., L. Lyles, F.H. Siddoway, and D.W. Fryrear. 1972. How to control wind erosion. USDA-ARS Agric. Inf. Bull. No. 354. U.S. Gov. Print. Office, Washington, DC.

Wright, S.F., J.L. Starr, and I.C. Paltineanu. 1999. Changes in aggregate stability and concentration of glomalin during tillage management transition. Soil Sci. Soc. Am. J. 63:1825–1829.

Zobeck, T.M., S. Van Pelt, J.E. Stout, and T.W. Popham. 2001. Validation of the revised wind erosion equation (RWEQ) for single events and discrete periods. p. 471–474. *In* J.C. Ascough and D.C. Flanagan (ed.) Soil Erosion for the 21st Century, Proc. Int. Symp. ASAE Pub. 701P00007.

20

Grazing Impacts on Soil Physical, Chemical, and Ecological Properties in Forage Production Systems

Miguel A. Taboada, Gerardo Rubio, and Enrique J. Chaneton

Extensive systems of animal husbandry (e.g., meat and dairy cattle, sheep, goats, and so on) are mainly based on the direct grazing of grasslands, pastures, fodder crops, and crop residues by livestock. Grazing effects on soil properties of forage production systems follow direct and indirect pathways. Direct effects relate to animal trampling and excretion, while indirect effects are mediated by changes in vegetation structure and function. Figure 20|1 shows a simplified model of a grazing system, integrating both direct and indirect effects of livestock on soil physical properties, as proposed by Greenwood and McKenzie (2001).

Grazing animals affect soil and vegetation properties through the action of treading, defoliation, and excretal returns. This may result in soil compaction and/or poaching damage to the pasture, which may recover as a function of the effectiveness of abiotic (e.g., wetting and drying cycles) or biotic (e.g., roots and worms activity) mechanisms of structural resilience. In addition, defoliation and excretal returns impact carbon and nutrient cycling in soil, which involve particular responses by soil organisms. Soil pores are the habitat for microbes and meso- and microfauna (Young and Ritz, 2005), which in turn are responsible for biotic mechanisms of soil structural resilience. In this chapter, we review the various effects of grazing on soil physical properties, carbon, and nutrient cycling in forage production systems. Where possible, we also discuss some management implications focusing on possible ways of minimizing negative grazing effects as well as recovering desirable soil properties for sustainable production.

Soil Physical Impacts of Livestock Grazing

Several review articles have been published over the last decades on the soil physical impacts of grazing (e.g., Gifford and Hawkins, 1978; Willatt and Pullar, 1983; Greenwood and McKenzie, 2001; Drewry, 2006), and most of their main conclusions are summarized in this section. Grazing impacts on soil physical properties are caused by defoliation and treading. Both actions cannot

Miguel A. Taboada, Soil Institute, CIRN, INTA, Division of Fertility and Fertilizers, Agronomy Faculty, University of Buenos Aires and CONICET, Nicolás Repetto and De Los Reseros, 1686 Hurlingham, Buenos Aires Province, Argentina; Gerardo Rubio, Division of Fertility and Fertilizers, Agronomy Faculty, University of Buenos Aires, and INBA-CONICET; Enrique J. Chaneton, Division of Ecology, Agronomy Faculty, University of Buenos Aires, and IFEVA-CONICET, San Martín Avenue 4453, C1417 DSE Buenos Aires, Argentina. *Corresponding author (mtaboada@cnia.inta.gov.ar; mtaboada@agro.uba.ar).

doi:10.2136/2011.soilmanagement.c20

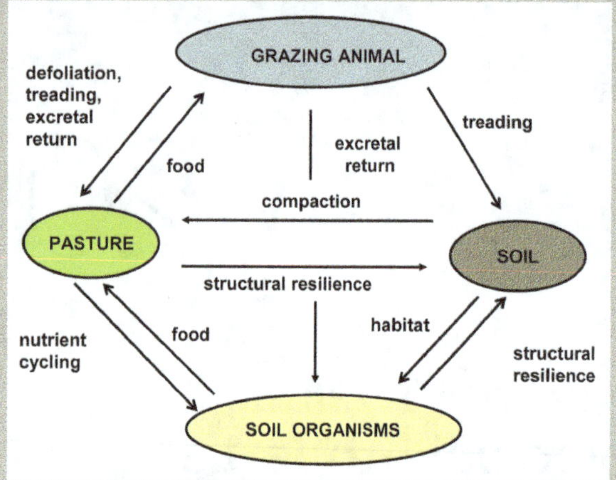

Fig. 20|1. Interactions among soil, vegetation, and livestock in grazing systems (modified from Greenwood and McKenzie, 2001).

be easily decoupled in field studies (Curll and Wilkins, 1983), but their various consequences deserve separate analyses.

Changes Imposed by Defoliation and Decreased Cover

The intensity of defoliation has greater effects on net herbage accumulation than treading (Curll and Wilkins, 1983). Grazing exerts more pressure on the most palatable plant species, which may decrease depending on grazing system and exclusion periods, changing grassland floristic composition, and vertical canopy structure (Sala et al., 1986). Defoliation increases the proportion of bare soil, and thus daily temperature fluctuation and maximum daily temperatures. Such effects are illustrated in Table 20|1 based on results from a natural grassland in the Flooding Pampa of Argentina (Lavado and Taboada, 1987). In this

study, an area grazed year-round by cattle for more than a century (average stocking rate 0.5 cattle ha^{-1}) was compared with an adjacent, 4-ha exclosure protected from grazing for 7 yr. Within the exclosure, accumulation of litter and standing dead and live plant biomass reduced increases of topsoil temperatures during the day. Such increases did occur in the grazed area, because of its low ground cover and short canopy height, which resulted from greater cover by short grasses and low-growing forbs as compared with the exclosure (Sala et al., 1986). Similar changes in topsoil temperatures were found in semiarid South African rangelands by Snyman and du Preez (2005). Increased topsoil temperatures caused by grazing led to higher soil water evaporation rates (Table 20|1). In the Flooding Pampa case, soil evaporation rates in the grazed area were 3.6 to 10.2 higher than in the ungrazed area, showing the effect of mulching by plant litter in soil water conservation.

High evaporation rates under grazing (Table 20|1) may promote the accumulation of salts in the topsoil because of the upward movement of water and soluble salts from salinized deep horizons (Lavado and Taboada, 1987; Alconada et al., 1993). An example of this process is shown in Fig. 20|2, after results from Lavado and Taboada (1987). Salts did not accumulate at the soil surface in the absence of grazing because a thick layer of litter reduced evaporative losses and thus decreased the upward flux of water and salts. As a result, only minor salinity fluctuations were detected in the A and BA horizons of a nitric soil (typic Natraquoll). The episodic salinization of

Table 20|1. Topsoil temperature and soil evaporation rate in two summer dates (modified from Lavado and Taboada, 1987).

Months	Soil temperature range		Evaporation rate			
	GE†	G†	GE†		G†	
			Mean	CV	Mean	CV
	———— °C ————		mm d^{-1}	%	mm d^{-1}	%
January	22.8–25.5	26.4–32.7	0.19	93.12	2.13**	43.89
February	17.7–23.9	21.5–30.4	0.39	81.65	1.79**	30.41

** Highly significant differences between treatments ($P < 0.01$).
† GE, grazing exclusion; G, grazed land.

the topsoil may substantially decrease grassland forage production. It can be controlled by increasing ground cover through changes in grazing regime, by introducing longer rest periods or changing from continuous to rotational grazing. In the Flooding Pampa, Alconada et al. (1993) showed that salt ascension by capillarity from groundwater may be effectively limited through the retention of runoff water by earth banks. Interestingly, this practice also caused the replacement of unpalatable halophytic grasses by more palatable and productive hygrophilous species (Alconada et al., 1993).

Changes Imposed by Trampling and Treading

The mechanical impact of animal hooves on soil surface can be a severe disturbance on topsoil structure through the external forces applied by treading and trampling (Gifford and Hawkins, 1978; Willatt and Pullar, 1983 Greenwood and McKenzie, 2001; Drewry, 2006). This mechanical impact changes the form and stability of soil aggregates, which results in changes in bulk density, pore size distribution, and soil strength, among other properties. Different factors may change aggregate form and stability in grazed soils. Figure 20|3 shows a proposed conceptual model that integrates those factors in a dynamic equilibrium affected by opposed disintegration and regeneration forces.

Influence of Animal Type, Soil Properties, and Vegetation

Disintegration forces are due to animal treading and

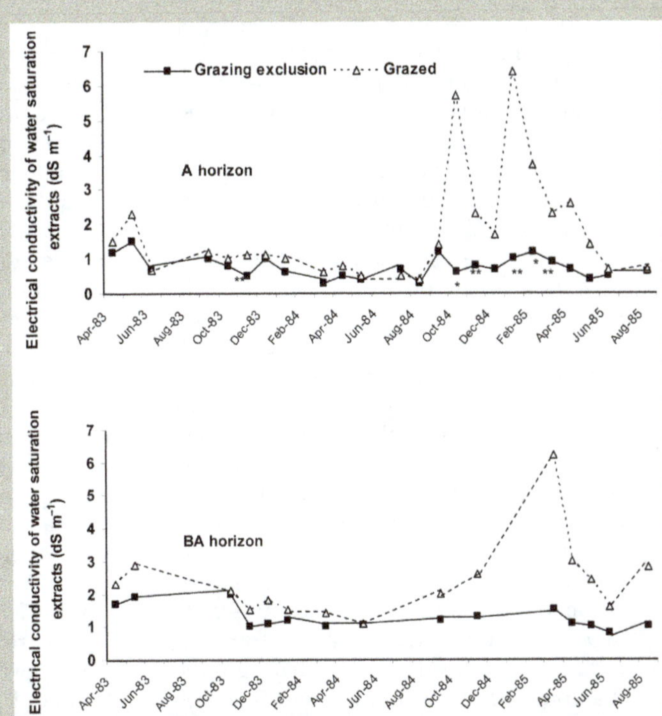

Fig. 20|2. Soil salinity measured as electrical conductivity of saturation extracts in A and BA horizons of a Typic Natraquoll in the Flooding Pampa, Argentina, under long-term grazing exclusion and continuous grazing (after Lavado and Taboada, 1987).

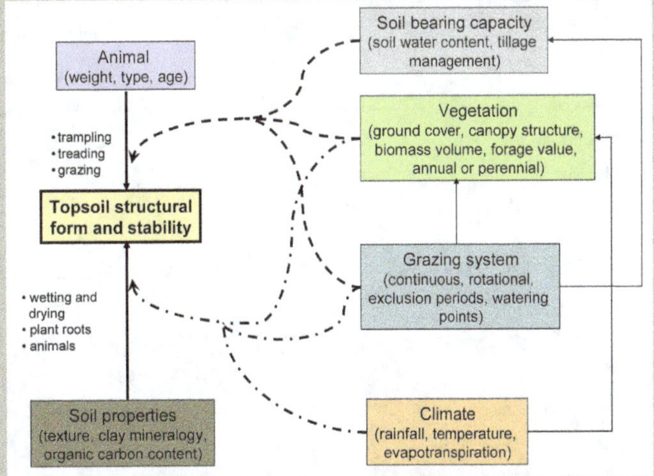

Fig. 20|3. Conceptual model describing factors causing variations in topsoil structural form and stability in grazed systems. Solid lines denote direct influences, dotted lines indirect influences.

trampling, and their impact depends on animal type, soil bearing capacity, vegetation, and grazing system. These forces are counteracted by regeneration forces. The abiotic or biotic nature of this regenerative process depends on soil type, and recovery times on climatic, vegetation, and grazing system factors. The pressure exerted by grazing animals on soil is a function of the animal's mass, foot size, and kinetic energy. Table 20|2 shows static pressures exerted by different kinds of livestock. These data show that pressures exerted by sheep and cattle average 66 kPa and 138 kPa, respectively, when standing. Despite sheep exerting lower static pressure than cows, herbivory and trampling by sheep may lead to landscape degradation in semiarid and desert areas (Rostagno 1989; Golodets and Boeken, 2006; Li et al., 2008). Most animal movement probably occurs during grazing, causing the amount of treading by each animal to depend on forage availability and grazing period. Cattle tread more on bare ground than on ground covered by tussocks, probably due to a preference to walk on even surfaces (Balph et al., 1989). The distance traveled by grazing animals is also affected by the distance to water points and quality of feed (Greenwood and McKenzie, 2001; Fensham and Fairfax, 2008).

Soil water content largely determines surface bearing capacity, and then controls structural damages by treading (Greenwood and McKenzie, 2001). Table 20|3 summarizes the most expected soil structural effects resulting from animal treading with different soil water conditions. For soils trampled when dry, structural deterioration results from aggregate crushing by animal hooves. This leads to the prevalence of smaller aggregates in the upper horizons (Warren et al., 1986; Taboada et al., 1999). When moist, animal trampling compresses the soil beneath the hoof (Scholefield et al., 1985) and collapses the larger soil pores (Warren et al., 1986; Taboada and Lavado, 1993). This results in increases in bulk density by compaction in the soil surface (Willatt and Pullar, 1983; Greenwood and McKenzie, 2001). Grazing compaction often results in reduced infiltration rates and low saturated hydraulic conductivity (Gifford and Hawkins, 1978; Willat and Pullar, 1983). Greenwood et al. (1997) concluded that the loss of porosity caused by grazing was related to a decrease in the number and/or continuity of pores greater than 1.2

Table 20|2. Comparative weight, foot area, and static pressure of grazing animals (modified from Greenwood and McKenzie, 2001).

	Mass	Total foot area	Static pressure
	kg	cm^2	kPa
Sheep	40–54	55–84	48–83
Cattle	306–612	264–460	98–192
Horses	400–700	736	54–95
Goats	40	55	60–73

Table 20|3. Expected damages produced by animal treading at different soil water conditions, affected soil depth, and changes in soil physical properties.

	Soil water conditions during treading			
	Dry	Moist	Saturated	Ponded
Expected damage	aggregate crushing	compaction	puddling and poaching	no effect
Depth	<5 cm	<10 cm	up to 20 cm	0
Bulk density	no effect	increase	increase/decrease	no effect
Macroporosity	no effect	decrease	decrease	no effect
Resistance	no effect	increase	decrease	no effect
Infiltration rate	may decrease	decrease	decrease	no effect
Saturated hydraulic conductivity	no effect	decrease	decrease	no effect
Structural stability	decrease	decrease	decrease	increase

mm. Bulk density and porosity damages by grazing can be attributed to treading rather than defoliation (Greenwood and McKenzie, 2001). Grazing compaction usually leaves little evidence of animal treading at the topsoil. Their effect may reach from only a few centimeters (Taboada and Lavado, 1988; Singleton and Addison, 1999; Martínez and Zinck, 2004) to about 20 cm of soil depth (Chanasyk and Naeth, 1995; Bell et al., 1997; Greenwood and McKenzie, 2001; Villamil et al., 2001; Drewry, 2006). Grazing consistently increased the topsoil strength when comparisons were made at the same water potentials (Greenwood and McKenzie, 2001).

Treading on saturated or nearly saturated soils causes remolding of the soil around the animal's hooves. This results in structural damages by poaching or kneading. When wet soils are trampled, there is plastic flow around the hoof (Mullins and Fraser, 1980; Scholefield et al., 1985). Repeated treading in these conditions produces deep hoofprints, which damage the sward (Mullins and Fraser, 1980; Scholefield and Hall, 1986). Poaching generally coincides with the presence of a layer of free surface water, and creates dense, unstable surface clods (Mulholland and Fullen, 1991). Poaching causes hoofprint depressions in the soil greater than 40 mm deep (Mullins and Fraser, 1980; Scholefield and Hall, 1986) and causes detrimental effects on soil macroporosity and water infiltration (Warren et al., 1986; Rubio and Lavado, 1990; Alconada et al., 1993). Many wetlands and lowlands are treaded when ponded, but literature on trampling effects under these conditions is scarce. In the Flooding Pampa, grazing cattle were found not to change bulk density but to increase macroporosity and aggregate stability (Taboada and Lavado, 1988, 1993; Taboada et al., 1999). The process of stability recovery was related to soil swelling, which originated with air entrapment during flooding periods (Taboada et al., 2001).

Soil infiltration rate is perhaps the most sensitive parameter for detecting soil physical damage by grazing. Gifford and Hawkins (1978) found grazing to decrease steady state infiltration by about 75% as compared with ungrazed treatments. Increased runoff associated with decreased infiltration may lead to increased erosion, with associated loss of organic matter and decreased plant available water. Infiltration in grazed pastures is decreased by the loss of macropores, as well as by vegetation removal. A ground cover of 70 to 75% was regarded in New South Wales, Australia, as the critical value below which runoff and soil loss increased markedly (Greenwood and McKenzie, 2001). Reductions in infiltration rates have also been measured where poaching has occurred (Mulholland and Fullen, 1991). Despite these changes in infiltration rate, as well as those reported in evaporation rate (Table 20|1), results from 15 articles reviewed by Greenwood and McKenzie (2001) showed no consistent trend of grazing effect on soil water. This was ascribed to the opposing effects of stocking rate on other factors that affect soil water, such as evaporation and infiltration rates on one side, and transpiration losses on the other side.

Other soil properties also affect bearing capacity. For instance, only small effects from grazing were observed when topsoils had high organic matter content and abundant plant root biomass (Taboada and Lavado, 1993; Taboada et al., 1999; Martínez and Zinck, 2004). Van Haveren (1983) found an interaction between soil texture and grazing intensity and showed that not all soils may be compacted by trampling because of the high sand content of some soils. Lagocki (1978) even found that bulk density decreased when trampling was performed on a soil with a high water table. However, studies comparing different grazing impacts in different soil types are not abundant.

Vegetation can also influence the expression of treading and trampling effects. Hamza and Anderson (2005) quoted several works showing that the impact of animal trampling can be reduced if the soil surface is covered by vegetation. For instance, 50% cover, 2-cm canopy height, or 1.0 t ha^1 dry matter over the soil were reported as thresholds for maximizing infiltration rates and preventing erosion. Increased treading generally decreases accumulated pasture biomass, although it is difficult to separate the effects on production of direct physical plant damage from the indirect soil compaction caused by treading. Brown (quoted by Greenwood and McKenzie, 2001) compared treading with and without the protective cover of standing pasture. Pasture trodden when 25 mm high was much less productive than pasture trodden when up to 127 mm high. Soil protection by vegetation is most important in semiarid and desert grasslands, where grazing decreases in infiltration rates are closely associated with bare ground areas (Rostagno, 1989; Li et al., 2008). Ground cover decreases by overgrazing and leads to desertification, which is coupled to litter removal and soil organic carbon decreases (Tongway et al., 2003; Li et al., 2008). This may change soil microtopography by remobilization of fine sediments and soil resources and wind erosion (Trimble and Mendel, 1995; Nash et al., 2003; Bisigato et al., 2008).

Influence of Stocking Rate and Grazing System

Soil compaction, as measured by bulk density, macropore volume, or surface strength, is generally increased by stocking rate (Willatt and Pullar, 1983). This leads to decreased soil water intake by infiltration rate (Gifford and Hawkins, 1978) and decreased hydraulic conductivity (Willatt and Pullar, 1983). Seventeen out of 23 studies reviewed by Greenwood and McKenzie (2001) found soil bulk density to increase with stocking rate, while 6 out of 9 found structural stability to decrease with stocking rate. These structural damages differ in their magnitude and depth reach as a result of water content during trampling and soil type. Lack of stocking rate effects on soil bulk density have been attributed either to dry soil conditions (i.e., high bearing capacity) or stocking rates not high enough to change bulk density (Taboada and Lavado, 1988, 1993; Taboada et al., 1999).

Management systems based on continuous grazing often decrease forage offer and quality (Greenwood and McKenzie, 2001; Yong-Zhong et al., 2005). However, even under continuous grazing, the distribution of treading is not uniform but concentrated in areas such as gateways and around troughs. Pietola et al. (2005) found that even at low intensity grazing reduced infiltration and hence increased susceptibility to erosion around drinking sites. Fensham and Fairfax (2008) compiled empirical data to identify threshold distances from water containing a 95% population of grazing animals. They found threshold distances of 3 km for sheep and 6 km for cattle.

Rotational grazing systems aim to avoid damages to the sward (Savory and Parsons, 1980) and are essential when vegetation has low regeneration potential (Müller et al., 2007). Livestock is allowed to graze during short time periods at moderate to high stocking rates, using two or more fields separated by fences (The Forage and Grazing Terminology Committee, 1991). Rotational grazing is considered highly successful for increasing forage offer and quality, but it has no clear impact on the recovery of soil physical properties. For instance, Rubio and Lavado (1990) observed in a sodic soil (Typic Natraqualf) that topsoil bulk density decreased under rotational grazing as compared with a continuously grazed soil, matching that found in ungrazed areas. In contrast, in the Brazilian savanna ("Cerrado") Leão et al. (2006) found that short-duration grazing was more restrictive to potential root growth, as evaluated by the least limiting water range (LLWR), than continuous grazing. In the same region, Pires da Silva et al. (2003) found that adopting an irrigated short-duration grazing system with high stocking rate may adversely affect the soil physical quality.

Grazing may affect soil physical properties in integrated livestock–grain cropping systems, in which crop residues, fodder and cover crops, and even weeds are part of the animal feeding regime during winter or summer. Tillage management changes soil bearing capacity, and hence the trampling compaction hazard. Literature on this subject is scant. In the Argentine Rolling Pampas, Díaz-Zorita et al. (2002) found that intensive animal trampling during winter after maize and soybean harvest increased bulk density at the soil surface (0–5 cm) in till relative to no-till systems, but no significant changes were observed at the 5- to 15-cm soil layer. They concluded that crop residues could be grazed without inducing significant increases in soil compaction if row cropping is managed with no-till farming practices (see also Fernández et al., 2010; García-Prechac et al., 2004). In the semiarid Pampas, Quiroga et al. (2009) found grazing to have a negative effect on soil bulk density when averaging data across tillage systems, while there was no effect on aggregate stability, and a positive one on the proportion of >8 mm aggregates. They concluded that in this region the introduction of grazing animals in no-till crop systems would not be detrimental to soil conditions and quality. Franzluebbers and Stuedemann (2008) found that the introduction of cattle to consume high-quality cover crop forage did not cause significant changes in bulk density, water-stable aggregation, and infiltration rate in a Typic Kanhapludult, GA. In coastal plains soils (Typic Kandiudult, AL) integrating winter-annual grazing in a cotton–peanut rotation using a conservation tillage system improved soil quality by reducing cone index, increasing infiltration and increasing soil organic carbon in surface (Siri-Prieto et al., 2007). Overall, these results suggest that it is possible the integration of livestock in cropping systems without substantially damaging soil physical structure.

Regeneration of Structural Forms and Stability

Soil structural damage by grazing can be restored by natural regenerative processes, which include wetting and drying cycles, growth and decay of plant roots, and the action of soil animals when stress factors are reduced or removed altogether (Greenwood and McKenzie, 2001; Drewry, 2006). Soil compaction by treading and subsequent natural recovery of soil physical properties have been shown to be cyclical (Taboada et al., 1999; Drewry et al., 2004). However, most studies were done in rotational grazing management timeframes, which do not always allow a complete natural recovery that often occurs with complete animal exclusion for months or years. Natural recovery of soil physical condition improves soil properties including hydraulic conductivity, macropore volume, and bulk density. Drewry (2006) distinguished between studies reporting natural recovery of soil physical condition for periods less than 1 yr and those reporting recovery for periods 1 yr or greater. Short-term recovery was reported by five studies that were performed in organic carbon rich (>4.5%) silty loam and clay loam topsoils (up to 10 cm). Soil macroporosity (>30 μm) and infiltration rate showed the highest improvement percentages after grazing cessation (up to 127%). Visual hoof damages from soil puddling and deformation were reduced to a half of the initial value within 87 to 165 d, depending on the characteristics of the soil. Physical deterioration of soil to about 5-cm depth can be naturally ameliorated by the burrowing activities of macroinvertebrates associated with dung deposition (Herrick and Lal, 2005). Sometimes, natural recovery forces are strong enough not to require cessation of grazing, as in flooded grassland soils in Argentina (Taboada and Lavado, 1993; Taboada et al., 1999, 2001).

Recovery generally demands many years in drier climates (Gifford and Hawkins, 1978; Greenwood and McKenzie, 2001). Gifford and Hawkins (1978) reported for rangelands in the USA that infiltration rates might be still increasing 13 yr after grazing stopped. Other authors also found long periods (up to several years) of recovery (Braunack and Walter, 1985; Steffens et al., 2008). These long recovery periods require grazing exclusion for several years, which is only practically possible under "cut-and-carry" or nil-grazing systems, which are difficult to implement under extensive grazing systems. Faster responses are however possible in temperate climates, because of the action of frequent wetting and drying cycles, freezing and thawing cycles, and vigorous pasture root growth.

Grazing Impacts on Soil Carbon Dynamics
Carbon Budget in Grazing Systems

More than two-thirds of the carbon stored in grasslands is located below ground in soil organic matter pools (Parton et al., 1987; Burke et al., 1989). Grassland carbon stocks are primarily determined by climatic factors; total carbon increases with precipitation as a result of increased primary production, and decreases with increasing temperature due to increased decomposition (Burke et al., 1989; Sala et al., 1996). Large grazers remove 20 to 75% of the aboveground net primary production (Milchunas and Lauenroth, 1993; Oesterheld et al., 1999). By diverting carbon away from the plant detrital pathway, grazing animals may reduce the energy supply to soil decomposers (Cebrian, 1999; Wardle and Bardgett, 2004). Thus, grazing exerts a major influence on grassland carbon cycling, affecting not only transfers among vegetation and soil compartments, but also ecosystem input and output flows (Fig. 20│4). In the long run, these alterations may have important consequences for the capacity of managed grasslands to store carbon and

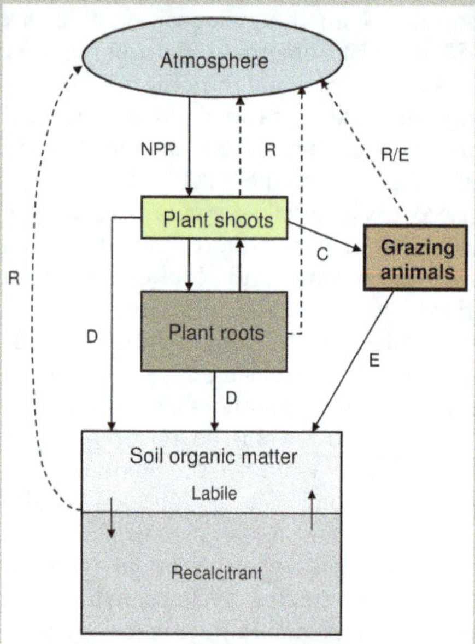

Fig. 20|4. Diagrammatic model of the carbon budget in a grazing system showing major carbon pools and transfer flows. In this schematic, microbial decomposers are included in the soil labile organic matter compartment. Solid arrows depict ecosystem inputs and transfers among compartments; broken arrows denote gaseous losses. NPP, net primary productivity; R, respiration; C, consumption; D, detritus production; E, excretion. Grazing may directly or indirectly control all these fluxes.

contribute to the regulation of the global carbon cycle (Sala et al., 1996; Steinfeld and Wassenaar, 2007).

In this section, we discuss the effect pathways whereby grazers alter the carbon budget of grasslands. We begin by discussing grazing impacts on carbon stocks and flows, and then move on to review the mechanisms for grazing effects on litter decomposition, a key process in the maintenance of soil fertility and forage production. In doing so, we consider how changes in plant functional composition influence soil functioning. Lastly, we address some management implications of vegetation–soil feedbacks for restoring carbon stocks in degraded grazing lands. To a large extent, these arguments apply also to nutrient cycling, which is the focus of the next section.

Grazing Impacts on Carbon Pools

The amount of CO_2 fixed by plants through net primary production (NPP) can follow different pathways (Fig. 20|4). Carbon is translocated from above- to belowground plant organs and back to shoots depending on season and grazing pressure. Herbivores consume a sizeable fraction of the aboveground NPP; the rest of the carbon is accumulated as plant biomass and eventually enters above- and belowground detrital routes, undergoing decomposition by soil microorganisms. Grazers also transfer carbon into the soil organic matter via waste deposition. Soil organic matter can be divided into different compartments according to their turnover rates (Parton et al., 1987, 1988); here, soil microbes are included in a "labile" carbon pool with rapid (1–25 yr) turnover times. A small fraction of recalcitrant detritus, largely of root origin, slowly accumulates in a "passive" soil pool with a slow turnover time (>200 yr), which represents a net carbon sink (Chapin et al., 2002). Most carbon losses from grasslands occur through plant, herbivore, and soil respiration (Fig. 20|4). Yet livestock also releases substantial amounts of methane (CH_4) through digestive fermentation (Steinfeld and Wassenaar, 2007).

A static view of the ecosystem allows one to assess the effects of grazing on carbon distribution among plant and soil compartments. Such studies typically involve comparisons between grazed grassland and ungrazed exclosures and are subjected to several sources of variation, including management, age of exclosure, site productivity, vegetation composition, soil type, and topography (Milchunas and Lauenroth, 1993; Stohlgren et al., 1999). Nevertheless, taken as a whole, exclosure studies offer a global picture of the direction and magnitude of grazing effects on ecosystem function (Milchunas and Lauenroth, 1993). Whereas in mesic grasslands grazing decreases the amount of carbon stored in aboveground vegetation, except where woody species encroachment shifts the dominant life form (Asner et al., 2004), its impact on root mass is far more variable and context dependent (Milchunas and Lauenroth, 1993). On the other hand, livestock grazing has been reported to decrease

(Marrs et al., 1989; Frank et al., 1995; Altesor et al., 2006), increase (Dormaar et al., 1994; Chaneton and Lavado, 1996; Schuman et al., 1999), or even produce no observable change in soil organic carbon (Henderson et al., 2004).

A few examples suffice to illustrate the variability of grazing effects on soil carbon pools. In mixed prairies of the North-Central USA Frank et al. (1995), found that moderately grazed sites contained 15% less soil carbon (107-cm depth) than grazing exclosures. The effect disappeared for heavily grazed sites, where increased cover by shallow-rooted grasses would have compensated for soil carbon losses (Frank et al., 1995). In similar grassland, Schuman et al. (1999) reported an increase of soil carbon in the rooting zone (0–30 cm) under heavy stocking rates. In the Flooding Pampa of Argentina, cattle grazing did not affect soil organic carbon (10-cm depth) in an upland site, but increased it by 28% in a lowland site (Chaneton and Lavado, 1996), suggesting that topographic position modulated grazing effect on carbon cycling. Moreover, grazing effects may depend on the soil pool being sampled. For several native grasslands in Uruguay, Altesor et al. (2006) found more organic carbon at the surface soil (0–5 cm) layer in grazed compared with ungrazed plots, although they also detected an 8% reduction in carbon stock for deeper layers (60–100 cm) of grazed sites. This pattern may suggest that increased root production of grazed vegetation (Doll, 1991; Altesor et al., 2006) may counteract grazing-induced reductions in aboveground detrital inputs to soil.

Lack of consistent grazing effects on soil carbon has also been commonplace in some grassland regions. Henderson et al. (2004) sampled nine locations across the Canadian Great Plains and detected no trend in root mass, subsurface litter, or soil particulate carbon in response to grazing. Stohlgren et al. (1999) found no definite, overall grazing effect on soil carbon for several Rocky Mountain grasslands, although negative and positive effects occurred within different management units. Neutral grazing effects may well be associated with the short time scale inherent to exclosure studies, given the slow turnover rate of major soil carbon pools (Milchunas and Lauenroth, 1993; Piñeiro et al., 2006). A different trend emerges for arid rangelands; there

reductions in soil carbon due to overgrazing appear to be the norm (Schlesinger et al., 1990; Abril and Bucher, 2001; Asner et al., 2004; Neff et al., 2005). Indeed, in a global analysis of grazing impacts across systems (200–800 mm yr^{-1}), Milchunas and Lauenroth (1993) found that soil organic matter changes were similarly divided between negative and positive, even though grazing decreased aboveground NPP irrespective of habitat.

Grazing Alteration of Major Carbon Flows

A dynamic view of input and output flows may help to understand how soil carbon stocks respond to grazing in different systems. Aboveground NPP increases along a precipitation gradient from arid to humid grasslands (Milchunas and Lauenroth, 1993; Oesterheld et al., 1999). Stocking rates and biomass consumption by livestock also increase with precipitation, but in an exponential fashion (Oesterheld et al., 1998, 1999). As a result, grazing impacts on vegetation composition are greater in humid than in dry grasslands, and this is correlated with increasing reductions in aboveground NPP by grazing (Milchunas and Lauenroth 1993). Trends in total root mass are less clear-cut; grazing effects are generally positive but seem unrelated to aboveground NPP changes across ecosystems (see Milchunas and Lauenroth 1993). Thus, grazing is expected to reduce the carbon flow from standing vegetation to surface litter and to the soil organic matter (Fig. 20|4), although larger quantities of plant carbon may be allocated belowground in grazed grassland (Doll, 1991; Schuman et al., 1999).

Microbial decomposition of litter entering the labile soil pool is the main process whereby carbon is released to the atmosphere (Fig. 20|4). Decomposition rates vary markedly between shoot and root litter substrates, the latter being less degradable (Semmartin et al., 2004, 2008). Grazing may accelerate or retard aboveground litter breakdown through various mechanisms (see below), with the direction of the effect depending on site productivity and plant community turnover (Bardgett and Wardle, 2003). In contrast, grazing effects on root decomposition are often negligible (Semmartin et al., 2008). This suggests that carbon release per unit of the largest litter pool may not be substantially affected by livestock grazing.

We now synthesize the above discussion in a hypothetical model of grazing effects on the amount of organic carbon stored in the rooting zone (i.e., the layer in which most roots are located). The model suggests that the relative effect of grazing on soil carbon may shift from negative in arid systems, to positive in mesic through humid habitats (Fig. 20|5). Yet, in mesic-to-humid systems, soil carbon may be either increased or decreased, depending on the balance between root production (input) and above- and belowground litter decomposition (output). Both of these processes will be influenced by the functional traits of plant species that come to dominate with grazing, and by site conditions (Bardgett and Wardle, 2003; Garibaldi et al., 2007). At the opposite end of the gradient, negative grazing effects on soil carbon prevail (Fig. 20|5). A large body of evidence suggests that reduced litter inputs from herbivory (Oesterheld et al., 1999; Abril and Bucher, 2001), increased erosion (Asner et al., 2004; Neff et al., 2005), and photodegradation of litter (Austin and Vivanco, 2006) may lead to significant reductions in soil carbon stocks in arid rangelands. Note that this model does not allow for within-site heterogeneity in soil carbon pools, e.g., "fertility islands" in arid lands (Schlesinger et al., 1990).

Long-term changes in input and output fluxes may not only alter topsoil carbon content, but may also be transmitted to deeper layers, affecting storage in less active pools, and the grassland net carbon balance (Chapin et al., 2002). To account for grazing impacts on whole-grassland carbon dynamics, a long-term (>100 yr) perspective is needed (Burke et al., 1989). Recently, Piñeiro et al. (2006) tackled this issue for the Río de la Plata grasslands in southern South America, using the CENTURY model (Parton et al., 1987, 1988), which simulates long-term dynamics of soil carbon and nitrogen pools. They evaluated the impact of livestock grazing since its introduction to the region some 400 years ago. They estimated changes in grassland carbon stocks in vegetation and soil pools by comparing model outputs between c.1600 and 1970.

Total carbon in vegetation and surface microbes was estimated to have decreased an average of 32% (n = 11 sites) after livestock introduction (1600: 1804 g C m^{-2} vs. 1970: 1223 g C m^{-2}), while carbon stored in soil pools declined by 21% (1600: 10,082 g C m^{-2} vs. 1970: 7967 g C m^{-2}). The largest decrease (32%) was in the "slow" organic carbon pool (turnover ~25 yr). As a result, grazing redistributed soil carbon to the "passive" pool from 39% to 47%. These changes partly reflected the imbalance between input and output carbon fluxes (Table 20|4). Both NPP inputs and respiration losses decreased with domestic grazing but total respiration decreased more. A greater proportion of NPP was released through herbivore respiration, which reduced transfers to soil microbes, decreasing soil respiration (Table 20|4). The increase in carbon losses through herbivore respiration, however, was smaller than the absolute decrease in carbon fixation by NPP (Piñeiro et al., 2006).

Most importantly, livestock introduction augmented gaseous N output (mainly as NH$_3$) from animal excreta.

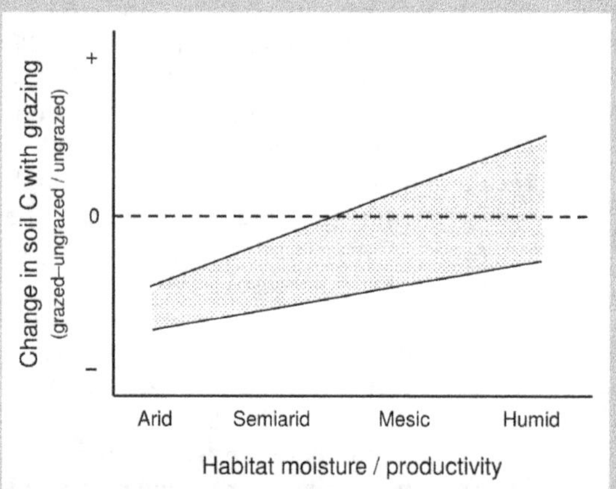

Fig. 20|5. A summary model of the relative effect of grazing on soil organic carbon in the rooting zone, along a habitat moisture gradient from dry to humid grasslands. Values above (positive) and below (negative) the dashed line indicate increases and decreases in soil carbon as a result of grazing, respectively. The stippled area represents cross-site variability in soil carbon response within habitat types.

Table 20|4. Major carbon flows (g C m^{-2} yr^{-1}) estimated before and after the introduction of livestock grazing in Río de la Plata grasslands (after Piñeiro et al., 2006).

	1600	1970	Change
			%
Net primary productivity	889	670	−24.6
Detrital inputs to soil	884	562	−36.4
Herbivore respiration	5	107	+2040
Soil respiration	884	571	−35.4

These N losses were associated with higher C:N ratios of all system compartments, suggesting that N availability, rather than carbon, constrained organic matter accumulation in soils (Piñeiro et al., 2006). These results illustrate the remarkable control that long-term livestock grazing may exert on the size of major elemental pools and flux rates in rangelands.

Grazing Effects on Soil Microbes and Decomposition

Human activity has been historically associated with large changes in herbivore loads. A pervasive consequence of managed grazing systems is the alteration of ecological feedbacks between above- and belowground processes (Wardle and Bardgett, 2004), which parallel grazing-induced changes in vegetation structure and composition (Milchunas and Lauenroth, 1993; Chaneton et al., 1996; Asner et al., 2004). Up to this point, we included decomposers in the labile soil carbon pool. We now discuss responses of soil organisms to livestock grazing to provide for a mechanistic understanding of grazing effects on soil carbon cycling.

Recent studies have found variable effects of grazing on litter decomposition. In these experiments, different litter types are incubated in a common environment or, less often, in grazed vs. ungrazed sites.

As a general trend, data show marked differences in decomposition rate among litter species, as well as strong effects of incubation site, but patterns do not always conform to a grazing effect. Litter from grazing-promoted species may either decompose faster or more slowly than litter from grazing-reduced species (Garibaldi et al., 2007). In a study comparing rangelands along a 900-mm rainfall gradient, Semmartin et al. (2004) found that species increasing with grazing in a humid site decomposed faster than those decreasing with grazing. However, this pattern was reversed for semiarid and arid grasslands. On the other hand, shoot decomposition rates were higher in grazed grassland than in exclosures, but this was not true for root-derived litter (Semmartin et al., 2008). This suggests that historical grazing effects on the soil environment for aerial litter decomposition may play a crucial role in carbon dynamics.

Large grazers can affect soil microorganisms and the processes they regulate through direct and indirect pathways involving changes in the quantity and quality of

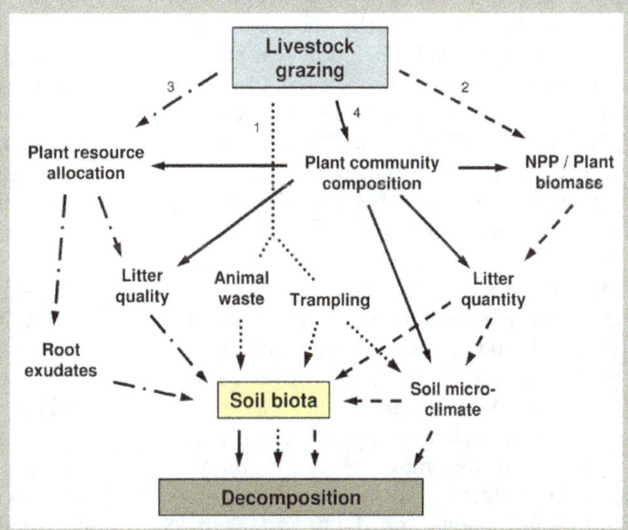

Fig. 20|6. Direct and indirect mechanisms whereby grazing affects soil biota and litter decomposition. Numbers and line patterns indicate different effect pathways: direct pathway, through excretion and trampling (1, dotted lines); indirect pathways, through changes in plant productivity (2, dashed lines), plant resource allocation (3, chained lines), and plant community composition (4, solid lines). Litter quantity and quality are key intermediary factors of indirect pathways, comprising both shoot and root residues.

resources entering the soil (Bardgett and Wardle, 2003). Figure 20|6 depicts several key, interrelated mechanisms whereby grazing influences the soil biota and carbon cycling rates during litter decomposition. There can be positive, negative, or neutral effects of grazing on decomposition depending on the relative importance of these mechanisms. First, grazing animals directly alter the quality of resource inputs to soil by returning soluble organic matter in feces (Fig. 20|6), which represent a nutrient-rich, albeit localized, carbon source that stimulates microbial biomass (Bardgett and Wardle, 2003). Trampling by livestock may also alter the physical environment for decomposers (see sections above). This includes disturbance of biological soil crusts, which help to stabilize surface soil carbon in arid lands (Neff et al., 2005).

Second, grazing controls the quantity of plant litter produced through its impact on above- and belowground NPP (Fig. 20|6). Long-term negative effects on NPP (see Milchunas and Lauenroth, 1993) may drive significant decreases in microbial biomass, perhaps to the extent of reducing carbon mineralization. Stimulation of aerial or root productivity (Doll, 1991) at moderate stocking rates, particularly in productive habitats, would cause the opposite effect by alleviating carbon limitation of soil microbes (Bardgett and Wardle, 2003). In addition, decreased plant cover by animal consumption and treading can drastically modify soil temperatures and moisture content (Lavado and Taboada, 1987) in ways that could affect microbial communities (Young and Ritz, 2005). Soil cover may further determine the extent of carbon release through physical degradation of litter by UV solar radiation, especially in arid zones (Austin and Vivanco, 2006).

A third pathway by which grazing affects soil decomposers is by modifying the patterns of resource allocation of individual plants (Bardgett and Wardle, 2003; Fig. 20|6). Shoot herbivory in high-productivity grasslands on fertile soils often increases leaf nutrient content, thus enhancing litter quality and decomposition (Semmartin et al., 2008). Across terrestrial ecosystems, plant nutritional quality is positively correlated to the percentage NPP consumed by herbivores, and also to the percentage detrital production decomposed annually (Cebrian and Lartigue, 2004). Thus, plant tissue quality may be regarded as an indicator of grazing control on plant biomass, detritus accumulation, and ecosystem carbon storage (see Cebrian, 1999). Further, herbivory may alter root exudation. Carbon allocated to root exudation may increase in grazed plants, which may then stimulate microbial biomass and activity in the rhizosphere, driving a feedback mechanism that accelerates nutrient return to plants and productivity (Hamilton and Frank, 2001; Bardgett and Wardle, 2003). However, the significance of this mechanism in managed grasslands remains to be tested.

Finally, changes in plant community composition associated with grazing may strongly affect soil microbial communities and decomposition (Bardgett and Wardle, 2003). Plant species composition may influence decomposers through various pathways (Fig. 20|6). This mechanism operates over successional time scales, and should be most important when compositional changes involve a shift in life-form dominance, such as an increase in forb species at the expense of perennial C_4 grasses (Garibaldi et al., 2007). It may lead to a reduction in average community litter quality when unpalatable, low-quality plants take over, or otherwise may increase litter quality inputs to soil when palatable, high-quality species remain dominant under grazing (Bardgett and Wardle, 2003; cf. Garibaldi et al., 2007). Invasion by exotic species plays a central role in these community changes. It appears that invasive plants generally accelerate carbon cycling, and may also increase plant and soil carbon stocks (Liao et al., 2008). However, the effect of woody invasions on soil organic carbon was found to shift from positive to negative along a decreasing precipitation gradient (Jackson et al., 2002). These authors showed that in drier grasslands carbon losses offset gains in vegetation due to woody species establishment.

Furthermore, grazing has been found to alter the functional composition of the soil microbial community. In some systems, decomposition is dominated by bacteria in grazed sites, whereas it becomes dominated by fungi in ungrazed or lightly grazed areas. Hence, a fast carbon cycle would prevail in grazed sites as more energy goes through the "bacterial channel," while a slow cycle would dominate in the absence of grazing, as more energy is processed by the "fungal channel" (Wardle, 2005).

Soil Carbon Stocks: Management Implications

Overall, there is increasing empirical evidence suggesting that grazing effects on soil decomposers and the processes they regulate may vary predictably with habitat moisture and productivity. Positive grazing effects on decomposition would be common in fertile systems with high herbivore loads, whereas negative effects may be most common in low fertility systems that sustain low grazing pressures (Bardgett and Wardle, 2003).

These contrasting ecosystem dynamics would also apply to early- vs. late-successional grasslands, which are alternatively dominated by fast-growing species with high litter quality and slow-growing plants with poor litter quality, respectively (Bardgett and Wardle 2003; Wardle, 2005). This broad trend may have important implications for the restoration of degraded rangelands, a major worldwide concern (Asner et al., 2004). Indeed, plant functional composition may determine the rate of carbon accumulation in soil, as recently shown by Fornara and Tilman (2008). These authors documented a 500% increase in soil carbon stored to 1-m depth after 12 yr of succession in mixed, C_4 grass–legume communities, while communities lacking either plant group were much less efficient at storing carbon. They suggest that complementary resource use between grasses and legumes leading to higher total root biomass accrual was the mechanism responsible for the observed effect (Fornara and Tilman 2008). Hence, restoration of soil fertility and basic ecosystem functions (i.e., forage production) in overgrazed rangelands may need to consider the functional composition of the successional vegetation.

Grazing Impacts on Soil Nutrient Cycling

Grazing animals strongly affect the dynamics of nutrients through the soil–plant–animal system in many ways (cf. Fig. 20|6). They return to soil around 60 to 95% of the nutrients contained in the forage they ingest. These nutrients are usually returned to the soil in a different chemical form compared with that of nutrients they absorbed by plants, thus significantly affecting the rates of nutrient cycling (Chaneton et al., 1996; Bardgett and Wardle, 2003). Grazers return nutrients as dung and urine forming localized excretal patches. Nutrient partitioning between dung and urine depends on the specific element and on other factors, notably the nutrient composition of the diet. Excretal patches cover only a small proportion of the pasture area, which increases the spatial heterogeneity of nutrient inputs to soil and availability to plants. Impacts of grazing are also evident in nutrient losses, which are concentrated in excretal patches and comprise both gaseous and leaching outputs. In this section, we discuss aspects specific to the influence of grazing animals on nutrient cycling. We focus on N and P, which are the main limiting nutrients to forage production in most grazing systems worldwide.

Nutrient Budgets in Grazing Systems

The demand for nutrients by grazing animals is relatively stable. For example, values of 25 g N kg^{-1} and 8 g P kg^{-1} retained by cattle adult body mass, 30 g N kg^{-1} and 8 g P kg^{-1} in the calf body mass, and 6 g N kg^{-1} and 1 g P kg^{-1} in the milk mass are generally accepted (ARC, 1980). However, it is very difficult to make generalizations on the amount of nutrients removed by grazers because they are largely determined by the total amount of ingested forage. The consumption of forage is intrinsically variable because it is regulated simultaneously by the availability of forage, the farming management and the rate of forage utilization, among other factors. Forage availability is highly variable because it depends on unpredictable factors such as rainfall. As a consequence, the total amount of nutrients consumed by grazing animals, and hence the role of grazing on nutrient budgets, may be difficult to predict accurately (Batabyal, 2007). For example, in a Flooding Pampa grassland under low stocking rate, Chaneton et al. (1996) found that the N added from rainfall (7 kg yr^{-1}) was sufficient to compensate for the N exported in animal body mass. In contrast, a continuous loss of P (0.4 kg ha^{-1} yr^{-1}) was estimated for the same nonfertilized grassland. Haygarth et al. (1998) compared the P budgets for two contrasting grassland farming systems: a dairy farm and an extensively managed hill sheep farm. They observed a high rate of soil P accumulation (26 kg ha^{-1} yr^{-1}) in the dairy farm, which received an annual input of 16 kg P per ha^{-1} from P fertilization. This surplus was related to the feed concentrates, which represented around 27 kg ha^{-1} yr^{-1}. The low stocking rate sheep farm, contrastingly, showed an almost equilibrated P budget (+0.28 kg ha^{-1} yr^{-1}).

In their study of Río de la Plata grasslands, Piñeiro et al. (2006) estimated that, in the long term, grazing altered the N cycling by accelerating returns via animal depositions and increasing N outputs by volatilization and leaching. As a result, soil N storage diminished 19% and N outputs increased 67% after four centuries of simulated domestic livestock grazing.

Nutrient Recycling by Grazing Animals
Proportion of Nutrients Returned to Soil

Although it is difficult to make quantitative generalizations on nutrient budgets in grazing systems, some common qualitative patterns emerge if we analyze the fate of nutrients after they are ingested by livestock. In ungrazed systems, nutrients taken up by plants are returned to the soil largely through the litter deposited on plant senescence. These nutrients may be re-used by plants and soil microorganisms. When livestock is introduced to the system, only a fraction of the plant nutrients are returned to the soil via the plant detrital pathway (cf. Fig. 20|4). Most nutrients go through the more complex path in the animal before they are returned to the soil. The amount of nutrients that are returned to the soil depends largely on the type of animal and the farming management (Mohamed Saleem, 1998). Depending on the grassland productivity and herbivore management, the rate of forage utilization will determine the grazing productivity and nutrient cycling.

Only a small fraction of the ingested nutrients are retained in animal products. Dahlin et al. (2005) reported that only 10 and 13% of the ingested N and P are retained by the animal in low-input grazed pastures, so that the percentage of ingested nutrients returned to the soil was about 90 and 87% for N and P, respectively. These figures are lower if dairy cows are considered. In such a case, a substantial part of the ingested and retained nutrients is exported in the milk. For example, Haynes and Williams (1993) calculated that as much as 10 and 25% of the ingested N and P are contained in the exported milk. If these proportions are coupled with the nutrients retained by the animal, the percentage of ingested nutrients returned to the soil lowers to 80 and 65%, for N and P, respectively.

The partitioning of nutrients between dung and urine differs broadly among nutrients and grazing systems. Some nutrients are excreted in a rather equivalent proportion between dung and urine, others preferentially in dung, and others in urine. Nitrogen is excreted in a rather equivalent proportion in both dung and urine. Sodium, Cl, and S are also partitioned significantly in both ways. Phosphorous is excreted mainly in dung, with reported values ranged around 95% of the total deposition (Haynes and Williams, 1993). Only a small amount of P is found in urine, although the proportion can increase if the diet is rich in this element. Calcium, Mg, Zn, Cu, Fe and Mn are also predominantly excreted in the solid feces. Potassium is the only nutrient predominantly excreted in urine (Haynes and Williams, 1993).

Forms of Nutrients Returned to Soil

Grazing may have a positive effect on the nutrient cycling of the soil. Compared with above- and belowground dead plant materials, animal excretions have higher concentrations of nutrients than the nondigested materials. More than half of the N in the urine is constituted by urea, which is rapidly hydrolyzed to ammonium in the excretal patch. This ammonium is highly susceptible to volatilization, causing gaseous N losses, which can be small (Chaneton et al., 1996) or large (Piñeiro et al., 2006), relative to internal nutrient flows, depending on the time-frame of analysis. The remaining proportion of urine-N is predominantly constituted by amino acids and peptides. In an elegant work, Clough et al. (1998) studied the fate of the ^{15}N labeled urine applied to four soils. Plant N uptake was the predominant sink of urine-N, accounting from 22 to 31% of the applied N. Nitrogen retained in the soil ranged from 20 to 24%, almost all in organic compounds. Leachate accounted from 13 to 32% whereas measured gas loss was less than 4% in all cases.

Phosphorous is returned to soil mainly in inorganic forms, dicalcium phosphate being the predominant P form in dung. Results from Sharpley and Moyer (2000) show consistently more inorganic P (63%) that organic P in dairy manure. Most of the inorganic P (around 80%) was water extractable. The total P in feces increases as the P content of the diet increases, but particularly the amount and proportion of water-soluble inorganic P compounds. For instance, Dou et al. (2002) found that increasing the proportion of P in the diet from 3.4 to 6.7 g P kg^{-1} feed dry matter caused an increase in the water-soluble fraction of fecal P of 2.9 to 10.5 g kg^{-1} fecal dry matter, representing 56 and 83% of acid digest P. The water-soluble fraction of fecal P is the most susceptible fraction to loss in the environment, which suggests that higher P contents in the diet lead to an increase in the susceptibility to environmental losses (Sharpley and Moyer, 2000). In such cases, Dou et al. (2002) suggest a measure of 2 g P kg^{-1} fecal dry matter, as inorganic P in a 1-h extract, to be a simple fecal P indicator that meets lactating cow requirements with minimal excess. The other fecal P fractions, including organic P and less soluble inorganic P, are less affected by dietary P concentration (Dou et al., 2002).

Organic P forms in the dung contain inositol hexaphosphates, among other compounds. The proportion of organic P in the animal excretions is commonly lower than the proportion of organic P in the ingested forage. This means that a significant mineralization of ingested organic P is produced during passage through the animal. Since plants absorb mainly soil P inorganic forms, grazing animals are expected to exert a positive effect on P cycling. Accordingly, Aarons et al. (2004) observed that dung P contributed only to increases in available inorganic P, with extractable organic P unchanged in the short term.

Grazing Effects on Plant Litter Quality

Litter quality can be significantly affected by changes in the grassland species composition promoted by grazing (Bardgett and Wardle, 2003). Published evidence reveals that the pattern of response is far from universal and depends largely on the functional groups involved in the species replacement that is promoted by grazing. In humid grassland, it was found that grazing promoted a change in the botanical composition with an increase of forb species and a decrease of the more palatable graminoids (Semmartin et al., 2004). The forbs produced litter with higher N and P contents and higher N mineralization rates than the graminoids, and this coincided with higher soil N availability in grazed than in ungrazed sites (Garibaldi et al., 2007). In contrast, Ritchie et al. (1998) observed that grazing decelerated the cycling of N and lowered soil N availability by reducing the proportion of highly palatable legumes. Therefore, the particular group of species that are diminished and promoted by grazing must be taken into account when analyzing the consequences of plant–herbivore interactions for the rates of nutrient cycling in grasslands.

Spatial Redistribution of Soil Nutrients

In their excellent review, Haynes and Williams (1993) found a range of 8 to 12 urinations per day for cattle and 18 to 20 for sheep. Each urination event ranged from 1.6 to 2.2 and 0.10 to 0.18 L for cattle and sheep, respectively. In regard to the number of defecations, these authors reported a range of 11 to 16 and 7 to 26 for cattle and sheep, respectively, with a range of 1.5 to 2.7 and 0.03 to 0.17 kg per defecation. These quantities are influenced by environmental and farming conditions. For instance, the number of urinations is largely affected by water intake and water content of the diet. The surface covered by a single urination event is 0.16 to 0.49 m^2 for cattle and 0.03 to 0.05 m^2 for sheep. In the case of dung, the area covered by a single defecation event ranges from 0.05 to 0.09 m^2 for cattle and 0.008 to 0.025 m^2 for sheep (Haynes and Williams 1993). This great range for sheep is because the fecal depositions are in the form of pellets that are often distributed over a wide area.

Excretion patches are distributed non-uniformly over the soil surface, which may determine two scales of spatial heterogeneity. At a broader scale, soil nutrient concentrations are particularly high at stock camps (i.e., areas within the paddock where animals congregate spontaneously), around water sources and fences, beneath trees, and areas where supplemental feed is provided. Dahlin et al. (2005) found an average transfer to stock camps of 32 and 22% (percentage of ingested nutrients) for N and P, respectively. This translocation increases the risk of nutrient losses, which is aggravated by the fact that those areas are usually trampled and, consequently, forage production and nutrient uptake is small. In such sense, stock camps are highly enriched with nutrients and constitute "'hot spots'" for nutrient losses, especially mobile nutrients, such as N. Less mobile nutrients, such as P, are retained by the soil and may become available to the herbage in subsequent growing seasons (Oenema et al., 2006). In the remaining areas of the paddock there is a pattern of spatial heterogeneity at a smaller scale. This is primarily due to the deposition of excreta in patches rather than evenly over the field. A grazed grassland paddock consists of a mixture of aggregates of grazed and ungrazed microsites. Excretal patches are old, fresh, isolated, overlapped, or mixed urine and dung depositions covering 40% of the total area (Lantinga et al., 1987). West et al. (1989) found that grazing markedly affected the pattern of spatial variability of N and P, which is highly dependant on the stocking rate. For example, Lavado et al. (1996) attributed the lack of spatial variability on soil N and P in a native grassland devoted to calf production to the low livestock density, but also to the size of the paddock (800 ha) and to the distance from the study site to the water source (1200 m).

Therefore, it is expected that nutrient cycling will display high spatial variability in grazed grasslands. This spatial heterogeneity creates technical and logistical problems that make the study of nutrient dynamics difficult. For this reason, modeling is a natural alternative for the study of nutrient cycling in grazed pastures (Hutchings et al., 2007).

Management Strategies to Increase Nutrient Use Efficiency

Rotational grazing is an effective practice to increase forage utilization efficiency, minimize daily variations in intake and quality feed, and allocate pasture to animals more efficiently, based on nutritional needs. With large pastures, the animals decide where, when, and how long to graze. Rotational grazing reduces the size of the pasture and transfers these decisions from animal to manager and usually results in more efficient use of the pasture resources and nutrient cycling. Rotational grazing contributes to minimizing camping and to spreading out animal depositions over large areas and thus reduces heterogeneity and nutrient losses (Oenema et al., 2006). The components of a good rotational grazing system are a balanced forage system, an electrical fencing system, and spatially distributed water and shade supply.

An interesting alternative to farms that use both stable and direct grazing is the prototype system for the realization of very high environmental standards that has been developed in the Netherlands (Oenema et al., 2006). The system consists of grazing for a restricted number of hours alternated with periods in the stable for supplementation. This practice reduces the fraction of dung and urine deposited on the soil, thus reducing field heterogeneity. The excreta collected in the stable can be distributed through appropriate application technology to the pasture at the right amount and spatial distribution.

Concluding Remarks

We have discussed how grazing affects soil physical properties, carbon, and nutrient cycling in forage production systems. These effects vary across the different climates, soil types, grazing systems, and livestock types. Nonetheless, some broad generalizations can be pointed out:

1. Grazing changes soil physical properties by defoliation and treading. Defoliation decreases ground cover, which increases topsoil temperature and evaporation rates, and causes salinization in salt-affected areas. Animal treading often results in shallow compaction when soil is trampled moist, and causes puddling when soil is trampled wet. Whereas soil infiltration rate is highly sensitive for detecting these changes, particularly in

bare ground areas, the effect of grazing on soil water content is hard to predict.

2. Structural damage by grazing can be restored by natural regenerative processes, which may take many years in dry climates. Faster responses are possible in temperate climates, because of the action of frequent wetting and drying cycles, freezing and thawing cycles, and vigorous pasture root growth.

3. The balance between soil carbon inputs and outputs in grazed systems varies widely among systems, making it difficult to make generalizations about the effect of grazing on soil carbon dynamics. However, the size of major carbon pools and fluxes in rangelands can be strongly controlled by grazing. Grazing alters litter decomposition and the functioning of the soil microbial community, and these effects are to a large extent mediated by plant community turnover. Several lines of evidence suggest that grazing effects on soil carbon and nutrient cycling may vary predictably along habitat moisture/fertility gradients.

4. The partitioning of nutrients between dung and urine differs among nutrients and grazing systems. Nitrogen is excreted in a rather equivalent proportion in dung and urine, whereas most P returns to soil in dung paths. Grazing alters N and P cycling by accelerating returns via animal depositions, and by increasing N outputs through volatilization and leaching.

5. Rotational grazing is an effective practice to increase forage utilization efficiency and reduce the fraction of dung and urine deposited on the soil, thus reducing field heterogeneity. However, rotational grazing has less clear consequences on the recovery of soil physical properties and carbon stocks.

6. Soil protection by vegetation is most important in semiarid and desert grasslands, where decreased infiltration rates by grazing are closely associated with bare ground areas. This objective can be attained by excluding grazing for several years (rest periods). However, restoration of soil organic matter in degraded arid rangelands would take exceedingly long (and economically unsustainable) periods of livestock exclusion.

Acknowledgments
We thank the Consejo Nacional de Investigaciones Científicas y Técnicas (CONICET), Agencia Nacional de Promoción Científica y Tecnológica (FONCYT grant scheme), and University of Buenos Aires for funding our research during the preparation of this chapter.

References

Aarons, S.R., H.M. Hosseini, L. Dorling, and C.J.P. Gourley. 2004. Dung decomposition in temperate dairy pastures. II. Contribution to plant-available soil phosphorus. Aust. J. Soil Res. 42:115–123.

Abril, A., and E.H. Bucher. 2001. Overgrazing and soil carbon dynamics in the western Chaco of Argentina. Appl. Soil Ecol. 16:243–249.

Alconada, M., O.E. Ansin, R.S. Lavado, V.A. Deregibus, G. Rubio, and F.H. Gutiérrez Boem. 1993. Effect of water retention of run-off water and grazing on soil and on vegetation of a temperate humid grassland. Agric. Water Manage. 23:233–246.

Altesor, A., G. Piñeiro, F. Lezama, R.B. Jackson, M. Sarasola, and J.M. Paruelo. 2006. Ecosystem changes associated with grazing in subhumid South American grasslands. J. Veg. Sci. 17:323–332.

ARC (Agricultural Research Council). 1980. The nutrient requirement of ruminant livestock. Commonwealth Agricultural Bureaux. Slough UK.

Asner, G.P., A.J. Elmore, L.P. Olander, R.E. Martin, and A.T. Harris. 2004. Grazing systems, ecosystem responses, and global change. Annu. Rev. Environ. Resour. 29:261–299.

Austin, A.T., and L. Vivanco. 2006. Plant litter decomposition in a semi-arid ecosystem controlled by photodegradation. Nature 442:555–558.

Balph, B.F., M.H. Balph, and J.C. Malechek. 1989. Cues cattle use to avoid stepping on crested wheatgrass tussocks. J. Range Manage. 42:376–377.

Bardgett, R.D., and D.A. Wardle. 2003. Herbivore-mediated linkages between aboveground and belowground communities. Ecology 84:2258–2268.

Batabyal, A. 2007. Some statistical properties of total forage consumption by grazing animals on a rangeland. Ecol. Econ. 64:5–8.

Bell, M.J., B.J. Bridge, G.R. Harch, and D.N. Orange. 1997. Physical rehabilitation of degraded Krasnozems using ley pasture. Aust. J. Soil Res. 35:1093–1113.

Bisigato, A.J., R.M.L. Laphitz, and A.L. Carrera. 2008. Non-linear relationships between grazing pressure and conservation of soil resources in Patagonian Monte shrublands. J. Arid Environ. 72:1464–1475.

Braunack, M.V., and J. Walter. 1985. Recovery of surface soil properties of ecological interest after sheep grazing in a semi-arid woodland. Aust. J. Ecol. 10:451–460.

Burke, I.C., C.M. Yonker, W.J. Parton, C.V. Cole, K. Flach, and D.S. Schimel. 1989. Texture, climate and, cultivation effects on soil organic matter content in U.S. grassland soils. Soil Sci. Soc. Am. J. 53:800–805.

Cebrian, J. 1999. Patterns in the fate of production in plant communities. Am. Nat. 154:449–468.

Cebrian, J., and J. Lartigue. 2004. Patterns of herbivory and decomposition in aquatic and terrestrial ecosystems. Ecol. Monogr. 74:237–259.

Chanasyk, D., and A. Naeth. 1995. Grazing impacts on bulk density and soil strength in the foothills fescue grasslands of Alberta, Canada. Can. J. Soil Sci. 75:551–557.

Chaneton, E.J., and R.S. Lavado. 1996. Soil nutrients and salinity after long-term grazing exclusion in a Flooding Pampa grassland. J. Range Manage. 49:182–187.

Chaneton, E.J., J.H. Lemcoff, and R.S. Lavado. 1996. Nitrogen and phosphorus cycling in grazed and ungrazed plots in a temperate subhumid grassland in Argentina. J. Appl. Ecol. 33:291–302.

Chapin, F.S., III, P.A. Matson, and H.A. Mooney. 2002. Principles of terrestrial ecosystem ecology. Springer, New York.

Clough, T.J., S.F. Ledgard, M.S. Sprosen, and M.J. Kear. 1998. Fate of 15N labelled urine on four soil types. Plant Soil 199:195–203.

Curll, M. L., and R.J. Wilkins. 1983. The comparative effects of defoliation, treading and excreta on a Lolium perenne-Trifolium repens pasture grazed by sheep. J. Agric. Sci. (Cambridge) 100:451–460.

Dahlin, A.S., U. Emanuelsson, and J.H. McAdam. 2005. Nutrient management in low input grazing-based systems of meat production. Soil Use Manage. 21:122–131.

Díaz-Zorita, M., G. Duarte, and J. Grove. 2002. A review of no-till systems and soil management for sustainable crop production in the subhumid and semiarid pampas of Argentina. Soil Tillage Res. 65:1–18.

Doll, U. 1991. C-14 translocation to the below ground subsystem in a temperate humid grassland (Argentina). p. 350–358. in B. McMichael, and H. Persson (ed.) Plant roots and their environment. Elsevier, Amsterdam.

Dormaar, J.F., B.W. Adams, and W.D. Willms. 1994. Effect of grazing and abandoned cultivation on a Stipa-Bouteloua community. J. Range Manage. 47:28–32.

Dou, Z., K.F. Knowlton, R.A. Kohn, Z. Wu, L.D. Satter, G. Zhang, J.D. Toth, and J.D. Ferguson. 2002. Phosphorus characteristics of dairy feces affected by diets. J. Environ. Qual. 21:2058–2065.

Drewry, J.J. 2006. Natural recovery of soil physical properties from treading damage of pastoral soils in New Zealand and Australia: A review. Agric. Ecosyst. Environ. 114:159–169.

Drewry, J.J., R.J. Paton, and R.M. Monaghan. 2004. Soil compaction and recovery cycle on a Southland dairy farm: Implications for soil monitoring. Aust. J. Soil Res. 42:851–856.

Fensham, R.J., and R.J. Fairfax. 2008. Water-remoteness for grazing relief in Australian arid-lands. Biol. Conserv. 141:1447–1460.

Fernández, P.L., C.R. Álvarez, V. Schindler, and M.A. Taboada. 2010. Changes in topsoil bulk density after grazing crop residues under no-till farming. Geoderma 159:24–30.

Fornara, D.A., and D. Tilman. 2008. Plant functional composition influence rates of soil carbon and nitrogen accumulation. J. Ecol. 96:314–322.

Frank, A.B., D.L. Tanaka, L. Hofman, and R.F. Follett. 1995. Soil carbon and nitrogen of Northern Great Plains grasslands as influenced by long-term grazing. J. Range Manage. 48:470–474.

Franzluebbers, A.J., and J.A. Stuedemann. 2008. Soil physical responses to cattle grazing cover crops under conventional and no tillage in the southern Piedmont, USA. Soil Tillage Res. 100:141–153.

García-Prechac, F., O. Ernst, G. Siri-Prieto, and J.A. Terra. 2004. Integrating no-till into crop-pasture rotations in Uruguay. Soil Tillage Res. 77:1–13.

Garibaldi, L.A., M. Semmartin, and E.J. Chaneton. 2007. Grazing-induced changes in plant composition affect litter quality and nutrient cycling in flooding Pampa grasslands. Oecologia 151:650–652.

Gifford, G.F., and R.H. Hawkins. 1978. Hydrologic impact of grazing on infiltration. Water Resour. Res. 14:305–313.

Golodets, C., and B. Boeken. 2006. Moderate sheep grazing in semiarid shrubland alters small-scale soil surface structure and patch properties. Catena 65:285–291.

Greenwood, K.L., and B.M. McKenzie. 2001. Grazing effects on soil physical properties and the consequences for pastures: A review. Aust. J. Exp. Agric. 41:1231–1250.

Greenwood, K.L., D.A. MacLeod, and K.J. Hutchinson. 1997. Long-term stocking rate effects on soil physical properties. Aust. J. Exp. Agric. 37:437–439.

Hamza, M.A., and W.K. Anderson. 2005. Soil compaction in cropping systems. A review of the nature, causes and possible solutions. Soil Tillage Res. 82:121–145.

Hamilton, E.W., and D.A. Frank. 2001. Can plants stimulate soil microbes and their own nutrient supply? Evidence from a grazing tolerant grass. Ecology 82:2397–2402.

Haygarth, P.M., P.J. Chapman, S.C. Jarvis, and R.V. Smith. 1998. Phosphorus budgets for two contrasting grassland farming systems in the UK. Soil Use Manage. 14:160–167.

Haynes, R.J., and P.H. Williams. 1993. Nutrient cycling and soil fertility in the grazed pasture ecosystem. Adv. Agron. 49:119–199.

Henderson, D.C., B.H. Ellert, and M.A. Naeth. 2004. Grazing and soil carbon along a gradient of Alberta rangelands. Rangeland Ecol. Manag. 57:402–410.

Herrick, J.E., and R. Lal. 2005. Soil physical property changes during dung decomposition in a tropical pasture. Soil Sci. Soc. Am. J. 59:908–912.

Hutchings, N.J., J.E. Olesen, B.M. Petersen, and J. Berntsen. 2007. Modelling spatial heterogeneity in grazed grassland and its effects on nitrogen cycling and greenhouse gas emissions. Agric. Ecosyst. Environ. 121:153–163.

Jackson, R.B., J.L. Banner, E.G. Jobbagy, W.T. Pockman, and D.H. Wall. 2002. Ecosystem carbon loss with woody plant invasion of grasslands. Nature 418:623–626.

Lagocki, H.F.R. 1978. Surface soil stability and controlled by a drainage criterion. p. 233–237. In W.W Emerson, R.D. Bond, and A.R. Dexter (ed.) Modification of soil structure. John Wiley & Sons, Hoboken.

Lantinga, E.A., J.A. Keuning, J. Groenwold, and P. Deenen. 1987. Distribution of excreted nitrogen by grazing cattle and its effect on sward quality, herbage production and utilization. p. 103–117. In H.G. van der Meer, R.J. Unwin, T.A. van Dijk, and G.C. Ennik (ed.) Animal manure on grassland and fodder crops. Fertilizer or waste? Martinus Nijhoff, Dordrecht.

Lavado, R.S., and M.A. Taboada. 1987. Soil salinization as an effect of grazing in a native grassland soil in the flooding Pampa of Argentina. Soil Use Manage. 4:143–148.

Lavado, R.S., J.O. Sierra, and P. Hashimoto. 1996. Impact of grazing on soil nutrients in a Pampean grassland. J. Range Manage. 49:452–457.

Leão, T.P., A.P. da Silva, M.C.M. Macedo, S. Imhoff, and V.P.B. Euclides. 2006. Least limiting water range: A potential indicator of changes in near-surface soil physical quality after the conversion of Brazilian Savanna into pasture. Soil Tillage Res. 88:279–285.

Li, C., X. Hao, M. Zhao, G. Han, and W.D. Willms. 2008. Influence of historic sheep grazing on vegetation and soil properties of a desert steppe in Inner Mongolia. Agric. Ecosyst. Environ. 128:109–116.

Liao, C., R. Peng, Y. Luo, X. Zhou, X. Wu, C. Fang, J. Chen, and B. Li. 2008. Altered ecosystem carbon and nitrogen cycles by plant invasion: A meta-analysis. New Phytol. 177:706–714.

Marrs, R.H., A. Rizand, and A.F. Harrison. 1989. The effects of removing sheep grazing on soil chemistry, above-ground nutrient distribution, and selected aspects of soil fertility in long-term experiments at Moor House National Nature Reserve. J. Appl. Ecol. 26:647–661.

Martínez, L.J., and J.A. Zinck. 2004. Temporal variations of soil compaction and deterioration of soil quality in pasture areas of Colombian Amazonia. Soil Tillage Res. 75:3–17.

Milchunas, D.G., and W.K. Lauenroth. 1993. Quantitative effects of grazing on vegetation and soils over a global range of environments. Ecol. Monogr. 63:327–366.

Mohamed Saleem, M.A. 1998. Nutrient balance pattern in African livestock systems. Agric. Ecosyst. Environ. 71:241–254.

Mulholland, B., and M.A. Fullen. 1991. Cattle trampling and soil compaction on loamy sands. Soil Use Manage. 7:189–193.

Müller, B., K. Frank, and C. Wissel. 2007. Relevance of rest periods in non-equilibrium rangeland systems–A modeling analysis. Agric. Syst. 92:295–317.

Mullins C.E. and A. Fraser. 1980. Use of the drop penetrometer on undisturbed and remolded soils at a range of soil water tensions. J. Soil Sci. 31:25–32.

Nash, M.S., E. Jackson, and W.G. Whitford. 2003. Soil microtopography on grazing gradients in Chihuahuan desert grasslands. J. Arid Environ. 55:181–192.

Neff, J.C., R.L. Reynolds, J. Belnap, and P. Lamothe. 2005. Multi-decadal impacts of grazing on soil physical and biogeochemical properties in southeast Utah. Ecol. Appl. 15:87–95.

Oenema, O., T.V. Vellinga, and H. Van Keulen. 2006. Nutrient management under grazing. p. 63–83. In J. D. Elgersma, J. Dijkstra, and S. Tamminga (ed.). Fresh herbage for dairy cattle. Springer, Dordrecht.

Oesterheld, M., C.M. DiBella, and H. Kerdiles. 1998. Relation between NOAA-AVHRR satellite data and stocking rate of rangelands. Ecol. Appl. 8:207–212.

Oesterheld, M., J. Loreti, M. Semmartin, and J.M. Paruelo. 1999. Grazing, fire, and climate effects on primary productivity of grasslands and savannas. p. 287–306. In L.R. Walker (ed.) Ecosystems of disturbed ground. Ecosystems of the world 16. Elsevier, Amsterdam.

Parton, W.J., D.S. Schimel, C.V. Cole, and D.S. Ojima. 1987. Analysis of factors controlling soil organic matter levels in Great Plains grasslands. Soil Sci. Soc. Am. J. 51:1173–1179.

Parton, W.J., J.W.B. Stewart, and C.V. Cole. 1988. Dynamics of C, N, P and S in grassland soils: A model. Biogeochemistry 5:109–131.

Pietola, L., R. Horn, and M. Yli-Halla. 2005. Effects of trampling by cattle on the hydraulic and mechanical properties of soil. Soil Tillage Res. 82:99–108.

Piñeiro, G., J.M. Paruelo, and M. Oesterheld. 2006. Potential long-term impacts of livestock introduction on carbon and nitrogen cycling in grasslands of Southern South America. Glob. Change Biol. 12:1267–1284.

Pires da Silva, A., S. Imhoff, and M. Corsi. 2003. Evaluation of soil compaction in an irrigated short-duration grazing system. Soil Tillage Res. 70:83–90.

Quiroga, A., R. Fernández, and E. Noellmeyer. 2009. Grazing effect on soil properties in conventional and no-till Systems. Soil Tillage Res. 105:164–170.

Ritchie, M.E., D. Tilman, and J.M.H. Knops. 1998. Herbivore effects on plant and nitrogen dynamics in oak savanna. Ecology 79:165–177.

Rostagno, C.M. 1989. Infiltration and sediment production as affected by soil surface conditions in a shrubland of Patagonia, Argentina. J. Range Manage. 42:382–385.

Rubio, G., and R.S. Lavado. 1990. Grazing management effects on the bulk density of a Natraqualf (In Spanish, with English abstract). Cienc. Suelo 8:79–82.

Sala, O.E., W.K. Lauenroth, and I.C. Burke. 1996. Carbon budgets of temperate grasslands and the effects of global change. p. 101–119. In A.I. Breymeyer, D.O. Hall, J.M. Melillo, and G.I. Ågren (ed.) Global change: Effects on coniferous forests and grasslands. John Wiley & Sons, Hoboken.

Sala, O.E., M. Oesterheld, A. Soriano, and R.J.C. León. 1986. Grazing effect upon plant community structure in subhumid grasslands of Argentina. Vegetatio 67:27–32.

Savory, A., and S.D. Parsons. 1980. The Savory grazing method. Rangelands 2:234–237.

Schlesinger, W.H., J.F. Reynolds, G.L. Cunningham, L.F. Huenneke, W.M. Jarrell, R.A. Virginia, and W.G. Whitford. 1990. Biological feedbacks in global desertification. Science 247:1043–1048.

Scholefield, D., and D.M. Hall. 1986. A recording penetrometer to measure the strength of soil relation to the stresses exerted by a walking cow. J. Soil Sci. 37:165–172.

Scholefield, D., P.M. Patto, and D.M. Hall. 1985. Laboratory research on the compressibility of four topsoils from grassland. Soil Tillage Res. 6:1–16.

Schuman, G.E., J.D. Reeder, J.T. Manley, R.H. Hart, and W.A. Manley. 1999. Impact of grazing management on the carbon and nitrogen balance of a mixed-grass rangeland. Ecol. Appl. 9:65–71.

Semmartin, M.M.R.A., R.A. Distel, A.S. Moretto, and C.M. Ghersa. 2004. Litter quality and nutrient cycling affected by grazing-induced species replacements along a precipitation gradient. Oikos 107:148–161.

Semmartin, M., L.A. Garibaldi, and E.J. Chaneton. 2008. Grazing history effects on above- and below-ground litter decomposition and nutrient cycling in two co-occurring grasses. Plant Soil 303:177–189.

Sharpley, A., and B. Moyer. 2000. Phosphorus forms in manure and compost and their release during simulated rainfall. J. Environ. Qual. 29:1462–1469.

Singleton, P.L., and B. Addison. 1999. Effect of cattle treading on physical properties of three soils used for dairy farming in the Waikato, North Island, New Zealand. Aust. J. Soil Res. 37:891–902.

Siri-Prieto, G., D. Wayne Reeves, and R.L. Raper. 2007. Tillage systems for a cotton-peanut rotation with winter-annual grazing: Impacts on soil carbon, nitrogen and physical properties. Soil Tillage Res. 96:260–268.

Snyman, H.A., and C.C. du Preez. 2005. Rangeland degradation in a semi-arid South Africa-II: Influence on soil quality. J. Arid Environ. 60:483–507.

Steffens, M., A. Kölbl, K.U. Totsche, and I. Kögel-Knabner. 2008. Grazing effects on soil chemical and physical properties in a semiarid steppe of Inner Mongolia (P.R. China). Geoderma 143:63–72.

Steinfeld, H., and T. Wassenaar. 2007. The role of livestock production in carbon and nitrogen cycles. Annu. Rev. Environ. Resour. 32:271–294.

Stohlgren, T.J., L.D. Schell, and B. Vanden Heuvel. 1999. How grazing and soil quality affect native and exotic plant diversity in Rocky Mountain grasslands. Ecol. Appl. 9:45–64.

Taboada, M.A., and R.S. Lavado. 1988. Grazing effects of the bulk density in a Natraquoll of Argentina. J. Range Manage. 41:502–505.

Taboada, M.A., and R.S. Lavado. 1993. Influence of trampling on soil porosity under alternate dry and ponded conditions. Soil Use Manage. 9:139–143.

Taboada, M.A., R.S. Lavado, H. Svartz, and A.M.L. Segat. 1999. Structural stability changes in a grazed grassland Natraquoll of the Flooding Pampa of Argentina. Wetlands 19:50–55.

Taboada, M.A., R.S. Lavado, G. Rubio, and D.J. Cosentino. 2001. Soil volumetric changes in natric soils caused by air entrapment following seasonal ponding and water table. Geoderma 101:49–64.

The Forage and Grazing Terminology Committee. 1991. Terminology for Grazing Lands and Grazing Animals. Pocahontas Press, Inc. Blacksburg, Virginia.

Tongway, D.J., A.D. Sparrow, and M.H. Friedel. 2003. Degradation and recovery processes in arid grazing lands of central Australia. Part 1: Soil and land resources. J. Arid Environ. 55:301–326.

Trimble, S.W., and A.C. Mendel. 1995. The cow as a geomorphic agent—A critical review. Geomorphology 13:233–253.

Van Haveren, B.P. 1983. Soil bulk density as influenced by grazing intensity and soil type on a shortgrass prairie site. J. Range Manage. 36:586–588.

Villamil, M.B., N.M. Amiotti, and N. Peinemann. 2001. Soil degradation related to overgrazing in the semi-arid southern caldenal area of Argentina. Soil Sci. 166:441–452.

Wardle, D.A., and R.D. Bardgett. 2004. Human-induced changes in large herbivorous mammal density: The consequences for decomposers. Front. Ecol. Environ. 2:145–153.

Wardle, D.A. 2005. How plant communities influence decomposer communities. p. 119–138. *In* R.D. Bardgett, M.B. Usher, and D.W. Hopkins (ed.) Biological diversity and function in soils. Cambridge Univ. Press, Cambridge, UK.

Warren, S.D., M.B. Nevill, W.H. Blackburn, and N.E. Garza. 1986. Soil response to trampling under intensive rotation grazing. Soil Sci. Soc. Am. J. 50:1336–1340.

West, C.P., A.J. Mallarino, W.F. Wedin, and D.B. Marx. 1989. Spatial variability of soil chemical properties in grazed pastures. Soil Sci. Soc. Am. J. 53:784–789.

Willatt, S.T., and D.M. Pullar. 1983. Changes in soil physical properties under grazed pastures. Aust. J. Soil Res. 22:343–348.

Yong-Zhong, S., L. Yu-Lin, C. Jian-Yuan, and Z. Wen-Zhi. 2005. Influences of continuous grazing and livestock exclusion on soil properties in a degraded sandy grassland, Inner Mongolia, northern China. Catena 59:267–278.

Young, I.M., and K. Ritz. 2005. The habitat of soil microbes. p. 31–41. *In* R.D. Bardgett, M.B. Usher, and D.W. Hopkins (ed.) Biological diversity and function in soils. Cambridge Univ. Press, Cambridge, UK.

21

The Use of Cover Crops to Manage Soil

T.C. Kaspar and J.W. Singer

Cover crops are used to manage soils for many different reasons and are known by many different names. Cover crops are literally "crops that cover the soil" and one of their first uses was to reduce soil erosion during fallow periods in annual cropping systems. Cover crops are also known as "green manures," "catch crops," or "living mulch." Green manure cover crops are usually legumes that fix N and are grown to provide N to the following cash crop. Catch crops are cover crops that are grown during fallow periods in cropping systems to take up nutrients, especially N, that would be lost if plants are not present. Lastly, living mulches are cover crops that are grown both during and after the cash crop growing season and are suppressed or managed to reduce their competition with the cash crop when it is growing. After the cash crop has matured and before it begins growing again, the living mulch is allowed to grow unhindered. One way to manage living mulches is to restrict them to the "fallow" spaces between crop rows. Orchards or vineyards are sometimes managed with living mulches, but it is also possible to incorporate living mulches into annual cropping systems. Thus, as can be seen from their many names and descriptions, cover crops can fulfill many soil management functions.

In terms of soil management, the basic premise for using cover crops is to reduce fallow periods and spaces in cropping systems. Natural ecosystems typically have some plants growing, covering the soil, transpiring water, taking up nutrients, fixing carbon, and supporting soil fauna for most of the time that the ground is not frozen. Agricultural cropping systems producing grain, oilseed, and fiber crops in temperate regions typically only have living plants for four to six months of the year and are fallow for the remaining six to eight months. Current planting and tilling practices often leave soil bare and exposed during fall, winter, and early spring. Some perennial cropping systems for nut or fruit crops (e.g., almonds and grapes) keep the spaces between rows fallow and tilled for extended periods. As a result of these fallow periods and fallow spaces in annual and perennial cropping systems, soil is left unprotected from erosive forces, nutrients and organic matter are lost or not replenished, runoff increases, soil fauna are stressed, and soil productivity diminishes. Thus, inserting cover crops into fallow periods or fallow spaces in cropping systems can accomplish multiple soil management goals. This discussion is not intended to be a comprehensive review and will focus on the general principles and evidence for using cover crops to manage soil erosion, runoff, soil nutrients, soil physical properties, soil water, soil organic carbon, soil chemical properties, and soil biology.

T.C. Kaspar (Tom.Kaspar@ars.usda.gov) and J.W. Singer (jeremy.singer@ars.usda.gov), USDA-ARS, National Laboratory for Agriculture and the Environment, 2110 University Boulevard, Ames, IA 50011.

doi:10.2136/2011.soilmanagement.c21

Erosion and Runoff

Reducing water erosion is one of the main reasons for growing cover crops (Langdale et al., 1991). Soils are generally more susceptible to erosion when they are not covered with the canopies of living plants or their plant residues. Annual crop plants, such as corn (*Zea mays* L.) and soybean [*Glycine max* (L.) Merr.], only provide significant canopy cover for four months or less each year. Additionally, crops such as soybean, cotton (*Gossypium hirsutum* L.), or corn harvested for silage often do not leave enough residues to fully protect the soil between harvest and development of the next crop canopy. Cover crops and their residues reduce erosion through the same mechanisms as the cash crops. However, when cover crops are grown in the fallow intervals between cash crops they extend the time the soil is covered with living plants and also supplement and anchor residues left by annual crops.

The impact of cover crops on erosion processes depends on how much they reduce the forces of soil detachment and transport. Cover crops reduce interrill erosion primarily because they increase the amount and duration of soil cover either with living plants or plant residues. Soil cover is the principal characteristic of cropping systems that affects the amount of interrill erosion. Because interrill erosion results from the detachment of soil particles by raindrop impact, living or dead plant material that intercepts raindrops and dissipates impact energy will reduce interrill erosion. Ram et al. (1960) observed that soil detachment from raindrop impact was reduced as cover crop canopy increased either because of plant density or plant growth. Later, Laflen et al. (1985) showed that the relationship between surface cover and erosion reduction is exponential with smaller decreases in erosion as surface cover approaches 100%. This explains why the relative benefit of incorporating cover crops into a cropping system depends to some extent on the quantity, duration, and distribution of residues and plant canopies that are normally present in the cropping system throughout the year (Mutchler and McDowell, 1990). For example, Mutchler and McDowell (1990) found that the reduction of erosion by a hairy vetch (*Vicia villosa* Roth)–wheat (*Triticum aestivum* L.) winter cover crop mixture

was much greater in tilled systems than in no-till systems and much greater in no-till following a soybean crop than following a soybean–wheat double crop. Kaspar et al. (2001), however, showed that oat (*Avena sativa* L.) or rye (*Secale cereale* L.) cover crops in no-till soybean reduced interrill erosion in two of three rainfall simulator trials even though residue cover did not increase significantly with cover crops and was already greater than 75% without cover crops. They hypothesized that the decreases in interrill erosion with cover crops was caused by reduced interrill transport of sediments. They observed that the anchoring of cover crop plants or residues to the soil by roots resulted in the formation of microdams, which probably resulted in sediment deposition. Lattanzi et al. (1974) also observed that increasing amounts of surface residues reduced interrill erosion by intercepting splash transport of sediment, slowing interrill flow velocity, and increasing water film depth behind residue microdams.

Whereas interrill erosion is largely dependent on raindrop impact to detach soil particles, rill erosion relies on the shear force of water flowing in concentrated flow paths to both detach and transport soil particles (Flanagan, 2002). Cover crops can reduce rill erosion by reducing the shear force of flowing water or by increasing the resistance of soil particles to detachment. One way cover crops reduce the shear force of runoff water is by reducing its volume through increased infiltration. This occurs because cover crops prevent surface sealing, increase storage capacity, and improve soil structure (Dabney, 1998). Additionally, cover crops or surface residues (Brown and Norton, 1994) can slow flow velocity at the surface by increasing hydraulic resistance. Lastly, because cover crop plants or residues are anchored to the surface by roots and because they hold other unanchored surface residues in place (Kaspar et al., 2001), flowing water cannot easily move residues and expose the soil surface to shear forces of water (Foster et al., 1982).

Cover crops also reduce both rill and interrill erosion by increasing soil resistance to detachment. Cover crops are known to increase soil organic matter near the soil surface (Wander et al., 1994), which in turn should result in larger, more stable aggregates that are less susceptible to detachment (Dabney, 1998). Additionally, cover crop roots can physically bind aggregates together, which makes them even more resistant to detachment by flowing water or raindrop impact. Mamo and Bubenzer (2001) confirmed that plant roots substantially reduced soil detachment and rill erodibility.

A number of field studies have measured significant erosion reductions with cover crops. In Missouri, Zhu et al. (1989) compared soil erosion of no-till soybean with winter cover crops with erosion of no-till soybean without cover crops under natural rainfall. Annual soil loss was decreased 87%, 95%, and 96% by chickweed (*Stellaria media* L.), Canada bluegrass (*Poa compressa* L.), and downy brome (*Bromus tectorum* L.) winter cover crops, respectively, compared with no cover crops. Apparently, the no-till soybean crop at this location did not produce enough residues to adequately protect the soil. In another Missouri study (Wendt and Burwell, 1985), a winter rye or winter wheat cover crop reduced the annual soil loss of no-till corn grown for silage from 22 Mg ha^{-1} to 0.9 Mg ha^{-1} and annual runoff from 245 to 122 mm. In this same study, however, no-till corn without silage removal had less soil erosion than no-till corn with silage removal and a cover crop, but both of these treatments had less erosion than corn grown with either moldboard plowing or field cultivating. In Mississippi, Mutchler and McDowell (1990) found that a wheat or hairy vetch cover crop following cotton reduced annual soil loss of conventional-tilled cotton from 74.2 Mg ha^{-1} to 20.4 Mg ha^{-1} and that of no-till cotton from 19.2 Mg ha^{-1} to 2.3 Mg ha^{-1} when cotton followed soybean. Lastly, a 3-yr rainfall simulator study in Iowa (Kaspar et al., 2001) showed that winter rye cover crops overseeded into no-till soybean in late summer reduced interrill erosion the following spring by 54% and rill erosion by

90% compared with no-till without cover crops. Oat cover crops, which winter-kill in Iowa, reduced interrill and rill erosion by 26% and 65%, respectively. Neither rye nor oat cover crops significantly increased surface cover, which was already greater than 75% for the no-till soybean without cover crops. Because of the high residue cover, the quantity of interrill erosion was relatively low even for no-till without cover crops. Additionally, the rye cover crop increased infiltration in only 1 of the 3 yr. In spite of this, there were substantial reductions in rill erosion. Kaspar et al. (2001) observed that the cover crops prevented soybean residues from moving or dislodging with surface water flow. As a result, the soil surface was not exposed to the shear force of flowing water and this was partly responsible for the cover crops' success in reducing rill erosion.

Phosphorus Losses in Runoff

Losses of P from agricultural systems to surface waters are largely dependent on the amount of surface runoff and sediment transport that occurs. Phosphorus is transported in runoff as soluble P and particulate P (Sharpley and Smith, 1991). Particulate P consists of P bound to soil sediment and P contained in organic matter. Sharpley and Smith (1991) summarized research on the effect of cover crops on P losses and found that reductions in total P losses, which consist mostly of particulate P, ranged from 54% to 94% (Table 21|1). This is not surprising considering that cover crops reduce runoff and sediment detachment and transport. They also pointed out, however, that the effects of cover crops on soluble P in runoff were more variable (Table 21|1). Including cover crops in a cropping system sometimes resulted in higher concentrations of soluble P in runoff

Table 21|1. Literature summary of percent reduction (–) or increase (+) in total P, soluble P concentration, or soluble P in runoff due to winter cover crops (adapted from Sharpley and Smith, 1991).

Reference	Location	Cover crop	Change in total P losses in runoff	Change in soluble P concentration in runoff	Change in soluble P in runoff
Angle et al. (1984)	Maryland	Barley	–92%	+460%	–13%
Langdale et al. (1985)	Georgia	Rye	–66%	+54%	+8%
Pesant et al. (1987)	Quebec	Alfalfa/timothy	–94%	–60%	–12%
Yoo et al. (1988)	Alabama	Wheat	–54%	0%	–50%

and did not always reduce the cumulative amount of soluble P in runoff compared with no cover crops. Several studies have shown that soluble P can be lost in runoff flowing over plant residues (Timmons et al., 1970; Bechmann et al., 2005). However, on an annual basis, plant water use and infiltration would be expected to increase with cover crops, which should reduce the volume of runoff. This would offset to some extent the higher soluble P concentrations of runoff in the presence of cover crops.

Soil Physical Properties

Soil structure in simple terms is the physical relationship between the solid, liquid, and gaseous phases of soil. The arrangement of soil particles into peds or aggregates determines the size and shape of soil voids or pores and this greatly influences the movement of water and gases in soil. As a result, soil structure can have a substantial impact on plant growth. Conversely, cover crops, like any plants, can alter soil physical and structural properties directly through formation of pores and aggregates by roots or indirectly through the input and decomposition of shoot and root residues. Obviously, a cover crop's impact on soil structure will vary depending on climate, soil type, soil texture, soil depth, tillage, cropping system, cover crop biomass, cover crop species, and cover crop frequency in the crop rotation. For example, Ball-Coelho et al. (2000) reported that in a 3-yr corn–soybean–winter wheat rotation, cover crops after winter wheat and corn had a small effect on stability of micro-aggregates and no effect on dry-aggregate size distribution or wet-aggregate stability of macroaggregates in the upper 0.075 m of a sandy soil with conventional, chisel, and no-till tillage systems. They concluded that because microaggregates were more stable than macroaggregates with cover cropping, the binding mechanisms probably involved humic materials or microbial products, which likely were not as important for macroaggregate stability (Degens et al., 1996). Alternately, Dapaah and Vyn (1998) reported an increase in wet aggregate stability in the upper 0.07 m of a sandy loam and loam soils after only one cycle of annual ryegrass (*Lolium multiflorum* Lam.), red clover (*Trifolium pratense* L.), and oilseed radish

(*Raphanus sativus* L.) cover crops. Annual ryegrass maintained greater wet aggregate stability than the other cover crops at samplings from May through September presumably because of more persistent aggregate binding that led to less aggregate breakdown (Dapaah and Vyn, 1998). Tisdall and Oades (1979) reported that perennial ryegrass (*Lolium perenne* L.) was more efficient than white clover (*Trifolium repens* L.) in stabilizing aggregates in a loam soil because perennial ryegrass supported a larger population of vesicular-arbuscular mycorrhizal hyphae and had greater hyphal and root length. Results from their electron micrographs revealed hyphae covered with an amorphous material, which they conjectured was a polysaccharide capable of binding clay particles. These studies demonstrate that cover crop impacts on soil aggregation vary with cover crop species, quantity of roots, soil type, and cropping systems.

Patrick et al. (1957) reported that after 25 yr of continuous cotton with tillage on a loam soil, 21.3% of the aggregates had diameters >0.21 mm when a hairy vetch cover crop was included in the cropping system compared with the 11.8% for a common vetch (*Vicia sativa* L.) cover crop treatment and 9.5% for a control without a cover crop. The authors concluded that hairy vetch improved aggregation better than common vetch because hairy vetch produced more biomass than common vetch. The hairy vetch treatment also had a lower bulk density, greater porosity, and greater water holding capacity than the no cover control in the surface 0.06 m of soil. Benoit et al. (1962) reported greater aggregation and hydraulic conductivity at the surface of a sandy loam soil after 3 yr of using a rye cover crop in a sweet corn and green bean (*Phaseolus vulgaris* L.) rotation with spring tillage. In one of their treatments they removed all the cover crop shoot material before spring tillage, demonstrating that much of the effect of the rye cover crop on soil structure was due mostly to the cover crop roots. To further investigate the effect of cover crop roots they made measurements below the plow layer (0.30–0.37 m) in the sixth year of the study and showed that cover crop roots decreased bulk density and increased capillary porosity and hydraulic conductivity relative to the no cover control. Williams and Weil (2004) also presented evidence that cover crop roots can improve subsequent cash crop root growth in silt loam soils with compacted plowpans by increasing macroporosity in the compacted soil layers. In their study, they used the minirhizotron camera technique to observe soybean roots growing through the plowpans in root channels created by decomposing cover crop roots of forage radish (*Raphanus sativus* L.).

Soil Water Status

Cover crops decrease soil water content through uptake and transpiration while living and can increase soil water content through increased surface residue cover and infiltration after termination (Wagger and Mengel, 1988; Unger and Vigil, 1998; Qi and Helmers, 2010). Thus, the relative impact of cover crops on soil water available to the following crop depends on cover crop management, timing and amount of precipitation, and total water holding capacity of the root-accessible portion of the soil profile (Frye et al., 1988). Cover crop transpiration and soil drying may be beneficial on heavy soils in wet springs because it may allow for earlier planting of the cash crop (Wagger and Mengel, 1988). In fields with subsurface drainage systems, cover crops can reduce drainage volume during the spring before cash crops are planted (Qi and Helmers, 2010), which can reduce losses of nitrate (Kanwar et al., 2005). Alternately, in dry years on coarse-textured soils with low water holding capacity or shallow rooting depth, water use by cover crops may reduce soil water available at planting and may ultimately reduce yields if rainfall does not replenish the water used by the cover crop (Campbell et al., 1984). Once a cover crop has been terminated, however, cover crop residues left on the soil surface can increase surface residue cover, reduce evaporation, and increase soil water contents (Wagger and Mengel, 1988).

Because cover crops increase transpiration when living and decrease evaporation when dead, there have been conflicting reports on the effect of cover crops on available soil water and cash crop yields. Campbell et al. (1984) observed that a rye cover crop terminated with herbicides after corn planting substantially reduced the soil water content of the upper 0.60 m of the soil profile and reduced corn grain yield on the Coastal Plain in South Carolina. Similarly, Ewing et al. (1991) reported lower soil water contents in the upper 0.15 m and lower corn yields following a crimson clover (*Trifolium incarnatum* L.) cover crop in North Carolina on a soil with a root-restricting soil layer. In both of these studies, the negative impact of water use by the cover crop was exaggerated by periods of little or no precipitation and the limited water holding capacity of the rooted soil volume. In contrast to these studies, Moschler et al. (1967) in Virginia observed greater corn yields and soil water contents when rye cover crop residues were left on the soil surface compared with no cover crop, removal of rye residues, or burying rye residues with tillage. Lastly, Clark et al. (1997) in Maryland found that rye, hairy vetch, and rye–hairy vetch mixture cover crops did not significantly deplete soil water contents in the upper 0.20 m of the soil. After cover crop termination, residues left on the soil surface conserved soil water later in the growing season and contributed to corn yield increases. They also showed that the greater residue cover produced by cover crops terminated closer to the time of corn planting conserved more soil water during corn growth than cover crops terminated earlier. Obviously, the impact of cover crops on soil water content is complicated. From a soil water management perspective, the decision of when to terminate a cover crop depends in part on availability of irrigation, probability of rainfall to replace the water used by the cover crop, and the need for cover crop residues to reduce evaporation from the soil surface. Additionally, unless irrigation is available, cover crops may not be suitable for cropping systems in semiarid regions where annual precipitation sometimes does not replace the water used by the previous cash crop or where the probability for replenishing water used by the cover crop is low (Unger and Vigil, 1998). Obviously, farmers will need location-based guidelines or decision-aide tools to assist them in managing soil water with cover crops.

Nitrogen

Cover crops can be utilized to manage N in agricultural soils by altering N cycling and availability. Cover crops grown during fallow periods in cropping systems change the annual patterns of N uptake and mineralization, reduce downward movement of NO_3, retrieve NO_3 from deep soil layers, and fix atmospheric N_2, if the cover crops are legumes. Ultimately, successful management of N using cover crops requires that N availability be synchronized so that inorganic N is readily available during periods of active uptake by cash crops and minimally available during periods when cash crops are not growing to reduce losses of N to air and water. Various aspects of the relationship between cover crops and N have been discussed in a number of previous reviews (Meisinger et al., 1991; Wagger et al., 1998; Dabney et al., 2001; Thorup-Kristensen et al., 2003). This discussion will examine how cover crops reduce losses of N from soil and affect N availability to cash crops.

Soil N, in the form of nitrate (NO_3), is soluble in water and can be lost from agricultural cropping systems with the downward movement of water through the soil profile. In many agricultural fields, percolating water and NO_3 are intercepted by agricultural drainage systems, which rapidly transport water and NO_3 offsite to surface waters. In fields without drainage systems, percolating water and NO_3 can eventually reach groundwater by continuing downward or reach surface waters by following subsurface flow pathways. The presence of living plants can dramatically reduce leaching losses of NO_3 in two ways: (i) by taking it up, which reduces its concentration in the soil solution, and (ii) by taking up water, which reduces the amount of water moving through the soil profile. Large leaching losses of NO_3 occur in many cropping systems, in part because there are extended fallow periods during each year when living plants are not removing NO_3 and water from the soil, usually from cash crop maturity in fall until crop canopy development the following spring. Cover crops reduce annual leaching losses of NO_3 because they extend the period of active N and water uptake to periods of the year when the cash crops are normally not present. For example, reductions in NO_3 leaching losses observed with winter cover crops range from 6 to 94% (Table 21|2).

The wide range in cover crop NO_3 leaching reductions reported resulted from differences in cover crop species, the amount of cover crop growth, the amount of N in the soil due to either fertilization or mineralization, and the amount of water moving through the soil. For example, non-legume cover crops usually reduce NO_3 leaching losses more than legumes (Tonitto et al., 2006). McCracken et al. (1994) in Kentucky observed that a rye cover crop reduced leaching losses by 94% whereas

Table 21|2. Literature summary of percent reduction in nitrate N leaching losses due to winter cover crops (adapted in part from Meisinger et al., 1991).

Reference	Location	Cover crop	Reduction in N leaching
Jones, 1942	Alabama	Oats	81%
Jones, 1942	Alabama	Hairy vetch	6%
Chapman et al. 1949	California	Mustard	80%
Chapman et al. 1949	California	Purple vetch	30%
Martinez and Guirard, 1990	France	Ryegrass	63%
Staver and Brinsfield, 1990	Maryland	Rye	77%
Staver and Brinsfield, 1998	Maryland	Rye	80%
McCracken et al., 1994	Kentucky	Rye	94%
McCracken et al., 1994	Kentucky	Hairy vetch	48%
Wyland et al., 1996	California	Rye	65–70%
Brandi-Dohrn et al., 1997	Oregon	Rye	32–42%
Ritter et al., 1998	Delaware	Rye	30%
Rasse et al., 2000	Michigan	Rye	28–68%
Strock et al., 2004	Minnesota	Rye	13%
Kladivko et al., 2004	Indiana	Winter wheat + less fertilizer	61%
Kaspar et al., 2007	Iowa	Rye	61%

a hairy vetch cover crop reduced leaching by 48%. In other studies, combinations of factors have affected the reductions in NO_3 leaching. In a Minnesota study (Strock et al., 2004), which showed a 13% average reduction in NO_3 leaching, the rye cover crop was planted only following the corn phase in a corn–soybean rotation, produced on average 1.4 Mg ha^{-1} of shoot dry matter, and in 1 of the 3 yr there was almost no drainage or NO_3 loss. In years of low winter precipitation, less water moves downward through soil and more NO_3 remains in the soil profile even without a cover crop (Fig. 21|1; Thorup-Kristensen et al., 2003). In contrast to the Minnesota study, Kaspar et al. (2007) in Iowa reported that over 4 yr a winter rye cover crop following both phases of a corn–soybean rotation reduced NO_3 loads in tile drainage water by an average of 61% and produced an average shoot biomass of 1.7 Mg ha^{-1}. In the Iowa study, cumulative annual drainage was always greater than 138 mm and significant NO_3 losses occurred every year. Other studies have shown that combining a winter cover crop with other N management practices can also effectively reduce NO_3 losses. A study in Indiana (Kladivko et al., 2004) reduced NO_3 loads by 61% with a reduction in fertilizer N rates and a winter wheat cover crop following corn. Thus, when a winter cover crop produces moderate growth and substantial water percolation occurs, cover crops can substantially reduce NO_3 losses to drainage water or deep percolation in annual cropping systems.

Living cover crops also have the potential to reduce gaseous losses of N as N_2, NO, and N_2O, from soil (Davidson et al., 2000). Of these, N_2O is also important as a greenhouse gas. A number of studies have shown that plant roots can effectively compete with soil microorganisms for available NO_3 and

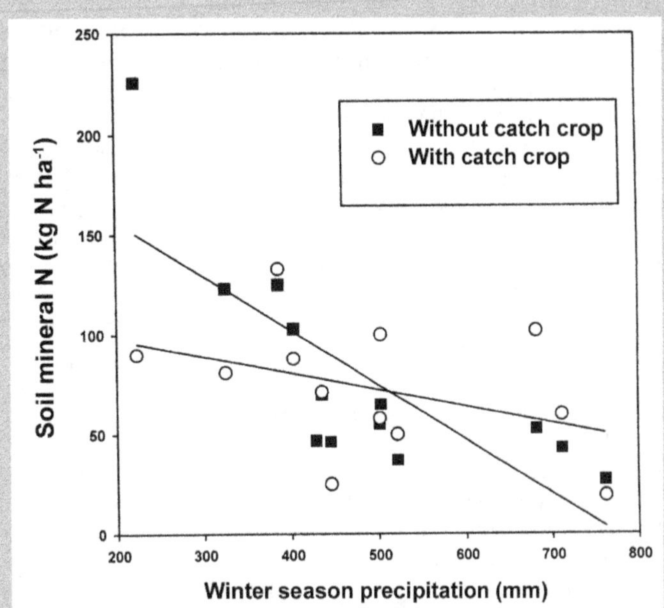

Fig. 21|1. Spring soil nitrate N content versus winter precipitation with and without cover crops from various experiments in Denmark (adapted from Thorup-Kristensen et. al., 2003).

reduce soil NO_3 concentrations, thus reducing its conversion by soil microorganisms to N_2O, NO, or N_2 during denitrification (Smith and Tiedje, 1979; Haider et al., 1987). Nitrous oxide and NO can also be lost during the aerobic nitrification process (Davidson et al., 2000). Because plants can also take up NH_4, they might also be expected to reduce gaseous losses by this aerobic process, but this has not been confirmed. In a controlled environment study, in which swine manure was injected into soil with or without a growing rye cover crop, Parkin et al. (2006) showed that N_2O emissions were significantly lower with a rye cover crop present. Thus, cover crops, like any living plants, have the potential to reduce gaseous losses of N through microbial reactions by reducing soil N availability (Davidson et al., 2000). After cover crops are dead, however, their residues can lose N through emissions of NH_3, N_2O, NO, or N_2. Quemada and Cabrera (1997) applied crimson clover residues to the surface of soil cores and measured maximum losses of NH_3 and N_2O of 6.0 and 2.6% of total residue N, respectively under very wet and warm conditions. Emissions of N_2O,

NO, or N_2 resulting from denitrification or nitrification of N mineralized from cover crop residues would be controlled by the same factors that control gaseous losses of fertilizer N: NO_3 or NH_4 availability, carbon availability, and the aeration status (Davidson et al., 2000; Rosecrance et al., 2000). For cover crop shoot residues, anaerobic conditions would be more likely to occur if residues are incorporated with tillage rather than left on the surface. Incorporation of legume cover crop residues with a low C to N ratio likely would increase net N mineralization, increase availability of both NH_4 and NO_3, and increase gaseous losses of N. The relative importance of denitrification or nitrification to the gaseous losses would depend on the aeration status of the soil (Davidson et al., 2000). Alternately, incorporation of grass cover crop shoot residues with a high C to N ratio likely would decrease net N mineralization, decrease availability of both NH_4 and NO_3, and decrease gaseous losses of N. For example, Rosecrance et al. (2000) observed that incorporated hairy vetch cover crop residues resulted in greater net N mineralization and greater N_2O losses than incorporated rye residues. On the other hand, soil incorporation of cover crop residues probably would make NH_3 emissions from residues less likely because NH_3 is soluble in water and NH_4 binds with soil particles.

Cover crops affect N availability to cash crops through uptake of inorganic N, fixation of N_2 by leguminous cover crops, and decomposition of cover crop residues. Interseeded cover crops or living mulches can directly reduce N availability for cash crops because they are growing and taking up N at the same time as the cash crop. Winter or off-season cover crops can also reduce soil N availability to the cash crop, if the N taken up by the cover crops still would have been present in the soil when the cash crop needed N. Nitrogen taken up by winter cover crops would not affect N availability for the cash crops, if the N taken up would have been lost through leaching or gaseous emissions or if it were replaced by mineralization or fertilization. For example, Thorup-Kristensen et al. (2003) used data from a number of experiments to show that NO_3–N remaining in the soil with or without cover crops was not different when winter precipitation exceeded 400 mm (Fig.

21|1). When winter precipitation was less than 400 mm, soil nitrate contents were much higher without a cover crop. Cover crop species also influences the amount of soil N taken up by the cover crop. In general, nonlegume cover crops take up more soil N than legume cover crops. Ranells and Wagger (1997a) observed that a rye cover crop recovered more [15]N (39%) than a crimson clover cover crop (4%). Shipley et al. (1992) reported that cereal rye and annual ryegrass took up 45 and 27%, respectively, of fall-applied ^{15}N, whereas hairy vetch and crimson clover only recovered 8%.

Cover crops can also have a positive impact on N availability by increasing the input of N into cropping systems through N_2 fixation by leguminous cover crops. Legumes will also take up N from the soil, but their growth is not limited by low soil N levels as are nonleguminous cover crops. Unfortunately, legumes generally do not grow rapidly in cool fall or winter weather and grow much better after the weather warms in the spring. Power and Zachariassen (1993) reported that there were considerable differences among eight legumes in N_2 fixation at soil temperatures of 10, 20, and 30°C and that for half of the legumes N_2 fixation at 10°C was much lower than that at the other temperatures. There is also considerable variation in legume cover crop growth and N_2 fixation among locations and cropping systems depending on climate, soils, planting date, and termination date. For example, Decker et al. (1994) evaluated three legume winter cover crops over seven environments in Maryland and on average the shoots of hairy vetch, Austrian winter pea (*Pisum sativum* L.), and crimson clover contained 152, 138, and 121 kg N ha^{-1}, respectively. In this same study, a winter wheat cover crop accumulated 39 kg N ha^{-1}, so we can probably assume that accumulation of N greater than 39 kg N ha^{-1} was probably the result of N_2 fixation. Hively and Cox (2001) in the colder environment of central New York, however, observed that four legume cover crop species produced only 2 to 35 kg N ha^{-1} over 2 yr. Whereas, Ebelhar et al. (1984) in Kentucky reported a 2-yr average of 209 kg N ha^{-1} in shoot biomass of a hairy vetch cover crop compared with 36 kg N ha^{-1} in a rye cover crop. Thus, in the appropriate cropping system and climate, legume winter cover crops can fix significant amounts of N.

In general, when fertilizer N is applied at planting it is more readily available and a greater proportion is recovered by annual crops than N contained in plant residues. For example, Harris et al. (1994) using ^{15}N showed that over 2 yr corn and barley crops took up 40% and 17% of the fertilizer and red clover ^{15}N, respectively. However, 47% of the red clover ^{15}N remained in the soil, whereas only 17% of the fertilizer ^{15}N was found in the soil. Availability to the cash crop of N taken up or fixed by a cover crop depends on the decomposition rate of cover crop residues. Residue decomposition is affected by many soil, plant, and environmental factors (Parr and Papendick, 1978). One of the main factors affecting decomposition of cover crop residues and release of N is the C to N ratio of the residues. Residues from grass or cereal cover crops have relatively high C to N ratios, decompose slowly, and in some cases immobilize NO_3 already present in the soil (Wagger, 1989b; Rosecrance et al., 2000). Legumes have lower C to N ratios than nonlegumes, decompose more quickly, and release or mineralize more N (Wagger, 1989a; Ruffo and Bollero, 2003). Using surface-applied ^{15}N-labeled cover crop residues Ranells and Wagger (1997a) showed that averaged over 2 yr corn recovered 14% of crimson clover N compared with only 4% of a rye cover crop N. The greater N availability of legume cover crops is also reflected in the N fertilizer response of corn following a rye or hairy vetch winter cover crop (Fig. 21|2; Miguez and Bollero, 2006). Miguez and Bollero (2006) showed that at the 0 kg N ha^{-1} fertilizer rate, the hairy vetch cover crop had greater corn yield than either the rye or no cover crop treatment indicating greater N availability. In contrast, the rye cover crop treatment had a lower corn yield than the no cover crop treatment at the 0 kg N ha^{-1} fertilizer rate, indicating that the rye cover crop probably immobilized soil N. In their study, increasing N fertilizer

eventually eliminated the yield difference between the rye and the no cover crop treatment, but not between the rye and hairy vetch treatments. They concluded that factors other than N availability were involved. Ranells and Wagger (1997b) have suggested planting grass and legume cover crop species in bicultures or mixtures to reduce the C to N ratio of the residues and increase the rate of residue decomposition and N release. Another approach to managing the C to N ratio of cover crops is to terminate the cover crop earlier in its life cycle. Wagger (1989a) found that terminating a rye, hairy vetch, or crimson clover cover crop 14 d earlier resulted in lower C to N ratios, lower concentrations of cellulose, hemicellulose, and lignin, faster decomposition, and more released N. In addition to changing the composition of cover crop residues, terminating a cover crop significantly before planting the cash crop allows more time for decomposition of cover crop residues and mineralization of residue N. Optimal synchronization of availability of mineralized residue N with uptake of N by the cash crop, however, is difficult to achieve.

Residue incorporation with tillage is another way to increase the rate of residue decomposition and N release. Tillage breaks or cuts cover crop residues into smaller pieces, covers residues with soil, and improves soil-to-residue contact. This

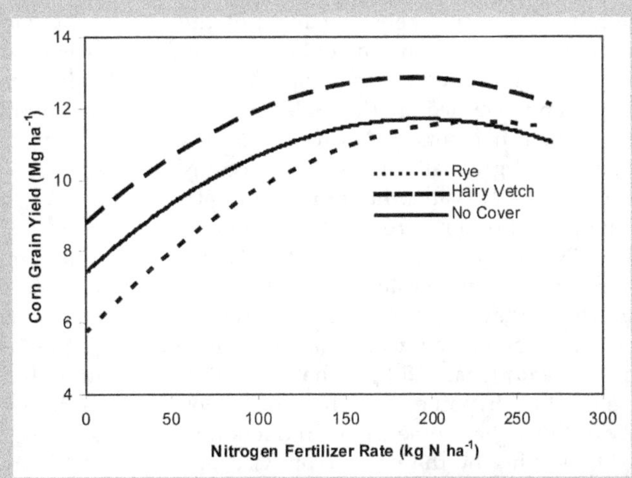

Fig. 21|2. Corn grain yield N fertilizer response curves for corn following no cover crop, a rye cover crop, or a hairy vetch cover crop in Illinois (adapted from Miguez and Bollero, 2006).

keeps residues wetter and more accessible to soil microorganisms. Varco et al. (1989) found that in the first year, corn recovered 32% of the N in ^{15}N-labeled hairy vetch residues when they were incorporated with tillage compared with only 20% when residues were left on the surface in no-till. Power et al. (1991) observed that with no added N fertilizer corn yielded substantially more when a hairy vetch cover crop was incorporated with tillage than when it was terminated with herbicide and remained on the surface. Some questions remain as to the eventual fate of all of the N in cover crop shoot residues when they are allowed to remain on the soil surface in no-till systems, but incorporation with tillage to increase availability also would negate some of the erosion control and water conservation benefits of cover crop residues.

Cover crops also impact long-term soil N availability by increasing total soil N through additions of fixed N or prevention of N losses. Few studies have looked at the cumulative effect of winter cover crops on total soil N when they have been planted every year for multiple years. Because only a portion of cover crop residue N is mineralized and taken up by the following cash crop, the remaining cover crop N may be susceptible to loss through leaching or gaseous emissions when it is mineralized (Harris et al., 1994). However, if cover crops are planted every year and fallow periods are avoided, it is likely that most of the cover crop N will continue to be taken up in later years either by cash crops or cover crops and recycled in the soil–plant system. Then, if inputs of N exceed outputs, this should gradually increase the total N content and N availability of the soil until a new equilibrium is reached. Garwood et al. (1999) calculated an estimated N balance for cropping systems over 8 yr that indicated that the treatments with winter rye cover crops had accumulated on average an additional 160 kg N ha^{-1} over that time. They hypothesized that the majority of this N had come from a reduction in NO$_3$ leaching by the cover crop. Gaseous losses of N, however, were not measured in this study and the change in soil organic matter was not significant. Kuo and Jellum (2000) measured an increase in soil organic N after 8 yr of rye, ryegrass, or hairy vetch cover crops in a corn silage system. Similarly, Sainju and Singh (2008) found that a hairy vetch cover crop increased total soil N over 3 yr in the 0.0- to 0.3-m soil layer. In many soils with high background levels of organic matter and total N, it is difficult to measure changes in total soil N due to changes in management practices. McCracken et al. (1989) was not able to measure a significant increase in total soil N, but did measure an increase in potential N mineralization and soil inorganic N levels in late May after 10 yr of hairy vetch cover crops. Hansen et al. (2000) observed that spring wheat required 27 kg N ha^{-1} less fertilizer to obtain the same yield after a perennial ryegrass winter cover crop was discontinued after 24 yr. When they terminated their study, this indirect evidence of increased N availability had persisted for 4 yr after the cover crop was discontinued. In some cases, however, improved soil N retention by long-term cover crops may not result in positive changes in soil N status, but instead may increase cash crop yield or grain N content, which would increase N outputs from the system (Ball Coelho et al., 2005). Thus, long-term cover crop use can change the N balance of cropping systems by reducing losses, by supplying fixed N, and by increasing total soil N. Eventually, as the system comes to equilibrium, N availability to the cash crop may increase. Further research examining N budgets, gaseous emissions, and changes in total soil N of cropping systems with cover crops are needed to better understand the cycling of N in these systems, which should improve soil productivity and management.

Soil Organic Carbon

Soil productivity is closely linked to soil organic matter (SOM) and its primary component soil organic carbon (SOC). Sequestration of C in SOM is also an important approach for reducing the concentration of CO$_2$ in the atmosphere (Lal et al., 1999). Soil organic matter, which includes soil humus and all the plant, animal, and microbial residues in the soil, is generally assumed to be 50 to 58% C by mass (Nelson and Sommers, 1996). In general, SOC increases when inputs of plant residue C to the soil are greater than C losses through decomposition, erosion, and leaching (Huggins and Fuchs, 1997; Paustian et al., 1997).

Because cover crops are normally grown during fallow periods of cropping systems, the addition of cover crops to a cropping system can increase total residue C inputs to soil and has the potential to increase SOC (Karlen and Cambardella, 1996; Lal et al., 1999; Jarecki and Lal, 2003). Similarly, the rate at which cover crop residues decompose also affects the balance between losses and inputs of C into soil. Cover crop residue decomposition depends primarily on temperature, water content, biochemical constituents, residue quantity, C to N ratio, and soil contact. Kuo et al. (1997) reported that SOC half-lives for rye, hairy vetch, and annual ryegrass were similar and averaged 31 d and 57 d in 2 yr when residues were buried 15 cm below the soil surface in mesh bags, before planting corn. They attributed the slower decomposition in 1 of the 2 yr to wetter soils and lower temperature. Although hairy vetch had lower shoot C to N ratios than ryegrass or rye, this did not affect the observed decay rate, which probably indicates that N was not limiting and that other factors such as lignin concentration of residues or environmental conditions limited the decomposition rate. In contrast, Ruffo and Bollero (2003) reported that hairy vetch decomposed more rapidly than rye and that the decomposition rate of both responded to water content and temperature.

Cover crops have been used successfully to increase soil C (Table 21|3) especially in locations with mild winters that allow substantial cover crop growth (Beale et al., 1955; Patrick et al., 1957; Utomo et al., 1990; Kuo et al., 1997; Nyakatawa et al., 2001; Sainju et al., 2002). In some of these studies, cover crop residues were incorporated with tillage (Beale et al., 1955; Patrick et al., 1957; Kuo et al., 1997; Sainju et al., 2002). As

mentioned earlier, however, soil incorporation or more extensive tillage can increase the relative rate of cover crop decomposition and reduce the retention of cover crop C in soil. Beale et al. (1955) reported that SOM was 28% higher after 10 yr of a vetch and rye cover crop with mulch tillage compared with cover crops and moldboard plow tillage. In spite of this evidence for increases in soil C and residue C inputs with cover crops, it is often difficult to measure a change in SOC in cropping systems to which cover crops have been added. This is partly because it is difficult to measure small changes in SOC in field soils with relatively high background SOC levels and large variations in SOC with depth and terrain (Kaspar et al., 2006). Additionally, cover crops may not produce large amounts of biomass in some locations or climates and cover crop biomass may be a relatively small percentage of the total biomass produced in some cropping systems like continuous corn. For example, Eckert (1991) in Ohio was not able to detect an increase in soil C with a rye cover crop in no-till continuous corn or corn–soybean rotations. Duiker and Hartwig (2004) reported similar SOC levels in a crownvetch (*Coronilla varia* L.) living mulch treatment and the control after 13 yr. They concluded that severe suppression of the crownvetch living mulch to reduce competition with the corn crop had also reduced the soil C inputs and the SOC benefits. Similarly, Utomo et al. (1990) observed no change in soil C with a rye cover crop in either no-till or conventional tillage, but measured an increase with a hairy vetch cover crop in no-till, which produced more biomass than rye in a 0 N treatment. Mendes et al. (1999) found that red clover or triticale (*×Triticosecale* Wittmack) winter cover crops did not increase soil C in a tilled vegetable production system.

Table 21|3. Cover crop increases in soil organic matter (SOM) or soil organic carbon (SOC).

Reference	Location	Cover crop	Depth of sample	Increase in SOM or SOC†
Beale et al. (1955)	South Carolina	Vetch and rye	0.13 m	31%
Patrick et al. (1957)	Louisiana	Hairy vetch	0.07 m	85%
Kuo et al. (1997)	Washington	Rye	0.30 m	7%†
Sainju et al. (2002)	Georgia	Rye	0.20 m	12%†
Villamil et al. (2006)	Illinois	Rye and hairy vetch	0.30 m	9%

† Soil organic carbon.

Soil Chemical Properties

Aside from the impacts of cover crops on C and N cycling in soil, little is known about the effect of cover crops on other nutrients or soil pH. Nyakatawa et al. (2001) reported no effect of a winter rye cover crop in a continuous cotton system on soil pH after two cycles compared with a winter fallow treatment. Eckert (1991) observed that a rye winter cover crop increased soil pH by a small amount in two of three cropping systems where corn followed rye and concluded that this was due to rye's assimilation of NH_4 and not a relocation of Ca. Eckert (1991) also reported that rye increased exchangeable K concentrations in the surface 0.05 m, presumably by accumulating K from the soil and depositing the K-containing shoot and root residues on the soil surface and in the upper 0.05 m. The rye cover crop did not significantly affect soil concentrations of C, P, Ca, or Mg in these studies.

Soil Biology

Cover crops increase the potential for macro- and microfaunal activity in soils because they increase the total inputs of organic material to soils, they increase the length of time each year that plants are growing and inputting C into the soil, and they moderate changes in soil temperature and water content by increasing surface cover. For example, Mele and Carter (1999) concluded that crop residues on the soil surface and no soil disturbance support higher densities and weight of some species of earthworms because of increased water content of surface layers, food sources at the surface, and retention of burrows. Reeleder et al. (2006) reported higher densities of earthworms, predominately *Aporrectodea turgida*, in 1 of 2 yr of their study after 8 yr of a rye cover crop. Averaged across tillage, the rye treatment had 33.3 worms m^{-2} compared with 12.8 worms m^{-2} in the no rye treatment. They postulated that earthworm populations may have been higher in the rye treatment because of increased availability of water. Also, in this same study populations of microarthropods were generally higher with a rye cover crop than without, but the total population of soil fungi was unaffected by cover crops. Ketterings et al. (1997)

artificially altered earthworm populations and determined that earthworms increased stability of aggregates >1.0 mm diameter following a cereal rye–hairy vetch cover crop. They also reported that earthworms exhibited a preference for organic materials high in N, increased the storage of C and N in aggregates, and facilitated the decomposition of coarse organic material deposited on the soil surface.

Cover crops can also increase the potential for microfaunal activity. Reddy et al. (2003) reported that after 3 yr with crimson clover or cereal rye cover crops soil had greater total bacterial and fungal propagule density and fluorescein diacetate hydrolytic activity (FDA) than the soil without a cover crop. The FDA assay is a measure of the soil enzyme esterase and is used as an indicator of microbial activity and biomass. In their study, the crimson clover cover crop had a greater stimulatory effect on soil biology than cereal rye. The authors speculated that the legume cover crop had more readily available amino acids and carbohydrates than the grass cover crops because of a lower C to N ratio. Lundquist et al. (1999) reported on the short-term (42 d) effects of cereal rye incorporation in contrasting vegetable management systems. Their results indicated that following rye incorporation, counts of active bacteria increased 24 to 52% in the first 7 d and populations of bacterial-feeding nematodes increased 400 to 600% between 7 and 14 d. Active fungal hyphal lengths and fungal-feeding nematodes were less responsive to rye incorporation during the 42-d period.

Considerable evidence exists regarding the control of weeds, plant pathogens, and nematodes by chemical substances released from cover crop plants or from their residues into the soil (Inderjit and Keating, 1999). Chemicals released during decomposition of cover crop residues may be released in their unaltered form or may be altered or transformed by soil microorganisms. As with many of the effects of cover crops on soil properties that have been discussed, these chemicals released into the soil are common to many plant species and not just cover crops. These substances are known as allelochemicals and they can have positive or negative effects on plants or soil organisms (Inderjit and Keating, 1999). These compounds are primarily secondary metabolites and include phenolics, terpenoids, alkaloids, steroids, polyacetylenes, and essential oils (Inderjit and Keating, 1999). Inderjit and Keating (1999), however, also reported that amino acids and organic acids have been shown to possess allelopathic potential, although clear effects on biological activity have not always been proven. In any case, cover crops offer unique opportunities for using allelochemicals to control weeds, pathogens, and nematodes in agricultural cropping systems because they can be grown during the fallow periods in these systems.

Cover crops that are members of the *Poaceae* and *Cruciferae* families have the potential to be used as biofumigants to control plant parasitic nematodes and soil fungi. Brassicas contain glucosinolates that undergo hydrolysis to produce compounds with broad biocidal activity (Brown and Morra, 1997). Kirkegaard and Sarwar (1998) quantified levels of glucosinolates in 65 brassica species and reported that total glucosinolates on a per unit area basis ranged from 0.8 to 45.3 mmol m^{-2} when sampled at the mid-flowering growth stage. They concluded that the variation in biomass and glucosinolate compounds and their concentrations in various plant parts provide opportunities for enhanced selection to increase their biofumigation potential. Sarwar et al. (1998) showed that the compounds released from Brassicas suppressed five cereal root pathogens in laboratory tests. Similarly, Potter et al. (1998) found that leaf tissue of a variety of Brassica species incorporated into soil was highly nematicidal. McBride et al. (2000) also found that incorporated cereal rye significantly suppressed nematode activity. They had hypothesized that low molecular weight organic acids were responsible for the nematode-suppressing qualities of rye and they detected organic acids in soil solution shortly after incorporation of fresh rye foliage. The rapid degradation of the organic acids in the soil, however, led them to conclude that these acids probably were not solely responsible for suppressing the nematodes.

Understanding the chemical mechanisms for weed suppression by cover crops is challenging. Barnes and Putnam (1986) in a greenhouse study reported that rye residues and their aqueous extracts lowered emergence and radicle elongation for several species and that shoot residues were more effective than root residues at inhibiting lettuce (*Lactuca sativa* L.) germination. Liebl et al. (1992) reported that cereal rye provided excellent weed control, with or without herbicides, and concluded that the weed suppression of the rye cover crop could have been caused by an allelopathic effect of decaying rye residues or the physical presence of the rye mulch on the soil surface or both. Similar to the range of allelochemical concentrations found among Brassica spp., Burgos et al. (1999) found that concentrations of the allelochemicals in rye DIBOA [2,4-dihydroxy-1,4-(2H)-benzoxazine-3-one] and BOA [2-(3H)-benzoxazolinone], both hydroxamic acids, ranged from 137 to 1469 µg g^{-1} dry tissue among eight cultivars grown in the field. They also reported data from a parallel greenhouse study that demonstrated that concentrations of hydroxamic acid peaked 60 d after planting rye. Although the allelochemicals produced by rye and other cover crops may provide beneficial suppression of weeds, these same chemicals may also be responsible for yield decreases of cash crops following cover crops. For example, both Johnson et al. (1998) and Raimbault et al. (1990) reported reductions in either grain or silage yields when corn was planted immediately after terminating a rye cover crop with herbicide. In later studies, Raimbault et al. (1991) showed that corn planted 14 d after rye cover crop termination produced 9% higher silage yields than corn planted immediately after rye was terminated. They hypothesized that one of many possible explanations could be that the extra time allowed any allelochemicals produced by the rye to be dissipated or denatured.

Summary

Inserting cover crops into fallow periods and spaces in cropping systems is a beneficial soil management practice. Cover crops can protect the soil from erosion, reduce losses of N and P, increase soil C, reduce runoff, inhibit pests, and support beneficial soil fauna. Although cover crops have the potential to maintain and enhance soil productivity and to reduce offsite impacts of N, P, sediment, and greenhouse gases (CO_2 and N_2O), they are not widely used in most agricultural systems (Singer et al.,

2007). Incorporating cover crops into cropping systems requires time, money, inputs, machinery, labor, and modifications of current practices without immediate financial return to the farmer. Improvements in soil productivity resulting from cover crops may require many years before benefits are detectable. Additionally, availability of cover crop management information, cover crop seed, and custom planting and spraying services need to be improved to facilitate adoption of cover crops in grain, oilseed, and fiber cropping systems. However, if farmers are given financial and productivity incentives to grow cover crops, farmers, crop consultants, and scientists should be able to overcome these management, supply, and service problems relatively quickly. Therefore, scientists need to demonstrate to farmers with long-term integrated studies that maintaining or enhancing soil productivity with cover crops provides long-term financial benefits. Additionally, scientists need to demonstrate to nonfarmers that providing incentives for widespread adoption of cover crops could provide valuable ecosystem services and improvement of water quality, while still maintaining current levels of food, biomass, and fiber production.

References

Angle, J.S., G. Mc Clung, M.S. Mc Intosh, P.M. Thomas, and D.C. Wolf. 1984. Nutrient losses in runoff from conventional and no-till corn watersheds. J. Environ. Qual. 13:431–435.

Ball-Coelho, B.R., R.C. Roy, and C.J. Swanton. 2000. Tillage and cover crop impacts on aggregation of a sandy soil. Can. J. Soil Sci. 80:363–366.

Ball Coelho, B.R., R.C. Roy, and A.J. Bruin. 2005. Long-term effects of late-summer overseeding of winter rye on corn grain yield and nitrogen balance. Can. J. Plant Sci. 85:543–554.

Barnes, J.P., and A.R. Putnam. 1986. Evidence for allelopathy by residues and aqueous extracts of rye (*Secale cereale*). Weed Sci. 34:384–390.

Beale, O.W., G.B. Nutt, and T.C. Peele. 1955. The effects of mulch tillage on runoff, erosion, soil properties, and crop yields. Soil Sci. Soc. Am. Proc. 19:244–247.

Bechmann, M.E., P.J.A. Kleinman, A.N. Sharpley, and L.S. Saporito. 2005. Freeze-thaw effects on phosphorus loss in runoff from manured and catch-cropped soils. J. Environ. Qual. 34:2301–2309.

Benoit, R.E., N.A. Willits, and W.J. Hanna. 1962. Effect of rye winter cover crop on soil structure. Agron. J. 54:419–420.

Brandi-Dohrn, F.M., M. Hess, J.S. Selker, R.P. Dick, S.M. Kauffman, and D.D. Hemphill, Jr. 1997. Nitrate leaching under a cereal rye cover crop. J. Environ. Qual. 26:181–188.

Brown, L.C., and L.D. Norton. 1994. Surface residue effects on soil erosion from ridges of different soils and formation. Trans. ASAE 37:1515–1524.

Brown, P.D., and M.J. Morra. 1997. Control of soil-borne plant pests using glucosinolate-containing plants. Adv. Agron. 61:167–231.

Burgos, N.R., R.E. Talbert, and J.D. Mattice. 1999. Cultivar and age differences in the production of allelochemicals by Secale cereale. Weed Sci. 47:481–485.

Campbell, R.B., D.L. Karlen, and R.E. Sojka. 1984. Conservation tillage for maize production in the U.S. Southeastern Coastal Plain. Soil Tillage Res. 4:511–529.

Chapman, H.D., G.F. Liebig, and D.S. Rayner. 1949. A lysimeter investigation of nitrogen gains and losses under various systems of covercropping and fertilization and a discussion of error sources. Hilgardia 19:57–95.

Clark, A.J., A.M. Decker, J.J. Meisinger, and M.S. McIntosh. 1997. Kill date of vetch, rye, and a vetch-rye mixture: II. Soil moisture and corn yield. Agron. J. 89:434–441.

Dabney, S.M. 1998. Cover crop impacts on watershed hydrology. J. Soil Water Conserv. 53:207–213.

Dabney, S.M., J.A. Delgado, and D.W. Reeves. 2001. Using winter cover crops to improve soil and water quality. Commun. Soil Sci. Plant Anal. 32:1221–1250.

Dapaah, H.K., and T.J. Vyn. 1998. Nitrogen fertilization and cover crop effects on soil structural stability and corn performance. Commun. Soil Sci. Plant Anal. 29:2557–2569.

Davidson, E.A., M. Keller, H.E. Erickson, L.V. Verchot, and E. Veldkamp. 2000. Testing a conceptual model of soil emissions of nitrous and nitric oxides. Bioscience 50:667–680.

Decker, A.M., A.J. Clark, J.J. Meisinger, F.R. Mulford, and M.S. McIntosh. 1994. Legume cover crop contributions to no-tillage corn production. Agron. J. 86:126–135.

Degens, B.P., G.P. Sparling, and L.K. Abbott. 1996. Increasing the length of hyphae in a sandy soil increases the amount of water-stable aggregates. Appl. Soil Ecol. 3:149–159.

Duiker, S.W., and N.L. Hartwig. 2004. Living mulches of legumes in imidazolinone-resistant corn. Agron. J. 96:1021–1028.

Ebelhar, S.A., W.W. Frye, and R.L. Blevins. 1984. Nitrogen from legume cover crops for no-tillage corn. Agron. J. 76:51–55.

Eckert, D.J. 1991. Chemical attributes of soils subjected to no-till cropping with rye cover crops. Soil Sci. Soc. Am. J. 55:405–409.

Ewing, R.P., M.G. Wagger, and H.P. Denton. 1991. Tillage and cover crop management effects on soil water and corn yield. Soil Sci. Soc. Am. J. 55:1081–1085.

Flanagan, D.C. 2002. Erosion. *In* R. Lal (ed.) Encyclopedia of soil science. Marcel Dekker, Inc., New York.

Foster, G.R., C.B. Johnson, and W.C. Moldenhauer. 1982. Critical slope lengths for unanchored cornstalk and wheat straw residue. Trans. ASAE 25:935–939, 947.

Frye, W.W., R.L. Blevins, M.S. Smith, S.J. Corak, and J.J. Varco. 1988. Role of annual legume cover crops in efficient use of water and nitrogen. p. 129–154. *In* W.L. Hargrove (ed.) Cropping strategies for the efficient use of water and nitrogen. ASA, CSSA, and SSSA, Madison, WI.

Garwood, T.W.D., D.B. Davies, and A.R. Hartley. 1999. The effect of winter cover crops on yield of the following spring crops and nitrogen balance in a calcareous loam. J. Agric. Sci. 132:1–11.

Haider, K., A. Mosier, and O. Heinemeyer. 1987. The effect of growing plants on denitrification at high soil nitrate concentrations. Soil Sci. Soc. Am. J. 51:97–102.

Hansen, E.M., J. Djurhuus, and K. Kristensen. 2000. Nitrate leaching as affected by introduction or discontinuation of cover crop use. J. Environ. Qual. 29:1110–1116.

Harris, G.H., O.B. Hesterman, E.A. Paul, S.E. Peters, and R.R. Janke. 1994. Fate of legume and fertilizer nitrogen-15 in a long-term cropping systems experiment. Agron. J. 86:910–915.

Hively, W.D., and W.J. Cox. 2001. Interseeding cover crops into soybean and subsequent corn yields. Agron. J. 93:308–313.

Huggins, D.R., and D.J. Fuchs. 1997. Long-term N management effects on corn yield and soil C of an Aquic Haplustoll in Minnesota. p. 121–128. *In* E.A. Paul et al. (ed.) Soil organic matter in temperate ecosystems: Long-term experiments in North America. CRC Press, Boca Raton, FL.

Inderjit, and K.I. Keating. 1999. Allelopathy: Principles, procedures, processes, and promises for biological control. Adv. Agron. 67:141–231.

Jarecki, M.K., and R. Lal. 2003. Crop management for soil carbon sequestration. Crit. Rev. Plant Sci. 22:471–502.

Johnson, T.J., T.C. Kaspar, K.A. Kohler, S.J. Corak, and S.D. Logsdon. 1998. Oat and rye overseeded into soybean as fall cover crops in the upper Midwest. J. Soil Water Conserv. 53:276–279.

Jones, R.J. 1942. Nitrogen losses from Alabama soils in lysimeters as influenced by various systems of green manure crop management. Agron. J. 34:574–585.

Kanwar, R.S., R.M. Cruse, M. Ghaffarzadeh, A. Bakhsh, D.L. Karlen, and T.B. Bailey. 2005. Corn-soybean and alternative cropping systems effects on NO_3–N leaching losses in subsurface drainage water. Appl. Eng. Agric. 21:181–188.

Karlen, D.L., and C.A. Cambardella. 1996. Conservation strategies for improving soil quality and organic matter storage. p. 395–420. *In* M.R. Carter and B.A. Stewart (ed.) Structure and organic matter storage in agricultural soils. Advances in Soil Science. CRC Press Inc., New York.

Kaspar, T.C., J.K. Radke, and J.M. Laflen. 2001. Small grain cover crops and wheel traffic effects on infiltration, runoff, and erosion. J. Soil Water Conserv. 56:160–164.

Kaspar, T.C., T.B. Parkin, D.B. Jaynes, C.A. Cambardella, D.W. Meek, and Y.S. Jung. 2006. Examining changes in soil organic carbon with oat and rye cover crops using terrain covariates. Soil Sci. Soc. Am. J. 70:1168–1177.

Kaspar, T.C., D.B. Jaynes, T.B. Parkin, and T.B. Moorman. 2007. Rye cover crop and gamagrass strip effects on NO_3 concentration and load in tile drainage. J. Environ. Qual. 36:1503–1511.

Ketterings, Q.M., J.M. Blair, and J.C.Y. Marinissen. 1997. Effects of earthworms on soil aggregate stability and carbon and nitrogen storage in a legume cover crop agroecosystem. Soil Biol. Biochem. 29:401–408.

Kirkegaard, J.A., and M. Sarwar. 1998. Biofumigation potential of brassicas. I. Variation in glucosinolate profiles of diverse field-grown brassicas. Plant Soil 201:71–89.

Kladivko, E.J., J.R. Frankenberger, D.B. Jaynes, D.W. Meek, B.J. Jenkinson, and N.R. Fausey. 2004. Nitrate leaching to subsurface drains as affected by drain spacing and changes in crop production system. J. Environ. Qual. 33:1803–1813.

Kuo, S., U.M. Sainju, and E.J. Jellum. 1997. Winter cover crop effects on soil organic carbon and carbohydrate in soil. Soil Sci. Soc. Am. J. 61:145–152.

Kuo, S., and E.J. Jellum. 2000. Long-term winter cover cropping effects on corn (*Zea mays* L.) production and soil nitrogen availability. Biol. Fertil. Soils 31:470–477.

Laflen, J.M., G.R. Foster, and C.A. Onstad. 1985. Simulation of individual-storm soil loss for modeling the impact of soil erosion on crop productivity. p. 285–295. *In* S.A. El-Swaify et al. (ed.) Soil erosion and conservation. Soil and Water Conserv. Soc., Ankeny, IA.

Lal, R., J.M. Kimble, R.F. Follett, and C.V. Cole. 1999. The potential of U.S. cropland to sequester carbon and mitigate the greenhouse effect. Ann Arbor Press, Chelsea, MI.

Langdale, G.W., R.A. Leonard, and A.W. Thomas. 1985. Conservation practice effects on phosphorus losses from southern piedmont watersheds. J. Soil Water Conserv. 40:157–161.

Langdale, G.W., and R.L. Blevins. D.L. Karlen, D.K. McCool, M.A. Nearing, E.L. Skidmore, A.W. Thomas, D.D, Tyler, and J.R. Williams. 1991. Cover crop effects on soil erosion by wind and water. p. 15–22. *In* W.L. Hargrove (ed.) Cover crops for clean water. Proc. Int. Conf., Jackson, TN. 9–11 April 1991. Soil and Water Conserv. Soc., Ankeny, IA.

Lattanzi, A.R., L.D. Meyer, and M.F. Baumgardner. 1974. Influences of mulch rate and slope steepness on interrill erosion. Soil Sci. Soc. Am. J. 38:946–950.

Liebl, R., F.W. Simmons, L.M. Wax, and E.W. Stoller. 1992. Effect of rye (*Secale cereale*) mulch on weed-control and soil-moisture in soybean (*Glycine max*). Weed Technol. 6:838–846.

Lundquist, E.J., L.E. Jackson, K.M. Scow, and C. Hsu. 1999. Changes in microbial biomass and community composition, and soil carbon and nitrogen pools after incorporation of rye into three California agricultural soils. Soil Biol. Biochem. 31:221–236.

Mamo, M., and G.D. Bubenzer. 2001. Detachment rate, soil erodibility, and soil strength as influenced by living plant roots. Part II: Field study. Trans. ASAE 44:1175–1181.

Martinez, J., and G. Guiraud. 1990. A lysimeter study of the effects of a ryegrass catch crop, during a winter-wheat maize rotation, on nitrate leaching and on the following crop. J. Soil Sci. 41:5–16.

McBride, R.G., R.L. Mikkelsen, and K.R. Barker. 2000. The role of low molecular weight organic acids from decomposing rye in inhibiting root-knot nematode populations in soil. Appl. Soil Ecol. 15:243–251.

McCracken, D.V., S.J. Corak, M.S. Smith, W.W. Frye, and R.L. Blevins. 1989. Residual effects of nitrogen fertilization and winter cover cropping on nitrogen availability. Soil Sci. Soc. Am. J. 53:1459–1464.

McCracken, D.V., M.S. Smith, J.H. Grove, R.L. Blevins, and C.T. MacKown. 1994. Nitrate leaching as influenced by cover cropping and nitrogen source. Soil Sci. Soc. Am. J. 58:1476–1483.

Meisinger, J.J., W.L. Hargrove, R.L. Mikkelsen, J.R. Williams, and V.W. Benson. 1991. Effects of cover crops on groundwater quality. p. 57–68. *In* W.L. Hargrove (ed.) Cover crops for clean water. Proc. Int. Conf., Jackson, TN. 9–11 April 1991. Soil and Water Conserv. Soc., Ankeny, IA.

Mele, P.M., and M.R. Carter. 1999. Impact of crop management factors in conservation tillage farming on earthworm density, age structure and species abundance in south-eastern Australia. Soil Tillage Res. 50:1–10.

Mendes, I.C., A.K. Bandick, R.P. Dick, and P.J. Bottomley. 1999. Microbial biomass and activities in soil aggregates affected by winter cover crops. Soil Sci. Soc. Am. J. 63:873–881.

Miguez, F.E., and G.A. Bollero. 2006. Winter cover crops in Illinois: Evaluation of ecophysiological characteristics of corn. Crop Sci. 46:1536–1545.

Moschler, W.W., G.M. Shear, D.L. Hallok, R.D. Sears, and G.D. Jones. 1967. Winter cover crops for sod-planted corn: Their selection and management. Agron. J. 59:547–551.

Mutchler, C.K., and L.L. McDowell. 1990. Soil loss from cotton with winter cover crops. Trans. ASAE 33:432–436.

Nelson, D.W., and L.E. Sommers. 1996. Total carbon, organic carbon, and organic matter. p. 961–1010. *In* D.L. Sparks et al. (ed.) Methods of soil analysis. Part 3. Chemical methods. SSSA Book Ser. 5. SSSA, Madison, WI.

Nyakatawa, E.Z., K.C. Reddy, and K.R. Sistani. 2001. Tillage, cover cropping, and poultry litter effects on selected soil chemical properties. Soil Tillage Res. 58:69–79.

Parkin, T.B., T.C. Kaspar, and J.W. Singer. 2006. Cover crop effects on the fate of N following soil application of swine manure. Plant Soil 289:141–152.

Parr, J.F., and R.I. Papendick. 1978. Factors affecting the decomposition of crop residues by microorganisms. p. 109–129. *In* W.R. Oschwald (ed.) Crop residue management systems. ASA, Madison, WI.

Patrick, W.H., Jr., C.B. Haddon, and J.A. Hendrix. 1957. The effect of longtime use of winter cover crops on certain physical properties of Commerce loam. Soil Sci. Soc. Am. J. 21:366–368.

Paustian, K., H.P. Collins, and E.A. Paul. 1997. Management controls on soil carbon. p. 15–49. *In* E.A. Paul et al. (ed.) Soil organic matter in temperate ecosystems: Long-term experiments in North America. CRC Press, Boca Raton, FL.

Pesant, A.R., J.L. Dionne, and J. Genest. 1987. Soil and nutrient losses in surface runoff from conventional and no-till corn systems. Can. J. Soil Sci. 67:835–843.

Potter, M.J., K. Davies, and A.J. Rathjen. 1998. Suppressive impact of glucosinolates in brassica vegetative tissues on root lesion nematode *Pratylenchus neglectus*. J. Chem. Ecol. 24:67–80.

Power, J.F., J.W. Doran, and P.T. Koerner. 1991. Hairy vetch as a winter cover crop for dryland corn production. J. Prod. Agric. 4:62–67.

Power, J.F., and J.A. Zachariassen. 1993. Relative nitrogen utilization by legume cover crop species at three soil temperatures. Agron. J. 85:134–140.

Qi, Z., and M.J. Helmers. 2010. Soil water dynamics under winter rye cover crop in central Iowa. Vadose Zone J. 9:53–60.

Quemada, M., and M.L. Cabrera. 1997. Temperature and moisture effects on C and N mineralization from surface applied clover residue. Plant Soil 189:127–137.

Raimbault, B.A., T.J. Vyn, and M. Tollenaar. 1990. Corn response to rye cover crop management and spring tillage systems. Agron. J. 82:1088–1093.

Raimbault, B.A., T.J. Vyn, and M. Tollenaar. 1991. Corn response to rye cover crop, tillage methods, and planter options. Agron. J. 83:287–290.

Ram, D.N., M.T. Vittum, and P.J. Zwerman. 1960. An evaluation of certain winter cover crops for the control of splash erosion. Agron. J. 52:479–482.

Ranells, N.N., and M.G. Wagger. 1997a. Nitrogen-15 recovery and release by rye and crimson clover cover crops. Soil Sci. Soc. Am. J. 61:943–948.

Ranells, N.N., and M.G. Wagger. 1997b. Grass-legume bicultures as winter annual cover crops. Agron. J. 89:659–665.

Rasse, D.P., J.T. Ritchie, W.R. Peterson, J. Wei, and A.J.M. Smucker. 2000. Rye cover crop and nitrogen fertilization effects on nitrate leaching in inbred maize fields. J. Environ. Qual. 29:298–304.

Reddy, K.N., R.M. Zablotowicz, M.A. Locke, and C.H. Koger. 2003. Cover crop, tillage, and herbicide effects on weeds, soil properties, microbial populations, and soybean yield. Weed Sci. 51:987–994.

Reeleder, R.D., J.J. Miller, B.R.B. Coelho, and R.C. Roy. 2006. Impacts of tillage, cover crop, and nitrogen on populations of earthworms, microarthropods, and soil fungi in a cultivated fragile soil. Appl. Soil Ecol. 33:243–257.

Ritter, W.F., R.W. Scarborough, and A.E.M. Chirnside. 1998. Winter cover crops as a best management practice for reducing nitrogen leaching. J. Contam. Hydrol. 34:1–15.

Rosecrance, R.C., G.W. McCarty, D.R. Shelton, and J.R. Teasdale. 2000. Denitrification and N mineralization from hairy vetch (*Vicia villosa* Roth) and rye (*Secale cereale* L.) cover crop monocultures and bicultures. Plant Soil 227:283–290.

Ruffo, M.L., and G.A. Bollero. 2003. Modeling rye and hairy vetch residue decomposition as a function of degree-days and decomposition-days. Agron. J. 95:900–907.

Sainju, U.M., B.P. Singh, and W.F. Whitehead. 2002. Long-term effects of tillage, cover crops, and nitrogen fertilization on organic carbon and nitrogen concentrations in sandy loam soils in Georgia, USA. Soil Tillage Res. 63:167–179.

Sainju, U.M., and B.P. Singh. 2008. Nitrogen storage with cover crops and nitrogen fertilization in tilled and nontilled soils. Agron. J. 100:619–627.

Sarwar, M., J.A. Kirkegaard, P.T.W. Wong, and J.M. Desmarchelier. 1998. Biofumigation potential of brassicas. III. In vitro toxicity of isothiocyanates to soil-borne fungal pathogens. Plant Soil 201:103–112.

Sharpley, A.N., and S.J. Smith. 1991. Effects of cover crops on surface water quality. p. 41–49. *In* W.L. Hargrove (ed.) Cover crops for clean water. Proc. Int. Conf., Jackson, TN. 9–11 April 1991. Soil and Water Conserv. Soc., Ankeny, IA.

Shipley, P.R., J.J. Messinger, and A.M. Decker. 1992. Conserving residual corn fertilizer nitrogen with winter cover crops. Agron. J. 84:869–876.

Singer, J.W., S.M. Nusser, and C.J. Alf. 2007. Are cover crops being used in the US corn belt? J. Soil Water Conserv. 62:353–358.

Smith, M.S., and J.M. Tiedje. 1979. The effect of roots on soil denitrification. Soil Sci. Soc. Am. J. 43:951–955.

Staver, K.W., and R.B. Brinsfield. 1990. Patterns of soil nitrate availability in corn production systems: implications for reducing groundwater contamination. J. Soil Water Conserv. 45:318–323.

Staver, K.W., and R.B. Brinsfield. 1998. Using cereal grain winter cover crops to reduce groundwater nitrate contamination in the mid-Atlantic coastal plain. J. Soil Water Conserv. 53:230–240.

Strock, J.S., P.M. Porter, and M.P. Russelle. 2004. Cover cropping to reduce nitrate loss through subsurface drainage in the northern U.S. Corn Belt. J. Environ. Qual. 33:1010–1016.

Thorup-Kristensen, K., J. Magid, and L.S. Jensen. 2003. Catch crops and green manures as biological tools in nitrogen management in temperate zones. Adv. Agron. 79:227–302.

Timmons, D.R., R.F. Holt, and J.J. Latterel. 1970. Leaching of crop residues as a source of nutrients in surface runoff water. Water Resour. Res. 6:1367–1375.

Tisdall, J.M., and J.M. Oades. 1979. Stabilization of soil aggregates by the root systems of ryegrass. Aust. J. Soil Res. 17:429–441.

Tonitto, C., M.B. David, and L.E. Drinkwater. 2006. Replacing bare fallows with cover crops in fertilizer-intensive cropping systems: A meta-analysis of crop yield and N dynamics. Agric. Ecosyst. Environ. 112:58–72.

Unger, P.W., and M.F. Vigil. 1998. Cover crop effects on soil water relationships. J. Soil Water Conserv. 53:200–207.

Utomo, M., W.W. Frye, and R.L. Blevins. 1990. Sustaining soil nitrogen for corn using hairy vetch cover crop. Agron. J. 82:979–983.

Varco, J.J., W.W. Frye, M.S. Smith, and C.T. MacKown. 1989. Tillage effects on nitrogen recovery by corn from a nitrogen-15 labeled legume cover crop. Soil Sci. Soc. Am. J. 53:822–827.

Villamil, M.B., G.A. Bollero, R.G. Darmody, F.W. Simmons, and D.G. Bullock. 2006. No-till corn/soybean systems including winter cover crops: Effects on soil properties. Soil Sci. Soc. Am. J. 70:1936–1944.

Wagger, M.G., and D.B. Mengel. 1988. The role of nonleguminous cover crops in the efficient use of water and nitrogen. p. 115–127. *In* W.L. Hargrove (ed.) Cropping strategies for efficient use of water and nitrogen. ASA, CSSA, and SSSA, Madison, WI.

Wagger, M.G. 1989a. Time of desiccation effects on plant composition and subsequent nitrogen release from several winter annual cover crops. Agron. J. 81:236–241.

Wagger, M.G. 1989b. Cover crop management and nitrogen rate in relation to growth and yield of no-till corn. Agron. J. 81:533–538.

Wagger, M.G., M.L. Cabrera, and N.N. Ranells. 1998. Nitrogen and carbon cycling in relation to cover crop residue quality. J. Soil Water Conserv. 53:214–218.

Wander, M.M., S.J. Traina, B.R. Stinner, and S.E. Peters. 1994. Organic and conventional management effects on biologically active soil organic matter pools. Soil Sci. Soc. Am. J. 58:1130–1139.

Wendt, R.C., and R.E. Burwell. 1985. Runoff and soil losses for conventional, reduced, and no-till corn. J. Soil Water Conserv. 40:450–454.

Williams, S.M., and R.R. Weil. 2004. Crop cover root channels may alleviate soil compaction effects on soybean crop. Soil Sci. Soc. Am. J. 68:1403–1409.

Wyland, L.J., L.E. Jackson, W.E. Chaney, K. Klonsky, S.T. Koike, and B. Kimple. 1996. Winter cover crops in a vegetable cropping system: Impacts on nitrate leaching, soil water, crop yield, pests and management costs. Agric. Ecosyst. Environ. 59:1–17.

Yoo, K.H., J.T. Touchton, and R.H. Walker. 1988. Runoff, sediment and nutrient losses from various tillage systems of cotton. Soil Tillage Res. 12:13–24.

Zhu, J.C., C.J. Gantzer, S.H. Anderson, E.E. Alberts, and P.R. Beuselinck. 1989. Runoff, soil, and dissolved nutrient losses from no-till soybean with winter cover crops. Soil Sci. Soc. Am. J. 53:1210–1214.

22

Intercropping and its Implications for Soil Management

S. Walker, C.J. Stigter, E. Ofori, and N. Kyei-Baffour

Intercropping has always been part of the traditional farming systems in the tropics. The local knowledge about the most appropriate mixes of crops has been passed from generation to generation. When people looked at the natural ecosystems, they saw many species growing together in each location, so it was natural to also plant more than one crop species in the field at one time or during one season. These polyculture cropping systems that have been traditionally used by people in developing countries in Africa and Latin America can provide a key solution to sustainable agriculture in the tropics (Francis and Adipala, 1994; Rusinamhodzi et al., 2006). Stern (1984) reported that about 80% of cultivated land in West Africa was under intercropping. Scientific research is continually being performed on a range of multicropping systems and Connolly et al. (2001) reported on studies in at least 50 countries and with various combinations of a total of 55 different species. The top four species in these studies were maize (*Zea mays* L.), cowpea (*Vigna unguiculata* L. Walp.), wheat (*Triticum aestivum* L.), and groundnut (*Arachis hypogaea* L.), 67% of the studies were performed on-station, and 49% of the studies only lasted 1 yr (Connolly et al., 2001).

Common intercropping systems include maize (*Zea mays* L.), millet [*Pennisetum glaucum* (L.) R. Br.], and sorghum [*Sorghum bicolor* (L.) Muench] as dominant cereal crops, intercropped with legume crops such as beans (*Phaseolus vulgaris* L.), cowpea, groundnut, pigeon-pea [*Cajanus cajan* (L.) Huth], and soybean [*Glycine max* (L.) Merr.] as the companion plant species. Maize–cowpea is promoted among smallholder farms in Zimbabwe (Ncube et al., 2007), Maize–pigeon-pea is practiced in Malawi (Gladwin et al., 2001) and northern Mozambique. According to Olufajo (1992), in the traditional soybean growing areas of Nigeria, soybean is commonly intercropped with cereals like maize, sorghum, and millet. According to Ofori and Stern (1987), cereal and legume intercropping is recognized as a common cropping system throughout tropical developing countries for both capital-intensive commercial and subsistence farming (Fig. 22|1a,b). In Latin America, crops produced by farmers include maize, beans, cassava (*Manihot esculenta* Crantz), and quinoa (*Chenopodium quinoa* Willd.) (Altieri, 1996). The smallholder farmer uses land intensively, with maize and beans often being grown simultaneously in the same field. An *Acacia senegal* (L.) Willd. agroforestry system that has been traditionally practiced in the Sudano-Sahelian region is a complex and

S. Walker, Dep. of Soil, Crop and Climate Sciences, Univ. of the Free State, Bloemfontein, South Africa (walkers@ufs.ac.ca); C.J. Stigter, Agromet Vision, Bondowoso, Indonesia and Bruchem, the Netherlands (cjstigter@usa.net); E. Ofori, Dep. of Agric. Eng., Kwame Nkrumah Univ. of Science and Technol., Kumasi, Ghana (oforiegh@yahoo.com); N. Kyei-Baffour, Dep. of Agric. Eng., Kwame Nkrumah Univ. of Science and Technol., Kumasi, Ghana (nkyeibaffour@hotmail.com).

doi:10.2136/2011.soilmanagement.c22

Fig. 22|1. (a) Maize and bean intercropping in Zimbabwe. (b) Maize and soybean intercropping in South Africa.

Fig. 22|2. (a) Cashew and coconut intercropping in Mozambique. (b) Citrus and bean intercropping in Kenya.

yet dynamically resilient system across a wide range of climates, soils, and topography (Raddad and Luukkanen, 2007).

More than half of the studies performed on intercropping focus on the factors affecting intercropping productivity and the economic benefit of using more than one crop on a single field in any season (Connolly et al., 2001). A range of treatments concerned with various management manipulations (fertilizer, spacing, weeding, irrigation, time of planting, etc.) and their effect on production have been investigated (Connolly et al., 2001). However, there seems to be little interest in investigating the mechanisms or processes that result in the changes (e.g., Carrubba et al., 2008) and many studies fail to measure any of the soil effects.

There is a wide range of mixtures in both space and time that fall into the intercropping group of systems. In general, it must be a system with more than one species growing on the same land during the same season with at least part of the growing cycles of the crops overlapping (Baldy and Stigter, 1997). The intercropping can be of various combinations of annual and perennial crops and/or tree species (Fig. 22|2a,b) and so often would include agroforestry, where trees play a vital role in providing beneficial effects. The arrangement of the component crops can either be systematic (in rows or strips or alleys) or mixed randomly and spaced without any particular arrangement (Andrews and Kassam, 1976). The component crops can also have different lengths of growing seasons or may be sown at different times and so may only be intercropped for part of the season.

The land equivalent ratio (LER) has been developed for use in evaluating the yield advantage of intercropping systems, particularly for the replacement series. Land equivalent ratio is defined as the area that a sole crop has to occupy to produce the same amount as its component in the intercrop (Willey and Osiru, 1972; Mead and Willey, 1980; Baldy and Stigter, 1997). Baldy and Stigter (1997) state that "it is the only method which permits an effective comparison of different yields from the same surface because each intercrop is compared with its sole stand." The LER compares the yield of each part of the intercrop to the yield of that same species grown alone as a sole crop. The equation is as follows:

$$LER_T = LER_a + LER_b + \ldots\ldots + LER_n$$

where LER_T is the total LER and the subscripts a, b, to n give partial LERs for each different species that is a component of the intercropping system. Each of these partial LERs can be calculated as follows:

$$LER_a = Y_{Ia}/Y_{Sa} \ldots\ldots \text{ etc.}$$

where Y_{Ia} is the mass yield per unit area of the first species "a" in the intercrop and Y_{Sa} is the mass yield per unit area of the same species when grown as a sole crop (Connolly et al., 2001; Tsubo et al., 2005). If the LER_T is greater than one (LER > 1), then it shows that intercropping has a yield advantage over the sole cropping of each crop individually. It also shows that if sole cropping were used, a larger piece of land would be required to produce the same total yield produced under the intercropping system (Tsubo et al., 2005). The LER for each species in an intercropping system can then be calculated and used to compare studies. Many times the advantage of the intercrop is due to financial advantage because one of the crops has a higher market value that the other. Then the LER can be calculated using the "monetary value" in place of the mass yield and again the comparison can be made according to an assessment of the advantages of one or other cropping system.

There are many sorts of combination used in intercropping layouts or designs in the field. There are also a variety of different ways of classifying the intercropping systems—by height or climate zone or reason or purpose for which the crops are grown, etc. Even if one considers that the intercrops are limited to those that are grown in the same field or land at the same time, and for only part of their growing seasons, there can still be many different combinations. The crops can be either annuals or perennials and they can be of similar or different growth cycles and growth habits. The arrangement in the field also has a range of possibilities: completely randomly mixed if seed is broadcast, trees randomly spaced in nature with the crops grown under them, rows with crops mixed in the same row, alternating rows or groups of rows of each crop with or without rows of trees, or as alternating strips of each crop (Baldy and Stigter, 1997). Compared with the sole crop there can also be a variety of densities—they can be a pairwise series or additive series or replacement series or a responsive series design (Connolly et al., 2001). So when one is discussing intercropping, one should take care to carefully describe the plant arrangement and planting timing systems before one tries to analyze the results and especially before one tries to extrapolate the results into a different situation.

This review will be limited to the influence of the combination of more than two crops and/or trees growing in the same land for at least part of each crop's growing season. The aim will be to discuss how intercropping influences soil water management and soil properties (both physical and chemical). The productivity of intercropping and its potential to build stable cropping systems will also be addressed.

Intercropping Effect on Soil Properties

Generally the effects of including a legume in a crop rotation or as part of an intercropping system are accepted to be via the N_2 fixed from the atmosphere by symbiotic microbes (Paynel et al., 2008). However, there appears to be evidence that the positive influence of the legume should be attributed to a combination of effects on all nutrients and the soil physical properties. The legume crop does influence the N and P in the soil, as well as the exchangeable cations and the soil organic carbon (SOC) (Mathuva et al., 1998; Rao and Mathuva, 2000; Fischler et al., 1999; Bronick and Lal, 2005; Snapp et al., 1998).

Trees hold nutrients in an efficient, closed cycling system. Their deeper roots tap nutrients from lower soil levels that are inaccessible to forage species, and N-fixing trees can raise the nutrient levels of pasture soils. Well-managed forage production provides improved nutrition for livestock growth and production. There may be supplemental nutrient-rich fodder production in the tree leaves. Fertilizer applied for forage is also used by trees, livestock manure recycles nutrients to trees and forage, and grazing controls weeds and reduces competition with grass for moisture, nutrients, and sunlight. Shade and reduced wind speed raise the moisture level of soils by reducing evaporation. There is reduced soil compaction due to root development. Organic matter input is increased; erosion is reduced, thereby improving soil structure and soil biological activity. Increased wildlife diversity may be expected and forage or understory litter protects the soil from water and wind erosion.

Fischler et al. (1999) found that there were sustained improvements in soil physical properties due to the crotalaria green manure combined into the maize–bean cropping system. They noted a small decrease in bulk densities and increase in porosity as well as increased water infiltration. However, it is difficult to partition the influence of the green manure or of the legume intercropping. The root systems of cereals are different from those of legumes. It is generally accepted that the presence of extensive fibrous roots will produce high levels of macro-aggregation in the soil (Harris et al., 1966). The roots produced by legumes are also associated with a high microbial activity and therefore increased microbial biomass in the soil (Haynes and Beare, 1997). This results in increased soil aggregation. Soil microbe composition may be more diverse under agroforestry than conventional sole and monocropping systems (Lacombe et al., 2008). This increased biodiversity can be an additional advantage of the trees being included in the intercropping system. The tree (poplar *Populus tremula* L., walnut genus *Juglans* L., and white ash *Fraxinus* L.)–soybean intercropping system in Quebec was shown to have a higher incident of arbuscular mycorrhizal fungi and a greater ratio of gram-positive to gram-negative bacteria and a higher microbial stability and diversity in the tree-based intercrops on heavy clay soils (Lacombe et al., 2008).

Some of the legume species used, particularly in hedgerows or with perennial legume crops, are also deep rooted. This assists in the recycling of nutrients and particularly in bringing them up from the deeper soil layers (Snapp et al., 1998). These deep-rooted legumes thus serve a dual role, in promoting deep uptake of nutrients, which are then more readily available for the other crops, as well as influencing the soil aggregation. But it is difficult to separate the multiple effects different organisms (roots, fungi, bacteria, microbes) have on aggregation (Bronick and Lal, 2005). Pigeon-pea is one of the deep-rooted legumes that provide

substantial biological N_2 fixation capacity under low-fertility conditions (Snapp et al., 1998). Pigeon-pea also has multiple uses as the leaves, stems, pods, and seeds are edible by humans and livestock and the stems are used for roofing or fences. Paynel et al. (2008) stated that in white clover (*Trifolium repens* L.)–ryegrass (*Lolium perenne* L.) systems the N is transferred mostly through rhizodeposition into the soil by the clover and then is followed by the uptake of the N by the ryegrass. As the rhizodeposition can occur by exudation from living tissue or by decomposition of the legume tissues after senescence, the N is available both during the growing season and in the following season.

There are not many studies that include detailed measurements of the roots and especially few for intercropping systems (Lose et al., 2003); however, the effects of the roots are often studied indirectly from the measured changes in soil water or nutrients or other plant parameters. Lose et al. (2003) investigated the root growth of maize, cassava, and various tree species under intercropping and alley cropping systems. They found that there was a reduction in the cassava root length density under the intercropping system but under the alley cropping there was less interference between the annuals and perennials. It appears that the hedgerow type systems have fine roots in the topsoil that compete with those of the cassava (Lose et al., 2003).

Under intercropping systems there is also an improvement in the SOC. For instance, after 11 yr of additions of gliricidia [*Gliricidia sepium* (Jacq.) Kunth] prunings, the gliricidia–maize intercropping system had 3 g kg^{-1} higher SOC than the sole maize system. The organic carbon was also detected deeper under the intercropping than the sole maize system (Makumba et al., 2006). Rao and Mathuva (2000) showed that a cowpea–maize rotation in semiarid Kenya maintained the SOC whereas other systems caused small reductions of 6 to 13% in SOC. The SOC pool can be increased by mixed cropping, cover crops, and crop rotations (Bronick and Lal, 2005). The SOC is usually not evenly distributed through the soil, so aggregation "hot spots" form. The increased SOC thus related to increased aggregation by being dependent on the deposition rates in the soil (Bronick and Lal., 2005), which will vary according to the other environmental conditions in different regions. But under a range of other organic matter technologies, including intercropping in Malawi, there were no measurable differences in the SOC (Snapp et al., 1998). In Niger, following pearl millet (*Pennisetum glaucum* (L.) R. Br.)–cowpea intercrop the soil organic matter was significantly higher (by 29%) compared with the sole crop of pearl millet (Reddy et al., 1992).

As legumes are more efficient at uptake of P and K, especially under fast-growing legumes, they may prove to be severe competition for the cereal crops (Fischler et al., 1999). This will be accentuated in soils that are of poor nutrient status, such as many African soils. There can be recycling of small quantities of nutrients through the litter and roots. For example, pigeon-pea and cowpea intercrops could recycle 30 kg N ha^{-1} and 2.2 kg P ha^{-1} under the semiarid Kenyan conditions (Rao and Mathuva, 2000). However, there may be little or no effect on subsequent crops as usually the legume grain is harvested and most of the legume stover will be removed from the field and stored as livestock fodder (Snapp et al., 1998; Ncube, 1997). If, however, the residues are left on the fields, some subsequent crops can benefit greatly in a wet season (Ncube et al., 2007). It is a pity that more soil information is not available from the many intercropping studies that have been conducted in Africa. Intercropping holds a variety of advantages to enhance the soil as a natural resource. The effects of the residual legume roots in the soil will have a longer-term effect on the soil fertility and cannot be removed for other uses in the farming system. In contrast, the N-rich aboveground biomass from the legumes is most often removed and so does not remain as a benefit to the soil resource base.

Soil Water Management

One of the greatest effects that intercropping can have is on soil water management through the growing season. To obtain an understanding of the water strategies of the intercrops, one has to investigate and understand the distribution of the roots in the soil profile. This will then help to explain the water extraction patterns and water use by the combined crops in the intercropping system. For instance, the cowpea rooting system can be well developed and more extensive even than other legumes (Miller, 1998; Rusinamhodzi et al., 2006). These more extensive rooting systems can then explore the soil to greater depths and volumes and so provide a means of promoting better utilization of natural resources. This was the case when the soil water balance was measured to a depth of 1200 mm and the maize–bean intercrop was able to extract more water than either of the sole crops of maize or beans (Walker and Ogindo, 2003). The intercrop's transpiration through the season was higher than that of either soil crop and this was attributed to the more prolific root system of the combined crops being able to explore the soil profile more fully (Ogindo, 2003; Ogindo and Walker, 2005). This is in contrast to the study reported on sorghum–cowpea intercrop where Shackel and Hall (1984) reported that the total quantities and patterns of soil water depletion were similar for the intercropping and both the sole cropping systems.

Intercrops have been shown to substantially improve water use efficiency (WUE) (Reddy and Willey, 1981; Mukhala, 1998; Morris and Garrity, 1993). The components used in calculating WUE are the yield and the water use. The yield under intercropping needs to include that from both contributing crops and is therefore probably best expressed in terms of the energy value. In this way the amount of energy stored in each of the seeds or reproductive parts of the crop is converted into an amount of energy that represents the bioproductivity of the crop (Beets, 1977; Mukhala et al., 1999). Then the potential energy produced from different crops can be added and a total energy yield from the intercrops can be compared with that from the individual or sole crop's energy produced (Tsubo et al., 2004).

The components of the water use can be divided into the sum of the transpiration (T) and the evaporation directly from the soil surface (Es). However, these components are rarely separated in agronomic studies. But if one is to evaluate the true productivity of the water used by the crops, one should only use the transpiration or green water as this is what is converted into biomass and yield (Walker and Ogindo, 2003). The method often used to separate the two components is that of the transpiration coefficient (Tanner, 1981; Walker, 1986; Walker et al., 2001;

Ogindo and Walker, 2004). In semiarid areas, the soil evaporation has been reported to be as high as 75% (Bennie et al., 1994); under bare soil conditions, it can be as high as 95% in a typical farmer's fields (Rockström, 1997). Under these high evaporative demand situations, the useful water for transpiration and thus conversion to biomass is extremely low. However, if intercropping were to be practiced, there would be a higher coverage of the soil by the second crop and a sort of green or living mulch would be formed. While it would use water for transpiration, it would also produce a crop and thus be useful water and at the same time decrease the bare soil surface evaporation. This would then also increase the WUE of the combined crop (Walker and Ogindo, 2003).

The process whereby the soil surface is shaded quicker under the intercropping system is due to the faster growth of the leaves due to competition. This has been shown from the leaf area measurements of the maize–bean intercropping system where the intercrop had a higher leaf area index by 40 d after planting (Walker and Ogindo, 2003). This earlier closure of the crop canopy reduced the amount of solar radiation reaching the soil surface. Using the Ritchie (1972) equation, the reduction in soil surface evaporation was calculated. There was a reduction of between 5 and 8% in the soil surface evaporation as a percentage of the rainfall, except for the one season when 161mm was received soon after planting (Walker and Ogindo, 2003). The reduction in soil evaporation due to an earlier and higher leaf area resulted in 2 to 7% more transpiration in the intercrop than in the sole cropping systems.

Productivity of Intercrops

In addition to providing sustainable cropping systems (Connolly et al., 2001), the intercropping of a cereal together with a legume enhances food production. In areas of Africa, the legume intercropped with maize has provided a stabilizing effect on the food security of the smallholder farmers (Snapp et al., 1998). The intercropping systems provide the best alternative for the semiarid areas by providing a more stable production that is less variable under the variable climate (Rose and Adiku, 2001).

In a multilocation study, the maize–soybean intercrop gave a larger combined yield and higher monetary return than either of the sole crops (Ahmed and Rao, 1982). The LER showed a yield advantage of between 42 and 64% for the intercrop with different amounts of N fertilizer (Ahmed and Rao, 1982). In Ethiopia, wheat and faba-bean (*Vicia faba* L.) were grown in an additive intercropping system that increased the LER by between 3 and 22%. The highest total grain yield was obtained when a full wheat crop was intercropped with a 37.5% faba-bean crop (Agegnehu et al., 2008). In northern Zambia, with maize–bean intercropping the LERs were all above 1.0 (overall mean 1.46), showing large yield advantages for this intercropping system (Siame et al., 1998). From the farmer's perspective it must be compared with that of the sole maize. If no inorganic N fertilizer was added, then the intercrop gave a slightly lower yield; however, with some applied N, the yield was above the sole maize production. Therefore, as the returns on N application were higher under the maize–bean intercropping system, it is more economically viable and should be recommended (Siame et al., 1998).

By using the LER one is able to compare the production of the intercrop and express it relative to the yield from a sole crop of the dominant species (Beets, 1982). In a row orientation trial at Bloemfontein (29S and 26E at 1411 msl), no statistical difference was found between the north-south and east-west rows. However, the intercrop of maize and beans had an advantage of 8% over the sole cropping systems (Tsubo et al., 2003). This means that if the maize were planted alone on a piece of land, it would need 8% more land surface area to produce the same yield as produced by the intercropping system. This clearly shows that the inclusion of the beans with the maize on one piece of land gave an advantage to the concurrent maize production and did not reduce it in any way. In this experiment the beans were planted at half the population they would be planted under a sole production system and when intercropped with maize they produced about 50% of the sole bean crop (Tsubo, 2000). In plant density trials, the yield of the maize–bean intercropping system gave between a 15 and 26% advantage on sole maize plantings. The medium density (8–6 plants m^{-2}) gave the highest advantage (Mukhala, 1998). The effects of sowing dates on the yields at the same site are not easily compared as the amount of rainfall received during each growing season is so vastly different and it is well known that the later plantings of maize give lower yield even under well-watered conditions. Overall, the maize–bean intercropping system was more effective and more efficient relative to the individual plantings, as all the sowing dates gave LER higher than 1.0 (Tsubo et al., 2004). In general, in these trials the maize yield was not decreased by the addition of beans as an intercrop, showing the advantage of this intercropping system (Tsubo et al., 2005).

The high production of the intercrop can be explained by the more efficient use of the limited natural resources, namely radiation and water. When considering the radiation penetration into the maize canopy, one finds some of the solar radiation will usually be transmitted to the soil surface. However, with a bean intercrop, this radiation will be intercepted by the bean crop. This provides more efficient radiation capture by the intercrop than sole cropping systems (Rose and Adiku, 2001). The radiation use efficiency (RUE) can be calculated from the intercepted photosynthetically active radiation and the aboveground biomass accumulation (Muchow, 1985; Sinclair and Muchow, 1999). The RUE was generally higher for the intercropping system than both the sole crops except under both north-south and east-west in the row orientation trial. The RUE was also reduced by the drought conditions during some seasons, which is a similar result to that found by Muchow (1985).

From many studies across Africa, it seems that the intercropping systems have a higher productivity than the sole cereal systems in semiarid areas—for example, maize–legume (Fisher, 1977a,b; Austin and Marais, 1987; Pilbeam et al., 1994; Alemseged et al., 1996a,b) and for sorghum–legume (Rees, 1986a,b,c; Lightfoot and Tayler, 1987a,b). These yield advantages have also been explained by the more efficient use of both the radiation and the water. Therefore, the C4 cereal–legume intercropping can be recommended for the semiarid areas of Africa as they produce more stable and reliable crops despite climate variability.

Benefits of Silvopastoral Systems

Silvopastoral systems provide strategies to help farmers be prepared and resilient to climate variability. There are many implications of merging forestry and agricultural systems, from economic and development considerations, to local land use, zoning, or cost sharing. The combinations of trees, crops, and/or animals should be intentionally designed to provide a service and then managed to achieve the desired benefits while maintaining the soil resource base. A well-planned system should take into consideration the following factors (Fig. 22|3a,b,c,d):

1. Spacing—the number and size of trees should be managed for continuous, even spacing that optimizes (i) radiation, (ii) growing space for the tree crop, and (ii) fodder production, and reduces competition (Lose et al., 2003; Raddad and Luukkanen, 2007);
2. Species selection—crop and tree selection, potential markets, soil type, climate conditions, and species compatibility;
3. Environment—how the pattern of trees affects wildlife habitat, ease of livestock handling, forage and tree growth and competition, as well as microclimate.

Criteria for tree crop selection (banana genus *Musa* L., cashew *Anacardium occidentale* L., citrus genus *Citrus*, coconut *Cocos nucifera* L., mango *Mangifera indica* L., plantain *Musa ×paradisiaca*, L., etc.) include: marketability, quality of crop or plant, growth characteristics, rooting characteristics, drought tolerance, capability of providing the desired products and environmental benefits, radiation penetration (Reynolds et al., 2007), and extent of ground protection by leaf litter. The forage component (beans, pulses, grasses, etc.) should be a perennial crop that is suitable for livestock grazing, compatible with the site, productive under partial shade, responsive to intensive management, tolerant to heavy utilization, deliver protection of the soil from erosion and drying up, impede development of weeds, enrich soil with N, and contribute to soil organic matter and humus content.

From the management point of view, livestock grazing should be intensively and

Fig. 22|3. (a) Pepper and tea grown as an intercrop in India. (b) Cardamom grown together with pepper as an intercrop in India. (c). Pineapple and cashew nut tree intercropping in Mozambique. (d) Coconut and maize grown along the river in the Philippines.

carefully managed so that system compatibility (soil, fauna, livestock, and tree plants) is maintained. The following management tools should be considered: tree harvesting (Nichols et al., 2001; Norgrove and Hauser, 2002; Raddad and Luukkanen, 2007; Reynolds et al., 2007), thinning, and pruning (Droppelmann and Berliner, 2003; Makumba et al., 2006); fertilizing to improve both forage and tree production (Lose et al., 2003; Wise et al., 2007), if necessary; planting legumes for N fixation and forage production (Ghosh, 2004; Nichols et al., 2001; TangYa et al., 2003); multipasture use, rotating grazing and developing walkways to facilitate it; and supplemental feeding to counter anticipated adverse conditions, to allow for pasture regeneration, or to avoid

system vulnerability periods (tree fruiting, wet soil trampling, etc.).

From the economic benefits point of view, integrating tree forage and livestock creates a land management system to produce marketable products while maintaining long-tem productivity. Economic risk is reduced because the system produces multiple products (dairy, beef, timber, charcoal, fruit, fodder, forage), most of which have established markets. Comprehensive land utilization in silvopastoral systems provides a relatively constant income from livestock sales and selective sale of timber and other products. As other economic benefits, production costs are reduced and marketing flexibility is enhanced by distributing management costs between timber and livestock components. Carbon sequestration in trees and pasture roots and soils, biodiversity and watershed protection, and agrotourism and agroecotourism are also potential benefits.

Intercropping as a Stable Base for Sustainable Agriculture

Sustainable agriculture is intricately entwined with all aspects of the environment including both soil and climate. Usually sustainability is considered to stand on five pillars and be a balance between human needs and the ability of the environment to supply them. There are many definitions available and their interaction with agroclimatology was reviewed by Sivakumar et al. (2000). The inclusion of "natural resources" is key to most of them and a common theme is that the agricultural system needs to remain productive over a long time period (i.e., years) while not affecting the natural environment. Many definitions include words like conserve, enhance, improve, maintain, manage, protect, provide, and use. So the natural resource base and environmental quality are important and need to be maintained in a healthy state without environmental degradation. But the basic human needs for food and fiber via agricultural production must also be maintained at a viable level of productivity (ASA, 1989).

The intercropping systems that combine cereal and legume production have been shown to meet these requirements. As many of the intercropping systems discussed increase the land use efficiency while producing food, they thereby enhance sustainable crop production (Agegnehu et al., 2008). In contrast to intensive monocropping systems that clearly have negative impacts on the natural resources, the intercropping systems have potential advantages such as higher overall productivity and enhanced ecological services (Malezieux et al., 2008). It has been shown that there are nutritional benefits from intercropping for the human diet (Fig. 22|4a,b). Both vitamin C and calcium are contributed by the beans, as none was measured in a porridge made only from maize meal (Mukhala et al., 1999). Maize and bean intercropping contributed more magnesium, phosphorus, and potassium to one's diet than a sole maize crop (Mukhala

et al., 1999). Therefore, if smallholders were encouraged to grow maize–bean intercrops, they would benefit both in terms of yield per unit area and nutrient availability and accessibility in their meals.

Gladwin et al. (2001) have promoted the idea that governments should help farmers to improve the returns from their available resources in the broadest sense of the word to address the household food insecurity in Africa. A simulation of a woman-headed household under normal circumstances in southern Malawi shows that the staple food of maize would be intercropped with cassava, groundnuts, beans, and pigeon-peas.

Fig. 22|4. (a) Maize, pumpkin, and melon grown together for household consumption in Zimbabwe. (b) Taro and banana intercropped for food production in India.

All the maize would be consumed and part of the other crops eaten and part sold, so they would benefit from the better nutrition. However, it is a very complex livelihood system with many options (Gladwin et al., 2001). One of the recommended strategies is to help women farmers increase the return from their land by combining small amounts of inorganic fertilizers with grain–legume intercrops or green manures (Gladwin et al., 2001). This would result in the cost of maize to a farmer growing her own maize with a little fertilizer being much less than the price she would have to pay when purchasing it (Gladwin et al., 2001).

Crews and Peoples (2004) have used a model to suggest that if one were able to increase the contribution of fixed N in the agricultural system, it would have the greatest impact. They suggest that the most important areas to address are increasing N_2 fixation by legumes in some places by fertilizing with deficient nutrients (especially phosphorus) and decreasing the subsoil constraints to promote deeper root penetration. From the discussion above, it has been shown that the cereal–legume intercrops do address these issues by fixing N and having deep rooting systems. Crews and Peoples (2004) continue by emphasizing that alternative cropping systems including different legumes should be implemented, along with future breeding research, to promote better N fixation. Further, one of the more important immediate strategies for increasing N in legume-based systems is to minimize the time that fields are left with a bare, exposed soil surface and no vegetation. Thus they promote cover crops and perennial legume systems to maintain the capacity to take up soil N.

Final Remarks

Intercropping systems are prevalent in many parts of the world as farmers in the temperate regions use alternating strips of various crops, e.g., maize and soybeans (Sullivan, 2003). However, multiple cropping is basically a traditional method of intensive farming in warm climates attempting to optimize the use of land, water, solar radiation, and other climatic factors, as well as nutrients, labor, and other socioeconomic factors such as those related to year-round food availability for the household, marketability, and prices. It is a system of farming that is suitable for averting risk of total crop failure in dryland farming, a system used in association with local or other varieties of food and cash crops (including trees) and proven to be suitable. Intercropping can be done with annual food crops, perennial crops, fodder crops, tree crops, and/or energy crops. It offers an alternative food production system in the face of the current food and energy crises; it offers solutions for food security in agricultural production in technologically advanced countries and also for smallholder, resource-poor farmers in developing countries. If the appropriate climatic conditions exist and sufficient water is available, intercropping can provide a year-round supply of cereals, fruits, fodder, biofuels, and vegetables.

Advantages of intercropping systems in comparison with monocropping systems include insurance against crop failure, higher total yield per unit area, optimum use of available resources (soil, sunlight, water, and labor), soil fertility sustenance, and greater economic return to the farmer. From a small-scale, resource-poor farmer perspective in developing countries, intercropping provides a strategy of risk aversion, livelihood sustainability, better nutrition, and crop diversification. In their selection of land use and cropping patterns, according to Rosset (1999), small-scale farmers are more likely to intercrop various crops on the same field, plant multiple times during the year, and integrate crops, livestock, and even aquaculture, to make much more intensive use of space and time. To the small-scale, resource-poor farmer, multiple cropping serves to enhance risk avoidance regarding income, yield, and ecological balance. From a commercial or large-scale agriculture perspective in developed countries, multiple cropping is practiced to gain the advantages of increased land productivity, maintenance of diversity, response to the effects of climate change, disease and pest control, and ecosystem health.

References

Agegnehu, G., A. Ghizaw, and W. Sinebo. 2008. Yield potential and land-use efficiency of wheat and faba bean mixed intercropping. Agron. Sustain. Dev. 28:257–263.

Ahmed, S., and M.R. Rao. 1982. Performance of maize-soyabean intercrop combination in the tropics: Results of a multi-location study. Field Crops Res. 5:147–161.

Alemseged, Y.B., G.W. King, L.R. Coppock, and J.C. Tothill. 1996a. A preliminary investigation of the potential for maize-legume intercropping in thee semi-arid area of Sidamo region, Ethiopia. I. Maize response. S. Afr. J. Plant Soil 13:120–124.

Alemseged, Y.B., G.W. King, L.R. Coppock, and J.C. Tothill. 1996b. A preliminary investigation of the potential for maize-legume intercropping in the semi-arid area of Sidamo region, Ethiopia. II. Legume response. S. Afr. J. Plant Soil 13:125–130.

Altieri, M. 1996. Indigenous knowledge re-valued in Andean agriculture. ILEIA Newsl. 12(1):7–8.

Andrews, D.J., and A.H. Kassam. 1976. The importance of multiple cropping in increasing world food supplies. p. 1–10. In R.I. Papendick, A. Sánchez, and G.B. Triplett (ed.) Multiple Cropping. ASA Spec. Publ. 27. ASA, Madison, WI

ASA. 1989. Decision reached on sustainable agriculture. Agron. News Jan., p. 15.

Austin, M.N., and J.N. Marais. 1987. Methods of presenting intercropping results and preliminary results with Zea mays and Phaseolus vulgaris. S. Afr. J. Plant Soil 4:1–6.

Baldy, C., and C.J. Stigter. 1997. Agrometeorology of multiple cropping in warm climates. Science Publ., Enfield, NH.

Beets, W.C. 1977. Multiple cropping of maize and soyabeans under a high level of crop management. Neth. J. Agric. Sci. 25:95–102.

Beets, W.C. 1982. Multiple cropping and tropical farming systems. Westview Press, Boulder, CO.

Bennie, A.T.P., J.E. Strydom, and H.S. Very. 1994. Storage and utilization of rain water in soils for stabilizing crop production in semi-arid areas. Report No. 227/1/94. (In Afrikaans with English abstract.) Water Res. Commission, Pretoria.

Bronick, C.J., and R. Lal. 2005. Soil structure and management: A review. Geoderma 124:3–22.

Carrubba, A., R.L. la Torre, F. Saiano, and P. Aiello. 2008. Sustainable production of fennel and dill by intercropping. Agron. Sustain. Dev. 28:247–256.

Connolly, J., H.C. Goma, and K. Rahim. 2001. The information content of indicators in intercropping research. Agric. Ecosyst. Environ. 87:191–207.

Crews, T.E., and M.B. Peoples. 2004. Legume versus fertilizer sources of nitrogen: Ecological tradeoffs and human needs. Agric. Ecosyst. Environ. 102:279–297.

Droppelmann, K., and P. Berliner. 2003. Runoff agroforestry—a technique to secure the livelihood of pastoralists in the Middle East. J. Arid Environ. 54:571–577.

Fisher, N.M. 1977a. Studies in mixed cropping. I. Seasonal differences in relative productivity of crop mixtures and pure stands in the Kenya highlands. Exp. Agric. 13:177–184.

Fisher, N.M. 1977b. Studies in mixed cropping. II. Population pressures in maize-bean mixtures. Exp. Agric. 13:185–191.

Fischler, M., C.S. Wortmann, and B. Feil. 1999. Crotalaria (C. ochroleuca G. Don.) as a green manure in maize-bean cropping systems in Uganda. Field Crops Res. 61:97–107.

Francis, A.C., and E. Adipala. 1994. Tropical intercropping systems. What is the future? Afr. Crop Sci. J. 2:131–133.

Ghosh, P.K. 2004. Growth, yield, competition and economics of groundnut/cereal fodder intercropping systems in the semi-arid tropics of India. Field Crops Res. 88:227–237.

Gladwin, C.H., A.M. Thomson, J.S. Peterson, and A.S. Anderson. 2001. Addressing food security in Africa via a multiple livelihood strategies of women farmers. Food Policy 26:177–207.

Harris, R.F., G. Chesters, and O.N. Allen. 1966. Dynamics of soil aggregation. Adv. Agron. 18:108–169.

Haynes, R.J., and M.H. Beare. 1997. Influence of six crop species on aggregate stability and some labile organic matter fractions. Soil Biol. Biochem. 29:1647–1653.

Lacombe, S., R.L. Bradley, C. Hamel, and C. Beaulieu. 2008. Do tree-based intercropping systems increase the diversity and stability of soil microbial communities? Agric. Ecosyst. Environ. 131:25–31.

Lightfoot, C.W.F., and R.S. Tayler. 1987a. Intercropping sorghum with cowpea in dryland farming systems in Botswana. I. Field experiments and relative advantages of intercropping. Exp. Agric. 23:425–434.

Lightfoot, C.W.F., and R.S. Tayler. 1987b. Intercropping sorghum with cowpea in dryland farming systems in Botswana. II. Comparative stability of alternative cropping systems. Exp. Agric. 23:435–442.

Lose, S.J., T.H. Hilger, D.E. Leihner, and J. Kroschel, J. 2003. Cassava, maize and tree root development as affected by various agroforestry and cropping systems in Bénin, West Africa. Agric. Ecosyst. Environ. 100:137–151.

Makumba, W., B. Janssen, O. Oenema, F.K. Akinnifesi, D. Mweta, and F. Kwesiga. 2006. The long-term effects of a gliricidia-maize intercropping system in Southern Malawi, on gliricidia and maize yields, and soil properties. Agric. Ecosyst. Environ. 116:85–92.

Malezieux, E., Y. Crozat, C. Dupraz, M. Laurans, D. Makowski, H. Ozier-Lafontaine, B. Rapidel, S. de Tourdonnet, and M. Valantin-Morison. 2008. Mixing plant species in cropping systems: Concepts, tools and models. A review. Agron. Sustain. Dev. 29:43–62.

Mathuva, M.N., M.R. Rao, P.C. Smithson, and R. Coe. 1998. Improving maize (Zea mays) yields in semiarid highlands of Kenya: Agroforestry or inorganic fertilizers? Field Crops Res. 55:57–72.

Mead, R., and R.W. Willey. 1980. The concept of a 'land equivalent ratio' and advantages in yields from intercropping. Exp. Agric. 16:217–228.

Miller, R. 1998. Legume cover crops for northern California. Small Farm Center UC Coop. Ext. UC Davis. Davis, CA.

Morris, R.A., and D.P. Garrity. 1993. Resource capture and utilization in intercropping: Water. Field Crops Res. 34:303–317.

Muchow, R.C. 1985. An analysis of the effects of water deficits on grain legumes grown in semi-arid tropical environment in terms of radiation interception and its efficiency of use. Field Crops Res. 11:309–323.

Mukhala, E. 1998. Radiation and water utilization efficiency by mono-culture and inter-crop to suit small-scale irrigation farming. Ph.D. diss. Univ. of the Orange Free State, Bloemfontein, South Africa.

Mukhala, E., J.M. de Jager, L.D. van Rensburg, and S. Walker. 1999. Dietary nutrient deficiency in small-scale farming communities in South Africa: Benefits of intercropping maize (Zea mays) and beans (Phaseolus vulgaris). Nutr. Res. 19:629–641.

Ncube, B. 1997. Understanding cropping systems in the semi-arid environments of Zimbabwe: Options for soil fertility management. PhD diss. Wageningen Univ., Wageningen, the Netherlands.

Ncube, B., S.J. Twomlow, M.T. van Wijk, J.P. Dimes, and K.E. Giller. 2007. Productivity and residual benefits of grain legumes to sorghum under semi-arid conditions in southwestern Zimbabwe. Plant Soil 299:1–15.

Nichols, J.D., M.E. Rosemeyer, F.L. Carpenter, and J. Kettler. 2001. Intercropping legume trees with native timber trees rapidly restores cover to eroded tropical pasture without fertilization. For. Ecol. Manage. 152:195–209.

Norgrove, L., and S. Hauser. 2002. Yield of plantain grown under different tree densities and 'slash and mulch' versus 'slash and burn' management in an agrisilvicultural system in southern Cameroon. Field Crops Res. 78:185–195.

Ofori, F., and W.R. Stern. 1987. Cereal-legume intercropping systems. Adv. Agron. 41:41–90.

Ogindo, H.O. 2003. Comparing the precipitation use efficiency of maize-bean intercropping with sole cropping in a semi-arid ecotope. Ph.D. diss. Univ. of the Free State, Bloemfontein, South Africa.

Ogindo, H.O., and S. Walker. 2004. The determination of transpiration efficiency coefficient for common bean. Phys. Chem. Earth 29:1083–1089.

Ogindo, H.O., and S. Walker. 2005. Comparison of measured changes in seasonal soil water content by rainfed maize-bean intercrop and component cropping systems in a semi-arid region of southern Africa. Phys. Chem. Earth 30:799–808.

Olufajo, O.O. 1992. Response of soybean intercropping with maize on a sub-humid tropical environment. Trop. Oilseed J. 1:27–33.

Paynel, F., F. Lesuffleur, J. Bigot, S. Diquélou, and J.-B. Cliquet. 2008. A study of ^{15}N transfer between legumes and grasses. Agron. Sustain. Dev. 28:281–290.

Pilbeam, C.J., J.R. Okalebo, L.P. Simmands, and K.W. Gathua. 1994. Analysis of maize-common bean intercrops in semi-arid Kenya. J. Agric. Sci. (Cambridge) 123:191–198.

Raddad, E.Y., and O. Luukkanen. 2007. The influence of different *Acacia senegal* agroforestry systems on soil water and crop yields in clay soils of the Blue Nile region, Sudan. Agric. Water Manage. 87:61–72.

Rao, M.R., and M.N. Mathuva. 2000. Legumes for improving maize yield and income in semi-arid Kenya. Agric. Ecosyst. Environ. 78:123–137.

Reddy, K.C., P. Visser, and P. Buckner. 1992. Pearl millet and cowpea yields in sole and intercrop systems, and their after-effects on soil and crop productivity. Field Crops Res. 28:315–326.

Reddy, M.S., and R.W. Willey. 1981. Growth and resource-use studies in an intercrop of pearl millet/groundnut. Field Crops Res. 4:13–24.

Rees, D.J. 1986a. Crop growth, development and yield in semi-arid conditions in Botswana. II. The effects of intercropping *Sorghum bicolor* with *Vigna unguiculata*. Exp. Agric. 22:169–177.

Rees, D.J. 1986b. The effects of population density and intercropping with cowpea on the water use and growth of sorghum in semi-arid conditions in Botswana. Agric. For. Meteorol. 37:293–308.

Rees, D.J. 1986c. The effects of population density, row spacing and intercropping on the interception and utilization of solar radiation by *Sorgham bicolor* and *Vigna unguiculata* in semi-arid conditions in Botswana. J. Appl. Ecol. 223:917–928.

Reynolds, P.E., J.A. Simpson, N.V. Thevathasan, and A.M. Gordon. 2007. Effects of tree competition on corn and soybean photosynthesis, growth, and yield in a temperate tree-based agroforestry intercropping system in southern Ontario, Canada. Ecol. Eng. 29:362–371.

Ritchie, J.T. 1972. Model for predicting evaporation from a row crop with incomplete cover. Water Resour. Res. 8:1204–1211.

Rockström, J. 1997. On-farm agrohydrological analysis of the Sahelian yield crisis: Rainfall partitioning, soil nutrients and water use efficiencey of pearl millet. Ph.D. diss. Univ. of Stockholm, Sweden.

Rose, C.W., and S. Adiku. 2001. Conceptual methodologies in agro-environmental systems. Soil Tillage Res. 58:141–149.

Rosset, P.M. 1999. Policy Brief No. 4: The multiple functions and benefits of small farm agriculture in the context of global trade negotiation. Available at http://www.foodfirst.org/en/node/246 (verified 10 Sept. 2010). Food First/Institute for Food and Development Policy, Oakland, CA.

Rusinamhodzi, L., H.K. Murwira, and J. Nyamangara. 2006. Cotton-cowpea intercropping and its N2 fixation capacity improves yield of a subsequent maize crop under Zimbabwean rain-fed conditions. Plant Soil 287:327–336.

Shackel, K.A., and A.E. Hall. 1984. Effect of intercropping on water relations of sorghum and cowpea. Field Crops Res. 8:381–387.

Siame, J., R.W. Willey, and S. Morse. 1998. The response of maize/*Phaseolus* intercropping to applied nitrogen on Oxisols in northern Zambia. Field Crops Res. 55:73–81.

Sinclair, T.R. and R.C. Muchow. 1999. Radiation use efficiency. Adv. Agron. 65:215–265.

Sivakumar, M.V.K., R. Gommes, and W. Baier. 2000. Agrometeorology and sustainable agriculture. Agric. For. Meteorol. 103:11–26.

Snapp S.S., P.L. Mafongoya, and S. Waddington. 1998. Organic matter technologies for integrated nutrient management in smallholder cropping systems of southern Africa. Agric. Ecosyst. Environ. 71:185–200.

Stern, K.G. 1984. Intercropping in tropical small-holder agriculture with special reference to West Africa. GTZ, Eschborn, Germany.

Sullivan, P. 2003. Intercropping principles and production practices. Natl. Sustainable Agric. Inf. Serv., Fayetteville, AR.

TangYa, Z. Yan-zhou, X. Jia-sui, and S. Hui. 2003. Incorporation of mulberry in contour hedgerows to increase overall benefits: A case study from Ningnan County, Sichuan Province, China. Agric. Syst. 76:775–785.

Tanner, C.B. 1981. Transpiration efficiency of potato. Agron. J. 73:59–64.

Tsubo, M. 2000. Radiation interception and use in a maize and bean intercropping system. Ph.D. diss. Univ. of the Orange Free State, Bloemfontein, South Africa.

Tsubo, M., E. Mukhala, H.O. Ogindo, and S. Walker. 2003. Productivity of maize-bean intercropping in a semi-arid region of South Africa. Water S.A. 29:381–388.

Tsubo, M., H.O. Ogindo, and S. Walker. 2004. Yield evaluation of maize-bean intercropping in a Semi-arid region of South Africa. Afr. Crop Sci. J. 12:351–358.

Tsubo, M., S. Walker, and H.O. Ogindo. 2005. A simulation model of cereal–legume intercropping systems for semi-arid regions I. Model development. Field Crops Res. 93:10–22.

Walker, C.K. 1986. Transpiration efficiency of field grown maize. Field Crops Res. 14:29–38.

Walker, S., and H.O. Ogindo. 2003. The water budget of rainfed maize and bean intercrop. Phys. Chem. Earth 28:919–926.

Walker, S., H.O. Ogindo, and M. Hensley. 2001. Measurement of a transpiration coefficient for common bean using a lysimeter. *In* Annual meetings abstracts [CD-ROM]. ASA, CSSA, and SSSA, Madison, WI.

Willey, R.W., and D.S.O. Osiru. 1972. Studies on mixtures of maize and beans (*Phaseolus vulgaris*) with particular reference to plant population. J. Agric. Sci. (Cambridge) 79:519–529.

Wise, R., O. Cacho, and R. Hean. 2007. Fertilizer effects on the sustainability and profitability of agroforestry in the presence of carbon payments. Environ. Model. Softw. 22:1372–1381.

23

Agroforestry

Thomas J. Sauer and Guillermo Hernandez-Ramirez

> "The diligent farmer plants trees, of which he himself will never see the fruit."
>
> Cicero

Humans have long exploited the climate-altering effects that trees provide through shade from a hot sun and shelter from strong winds (Hall et al., 1958). Behavior that initially produced greater physical comfort evolved into purposeful planting, selection, and tending to increase and expand the multiple benefits trees can provide including food, fiber, fuel, and medicinal products. Agroforestry systems (AFS) integrate woody perennial plants with agricultural crops or animal production on the same land area. A fundamental advantage of AFS is that the combination of trees with understory plants or animals has greater potential for production of food, forage, and fiber than any one element alone. A numerical scale to express this multiple-product concept as a land equivalent ratio was developed for AFS by Keesman et al. (2007). Agroforestry systems have great potential to increase per unit land area productivity as the trees exploit resources (light, water, and nutrients) through their multilayered architecture, deeper rooting, and extended growing seasons that may not be as readily captured by annual crops. The inherent benefits of agroforestry also include enhanced ecosystem services, increased ecological and economic diversity, and the ability to protect or restore vulnerable or degraded soils. These multiple benefits illustrate AFS's great potential to contribute to achieving the Millennium Development Goals to reduce hunger, poverty, disease, and environmental degradation (Garrity, 2004).

Due to innate variation in climate, soil characteristics, and socioeconomic conditions, there is a rich diversity of AFS around the world. Agroforestry systems in the tropics and subtropics are often designed to mimic the highly productive natural forest ecosystems there and may involve multiple species with vertical stratification within the canopy. Much research has been devoted to the cultural and production aspects of tropical and subtropical AFS, producing numerous research articles and several technical books (e.g., Young, 1989; Nair, 1993; Schroth and Sinclair, 2003; Nair et al., 2004; Batish et al., 2008). Agroforestry concepts have been more slowly adapted to temperate regions. As a result, considerably less information is available on temperate AFS (Byington, 1990; Long, 1993; Gordon and Newman, 1997; Garrett et al., 2000). The low adoption of agroforestry practices in temperate regions can be attributed to several factors including concern

T.J. Sauer, USDA-ARS, National Laboratory for Agriculture and the Environment, 2110 University Boulevard, Ames, IA 50011-3120 (Tom.Sauer@ars.usda.gov); G. Hernandez-Ramirez, The New Zealand Institute for Plant and Food Research Limited, Soil Water and Environment Group, Plant and Food , Research Lincoln Gerald Street, Lincoln, 7608, New Zealand (Guillermo.Hernandez@plantandfood.co.nz).

doi:10.2136/2011.soilmanagement.c23

regarding tree–crop competition (especially for nutrients and water in drier climates), readily available and relatively inexpensive external inputs (e.g., fuel, fertilizer, and mechanization) that are often unavailable or too expensive in the tropics, and a cultural reluctance to "farm with trees."

While cropping and forestry components of the different AFS have received considerable attention, soil management aspects have not been developed to a comparable level. Young (1997) and Schroth and Sinclair (2003) are the only books with a focus on soil management aspects of AFS. Many principles of soil management were developed specifically for arable cropping systems on soils formed predominantly under grasslands. Traditionally, forest soil science has been considered a specialty area of soil science due to distinct differences between soils formed in forests or grasslands (Jenny, 1941). Forest soils stand apart due to the influence of the forest vegetation including the forest litter, tree roots, and associated organisms on forest soil processes and properties (Wilde, 1958; Armson, 1977; Pritchett and Fisher, 1987). In forest ecosystems, litterfall on the soil surface is the primary annual organic input and it is the decomposition and mixing of the litter layer by micro- and macrofauna that has profound implications for C and nutrient cycling and the physical characteristics of forest soils (Dickinson and Pugh, 1974; Cadisch and Giller, 1997; Berg and Laskowski, 2006). Soil management in AFS requires an integration of the features and processes of soils with simultaneous crop and tree culture.

The ability of trees to modify the local microclimate creates a special ability for AFS to adapt to climate change (Lin, 2007; Calfapietra et al., 2010). Long-term shifts in temperature and precipitation patterns may result in "normal" conditions that are outside the optimal range for crops currently under cultivation in an area. Another climate-related environmental stress is with regard to the extremes of episodic events. Agroforestry systems are inherently more resilient to climate change and extremes than traditional arable cropping systems in two important ways. First, AFS involve multiple species and perennial vegetation, thereby providing greater plant diversity and less vulnerability to climate stress than is provided by monocropping and annual

species. Second, as mentioned previously, the perennial woody vegetation itself serves to modify the local microclimate by influencing sunlight interception and airflow patterns, offering protection for the understory species from extremes in temperature and damaging winds (Stigter, 1988; Brenner, 1996; Cleugh and Hughes, 2002). Deeper rooting perennial vegetation and tighter within-system nutrient cycling also afford greater resilience to drought and the efficiency of nutrient use further enhances the potential utility of AFS to adapt to the uncertainties of climate (Wallace, 1996; Kho, 2008).

Although AFS are inherently productive and resilient to environmental stresses, sharply increasing global demand for food, fuel, and fiber require even greater intensification of production of each of these commodities on a per unit land area basis. Effective soil management of AFS will require skillful and timely application of existing techniques and development of new techniques to optimize and sustain production of all components of the AFS. We will begin our discussion with a brief overview of five broad categories of AFS: (i) riparian forest buffers, (ii) alley cropping, (iii) silvopasture, (iv) field windbreaks or shelterbelts, and (v) forest farming (Fig. 23|1). This will be followed by a discussion of temperate zone AFS that focuses on soil physical properties, nutrient cycling and pH, and soil biology and ecology. Discussion of unique features of subtropical and tropical AFS will follow with a similar treatment of the properties and processes of AFS in these regions. A summary concludes the discussion by reviewing the major points and unifying principles of soil management in AFS.

An Overview of Agroforestry Practices

Natural riparian corridors occur when trees are distributed in a narrow strip along streams, rivers, or lakes. Planted riparian forest buffers are often a restoration of the natural vegetative cover and are designed to filter nutrients and sediment from overland flow before the runoff enters surface water bodies (Peterjohn and Correll, 1984; Lowrance et al., 1984; Snyder et al., 1998) and/or reduce nutrient fluxes through shallow groundwater (Hubbard and Lowrance,

Fig. 23|1. Examples of agroforestry practices. Clockwise from top left, alley cropping, riparian buffer, field windbreak, silvopasture, and forest farming. Photo credits (clockwise from top left): courtesy of The Center for Agroforestry at the University of Missouri, photo by Lynn Betts, USDA Natural Resources Conservation Service, photo by Erwin Cole, USDA Natural Resources Conservation Service, photo by Todd Groh taken from Nowak et al. (2002), and photo by Scott Josiah, Nebraska Forest Service.

1996). As riparian forests are located along hydrologic flow paths, runoff water and shallow groundwater and the nutrients they convey are available for uptake by the trees and understory vegetation (Lowrance et al., 1984). Much focus on the nutrient cycling processes in riparian forests has been on N cycling and particularly nitrate (NO_3) removal from shallow groundwater (Groffman et al., 1996; Hill, 1996). Riparian plantings are often designed with species and their placement optimized to intercept surface runoff, increase infiltration of the runoff, and encourage plant uptake of water from the vadose zone and groundwater. The primary objective is to slow surface water contribution to the stream and filter out eroded sediment and nutrients to improve stream water quality.

Riparian forests are often highly productive due to the readily available and often nutrient-enriched water in the riparian root zone. Due to their position in the landscape, species within natural riparian ecosystems are generally adaptable to the hydrologic extremes of flood and drought with the ability to tolerate temporary submersion or extend roots to extract water from receding aquifers. For this reason, riparian forests may be more resilient than some other ecosystems to climate-induced stress and variability. Decay of the dead wood and grass and leaf litter will return nutrients and C to the soil. This process and the runoff- and sediment-trapping features of riparian corridors can create concern that these areas may become excessively nutrient enriched unless some management that includes nutrient removal (i.e., biomass harvesting) is employed.

Alley cropping (also referred to as hedge rows or agroforestry intercropping) involves widely spaced trees (in rows, in some other geometric pattern, or in a sparse stand resulting from thinning an existing forest) with the area between the trees (alleyways) used for agricultural (grains, legumes, or forages) or horticultural (shrubs, berries, or vines) crops (Tang et al., 1990). Successful alley cropping requires careful management to assure a balance between competition for and efficient utilization of light, water, and nutrients by the trees and crops (Gillespie et al., 2000; Livesley et al., 2004). Alley cropping systems do allow a high degree of light, water, and nutrient-use manipulation in both space and time (season). For instance, trimming of tree branches increases light penetration to the understory crop while the decomposing prunings provide a slow-release source of nutrients for the crops and a mulch layer to reduce evaporation (Tang et al., 1990; Palm, 1995). Such AFS systems often require intensive management but can be highly efficient, require low external inputs, and can be economically more profitable than conventional systems (Lu, 2006). Alley cropping is a popular AFS for some high-value crops including coffee [*Coffea* L. (Rubiaceae)] and tea [*Camellia sinensis* (L.) Kuntze] and even lower-value row crops or forages.

The architecture and management of silvopasture AFS can be very similar to alley cropping AFS, especially for the tree overstory, but with an understory of forages for consumption by grazing animals instead of growing crops (Sharrow, 1999; Garrett et al., 2004; Schnabel and Ferreira, 2004; Mosquera-Losada and Giguerio, 2005). Silvopastures are a managed analog to natural savanna ecosystems with widely spaced trees and grasses in the open spaces that are often found in a vegetative transition zone between forests and grasslands (Dyksterhuis, 1957). Silvopastures may also represent a transition in land use when trees are planted into existing pasture and grazed for some years until light levels beneath the trees are insufficient for forage growth or when forages are planted beneath a recently thinned forest. Like alley cropping, management of silvopastures requires a balance of resource utilization between trees and forage with the added grazing animal element. Influence of the tree canopy on light penetration, water use, and temperature are key factors affecting forage production and quality and the ability to support grazing animals (Lin et al., 1999; Silva-Pando et al., 2002). Effective management or favorable rainfall and temperature patterns will produce more forage and encourage greater stocking density, which may not be sustainable during subsequent suboptimal growing seasons and has important implications for soil quality.

Field windbreaks or shelterbelts are AFS designed specifically for changing the local microclimate primarily by reducing wind speed in their lee (van Eimern, 1964; Bird et al., 1992; Brandle et al., 2004). Shelterbelt plantings consisting of single to multiple rows of trees and/or shrubs have been frequently employed in semiarid areas with extensive plantings in the steppes of Russia (Vyssotsky, 1935) and during the 1930s Dust Bowl in the American Great Plains (U.S. Forest Service, 1935; Droze, 1977). The tree rows reduce wind speed to a distance downwind of approximately 20 times the tree height with multiple beneficial effects on the local microclimate. In general, crop growth in the lee of a shelterbelt is increased due to less evaporation, more plant-available water, and less mechanical stress (Plate, 1971; Rosenberg, 1979; Kort, 1988; Brenner, 1996; Cleugh and Hughes, 2002; Peri and Bloomberg, 2002). Shelterbelts have also been shown to significantly reduce wind erosion (Gupta et al., 1983). Other benefits include trapping snow, increasing wildlife habitat, and improving aesthetics (Cook and Cable, 1995).

Forest farming refers to the cultivation of usually higher-value specialty crops beneath a tree canopy. Forest farming is often used for the cultivation of shade tolerant ornamentals (flowers, ferns, bushes, and decorative florals), medicinals and botanicals (herbs, teas, and natural health products), or food products (mushrooms, fruits, berries, and nuts). A type of forest farming more common in the tropics is the home garden where plants for food products are grown in the understory next to a dwelling. Smith (1953) and Sholto Douglas and de J. Hart (1978) are two classic references for forest farming, both written from a strong ecological perspective for sustainable food production. These writers focused primarily on management of the forest canopy as a food source: fruits and nuts for humans and fodder for livestock. More recently, greater attention has been focused on management of understory species (e.g., ginseng, *Panax* spp.) with the overstory species now having the primary role of providing a managed microclimate (Hill and Buck, 2000; Rao et al., 2004).

Soil Management in Temperate Agroforestry Practices

Although the next three sections will deal with soil physical properties, nutrient cycling and pH, and soil biology and ecology individually they are, in fact, intimately linked. This linkage is perhaps most easily and directly observed with regard to the recycling of organic matter in the soil and the effect of decomposition pathways on nutrient cycling and the amount and quality of soil organic carbon (SOC). As AFS integrate multiple plant species and soil management practices, this discussion will by necessity include concepts covered in far greater detail in other chapters of this volume. Relevant fundamental principles will therefore not be repeated here, where instead the aim is to briefly introduce and synthesize concepts and then relate the principles, processes, and practices of soil management to AFS.

Soil Physical Properties

The size, shape, and arrangement of soil particles and the gathering of these particles into aggregates have profound effects on the transport of water and energy (heat) in soils. Equally important are the voids among the aggregates through which the dynamic transport processes of two vital fluids, air and water, occur. The form and strength of soil structure is very much influenced by shrinking–swelling processes due to wetting–drying and freezing–thawing cycles (Horn and Smucker, 2005) and biological processes associated with root growth and secretions and the activities of micro- and macrofauna (Angers and Caron, 1998). The amount, size, and connectedness of pores between and within the structural aggregates have a tremendous impact on plant growth as it is through these pores that water (and the nutrients it contains) is absorbed by roots. Well-structured soils are also best able to balance the drainage

of excess water and retention of water for plant uptake with the maintenance of a sufficiently oxygenated void space (Gliński and Stępniewski, 1985; Kirkham, 2005). Processes that contribute to good soil structure often lead to increasing SOC content, which has a strong correlation with the amount of plant-available water a soil is capable of storing (Hudson, 1994). Due to the marked difference between the thermal properties of air and water, soil thermal properties are directly correlated with soil water content. Heat and water transport are therefore intimately coupled (Parlange et al., 1998) with important implications for all chemical reactions and biological activity in soils.

Many AFS offer great potential to improve soil structure due to their diversity of plant species and their contrasting growth habits across spatial and temporal scales. Agroforestry systems have produced significantly lower soil bulk density within a multispecies riparian buffer in Iowa (Bharati et al., 2002), beneath a two-row red-cedar (*Juniperus virginiana* L.)–Scotch pine (*Pinus sylvestris* L.) field windbreak in Nebraska (Sauer et al., 2007), and in alley cropping and silvopasture systems in Missouri (Seobi et al., 2005; Kumar et al., 2010). Udawatta and Anderson (2008) reported 2.6 times greater macropores in soil beneath the oak (*Quercus* spp.)–grass agroforestry plantings as compared with the cropped field for the same site as Seobi et al. (2005). The changes in soil pore structure in the AFS at this site associated with differences in bulk density including greater porosity, order-of-magnitude higher saturated hydraulic conductivity, and increased potential water storage were observed already 6 yr after establishment (Seobi et al., 2005; Udawatta et al., 2006). Increased soil aggregate stability beneath a similar *Quercus* spp.–grass AFS in Missouri (Udawatta et al., 2008) also suggests that these changes in soil structure and porosity associated with the AFS are resilient. Although Karki et al. (2009) found a lower percentage of water-stable aggregates in a silvopasture in Georgia, soil penetration resistance was lower in the silvopasture compared with an open pasture.

Infiltration beneath silver maple (*Acer saccharinum* L.) in a multispecies riparian buffer in Iowa was significantly greater than adjacent grass, crop, and pasture sites and was attributed to greater sand content, macropores from decayed roots, and soil faunal activity (Bharati et al., 2002). Sharrow (2007) however, found 13% higher bulk density and 7% lower total porosity in a Douglas-fir (*Pseudotsuga menziesii* Mirb. Franco)–subterranean clover (*Trifolium subterraneum* L.)–tall fescue (*Festuca arundinacea* Schreb.) silvopasture in Oregon as compared with adjacent ungrazed forest. The silvopasture soil also had a 38% lower infiltration rate but had the same amount of available water at field capacity in the top 6 cm. Cessation of grazing for 2 yr reduced the differences between silvopasture and forest soils with little detrimental effect on forage or tree production. Anderson et al. (2009) reported no significant differences in ponded infiltration between agroforestry buffer strip and no-till row crop areas in an alley cropped watershed in Missouri. Greater water depletion during the growing season was observed in the buffer strip soil that enabled increased recharge during storm events resulting in more water storage and less surface runoff.

AFS effects on soil physical properties and the soil moisture regime are important factors influencing their performance and management. Jose et al. (2000a) and Reynolds et al. (2006) evaluated the tree–crop competition for water in alley cropping systems in Indiana and Ontario and determined that management strategies needed to address tree water use and shading if losses in crop productivity were to be avoided. In contrast, Balandier et al. (2008) found that even though 10 yr-old wild cherry (*Prunus avium* L.) in a silvopastoral AFS in central France had a different rooting pattern than the mixed grass and legume forage, the wild cherry experienced severe competition for water. Carlson et al. (1994) also reported tree water stress for a Douglas fir–subterranean clover–tall fescue silvopasture in Oregon. These results in more water-limited climates contrast with the findings of Gyenge et al. (2008) who found that the deeper roots of ponderosa pine (*Pinus ponderosa* Dougl. ex Laws) were able to extract underutilized deep-water reserves in a silvopasture in Patagonia. Clearly, understanding of the relevant processes of soil water storage and plant water use and careful management practices are necessary to avoid adverse impacts on both crop and tree growth and productivity in AFS.

Nutrient Cycling and Soil pH

The observed effects of AFS on soil structure and porosity and the associated changes in the soil water regime have important implications for the biological processes associated with nutrient and organic matter cycling. Litterfall is the primary annual organic input to soils in forests as compared with crop and grassland ecosystems where the primary organic input is from root decomposition (Anderson, 1987; Gale and Cambardella, 2000; Berg and Laskowski, 2006; Kong and Six, 2010). The biochemical makeup of these organic materials (e.g., lignin content and C to N ratio) and their mode and rate of decomposition directly affect the recycling of nutrients and nutrient losses as gaseous emissions or via surface runoff and groundwater flow. Forest ecosystems are generally considered to be more conservative in nutrient cycling with large total nutrient pools but low amounts of available or mobile nutrients located on the forest floor and in shallow soil layers. Forest soils are typified by a thin, organic-rich O horizon over an A and deeper horizons with lower nutrient concentrations due to significant losses of soluble organic N, P, and S compounds and the cations Ca, Mg, and K through leaching (Anderson, 1987). The more extensive tree roots are capable of extracting nutrients from deeper soil layers and nutrient uptake patterns of some species can acidify the soil, thereby accelerating weathering of minerals and furthering nutrient release (Arnold, 1992; Binkley and Giardina, 1998). By contrast, arable soils are typically much more intensively managed with frequent (often annual for N) external nutrient additions, rapid organic matter decomposition, and seasonally large but highly transient pools of plant-available nutrients.

Nair (1993) lists several soil fertility-related benefits that trees can provide in AFS including N fixation, access to deeper sources of nutrients, enhancement of dry and wet atmospheric deposition of nutrients, and release of root exudates, all of which can contribute to increased nutrient use efficiency. Differences in the amounts, properties, and decomposition of tree-, crop-, and animal-derived (silvopasture) organic inputs, while resulting in a more complex system of nutrient cycling, also create a more diverse and resilient system with potential for synergism as well as competition. Wedderburn and Carter (1999) compared the decomposition of litter from four functional tree types (deciduous N-fixer, evergreen N-fixer, deciduous, and evergreen) for silvopastoral AFS and found initial lignin content and lignin to N ratio controlled the rate of litter decay. They concluded that litter properties could be independent of tree functional group but chemical differences between species were more important than seasonal changes in litter quantity or quality. Forage legumes and animal manures have also been evaluated, especially for supplying N, in silvopasture AFS. Blazier et al. (2008) reported that subterranean clover integrated into a loblolly pine (*Pinus taeda* L.)–Bahia grass (*Paspalum notatum* Flüggé) silvopasture in Louisiana helped retain more P from applied fertilizer or poultry litter in the surface soil layer and resulted in enhanced pine growth. A 10 Mg litter ha^{-1} rate of litter application did result in P accumulation in the surface soil and likely increased N and P leaching potential. Karki et al. (2009) also found benefits of integrating a legume into a silvopasture AFS in Georgia. Crimson clover (*Trifolium incarnatum* L. 'Dixie') overseeded into Bahia grass under longleaf pine (*Pinus palustris* Mill.) resulted in improved forage productivity and forage and soil quality during the hay production period of pasture to silvopasture conversion.

Jose et al. (2000b) found faster release of both N and P from fine roots as compared with leaves from black walnut (*Juglans nigra* L.) and red oak (*Quercus rubra* L.) in an alley cropping AFS in Indiana. Plant competition for fertilizer N was considered minimal, however, as N uptake by the black walnut and corn (*Zea mays* L.) crop were not synchronized. Competition for N mineralized from leaf and root tissues could occur but would depend on soil N status and water content. Significant changes in available nutrients, exchangeable acidity, and pH were observed in the surface soil layer beneath a red-cedar–Scotch pine field windbreak in Nebraska 35 yr after the trees were planted (Sauer et al., 2007). Available Ca and Mg were significantly greater and P lower beneath the trees compared with the adjacent cropped fields. Eastern red-cedar

(*Juniperus virginiana* L.) leaves are known to contain high concentrations of Ca (Read and Walker, 1950) so the increase in Ca is likely due to Ca uptake by the trees followed by litterfall and decomposition with nutrient incorporation into the surface soil. The comparatively lower P concentration beneath the trees is likely due to P increases in the cultivated field that received several manure applications. Soil pH in water varied from 4.3 to 7.3 and was highly correlated with tree species with the low pH values (and exchangeable acidity and cation exchange capacity) observed near the Scotch pine trees. Soil acidification with various pine species has been observed previously (Coile, 1933; Arnold, 1992; Sariyildiz et al., 2005) while eastern red-cedar has been found to increase soil pH (Coile, 1933; Read and Walker, 1950). It is clear that different tree species can have profound, localized effects on soil chemical properties and nutrient cycling and distribution in AFS.

Forest riparian buffers have often been designed and planted for the principal purpose of extracting nutrients, especially N in the form of NO_3, and P, from overland flow and shallow groundwater. Peterjohn and Correll (1984) reported significant removal of both N and P in surface and subsurface flows across a riparian forest in Maryland. Entry and Emmingham (1996) measured nutrients stored in the litter and surface mineral soil in forest and grass buffers in Oregon and found substantially greater amounts of macronutrients (P, K, Ca, Mg, and Fe) and Mn in the surface litter and soil of the forest but smaller quantities of the Zn, B, and Cu. Continuous nutrient accumulation in both litter and soil could saturate the storage capacity, requiring biomass removal (tree harvest) to promote greater uptake and continued removal of nutrients transported from agricultural fields. Poplar (*Populus ×euroamericana* 'Eugenei') trees within a multispecies riparian buffer in Iowa immobilized significant N (37 kg ha^{-1} yr^{-1}), thereby slowing or preventing N losses to water resources or the atmosphere (Tufekcioglu et al., 2003). This substantial quantity of N could be removed from the system via tree harvest or could re-enter the terrestrial nutrient cycle following death of the tree and decomposition of the woody biomass.

Soil Biology and Ecology

Studies of soil biological processes face numerous challenges including aspects of soil heterogeneity, the large diversity of organisms, and the abundance of interacting processes occurring in the dynamic soil environment (Andrén et al., 2008). Soil biological processes are especially important in AFS as biotic systems respond to the physical environment (temperature and moisture regimes) created by the tree–crop canopies and they often control the rate and direction of C and nutrient transformations. Studies of biological aspects of forest litter decomposition abound (e.g., Andersson et al., 2004; Berg and Laskowski, 2006; Kanerva and Smolander, 2007; Niemi et al., 2007) including potential impacts of climate change (Cotrufo et al., 1994; Arp et al., 1997; Oren et al., 2001; Busse et al., 2009). Many of the principles relating to C and nutrient cycling discovered for forest ecosystems have direct application to AFS. Recent attention on C cycling in AFS has focused on global climate change and the C sequestration potential of the different systems as affected by local climate and soil properties (Nair, 1993; Schroeder, 1994; Kort and Turnock, 1999; Nair and Nair, 2003; Schoeneberger, 2008). Two other areas of intense interest concern the use of soil enzyme activity as an indicator of soil quality and the role of mycorrhiza in enhancing nutrient use efficiency and water uptake (Haselwandter and Bowen, 1996; Ingleby et al., 2007; Trasar-Cepeda et al., 2008).

Udawatta et al. (2008 2009) reported that enzyme activities (fluorescein diacetate hydrolase, β-glucosidase, dehydrogenase, and glucosaminidase) increased in the tree rows of a pin oak (*Quercus palustris* Münchh.) alley cropping AFS in Missouri. The tree strips also had increased soil C and N as compared with the adjacent corn–soybean [*Glycine max* (L.) Merr.] fields. Mungai et al. (2005) studied soil enzyme activities and microbial functional diversity in a pecan [*Carya illinoinensis* (Wangenh.) K. Koch]–bluegrass (*Poa trivialis* L.) and a silver maple–corn–soybean alley cropping AFS in Missouri and found differing results for the two sites. They concluded that functionally different microbial populations may occur under the pecan trees that may affect nutrient availability in the cropped alleys. Ingleby et al. (2007) reported that trees and crops can share the same arbuscular mycorrhizal fungi (AMF) but it may take years for the colonization to benefit the growth of the crops. Lacombe et al. (2009) also reported contrasting results of microbial diversity and stability for two alley cropping sites in Quebec and Ontario and recommended that further research was needed to assess the role of tree roots in maintaining AMF and other beneficial organisms.

Szajdak et al. (2002) analyzed soils beneath a mixed species shelterbelt in eastern Poland to discern the influence of the shelterbelt on N transformations and the chemical structure of the humic acids in the soil organic matter. They reported that with increasing distance into the shelterbelt inorganic and organic N decreased as did the chemical maturity of the humic acids and the amino acids bound to them. Ivannikova et al. (2008) also found significant variation of soil properties, including biological activity, with distance and depth beneath a >100-yr-old shelterbelt in southeastern Russian. Data from different locations within the local microtopography (depressions and elevations of 10–30 cm) exhibited distinct patterns of biological activity as measured by CO_2 evolution during soil incubations. Sauer et al. (2007) reported significantly greater SOC in the surface 15 cm of soil beneath a 35-yr-old shelterbelt in Nebraska as compared with the adjacent cropped fields (Fig. 23|2). The observed increase in SOC represents an annual accrual of 10.6 g m^{-2} yr^{-1} and stable C isotope analysis indicates that fine particulate organic matter (POM) accounted for 21% of the SOC beneath the trees and 79% of the fine POM was tree-derived (Hernandez-Ramirez et al., 2011). Haile et al. (2008; 2010) also used stable C isotope signatures to determine that the majority of SOC in deeper soil layers down to 1.25 m were derived from tree sources in four slash pine (*Pinus elliottii* Engelm.)–Bahia grass silvopastoral sites representing Spodosols and Ultisols in Florida. Minimizing site disturbance and the increased diversity of plant species in AFS have been credited with reducing C losses and increasing the stability of SOC stocks.

Subtropical and Tropical Agroforestry Practices

General Differences from Temperate AFS

Similar to AFS in temperate regions, subtropical and tropical AFS can make substantial contributions to enhancing soil quality. Several significant differences in AFS structure and functioning can be noted across biomes as biophysical conditions among these biomes also differ. Solar radiation input is typically abundant, vertically incoming, and well distributed throughout the year in the tropics and subtropics compared with temperate regions. This basic difference entails the need to optimize spatial arrangements of plant canopies in multistrata configuration in subtropical and tropical AFS to maximize radiation capture, and hence maximum net ecosystem productivity (Budowski, 1993; Nygren et al., 1993; Mafongoya et al., 2006). Several subtropical and tropical crops [e.g., coffee, cacao (*Theobroma cacao* L.), ginger (*Zingiber officinale* Roscoe), black pepper (*Piper nigrum* L.), pearl millet [*Pennisetum glaucum* (L.) R. Br.], and vanilla (*Vanilla planifolia* Andrews) are also typically grown in association with shading trees (e.g., *Erythrina* spp., *Inga* spp., *Cordia alliodora* Ruiz & Pav., *Acacia mangium* Willd., *Azadirachta indica* A

Juss.) as they effectively tolerate shading and need protection from both excessive solar radiation and associated fluctuations in environmental conditions (e.g., temperature) to express their optimum yield potential (Muschler et al., 1993; Nygren et al., 1993; Kapp and Beer, 1995; Budowski and Russo, 1997; Beer et al., 1998; Nair et al., 1999; Somarriba et al., 2001). An example of the beneficial microclimate modification by tree overstories within subtropical and tropical AFS was documented in the Western Sahel by Payne et al. (1998). They observed soil temperature reductions up to 6°C at 5-cm depth and important enhancements in crop yield for pearl millet, corn, and sorghum [*Sorghum bicolor* (L.) Moench] when plants were grown under acacia trees [*Faidherbia albida* (Delile.) A. Chev.]. Crop growth failed with no tree shading under these extreme Sahel conditions.

Since the degree of shade tolerance differs across crops or forages and local environmental conditions vary too, the optimum AFS canopy structure seems to be unique for every case. Cusack and Montagnini (2004) studied the gradual transition from degraded pastureland to silvopastoral systems in Central America and concluded that intermediate tree canopy openness resulted in maximum regeneration and growth. Soto-Pinto et al. (2000) indicated

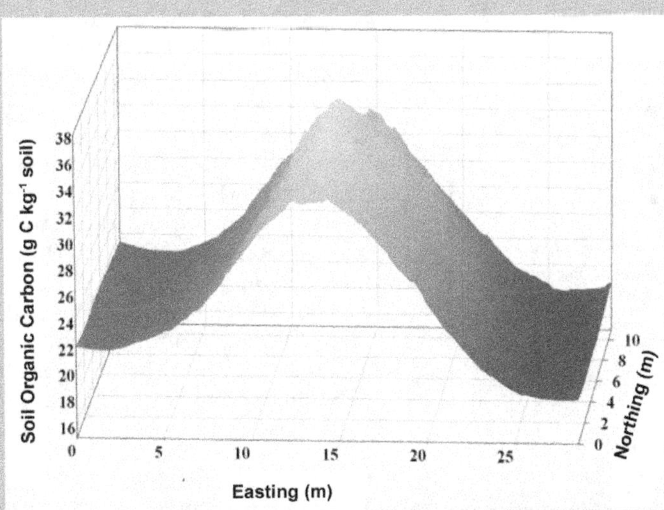

Fig. 23|2. Spatial distribution of soil organic carbon (SOC) storage from a sampling grid beneath and adjacent to a two-row shelterbelt in eastern Nebraska (Sauer et al., 2007). The uncultivated tree zone was from 5.5 to 23 m with the tree rows centered at approximately 11 and 14.5 m.

that maximum yield in coffee (a plant with C3 photosynthetic pathway) occurred with 38 to 48% of shade cover (measured by photographic techniques) in southern Mexico (Fig. 23|3). Conversely, sparse tree canopies seem more desirable for growing C4 crops. Rao et al. (1998) presented yield data for corn grown with and with no trees suggesting a direct association between light availability and productivity of the crop component in AFS (Fig. 23|4). Earlier data by Salazar and Palm (1987) also supports decreasing yield in certain crops with proximity to the tree rows and as a result of excessive shading.

Another biophysical difference across biomes impacting AFS structures and functioning is the typically low soil fertility in tropics and subtropics relative to soils in temperate regions (Tiessen et al., 1994; Sanchez et al., 1997). Low-fertility acid soils cover 41, 27, and 26% of tropical America, Africa, and Asia, respectively (Sanchez, 1976; Mafongoya et al., 2006). Tropical acid soils partly depend on nutrient recycling from deep soil layers to maintain fertility in the surface soil, thus nutrient uptake by deep tree roots in tropical and subtropical AFS can contribute to this process (Szott et al., 1999; Chikowo et al., 2003; Mafongoya et al., 2006). This nutrient capture and recycling by tree components in AFS is completed through nutrient incorporation into biomass production, deposition on soil surface through litterfall, and decomposition (Rao et al., 1998; Nair et al., 1999). This organic matter cycling is basic for sustaining soil productivity particularly in tropical environments (Ohu et al., 1994; Mafongoya et al., 2006). In addition, soil organic matter turnover is generally much faster in subtropical and tropical AFS due to higher temperatures as observed by Oelbermann et al. (2004a, 2006a, 2006b) when comparing alley cropping systems in Costa Rica and southern Canada. They also suggested the need of increasing organic matter inputs in tropical AFS to support stable pools of both soil organic matter and associated nutrients. These multiple reports collectively indicate both much more dynamic and more growth-limiting nutrient levels in subtropical and tropical AFS than in temperate regions.

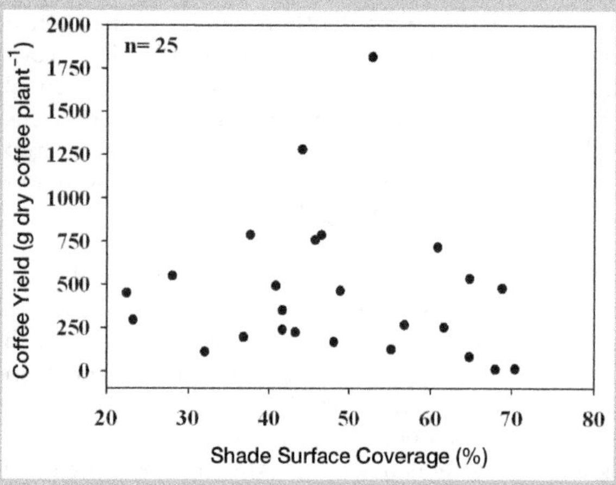

Fig. 23|3. Shading effects on coffee yields when maintaining a constant coffee population density (2200 plant ha⁻¹). Adapted from Soto-Pinto et al. (2000).

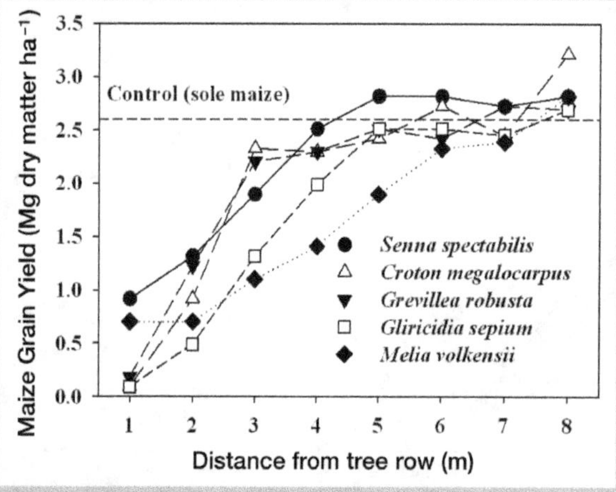

Fig. 23|4. Corn yield in alley cropping systems as a response to presence of trees (2 yr after establishment) and distance from tree rows for five different tree species at Machakos, Kenya. Adapted from Rao et al. (1998).

Types of AFS and Practices in the Subtropics and Tropics

The ample diversity of AFS in subtropics and tropics has been documented by Nair (1985), Lal

(1991), Montagnini (1992), Budowski (1993), and Brenner (1996). Although all AFS are tree-based systems, multiple differences in spatiotemporal arrangements, number of components, and level of interactions can be noted in subtropical and tropical AFS inventories. Practices range from simple AFS such as living fences of *Erythrina* spp. (Budowski, 1987; Russo, 1990a; Budowski and Russo, 1993) to much more complex AFS such as silvopastoral systems (Russo, 1990b; Montagnini, 1992; Montagnini et al., 2003) with high level of spatiotemporal interactions. Agroforestry technologies such as fodder banks are not considered AFS due to the absence of any direct interaction among system components (Nair, 1985); however, they are frequently mentioned in classification studies of agroforestry practices as they can provide similar products and services (animal food, fuelwood, timber, fiber, etc.). Typical examples of AFS and agroforestry practices in subtropics and tropics with potentially beneficial impacts on soil management are alley cropping, rotation or shifting, browsing or grazing, taungya, and orchards or home gardens.

Alley cropping is a simultaneous, spatially zoned system in which crops (typically annual crops) or pasture (mainly for mechanized forage harvest) are cultivated in wide alleys between single or multiple rows of trees or bushes as discussed above. These AFS in the tropics and subtropics typically involve a tree legume with the purpose of supplying N-rich mulch (Lal, 1991). With some exceptions, this N contribution to the overall system has consistently resulted in enhanced crop productivity. Okogun et al. (2000) found increased corn yield as a response to alley cropping with *Albizia lebbeck* (L.) Benth, *Gliricidia sepium* (Jacq.) Kunth, or *Leucaena leucocephala* Lam. de Wit. Similarly, several earlier studies suggested increasing crop productivity for alley cropping systems such as *L. leucocephala, Inga edulis* Mart., or *Erythrina poeppigiana* (Walp.) O.F. Cook with corn, and *Acioa barteri* (Hook. f. ex Oliv.) Engl. with cassava (*Manihot esculenta* Crantz) (Siaw et al., 1991). Collectively, these results support enhanced N availability status in alley cropping systems, particularly under low natural soil fertility conditions. An additional benefit can be obtained when alley cropping systems with *L. leucocephala* hedgerows are established in

steep tropical lands resulting in the formation of terraces (Lal, 1991) leading to more intensive land use capacity than on natural slopes. A typical management practice in alley cropping systems is pruning of the trees or bushes to facilitate crop growth as well as a way to enhance nutrient recycling.

One of the most traditional forms of AFS is shifting cultivation that consists of annual crop species—e.g., rice (*Oryza sativa* L.), corn, beans (*Phaseolus vulgaris* L.)—for 1 to 3 yr followed by natural regeneration of trees and woody species for 5 to 40 yr. This sequential system has been traditionally implemented by small farmers by applying slash-and-burn practices that sharply increase the dynamics and release of nutrients (Rao et al., 1998). This practice increases soil organic matter mineralization and nitrification rates during the first few years of cultivation (Montagnini and Buschbacher, 1989). Jordan (1992) showed a sharp increase in cation (Ca, Mg, and K) and NO_3 availability after cutting and burning of Amazon forest in Venezuela; however, pronounced nutrient depletions can occur after 2 to 3 yr as reported by Jordan (1992) and Tiessen et al. (1994). Leaching, runoff, and nutrient removal in harvests can account for a large portion of these nutrient losses caused by shifting cultivation. As pointed out by Sanchez (1976), shifting cultivation was traditionally viable only based on availability of sufficient land area; however, the increasing pressure for land use has restricted land availability, and hence, it has shortened (or completely eliminated) the fallow time between cropping periods, thereby impeding the restoration of natural soil conditions. Therefore, enhanced shifting cultivation systems such as improved fallows (3–4 yr) in sequence with annual crops have been proposed (Sanchez et al., 1997; Nair et al., 1999). These enhanced shifting cultivation systems substitute regeneration of natural vegetation by planting of selected tree species (typically legumes) during the fallow period (Mafongoya et al., 2006) and replace the practice of burning by mulching (Rao et al., 1998). As suggested by Rao et al. (1998), although nutrient availability can be limited shortly after mulching relative to burning, mulching assures higher soil organic matter contents, long-term steady nutrient release, reduced nutrient losses, and enhanced soil biological activity.

As a land management system and as a result of its unique spatial-temporal arrangement, taungya systems may be considered another alternative AFS for traditional shifting cultivation. With origins in Southeast Asia, taungya can be described as the planting of annual crops in the early stages of a forest plantation (Gajaseni, 1992). A typical example is teak (*Tectona grandis* L. f.) plantations in Thailand or Myanmar with annual crops—e.g., dryland rice, corn, pineapple [*Ananas comosus* (L.) Merr.], pepper, peanuts (*Arachis hypogaea* L.), cassava, and soybeans—cultivated between tree rows within 2 to 3 yr after tree planting. Since small farmers are involved in crop cultivation and in tree culture, this system can reduce the destruction of natural forest typically caused by traditional shifting cultivation. In the long term, however, sustainability of taungya systems has been questioned. Bruijnzeel (1992) reported a wide nutrient imbalance (i.e., P, Ca, K, Mg) in taungya experiments with *Pinus caribaea* Morelet and *Gmelina arborea* Roxb. in eastern Amazonia Brazil, and conifers *Agathis dammara* (Lamb.) Rich. in Indonesia and *T. grandis* in the Western Plains of Venezuela. This assessment estimated elevated nutrient losses (i.e., P, Ca, K, Mg) caused by timber extraction ranging from 33 to 82% across nutrients, locations, and types of sequential rotations. Leaching and soil erosion would further enlarge this nutrient deficit under repeated taungya systems. Bruijnzeel (1992) indicates the need for supplementary fertilization to the trees to sustain productivity after two successive forest rotations in low-fertility tropical soils.

Multiple-purpose trees intentionally grown in pasture with the primary purpose of animal browsing and/or grazing is also an extensively practiced AFS in the subtropics and tropics. Additional critical attributes of these silvopastoral systems are the appropriate compatibility of tree species with the grass component as well as the ability of the tree component to supply shelter to grazing animals (Ibrahim and Camargo, 2001). Successful examples of native tree species growing in silvopastoral systems of Costa Rica are *Alnus acuminata* Kunth (Russo, 1990a), *Hieronyma alchorneoides* Allemão and *Dipteryx panamensis* (Pittier) Record & Mell (Montagnini and Sancho-Mora, 1990; Montagnini et al., 2003), and *Cordia alliodora* (Ruis &. Pav.) Oken (Camargo et al., 2000). Since this type of AFS compresses more components (i.e., grass–tree–animal) and unique processes such as animal manure production take place, the level of interactions in silvopastoral systems is typically much higher (Montagnini, 1992). This increased level of interactions is expected to contribute additional resilience to the overall system (Sanchez, 1999) compared with pasture in monoculture.

Soil Physical Properties

Implementation of AFS in the subtropics and tropics can effectively enhance numerous soil properties. Data by Nyamadzawo et al. (2007) suggests enhanced soil structure and increasing infiltration rates in years following improved fallows compared with both natural fallow and continuous corn in tropical Africa. They also indicated that if improved fallows are followed by corn cultivation with no-tillage management, these enhancements in soil structure and infiltration persist over 2 yr during the post-fallow period. Reports by Hulugalle and Kang (1990) and Torquebiau and Kwesiga (1996) support significant reductions in soil bulk density and penetration resistance and increases in infiltration in response to alley cropping with *G. sepium* and improved fallow with *Sesbania sesban* L. Merr., respectively. Data by Hulugalle and Kang (1990) and Hulugalle and Ndi (1993) also support overall improved soil physical properties from alley cropping related to amelioration of surface seal formation following intensive rainfall events. After examining several cropping systems and soils, Dalland et al. (1993), Mapa and Gunasena (1995), and Buresh and Tian (1998) consistently associated enhancements in soil bulk density and infiltration with concomitant increases in soil aggregate stability and organic matter accumulation. Reports by Schroth et al. (1996), Torquebiau and Kwesiga (1996), Schroth (1999), and Schroth and Zech (1995) collectively suggest increases in soil organic matter, root growth and turnover, litterfall, and macrofauna activity as leading to enhanced soil physical conditions under AFS. Van Noordwijk et al. (1991) established the key role of tree roots in enhancing soil pore connectivity. Macropore formation by tree roots and associated increases in macrofauna activity can be an effective contribution of the tree component to enhance soil physical properties in AFS.

Land use systems involving multistrata canopy and mulching are also expected to show reductions in raindrop impact and run-off water (amount and velocity) as well as enhanced rainwater redistribution (Wallace, 1996). Reductions in ambient temperatures by tree shading in AFS can decrease both soil water evaporation and surface desiccation (Lal, 1991; Rao et al., 1998), typically leading to a more uniform soil water availability in semiarid subtropical and tropical regions. Collectively, these multiple favorable effects of AFS on soil physical properties also have major implications for minimizing soil losses by wind and/or water erosion. As documented by Lal (1991), Mapa and Gunasena (1995), Alegre and Rao (1996), and Mafongoya et al. (2006), AFS typically diminish soil erosion risks compared with crop-based farming systems. Soil erosion may also be decreased in AFS by combining tree species with different growth habits that rapidly create a dense, uniform ground coverage in early stages of the system (Juo et al., 1995; Buresh and Tian, 1998).

Nutrient Cycling and Soil pH

A beneficial contribution of the tree component in AFS in the tropics and subtropics to nutrient cycling and productivity is the increase in N supply via biological fixation of atmospheric N_2. Biological N_2 fixation is typically followed by incorporation of N into tree biomass (shoots + roots) and the subsequent recycling via soil organic matter decomposition (Haggar et al., 1993; Nair et al., 1999). As reviewed by Giller and Wilson (1991), numerous studies indicated the high potential of N-fixing trees (including legumes and several nonlegumes) in supplying additional N to other components in AFS in tropical and subtropical regions. Improved fallows based on tree legumes can accrue up to 200 kg N ha^{-1} in tropical soils (Giller et al., 1997). Similarly, Dalland et al. (1993) documented enhanced N supply to corn due to alley cropping particularly with *L. leucocephala* in tropical Africa. However, literature also indicates that although large amounts of N can be fixed in subtropical and tropical AFS, most of this N does not become readily available in the short term (Buresh and Tian, 1998; Mafongoya et al., 1998). Data by Oelbermann et

al. (2004a, 2004b, 2006b) in Central America suggest low N mineralization rates in alley cropping systems that may restrict the success of the overall AFS. Mafongoya et al. (1998) indicate only up to one-fifth of recent N additions via tree pruning, leaf drop, or litter may be readily released and taken up within the next crop growing season. Although N losses due to leaching and gaseous emissions cannot be neglected (Chikowo et al., 2004), the remaining 80% of the total N added via fixation and biomass cycling can be categorized as slow-release N. This relatively low N availability to plants can occur due to soil microbial N immobilization in early stages of the mineralization process and the recalcitrant N characteristics in the biomass of some tree legumes (Haggar et al., 1993; Palm, 1995). After examining biomass characteristics in tropical tree legumes, Buresh and Tian (1998) and Mafongoya et al. (1998) suggested adopting the ratio of lignin + polyphenol to N in tree foliage as an indicator of N mineralization rates in recently added litter in AFS as narrow lignin + polyphenol to N ratios strongly correlated with increasing N mineralization rates. In addition to understanding biomass attributes that precondition N release, the extent to which low N mineralization rates can limit AFS productivity also depends on the synchrony between N availability and crop N demand (Palm, 1995; Mugendi et al., 1999). There is a critical need for additional mechanistic understanding regarding soil N availability and its management in AFS.

An additional benefit to soil fertility of tree components in AFS is the nutrient uptake by deep-rooting trees from subsoil layers where nutrients are not accessible to most annual crops, and the subsequent nutrient redistribution to the topsoil via biomass production and decomposition (van Noordwijk et al., 1996; Nair et al., 1999; Buresh et al., 2004). The effective retrieval of deep soil NO_3 by trees in AFS and the resulting enhancement in overall N efficiency are well supported by multiple reports (Birch, 1964; Hartemink et al., 1996; Shepherd et al., 1996; Mekonnen et al., 1997; Jama et al., 1998b; Chikowo et al., 2003; see also Fig. 23I5). This NO_3 capturing effect by deep-rooted tree species may potentially also mitigate NO_3 leaching (Birch, 1964; Shepherd et al., 1996) and associated groundwater contamination. In addition to these reports about enhancement in N supply to

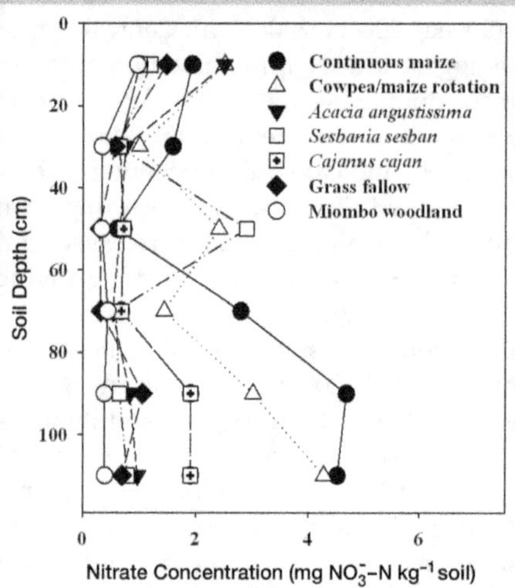

Fig. 23|5. Soil NO$_3$ profiles for corn grown in diverse cropping systems in eastern Africa. Adapted from Chikowo et al. (2003).

crops in AFS, favorable retrieval and cycling effects by deep tree roots have been also observed for mobile basic cations. Dalland et al. (1993) reported increasing soil Mg and K content in response to mulching in alley cropping with *L. leucocephala*. Similarly, Sanchez (1999) indicated enhanced K status in corn cultivation after a *S. sesban* fallow. These results collectively suggest that utilization of soil nutrients (and water) in AFS is greater than in monocultures as AFS typically combine shallow-rooted species with deep-rooted species. However, as pointed out by Lal (1991), very low-fertility acid soils may have insufficient nutrients available in deep soil layers to be recycled by tree roots. In addition, van Noordwijk et al. (1996) suggested shallow roots of trees in AFS may compete for nutrients and water (particularly under limiting soil moisture) with the crop and/or pasture components of AFS perhaps limiting uptake by crop roots. Understanding and managing competition for nutrients and water constitutes a key feature for sustainability and success in AFS, in particular for simultaneous AFS such as alley cropping (Sanchez, 1995; Rao et al., 1998).

Soil Al toxicity coupled with low P availability constitutes a fundamental biophysical constraint for crop productivity in the subtropics and tropics (Sanchez et al., 1997). The potential role of tree components in AFS for both Al toxicity alleviation and increasing P availability remains unclear as existing reports are inconsistent across geographic locations and AFS types. Buresh and Tian (1998) and Nair et al. (1999) indicated that AFS typically cycle insufficient P to sustain the overall system productivity. Similarly, Lal (1991) suggested that since AFS typically enhance soil N availability, nutrients such as P and Zn may become growth limiting. Conversely, although total P remained unaffected, data by Maroko et al. (1999) revealed that both natural and improved fallows in tropical soils can effectively increase preferential P allocation into labile fractions of soil organic matter. Soil P availability would be enhanced under these conditions assuming that these labile organic matter fractions can act as a source of readily available forms of P (Rao et al., 1998). Both Rao et al. (1998) and Mafongoya et al. (2006) suggested that abundant production of Al-binding organic acids by trees in AFS may result in soil Al detoxification and the associated enhancement in soil P availability. Although not clearly understood, reports by Pande and Tarafdar (2004) and Satter et al. (2006) indicated that mycorrhizal infections in tree roots can potentially contribute to enhanced soil P availability in AFS.

Strategic use of fertilizers has been proposed to alleviate pronounced nutrient deficiencies in tropical and subtropical AFS. Reports by Muschler et al. (1993), Szott and Kass (1993), Jama et al. (1998a), and Khanna (1998) across a variety of ecophysical conditions and management systems suggest the need for rational use of P fertilizers in AFS. Selection of tree species to be included in AFS can also critically impact nutrient management plans as different tree species may comparatively have both different nutrient requirements and diverse effects on soil fertility (Juo et al., 1995; Buresh and Tian, 1998; Rao et al., 1998; Montagnini et al., 2003; Mafongoya et al., 2006).

Soil Biology and Ecology

Biological activity is essential for maintaining soil fertility (i.e., nutrient turnover and availability) in sustainable cropping systems (Sanginga et al., 1992; Buresh and Tian, 1998; Rao et al., 1998; Mafongoya et al., 2006). These studies also suggest the lack of a comprehensive understanding of soil biological processes in tropical and subtropical environments. However, they also indicated that critical soil biological attributes such as macrofauna activity can be considerably enhanced by establishing AFS. Buresh and Tian (1998) found two to three times greater earthworm populations with different improved fallows than with continuous corn cultivation in western Africa. Budowski and Russo (1997) also indicated a greater earthworm population if *Erythrina* spp. is grown as shading trees in croplands in Central America. Likewise, Rao et al. (1998) presented data for macrofaunal biomass indicating five times more macrofaunal biomass with improved fallow as compared with corn monoculture. In their study, earthworm biomass was 10 times higher in the improved fallow. Sileshi and Mafongoya (2006) also found increasing numbers of several macrofaunal litter transformers as a response to AFS establishment. Similarly, Adejuyigbe et al. (1999) reported two- to six-fold higher soil microarthropod population densities (i.e., *Acari, Collembola*) in soil under fallows (i.e., planted and natural) compared with continuous cropping (i.e., corn and cassava) in southwestern Nigeria. They associated these enhancements in macrofaunal counts with both increasing lignin contents in tree litterfall resulting in relatively slower litter decomposition as well as greater soil water content. Collectively, these results support the contribution of the tree component in AFS to preserve and potentially enhance agricultural soils by restoring macrofaunal population and activity. As an additional ecosystem service, AFS such as living fences and silvopastures can also increase opportunities for biodiversity conservation through improved interconnectivity (i.e., biological corridors for wildlife) among surrounding natural ecosystems (León and Harvey, 2006).

Underlying mechanisms for improving soil biological and ecological processes in AFS may include increasing soil organic matter and microclimate modifications, particularly via shading. Favorable conditions for biological activity in AFS are directly promoted by: minimal soil disruption, mixture of plant species for enhanced biodiversity, permanent vegetative ground cover (with rapid regrowth), and litter management. Increased biomass input would typically lead to increased soil organic matter in AFS. In addition, microclimate modifications in AFS (i.e., via tree shading that buffers extreme temperature fluctuations) can also reduce soil organic matter decomposition rates. Many studies support these trends across a wide variety of ecosystems and AFS. Compared with continuous corn cultivation, data by Nyamadzawo et al. (2008) shows 28% greater C retention (to 20-cm depth) after 2 yr following their improved fallow management in eastern Africa. Rao et al. (1998) reported higher C accretion rates as a function of increasing plant residue inputs in coppicing improved fallow systems. After comparing numerous AFS in southern Mexico, Roncal-García et al. (2008) found increasing C accumulation to be associated with greater biodiversity and degree of complexity (i.e., number of tree species and morphology). Mapa and Gunasena (1995) in Sri Lanka and Oelbermann et al. (2006b) in Central America also reported higher C accumulation in response to alley cropping implementation.

Summary

Knowledge integrated in this chapter about the impacts of AFS on soil management in temperate, subtropical, and tropical biomes support the beneficial, holistic role of tree components in agricultural land use systems. Compared with annual monocultures, AFS can enhance several soil physical properties, improving soil resilience and reducing soil erosion losses. Likewise, in AFS, soil fertility and nutrient use efficiency of companion crops can be improved by trees through the release of nutrients from leaf, root, and woody components as well as via biological N_2 fixation and cycling (if N-fixing trees are included), and uptake and recycling of various nutrients from deep subsoil horizons. Contribution of AFS to biological diversity and activity, typically through sheltering effects coupled with both increases in amounts of SOC and enhancement of food web dynamics, can also be substantial. These various prospective advantages may reflect underlying mechanisms in the functioning of AFS oriented to optimize the utilization of resources (e.g., light, water, nutrients) in both time and space. Current research should increase the focus on identifying the best spatiotemporal combinations of system components (e.g., trees, crops, pastures, animals) to make AFS functioning and structure more efficient, with the aims of attaining optimum productivity and profitability with maximum environmental services and reduced economic risks. When assessing AFS performance as a whole, careful balance between crop productivity goals and benefits from the tree component needs to take into account potentially hidden beneficial, long-term contributions of trees to the overall system.

References

Adejuyigbe, C.O., G. Tian, and G.O. Adeoye. 1999. Soil microarthropod populations under natural and planted fallows in southwestern Nigeria. Agrofor. Syst. 47:263–272.

Alegre, J.C., and M.R. Rao. 1996. Soil and water conservation by contour hedging in the humid tropics of Peru. Agric. Ecosyst. Environ. 57:17–25.

Anderson, D.W. 1987. Pedogenesis in the grassland and adjacent forests of the Great Plains. p. 53–93 In B.A. Stewart (ed.) Adv. Soil Sci. Vol. 7. Springer-Verlag, New York.

Anderson, S.H., R.P. Udawatta, T. Seobi, and H.E. Garrett. 2009. Soil water content and infiltration in agroforestry buffer strips. Agrofor. Syst. 75:5–16.

Andersson, M.A.K., A. Kjøller, and S. Struwe. 2004. Microbial enzyme activities in leaf litter, humus and mineral soil layers of European forests. Soil Biol. Biochem. 36:1527–1537.

Andrén, O., H. Kirchmann, T. Kätterer, J. Magid, E.A. Paul, and D.C. Coleman. 2008. Visions of a more precise soil biology. Eur. J. Soil Sci. 59:380–390.

Angers, D.A., and J. Caron. 1998. Plant-induced changes in soil structure: Processes and feedbacks. Biogeochemistry 42:55–72.

Armson, K.A. 1977. Forest soils: Properties and processes. Univ. of Toronto Press, Toronto.

Arnold, G. 1992. Soil acidification as caused by the nitrogen uptake pattern of Scots pine (Pinus sylvestris). Plant Soil 142:41–51.

Arp, W.J., P.J. Kuikman, and A. Gorissen. 1997. Climate change: The potential to affect ecosystem functions through changes in amount and quality of litter. p. 187–200 In G. Cadisch and K.E. Giller (ed.) Driven by nature: Plant litter quality and decomposition. CABI, Wallingford, UK.

Balandier, P., F.-X. de Montard, and T. Curt. 2008. Root competition for water between trees and grass in a silvopastoral plot of 10 year old Prunus avium. p. 253–270 In D.R. Batish et al. (ed.) Ecological basis of agroforestry. CRC Press, Boca Raton.

Batish, D.R., R.K. Kohli, S. Jose, and H.P. Singh (ed.). 2008. Ecological basis of agroforestry. CRC Press, Boca Raton.

Beer, J., R. Muschler, D. Kass, and E. Somarriba. 1998. Shade management in coffee and cacao plantations. Agrofor. Syst. 38:139–164.

Berg, B., and R. Laskowski. 2006. Litter decomposition: A guide to carbon and nutrient turnover. Academic Press, Amsterdam.

Bharati, L., K.-H. Lee, T.M. Isenhart, and R.C. Schultz. 2002. Soil-water infiltration under crops, pasture, and established riparian buffer in Midwestern USA. Agrofor. Syst. 56:249–257.

Binkley, D., and C. Giardina. 1998. Why do tree species affect soils? The warp and woof of tree-soil interactions. Biogeochemistry 42:89–106.

Birch, H.F. 1964. Mineralization of plant nitrogen following alternative wet and dry conditions. Plant Soil 20:43–49.

Bird, P.R., D. Bicknell, P.A. Bulman, S.J.A. Burke, J.F. Leys, J.N. Parker, F.J. van der Sommen, and P. Voller. 1992. The role of shelter in Australia for protecting soils, plants and livestock. Agrofor. Syst. 20:59–86.

Blazier, M.A., L.A. Gaston, T.R. Clason, K.W. Farrish, B.P. Oswald, and H.A. Evans. 2008. Nutrient dynamics and tree growth of silvopastoral systems: Impact of poultry litter. J. Environ. Qual. 37:1546–1558.

Brandle, J.R., L. Hodges, and X.H. Zhou. 2004. Windbreaks in North American agricultural systems. Agrofor. Syst. 61:65–78.

Brenner, A.J. 1996. Microclimate modifications in agroforestry. p. 159–187 In C.K. Ong and P. Huxley (ed.) Tree-crop interactions. CABI, Wallingford, UK.

Bruijnzeel, L.A. 1992. Sustainability of fast-growing plantation species in the humid tropics with particular reference to nutrients. p. 51–67. In C.F. Jordan et al. (ed.) Taungya: Forest plantations with agriculture in Southeast Asia. CABI, Wallingford, UK.

Budowski, G. 1987. Living fences in tropical America, a widespread agroforestry practice. p. 169–178. In H.L. Gholz (ed.) Agroforestry: Realities, possibilities and potentials. Springer, Heidelberg.

Budowski, G. 1993. The scope and potential of agroforestry in Central-America. Agrofor. Syst. 23:121–131.

Budowski, G., and R.O. Russo. 1993. Live fence posts in Costa Rica: A compilation of the farmer's beliefs and technologies. J. Sustain. Agric. 3:65–87.

Budowski, G., and R.O. Russo. 1997. Nitrogen-fixing trees and nitrogen fixation in sustainable agriculture: Research and challenges. Soil Biol. Biochem. 29:767–770.

Buresh, R.J., E.C. Rowe, S.J. Livesley, G. Cadisch, and P. Mafongoya. 2004. Opportunities for capture of deep soil nutrients. p. 109–125. In M. van Noordwijk et al. (ed.) Belowground interactions in tropical agroecosystems: Concepts and models with multiple plant components. CABI, Wallingford, UK.

Buresh, R.J., and G. Tian. 1998. Soil improvement by trees in sub-Saharan Africa. Agrofor. Syst. 38:51–76.

Busse, M.D., F.G. Sanchez, A.W. Ratcliff, J.R. Butnor, E.A. Carter, and R.F. Powers. 2009. Soil carbon sequestration and changes in fungal and bacterial biomass following incorporation of forest residues. Soil Biol. Biochem. 41:220–227.

Byington, E.K. 1990. Agroforestry in the temperate zone. p. 228–289 In K.G. MacDicken and N.T. Vergara (ed.) Agroforestry: Classification and management. John Wiley & Sons, New York.

Cadisch, G., and K.E. Giller. 1997. Driven by nature: Plant litter quality and decomposition. CABI, Wallingford, UK.

Calfapietra, C., B. Gielen, D. Karnosky, R. Ceulemans, and G. Scarascia Mugnozza. 2010. Response and potential of agroforestry crops under global change. Environ. Pollut. 158:1095–1104.

Camargo, C., M. Ibrahim, E. Somarriba, B. Finegan, and D. Current. 2000. Factores ecológicos y socioeconómicos que influyen en la regeneración natural de laurel (Cordia alliodora) en sistemas silvopastoriles del trópico húmedo y sub-húmedo de Costa Rica. (In Spanish.) Agroforestería en las Américas 7:46–52.

Carlson, D.H., S.H. Sharrow, W.H. Emmingham, and D.P. Lavender. 1994. Plant-soil-water relations in forestry and silvopastoral systems in Oregon. Agrofor. Syst. 25:1–12.

Chikowo, R., P. Mapfumo, P. Nyamugafata, and K.E. Giller. 2004. Mineral N dynamics, leaching and nitrous oxide losses under maize following two-year improved-fallows on a sandy loam soil in Zimbabwe. Plant Soil 259:315–330.

Chikowo, R., P. Mapfumo, P. Nyamugafata, G. Nyamadzawo, and K.E. Giller. 2003. Nitrate-N dynamics following improved fallows and maize root development in a Zimbabwean sandy clay loam. Agrofor. Syst. 59:187–195.

Cleugh, H.A., and D.E. Hughes. 2002. Impact of shelter on crop microclimates; a synthesis of results from wind tunnel and field experiments. Aust. J. Exp. Agric. 42:679–701.

Coile, T.S. 1933. Soil reaction and forest types in the Duke Forest. Ecology 14:323–333.

Cook, P.S., and T.T. Cable. 1995. The scenic beauty of shelterbelts on the Great Plains. Landsc. Urban Plan. 32:63–69.

Cotrufo, M.F., P. Ineson, and A.P. Rowland. 1994. Decomposition of tree leaf litters grown under elevated CO_2: Effect of litter quality. Plant Soil 163:121–130.

Cusack, D., and F. Montagnini. 2004. The role of native species plantations in recovery of understory woody diversity in degraded pasturelands of Costa Rica. For. Ecol. Manage. 188:1–15.

Dalland, A., P.I. Våje, R.B. Mathews, and B.R. Singh. 1993. The potential of alley cropping in improvement of cultivation systems in the high rainfall areas of Zambia. III. Effects on soil chemical and physical properties. Agrofor. Syst. 21:117–132.

Dickinson, C.H., and G.J.F. Pugh. 1974. Biology of plant litter decomposition. Vol. 2. Academic Press, London.

Droze, W.H. 1977. Trees, prairies, and people: A history of tree planting in the Plains States. Texas Women's Univ., Denton.

Dyksterhuis, E.J. 1957. The savannah concept and its use. Ecology 38:435–442.

Entry, J.A., and W.H. Emmingham. 1996. Nutrient content and extractability in riparian soils supporting forests and grasslands. Appl. Soil Ecol. 4:119–124.

Gajaseni, J. 1992. Overview of taungya. p. 3–8. In C.F. Jordan et al. (ed.) Taungya: Forest plantations with agriculture in Southeast Asia. CABI, Wallingford, UK.

Gale, W.J., and C.A. Cambardella. 2000. Carbon dynamics of surface residue- and root-derived organic matter under simulated no-till. Soil Sci. Soc. Am. J. 64:190–195.

Garrett, H.E., M.S. Kerley, K.P. Ladyman, W.D. Walter, L.D. Godsey, J.W. van Sambeek, and D.K. Brauer. 2004. Hardwood silvopasture management in North America. Agrofor. Syst. 61:21–33.

Garrett, H.E., W.J. Rietveld, and R.F. Fisher (ed.). 2000. North American agroforestry: An integrated science and practice. ASA, Madison, WI.

Garrity, D.P. 2004. Agroforestry and the achievement of the millennium development goals. Agrofor. Syst. 61:5–17.

Giller, K.E., G. Cadisch, C. Ehaliotis, E. Adams, W.D. Sakala, and P.L. Mafongoya. 1997. Building soil nitrogen capital in Africa. p. 151–192. In R.J. Buresh et al. (ed.) Replenishing soil fertility in Africa. SSSA Spec. Publ. 51. SSSA, Madison, WI.

Giller, K.E., and K.J. Wilson. 1991. Nitrogen fixation in tropical cropping systems. CABI, Wallingford, UK.

Gillespie, A.R., S. Jose, D.B. Mengel, W.L. Hoover, P.E. Pope, J.R. Seifert, D.J. Biehle, T. Stall, and T.J. Benjamin. 2000. Defining competition vectors in a temperate alley cropping system in the Midwestern USA. I. Production physiology. Agrofor. Syst. 48:25–40.

Gliński, J., and W. Stępniewski. 1985. Soil aeration and its role for plants. CRC Press, Boca Raton, FL.

Gordon, A.M., and S.M. Newman (ed.). 1997. Temperate agroforestry systems. CABI, Wallingford, UK.

Groffman, P.M., G. Howard, A.J. Gold, and W.M. Nelson. 1996. Microbial nitrate processing in shallow groundwater in a riparian forest. J. Environ. Qual. 25:1309–1316.

Gupta, J.P., G.G.S.N. Rao, G.N. Gupta, and B.V. Ramana Rao. 1983. Soil drying and wind erosion as affected by different types of shelterbelts planted in the desert region of Western Rajasthan, India. J. Arid Environ. 6:53–58.

Gyenge, J.E., M.E. Fernández, and T.M. Schlichter. 2008. Tree-grass interactions and water use in silvopastoral systems in N.W. Patagonia. p. 171–180 In D.R. Batish et al. (ed.) Ecological basis of agroforestry. CRC Press, Boca Raton.

Haggar, J.P., E.U. Tanner, J.W. Beer, and D.C.L. Kass. 1993. Nitrogen dynamics of tropical agroforestry and annual cropping systems. Soil Biol. Biochem. 25:1363–1378.

Haile, S.G., P.K.R. Nair, and V.D. Nair. 2008. Carbon storage of different soil-size fractions in Florida silvopastoral systems. J. Environ. Qual. 37:1789–1797.

Haile, S.G., V.D. Nair, and P.K.R. Nair. 2010. Contribution of trees to carbon storage in soils of silvopastoral systems in Florida, USA. Glob. Change Biol. 16:427–438.

Hall, N., V.S. Russell, and C.D. Hamilton. 1958. Trees and microclimate. p. 259–264 In Climatology and Micrometeorology: Proc. of the Canberra Symposium. UNESCO, Paris.

Hartemink, A.E., R.J. Buresh, B. Jama, and B.H. Janssen. 1996. Soil nitrate and water dynamics in sesbania fallows, weed fallows, and maize. Soil Sci. Soc. Am. J. 60:568–574.

Haselwandter, K., and G.D. Bowen. 1996. Mycorrhizal relations in trees for agroforestry and land rehabilitation. For. Ecol. Manage. 81:1–17.

Hernandez-Ramirez, G., T.J. Sauer, C.A. Cambardella, J.R. Brandle, and D.E. James. 2011. Carbon sources and dynamics in afforested and cultivated corn belt soils. Soil Sci. Soc. Am. J. 75:216–225.

Hill, A.R. 1996. Nitrate removal in stream riparian zones. J. Environ. Qual. 25:743–755.

Hill, D.B., and L.E. Buck. 2000. Forest farming practices. p. 283–320 In H.E. Garrett et al. (ed.) North American agroforestry: An integrated science and practice. ASA, Madison, WI.

Horn, R., and A. Smucker. 2005. Structure formation and its consequences for gas and water transport in unsaturated arable and forest soils. Soil Tillage Res. 82:5–14.

Hubbard, R.K., and R.R. Lowrance. 1996. Solute transport and filtering through a riparian forest. Trans. ASAE 39:477–488.

Hudson, B.D. 1994. Soil organic matter and available water capacity. J. Soil Water Conserv. 49:189–194.

Hulugalle, N.R., and B.T. Kang. 1990. Effect of hedgerow species in alley cropping systems on surface soil physical properties of an Oxic Paleustalf in southwestern Nigeria. J. Agric. Sci. 114:301–307.

Hulugalle, N.R., and J.N. Ndi. 1993. Effects of no-tillage and alley cropping on soil properties and crop yields in a typic kandiudult of southern Cameroon. Agrofor. Syst. 22:207–220.

Ibrahim, M., and J.C. Camargo. 2001. Produtividade e servicos ambientais de sistemas silvipastoris: Experiencias do CATIE. (In Portuguese.) p. 331–347. In M.M. Carvalho (ed.) Sistemas agroflorestais pecuarios: Opcoes de sustentabilidade para áreas tropicais e subtropicais. Ministerio da Agricultura Pecúaria e Abastecimento, Embrapa Gado de Leite, FAO, Juiz de Fora, Brazil.

Ingleby, K., J. Wilson, R.C. Munro, and S. Cavers. 2007. Mycorrhizas in agroforestry: Spread and sharing of arbuscular mycorrhizal fungi between trees and crops: Complementary use of molecular and microscopic approaches. Plant Soil 294:125–136.

Ivannikova, L.A., V.M. Alifanov, and L.A. Gugalinskaya. 2008. Biological activity of chernozem on different elements of microtopography. Eurasian Soil Sci. 41:1456–1462.

Jama, B., R.J. Buresh, and F.M. Place. 1998a. Sesbania tree fallows on phosphorus-deficient sites: Maize yield and financial benefit. Agron. J. 90:717–726.

Jama, B., J.K. Ndufa, R.J. Buresh, and K.D. Shepherd. 1998b. Vertical distribution of roots and soil nitrate: Tree species and phosphorus effects. Soil Sci. Soc. Am. J. 62:280–286.

Jenny, H. 1941. Factors of soil formation: A system of quantitative pedology. McGraw-Hill, New York.

Jordan, C. 1992. Nutrient cycling in taungya. p. 44–50. *In* C.F. Jordan et al. (ed.) Taungya: Forest plantations with agriculture in Southeast Asia. CABI, Wallingford, UK.

Jose, S., A.R. Gillespie, J.R. Seifert, and D.J. Biehle. 2000a. Defining competition vectors in a temperate alley cropping system in the Midwestern USA: 2. Competition for water. Agrofor. Syst. 48:41–59.

Jose, S., A.R. Gillespie, J.R. Seifert, D.B. Mengel, and P.E. Pope. 2000b. Defining competition vectors in a temperate alley cropping system in the Midwestern USA: 3. Competition for nitrogen and litter decomposition dynamics. Agrofor. Syst. 48:61–77.

Juo, A.S.R., K. Franzluebbers, A. Dabiri, and B. Ikhile. 1995. Changes in soil properties during long-term fallows and continuous cultivation after forest clearing in Nigeria. Agric. Ecosyst. Environ. 56:9–18.

Kanerva, S., and A. Smolander. 2007. Microbioal activities in forest floor layers under silver birch, Norway spruce and Scots pine. Soil Biol. Biochem. 39:1459–1467.

Kapp, G.B., and J. Beer. 1995. A comparison of agrisilvicultural systems with plantation forestry in the Atlantic Lowlands of Costa Rica. Agrofor. Syst. 32:207–223.

Karki, U., M.S. Goodman, and S.E. Sladden. 2009. Nitrogen source influences on forage and soil in young southern-pine silvopasture. Agric. Ecosyst. Environ. 131:70–76.

Keesman, K.J., W. van der Werf, and H. van Keulen. 2007. Production ecology of agroforestry systems: A minimal mechanistic model and analytical derivation of the land equivalent ratio. Math. Biosci. 209:608–623.

Khanna, P.K. 1998. Nutrient cycling under mixed-species tree systems in Southeast Asia. Agrofor. Syst. 38:99–120.

Kho, R.M. 2008. Approaches to tree-environment-crop interactions. p. 51–72 *In* Batish et al (ed.) Ecological basis of agroforestry. CRC Press, Boca Raton, FL.

Kirkham, M.B. 2005. Principles of soil and plant water relations. Elsevier Academic Press, Burlington, MA.

Kong, A.Y.Y., and J. Six. 2010. Tracing root vs. residue carbon into soils from conventional and alternative cropping systems. Soil Sci. Soc. Am. J. 74:1201–1210.

Kort, J. 1988. Benefits of windbreaks to field and forage crops. Agric. Ecosyst. Environ. 22/23:165–190.

Kort, J., and R. Turnock. 1999. Carbon reservoir and biomass in Canadian prairie shelterbelts. Agrofor. Syst. 44:175–186.

Kumar, S., S.H. Anderson, and R.P. Udawatta. 2010. Agroforestry and grass buffer influences on macropores measured by computed tomography under grazed pasture systems. Soil Sci. Soc. Am. J. 74:203–212.

Lacombe, S., R.L. Bradley, C. Hamel, and C. Beaulieu. 2009. Do tree-based intercropping systems increase the diversity and stability of soil microbial communities? Agric. Ecosyst. Environ. 131:25–31.

Lal, R. 1991. Myths and scientific realities of agroforestry as a strategy for sustainable management for soils in the tropics. Adv. Soil Sci. 15:91–137.

León, M.C., and C.A. Harvey. 2006. Live fences and landscape connectivity in a neotropical agricultural landscape. Agrofor. Syst. 68:15–26.

Lin, B.B. 2007. Agroforestry management as an adaptive strategy against potential microclimate extremes in coffee agriculture. Agric. For. Meteorol. 144:85–94.

Lin, C.H., R.L. McGraw, M.F. George, and H.E. Garrett. 1999. Shade effects on forage crops with potential in temperate agroforestry practices. Agrofor. Syst. 44:109–119.

Livesley, S.J., P.J. Gregory, and R.J. Buresh. 2004. Competition in tree agroforestry systems. 3. Soil water distribution and dynamics. Plant Soil 264:129–139.

Long, A.J. 1993. Agroforestry in the temperate zone. p. 443–468 *In* P.K.R. Nair (ed.) An introduction to agroforestry. Kluwer, Dordrecht, the Netherlands.

Lowrance, R., R. Todd, J. Fail, Jr., O. Hendrickson, Jr., R. Leonard, and L. Asmussen. 1984. Riparian forests as nutrient filters in agricultural watershed. Bioscience 34:374–377.

Lu, J. 2006. Energy balance and economic benefits of two agroforestry systems in northern and southern China. Agric. Ecosyst. Environ. 116:255–262.

Mafongoya, P.L., K.E. Giller, and C.A. Palm. 1998. Decomposition and nitrogen release patterns of tree prunings and litter. Agrofor. Syst. 38:77–79.

Mafongoya, P.L., E. Kuntashula, and G. Sileshi. 2006. Managing soil fertility and nutrient cycles through fertilizer trees in Southern Africa. p. 273–289. *In* N. Uphoff et al. (ed.) Biological approaches to sustainable soil systems. CRC Press, Boca Raton, FL.

Mapa, R.B., and H.P.M. Gunasena. 1995. Effect of alley cropping on soil aggregate stability of a tropical alfisol. Agrofor. Syst. 32:237–245.

Maroko, J.B., R.J. Buresh, and P.C. Smithson. 1999. Soil phosphorus fractions in unfertilized fallow-maize systems on two tropical soils. Soil Sci. Soc. Am. J. 63:320–326.

Mekonnen, K., R.J. Buresh, and B. Jama. 1997. Root and inorganic nitrogen distributions in sesbania fallow, natural fallow, and maize fields. Plant Soil 188:319–327.

Montagnini, F. 1992. Sistemas agroforestales: Principios y aplicaciones en los tropicos. (In Spanish.) OTS, San Jose, Costa Rica.

Montagnini, F., and R. Buschbacher. 1989. Nitrification rates in two undisturbed tropical rain forests and three slash-and-burn sites of the Venezuelan Amazon. Biotropica 21:9–14.

Montagnini, F., and F. Sancho-Mora. 1990. Impacts of native trees on tropical soils: A study in the Atlantic lowlands of Costa Rica. Ambio 19:386–390.

Montagnini, F., L. Ugalde, and C. Navarro. 2003. Growth characteristics of some native tree species used in silvopastoral systems in the humid lowlands of Costa Rica. Agrofor. Syst. 59:163–170.

Mosquera-Losada, M.R., and A. Giguerio. 2005. Silvopastoralism and sustainable land management. CABI, Wallingford, UK.

Mugendi, D.N., P.K.R. Nair, J.N. Mugwe, M.K. O'Neill, M.J. Swift, and P.L. Woomer. 1999. Alley cropping of maize with calliandra and leucaena in the subhumid highlands of Kenya. Part 2: Biomass decomposition, N mineralization, and N uptake by maize. Agrofor. Syst. 46:51–64.

Mungai, N.W., P.P. Motavalli, R.J. Kremer, and K.A. Nelson. 2005. Spatial variation of soil enzyme activities and microbial functional diversity in temperate alley cropping systems. Biol. Fertil. Soils 42:129–136.

Muschler, R.G., P.K.R. Nair, and L. Meléndez. 1993. Crown development and biomass production of pollarded *Erythrina berteroana, E. fusca* and *Gliricidia sepium* in the humid tropical lowlands of Costa Rica. Agrofor. Syst. 24:123–143.

Nair, P.K.R. 1985. Classification of agroforestry systems. Agrofor. Syst. 3:97–128.

Nair, P.K.R. 1993. An introduction to agroforestry. Kluwer Acad. Publ., Dordrecht, the Netherlands.

Nair, P.K.R., R.J. Buresh, D.N. Mugendi, and C.R. Latt. 1999. Nutrient cycling in tropical agroforestry systems: Myths and science. p. 1–31. *In* L.E. Buck et al. (ed.) Agroforestry in sustainable agricultural systems. CRC Press, Boca Raton, FL.

Nair, P.K.R., and V.D. Nair. 2003. Carbon storage in North American agroforestry systems. p. 333–346 *In* J.M. Kimble et al. (ed.) The potential of U.S. forest soils to sequester carbon and mitigate the greenhouse effect. CRC Press, Boca Raton, FL.

Nair, P.K.R., M.R. Rao, and L.E. Buck (ed.). 2004. New vistas in agroforestry: A compendium for the 1st World Congress of Agroforestry. Orlando, FL. 27 June–2 July 2004. Kluwer Acad. Publ., Dordrecht, the Netherlands.

Niemi, R.M., M. Vepsäläinen, K. Erkomaa, and H. Ilvesniemi. 2007. Microbial activity during summer in humus layers under *Pinus sylvestris* and *Alnus incana*. For. Ecol. Manage. 242:314–323.

Nowak, J., A. Blount, and S. Workman. 2002. Integrated timber, forage and livestock production—Benefits of silvopasture. Circular 1430. University of Florida, Institute of Food and Agricultural Sciences and Florida Cooperative Extension Service.

Nyamadzawo, G., R. Chikowo, P. Nyamugafata, J. Nyamangara, and K.E. Giller. 2007. Improved legume tree fallows and tillage effects on structural stability and infiltration rates of a kaolinitic sandy soil from Central Zimbabwe. Soil Tillage Res. 96:182–194.

Nyamadzawo, G., R. Chikowo, P. Nyamugafata, J. Nyamangara, and K.E. Giller. 2008. Soil organic carbon dynamics of improved fallow-maize rotation systems under conventional and no-tillage in Central Zimbabwe. Nutr. Cycling Agroecosyst. 81:85–93.

Nygren, P., F. Maraux, and G.A. Sanchez. 1993. Solar-radiation transmission in the canopy of Erythrina-poeppigiana (Walkers) Cook, O.F. (In Portuguese.) Pesquisa Agropecu. Bras. 28:167–176.

Oelbermann, M., R.P. Voroney, and A.M. Gordon. 2004a. Carbon sequestration in tropical and temperate agroforestry systems: A review with examples from Costa Rica and southern Canada. Agric. Ecosyst. Environ. 104:359–377.

Oelbermann, M., R.P. Voroney, D.C.L. Kass, and A.M. Schlönvoigt. 2006a. Soil carbon and nitrogen dynamics using stable isotopes in 19-and 10-year-old tropical agroforestry systems. Geoderma 130:356–367.

Oelbermann, M., R.P. Voroney, A.M. Schlönvoigt, and D.C.L. Kass. 2004b. Decomposition of *Erythrina poeppigiana* leaves in 3-, 9-, and 18-year-old alley cropping systems in Costa Rica. Agrofor. Syst. 63:27–32.

Oelbermann, M., R.P. Voroney, N.V. Thevathasan, A.M. Gordon, D.C.L. Kass, and A.M. Schlönvoigt. 2006b. Soil carbon dynamics and residue stabilization in Costa Rica and southern Canadian alley cropping system. Agrofor. Syst. 68:27–36.

Ohu, J.O., E.T. Ekwue, and O.A. Folorunso. 1994. The effect of addition of organic matter on the compaction of a Vertisol from Northern Nigeria. Soil Technol. 7:155–162.

Okogun, J.A., N. Sanginga, and K. Mulongoy. 2000. Nitrogen contribution of five leguminous trees and shrubs to alley cropped maize in Ibadan, Nigeria. Agrofor. Syst. 50:123–136.

Oren, R., D.S. Ellsworth, K.H. Johnsen, N. Phillips, B.E. Ewers, C. Maier, K.V.R. Schäfer, H. McCarthy, G. Hendrey, S.G. McNulty, and G.G. Katul. 2001. Soil fertility limits carbon sequestration by forest ecosystems in a CO_2-enriched atmosphere. Nature 411:469–472.

Palm, C.A. 1995. Contribution of agroforestry trees to nutrient requirements in intercropped plants. Agrofor. Syst. 30:105–124.

Pande, M., and J.C. Tarafdar. 2004. Arbuscular mycorrhizal fungal diversity in neem-based agroforestry systems in Rajasthan. Appl. Soil Ecol. 26:233–241.

Parlange, M.B., A.T. Cahill, D.R. Nielsen, J.W. Hopmans, and O. Wendroth. 1998. Review of heat and water movement in field soils. Soil Tillage Res. 47:5–10.

Payne, W.A., J.H. Williams, M.M. Keitella, and R.D. Stern. 1998. Crop diversification in the Sahel through use of environmental changes near *Faidherbia albida* (Del.) A. Chev. Crop Sci. 38:1585–1591.

Peri, P.L., and M. Bloomberg. 2002. Windbreaks in southern Patagonia, Argentina: A review of research on growth models, windspeed reduction, and effects on crops. Agrofor. Syst. 56:129–144.

Peterjohn, W.T., and D.L. Correll. 1984. Nutrient dynamics in an agricultural watershed: Observations on the role of a riparian forest. Ecology 65:1466–1475.

Plate, E.J. 1971. The aerodynamics of shelter belts. Agric. Meteorol. 8:203–222.

Pritchett, W.L., and R.F. Fisher. 1987. Properties and management of forest soils. 2nd ed. John Wiley & Sons, New York.

Rao, M.R., P.K.R. Nair, and C.K. Ong. 1998. Biophysical interactions in tropical agroforestry systems. Agrofor. Syst. 38:3–50.

Rao, M.R., M.C. Palada, and B.N. Becker. 2004. Medicinal and aromatic plants in agroforestry systems. Agrofor. Syst. 61:107–122.

Read, R.A., and L.C. Walker. 1950. Influence of eastern redcedar on soil in Connecticut pine plantations. J. For. 48:337–339.

Reynolds, P.E., J.A. Simpson, N.V. Thevathasan, and A.M. Gordon. 2006. Effects of tree competition on corn and soybean photosynthesis, growth, and yield in a temperate tree-based agroforestry intercropping system in southern Ontario, Canada. Ecol. Eng. 29:362–371.

Roncal-García, S., L. Soto-Pinto, J. Castellanos-Albores, N. Ramírez-Marcial, and B. de Jong. 2008. Sistemas agroforestales y almacenamiento de carbono en comunidades indígenas de Chiapas, México. (In Spanish, with English abstract.) Interciencia 33:200–206.

Rosenberg, N.J. 1979. Windbreaks for reducing moisture stress. p. 394–408 *In* B.J. Barfield and J. F. Gerber (ed.) Modification of the aerial environment of plants. Am. Soc. of Agric. Eng., St. Joseph, MI.

Russo, R.O. 1990a. Erythrina (Leguminosae: Papilionoideae): A versatile genus for agroforestry systems in the tropics. J. Sustain. Agric. 1:89–109.

Russo, R.O. 1990b. Evaluating *Alnus acuminata* as a component in agroforestry systems. Agrofor. Syst. 10:241–252.

Salazar, A., and C. Palm. 1987. Screening of leguminous trees for alley cropping on acid soils of the humid tropics. p. 61–67. *In* D. Withington et al. (ed.) Gliricidia sepium (*Jacq.*) Walp.: Management and improvement. NFTA Spec. Publ. 87–01. Nitrogen Fixing Tree Assoc., Waimanalo, HI

Sanchez, P.A. 1976. Properties and management of soils in the tropics. John Wiley & Sons, New York.

Sanchez, P.A. 1995. Science in agroforestry. Agrofor. Syst. 30:5–55.

Sanchez, P.A. 1999. Improved fallows come of age in the tropics. Agrofor. Syst. 30:5–55.

Sanchez, P.A., K.D. Shepherd, M.J. Soule, F.M. Place, R.J. Buresh, and A.-M.N. Izak. 1997. Soil fertility replenishment in Africa: An investment in natural resource capital. p. 1–46. *In* R.J. Buresh et al. (ed.) Replenishing soil fertility in Africa. SSSA Spec. Publ. 51. SSSA, Madison, WI.

Sanginga, N., K. Mulongoy, and M.J. Swift. 1992. Contribution of soil organisms to the sustainability and productivity cropping systems in the tropics. Agric. Ecosyst. Environ. 41:135–152.

Sariyildiz, T., J.M. Anderson, and M. Kucuk. 2005. Effects of tree species and topography on soil chemistry, litter quality, and decomposition in Northeast Turkey. Soil Biol. Biochem. 37:1695–1706.

Satter, M.A., M.M. Hanafi, T.M.M. Mahmud, and H. Azizah. 2006. Role of arbuscular mycorrhiza and phosphorus in Acacia mangium-peanut agroforestry system for rejuvenation of tin tailings. J. Sustain. Agric. 28:55–68.

Sauer, T.J., C.A. Cambardella, and J.R. Brandle. 2007. Soil carbon and tree litter dynamics in a red cedar-scotch pine shelterbelt. Agrofor. Syst. 71:163–174.

Schnabel, S.W., and A. Ferreira (ed.). 2004. Sustainability of agrosilvopastoral systems: Dehesas,

Montados. Advance in Geoecology 37. Catena Verlag, Reiskirchen, Germany.

Schoeneberger, M.M. 2008. Agroforestry: Working trees for sequestering carbon on agricultural lands. Agroforest. Syst. 75:27–37. doi:10.1007/s10457-008-9123-8.

Schroeder, P. 1994. Carbon storage benefits of agroforestry systems. Agrofor. Syst. 27:89–97.

Schroth, G. 1999. A review of belowground interactions in agroforestry, focussing on mechanisms and management options. Agrofor. Syst. 43:5–34.

Schroth, G., D. Kolbe, B. Pity, and W. Zech. 1996. Root system characteristics with agroforestry relevance of nine leguminous tree species and a spontaneous fallow in a semi-deciduous rainforest area of West Africa. For. Ecol. Manage. 84:199–208.

Schroth, G., and F.L. Sinclair (ed.). 2003. Trees, crops and soil fertility. CABI, Wallingford, UK.

Schroth, G., and W. Zech. 1995. Above- and belowground biomass dynamics in a sole cropping and an alley cropping system with *Gliricidia sepium* in the semi-deciduous rainforest zone of West Africa. Agrofor. Syst. 31:181–198.

Seobi, T., S.H. Anderson, R.P. Udawatta, and C.J. Gantzer. 2005. Influence of grass and agroforestry buffer strips on soil hydraulic properties for an albaqualf. Soil Sci. Soc. Am. J. 69:893–901.

Sharrow, S.H. 1999. Silvopastoralism: Competition and facilitation between trees, livestock, and improved grass-clover pastures on temperate rainfed lands. p. 111–130 *In* Buck L.E. et al. (ed.) Agroforestry in sustainable agricultural systems. Lewis Publ., Boca Raton, FL.

Sharrow, S.H. 2007. Soil compaction by grazing livestock in silvopastures as evidenced by changes in soil physical properties. Agrofor. Syst. 71:215–223.

Shepherd, K.D., E. Ohlsson, J.R. Okalebo, and J.K. Ndufa. 1996. Potential impact of agroforestry on soil nutrient balances at the farm scale in the East African highlands. Fert. Res. 44:87–99.

Sholto Douglas, J., and R.A. de J. Hart. 1978. Forest farming: Towards a solution to problems of world hunger and conservation. Rodale Press, Emmaus, PA.

Siaw, D.E.K.A., B.T. Kang, and D.U.U. Okali. 1991. Alley cropping with *Leucaena leucocephala* (Lam) De Wit and *Acioa barteri* (Hook.f.) Engl. Agrofor. Syst. 14:219–231.

Sileshi, G., and P.L. Mafongoya. 2006. Variation in macrofaunal communities under contrasting land use systems in eastern Zambia. Appl. Soil Ecol. 33:49–60.

Silva-Pando, F.J., M.P. Gonzalez-Hernandez, and M.J. Rozados-Lorenzo. 2002. Pasture production in a silvopastoral system in relation with microclimate variables in the Atlantic coast of Spain. Agrofor. Syst. 56:203–211.

Smith, J.R. 1953. Tree crops: A permanent agriculture. The Devin-Adair Co., New York.

Snyder, N.J., S. Mostaghimi, D.F. Berry, R.B. Reneau, S. Hong, P.W. McClellan, and E.P. Smith. 1998. Impact of riparian forest buffers on agricultural nonpoint source pollution. J. Am. Water Resour. Assoc. 34:385–395.

Somarriba, E., J. Beer, and R.G. Muschler. 2001. Research methods for multistrata agroforestry systems with coffee and cacao: Recommendations from two decades of research at CATIE. Agrofor. Syst. 53:195–203.

Soto-Pinto, L., I. Perfecto, J. Castillo-Hernandez, and J. Caballero-Nieto. 2000. Shade effect on coffee production at the northern Tzeltal zone of the state of Chiapas, Mexico. Agric. Ecosyst. Environ. 80:61–69.

Stigter, C.J. 1988. Microclimate management and manipulation in agroforestry. p. 145–168. *In* K.F. Wiersum (ed.) Viewpoints on agroforestry. 2nd renewed ed. Agric. Univ., Wageningen, the Netherlands.

Szajdak, L., V. Maryganova, T. Meysner, and L. Tychinskaya. 2002. Effect of shelterbelt on two kinds of soils on the transformation of organic matter. Environ. Int. 28:383–392.

Szott, L.T., and D.C.L. Kass. 1993. Fertilizers in agroforestry systems. Agrofor. Syst. 23:157–176.

Szott, L.T., C.A. Palm, and R.J. Buresh. 1999. Ecosystem fertility and fallow function in the humid and subhumid tropics. Agrofor. Syst. 47:163–196.

Tang, B.T., L. Reynolds, and A.N. Atta-Krah. 1990. Alley farming. Adv. Agron. 43:315–359.

Tiessen, H., E. Cuevas, and P. Chacón. 1994. The role of soil organic matter in sustaining soil fertility. Nature 371:783–785.

Torquebiau, E.F., and F. Kwesiga. 1996. Root development in a *Sesbania sesban* fallow-maize system in Eastern Zambia. Agrofor. Syst. 34:193–211.

Trasar-Cepeda, C., M.C. Leirós, and F. Gil-Sotres. 2008. Hydrolytic enzyme activities in agricultural and forest soils. Some implications for their use as indicators of soil quality. Soil Biol. Biochem. 40:2146–2155.

Tufekcioglu, A., J.W. Raich, T.M. Isenhart, and R.C. Schultz. 2003. Biomass, carbon and nitrogen dynamics of multi-species riparian buffers within an agricultural watershed in Iowa, USA. Agrofor. Syst. 57:187–198.

Udawatta, R.P., and S.H. Anderson. 2008. CT-measured pore characteristics of surface and subsurface soils influenced by agroforestry and grass buffers. Geoderma 145:381–389.

Udawatta, R.P., S.H. Anderson, C.J. Gantzer, and H.E. Garrett. 2006. Agroforestry and grass buffer influence on macropore characteristics: A computed tomography analysis. Soil Sci. Soc. Am. J. 70:1763–1773.

Udawatta, R.P., R.J. Kremer, B.W. Adamson, and S.H. Anderson. 2008. Variations in soil aggregate stability and enzyme activities in a temperate agroforestry practice. Appl. Soil Ecol. 39:153–160.

Udawatta, R.P., R.J. Kremer, H.E. Garrett, and S.H. Anderson. 2009. Soil enzyme activities and physical properties in a watershed managed under agroforestry and row-crop systems. Agric. Ecosyst. Environ. 131:98–104.

U.S. Forest Service. 1935. Possibilities of shelterbelt planting in the plains region. Lake States Forest Expt. Sta., Special Pub. U.S. Gov. Print. Office, Washington, DC.

van Eimern, J. (ed.). 1964. Windbreaks and shelterbelts. Tech. Note No. 59. World Meteorological Organ., Geneva. Switzerland.

van Noordwijk, M., G. Lawson, J.J.R. Groot, and K. Hairiah. 1996. Root distribution in relation to nutrients and competition. p. 319–364. *In* C.K. Ong and P.A. Huxley (ed.) Tree–crop interactions: A physiological approach. CABI, Wallingford, UK.

van Noordwijk, M., M. Widianto, M. Heinen, and K. Hairiah. 1991. Old tree roots channels in acid soils in the humid tropics: Important for crop root penetration, water infiltration and nitrogen management. Plant Soil 134:37–44.

Vyssotsky, G.N. 1935. Shelterbelts in the steppes of Russia. J. For. 33:781–788.

Wallace, J.S. 1996. The water balance of mixed tree-crop systems. p. 73–158. *In* C.K. Ong and P.A. Huxley (ed.) Tree–crop interactions: A physiological approach. CABI, Wallingford, UK.

Wedderburn, M.E., and J. Carter. 1999. Litter decomposition by four functional tree types for use in silvopastoral systems. Soil Biol. Biochem. 31:455–461.

Wilde, S.A. 1958. Forest soils. The Ronald Press Co., New York.

Young, A. 1989. Agroforestry for soil conservation. CABI, Wallingford, UK.

Young, A. 1997. Agroforestry for soil management. 2nd ed. CABI, Wallingford, UK.

24

Soil Management Implications of Producing Biofuel Feedstock

Jane M.F. Johnson, David W. Archer, Douglas L. Karlen,
Sharon L. Weyers, and Wally W. Wilhelm

The use of plant biomass for energy has existed since humans mastered the use of fire, although utilization beyond the open fire has evolved. The concept of using recent biomass as a major energy feedstock is being revisited, driven by high consumer demand (growing population), declining domestic oil supplies, increasing cost of fossil fuels, and a desire to curb the emission of greenhouse gases (Johnson et al., 2007b). In general terms, agriculture and forestry are the economic sectors commercially producing a wide array of bioenergy feedstocks (e.g., grains, herbaceous annuals, herbaceous perennials, and woody perennials). For this review, biomass feedstock is any nongrain, plant-derived feedstock. These commodities can serve as feedstock for cellulosic ethanol or other thermochemical platforms such as gasification or pyrolysis.

The type of bioenergy feedstock produced and the desired energy product can alter the management implications, which likely will vary by region. It is also likely that a given farm operation may produce multiple feedstocks, including corn and soybean grain, perennial grasses, and crop residues. The potential risks and benefits of growing and using feedstocks vary considerably (Johnson et al., 2007b). The challenge of establishing a perennial biomass system depends on prior management. Conversion of highly diverse grassland systems to low-diversity or mono-culture perennial systems could reduce the environmental benefits of these lands. Conversely, converting from high-input, annual crop species to perennial species could reduce input requirements (fertilizer, fuel, pesticides) and reduce erosion risks, and thus have positive environmental impacts (Mann and Tolbert, 2000). Agronomic, environmental, and economic issues need to be addressed for the wide range of feedstocks and feedstock combinations to assure sustainability.

Agronomic management of the major cash crops, i.e., corn (*Zea mays* L.), soybean (*Glycine max* L. [Merr.]), wheat (*Triticum aestivum* L.), and rice (*Oryza sativa* L.), has been studied for many decades. However, changes in traditional management strategies and practices are necessary when crop residues such as stover or straw are harvested routinely. Answers to the following questions will provide the framework for making these changes: (i) how much biomass can be harvested without exacerbating soil erosion or loss of soil organic matter (SOM) and soil organic carbon (SOC),

J.M.F. Johnson (Jane.johnson@ars.usda.gov) and S.L. Weyers, USDA-Agricultural Research Service, 803 Iowa Ave., Morris, MN 56267; D.W. Archer, USDA-Agricultural Research Service, 1701 10th Ave. SW, P.O. Box 459, Mandan, ND 58554; D.L. Karlen, USDA-Agricultural Research Service National Laboratory for Agriculture and the Environment, 2110 University Blvd., Ames, IA 50011; W.W. Wilhelm, deceased, formerly USDA-ARS, University of Nebraska, Lincoln. The U.S. Department of Agriculture offers its programs to all eligible persons regardless of race, color, age, sex, or national origin, and is an equal opportunity employer.

doi:10.2136/2011.soilmanagement.c24

(ii) how will harvesting biomass impact fertility management, (iii) how often should the biomass be harvested, (iv) how can perennial biomass crops be integrated into a current farming system, (v) can tillage be reduced when residues are harvested, and (vi) how will the residue be harvested, stored, and transported? Furthermore, utilization of new or alternative crops has agronomic unknowns such as pesticide and fertility management, and economic questions such as market demand. The risks to farmers can be high due to lack of agronomic experience, uncertain markets, and lack of crop insurance or subsidies.

Many aspects of crop residue management were reviewed in the ASA Special Publication Number 31, edited by W. R. Oschwald et al. (1978). Although 30 years have passed since it was published, many of the principles addressed are still relevant: conservation, soil erosion control, soil chemistry, disease, and weed control. This review focuses on preventing soil erosion, maintaining or building SOM, and managing nutrients and water for agricultural

Fig. 24|1. A visual assessment of soil quality impacts of harvesting corn stover. Notice the soil erosion in the foreground following intense spring rains where corn stover had been harvested during the previous autumn (ARS Photo d1235–30).

commodity crops, herbaceous annuals, and perennials. It also includes a discussion of economic influences on management in biofuel feedstock production.

Soil Erosion and Soil Organic Carbon

Many soil conservation improvements have been realized by using the tolerable annual soil loss (T) concept, which was incorporated into the United States' conservation policies (USDA-NRCS, 2006). Soil erosion predictive models, the Revised Universal Soil Loss Equation version 2 (RUSLE2) and the Wind Erosion Equation (WEQ), have been widely used by the USDA-NRCS as planning tools to control soil erosion to T or below (Cox, 2008). However, these tools alone do not address loss of SOC. The USDA-NRCS estimated that managing soil to maintain SOC could save an additional 1.2 billion Mg of soil and 8.2 billion U.S. dollars annually (http://soils.usda.gov/sqi/concepts/soil_organic_matter/som_manage.html, verified 16 Sept. 2010). Safeguarding soil productivity in an era of biomass feedstock harvest and competing demands for agricultural production is paramount (Lal, 2004; Wilhelm et al., 2004; Graham et al., 2007; Johnson et al., 2007b).

No-tillage farming retains all unharvested crop residues on the soil surface, dramatically reducing soil erosion. Both the Perlack et al. (2005) and Graham et al. (2007) assessments of harvestable biomass assume universal conversion to no-tillage farming systems. Unfortunately, it is unlikely that no-tillage farming will be adopted universally. In northern states such as Minnesota, less than 5% of the cropland is managed using no-tillage; although about 50% of the acreage has some form of conservation tillage (chisel plow, ridge tillage, or mulch tillage) (CTIC, 2002). The low adoption rate in the northern tier states is related to the short growing season and cool, wet springs. Tillage aids in drying and warming soil, especially in early spring. In drier climates, retaining residue on the soil surface by eliminating tillage is a strategy to reduce evaporation (Al-Darby et al., 1989) and to improve crop productivity (Wilhelm et al., 1986). Unfortunately, harvesting biomass reduces the effectiveness of conservation- or no-tillage to retard erosion (Fig. 24|1).

Policy and conservation guidelines for minimizing erosion already exist and are applicable to biomass harvest; however, these guidelines may not be sufficient to maintain SOC. Wilhelm et al. (2007) found that the residue requirements for maintaining SOC exceeded those needed to limit erosion at or below T (Fig. 24|2). Thus, developing harvest recommendations constrained only for erosion control risks loss of SOC. Part of the challenge in developing guidelines that address both erosion and SOC is that mechanisms controlling soil erosion differ from managing SOC (Wilhelm et al., 2004). Reducing erosion is a function of percent soil coverage (Stocking, 1988; Bilbro and Fryrear, 1994), but maintaining SOC requires biomass inputs equal biomass outputs.

Using empirical data and linear regression between C inputs and change in SOC, Johnson et al. (2006a) proposed the term "minimum source C (MSC)," which is the annual C inputs necessary to maintain SOC content. Based on literature data for several crops and tillage practices, Johnson et al. (2006a) estimated an average MSC at 2.21 ± 1.1 Mg C ha^{-1} yr^{-1} (n = 21). Since the Johnson et al. (2006a) review, several other studies reported empirical MSC estimates ranging from 0.032 to 8.7 Mg C ha^{-1} yr^{-1} (Table 24|1). Wheat tends to have lower MSC compared with corn or soybean systems, perhaps because wheat-based systems are more common in cooler, dryer climates. Banowetz et al. (2008) using the USDA-NRCS Soil Conditioning Index (SCI) predicted a MSC of 2.0 Mg C ha^{-1} for cereal crops and seed grass in Washington. Theoretically, once MSC for a given management system is known, the

amount of sustainably harvestable biomass can be predicted. However, not all empirical studies of SOC demonstrate a correlation with C inputs (Dexter et al., 1982; Johnson and Chamber, 1996; Nicholson et al., 1997; Sainju et al., 2006b; Huggins et al., 2007); in such cases, MSC cannot be estimated. A lack of correlation to C inputs indicates that either the rate of humification and/or the rate of mineralization were changing (Bayer et al., 2006). Based on the wide range in MSC reported and the examples where MSC cannot be estimated we suggest that additional research and modeling efforts are needed to

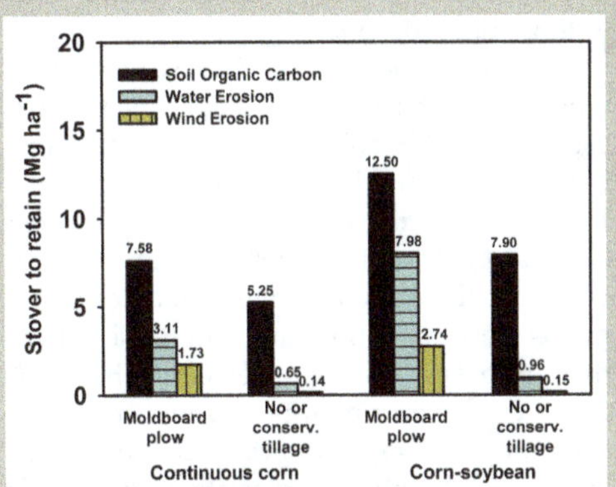

Fig. 24|2. Tillage and crop rotation effects on the annual average amount of corn stover required for protecting soil resources against wind or water erosion and to sustain soil carbon (organic matter) levels (Wilhelm et al., 2007).

Table 24|1. Recent empirical estimates on the amount of annual aboveground non-grain C inputs required for maintaining soil organic C levels.

Location	Crop†	Primary tillage‡	Soil type§	C	Citation
				Mg ha^{-1} yr^{-1}	
SD	M	CP	L	3.21	Pikul et al. (2008)
NE	M, S	D	SiL	2.4	Varvel and Wilhelm (2008a)
MN	M, S	NT	CL	8.7	Huggins et al. (2007)
MT	W	NT	CL	0.82	Sainju et al. (2006a)
CA	W, M, T	CT	SiL, SiCL	2.6	Kong et al. (2005)
India	W, S	NR	SaL	0.032	Kundu et al. (2007)
Brazil	O, M, V, C	CT	SaCL	6.2	Bayer et al. (2006)
Brazil	O, M, V, C	NT	SaCL	2.7	Bayer et al. (2006)

† C, cowpea [*Vigna unguiculata* (L.) Walp.]; M, maize (corn); O, oat (*Avena strigosa* Schreb.); S, soybean; T, tomato (*Lycopersicon esculentum* Mill.); V, vetch (*Vicia sativa* L.); W, wheat.

‡ CP, chisel plow; CT, conventional tillage, details not provided; D, disk; NR, not reported; NT, no-tillage.

§ Si, silt; Sa, sandy; L, loam; C, clay.

more accurately predict harvest rates that will not degrade SOC.

Currently, the USDA-NRCS SCI tool provides a user-friendly, accessible means for estimating if harvesting a chosen amount of biomass would likely cause a loss of SOC. The SCI has three primary components: (i) plant biomass input or removal rate, (ii) effect of tillage and management on organic matter decomposition, and (iii) erosion prediction based on management. The tool does not give an amount of C sequestered or lost, but provides guidance on the direction of change. Additional scientific review and validation is needed to expand its validity to a wider variety of agricultural regions and management practices including irrigated systems and nontraditional cropping systems (Cox, 2008). The data tables within SCI are updated as new information becomes available (USDA-NRCS, 2002). Other models, such as CQSTR (Rickman et al., 2001), EPIC (Izaurralde et al., 2006), CENTURY (Parton et al., 1988), and DAISY (Bruun et al., 2003) offer additional tools for predicting the impact of biomass harvest on SOC, but are more complicated, requiring more input data and more knowledge for interpretation.

Management strategies for preventing SOC loss and controlling wind and water erosion should be included whenever biomass feedstocks (annual or perennial) are harvested. Strategies may include one or more of the following management modifications: (i) reducing or eliminating soil tillage, (ii) maintaining adequate soil cover, (iii) adding perennial crops into the rotation, and (iv) using green manures, cover crops, and/or living mulches (Syers, 1997; Thorup-Kristensen et al., 2003; Johnson et al., 2007b; Johnson et al., 2007c). Reducing or eliminating tillage is well known for keeping crop residue on the soil surface (e.g., Siemens and Oschwald, 1978). Herbaceous perennials typically have extensive root systems, which can increase SOC while helping protect the soil from erosive forces (McLaughlin and Walsh, 1998; Frank et al., 2004; Liebig et al., 2005; Lee et al., 2007; Liebig et al., 2008). Thus, inclusion of a perennial into a rotation is a strategy to offset SOC loss during rotation phases that return less C into the soil. Green manures have been used as biological tools for N management, erosion control, and to increase SOC (Thorup-Kristensen et al., 2003). Cover crops, which are more commonly used in warmer, wetter climates (Dabney et al., 2001), protect the soil from erosive forces, suppress weeds, help manage N fertility, and increase C input by extending the growing season (Reicosky and Forcella, 1998; Dabney et al., 2001; Thorup-Kristensen et al., 2003; Singer, 2005; Baker et al., 2007).

New and Alternative Crops

Many perennial grass species can serve as biomass feedstocks (Jasinskas et al., 2008; Monti et al., 2008; Mulkey et al., 2008). However, greater emphasis has been placed on developing switchgrass (*Panicum virgatum* L.) and miscanthus (*Miscanthus* ×*giganteus* J. M. Greef & Deuter ex Hodk. & Renvoize), for feedstock production (Wright, 1994; Heaton et al., 2008). Switchgrass has a long history as a forage crop (Moser and Vogel, 1995; Wolf and Fiske, 1996; Vogel et al., 1999). Therefore, multiple cultivars (Cassida et al., 2005b; Adler et al., 2006) and agronomic recommendations for switchgrass are available in several regions (Wolf and Fiske, 1996; Teel et al., 2003; Lee and Boe, 2005; Nyoka, 2007). Research is ongoing to improve breeding, production, and management to improve switchgrass and other grasses (McLaughlin and Adams Kszos, 2005; Heaton et al., 2008; Lemus et al., 2008).

Additional bioenergy resources are also in development. Several oilseed crops are being explored for biodiesel production, for industrial products (e.g., lubricants and plasticizers), and even for jet fuel (Table 24|2), depending on the conversion process utilized. Many of the oilseed crops have oil contents as high as or higher than soybean, and can be grown on marginal land with fewer inputs than soybean. One example is cuphea (*Cuphea* spp.), which is only semidomesticated, with an indeterminate growth habit, and shattering problems that need to be resolved through improved plant breeding (Knapp and Crane, 2000; Gesch et al., 2006). Currently, alternative oilseed crop expansion in the United States is hampered by lack of markets, subsidies, and crop insurance (Johnson et al., 2007b). In addition, agronomic management tools (e.g., nutrient and pesticide recommendations, rotations, harvest strategies) are needed.

Table 24|2. Alternative oilseed crops in production or development.

Common name	Scientific name	Potential use	Commercially grown	Citation
Oilseed rape	*Brassica napus* L.	biodiesel	yes	Demirbas (2006)
Crambe	*Crambe abyssinica* Hochst. ex R. E. Fr.	lubricants, plasticizers	yes	Carlson et al. (1996)
Lesquerella	*Lesquerella fendleri* (A. Gray) S. Watson	lubricants, plasticizers	no	Carlson et al. (1996); Dierig et al. (1996)
Camelina	*Camelina sativa* (L.) Crantz	industrial use	yes	Putnam et al. (1993)
Pennycress	*Thlaspi arvense* L.	cover crop, biodiesel	no	Johnson et al. (2007b)
Castor-bean	*Ricinus cummunis* L.	lubricant	yes	Brigham (1993); Goodrum and Geller (2005)
Cuphea	*Cuphea* spp.	biodiesel, jet fuel, lubricant	yes	Gesch et al. (2006)

Nutrient Management

Adaptation of current cropping systems for biomass harvest and development of dedicated biomass crops raise nutrient management questions such as: (i) amount of various nutrients actually removed, (ii) impact of removal on soil fertility, (iii) management strategies to replace or reduce nutrient removal, and (iv) economics of nutrient replacement (discussed in economic section below). Harvesting annual crop residue in addition to grain removes additional nutrients; thus nutrient management may need to be adjusted. Likewise, nutrient management in perennial systems will need to be modified to reflect change in commodity from forage, such as alfalfa (*Medicago* spp.) or mixed hay to biomass feedstock. Historically, these species have been grown in pastures for grazing, harvested for hay or silage production, grown in Conservation Reserve Program (CRP) land, or grown in other conservation areas such as grassy waterways. Perennials in conservation programs were rarely, if ever, harvested, while forage crops are managed for high feed value (protein and energy content) and generally harvested several times each year. As a biomass feedstock, energy or carbohydrate content is more important than protein. This difference in end use means that all aspects of nutrient management likely will be substantially different.

Nutrient concentration varies among crop species, plant organ, and physiological stage (Lewandowski and Kicherem, 1997; Fageria, 2004; Johnson et al., 2007a; Monti et al., 2008). The concentration averaged across all feedstocks and organs was 9.1 ± 6.8 g N kg^{-1}, 0.9 ± 0.5 g P kg^{-1}, and 8.7 ± 7.3 g K kg^{-1} (Table 24|3). Annuals crops did not appear to have different concentrations of N, P, or K compared with perennial species. Corn stems and cobs had lower nutrient concentration than corn leaves. Among these studies, cobs and switchgrass stems had the lowest concentration of N, switchgrass and miscanthus stems had the lowest concentration of P, while giant reed grass stems (*Arundo donax* L.) had the lowest concentration of K. Feedstock likely will be harvested when it is relatively dry, thus nutrient concentrations will be lower than if the same feedstock were harvested for animal feed (Lewandowski and Kicherem, 1997; Reynolds et al., 2000; Hoskinson et al., 2007). Other macronutrients (Ca, Mg, and S) and micronutrients (B, Cu, Fe, Mn, and Zn) are also removed when biomass feedstocks are harvested (Fageria, 2004; Hoskinson et al., 2007; Monti et al., 2008). Therefore, producers will need to monitor soil fertility, scout crops for deficiency symptoms, and apply appropriate fertilizer mixtures as necessary.

Nutrient removal depends both on concentration and harvest rate. Therefore, if a feedstock removes 7.1 g N kg^{-1} dry biomass and 5 Mg ha^{-1} is harvested annually, then 35.5 kg N ha^{-1} would be removed; similarly, 10 Mg feedstock ha^{-1} removes 71 kg N ha^{-1}. In the Pacific Northwest, switchgrass yielding 14 to 20 Mg ha^{-1} exports about 210 kg N ha^{-1}, 40 kg P ha^{-1}, and 350 kg K ha^{-1} with harvested biomass (Hal Collins, personal communication). Duffy and Nanhou (2001) estimated 0.42 kg P Mg^{-1} and 9.4 kg K Mg^{-1} were removed with every Mg of switchgrass harvested in the fall, which corresponds to 5.2 kg P ha^{-1} and 125 kg K ha^{-1} removed at a harvest rate of 13.4 Mg ha^{-1}.

Table 24|3. Plant concentration of N, P and K of potential annual and herbaceous perennial biomass feedstocks.

Crop organ	N	P	K	Comment	Citation
		g kg⁻¹			
Annual crops					
Barley straw	6.4	nr†	nr		Cookson et al. (1998)
Barley straw	6.1	nr	nr		Andren and Paustian (1987)
Barley straw	6.0	nr	nr		Christensen (1986)
Barley straw	7.5	1.1	12.5		Lindstrom (1986)
Barley straw	7.9	nr	nr		Mitchell et al. (2001)
Barley straw	6.3	nr	nr		Velthof et al. (2002)
Barley straw	4.8	nr	nr	avg. 3 yr and treatments	Halvorson and Reule (2007)
Corn cob	3.3	nr	nr		Burgess et al. (2002)
Corn cob	3.8	nr	nr		Halvorson and Johnson (2009)
Corn cob	10.0	nr	nr		Yu et al. (2008)
Corn leaf	10.3	nr	nr		Burgess et al. (2002)
Corn leaf	13.6	nr	nr		Johnson et al. (2007a)
Corn stem	5.7	nr	nr		Burgess et al. (2002)
Corn stover	5.9	nr	nr		Johnson et al. (2007a)
Corn stover	3.4	1.0	3.4		Breakwell and Turco (1989)
Corn stover	10.0	1.33	20.6		Tian et al. (1992)
Corn stover	5.7	nr	nr	avg. of treatments	Al-Kaisi et al. (2005)
Corn stover	6.7	nr	nr		Burgess et al. (2002)
Corn stover	7.7	0.9	11.7	130 d after planting	Fageria (2004)
Corn stover	7.5	0.7	10.0	avg. of cutting heights	Hoskinson et al. (2007)
Corn stover	11.1	1.8	13.3		Lindstrom (1986)
Corn stover	8.0	2.0	nr		Manlay et al. (2002)
Corn stover	7.3	nr	nr		Velthof et al. (2002)
Millet straw‡	2.7	0.7	nr		Manlay et al. (2002)
Millet straw	10.0	1.0	12.8		Fatondji et al. (2006)
Millet straw	13.8	nr	nr		Sarr et al. (2008)
Rice hulls	3.1	0.8	3.6		Linquist et al. (2007)
Rice leaf	26.0	nr	nr		Abiven et al. (2005)
Rice leaf	14.6	nr	nr	avg. 10 cultivars	Ying et al. (1998)
Rice stem	10.0	nr	nr		Abiven et al. (2005)
Rice stem	6.2	nr	nr	avg. 10 cultivars	Ying et al. (1998)
Rice straw	8.4	0.53	23.4		Tian et al. (1992)
Rice straw	6.5	nr	nr		Tirol-Padre et al. (2005)
Rice straw	5.0	0.5	31.0		Kaewpradit et al. (2008)
Rice straw	4.1	1.0	nr		Manlay et al. (2002)
Sorghum leaf	26.0	nr	nr		Abiven et al. (2005)
Sorghum leaf	13.5	1.3	10.2	avg. among varieties	Monti et al. (2008)
Sorghum straw	6.2	nr	nr	avg. among treatments	Franzluebbers et al. (1995)
Sorghum straw	7.0	0.5	nr		Saffigna et al. (1989)
Sorghum stem	9.0	nr	nr		Abiven et al. (2005)
Sorghum stem	3.5	0.7	12.8	avg. among varieties	Monti et al. (2008)
Soybean leaf	34.0	nr	nr	avg. 3 yr, Oct. harvest	Rao et al. (2005)
Soybean leaf	44.0	nr	nr		Abiven et al. (2005)
Soybean leaf	15.8	nr	nr		Johnson et al. (2007a)
Soybean straw	9.7	nr	nr		Al-Kaisi et al. (2005)
Soybean straw	12.8	1.5	15.2		Fageria (2004)
Soybean straw	14.8	nr	nr	avg. among treatments	Franzluebbers et al. (1995)
Soybean straw	22.5	2.2	10.5		Lindstrom (1986)
Soybean stem	8.8	nr	nr	avg. among treatments	Franzluebbers et al. (1995)
Soybean stem	7.0	nr	nr	avg. 3 yr, Oct. harvest	Rao et al. (2005)

Crop organ	N	P	K	Comment	Citation
		g kg⁻¹			
Soybean stem	20.0	nr	nr		Abiven et al. (2005)
Soybean stem	4.4	nr	nr		Johnson et al. (2007a)
Wheat leaf	7.0	nr	nr		Abiven et al. (2005)
Wheat straw	5.9	nr	nr		Cookson et al. (1998)
Wheat straw	9.4	1.53	5.88		Borie et al. (2002)
Wheat straw	6.0	nr	nr	avg. among treatments	Franzluebbers et al. (1995)
Wheat straw	10.6	nr	nr		Jawson and Elliott (1986)
Wheat straw	6.7	0.7	9.7		Lindstrom (1986)
Wheat straw	8.2	nr	nr		Mitchell et al. (2001)
Wheat straw	3.7	nr	nr		Tirol-Padre et al. (2005)
Wheat straw	6.2	nr	nr		Velthof et al. (2002)
Wheat stem	4.0	nr	nr		Abiven et al. (2005)
Mean ± Std. dev.	9.9 ± 7.5	1.1 ± 0.5	13.1 ± 7.6		
Perennial species					
Cardoon leaf	9.6	1.5	4.7	4-yr stand, after frost	Monti et al. (2008)
Cardoon stem	3.0	1.4	6.5	4-yr stand, after frost	Monti et al. (2008)
Reed Canary grass straw	11.7	1.4	3.0	3-yr stand, avg. 5 treatments	Katterer et al. (1998)
Giant reed grass stem	5.2	0.3	5.6	4-yr stand, after frost	Monti et al. (2008)
Giant reed grass leaf	15.7	0.8	5.1	4-yr stand, after frost	Monti et al. (2008)
Miscanthus leaf	6.3	0.4	3.3	4-yr stand, after frost	Monti et al. (2008)
Miscanthus straw	15.1	nr	nr	year 1 fall harvest	Clifton-Brown and Lewandowski (2002)
Miscanthus straw	13.4	nr	nr	year 1 spring harvest	Clifton-Brown and Lewandowski (2002)
Miscanthus straw	7.3	nr	nr	year 2 (avg. fall and spring harvest)	Clifton-Brown and Lewandowski (2002)
Miscanthus straw	4.3	nr	nr	year 3 (avg. fall and spring harvest)	Clifton-Brown and Lewandowski (2002)
Miscanthus stem	1.6	0.1	3.6	4-yr stand, after frost	Monti et al. (2008)
Switchgrass straw	13.6	nr	nr	late summer	Johnson et al. (2007a)
Switchgrass straw	7.9	0.7	1.8	4-yr stand, after frost	Monti et al. (2008)
Switchgrass straw	5.3	nr	nr	Nov., avg. treatment and year	Lemus et al. (2008)
Switchgrass straw	11.1	nr	nr	Jul., avg. treatment and year	Lemus et al. (2008)
Switchgrass straw	4.5	0.9	3.4	late frost	Adler et al. (2006)
Switchgrass straw	4.1	0.5	0.6	early spring	Adler et al. (2006)
Switchgrass straw	5.4	0.4	9.4	fall harvest	Duffy and Nanhou (2001); Lemus et al. (2002)
Switchgrass straw	nr	0.08	0.6	spring harvest	Duffy and Nanhou (2001)
Switchgrass straw	11.2	nr	nr	Sept. harvest, avg. 3 varieties	Bransby et al. (1998)
Switchgrass straw	6.0	0.9	nr	late fall harvest, avg. among genotypes	Cassida et al. (2005a)
Switchgrass straw	6.3	nr	nr	late fall, avg. among varieties and N treatment	Madakadze et al. (1999)
Switchgrass straw	7.3	nr	nr	summer harvest, avg. 5 yr	Reynolds et al. (2000)
Switchgrass straw	3.6	nr	nr	fall harvest, avg. tillage treatment and yr	Reynolds et al. (2000)
Switchgrass straw	18.1	nr	nr	boot to inflorescence	Vogel et al. (2002)
Switchgrass straw	8.9	nr	nr	post anthesis, avg. sites	Vogel et al. (2002)
Switchgrass straw	5.0	nr	nr	after killing frost	Vogel et al. (2002)
Switchgrass stem	4.1	nr	nr	late summer	Johnson et al. (2007a)
Switchgrass stem	3.2	0.3	3.1	4-yr stand, after frost	Monti et al. (2008)
Perennial mean ± SD	7.5 ± 4.5	0.73 ± 0.5	3.9 ± 2.4		

† nr, not reported.

‡ Millet (*Panicum miliaceum* L.).

Nutrient removal was reduced by delaying harvest until spring, which dropped P and K concentration to 0.08 kg P Mg^{-1} and 0.60 kg K Mg^{-1}. The portion of the plant harvested also impacts nutrient removal rates. For example, harvesting essentially all corn stover with cobs removed 42 kg N ha^{-1}, while harvesting the bottom 50% (by height) of the plant removed only 13.8 kg N ha^{-1} (Hoskinson et al., 2007). Harvesting the whole plant removed 34.3 kg K ha^{-1} and 4.0 kg P ha^{-1}, while harvesting only the bottom 50% removed 33.5 kg K ha^{-1} and 1.0 kg P ha^{-1} (Hoskinson et al., 2007). However, harvest of only the bottom 50% removed less N, P, and K because this portion has only about one-third the amount of dry matter and about twice the water relative to the top 50%, which included the ear shank and cob. Harvesting only cobs would remove 3 to 10 kg N ha^{-1} (Table 24|3).

Harvesting cellulosic feedstock removes plant nutrients and has the potential to reduce soil fertility (Apland et al., 1981; Smil, 1999; Lal, 2008). It is straightforward to determine nutrient removal from plant concentration and harvest rate, but more difficult to predict subsequent impacts on nutrient cycling and availability. Furthermore, response can vary by nutrient. For example, harvesting stover reduced plant available K and Mg after 32 yr, but did not significantly reduce other macro- (N, P, Ca) or micronutrient (Mn, Zn, Fe) concentrations in soil (Moebius-Clune et al., 2008). Removal of rice and barley (*Hordeum vulgare* L.) residues reduced both ammonification and nitrification (Kushwaha et al., 2000). Similarly, harvesting corn stover reduced potentially mineralizable N (Kapkiyai et al., 1999; Salinas-Garcia et al., 2001). In contrast, retaining stover in a no-tillage system reduced available N, presumably due to a reduction in soil temperature (Andraski and Bundy, 2008). The highest concentration of soil P and K were reported with 100% stover harvest and 0% harvest compared with partial harvest for no-tillage corn in Mexico (Roldan et al., 2003).

Inorganic or organic amendments can be used to balance nutrient removal and soil fertility. Traditional amendments such as inorganic fertilizer (Apland et al., 1981; Smil, 1999), animal manures, and composts (Kapkiyai et al., 1999), or unconventional amendments, such as pyrolysis char, gasification ash, or other energy-processing residues (Reijnders, 2006) can be applied to replace removed nutrients and improve soil fertility. For example, in a Kenyan study, application of only dry manure (10 Mg ha^{-1}) improved yield compared with applying comparable amount of inorganic N and P (Kapkiyai et al., 1999). In this same study, yield also was increased when manure was applied in addition to the inorganic P. Inclusion of legumes in a no-tillage system with 66% stover removal increased the activity of nutrient cycling microbial enzymes (Roldan et al., 2003). Legume intercropping with reed Canary grass (*Phalaris arundinacea* L.) and awnless brome (*Bromus inermis* Leyss.) effectively maintained yield in the absence of N fertilizer (Jasinskas et al., 2008).

Nutrient Management in Perennial Systems

Perennial systems annually translocate and cycle many plant nutrients; thus they require different nutrient management compared with annual crop species. Fertilizer recommendations likely differ between establishment year and subsequent years. For example, application of N fertilizer is not recommended during the establishment year of warm-season grasses, to reduce weed competition (Duffy and Nanhou, 2001; Nyoka, 2007). Perennial grasses may have a high affinity for N recovery, which is why they have been used in grass filter strips to reduce environmental contamination (Bransby et al., 1998). A typical characteristic of perennial grasses is to utilize and even scavenge N, thus it is anticipated perennial grasses will respond to N fertilization.

Switchgrass yield increased in response to N fertilizer applied up to 225 kg N ha^{-1} (Vogel et al., 2002; Heaton et al., 2004; Lemus et al., 2008). Recommended annual N application rates range from 50 to 224 kg ha^{-1} (McLaughlin and Adams Kszos, 2005; Nyoka, 2007; Khanna et al., 2008; Lemus et al., 2008). Switchgrass yields of 10.5 to 12.6 Mg ha^{-1} required 120 kg N ha^{-1} to replace N removed during harvest, when harvested at peak biomass (Vogel et al., 2002). Comparably, switchgrass required 50 to 140 kg N ha^{-1}, up to 36 kg P ha^{-1}, and up to 105 kg K ha^{-1} to maintain yields from 9 to 15 Mg ha^{-1}

(Khanna et al., 2008). Delaying biomass harvest reduced both N concentration (Table 24|3) and amount of biomass removed, which may translate into lower N inputs required (Vogel et al., 2002). Improved fertilizer recommendations for switchgrass are still needed to account for regional and cultivar variability (Sanderson et al., 1996; Cassida et al., 2005b; Parrish and Fike, 2005).

Limited information is available on nutrient management for other perennial grasses. Miscanthus produced average yields of 22 Mg ha^{-1} compared with 10 Mg ha^{-1} for switchgrass in several studies compared by Heaton et al. (2004). Nitrogen use efficiency decreased but biomass yield increased with N application for Kentucky bluegrass, smooth bromegrass (*Bromus inermis* Leyss.), and orchard grass (*Dactylis glomerata* L.) (Zemenchik and Albrecht, 2002). Seed yield of meadow bromegrass (*Bromus riparius* Rehmann) increased with fertilizer application (50 or 100 kg N ha^{-1}), and harvesting the straw stimulated seed at both fertilizer rates (Loeppky and Coulman, 2002). Mulkey et al. (2008) reported a positive N response for switchgrass, big bluestem (*Andropogon gerardii* Vitman) and Indian grass [*Sorghastrum nutans* (L.) Nash]. In this mixed stand, big bluestem outcompeted switchgrass, and Indian grass virtually disappeared from the stand within 3 yr regardless of N treatment, demonstrating that stand composition can change over time.

Water Management

Water quantity and quality issues related to bioenergy production relate both to growing bioenergy feedstock and feedstock conversion. Water requirements vary among conversion platforms, with very little water needed for using biomass as a substitute for natural gas or coal compared with fermentation. It takes about 3 L of water to produce 1 L of ethanol by fermentation from grain (Owens, 2007) and 1.9 to 6 L water to produce 1 L of ethanol by cellulosic fermentation (Aden, 2007). Assuming a conservative water use efficiency of 10 kg ha^{-1} mm^{-1} (Hatfield et al., 2001), it takes about 14 million L ha^{-1} to grow corn yielding 14 Mg ha^{-1}. To convert the 14 Mg ha^{-1} corn grain into 5500 L ethanol it takes about 16,500 L ha^{-1} water using a corn grain ethanol conversion rate of 0.396 L kg^{-1} (Shapouri et al., 2003). Clearly, more water is needed to raise corn compared with the amount of water needed to operate an ethanol fermentation plant. In regions where water supplies are limited, nonagricultural (e.g., human consumption, wildlife, recreation) water demands compete with water available for irrigation (Postel and Richter, 2003). A vigorous, factual, public discussion and debate of the pros and cons of using water to irrigate biofuel feedstock is appropriate for policy development, but beyond the scope of this discussion.

Evapotranspiration dynamics differ among species and can be altered by management (Hatfield et al., 2001), and will occur regardless if the crop was raised for grain or for grain and residue. However, more water is lost from the soil when crop residues are removed (Al-Darby et al., 1989), which is especially important in semiarid or arid lands. Residues increase infiltration and decrease evaporation, generally resulting in a net increase in soil moisture (Smika and Unger, 1986; Blevins and Frye, 1993; Wells et al., 2003; Govaerts et al., 2007). Therefore, additional water inputs may be needed when residues in addition to grain are harvested.

In general, water holding capacity is increased with increasing SOM (Hudson, 1994); therefore, if biomass harvest causes a reduction in SOM, water holding capacity could also be reduced. Maintaining soil cover tends to increase water use efficiency by (i) reducing the potential for soil crusting and erosion, (ii) improving water infiltration, and (iii) reducing evaporation (Aase and Pikul, 1995). Changes at the soil surface due to residue management can have hydrological impacts. For example, Tomer et al. (2005) reported incorporation of residue increased the overland flow component of stream discharge from small watersheds by nearly 50% in a 25-yr study. Alas, research data on the impact of widespread biomass feedstock harvest on watershed hydrology is lacking (Uhlenbrook, 2007).

Intensive agriculture has been identified as a significant source of nutrients and pesticides in surface and groundwater. Concerns have been raised about biomass feedstock production exacerbating pollution of surface and groundwater resources (Nyakatawa et al., 2006; Simpson et al.,

2008). Conversion of perennial systems into row crops presents a much greater environmental risk than converting row crops into perennial biomass systems (Nelson et al., 2006). Compensating measures such as including cover crops are recommended when crop straws are harvested (Dabney et al., 2001). However, in arid or semiarid regions water use by cover crops may be detrimental (Unger and Vigil, 1998) so other measures would be needed. Eliminating tillage can reduce pesticide leaching (Gish et al., 1998) and nitrate leaching (Mkhabela et al., 2008). Managing the water table through subsurface irrigation and drainage such that water and nitrates are kept in the soil profile is a strategy to reduce nitrate loading in surface- and groundwater (Elmi et al., 2004). Expansion or addition of riparian buffer strips alongside biomass harvest area would reduce surface runoff reaching waterways (Bharati et al., 2002). Water quality must be safeguarded through improved water management whereby nutrient and pesticide loading to surface and groundwater is reduced and ideally eliminated.

Harvest Strategies

Strategies will vary by the biomass feedstock (e.g., cobs, stover, straw, and perennial grasses) harvested. Baling small grain straw is a well-established practice, historically associated with animal husbandry. Thus, for many producers, utilizing the straw as a biomass feedstock would not require learning additional skills or acquiring additional equipment. Small grain is harvested earlier in the season, when there may be more time available for the producer to harvest the straw. Until a one-pass harvest system is commercially available, corn stover or small grain straw requires additional harvest operations such as baling (Petrolia, 2008). Completing corn harvest in a timely manner is an especially important driving force behind developing corn harvest equipment that can remove and separate grain and biomass feedstock (cobs or stover) in one pass (Fig. 24|3) (Hoskinson et al., 2007). A one-pass harvest system reduces soil contamination of feedstock, improves harvest timeliness, and reduces costs.

Stover Harvest

Hoskinson et al. (2007) used a prototype one-pass system to evaluate four different harvesting scenarios for corn grain and biomass components in both a continuous corn and corn–soybean rotation in Iowa. The four biomass harvest scenarios were (i) the top 50% of the plant by height, (ii) the bottom 50% of the plant, (iii) all harvestable stover, or (iv) no stover. Corn grain and stover were harvested in 2005, 2006, and 2007 at the continuous corn site, but only in 2005 and 2007 at the rotated site (Table 24|4). Information was collected on feedstock quantity for each of the four harvest treatments. These data were compared with guidelines developed by Wilhelm et al. (2007) (Fig. 24|2) to determine the potential impact on SOC levels. Harvesting all recoverable aboveground stover (whole plant treatment) from continuous corn exceeded the recommended harvest rates by 3.50, 5.87, and 3.17 Mg ha^{-1} in 2005, 2006, and 2007, respectively. All four

Fig. 24|3. Corn grain and stover being collected using a one-pass harvesting system near Ames, IA.

Table 24|4. Corn grain (15.5% moisture) and dry stover yield from continuous and rotated sites near Ames, IA. Both sites were managed using fall chisel plowing and field cultivation in the spring to prepare the seedbed each year.

Stover fraction	2005	2006	2007	2005	2006	2007
	Continuous corn					
	——— Mg grain ha^{-1} ———			——— Mg stover ha^{-1} ———		
Whole plant	10.41	9.41	12.04	4.71	6.14	5.61
Cob and top 50%	10.35	9.53	12.04	2.91	4.98	4.42
Bottom 50%	10.35	9.22	11.85	1.26	1.48	0.70
Grain only	10.41	8.97	11.41	–	–	–
Annual avg./avg. total stover†	10.38	9.28	11.84	8.77	7.84	10.0
	Corn–soybean rotation					
Whole plant	12.73	–	13.17	7.11	–	5.67
Cob and top 50%	12.35	–	12.86	4.62	–	4.93
Bottom 50%	11.66	–	13.29	1.48	–	1.55
Grain only	12.42	–	12.98	–	–	–
Annual avg/avg. total stover†	12.29		13.08	10.39		11.05

† Assuming a 1:1 dry grain to dry stover ratio, this is amount of biomass produced.

removal treatments used at the rotated site exceeded the guideline of 12.50 Mg ha^{-1} of stover necessary to sustain SOM and the 8 Mg ha^{-1} necessary for water erosion control. Therefore, no removal of biomass would be recommended, which was confirmed by the rill erosion occurring after an extremely intense rainfall event in spring 2008 (Fig. 24|1).

An alternative to annually limiting biomass harvest based on annual SOC guidelines (Wilhelm et al., 2007) would be to rotate harvest in time, such that on average the SOC guidelines are met. For example, assume two years of continuous corn, yielding 10 Mg dry stover ha^{-1} yr^{-1}. If during year one 7 Mg ha^{-1} were harvested and 3 Mg ha^{-1} left in the field and during year two no stover were harvested, then 13 Mg ha^{-1} stover or an average of 6.5 Mg ha^{-1} yr^{-1} would be returned. Therefore, in continuous corn, alternating stover harvest would provide on average enough C inputs as well as sufficient ground cover to maintain SOC and prevent erosion (Fig. 24|2). Stover should be harvested less often, perhaps only once every 3 yr, if more aggressive tillage is used, or every other corn year in a corn–soybean rotation. From a manager's perspective, especially in the absence of commercially available harvest equipment, it may be more desirable to rotate harvest in time rather than trying to harvest only a small percent of a given standing crop.

Cob Harvest

The low density of corn stover and other herbaceous material presents major transportation and handling challenges (Perlack and Turhollow, 2002). Furthermore, excessive stover harvest from any area within a field could adversely affect water entry, retention, runoff, nutrient cycling, productivity, and many other critical soil functions (Wilhelm et al., 2007). Therefore, the ethanol industry is exploring the feasibility of using corn cobs as a biofuel feedstock (e.g., http://www.poet.com/innovation/cellulosic/ and http://www.cvec.com/ [verified 16 Sept. 2010]). Corn cobs have several advantages compared with using all of the stover for either thermochemical or biochemical (fermentation) conversion platforms. Crofcheck and Montross (2004) found that cobs had higher glucose concentration during enzyme hydrolysis than other stover fractions with and without pretreatment. This suggests corn cobs could have very high quality as a cellulosic fermentation feedstock. Total energy content of corn stover fractions (cobs, husk, leaves, stalks, and grain) ranged from 16.7 to 20.9 kJ g^{-1} at physiological maturity, with cobs being at the higher end of this range (Pordesimo et al., 2005). Others (Treier et al., 2006; Yu et al., 2008) reported similar heat values (18.4–18.7 kJ g^{-1}) with low ash, N, and S relative to coal. Low N and S concentrations also reduce the potential production of NOx and SO$_2$ pollutants. Harvesting only the corn cobs also helps address the biomass density issue because of their relatively uniform size and shape that exists after coarse grinding. Collection of only cobs would also leave most of the corn stover in the field (Pordesimo et al., 2004), which has multiple benefits with regard to minimizing soil erosion (Lindstrom, 1986; Erenstein, 2002) by protecting the soil from erosive forces (Wilson et al., 2004) and sustaining soil carbon (Johnson et al., 2006a; Wilhelm et al., 2007). A multistate survey conducted by the USDA-Agricultural Resource Service, Renewable Energy Assessment Project (REAP) team found that on a weight basis, cobs accounted for 15 to 20% of the aboveground nongrain biomass under a wide variety of conditions (Wilhelm et al., 2010). This range is consistent with data for irrigated corn in Colorado and Texas (Halvorson and Johnson, 2009) and in the western Corn Belt (Varvel and Wilhelm, 2008b). Cobs have potential for feedstock, especially if only limited stover is available for harvest after meeting SOC and soil coverage needs.

Other Feedstock Harvest

In the Corn Belt, corn cob and stover are likely primary feedstocks, but in other regions, the primary feedstock may be small grain straw (Banowetz et al., 2008), perennial grasses (McLaughlin and Adams Kszos, 2005), or wood products. Integrating use of primary with secondary feedstocks could increase temporal and spatial diversity within the landscape. Multiple feedstocks may encourage growing more perennial species, including prairie mixtures.

Feedstocks with temporally diverse harvest windows are desirable from a time management standpoint. Multiple feedstocks with staggered harvest dates may reduce storage capacity needs, but must fit within the existing management. For example, harvesting perennial grasses after a killing frost may interfere with harvesting corn and soybean, due to labor constraints. Harvesting perennial grasses in very early spring before nesting would provide winter cover for wildlife, but in northern regions (e.g., Minnesota) late snows can delay or prevent spring harvest (Fig. 24|4). Managing harvest of perennial grasses will also require understanding the potential tradeoff in harvest timing between maximizing biomass yield versus timing harvest to minimize nutrient removal and maintaining stand integrity (Vogel et al., 2002; Mulkey et al., 2006; Lee et al., 2007).

As the bioeconomy develops, efficient, timely, and commercially available harvest systems will evolve. The amount of biomass that is needed on the land to control erosion and maintain SOC are considerations for determining what is harvested, how and when biomass is harvested, and how frequently. Expanding crop rotations, incorporating cover crops, reducing tillage, and rotating harvest in time are all strategies that can be used to sustain soil while sustaining economic viability of biofuel feedstock.

Fig. 24|4. Standing switchgrass in western Minnesota with early April snow accumulated, delaying spring harvest.

Economic Drivers

Biomass prices can have a critical impact on how crops are managed. In order for biomass harvest to be economically feasible, biomass prices must offset any additional costs incurred. Additional costs may include biomass harvest, storage, and transportation; nutrient replacement; impacts on current and future crop productivity; and implementing changes in production practices to facilitate biomass production (e.g., changes in tillage or rotation). As biomass prices increase relative to other crops, there are increased economic incentives for greater biomass removal. This includes incentives to harvest biomass on more acreage and incentives to increase production per unit land area.

Collection costs per unit of biomass tend to decline at higher removal rates and higher yields (Duffy and Nanhou, 2002; Graham et al., 2007). This provides an economic incentive for a farmer to harvest the highest yielding biomass crops at the highest practical rates, increasing production per unit land area. With annual crop residues, this might be somewhat tempered by high water content of stover, nutrient replacement requirements at higher removal rates, and

the degree to which biomass harvest activities interfere with grain harvest (Hoskinson et al., 2007). However, the importance of these effects depends on the harvest technology used. For example, high moisture content of the stover is not an important issue when wet harvest and storage is used instead of dry harvest and storage (Shinners et al., 2007). The importance of these effects also depends on the biomass price. Higher biomass prices increase the value of biomass production, so that it may be profitable to incur additional production costs, including paying higher nutrient replacement costs and making expenditures or production changes to reduce interference with grain harvest activities (Apland et al., 1981; Bender et al., 1984).

Biomass harvest removes plant nutrients (Table 24|3), and replacing these nutrients can represent a significant production expense. For whole stover removal, the replacement cost of both macro- and micronutrients was estimated as high as $118 ha^{-1} beyond the cost of producing grain (Table 24|5). Input costs of fertilizers vary among production systems. For instance, total operating costs over 20 yr were substantially higher for miscanthus production in Illinois than for switchgrass, $11,748 ha^{-1} compared

Table 24|5. Average nutrient removal for five site-years and 2008 replacement cost for various fractions of corn stover harvested as a potential biofuel feedstock near Ames, IA (Karlen et al., unpublished data).

Stover fraction	N		P		K	
	kg ha^{-1}	$ ha^{-1}†	kg ha^{-1}	$ ha^{-1}	kg ha^{-1}	$ ha^{-1}
Whole plant	39	47.93	3.2	22.20	34	37.67
Cob and top 50%	29	35.64	3.1	21.50	28	31.02
Bottom 50%	7	8.60	0.7	4.86	9	9.97
	Ca		Mg		Cu	
	kg ha^{-1}	$ ha^{-1}	kg ha^{-1}	$ ha^{-1}	g ha^{-1}	$ ha^{-1}
Whole plant	23	1.55	17	3.17	20	0.46
Cob and top 50%	14	0.94	10	1.87	20	0.46
Bottom 50%	6	0.40	5	0.93	2	0.05
	Fe		Mn		Zn	
	g ha^{-1}	$ ha^{-1}	g ha^{-1}	$ ha^{-1}	g ha^{-1}	$ ha^{-1}
Whole plant	523	2.93	128	0.92	88	0.77
Cob and top 50%	299	1.67	69	0.50	69	0.61
Bottom 50%	162	0.91	29	0.21	17	0.15

† Prices based on cost to growers on 23 July 2008: N, $1.229 kg^{-1}; P, $6.936 kg^{-1}; K, $1.108 kg^{-1}; Ca, $0.0672 kg^{-1}; Mg, $0.1867 kg^{-1}; Cu, $0.0229 kg^{-1}; Fe, $0.0056 g^{-1}; Mn, $0.0072 g^{-1}; Zn, $0.0088 g^{-1}. N-P-K were calculated for using anhydrous ammonia for N, diammonium phosphate for P$_2$O$_5$, and muriate (KCl) for K; Ca was based on calcitic limestone 40% Ca; Mg was based on dolomitic limestone 9% Mg; Cu as copper sulfate; Fe as iron sulfate; Mn as Manganese oxide; and Zn as zinc oxide.

with $2775 ha^{-1} at the farm gate, with $571 ha^{-1} and $484 ha^{-1} for N, P, K, and lime, respectively (Khanna et al., 2008). Replacement costs of P and K for switchgrass grown in Iowa were estimated at $7.78 and $47.32 ha^{-1} (Duffy and Nanhou, 2001), which would be considerably higher using 2008 dollars. One promising aspect of energy production would be to recover N before gasification in a coupled gasification and ethanol production process (Anex et al., 2007). Potentially 28 to 78% of the N fertilizer applied in corn stover and switchgrass biomass systems could be recovered by this process. Once recovered, the nutrients could be transported back to the farm.

Feedstock removal rates are limited by the need to leave sufficient nongrain biomass on the land to control soil erosion and to maintain soil quality (Johnson et al., 2006a; Johnson et al., 2006b; Wilhelm et al., 2007). While these limits might be imposed by regulation, producers recognize the economic incentives to limit removal rates. Excessive biomass removal can lead to increased soil erosion and reductions in SOC, which have environmental consequences as well as impacts on future crop productivity (Wilhelm et al., 2004). If producers perceive negative impacts on crop productivity, this provides an incentive against excessive removal. While increasing biomass price produces a short-term incentive to increase biomass harvest, it also produces an incentive to maintain future productivity. At some point, it may become profitable to change management in ways that allow for a higher harvest level. Several analyses of potential biomass supply assumed that producers would switch to less-intensive tillage systems to increase the amount of crop residues that could be harvested (Gallagher et al., 2003; Sheehan et al., 2004; Perlack et al., 2005; Graham et al., 2007). Other potential production changes include the use of cover crops (Anex et al., 2007) and changes in crop varieties or crop rotations to increase crop residue cover or biomass production, or to make biomass harvest more timely. Potential examples include changing from a corn–soybean to a continuous corn rotation to increase average biomass production (Sheehan et al., 2004), inclusion of perennial biomass crops in annual cropping systems (Anex et al., 2007), long-term conversion from annual

crops to perennial crops (Varvel et al., 2008), or inclusion of wheat in rotation with corn to allow early fall straw harvest instead of late fall corn stover harvest (Apland et al., 1981). Policy analyses helped identify the production shifts that might occur across broad areas and a limited set of production alternatives (De La Torre Ugarte et al., 2003; Schneider and McCarl, 2003). However, the relevant price levels and resulting production shifts are likely to be producer and site specific. Producers may be able to harvest biomass feedstock (such as corn stover or wheat straw) with only minor changes to their production systems, while incorporating dedicated biomass crops (perennial woody or herbaceous) could require additional equipment and facilities. Substantial management changes and associated costs with perennial species feedstocks are compounded by multiple-year commitments and a delay from planting to first harvest, dramatically increasing associated risks. Therefore, larger price incentives or market assurances may be required to stimulate a production shift (Larson et al., 2008).

Integration

The rapidly emerging soil management technologies being developed to use corn stover and other lignocellulosic materials to produce biofuel and other bioproducts offer an opportunity to increase the net environmental benefits of agriculture by using crop rotations that are more temporally and spatially diverse than current rotations. Implementing diverse rotations, especially rotations that include perennial biomass feedstocks and integrating multiple feedstocks, provides multiple environmental, social, and economic benefits. Using strategies to minimize negative impacts and maximize potential benefits will help avoid many of the concerns raised by Doornbosch and Steenblik (2007), Ernsting and Boswell (2007), Fargione et al. (2008), and Searchinger et al. (2008). These authors raised concerns that putting new lands (e.g., rainforest) into production, especially production of annual species, will release more greenhouse gas than can be offset by using biomass for energy. Their arguments strengthen the case for developing a bioeconomy that includes sustainability as an a priori criterion. This would support policy and management decisions that discourage land use change, and encourage the use of practices that protect soil, water, and air resources.

Implementation of more diverse rotations in time and space for producing bioenergy feedstocks and solving many of the current "externalities" associated with agriculture has become more feasible with the advent of global positioning systems (GPS), geographic information systems (GIS), remote sensing, and related technologies for precision—or site-specific—mapping and management of crops. These technologies can be used to manage fertilizer and pesticide inputs (Giles and Slaughter, 1997; Tian et al., 1999; Ferguson et al., 2002; Khosla et al., 2002; Robert, 2002), guide the placement of drainage tile and terraces (Zhang et al., 2002), encourage in-field and field-edge conservation practices (Berry et al., 2003; Dinnes, 2004), and support placement of specific crops, rotations, and tillage practices within individual fields (Kitchen et al., 2005). New technologies can also help optimize water quality across watersheds and ecoregions (Hatch et al., 2001).

Concluding Remarks

Corn stover and corn cobs represent near-term herbaceous feedstocks for cellulosic ethanol production and/or thermochemical platforms. As the bioeconomy matures, other feedstocks (cellulosic and oilseed) likely will develop and expand. The feedstock market and production (all-inclusive planting, harvest, transport, and storage) will develop and evolve. Dominant cellulosic feedstock will vary by region and season. Multiple feedstocks with temporally diverse harvest will improve time management and reduce on- or off-farm storage needs. Routine broadscale biomass harvest needs to be managed such that it does not exacerbate soil erosion or loss of SOC. Annual harvest rates should maintain sufficient cover to control erosion; harvestable biomass decreases as tillage intensity increases. In general, more biomass needs to be returned to maintain SOC than to control erosion. Sufficient biomass inputs can be achieved by limiting the amount of biomass harvested, for example harvesting only cobs

or the top portion of the corn stalk. Another strategy is to limit the frequency biomass is harvested on a given field. Management practices such as reducing or eliminating tillage, adding cover crops, and including perennials are all means of increasing the amount of harvestable biomass while maintaining SOC and soil quality. These management strategies also help safeguard water quality. Additional management strategies for protecting water quality include managing irrigation and drainage to reduce nutrient loading, and adding or expanding riparian buffers to reduce surface runoff. Furthermore, soil cover can increase water use efficiency by reducing the potential for soil crusting and erosion, improving water infiltration, and by reducing evaporation. Nutrient management will need to be altered to compensate for the additional nutrient removal and adapted for perennial crops. Limiting N availability during establishment of perennial grasses can reduce weed pressure. Delaying perennial harvest until after senescence or until spring can reduce the amount of nutrients removed in the biomass. Soil fertility should be monitored by soil testing and crops scouted for deficiency symptoms, including micronutrients, so nutrient management can be adjusted as necessary to support biomass harvest. Overall, protecting soil, water, and air resources through improved soil and crop management is essential to developing the bioenergy industry in a manner that ensures long-term environmental and economic sustainability.

Acknowledgments

This work contributes to the objectives of the USDA-ARS-Renewable Energy Assessment Project (REAP) cross-location project. The authors wish to thank B. Burmeister for careful editing, but take full responsibility for any errors. The authors dedicate this work to Wally Wilhelm for his inspiration and commitment to sustainable bioenergy.

References

Aase, J.K., and J.L. Pikul, Jr. 1995. Crop and soil response to long-term no tillage practices in the Northern Great Plains. Agron. J. 87:652–656.

Abiven, S., S. Recous, V. Reyes, and R. Oliver. 2005. Mineralisation of C and N from root, stem and leaf residues in soil and role of their biochemical quality. Biol. Fertil. Soils 42:119–128.

Aden, A. 2007. Water use for current and future ethanol production. Southwest Hydrol. 6(5):22–23.

Adler, P.R., M.A. Sanderson, A.A. Boateng, P.J. Weimer, and H.-J.G. Jung. 2006. Biomass yield and biofuel quality of switchgrass harvested in fall or spring. Agron. J. 98:1518–1525.

Al-Darby, A.M., M.A. Mustafa, A.M. Al-Omran, and M.O. Mahjoub. 1989. Effect of wheat residue and evaporative demands on intermittent evaporation. Soil Tillage Res. 15:105–116.

Al-Kaisi, M.M., X. Yin, and M.A. Licht. 2005. Soil carbon and nitrogen changes as affected by tillage system and crop biomass in a corn-soybean rotation. Appl. Soil Ecol. 30:174–191.

Andraski, T.W., and L.G. Bundy. 2008. Corn residue and nitrogen source effects on nitrogen availability in no-till corn. Agron. J. 100:1274–1279.

Andren, O., and K. Paustian. 1987. Barley straw decomposition in the field: A comparison of models. Ecology 68:163–210.

Anex, R.P., L.R. Lynd, M.S. Laser, A.H. Heggenstaller, and M. Liebman. 2007. Potential for enhanced nutrient cycling through coupling of agricultural and bioenergy systems. Crop Sci. 47:1327–1335.

Apland, J., B.A. McCarl, and T.G. Baker. 1981. Crop residue supply for energy generation: A prototype application to Midwestern U.S.A. grain farms. Energy Agric. 1:55–70.

Baker, J.M., T.E. Ochsner, R.T. Venterea, and T.J. Griffis. 2007. Tillage and soil carbon sequestration–What do we really know? Agric. Ecosyst. Environ. 118:1–5.

Banowetz, G.M., A. Boateng, J.J. Steiner, S.M. Griffith, V. Sethi, and H. El-Nashaar. 2008. Assessment of straw biomass feedstock resources in the Pacific Northwest. Biomass Bioenergy 32:629–634.

Bayer, C., T. Lovato, J. Dieckow, J.A. Zanatta, and J. Mielniczuk. 2006. A method for estimating coefficients of soil organic matter dynamics based on long-term experiments. Soil Tillage Res. 91:217–226.

Bender, D.A., R.M. Peart, D.H. Doster, J.R. Barrett, and M.O. Bagby. 1984. Energy crop evaluation by linear programming. Energy Agric. 3:199–210.

Berry, J.K., J.A. Delgado, R. Khosla, and F.J. Pierce. 2003. Precision conservation for environmental sustainability. J. Soil Water Conserv. 58(6):332–339.

Bharati, L., K.H. Lee, T.M. Isenhart, and R.C. Schultz. 2002. Soil-water infiltration under crops, pasture, and established riparian buffer in Midwestern USA. Agrofor. Syst. 56:249–257.

Bilbro, J.D., and D.W. Fryrear. 1994. Wind erosion losses as related to plant silhouette and soil cover. Agron. J. 86:550–553.

Blevins, R.L., and W.W. Frye. 1993. Conservation tillage: An ecological approach to soil management. Adv. Agron. 51:33–78.

Borie, F., Y. Redel, R. Rubio, J.L. Rouanet, and J.M. Barea. 2002. Interactions between crop residues application and mycorrhizal developments and some soil-root interface properties and mineral acquisition by plants in an acidic soil. Biol. Fertil. Soils 36:151–160.

Bransby, D.I., S.B. McLaughlin, and D.J. Parrish. 1998. A review of carbon and nitrogen balances in switchgrass grown for energy. Biomass Bioenergy 14:379–384.

Breakwell, D.P., and R.F. Turco. 1989. Nutrient and phytotoxic contributions of residue to soil in no-till continuous-corn ecosystems. Biol. Fertil. Soils 8:328–334.

Brigham, R.D. 1993. Castor: Return of an old crop. p. 380–383. In J. Janick and J. E. Simon (ed.) New crops. John Wiley & Sons, New York.

Bruun, S., B.T. Christensen, E.M. Hansen, J. Magid, and L.S. Jensen. 2003. Calibration and validation of the soil organic matter dynamics of the Daisy model with data from the Askov long-term experiments. Soil Biol. Biochem. 35:67–76.

Burgess, M.S., G.R. Mehuys, and C.A. Madramootoo. 2002. Decomposition of grain-corn residues (*Zea mays* L.): A litterbag study under three tillage systems. Can. J. Soil Sci. 82:127–138.

Carlson, K.D., J.C. Gardner, V.L. Anderson, and J.J. Hanzel. 1996. Crambe: New crop success. p. 306–322. In J. Janick (ed.) Progress in new crops. ASHS Press, Alexandria, VA.

Cassida, K.A., J.P. Muir, M.A. Hussey, J.C. Read, B.C. Venuto, and W.R. Ocumpaugh. 2005a. Biofuel component concentrations and yields of switchgrass in south central U.S. environments. Crop Sci. 45:682–692.

Cassida, K.A., J.P. Muir, M.A. Hussey, J.C. Read, B.C. Venuto, and W.R. Ocumpaugh. 2005b. Biomass yield and stand characteristics of switchgrass in south central U.S. environments. Crop Sci. 45:673–681.

Christensen, B.T. 1986. Barley straw decomposition under field conditions: Effects of placement and initial nitrogen content on weight loss and nitrogen dynamics. Soil Biol. Biochem. 18:523–529.

Clifton-Brown, J.C., and I. Lewandowski. 2002. Screening Miscanthus genotypes in field trials to optimise biomass yield and quality in Southern Germany. Eur. J. Agron. 16:97–110.

Cookson, W.R., M.H. Beare, and P.E. Wilson. 1998. Effects of prior crop residue management on microbial properties and crop residue decomposition. Appl. Soil Ecol. 7:179–188.

Cox, C.A. 2008. Beyond T: Guiding sustainable soil management. J. Soil Water Conserv. 63(5):162A–164A.

Crofcheck, C.L., and M.D. Montross. 2004. Effect of stover fraction on glucose production using enzymatic hydrolysis. Trans. ASABE 47:841–844.

CTIC. 2002. National Crop Residue Management Survey. Available at http://www.ctic.purdue.edu/CRM/ (posted 2002; verified 6 Oct. 2010). Conserv. Technol. Inf. Cent., West Lafayette, IN.

Dabney, S.M., J.A. Delgado, and D.W. Reeves. 2001. Using winter cover crops to improve soil and water quality. Commun. Soil Sci. Plant Anal. 32:1221–1250.

De La Torre Ugarte, D.G., M.E. Walsh, H. Shapouri, and S.P. Slinsky. 2003. The economic impacts of bioenergy crop production on U.S. agriculture. Agric. Econ. Rep. No. 816. USDA-ERS, Washington, DC.

Demirbas, A. 2006. Biodiesel production via non-catalytic SCF method and biodiesel fuel characteristics. Energy Convers. Manage. 47:2271–2282.

Dexter, A.R., D. Hein, and J.S. Hewett. 1982. Macro-structure of the surface layer of self-mulching clay in relation to cereal stubble management. Soil Tillage Res. 2:251–264.

Dierig, D.A., T.A. Coffelt, F.S. Nakayama, and A.E. Thompson. 1996. Lesquerell and vernonia: Oilseeds for arid lands. p. 347–354. In J. Janick (ed.) Progress in new crops. ASHS Press, Alexandria, VA.

Dinnes, D.L. 2004. Assessments of practices to reduce nitrogen and phosphorus nonpoint source pollution of Iowa's surface waters. Available at http://www.iowadnr.gov/water/nutrients/files/nps_assessments.pdf (verified 16 Sept. 2010). Iowa Dep. of Nat. Resour., Des Moines, IA and USDA-ARS Natl. Soil Tilth Lab., Ames, IA.

Doornbosch, R., and R. Steenblik. 2007. Biofuels: Is the cure worse than the disease? SG/SD/RT(2007)3. OECD, Paris.

Duffy, M.D., and V.Y. Nanhou. 2001. Costs of producing switchgrass for biomass in Iowa. PM 1866. Available at http://www.extension.iastate.edu/Publications/PM1866.pdf (posted April 2001; verified 16 Sept. 2010) Iowa State Univ. Ext., Ames.

Duffy, M.D., and V.Y. Nanhou. 2002. Costs of producing switchgrass for biomass in Southern Iowa. p. 267–275.

In J. Janick and A. Whipkey (ed.) Trends in new crops and new uses. ASHS Press, Alexandria, VA.

Elmi, A.A., C. Madramootoo, M. Egeh, and C. Hamel. 2004. Water and fertilizer nitrogen management to minimize nitrate pollution from a cropped soil in southwestern Quebec, Canada. Water Air Soil Pollut. 151:117–134.

Erenstein, O. 2002. Crop residue mulching in tropical and semi-tropical countries: An evaluation of residue availability and other technological implications. Soil Tillage Res. 67:115–133.

Ernsting, A., and A. Boswell. 2007. Agrofuels: Towards a reality check in nine key areas. Available at http://www.corporateeurope.org/docs/AgrofuelsReality-Check.pdf (verified 16 Sept. 2010). Biofuelwatch.

Fageria, N.K. 2004. Dry matter yield and shoot nutrient concentrations of upland rice, common bean, corn, and soybean grown in rotation on an Oxisol. Commun. Soil Sci. Plant Anal. 35:961–974.

Fargione, J., J. Hill, D. Tilman, S. Polasky, and P. Hawthorne. 2008. Land clearing and the biofuel carbon debt. Science 319:1235–1238.

Fatondji, D., C. Martius, C. Bielders, P. Vlek, A. Bationo, and B. Gerard. 2006. Effect of planting technique and amendment type on pearl millet yield, nutrient uptake, and water use on degraded land in Niger. Nutr. Cycling Agroecosyst. 76:203–217.

Ferguson, R.B., G.W. Hergert, J.S. Schepers, C.A. Gotway, J.E. Cahoon, and T.A. Peterson. 2002. Site-specific nitrogen management of irrigated maize: Yield and soil residual nitrate effects. Soil Sci. Soc. Am. J. 66:544–553.

Frank, A.B., J.D. Berdahl, J.D. Hanson, M.A. Liebig, and H.A. Johnson. 2004. Biomass and carbon partitioning in switchgrass. Crop Sci. 44:1391–1396.

Franzluebbers, A.J., F.M. Hons, and V.A. Saladino. 1995. Sorghum, wheat and soybean production as affected by long-term tillage, crop sequence and N fertilization. Plant Soil 173:55–65.

Gallagher, P.W., M. Dikeman, J. Fritz, E. Wailes, W. Gauthier, and H. Shapouri. 2003. Supply and social cost estimates for biomass from crop residues in the United States. Environ. Resour. Econ. 24:335–358.

Gesch, R.W., F. Forcella, A.E. Olness, D.W. Archer, and A. Hebard. 2006. Agricultural management of cuphea and potential for commercial production in the United States. Ind. Crops Prod. 24:300–306.

Giles, D.K., and D.C. Slaughter. 1997. Precision band sprayer with machine-vision guidance and adjustable yaw nozzles. Trans. ASAE 40(1):29–36.

Gish, T.J., D. Gimenez, and W.J. Rawls. 1998. Impact of roots on ground water quality. Plant Soil 200:47–54.

Goodrum, J.W., and D.P. Geller. 2005. Influence of fatty acid methyl esters from hydroxylated vegetable oils on diesel fuel lubricity. Bioresour. Technol. 96:851–855.

Govaerts, B., M. Fuentes, M. Mezzalama, J.M. Nicol, J. Deckers, J.D. Etchevers, B. Figueroa-Sandoval, and K.D. Sayre. 2007. Infiltration, soil moisture, root rot and nematode populations after 12 years of different tillage, residue and crop rotation managements. Soil Tillage Res. 94:209–219.

Graham, R.L., R. Nelson, J. Sheehan, R.D. Perlack, and L.L. Wright. 2007. Current and potential U.S. corn stover supplies. Agron. J. 99:1–11.

Halvorson, A.D., and J.M.F. Johnson. 2009. Irrigated corn cob production in the Central Great Plains. Agron. J. 101:390–399.

Halvorson, A.D., and C.A. Reule. 2007. Irrigated, no-till corn and barley response to nitrogen in northern Colorado. Agron. J. 99:1521–1529.

Hatch, L.K., A. Mallawatantri, D. Wheeler, A. Gleason, D. Mulla, J. Perry, K.W. Easter, R. Smith, L. Gerlach, and P. Brezonik. 2001. Land management at the major watershed–agroecoregion intersection. J. Soil Water Conserv. 56(1):44–51.

Hatfield, J.L., T.J. Sauer, and J.H. Prueger. 2001. Managing soils to achieve greater water use efficiency: A review. Agron. J. 93:271–280.

Heaton, E., T. Voigt, and S.P. Long. 2004. A quantitative review comparing the yields of two candidate C4 perennial biomass crops in relation to nitrogen, temperature and water. Biomass Bioenergy 27:21–30.

Heaton, E.A., F.G. Dohleman, and S.P. Long. 2008. Meeting U.S. biofuel goals with less land: The potential of Miscanthus. Global Change Biol. 14:2000–2014.

Hoskinson, R.L., D.L. Karlen, S.J. Birrell, C.W. Radtke, and W.W. Wilhelm. 2007. Engineering, nutrient removal, and feedstock conversion evaluations of four corn stover harvest scenarios. Biomass Bioenergy 31:126–136.

Hudson, B.D. 1994. Soil organic matter and available water capacity. J. Soil Water Conserv. 49:189–194.

Huggins, D.R., R.R. Allmaras, C.E. Clapp, J.A. Lamb, and G.W. Randall. 2007. Corn-soybean sequence and tillage effects on soil carbon dynamics and storage. Soil Sci. Soc. Am. J. 71:145–154.

Izaurralde, R.C., J.R. Williams, W.B. McGill, N.J. Rosenberg, and M.C.Q. Jakas. 2006. Simulating soil C dynamics with EPIC: Model description and testing against long-term data. Ecol. Modell. 192:362–384.

Jasinskas, A., A. Zaltauskas, and A. Kryzeviciene. 2008. The investigation of growing and using of tall perennial grasses as energy crops. Biomass Bioenergy 32:981–987.

Jawson, M.D., and L.F. Elliott. 1986. Carbon and nitrogen transformation wheat straw and root decomposition. Soil Biol. Biochem. 18:15–22.

Johnson, J.M.F., R.R. Allmaras, and D.C. Reicosky. 2006a. Estimating source carbon from crop residues, roots and rhizodeposits using the national grain-yield database. Agron. J. 98:622–636.

Johnson, J.M.F., N.W. Barbour, and S.L. Weyers. 2007a. Chemical composition of crop biomass impacts its decomposition. Soil Sci. Soc. Am. J. 71:155–162.

Johnson, J.M.F., M.D. Coleman, R.W. Gesch, A.A. Jaradat, R. Mitchell, D.C. Reicosky, and W.W. Wilhelm. 2007b. Biomass-bioenergy crops in the United States: A changing paradigm. The Americas J. Plant Sci. Biotechnol. 1:1–28.

Johnson, J.M.F., A.J. Franzluebbers, S.L. Weyers, and D.C. Reicosky. 2007c. Agricultural opportunities to mitigate greenhouse gas emissions. Environ. Pollut. 150:107–124.

Johnson, J.M.F., D.C. Reicosky, R.R. Allmaras, D. Archer, and W.W. Wilhelm. 2006b. A matter of balance: Conservation and renewable energy. J. Soil Water Conserv. 61(4):120A–125A.

Johnson, P.A., and B.J. Chamber. 1996. Effects of husbandry on soil organic matter. Soil Use Manage. 13:102–103.

Kaewpradit, W., B. Toomsan, P. Vityakon, V. Limpinuntana, P. Saenjan, S. Jogloy, A. Patanothai, and G. Cadisch. 2008. Regulating mineral N release and greenhouse gas emissions by mixing groundnut residues and rice straw under field conditions. Eur. J. Soil Sci. 59:640–652.

Kapkiyai, J.J., N.K. Karanja, J.N. Qureshi, P.C. Smithson, and P.L. Woomer. 1999. Soil organic matter and nutrient dynamics in a Kenyan nitisol under long-term fertilizer and organic input management. Soil Biol. Biochem. 31:1773–1782.

Katterer, T., O. Andren, and R. Pettersson. 1998. Growth and nitrogen dynamics of reed canarygrass (*Phalaris arundinacea* L.) subjected to daily fertilization and irrigation in the field. Field Crops Res. 55:153–164.

Khanna, M., B. Dhungana, and J. Clifton-Brown. 2008. Costs of producing miscanthus and switchgrass for bioenergy in Illinois. Biomass Bioenergy 32:482–493.

Khosla, R., K. Fleming, J.A. Delgado, T.M. Shaver, and D.G. Westfall. 2002. Use of site-specific management zones to improve nitrogen management for precision agriculture. J. Soil Water Conserv. 57:513–518.

Kitchen, N.R., K.A. Sudduth, D.B. Myers, R.E. Massey, E.J. Sadler, R.N. Lerch, J.W. Hummel, and H.L. Palm. 2005. Development of a conservation-oriented precision agriculture system: Crop production assessment and plan implementation. J. Soil Water Conserv. 60(6):421–430.

Knapp, S.J., and J.M. Crane. 2000. Registration of reduced shattering cuphea germplasm PSR23. Crop Sci. 41:299–300.

Kong, A.Y.Y., J. Six, D.C. Bryant, R.F. Denison, and C. van Kessel. 2005. The relationship between carbon input, aggregation, and soil organic carbon stabilization in sustainable cropping systems. Soil Sci. Soc. Am. J. 69:1078–1085.

Kundu, S., R. Bhattacharyya, V. Prakash, B.N. Ghosh, and H.S. Gupta. 2007. Carbon sequestration and relationship between carbon addition and storage under rainfed soybean-wheat rotation in a sandy loam soil of the Indian Himalayas. Soil Tillage Res. 92:87–95.

Kushwaha, C.P., S.K. Tripathi, and K.P. Singh. 2000. Variations in soil microbial biomass and N availability due to residue and tillage management in a dryland rice agroecosystem. Soil Tillage Res. 56:153–166.

Lal, R. 2004. Is crop residue a waste? J. Soil Water Conserv. 59:136–139.

Lal, R. 2008. Crop residues as soil amendments and feedstock for bioethanol production. Waste Manag. 28:747–758.

Larson, J.A., B.C. English, and L. He. 2008. Risk and return for bioenergy crops under alternative contracting arrangements. Southern Agric. Econ. Assoc. Ann. Meet., Dallas, TX. 2–6 Feb. 2008. Available at http://ageconsearch.umn.edu/bitstream/6842/2/sp08la02.pdf (verified 17 Sept. 2010).

Lee, D.K., and A. Boe. 2005. Biomass production of switchgrass in Central South Dakota. Crop Sci. 45:2583–2590.

Lee, D.K., V.N. Owens, and J.J. Doolittle. 2007. Switchgrass and soil carbon sequestration response to ammonium nitrate, manure, and harvest frequency on Conservation Reserve Program land. Agron. J. 99:462–468.

Lemus, R., E.C. Brummer, K.J. Moore, N.E. Molstad, C.L. Burras, and M.F. Barker. 2002. Biomass yield and quality of 20 switchgrass populations in southern Iowa, USA. Biomass Bioenergy 23:433–442.

Lemus, R., E.C. Brummer, C.L. Burras, K.J. Moore, M.F. Barker, and N.E. Molstad. 2008. Effects of nitrogen fertilization on biomass yield and quality in large fields of established switchgrass in southern Iowa, USA. Biomass Bioenergy 32:1187–1194.

Lewandowski, I., and A. Kicherem. 1997. Combustion quality of biomass: Practical relevance and experiments to modify the biomass quality of *Miscanthus x giganteus*. Eur. J. Agron. 6:163–177.

Liebig, M., M. Schmer, K. Vogel and R. Mitchell. 2008. Soil carbon storage by switchgrass grown for bioenergy. BioEnergy Res. 1:215–222.

Liebig, M.A., H.A. Johnson, J.D. Hanson, and A.B. Frank. 2005. Soil carbon under switchgrass stands and cultivated cropland. Biomass Bioenergy 28:347–354.

Lindstrom, M.J. 1986. Effects of residue harvesting on water runoff, soil erosion and nutrient loss. Agric. Ecosyst. Environ. 16:103–112.

Linquist, B.A., V. Phengsouvanna, and P. Sengxue. 2007. Benefits of organic residues and chemical fertilizer to productivity of rain-fed lowland rice and to soil nutrient balances. Nutr. Cycling Agroecosyst. 79:59–72.

Loeppky, H.A., and B.E. Coulman. 2002. Crop residue removal and nitrogen fertilization affects seed production in meadow bromegrass. Agron. J. 94:450–454.

Madakadze, I.C., K.A. Stewart, P.R. Peterson, B.E. Coulman, and D.L. Smith. 1999. Cutting frequency and nitrogen fertilization effects on yield and nitrogen concentration of switchgrass in a short season area. Crop Sci. 39:552–557.

Manlay, R.J., J.L. Chotte, D. Masse, J.Y. Laurent, and C. Feller. 2002. Carbon, nitrogen and phosphorus allocation in agro-ecosystems of a West African savanna III. Plant and soil components under continuous cultivation. Agric. Ecosyst. Environ. 88:249–269.

Mann, L., and V. Tolbert. 2000. Soil sustainability in renewable biomass plantings. Ambio 29:492–498.

McLaughlin, S.B., and L. Adams Kszos. 2005. Development of switchgrass (*Panicum virgatum*) as a bioenergy feedstock in the United States. Biomass Bioenergy 28:515–535.

McLaughlin, S., and M. Walsh. 1998. Evaluating environmental consequences of producing herbaceous crops for bioenergy Biomass Bioenergy 14:317–324.

Mitchell, R., J. Webb, and R. Harrison. 2001. Crop residues can affect N leaching over at least two winters. Eur. J. Agron. 15:17–29.

Mkhabela, M.S., A. Madani, R. Gordon, D. Burton, D. Cudmore, A. Elmi, and W. Hart. 2008. Gaseous and leaching nitrogen losses from no-tillage and conventional tillage systems following surface application of cattle manure. Soil Tillage Res. 98:187–199.

Moebius-Clune, B.N., H.M. van Es, O.J. Idowu, R.R. Schindelbeck, D.J. Moebius-Clune, D.W. Wolfe, G.S. Abawi, J.E. Thies, B.K. Gugino, and R. Lucey. 2008. Long-term effects of harvesting maize stover and tillage on soil quality. Soil Sci. Soc. Am. J. 72:960–969.

Monti, A., N. Di Virgilio, and G. Venturi. 2008. Mineral composition and ash content of six major energy crops. Biomass Bioenergy 32:216–223.

Moser, L.E., and K.P. Vogel. 1995. Switchgrass, big bluestem, and indiangrass. p. 409–420. *In* R.F. Barnes et al. (ed.) Forages, Vol. 1: An Introduction to grassland agriculture. Iowa State Univ. Press, Ames, IA.

Mulkey, V.R., V.N. Owens, and D.K. Lee. 2006. Management of switchgrass-dominated conservation reserve program lands for biomass production in South Dakota. Crop Sci. 46:712–720.

Mulkey, V.R., V.N. Owens, and D.K. Lee. 2008. Management of warm-season grass mixtures for biomass production in South Dakota USA. Bioresour. Technol. 99:609–617.

Nelson, R.G., J.C. Ascough II, and M.R. Langemeier. 2006. Environmental and economic analysis of switchgrass production for water quality improvement in northeast Kansas. J. Environ. Manage. 79:336–347.

Nicholson, F.A., B.J. Chambers, A.R. Mills, and P.J. Strachan. 1997. Effects of repeated straw incorporation on crop fertiliser nitrogen requirements, soil mineral nitrogen and nitrate leaching losses. Soil Use Manage. 13:136–142.

Nyakatawa, E.Z., D.A. Mays, V.R. Tolbert, T.H. Green, and L. Bingham. 2006. Runoff, sediment, nitrogen, and phosphorus losses from agricultural land converted to sweetgum and switchgrass bioenergy feedstock production in north Alabama. Biomass Bioenergy 30:655–664.

Nyoka, B. 2007. Management guide for feedstock production from switchgrass in the northern Great Plains. Available at http://ncsungrant.sdstate.org/uploads/publications/SGINC2-07.pdf (verified 17 Sept. 2010) North Central Sun Grant Center, South Dakota State Univ., Brookings.

Oschwald, W.R., M. Stelly, D.M. Kral, and J.H. Nauseef (ed.). 1978. Crop residue management systems. ASA Spec. Pub. 31. ASA, CSSA, and SSSA, Madison, WI.

Owens, S.R. 2007. Reduce Cooling Tower Water Consumption by 20 Percent. Available at http://ethanolproducer.com/article.jsp?article_id=3030 (posted June, 2007; verified 17 Sept. 2010). Ethanol Producer Magazine, Grand Forks, ND.

Parrish, D.J., and J.H. Fike. 2005. The biology and agronomy of switchgrass for biofuels. Crit. Rev. Plant Sci. 24:423–459.

Parton, W.J., J.W.B. Stewart, and C.V. Cole. 1988. Dynamics of C, N, P and S in grassland soils: A model. Biogeochemistry 5:109–131.

Perlack, R.D., L.L. Wright, A. Turhollow, R.L. Graham, B. Stokes, and D.C. Erbach. 2005. Biomass as feedstock for a bioenergy and bioproducts industry: The technical feasibility of a billion-ton annual supply. Available at http://www.eere.energy.gov/biomass/pdfs/final_billionton_vision_report2.pdf (posted 15 Jul. 2005; verified 17 Sept. 2010). USDOE and USDA.

Perlack, R.D., and A. Turhollow. 2002. Assessment options for the collection, handling and storage of corn stover. Available at http://bioenergy.ornl.gov/pdfs/ornltm-200244.pdf (posted Sept. 2002; verified 17 Sept. 2010). Oak Ridge Natl. Lab., USDOE.

Petrolia, D.R. 2008. The economics of harvesting and transporting corn stover for conversion to fuel ethanol: A case study for Minnesota. Biomass Bioenergy 32:603–612.

Pikul, J.L.J., J.M.F. Johnson, T.E. Schumacher, M. Vigil, and W.E. Riedell. 2008. Change in surface soil carbon under rotated corn in eastern South Dakota. Soil Sci. Soc. Am. J. 72:1738–1744.

Pordesimo, L.O., W.C. Edens, and S. Sokhansanj. 2004. Distribution of aboveground biomass in corn stover. Biomass Bioenergy 26:337–343.

Pordesimo, L.O., B.R. Hames, S. Sokhansanj, and W.C. Edens. 2005. Variation in corn stover composition and energy content with crop maturity. Biomass Bioenergy 28:366–374.

Postel, S., and B. Richter. 2003. Rivers for life: Managing water for people and nature. Island Press, Washington, DC.

Putnam, D.H., J.T. Budin, L.A. Field, and W.M. Breene. 1993. Camelina: A promising low-input oilseed. p. 314–322. In J. Janick and J. E. Simon (ed.) New crops. John Wiley & Sons, New York.

Rao, S.C., H.S. Mayeux, and B.K. Northup. 2005. Performance of forage soybean in the southern Great Plains. Crop Sci. 45:1973–1977.

Reicosky, D.C., and F. Forcella. 1998. Cover crops and soil quality interactions in agroecosystems. J. Soil Water Conserv. 53:224–229.

Reijnders, L. 2006. Conditions for the sustainability of biomass based fuel use. Energy Policy 34:863–876.

Reynolds, J.H., C.L. Walker, and M.J. Kirchner. 2000. Nitrogen removal in switchgrass biomass under two harvest systems. Biomass Bioenergy 19:281–286.

Rickman, R.W., C.L. Douglas, S.L. Albrecht, L.G. Bundy, and J.L. Berc. 2001. CQESTR: A model to estimate carbon sequestration in agricultural soils. J. Soil Water Conserv. 56:237–242.

Robert, P.C. 2002. Precision agriculture: A challenge for crop nutrition management. Plant Soil 247:143–149.

Roldan, A., F. Caravaca, M.T. Hernandez, C. Garcia, C. Sanchez-Brito, M. Velasquez, and M. Tiscareno. 2003. No-tillage, crop residue additions, and legume cover cropping effects on soil quality characteristics under maize in Patzcuaro watershed. Soil Tillage Res. 72:65–73.

Saffigna, P.G., D.S. Powlson, P.C. Brookes, and G.A. Thomas. 1989. Influence of sorghum residues and tillage on soil organic matter and soil microbial biomass in an Australian Vertisol. Soil Biol. Biochem. 21:759–765.

Sainju, U.M., A. Lenssen, T. Caesar-Thonthat, and J. Waddell. 2006a. Carbon sequestration in dryland soils and plant residue as influenced by tillage and crop rotation. J. Environ. Qual. 35:1341–1347.

Sainju, U.M., A. Lenssen, T. Caesar-Thonthat, and J. Waddell. 2006b. Tillage and crop rotation effects on dryland soil and residue carbon and nitrogen. Soil Sci. Soc. Am. J. 70:668–678.

Salinas-Garcia, J.R., A.D. Baez-Gonzalez, M. Tiscareno-Lopez, and E. Rosales-Robles. 2001. Residue removal and tillage interaction effects on soil properties under rain-fed corn production in Central Mexico. Soil Tillage Res. 59:67–79.

Sanderson, M.A., R.L. Reed, S.B. McLaughlin, S.D. Wullschleger, B.V. Conger, D.J. Parrish, D.D. Wolf, C. Taliaferro, A.A. Hopkins, W.R. Ocumpaugh, M.A. Hussey, J.C. Read, and C.R. Tischler. 1996. Switchgrass as a sustainable bioenergy crop. Bioresour. Technol. 56:83–93.

Sarr, P.S., M. Khouma, M. Sene, A. Guisse, A.N. Badiane, and T. Yamakawa. 2008. Effect of pearl millet–cowpea cropping systems on nitrogen recovery, nitrogen use efficiency and biological fixation using the ^{15}N tracer technique. Soil Sci. Plant Nutr. 54:142–147.

Schneider, U.A., and B.A. McCarl. 2003. Economic potential of biomass based fuels for greenhouse gas emission mitigation. Environ. Resour. Econ. 24:291–312.

Searchinger, T., R. Heimlich, R.A. Houghton, F. Dong, A. Elobeid, J. Fabiosa, S. Tokgoz, D. Hayes, and T.-H. Yu. 2008. Use of U.S. croplands for biofuels increases greenhouse gases through emissions from land-use change. Science 319:1238–1240.

Shapouri, H., J.A. Duffield, and M. Wang. 2003. The energy balance of corn ethanol revisited. Trans. ASAE 46:959–968.

Sheehan, J., A. Aden, K. Paustian, K. Killian, J. Brenner, M. Walsh, and R. Nelson. 2004. Energy and environmental aspects of using corn stover for fuel ethanol. J. Ind. Ecol. 7:117–146.

Shinners, K.J., B.N. Binversie, R.E. Muck, and P.J. Weimer. 2007. Comparison of wet and dry corn stover harvest and storage. Biomass Bioenergy 31:211–221.

Siemens, J.C., and W.R. Oschwald. 1978. Corn-soybean tillage systems: Erosion control, effects on crop production, costs. Trans. ASAE 21:293–302.

Simpson, T.W., A.N. Sharpley, R.W. Howarth, H.W. Paerl, and K.R. Mankin. 2008. The new gold rush: Fueling ethanol production while protecting water quality. J. Environ. Qual. 37:318–324.

Singer, J.W. 2005. Small grain cover crops for corn and soybean. Available at http://www.extension.iastate.edu/Publications/PM1999.pdf (verified 17 Sept. 2010). Iowa State Univ., Ames.

Smika, D.E., and P.W. Unger. 1986. Effect of surface residues on soil water storage. Adv. Soil Sci. 5:111–138.

Smil, V. 1999. Crop Residues: Agriculture's largest harvest. Bioscience 49:299–308.

Stocking, M.A. 1988. Assessing vegetative cover and management effects. p. 211–234. In R. Lal (ed.) Soil erosion research methods. Soil Water Conserv. Soc., Ankeny, IA.

Syers, J.K. 1997. Managing soils for long-term productivity. Philos. Trans. R. Soc. Lond. B Biol. Sci. 352:1011–1021.

Teel, A., S. Barnhart, and G. Miller. 2003. Management guide for the production of switchgrass for biomass fuel in Southern Iowa. PM 1710. Available at http://www.extension.iastate.edu/Publications/PM1710.pdf (posted May 2003; verified 17 Sept. 2010). Iowa State Univ., Ames.

Thorup-Kristensen, K., J. Magid, and L.S. Jensen. 2003. Catch crops and green manures as biological tools in nitrogen management in temperate zones. Adv. Agron. 79:227–302.

Tian, G., B.T. Kang, and L. Brussand. 1992. Biological effects of plant residues with contrasting chemical composition under humid tropical conditions—Decomposition and nutrient release. Soil Biol. Biochem. 24:1051–1060.

Tian, L., J.F. Reid, and J.W. Hummel. 1999. Development of a precision sprayer for site-specific weed management. Trans. ASAE 42:893–900.

Tirol-Padre, A., K. Tsuchiya, K. Inubushi, and J.K. Ladha. 2005. Enhancing Soil Quality through Residue Management in a Rice-Wheat System in Fukuoka, Japan. Soil Sci. Plant Nutr. 51:849–860.

Tomer, M.D., D.W. Meek, and L.A. Kramer. 2005. Agricultural practices influence flow regimes of headwater streams in western Iowa. J. Environ. Qual. 34:1547–1558.

Treier, K.E., M.H. Wicks, and D.M. Keener. 2006. Analysis of shelled corn as an agri-fuel-direct combustion vs. ethanol. ASABE Paper No. 066027. p. 1–12 *In* ASABE Annu. Intl. Meet. Am. Soc. of Agric. and Biol. Eng., Portland, OR. 9–12 July 2006. ASABE, St. Joseph, MI,

Uhlenbrook, S. 2007. Biofuel and water cycle dynamics: What are the related challenges for hydrological processes research? Hydrol. Processes 21:3647–3650.

Unger, P.W., and M.F. Vigil. 1998. Cover crop effects on soil water relationships. J. Soil Water Conserv. 53:200–207.

USDA-NRCS. 2002. National Agronomy Manual. 190 V-NAM. 3rd ed. USDA, Washington, DC.

USDA-NRCS. 2006. Soil conservation brief: Soil quality. No. 0601. Available at http://www.nrcs.usda.gov/feature/outlook/Soil%20Quality.pdf (posted Feb. 2006; verified 17 Sept. 2010). USDA-National Resource Conservation Service.

Varvel, G.E., K.P. Vogel, R.B. Mitchell, R.F. Follett, and J.M. Kimble. 2008. Comparison of corn and switchgrass on marginal soils for bioenergy. Biomass Bioenergy 32:18–21.

Varvel, G.E., and W.W. Wilhelm. 2008a. Soil carbon levels in irrigated western Corn Belt rotations. Agron. J. 100:1180–1184.

Varvel, G.E., and W.W. Wilhelm. 2008b. Cob biomass production in the western Corn Belt. BioEnergy Res. 1:223–228.

Velthof, G.L., P.J. Kuikman, and O. Oenema. 2002. Nitrous oxide emission from soils amended with crop residues. Nutr. Cycling Agroecosyst. 62:249–261.

Vogel, K.P., J.J. Brejda, D.T. Walters, and D.R. Buxton. 2002. Switchgrass biomass production in the Midwest USA: Harvest and nitrogen management. Agron. J. 94:413–420.

Vogel, K.P., J.F. Pedersen, S.D. Masterson, and J.J. Toy. 1999. Evaluation of a filter bag system for NDF, ADF, and IVDMD forage analysis. Crop Sci. 39:276–279.

Wells, R.R., D.A. DiCarlo, T.S. Steenhuis, J.Y. Parlange, M.J.M. Römkens, and S.N. Prasad. 2003. Infiltration and surface geometry features of a swelling soil following successive simulated rainstorms. Soil Sci. Soc. Am. J. 67:1344–1351.

Wilhelm, W.W., J.W. Doran, and J.F. Power. 1986. Corn and soybean yield response to crop management under no-tillage production systems. Agron. J. 7:184–189.

Wilhelm, W.W., J.M.F. Johnson, J.L. Hatfield, W.B. Voorhees, and D.R. Linden. 2004. Crop and soil productivity response to corn residue removal: A literature review. Agron. J. 96:1–17.

Wilhelm, W.W., J.M.F. Johnson, D.L. Karlen, and D.T. Lightle. 2007. Corn stover to sustain soil organic carbon further constrains biomass supply. Agron. J. 99:1665–1667.

Wilhelm, W.W., J.M.F. Johnson, D. Lightle, N.W. Barbour, D.L. Karlen, D.A. Laird, J. Baker, T.E. Ochsner, J.M. Novak, A.D. Halvorson, D.W. Archer, and F. Arriaga. 2010. Vertical distribution of corn stover dry mass grown at several U.S. locations. BioEnergy Res. doi:10.1007/s12155-010-9097-z.

Wilson, G.V., S.M. Dabney, K.C. McGregor, and B.D. Barkoll. 2004. Tillage and residue effects on runoff and erosion dynamics. Trans. ASAE 47:119–128.

Wolf, D.D., and D.A. Fiske. 1996. Planting and managing switchgrass for forage, wildlife, and conservation. Publication No. 418–013. Available at http://pubs.ext.vt.edu/418/418-013/418-013.pdf (posted June, 1996; verified 17 Sept. 2010). Virginia Polytechnic Inst. and State Univ., Blacksburg.

Wright, L.L. 1994. Production technology status of woody and herbaceous crops. Biomass Bioenergy 3:191–210.

Ying, J., S. Peng, G. Yang, N. Zhou, R.M. Visperas, and K.G. Cassman. 1998. Comparison of high-yield rice in tropical and subtropical environments: II. Nitrogen accumulation and utilization efficiency. Field Crops Res. 57:85–93.

Yu, F., R. Ruan, and P. Steele. 2008. Consecutive reaction model for the pyrolysis of corn cob. Trans. ASABE 51:1023–1028.

Zemenchik, R.A., and K.A. Albrecht. 2002. Nitrogen use efficiency and apparent nitrogen recovery of Kentucky bluegrass, smooth bromegrass, and orchardgrass. Agron. J. 94:421–428.

Zhang, N., M. Wang, and N. Wang. 2002. Precision agriculture—a worldwide overview. Comput. Electron. Agric. 36:113–132.

25

Emerging Challenges in Soil Management

Jerry L. Hatfield and Thomas J. Sauer

Soil provides a number of important functions, including agronomic and biomass productivity, purifying water, biodegradation of pollutants, regulating greenhouse gases (GHGs) emission from the soil surface, and providing a foundation for civil and urban structures. This is a wide range of functions, but the most critical function for humanity is providing the fundamental resource for the production of the food, feed, fiber, and fuel required to sustain the ever-increasing human population. In the history of agriculture, there has never been a time in which the importance of proper management of the soil resource has been more critical. Jenny (1980) pointed out that the functions of soil represent the diverse capabilities of soil and the multiple purposes of soil in answering a variety of human needs, from food to shelter.

The increase of the world's population from 6 to 9 billion over the next 30 to 40 years requires that the soil resource be carefully managed to increase the world's food supply. This has to be done in the face of an ever-degrading soil resource and an increasing variation in climatic resources and uncertainty in precipitation. The management of the soil resource to achieve this level of production will require that we better understand how soil responds to various practices. Within this volume we have assembled information to help establish a knowledge base to guide how we can cope with these challenges in providing a secure food supply in the face of a changing climate.

There have been continual assessments in the past few years that indicate soil condition continues to decline in terms of the organic matter content, soil structure, and potential rooting depth. Lal (1993) used the following expression to provide a framework for the assessment of soil quality, S_q, by combining the following attributes, $S_q = f(P, S, R, e, N, B)_t$, where P is productivity, S is an index of structural properties, R is rooting depth, e is charge density, N is nutrient reserve, B is a measure of biological diversity, and t is time. If we dissect this relationship, it is evident that soil quality is a function of a number of different parameters, ranging from the basic soil structure to the productivity of the agricultural system in terms of both yield and adding organic material back to the soil. This framework provides a ready reference as to the role of soil management practices in influencing the components affecting soil quality. An important aspect of Lal's approach is the incorporation of the nutrient component in the soil. Management of nutrients to achieve optimal productivity of different agricultural systems is part of the foundation of improving our capabilities of producing the food needed

J.L. Hatfield, Laboratory Director and Plant Physiologist, National Laboratory for Agriculture and the Environment, 2110 University Blvd., Ames, IA 50011 (jerry.hatfield@ars.usda.gov); T.J. Sauer, USDA-ARS, National Laboratory for Agriculture and the Environment, 2110 University Boulevard, Ames, IA 50011-3120 (Tom.Sauer@ars.usda.gov).

doi:10.2136/2011.soilmanagement.c25

to supply the increasing population. In this book we have provided an exploration of the current state of knowledge on critical nutrients and the emerging challenges in nutrient management.

One of the major challenges will be how to balance the need for the increased production of food, feed, fuel, and fiber with the need for enhanced environmental quality. Water and air quality continue to be major problems linked with soil management. One of the functions of soil is the purification of water, but there is increasing concern regarding the linkage between agricultural practices and movement of nitrate nitrogen and phosphorus into nearby water bodies. Manure is a valuable source of nutrients for agricultural systems and offers the potential to enhance the soil from a physical, chemical, and biological perspective. Improved manure management systems coupled with practices to decrease tillage represent an area in which soil management will have to be refined and the knowledge base expanded. In spite of the offsite movement of nutrients and agricultural chemicals, the loss of sediment from fields should still be of concern to everyone because of its long-term impacts on the soil resource. Water and wind erosion processes need to be understood for soils around the world to reduce the impact of these two major forces on the soil and also to identify management practices that could be implemented to reduce these erosion losses. A fundamental change in philosophy away from "tolerable" soil loss towards soil improvement is desperately needed if future generations are to have access to land resources to meet their needs. The continued loss of soil by erosion and tillage is increasing the spatial variation across agricultural landscapes and threatens to impact the potential for efficient production. Soil management practices that either lessen the impact of this spatial variation or enhance the soil in areas degraded by erosion or tillage will provide a lasting benefit to the soil and ultimately to humanity. However, the effectiveness of these practices needs to be evaluated and based on sound scientific principles so that the information will be transferable among regions. In this volume, our goal was to provide both an understanding of the basis for soil management practices and to provide an overview of different practices that could be adapted.

There are rapidly changing practices in terms of cropping systems being proposed. The development of biofuels based on either corn stover removal or planting of lands to perennial grasses for harvest of cellulosic materials for ethanol production or direct biomass burning creates a potential new stress on the soil in terms of water use, nutrient management, and protection of the soil resource. These are not easy questions to address because the problem changes in scope from being only a single field problem to a regional landscape problem encompassing a range of soils, topographies, and microclimates. Sound soil management practices that can be implemented as part of these production systems will have to account for how these systems utilize soil water, extract nutrients, and change the overall biology of the soil system. The challenge will be not only to increase production but also to increase the diversity of production that can be used for a variety of purposes.

Transitioning from oil to biofuels as part of the renewable fuel standard brings into discussion the role of soil and agricultural systems as part of the global carbon balance. The global carbon balance is part of the overall effort to understand the role of agricultural practices in both the mitigation and sequestration processes for GHGs. Carbon storage in soil has been considered a positive for the GHG uptake and enhances soil quality because of the role that increased carbon has on the organic matter content of the soil and the resultant improvement in soil water-holding capacity, soil biological processes, and nutrient cycling. These processes extend beyond carbon and include methane and nitrous oxide as critical GHGs. Improving our understanding of the role soil management plays in altering the GHG exchange will help drive whole new approaches in GHG mitigation strategies. One aspect of cropping systems that offers potential for improving the carbon capture is through the use of cover crops. Effective management and adaptation of cover crop systems across a wider range of soils and cropping systems will offer the potential of multiple benefits for carbon, water, and nutrient management with the challenge to develop and implement cover crop systems that are cost-effective and enhance the environment.

The role soil has on supplying adequate and timely nutrients, water, and oxygen to plants is critical to the development of cropping systems that operate at the highest level of efficiency. To meet the production needs for the next 50 years will require that we focus our energies on improved soil management practices leading to increased efficiency of water and nutrient use. The next revolution in agriculture won't be a result of any single factor but rather resilient production systems capable of coping with any number of stresses. Too often we focus on soil management impacts on the cropping systems and fail to address the critical aspects of pest management. The indirect impacts of pests on production systems are often overlooked and deserve greater attention as we move forward in understanding the role soil management has on production systems. Weeds, insects, and diseases are critical components in agricultural systems and are deserving of greater attention in future evaluations of soil management practices.

In assembling this volume, we selected chapter authors who would challenge our current thinking and provide insights into future efforts to enhance soil management practices. There are many gaps in our current knowledge, and we need to focus our efforts and discussion on how we can expand our knowledge base to fill those gaps, but also to provide educational tools to help scientists, consultants, and students understand the potential in improving soil management practices that enhance the soil, increase production efficiency, improve environmental quality, lessen pest pressure, and increase the capability of ensuring a stable and abundant supply of food, feed, fuel, and fiber for future generations. If we are indeed in the midst of a soil science renaissance (Hartemink and McBratney, 2008), there could be no better outcome than dramatic, new, and innovative improvements in soil management practices that protect and sustain this critical resource for future generations

References

Hartemink, A.E, and A. McBratney. 2008. A soil science renaissance. Geoderma 148:123–129.

Jenny, H. 1980. The soil resources. Springer-Verlag, New York.

Lal, R. 1993. Tillage effects on soil degradation, soil resilience, soil quality and sustainability. Soil Tillage Res. 27:1–8.

Page numbers followed by f and t indicate figures and tables.